6 Edition
Revised & Updated

A Handbook on

Electrical Engineering

Contains well illustrated formulae & key theory concepts

for

ESE, GATE, PSUs

& OTHER COMPETITIVE EXAMS

MADE EASY PUBLICATIONS

MADE EASY Publications Pvt. Ltd.

Corporate Office: 44-A/4, Kalu Sarai (Near Hauz Khas Metro Station), New Delhi-110016

Contact: 9021300500

E-mail: infomep@madeeasy.in

Visit us at: www.madeeasypublications.org

A Handbook on Electrical Engineering

First Edition: 2012
Second Edition: 2014
Third Edition: 2015
Fourth Edition: 2019
Reprint: 2020
Reprint: 2021
Fifth Edition: 2022

Sixth Edition: 2023

Director's Message

B. Singh (Ex. IES)

During the current age of international competition in Science and Technology, the Indian participation through skilled technical professionals have been challenging to the world. Constant efforts and desire to achieve top positions are still required.

I feel every candidate has ability to succeed but competitive environment and quality guidance is required to achieve high level goals. At MADE EASY, we help you discover your hidden talent and success quotient to achieve your ultimate goals. In my opinion CSE, ESE, GATE & PSUs exams are tools to enter in to the main stream of Nation serving. The real application of knowledge and talent starts, after you enter in to the working system. Here at MADE EASY you are also trained to become winner in your life and achieve job satisfaction.

MADE EASY alumni have shared their winning stories of success and expressed their gratitude towards quality guidance of MADE EASY. Our students have not only secured All India First Ranks in ESE, GATE and PSUs entrance examinations but also secured top positions in their career profiles. Now, I invite you to become an alumnus of MADE EASY to explore and achieve ultimate goal of your life. I promise to provide you quality guidance with competitive environment which is far advanced and ahead than the reach of other institutions. You will get the guidance, support and inspiration that you need to reach the peak of your career.

I have a true desire to serve the Society and the Nation easing path of the education for the people of India.

After a long experience of teaching Electrical Engineering over a period of time, MADE EASY team realised that there is a need of a good *Handbook* which can provide the crux of Electrical Engineering in a concise form to the student to brush up the formulae and important concepts required for ESE, GATE, PSUs and other competitive examinations. This *handbook* contains all the formulae and important theoretical aspects of Electrical Engineering. It provides much needed revision aid and study guidance before examinations.

B. Singh (Ex. IES)
CMD, MADE EASY Group

A Handbook on

Electrical Engineering

Contents

Contents

Contents

■■■■

1 Power Systems

I Supply System

Basic Structure of Power System

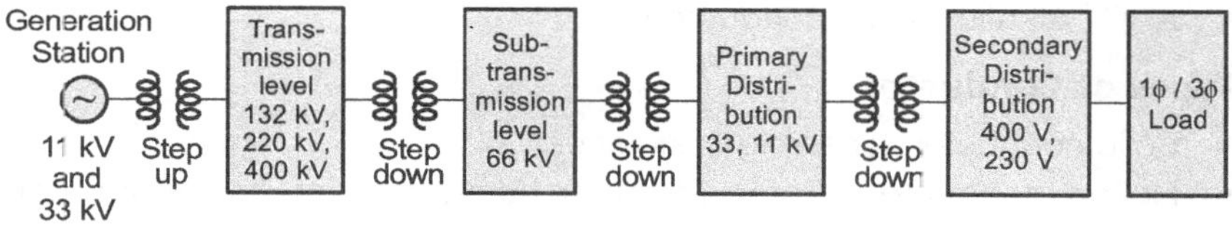

- Generating stations are interconnected by the lines.
- Transmission lines, when interconnected with each other, becomes transmission networks.
- The combined transmission and distribution network is known as the "power grid".

Effect of System Voltage on Transmission of Power

- Power loss in the line is inversely proportional to the system voltage and power factor both.
- Percentage voltage drop in resistance decreases with the increase in the system voltage.
- Weight of the conductor material for the line will decreases with the increase in supply voltage and power factor.
- Efficiency of transmission, increases with the increase of supply voltage and power factor.
- Higher supply voltages also enhances the system stability.
- The problems encountered with high voltages are the insulation of the equipment, corona, radio and television interference.
- The voltage level of a system is therefore governed by the amount of power to be transmitted and the length of the line.

Voltage Level

- Low voltage → 230 V (1-ϕ), 400 V (3-ϕ)
- High voltage → 11 kV, 33 kV

- Extra high voltage: 66 kV, 132 kV, 220 kV.
- Modern EHV: 400 kV
- Ultra high voltage: 765 kV and above.

Conductor Used for Transmission Line

- Copper conductor
- ACSR : Aluminium conductor steel reinforced.
- ACAR : Aluminium conductor alloy reinforced.
- AAAC : All Aluminium alloy conductor.
- Expanded ACSR conductor: Normally used for EHV lines.
- AAC : All Alluminium conductors.

Types of Conductor

- Solid conductor: It has high skin effect.
- Hollow conductor: Preferred under heavy current, i.e., more than 1000 Amp.
- Stranded conductor.
- Composite standard conductor: used for voltage ≤ 220 kV.
- Bundle conductor: Used for voltage > 275 kV.

Advantage of Bundle Conductor

- Self distance (GMR) increased without change in mutual distance.
- Voltage gradient reduced so corona loss reduce.
- It reduces the interference with nearby communication line.
- Inductance (L) of transmission line reduces and capacitance (C) increases.
- Surge impedance, i.e., $Z_s = \sqrt{\frac{L}{C}}$ decreases.
- Power system stability increases.

Insulators

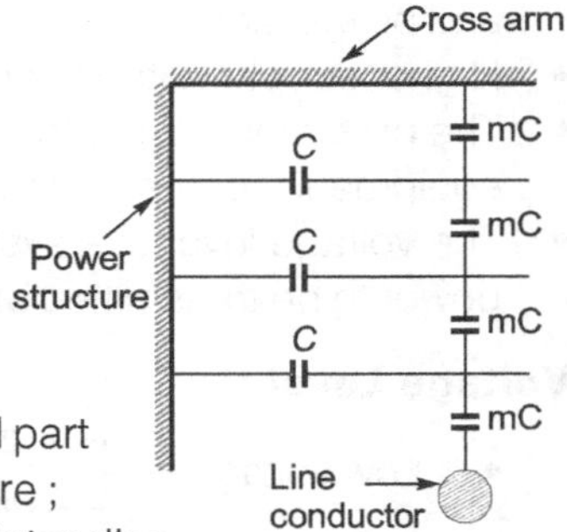

Over head line insulators provide the required insulation to the line conductors from each other and from the supporting structures electrically. Most commonly used materials are porcelain, toughened glass and steatite,

where, $C \rightarrow$ Capacitance between metal part of the insulator and tower structure ;

$mC \rightarrow$ Capacitance of each insulator disc.

$mC > C$

Note:

☑ The stress experienced by the disc near the power conductor is more than the stress experience by the disc near the cross-arm.

String Efficiency

$$\text{String efficiency} = \frac{\text{Voltage across the whole string}}{n \times (\text{Voltage across the unit adjacent to line conductor})}$$

where, $n \rightarrow$ Number of insulator discs in the string

String efficiency also defined as

$$\%\eta = \frac{\text{Flashover voltage of the string}}{n \times \text{flashover voltage of one string}} \times 100$$

Note:

☑ As the number of disc increases string efficiency decreases.

Methods of Equilising Potential Across Each Disc

- Increase the length of cross arm.
- Capacitance grading or grading of units.
- Use of grading rings or static shielding.

Remember:

☑ For static shielding the capacitance from the shield to the K_{th} link from the top

$$C_K = \frac{KC}{n-K} \qquad (\because n = \text{number of disc})$$

Types of Insulator

- **Pin type insulator:** Pin type insulator operate satisfactory upto 25 kV.
- **Multipine type insulator:** Operates upto 33 kV
- **Suspension type insulator:** A suspension insulator is designed to operate at 11 kV.
- **Strain type insulator:** Strain type insulator mechanically strong. It is used when direction of transmission line changes across river crossing and at the dead end of the transmission line.
- **Shackle type:** Shackle type insulator are used in low tension cable. These insulator can be operated either horizontally or vertically.

II Line Parameters

Transmission line is a carrier on which bulk amount of power from a remote generating station to the operative areas is being carried out.

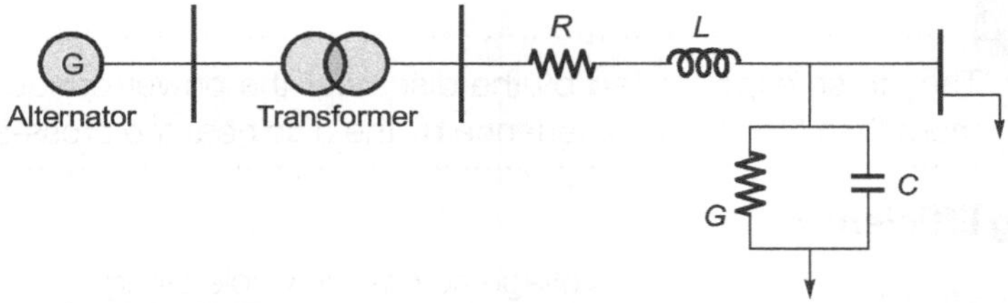

Transmission line is

- series combination of resistance (R) and inductance (L) and
- Parallel combination of shunt conductance (G) and capacitance (C).

Note:

- The line parameter of transmission line is calculated in per unity or per km and are constant for entire line length.
- The shunt conductance is caused by leakage current.
- In transmission line if $G = 0$ means leakage current is assume to be zero.
- Power loss in the conductor is only due to series resistance.
- Power transmission capacity of the line is mainly governed by the series inductance.

- **Resistance of a conductor,** $R_{\text{eff}} = \dfrac{\text{Power loss in conductor}}{I^2}$ ohms

  where, $R_{\text{eff}} \rightarrow$ Effective resistance of the conductor
- **D.C. Resistance of a Conductor,** $R_{dc} = \rho \dfrac{l}{A}$ ohms

  where, $\rho \rightarrow$ Resistivity of conductor, Ω-m ; $l \rightarrow$ Length of conductor, metre ; $A \rightarrow$ Cross-sectional area, m^2

Note:

☑ The effective resistance is equal to the dc resistance of the conductor only if the current is uniformly distributed throughout the cross-sectional area of the conductor (i.e. for DC only).

Skin Effect

If DC is passed in a conductor, the current density is uniform over the cross-section of the conductor but when an alternating current flows through a conductor, the distribution tends to become non uniform. There is a tendency of the current to crowd near the surface of the conductor. This phenomenon is called "skin effect". The effective conductor resistance increases in AC as compared to DC which causes larger power loss.

Remember:

☑ Skin effect increases with increase in frequency, conductor diameter and permeability.

Proximity Effect

When two or more conductors are in proximity, their electromagnetic field interact with each other, with the result that the current in each of them is redistributed such that the greater current density is concentrated in that part of the strand most remote from the interfering conductor. In each case, a reduced current rating results from the apparent increase of resistance.

Remember:

☑ Proximity effect is more pronounced in case of cable, large conductors and high frequencies.

Magnetic Flux Density

Biot-savart's law

- Magnetic flux at any point produced by a current carrying element

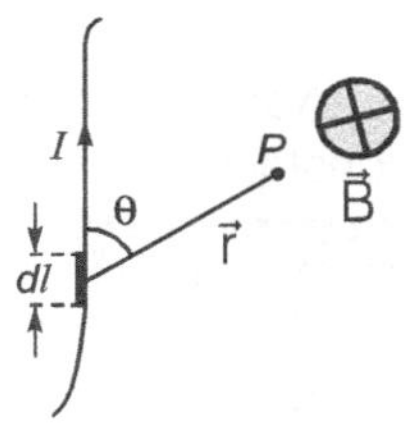

$$d\vec{B} = \frac{\mu}{4\pi}\frac{I\vec{dl}\times(\vec{r})}{r^3}$$

where, $dB \rightarrow$ Infinitesimal flux density at point P
$I \rightarrow$ Current in element ; $dl \rightarrow$ Length of element
$\theta \rightarrow$ Angle between current direction and radius vector to P
$r \rightarrow$ Radius vector ; $\mu \rightarrow$ Permeability of medium

- Magnetic flux density B at any point to an infinite conductor.

$$B = \frac{\mu I}{2\pi R}$$

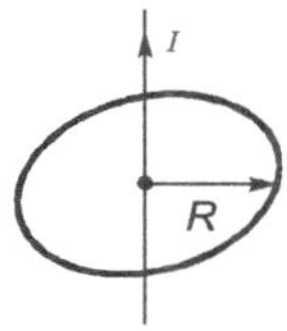

where, R = Radial distance of the point from the conductor.

Note:

The direction of the flux density is normal to the plane containing the conductor and radius vector R.

Amperes's law , $\oint H.dl = I_{\text{enclosed}}$

where, $H \rightarrow$ Magnetic field intensity

$I \rightarrow$ R.M.S. value of current enclosed by an amperian loop.

Relation Between Magnetic Flux Density and Magnetic Field Intensity

$$\boxed{B = \mu H,\ \mu = \mu_0\, \mu_r}$$

where, $\mu_0 \rightarrow 4\pi \times 10^{-7}$ H/m = Permeability of free space

$\mu_r \rightarrow$ Relative permeability of the medium = 1 (for non magnetic material)

Inductance

Inductance of an inductor is the ratio of its total magnetic flux linkages to the current I through the inductor.

$$\boxed{L = \frac{N\Psi_m}{I} = \frac{\lambda}{I}\ \text{Henry}}$$

where, $\Psi_m \rightarrow$ Magnetic flux linkages through a single turn

$N \rightarrow$ Total number of turns ; $\lambda \rightarrow$ Total magnetic flux linkages

Above formulae is valid for a medium in which the permeability is constant.

Remember:

The permeability of ferrous medium is not constant. For such cases the inductance is defined as the ratio of infinitesimal change in flux linkage to the infinitesimal change in current producing it

$$L = \frac{d\lambda}{dI}\ \text{Henry}$$

- Flux linkages within the conductor : $\Psi_{int} = \frac{\mu I}{8\pi}$ Wb-T/m

  where, $\Psi_{int} \rightarrow$ Total internal flux linkages ; $I \rightarrow$ R.M.S. value of current.

$$\Psi_{int} = 0.5I \times 10^{-7}\ \text{Wb-T/m}$$

- Inductance of the conductor, contributed by flux within the conductor:

$$\boxed{L_{int} = 0.5 \times 10^{-7}\ \text{H/m}} \quad \text{as} \quad \boxed{L_{int} = \frac{\Psi_{int}}{I}}$$

- Flux linkages outside the conductor

$$\Psi_{12} = \frac{\mu I}{2\pi} \ln\left(\frac{D_2}{D_1}\right)\ \text{Wb-T/m}$$

for $\mu_r = 1$: $\Psi_{12} = 2 \times 10^{-7} \ln\left(\frac{D_2}{D_1}\right)$ Wb-T/m

where $\Psi_{12} \rightarrow$ Total flux linkages between points 1 and 2

- Inductance of the conductor, contributed by flux between points 1 and 2:

$$\boxed{L_{12} = 2 \times 10^{-7} \ln\left(\frac{D_2}{D_1}\right) \text{ H/m}}$$

- Inductance of a single phase two wire line: $\boxed{L = 4 \times 10^{-7} \ln\left(\frac{D}{r'}\right)}$ H/m

where, $D \rightarrow$ Distance between two solid conductors of same radii r
$r' \rightarrow$ Radius of fictitious conductor = $0.7788\,r$

Inductance of Composite Conductor Lines

Conductor *M* consists of *m* similar parallel sub-conductors and conductor *N* consists of n similar parallel sub-conductors.

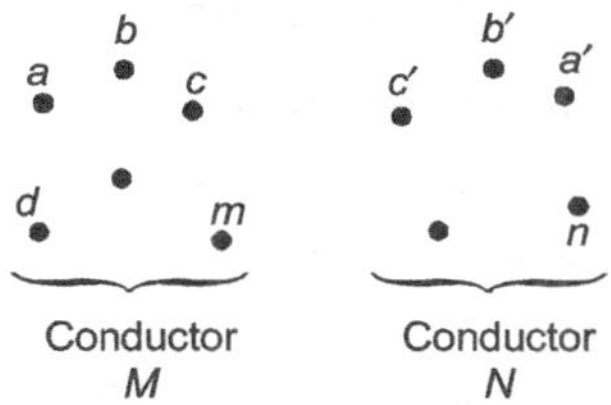

(Single phase line having composite conductors)

If line current is I, then each strand of conductor *M* carries a current I/m and each strand of conductor *N* carries a current of $-I/n$ (the conductor *N* being the return conductor).

$$L_M = 2 \times 10^{-7} \times \ln \frac{\left[(D_{aa'} D_{ab'} \ldots D_{an})(D_{ba'} D_{bb'} \ldots D_{bn}) \ldots (D_{ma'} D_{mb'} \ldots D_{mn})\right]^{1/mn}}{\left[(D_{aa}\, D_{ab} \ldots D_{am})(D_{ba}\, D_{bb} \ldots D_{bm}) \ldots (D_{ma}\, D_{mb} \ldots D_{mm})\right]^{1/m^2}} \text{ H/m}$$

where, $L_M \rightarrow$ inductance of conductor *M*

$$\boxed{L = 2 \times 10^{-7} \ln\left(\frac{\text{GMD}}{\text{GMR}}\right)}$$

Remember:

- GMD = mn^{th} root of the product of *mn* distances (known as the geometric mean distance between conductor *M* and conductor *N* and denoted by D_m).
- GMR = $(m^2)^{th}$ root of the product of m^2 distances these being the distances from each sub-conductor of conductor M to every other sub-conductor of conductor *M* (including $D_{aa}, D_{bb} \ldots D_{mm}$).
- GMR = Geometric mean radius (denoted by D_s), which depends on radius of conductors and does not effected by unequal spacing of conductors.
- D_{aa} = 0.7788 times the radius of sub-conductor '*a*'.

Inductance of 3-ϕ Line With Equivalent Spacing.

Assuming balanced currents, i.e., $(I_a + I_b + I_c = 0)$

$$L_a = 2 \times 10^{-7} \ln\left(\frac{D}{r'}\right) \text{ H/m}$$

where, $L_a \rightarrow$ Inductance of phase
$D \rightarrow$ Distance between any two phases
$r' \rightarrow 0.7788r$ = Radius of fictitious conductor
$\rightarrow$ 0.7788 times the radius of conductor
$L_a \rightarrow L_b = L_c$ (Because of symmetry)

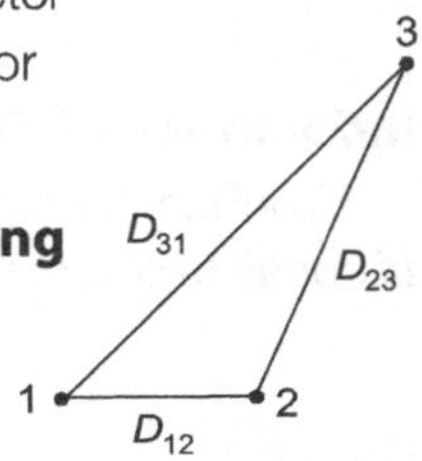

Inductance of 3-ϕ line with unsymmetrical spacing

In this case the lines are transposed.

Transposition of transmission line

- Whenever 3ϕ unsymmetrical line running parallel and neighbour to the communication line it cause interference in the communication line.
- In order to eliminate the communication interference transposition of line is recommended.
- Change the position of power conductor at regular interval with equidistance for a given line length, so that the position of power conductor is replaced by its successive phase conductor.

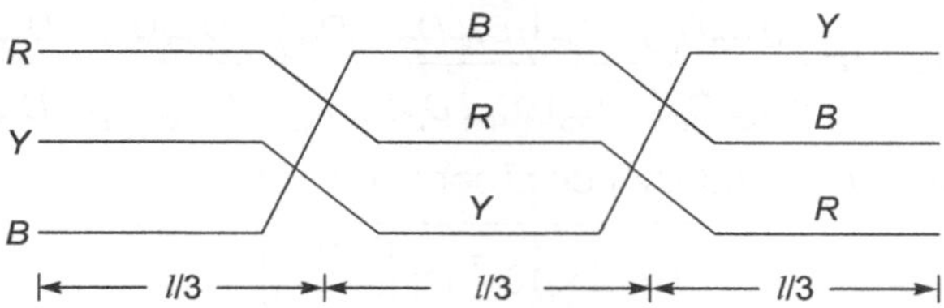

Advantages of Transposition

- Net resultant flux ϕ_r which link with communication line become zero.
- GMD/phase equal.
- L/phase equal.
- I/phase equal.
- Flux per phase equal.

Note:

☑ Transposition of transmission line is an old technique. The radio interference is eliminated by completely insulating any one of the phases.

Inductance of Phase-1

$$L_1 = 2\times10^{-7}\ln\left(\frac{D_{eq}}{r'}\right)\ \text{H/m}$$

where, $L_1 \to$ inductance of phase 1

$D_{eq} \to \sqrt[3]{D_{12}D_{23}D_{31}}$ = Equivalent spacing

= Geometric mean of the distance of the line.

Inductance of Bundled Conductor Lines

- *For a two conductor (duplex) arrangement*

$$D_S^b = \sqrt[4]{(D_S.d)^2} = \sqrt{D_s\cdot d}$$

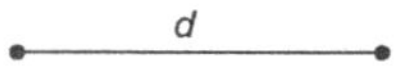

- *For a three conductor (tripex) arrangement*

$$D_S^b = \sqrt[9]{(D_s\cdot d\cdot d)^3} = \sqrt[3]{D_s\cdot d^2}$$

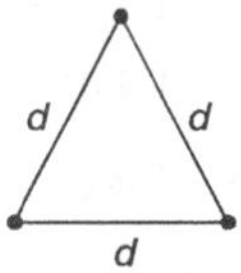

- *For a four-conductor (quadruplex) arrangement*

$$D_S^b = \sqrt[16]{(D_s\cdot d\cdot d\cdot d\sqrt{2})^4}$$

$$= 1.09\times\sqrt[4]{D_s\cdot d^3}$$

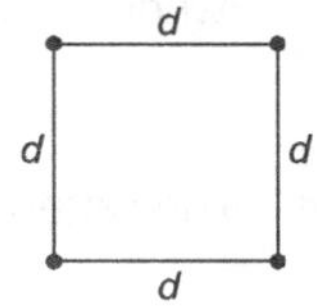

where, D_S^b = Geometric mean radius of bundled conductor

D_s = Geometric mean radius of each sub-conductor of bundle

d = Spacing between the sub-conductors of a bundle

Remember:

☑ Inductance of bundled conductor line is less than the inductance of the line with one conductor per phase.

☑ GMD depends only on the distance between the conductors and is independent of radius of conductors.

Inductance of Double Circuit 3-ϕ Line

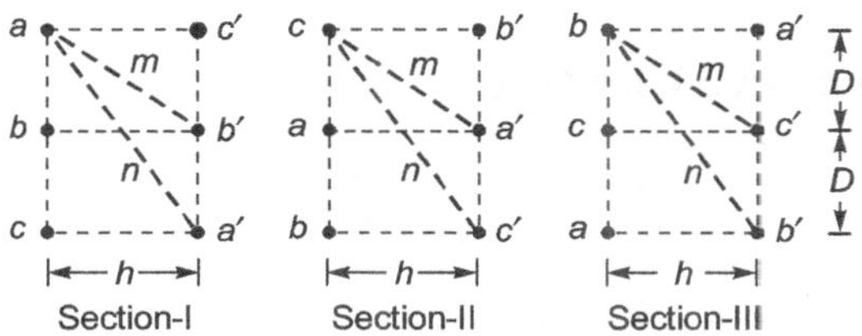

Inductance per phase per metre length

$$L = 2\times10^{-7}\ \ln\left[2^{1/6}\left(\frac{D}{r'}\right)^{1/6}\cdot\left(\frac{m}{n}\right)^{1/3}\right]\ \text{H/phase/m}$$

Mutual Inductance

Mutual inductance is defined as the flux linkages of one circuit due to the current in the second circuit per-ampere of current in the second circuit. If the current I_2 produces λ_{12} flux linkages with circuit 1. The mutual inductance is

$$\boxed{M_{12} = \frac{\lambda_{12}}{I_2}} \text{ Henry}$$

Electrical Field and Potential Difference

- The lines of electric flux originate on the positive charges on one conductor and terminate on the negative charges on the other conductor.
- If a long straight cylindrical conductor has a uniform charge throughout its length and is isolated from other charges.
- Electric field intensity E at any point, $E = \dfrac{q}{2\pi \in x}$ V/m

  where, $q \rightarrow$ Charge on conductor per unit length
  $\in \rightarrow$ Permittivity of the medium
  $x \rightarrow$ Distance from conductor to the point under consideration.
- The potential difference between two points

$$V_{xy} = \frac{q}{2\pi \in} \ln\left(\frac{D_y}{D_x}\right) \text{ Volts}$$

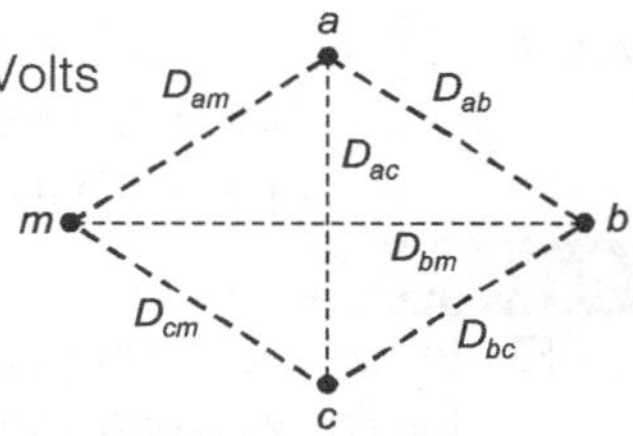

(An array of *m* charged conductors)

where, $D_x, D_y \rightarrow$ Distance of point x and y from charge q
$q \rightarrow$ Charge per unit length

- The potential difference between two conductor of an array of parallel conductors

$$V_{ab} = \frac{1}{2\pi \in}\left[q_a \ln\frac{D_{ab}}{r_a} + q_b \ln\frac{r_b}{D_{ba}} + q_c \ln\frac{D_{cb}}{D_{ca}} + \cdots + q_m \ln\frac{D_{mb}}{D_{ma}}\right]$$

Capacitance

Capacitance of Two Wire Line

$$\boxed{C_{ab} = \frac{\pi\varepsilon}{\ln\left(\frac{D}{r}\right)}} \text{ F/m}$$

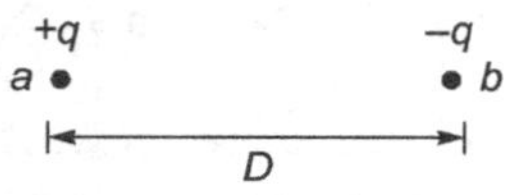

where, $C_{ab} \rightarrow$ Line to line capacitance
$q \rightarrow$ Charge per unit length; $r \rightarrow$ Radius of conductor a and b

If the conductor have different radii , $r = \sqrt{r_a \cdot r_b}$

where, $r_a, r_b \rightarrow$ Radius of conductor 'a' and conductor 'b' respectively.

Line to neutral capacitance

$$\boxed{C_{an} = C_{bn} = 2C_{ab}}$$

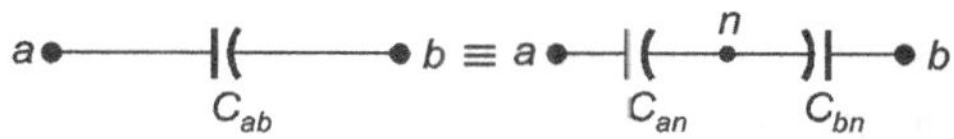

Charging Current

- The current caused by the alternate charging and discharging of the line due to alternating voltage is called **charging current** of the line.

Note:

☑ Charging current flows in a line even when the line is open circuited and affects the voltage drop, efficiency and power factor of the line.

Charging Current for 1-ϕ line

$$I_C = j\omega C_{an} V_{an} \text{ A/m / Phase}$$

where, $V_{an} \rightarrow$ Potential difference between conductor a and neutral
$\omega \rightarrow$ Frequency of alternating voltage

Capacitance of 3-ϕ line with equilateral spacing

$$\boxed{C_n = \frac{0.02412}{\log\left(\frac{D}{r}\right)}} \ \mu\text{F/km} \quad \text{or} \quad \boxed{C_n = \frac{2\pi\in}{\ln\left(\frac{D}{r}\right)}}$$

where, $C_n \rightarrow$ Line to neutral capacitance ; $D \rightarrow$ Spacing between conductors ; $r \rightarrow$ Radius of each conductor

Charging current per phase, $I_C = j\omega C_n V_{an}$

Capacitance of 3-ϕ line with asymmetrical spacing

$$\boxed{C_n = \frac{2\pi\in}{\ln\left(\frac{D_{eq}}{r}\right)}} \ \mu\text{F/km}$$

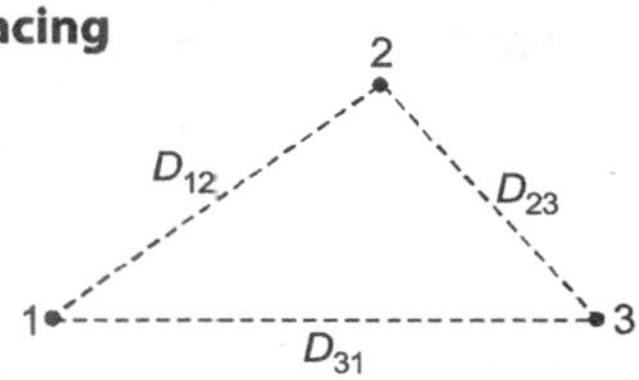

where, $\boxed{D_{eq} = \sqrt[3]{D_{12} D_{23} D_{31}}}$

Capacitance of bundled conductor lines

$$\boxed{C_n = \frac{2\pi\in}{\ln\left(\frac{D_{eq}}{D_{sc}^b}\right)}} \ \mu\text{F/km}$$

Self GMD of Bundle Conductors

- For a two conductor bundle, $D_{sc}^b = \sqrt{rd}$
- For a three conductor bundle, $D_{sc}^b = \sqrt[9]{(r \times d \times d)^3} = \sqrt[3]{rd^2}$
- For a four conductor bundle, $D_{sc}^b = \sqrt[16]{(r \times d \times d \times \sqrt{2}d)^4} = 1.09\sqrt[4]{rd^3}$

Effect of Ground on Line Capacitance (Method of Images)

The presence of ground alters the electric field of a line and hence affect the line capacitance.

For 1-ϕ line:

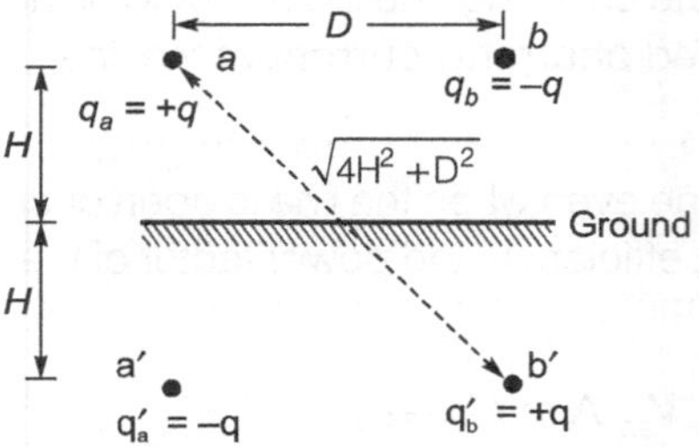

$$C_{ab} = \frac{\pi\varepsilon}{\ln\left(\dfrac{D}{r\left(1+\dfrac{D^2}{4H^2}\right)^{0.5}}\right)} \text{ F/m}$$

where, $H \rightarrow$ Height of conductor from ground

$$C_{an} = 2C_{ab} = \frac{2\pi\varepsilon}{\ln\left(\dfrac{D}{r'}\right)} \text{ F/m} \; ; \qquad r' = r\sqrt{1+\left(\frac{D}{2H}\right)^2}$$

For 3-ϕ line

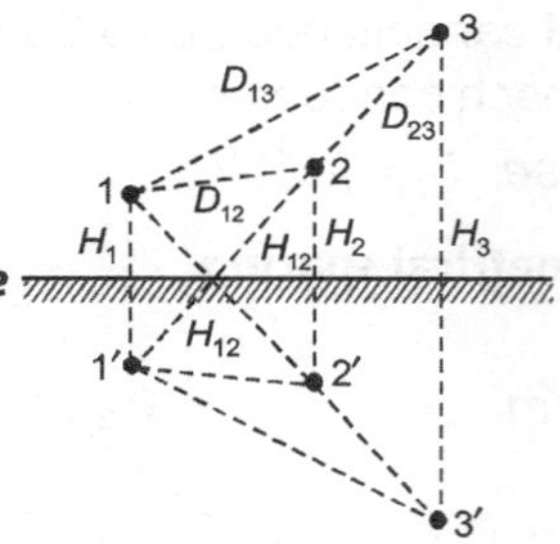

Conductors of a 3-phase line with image charges

$$C_n = \frac{2\pi\varepsilon}{\ln\left(\dfrac{D_{eq}}{r}\right) - \ln\left[\dfrac{\sqrt[3]{H_{12}H_{23}H_{31}}}{\sqrt[3]{H_1H_2H_3}}\right]}$$

where, H_1, H_2, H_3, and H_{12}, H_{23}, H_{31} are shown in figure.
C_n is in μF/km.

Note:

☑ Presence of ground increases the line capacitance by small amount.

III Performance of Transmission Line

Type of Transmission Line	Classification based on $f * l$	Classification based on length if line is power line i.e., f = 50 Hz	Based on operating voltage	Effect of capacitance
Short line	fl < 4000 Hz km	l < 80 km	0-20 kV	Neglected
Medium line	4000 < fl < 12000 Hz km	80 km < l < 240 km	20-100 kV	Capacitor is lumped and constant
Long line	fl > 12000 Hz km	l > 240 km	> 100 kV	Capacitance is uniformly distributed

where, $f \rightarrow$ Operating frequency ; $l \rightarrow$ Length of transmission line

Short transmission lines

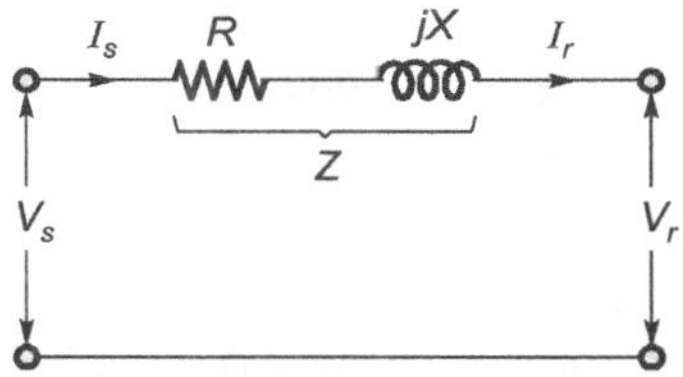

$$\boxed{I_s = I_r}, \quad \boxed{V_s = V_r + I_r Z}$$

where, $V_s \rightarrow$ Sending end voltage ; $V_r \rightarrow$ Receiving end voltage
$I_s \rightarrow$ Sending end current ; $I_r \rightarrow$ Receiving end current

In Matrix form:

$$\begin{bmatrix} V_s \\ I_s \end{bmatrix} = \begin{bmatrix} A & B \\ C & D \end{bmatrix} \begin{bmatrix} V_r \\ I_r \end{bmatrix} \quad ; \quad \begin{bmatrix} V_s \\ I_s \end{bmatrix} = \begin{bmatrix} 1 & Z \\ 0 & 1 \end{bmatrix} \begin{bmatrix} V_r \\ I_r \end{bmatrix}$$

So, $A = 1, B = Z, C = 0, D = 1$

$$\boxed{V_s \simeq V_r + I_r R \cos\phi_r + I_r X \sin\phi_r}$$

Note:

☑ The performance of transmission line is determine by efficiency and voltage regulation.

☑ For rotating machine speed regulation determine and for static machine voltage regulation determine.

☑ For satisfactory performance lower the regulation and higher the efficiency.

Efficiency : $\%\eta = \dfrac{P_R}{P_S} \times 100 = \dfrac{P_S - P_{Loss}}{P_S} \times 100 = \dfrac{P_R}{P_R + P_{Loss}} \times 100$

Voltage Regulation

It is the change in receiving end voltage from no load to full load while keeping the sending end voltage constant and made supply frequency constant.

$$\% \text{ Voltage regulation} = \frac{|V_S| - |V_R|}{|V_R|} \times 100 \quad \text{(OR)} \quad \frac{|V_{R,NL}| - |V_{R,FL}|}{|V_{R,FL}|} \times 100$$

where, $V_{R,NL}$ = Receiving end voltage under no load condition
$V_{R,FL}$ = Receiving end voltage under full load condition

Regulation of Short Transmission Line

$$\text{Regulation} = \frac{I_r R\cos\phi_r \pm I_r X\sin\phi_r}{V_R}$$

where, + → For lagging power factor; – → For leading power factor

Note:

- Regulation is always positive for lagging power factor.
- Regulation may be positive, negative or zero for leading power factor.
- In short line sending end power factor always less than receiving end power factor.
- Short line is always symmetrical and reciprocal.
- Zero Pf is possible only for leading load.

Remember:

Regulation of the Transmission Line	Required Condition	Occurs at
Zero regulation	$\phi_R = \left(\frac{\pi}{2} - \theta\right) = \tan^{-1}\left(\frac{R}{X_L}\right)$	Leading P.f.
Maximum regulation	$\phi_R = \theta = \tan^{-1}\left(\frac{X_L}{R}\right)$	Lagging P.f.
Negative regulation	$\cos\phi_R = 0$ leading	Leading P.f.

where, ϕ_R = Phase angle of load; θ = Impedance angle of load
$X = X_L$ = Inductance of transmission line

Medium Length Transmission Line

ABCD parameter in matrix form

For nominal T - circuit	For nominal π - circuit
$\begin{bmatrix} V_s \\ I_s \end{bmatrix} = \begin{bmatrix} 1+\frac{YZ}{2} & Z\left(1+\frac{YZ}{4}\right) \\ Y & 1+\frac{YZ}{2} \end{bmatrix} \begin{bmatrix} V_r \\ I_r \end{bmatrix}$	$\begin{bmatrix} V_s \\ I_s \end{bmatrix} = \begin{bmatrix} 1+\frac{YZ}{2} & Z \\ Y\left(1+\frac{YZ}{4}\right) & 1+\frac{YZ}{2} \end{bmatrix} \begin{bmatrix} V_r \\ I_r \end{bmatrix}$

Note:

☑ For a fix receiving end voltage the sending end voltage which is calculated in nominal-π model will be slightly high when compare to nominal-T so regulation in nominal-π is slightly high when compare to nominal-T.

Ferranti Effect

When receiving end of the transmission line is operating under no load condition or lightly load condition, sending end voltage V_s is less than receiving end voltage V_r.

Note:

☑ Ferranti effect is generally experienced in medium and long transmission line operating under no load condition or lightly load condition.

Long Transmission Line

ABCD parameter in matrix form

$$\begin{bmatrix} V_s \\ I_s \end{bmatrix} = \begin{bmatrix} \cosh\gamma l & Z_c \sinh\gamma l \\ \frac{1}{Z_c}\sinh\gamma l & \cosh\gamma l \end{bmatrix} \begin{bmatrix} V_r \\ I_r \end{bmatrix}$$

Propogation constant, $\boxed{\gamma = \sqrt{ZY}}$

also, $\boxed{\gamma = \alpha + j\beta}$ and for loss less line $\boxed{\beta = \omega\sqrt{LC}}$

where, $l \rightarrow$ Length of transmission line ;

$\alpha \rightarrow$ Attenuation constant in Neper/sec.

$\beta \rightarrow$ Phase constant in rad/km

$$\boxed{\sin\gamma l = \sqrt{YZ}\left[1 + \frac{YZ}{3!} + \frac{(YZ)^2}{5!} + \cdots\right]} \; ; \; \boxed{\cos\gamma l = 1 + \frac{YZ}{2!} + \frac{(YZ)^2}{4!} + \cdots}$$

Wavelength (λ)

The distance corresponding to which there is a phase changes of 2π or 360°.

$$\boxed{\lambda = \frac{2\pi}{\beta}}$$

Velocity of wave Propagation

$$\boxed{v_c = f\lambda}$$

where, $f \rightarrow$ Frequency ; $\lambda \rightarrow$ Wavelength

Equivalent π Circuit and ABCD Parameters

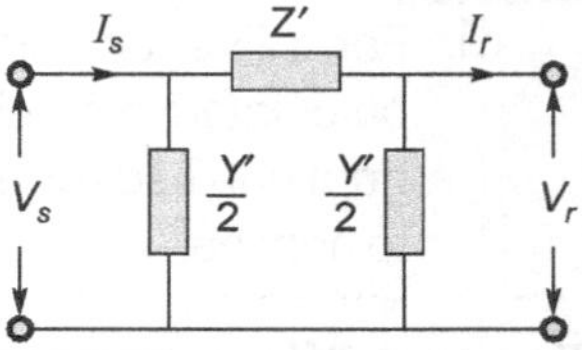

$$\begin{bmatrix} V_s \\ I_s \end{bmatrix} = \begin{bmatrix} 1+\dfrac{Y'Z'}{2} & Z \\ Y'\left(1+\dfrac{Y'Z'}{4}\right) & 1+\dfrac{Y'Z'}{2} \end{bmatrix} \begin{bmatrix} V_r \\ I_r \end{bmatrix}$$

where,

$$Z' = Z_c(\sinh\gamma l) = Z\cdot\frac{(\sinh\gamma l)}{\gamma l}$$

$$Y' = \frac{2}{Z_c}\tanh\left(\frac{\gamma l}{2}\right) = Y\frac{\tanh(\gamma l/2)}{(\gamma l/2)}$$

Equivalent T Circuit and ABCD Parameter

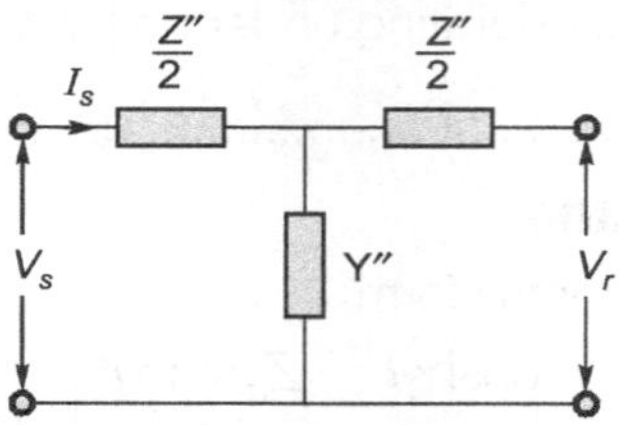

$$\begin{bmatrix} V_s \\ I_s \end{bmatrix} = \begin{bmatrix} 1+\dfrac{Y''Z''}{2} & Z''\left(1+\dfrac{Z''Y''}{4}\right) \\ Y'' & 1+\dfrac{Y''Z''}{2} \end{bmatrix} \begin{bmatrix} V_r \\ I_r \end{bmatrix}$$

where,

$$\frac{Z''}{2} = Z_c\left(\frac{\cosh\gamma l - 1}{\sinh\gamma l}\right) = \frac{Z}{2}\frac{\tanh(\gamma l/2)}{(\gamma l/2)}$$

$$Y'' = \frac{1}{Z_c}\sinh\gamma l = Y\frac{\sin\gamma l}{\gamma l}$$

Note:

☑ The equivalence is valid for only one frequency and only for the terminal conditions.

Power of Transmission Lines

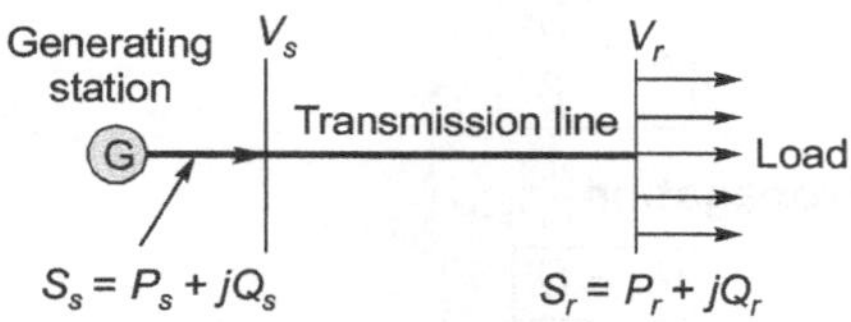

(A two bus power system)

Let $V_s = |V_s|\angle\delta, \quad V_r = |V_r|\angle 0$

$D = A = |A|\angle\alpha, \ B = |B|\angle\beta$

Parameters	Formula
Complex power per phase at the receiving end	$S_R = \frac{\lvert V_s\rvert\lvert V_r\rvert}{\lvert B\rvert}\angle(\beta-\delta) - \frac{\lvert A\rvert\lvert V_r\rvert^2}{\lvert B\rvert}\angle(\beta-\alpha)$
Active power at the receiving end of a transmission line	$P_R = \frac{\lvert V_S\rvert\lvert V_R\rvert}{\lvert B\rvert}\cos(\beta-\delta) - \frac{\lvert A\rvert\lvert V_R\rvert^2}{\lvert B\rvert}\cos(\beta-\alpha)$
Reactive power at the receiving end of a transmission line	$Q_R = \frac{\lvert V_S\rvert\lvert V_R\rvert}{\lvert B\rvert}\sin(\beta-\delta) - \frac{\lvert A\rvert\lvert V_R\rvert^2}{\lvert B\rvert}\sin(\beta-\alpha)$
Active power at the sending end of a transmission line	$P_S = \frac{\lvert D\rvert\lvert V_S\rvert^2}{\lvert B\rvert}\cos(\beta-\alpha) - \frac{\lvert V_S\rvert\lvert V_R\rvert}{\lvert B\rvert}\cos(\beta+\delta)$
Reactive power at the sending end of a transmission line	$Q_S = \frac{\lvert D\rvert\lvert V_S\rvert^2}{\lvert B\rvert}\sin(\beta-\alpha) - \frac{\lvert V_S\rvert\lvert V_R\rvert}{\lvert B\rvert}\sin(\beta+\delta)$

Note:

☑ Maximum power at receiving end condition : $\beta = \delta$

Surge Impedance

- The impedance of transmission line with losses, is known as characteristic impedance (Z_c).
- The impedance of transmission line without losses is known as surge impedance (Z_s) or natural impedance (Z_n)

$$Z_c = \sqrt{\frac{Z}{Y}} = \sqrt{\frac{R + j\omega L}{G + j\omega C}} \quad ; \quad Z_c = \sqrt{\frac{B}{C}} = \sqrt{Z_{O.C.}\, Z_{S.C.}}$$

where, Y, $Z \rightarrow$ Shunt admittance and series impedance per unit length respectively ; Z_{OC} = Open circuit impedance ;

$Z_{SC} \rightarrow$ Short circuit impedance

For lossless transmission line $R = 0$, $G = 0$

$$Z_s = \sqrt{\frac{L}{C}}$$

Note:

☑ Surge impedance does not depend upon length of the line.

Remember:

☑ $Z_c = 400\ \Omega$ for overhead transmission line

☑ $Z_c = 40\ \Omega$ for cable

Flat Line (or) Infinite Line

- A lossless transmission line terminated with surge impedance at the receiving end is known as infinite or flat line.

Surge Impedance Loading (SIL)

- The power delivered by the line to a purely resistive load equal to its surge impedance.
- The load at which the inductive and capacitive reactive voltamperes are equal and opposite, is called surge impedance loading of the line.

$$SIL = \frac{V_s V_R}{Z_s}$$

Note:

- SIL is independent of the distance and depends on the voltage.
- SIL is always less than the rated capacity of a line.
- If load on line is < SIL then power factor will be leading.
- If load on line is > SIL then power factor will be lagging.
- If load on line is = SIL then power factor will be unity.

Transmission Line Connection

Transmission Line Connected Either in Series or Cascaded

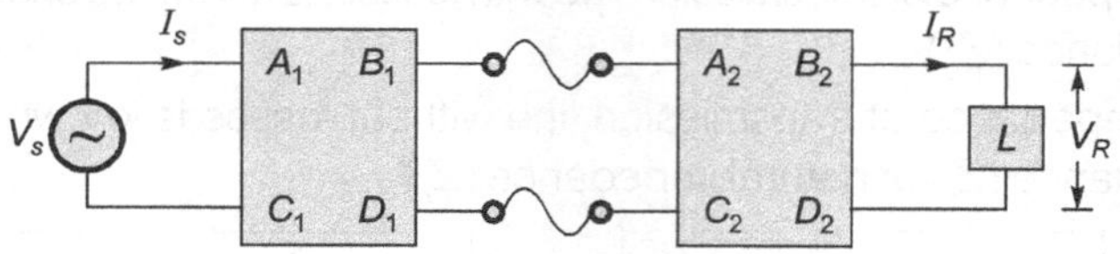

$$\begin{bmatrix} V_s \\ I_s \end{bmatrix} = \begin{bmatrix} A_1 & B_1 \\ C_1 & D_1 \end{bmatrix} \begin{bmatrix} A_2 & B_2 \\ C_2 & D_2 \end{bmatrix} \begin{bmatrix} V_R \\ I_R \end{bmatrix}$$

$$\begin{bmatrix} V_s \\ I_s \end{bmatrix} = \begin{bmatrix} A_1 A_2 + B_1 C_2 & A_1 B_2 + B_1 D_2 \\ C_1 A_2 + D_1 C_2 & C_1 B_2 + D_1 D_2 \end{bmatrix} \begin{bmatrix} V_R \\ I_R \end{bmatrix}$$

Transmission Line Connected in Parallel

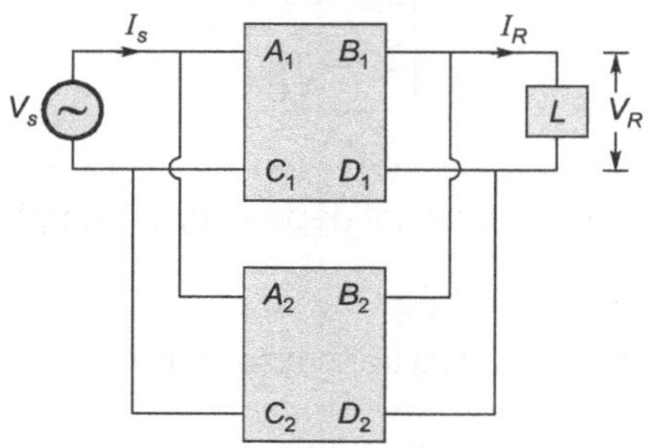

$$A = \frac{A_1 B_2 + A_2 B_1}{B_1 + B_2} \quad ; \quad B = \frac{B_1 B_2}{B_1 + B_2}$$

$$C = (C_1 + C_2) + \frac{(D_1 - D_2)}{B_1 + B_2} \quad ; \quad D = \frac{D_1 B_2 + D_2 B_1}{B_1 + B_2}$$

Reflection and Refraction of Waves

Line Terminated by an Impedance (Z)

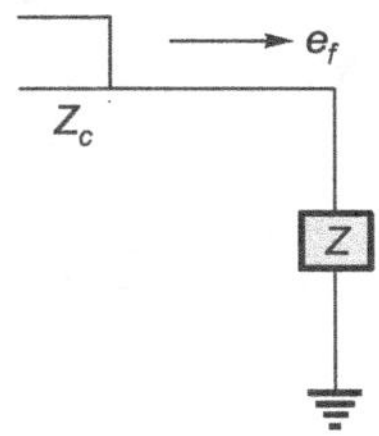

$$e_t = \frac{2Z}{Z + Z_c} \cdot e_f$$

$$e_r = \frac{Z - Z_c}{Z + Z_c} \cdot e_f$$

where, $e_f \rightarrow$ Incident or forward voltage wave
$e_t \rightarrow$ Refracted or transmitted voltage wave
$e_r \rightarrow$ Reflected voltage

$\frac{2Z}{Z + Z_c} \rightarrow$ Refraction coefficient of a line with surge impedance Z_c terminated by an impedance Z.

$\frac{Z - Z_c}{Z + Z_c} \rightarrow$ Reflection coefficient.

- Transmitted current, $i_t = \frac{2}{Z + Z_c} \cdot e_f$
- Reflected current, $i_r = \frac{Z_c - Z}{Z_c + Z} \cdot i_f$

Conditions

- If line is terminated by a resistance, equal, to the surge impedance.
  then, $e_t = e_f$, $i_t = i_f$, $e_r = 0$ and $i_r = 0$
  The incident wave continues as it is and there is no reflection.
- If line is open circuited line
  then, $e_t = 2e_f$, $i_t = 0$, $e_r = e_f$, $i_r = -i_f$
  An open circuit at the end of a line demands, that the current at that point be zero at all times.
- If line is short circuited line
  then, $e_t = 0$, $i_t = 2i_f$, $e_r = -e_f$, $i_r = i_f$

Equivalent Circuit of Travelling Waves

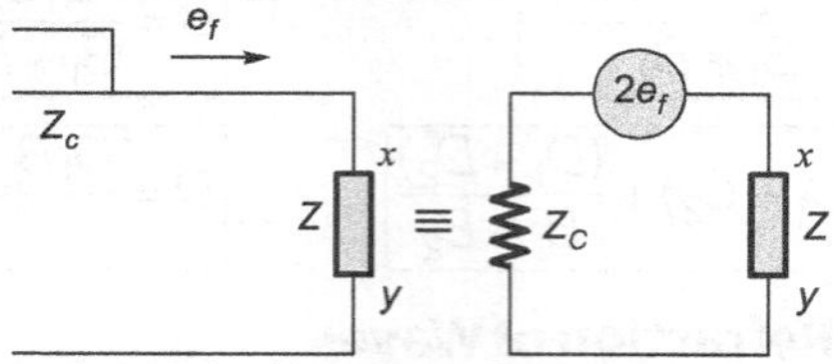

Actual circuit Equivalent circuit

Forked line

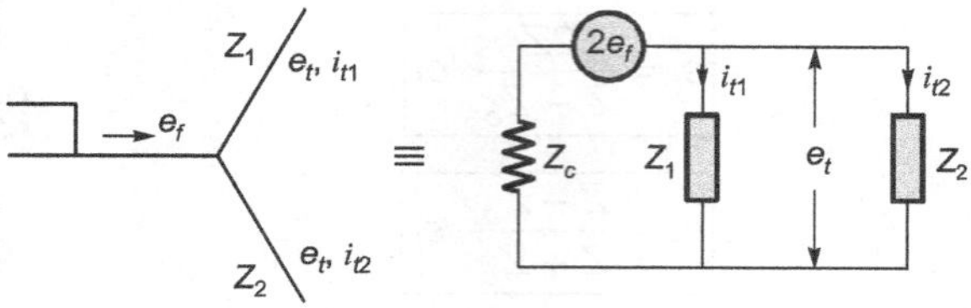

- Refracted or Transmitted voltage, $e_t = \dfrac{2e_f}{Z_c + \dfrac{Z_1 Z_2}{Z_1 + Z_2}}\left(\dfrac{Z_1 Z_2}{Z_1 + Z_2}\right)$
- Refracted or Transmitted currents, $i_{t_1} = \dfrac{e_t}{Z_1}$ and $i_{t_2} = \dfrac{e_t}{Z_2}$

Line terminated by inductance

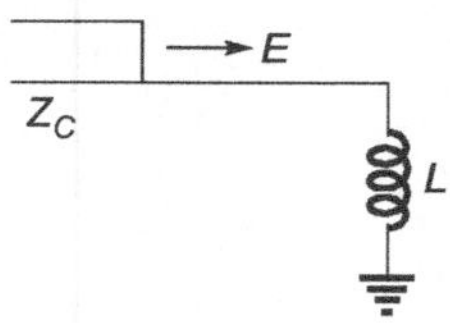

Transmitted current	$i_t = \dfrac{2E}{Z_c}\left[1 - \exp\left(-\dfrac{Z_c t}{L}\right)\right]$ where, $E \rightarrow$ A step wave travelling on a line of surge impedance Z_c
Transmitted voltage	$e_t = 2E \cdot \exp\left(\dfrac{-Z_c t}{L}\right)$
Reflected voltage	$e_r = E\left[2\exp\left(-\dfrac{Z_c t}{L}\right) - 1\right]$
Reflected current	$i_r = \dfrac{E}{Z_c}\left[1 - 2\exp\left(-\dfrac{Z_c t}{L}\right)\right]$

Line Terminated by Capacitance

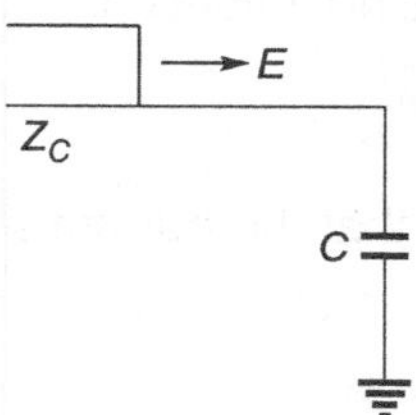

Transmitted current	$i_t = \dfrac{2E}{Z_c}\exp\left(-\dfrac{t}{CZ_c}\right)$
Transmitted voltage	$e_t = 2E\left[1 - \exp\left(-\dfrac{t}{CZ_c}\right)\right]$
Reflected voltage	$e_r = E\left[1 - 2 \cdot \exp\left(-\dfrac{t}{CZ_c}\right)\right]$
Reflected current	$i_r = \dfrac{E}{Z_c}\left[2 \cdot \exp\left(-\dfrac{t}{CZ_c}\right) - 1\right]$

IV Concept of Corona

Corona

- The complete disruption in dielectric strength or insulation of insulating material (air) near the surface of the conductor at certain point.
- Corona occurs only when the electric field intensity is greater than dielectric strength of air.

  In general corona occurs around the power conductor due to two reasons:
  - Electrical power transmission is at higher operating voltages.
  - Number of free electrons in the space surrounding the power conductor are more due to radio activity.

Critical Disruptive Voltage (V_d)

- The voltage at which self field intensified localised ionization of air takes place.

Disruptive Critical Voltage

For 1-ϕ line :

$$V_d = \frac{6\times 10^6}{\sqrt{2}} r\delta m_0 \ln\frac{D}{r} \text{ Volts}$$

where, $D \to$ Spacing between conductors, in metres.
$r \to$ Radius of conductor, in metres.
$V_d \to$ Critical disruptive voltage, in volts

For 3-ϕ line :

$$V_d = \frac{3\times 10^6}{\sqrt{2}} r\delta m_0 \ln\left(\frac{D_{eq}}{r}\right) \text{ Volts}$$

where, $D_{eq} \to$ Equivalent spacing between conductor, in metres.

Visual Critical Voltage

- The voltage at which the visual corona begins.
- Visual glow of corona occurs at a voltage higher than the critical disruptive voltage.

Visual Critical Voltage

For 1-ϕ line :

$$V_v = \frac{6\times 10^6}{\sqrt{2}} r\delta m_v \left(1+\frac{0.03}{\sqrt{\delta r}}\right) \ln\frac{D}{r} \text{ Volts}$$

For 3-ϕ line :

$$V_v = \frac{3\times 10^6}{\sqrt{2}} r\delta m_v \left(1+\frac{0.03}{\sqrt{\delta r}}\right) \ln\frac{D_{eq}}{r} \text{ Volts}$$

where, $m_v \to$ Irregularity factor ; $V_v \to$ Visual critical voltage

Critical Voltage & Intensity

Relative density of air (δ)

$$\delta = \frac{p}{76}\left(\frac{273+\theta_0}{273+\theta}\right)$$

where, $p \rightarrow$ Barometric pressure, cm ; $\theta \rightarrow$ Temperature, °C

Assume relative density at θ_0 °C and 76 cm of Hg as 1

Critical Intensity

$$E_0 = \left(\frac{3\times10^6}{\sqrt{2}}\right)\delta\cdot m_0 \text{ V/m}$$

where, $E_0 \rightarrow$ Rms value of critical intensity

$\delta \rightarrow$ Relative density of air ; $m_0 \rightarrow$ Roughness factor

Note:

- ☑ For smooth conductors of large diameter in air at normal temperature and pressure corona begins at peak value of critical field intensity 30 kV/cm.
- ☑ m_0 = 1.0 for smooth conductors and 0.93 to 0.98 for rough conductors exposed to atmosphere.

Visual Critical Intensity

$$E_v = \frac{3\times10^6}{\sqrt{2}}\cdot\delta m_v\left(1+\frac{0.03}{\sqrt{\delta r}}\right) \text{ V/m}$$ where, $m_v \rightarrow$ Irregularity Factor

Corona Loss

Peek's Formula (under fair weather conditions)

$$P_c = 243.5\frac{(f+25)}{\delta}\sqrt{\frac{r}{D}}\cdot(V-V_d)^2\times10^{-5} \text{ kW/km/phase}$$

where, $V \rightarrow$ Phase voltage, in kV (rms value)

$V_d \rightarrow$ Disruptive critical voltage in kV(rms value)

$r \rightarrow$ Radius of conductor, in metres ; $f \rightarrow$ System frequency

Note:

- ☑ Under stormy weather conditions use Vd as 80% of its fair weather value.
- ☑ $P_c \propto (f+25)$, i.e., corona loss increases with the increases in supply frequency.

Disadvantages of Corona

- There is certain real power loss in the form of corona loss.
- Corona causes an interference with neighbouring communication line.

Advantage of Corona

- The corona acts as a safety wall to the transmission line conductors against lightening stroke (peak magnitude of lightning surge will dissipate in the form of corona loss).

Method to Reduce Corona

- By the use large size of conductor.
- Use bundle conductor.
- Use hallow conductor.

Remember:

☑ In 400/220 kV substation bus bar are in the form of hallow tube.

☑ DC corona loss $= \frac{1}{3} AC$ corona loss (for same line voltage).

☑ Breakdown strength of air at NTP = 21.1 kV/cm/rms

V Mechanical Design of Overhead Lines

SAG and Tension

SAG

- The vertical distance between the conductor at the mid point and the line joining the two adjacent level support.

Catenary

- A line conductor of uniform cross-section and material, perfectly flexible but stretched inelastic between 2 support hanging freely under its own weight is represented by a curve known as catenary.

Sag Calculation

Supports at Same Level

Sag, $S = \frac{Wl^2}{8T}$ m

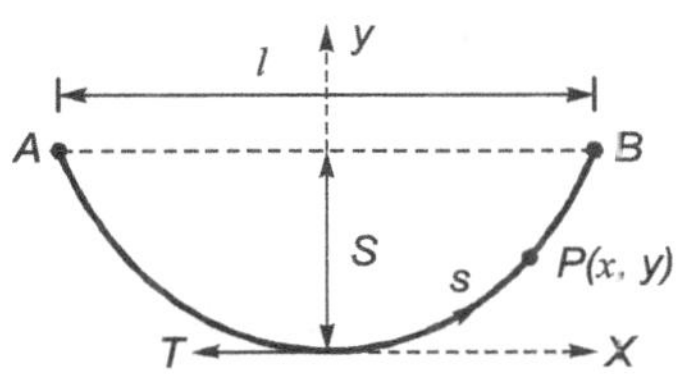

where, $l \rightarrow$ Length of span, metres
$S \rightarrow$ Sag at mid span, metres

$T \rightarrow$ Conductor tension (assumed constant over the whole span), N

$W \rightarrow$ Conductor weight, N/m

Spacing Between Conductors (without sparking)

$$\text{Spacing} = \sqrt{S} + \frac{V}{150} \text{ metres}$$

where, $S \rightarrow$ sag in metres; $V \rightarrow$ line voltage in kV

Effect of Ice and Wind

Weight of Ice Per Metre Length of Conductor

$$W_i = 2.8 \times 10^4 t(d+t) \text{ N/m}$$

where, d = diameter of conductor, metres

t = radial thickness of ice, metres

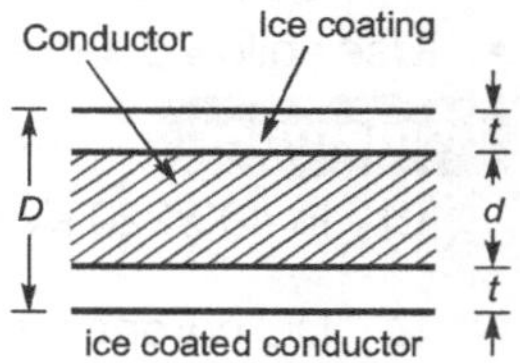

Wind load

$$F_W = P \times D \text{ N/m}$$

where, P = Wind pressure, Newton per square metre of projected area.

Total force acting on conductor per metre length

$$\boxed{F_t = \sqrt{(W+W_i)^2 + F_W^2} \text{ N/m}}$$

Sag under worst condition

$$S = \frac{F_t\, l^2}{8T} \text{ in new plane}$$

where, $F_t \rightarrow$ Total force per meter ; $T \rightarrow$ Limiting tension

Vertical sag

$$\boxed{\text{Vertical sag} = S \cos\gamma}$$

$$\because \quad \boxed{\tan\gamma = \frac{F_W}{W+W_i}}$$

Total length of conductor

$$Z = l\left(\frac{F_t^2 l^2}{24T^2}\right)$$

where, $Z \rightarrow$ Total length of the conductor

$F_t \rightarrow$ Total force acting on conductor per metre length

3. Supports at different levels

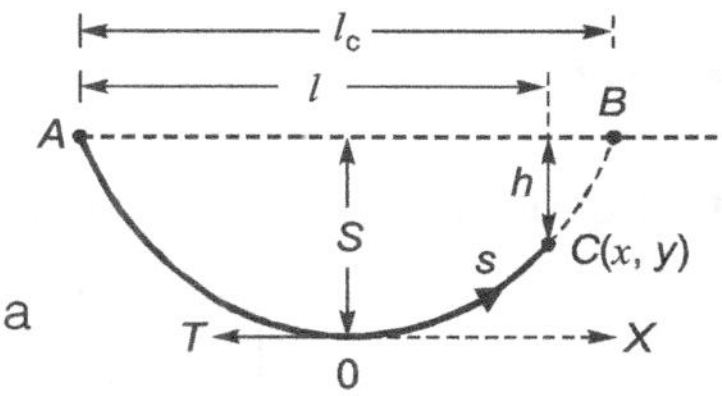

$$S = \frac{Wl_c^2}{8T} \quad \text{and} \quad l_c = l + \frac{2Th}{Wl}$$

where, l_c = Span of complete parabola

Remember:

☑ The formulas are also valid if two supports A and C fall on the same side of origin $\left(\text{i.e. if } l < \frac{l_c}{2}\right)$.

VI Balanced and Unbalanced Faults

Per Unit System

- The ratio of the quantity (in some unit) to its base value (in same unit).

$$\text{Per unit value} = \frac{\text{Actual value (in some units)}}{\text{Base or reference value (in same units)}}$$

Note:

☑ In transformer primary side p.u. reactance is always equal to secondary side p.u. reactance.

Remember:

- Out of 4-system quantities (kVA, kV, current and impedance) only two are independent. It is convenient to select the base value of kVA and kV and calculate the base values of other two.
- Ratio of 2-base unit give another base quantity

$$\frac{V_{base}}{I_{base}} = Z_{\text{base}}$$

- In transformer primary side p.u. reactance is always equal to secondary side p.u. reactance.

- Base current = $\dfrac{\text{Base kVA}}{\sqrt{3} \times \text{Base kV}}$ A (in 3-ϕ system)

- Base impedance = $\dfrac{\text{Line to neutral value of base voltage}}{\text{Base current}}$

$$\text{Base impedance} = \frac{(\text{Base kV})^2}{\text{Base MVA}}\ \Omega$$

- Per unit impedance, $Z_{pu} = \dfrac{Z_{\text{actual}}}{Z_{\text{base}}}$ or $X_{pu} = \dfrac{X_{\text{actual}}}{Z_{\text{base}}}$
- Per unit impedance referred to new base

$$= \begin{bmatrix}\text{Per unit impedance} \\ \text{referred to old base}\end{bmatrix}\left[\frac{\text{Base kV}_{\text{old}}}{\text{Base kV}_{\text{new}}}\right]^2\left[\frac{\text{Base kVA}_{\text{new}}}{\text{Base kVA}_{\text{old}}}\right]$$

- Short circuit kVA = Rated or base kVA $\times \dfrac{100}{\%Z}$

$$EI_{sc} = EI \times \frac{100}{\%Z}$$

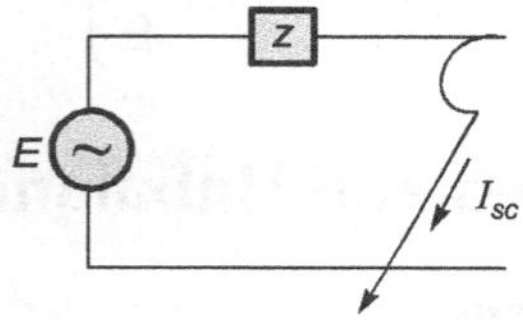

where, $E \rightarrow$ Rated voltage ; $I \rightarrow$ Rated current; $Z \rightarrow$ Internal impedance

Symmetrical Components

- A positive sequence, set of three symmetrical voltages (i.e. all numerically equal and all displaced from each other by 120°) having the same sequence '*abc*' as the original set and denoted by V_{a1}, V_{b1} and V_{c1}.
- A negative sequence, set of three symmetrical voltages having the phase sequence opposite to that of the original set and denoted by V_{a2}, V_{b2} and V_{c2}.
- A zero sequence, set of three voltages, all equal in magnitude and in phase with each other and denoted by V_{a0}, V_{b0} and V_{c0}.

$$V_a = V_{a1} + V_{a2} + V_{a0}$$
$$V_b = V_{b1} + V_{b2} + V_{b0}$$
$$V_c = V_{c1} + V_{c2} + V_{c0}$$

Sequence Component Vectors

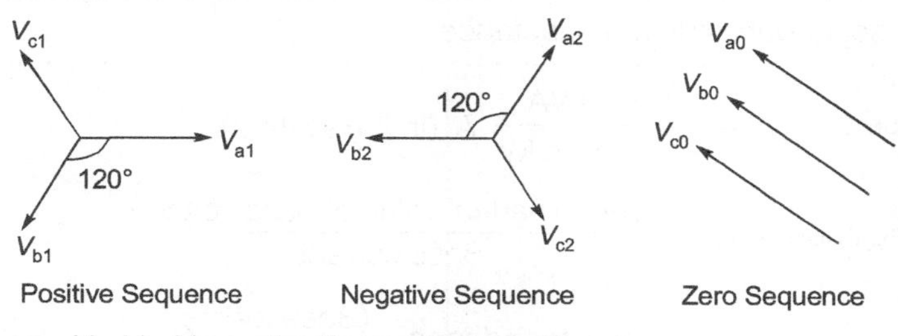

where, V_a, V_b, V_c are the three-phase voltages

$$V_{b1} = \alpha^2 V_{a1} \qquad V_{c1} = \alpha V_{a1}$$
$$V_{b2} = \alpha V_{a2} \qquad V_{c2} = \alpha^2 V_{a2}$$
$$V_{b0} = V_{a0} \qquad V_{c0} = V_{a0}$$

where, Operator $\alpha = 1\angle 120°$, $\alpha^2 = 1\angle -120°$

Phase voltages in matrix form

$$\begin{bmatrix} V_a \\ V_b \\ V_c \end{bmatrix} = \begin{bmatrix} 1 & 1 & 1 \\ 1 & \alpha^2 & \alpha \\ 1 & \alpha & \alpha^2 \end{bmatrix} \begin{bmatrix} V_{a_0} \\ V_{a_1} \\ V_{a_2} \end{bmatrix}$$

$$[V_{abc}] = [A][V_{012}] \qquad [V_{012}] = [A]^{-1}[V_{abc}]$$

where, $A = \begin{bmatrix} 1 & 1 & 1 \\ 1 & \alpha^2 & \alpha \\ 1 & \alpha & \alpha^2 \end{bmatrix}$ and $A^{-1} = \frac{1}{3}\begin{bmatrix} 1 & 1 & 1 \\ 1 & \alpha & \alpha^2 \\ 1 & \alpha^2 & \alpha \end{bmatrix}$

Complex Power

Total complex power in a 3-ϕ circuit

$$S = [V_{abc}]^T [I_{abc}]^* = 3[V_{012}]^T [I_{012}]^*$$

where, $[V_{012}] \rightarrow$ Symmetrical component matrix

$[V_{abc}] \rightarrow$ Unsymmetrical phasor matrix

$$\boxed{[Z_{012}] = [A]^{-1}[Z_{abc}][A]}$$

Remember:

☑ For almost all power system components the matrix $[Z_{abc}]$ is not diagonal but possess certain symmetries. These symmetries are such that $[Z_{012}]$ is diagonal either exactly or approximately.

Sequence Impedances

Sequence Impedances of Star Connected Loads

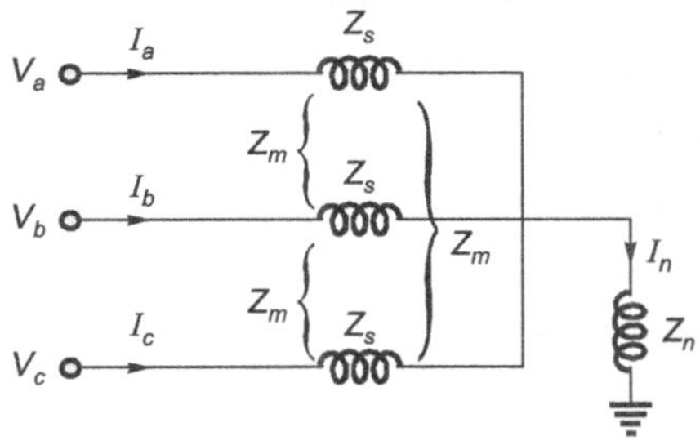

The load has self and mutual impedances Z_s and Z_m respectively and star point is grounded through impedance Z_n.

$$[z_{abc}] = \begin{bmatrix} Z_s + z_n & Z_m + Z_n & Z_m + Z_n \\ Z_m + Z_n & Z_s + Z_n & Z_m + Z_n \\ Z_m + Z_n & Z_m + Z_n & Z_s + Z_n \end{bmatrix}$$

$$[Z_{012}] = \begin{bmatrix} Z_s + 2\,Z_m + 3Z_n & 0 & 0 \\ 0 & Z_s - Z_m & 0 \\ 0 & 0 & Z_s - Z_m \end{bmatrix}$$

Remember:

- For a balanced load, the three sequences are independent. The current of one sequence produces voltage of only that sequence.
- Current flowing through Z_n is $3I_{a0}$. Thus in all zero sequence networks the neutral impedance Z_n appears as $3Z_n$.
- Positive and negative sequence impedance are equal, i.e., $Z_1 = Z_2 = j(X_s - X_m)$ for $Z_0 = j(X_s + 2X_m)$, for full transposed line.
- The positive and negative sequence impedances of linear and symmetrical static circuits do not depends on the phase sequence and are, therefore, equal.

Sequence Impedances of Transformers

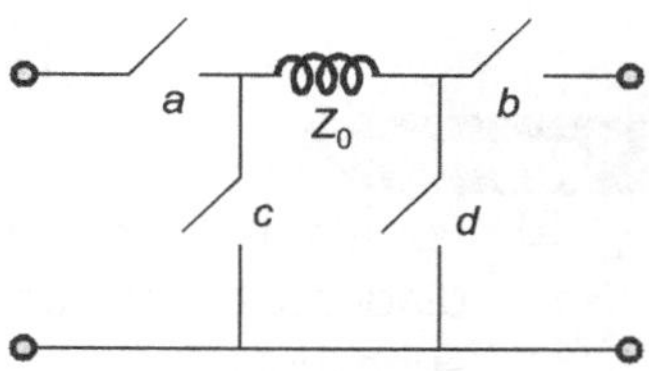

Z_0 = Zero sequence impedance of transformer.

⇒ Switch '*a*' is closed when primary winding is star connected with neutral grounded.

⇒ Switch '*c*' is closed when primary winding is Δ connected.

⇒ Keep both '*a*' and '*c*' open when primary winding is star connected with neutral isolated (not grounded).

⇒ Switch '*b*' is closed when secondary winding is star connect with neutral grounded.

⇒ Switch '*d*' is closed when secondary winding is Δ connected.

⇒ Keep both '*b*' and '*d*' open when secondary winding is star connected with neutral isolated.

Remember:

- Only positive sequence network contains voltage.
- For a synchronous generator, $Z_1 > Z_2 > Z_0$ (sequence impedances)
- For transmission line, cables, transformer, static loads, positive and negative sequence are the equal.

Unsymmetrical Faults

Single line to ground fault (SLG fault)

Terminal conditions (at the fault point)

$V_a = I_a Z_f,\ I_b = I_c = 0$

$$I_{a0} = I_{a1} = I_{a2} = \frac{I_a}{3}$$

where, I_{a0}, I_{a1}, I_{a2} are the sequence currents at the fault point.

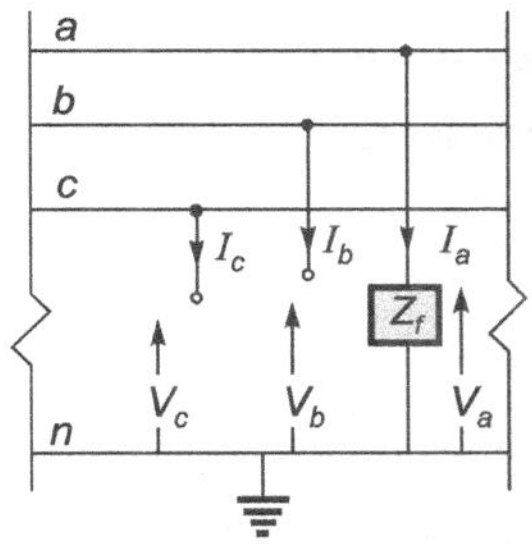

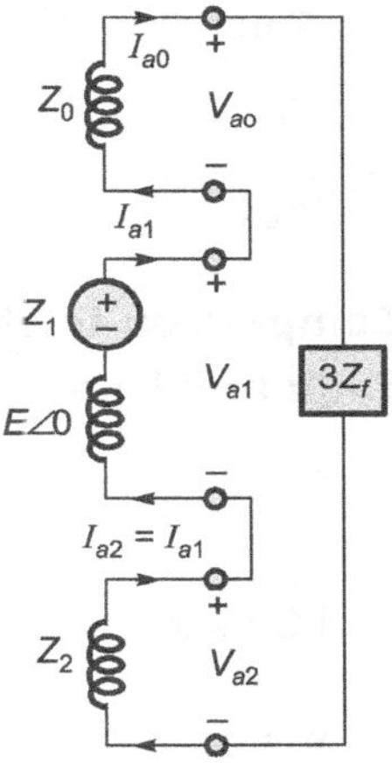

Fault current

$$I_f = I_a = I_{a0} + I_{a1} + I_{a2} = \frac{3E}{Z_1 + Z_2 + Z_0 + 3Z_f}\ ;\ I_{fLG} = 3I_{a0}$$

∵ Z_f: Fault impedance

Note:

☑ For solidly grounded, $Z_f = 0$.

Line to line fault (LL fault)

Terminal conditions

$I_a = 0,\ I_b = -I_c$

$V_b = V_c + Z_f I_B$

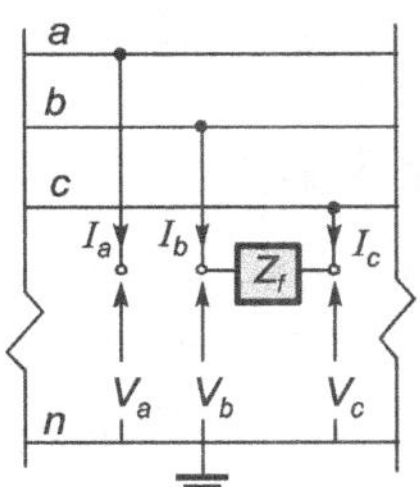

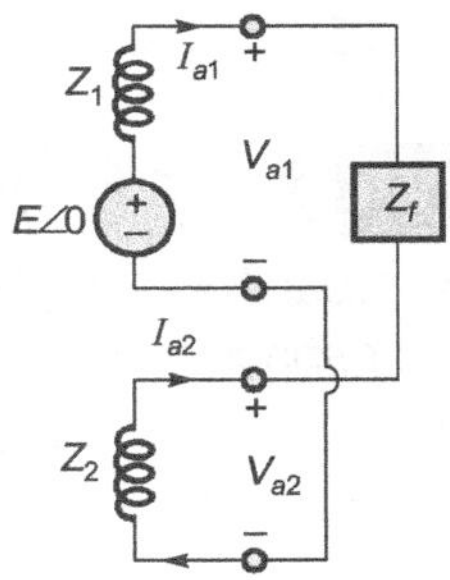

Fault current

$$I_f = I_b = -I_c = \frac{-j\sqrt{3}E}{Z_1 + Z_2 + Z_f}$$

$$I_{f_{LL}} = \sqrt{3}I_{a_1} = \sqrt{3}I_{a_2}$$

Double line to ground fault (LLG fault)

Terminal conditions

$I_a = 0,\ V_b = V_c = (I_b + I_c)Z_f$

$$I_{a1} = \frac{E}{Z_1 + \dfrac{Z_2(Z_0 + 3Z_f)}{Z_2 + Z_0 + 3Z_f}}$$

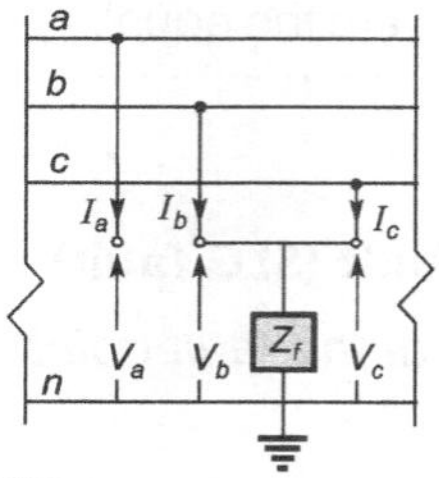

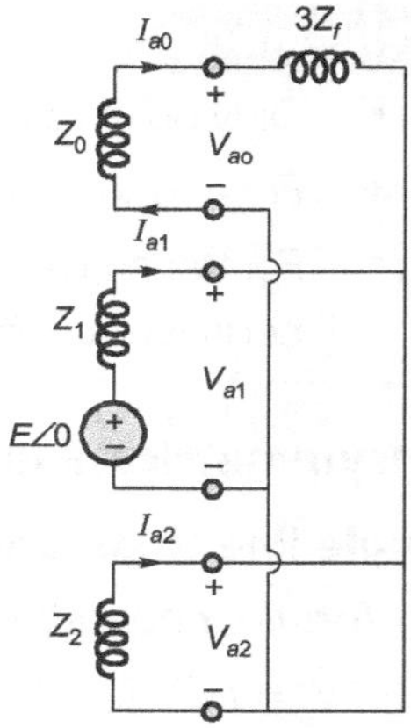

Fault Current

$I_f = I_b + I_c$

$I_{f_{LLG}} = 3I_{a_0}$

Comparison to Three-Phase Fault and Line-to-Ground

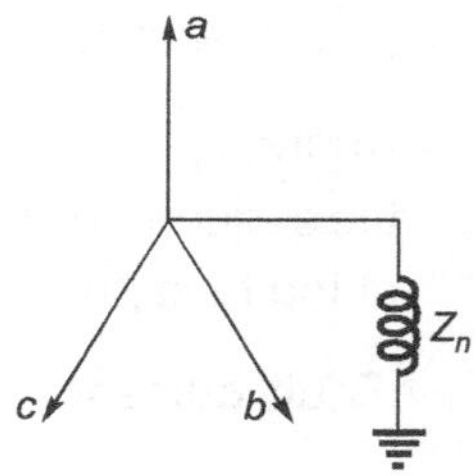

3ϕ, fault, $I_{f,3\phi} = \dfrac{E_a}{Z_1}$

LG fault, $I_{f,\text{LG}} = \dfrac{3E_a}{Z_1 + Z_2 + Z_0 + 3Z_n}$

where, E_a = Generator voltage

If $Z_n < \frac{1}{3}(Z_1 - Z_0)$ then $I_{f,\text{LG}} > I_{f,3\phi}$ $(Z_1 \approx Z_2)$

If $Z_n = \frac{1}{3}(Z_1 - Z_0)$ then $I_{f,\text{LG}} = I_{f,3\phi}$

If $Z_n > \frac{1}{3}(Z_1 - Z_0)$ then $I_{f,\text{LG}} < I_{f,3\phi}$

Note:

☑ For solidly grounded, $(Z_n = 0)$, $I_{fLG} \approx 1.5 I_{f,3\phi}$

VII Load Flow Study

- Load flow study help us to determine of the best location as well as optimal capacity of the proposed generating stations, substation and new lines.
- $I_{\text{BUS}} = Y_{\text{BUS}} \times V_{\text{BUS}}$ (Y_{BUS} = Bus admittance matrix)

where,
$$Y_{\text{BUS}} = \begin{bmatrix} Y_{11} & Y_{12} \ldots\ldots Y_{1n} \\ Y_{21} & Y_{22} \ldots\ldots Y_{2n} \\ \vdots & \vdots \\ Y_{n1} & Y_{n2} \ldots\ldots Y_{nn} \end{bmatrix}_{n \times n}$$

Assuming number of buses, $n = 4$, we have

$$Y_{11} = y_{11} + y_{12} + y_{13} + y_{14}$$
$$= \text{Total admittance connected to bus-1}$$
$$Y_{12} = -y_{12} = \text{Negative value of series admittance connected between bus-1 and bus-2}$$
$$Y_{13} = -y_{13} = \text{Negative value of series admittance connected between bus-1 and bus3}$$
$$Y_{14} = -y_{14} = \text{Negative value of series admittance connected between bus-1 and bus-4}$$

Similarly, other elements of Y_{BUS} are found out.

- Static load flow equations (SLFE) can be written as

$$P_i = \text{Real}\{S_i^*\} = |V_i| \sum_{k=1}^{n} |V_k||Y_{ik}|\cos(\theta_{ik} + \delta_k - \delta_i) \quad (i = 1, 2, \ldots, n)$$

and $$Q_i = -I_{\text{mag}}\{S_i^*\} = -|V_i| \sum_{k=1}^{n} |V_k||Y_{ik}|\sin(\theta_{ik} + \delta_k - \delta_i)$$

- Voltage at an ith bus is given by

$$V_i = \frac{1}{Y_{ii}}\left[\frac{P_i - jQ_i}{V_i^*} - \sum_{\substack{k=1 \\ k \neq i}}^{n} (Y_{ik} V_k)\right] \quad (\text{where, } i = 1, 2, \ldots, n)$$

Properties of Y_{Bus} Matrix

- Y_{Bus} is a sparsity matrix and also a symmetric matrix.
- Diagonal element of Y_{Bus} gives the self-admittances.
- OFF-diagonal elements of Y_{Bus} gives transfer admittances.
- Diagonal element = Total admittance connected to respective bus.
- Transfer element = Negative value of series admittance connected between two bus.
- The inverse of Y_{Bus} matrix is a full matrix, i.e., Z_{Bus}.

Properties of Z_{Bus} Matrix

- Z_{Bus} matrix is generally used for fault studies.
- The diagonal element are short circuit driving point impedance.
- OFF diagonal elements are short circuit transfer impedance.

Note:

☑ When a shunt element is connected to the existing system then only diagonals at Y_{Bus} will be affected.

Comparison between Gauss-Siedel and Newton-Raphson Method

Gauss Seidal Method	Newton Raphson Method
1. Load-flow equation are non-linear in rectangular coordinate.	1. Load-flow equation are non-linear in polar coordinate.
2. It has linear convergence characteristics.	2. It has quadratic convergence characteristics.
3. Less memory required.	3. More memory required.
4. Computation time per itera-tion is less.	4. Computation time per itera-tion is more.
5. Require large number of iteration.	5. Require less number of iteration.
6. Number of iteration is directly proportional to number of buses.	6. Independent upon bus.
7. As the selection of slack bus is changed, the convergent criteria also change.	7. Independent upon selection of slack bus.
8. Unreliable convergent, less accurate and used for smaller system.	8. Reliable convergent, more accurate, it can be use for large power system.

Remember:

- The accuracy of any iterative technique depends upon the tolerance (t) specified, it does not depends on the format of improvement and hence with same tolerance both NR and FDLF will give same result.
- Due to assumptions used in FDLF method, fast convergence characteristics of NR-method is lost in FDLF method, i.e., with same tolerance, FDLF method will give same result as obtained with NR-method, but FDLF method will use less time per iteration, more number of iterations and hence more overall time as compared to NR-method.

Note:

- In special sitatuions where extremly fast although approximate result is required, then FDLF method with fixed number of iterations will be used.
- FDLF method gives approximate result because it is used with fixed number of iterations; and NOT because it uses several assumptions are used in this method.

VIII Power System Stability

- The ability of alternators to maintain synchronism after the disturbance, which are working parallel.

- Power system stability problem is appearing in power system due to large variation in inertia of electrical machine.

Stability Limit

- The maximum amount of power that can be transfer to the point of disturbance, to maintain the stability is known as stability limit.

Note:
- Steady state stability limit is always more than transient stability limit.
- Transient stability limit can be improved maximum upto steady state stability limit.
- A system with high steady state stability limit is not guarantee for high transient stability limit, however a system with high transient stability limit is guarantee for high steady state stability limit.

Generator Power Output

Complex Power Output of Generator

$$S = \frac{|E||V|}{X_d}\sin\delta + j\left[\frac{|E||V|}{X_d}\cos\delta - \frac{|V|^2}{X_d}\right]$$

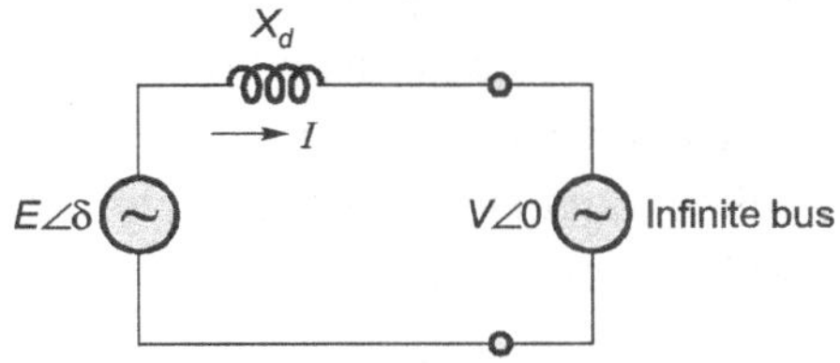

where, $X_d \rightarrow$ Direct axis synchronous reactance of synchronous machine

$E \rightarrow |E|\angle\delta$ = voltage behind direct axis synchronous reactance of generator

$V \rightarrow$ Terminal Voltage of generator

Real Power Output of Generator : $P_e = \frac{|E||V|}{X_d}\sin\delta = P_m \sin\delta$

Reactive Power Output of Generator : $Q_e = \frac{|E||V|}{X_d}\cos\delta - \frac{|V|^2}{X_d}$

Synchronising Power Coefficient : $P_r = \frac{|E||V|}{X_d}\cos\delta$

Note:

Maximum value of δ for successful operation is 90°.

Remember:

- Power can be transferred only if transfer reactance (X) is present. In case $X = 0$ then transfer of power will not be possible.

Swing Equation

- The equation describing the relative motion of the rotor (load angle δ) w.r.t. to the stator field as a function of time is called "swing equation".

$$M\frac{d^2\delta}{dt^2} = P_a = P_m - P_e \text{ ... Swing equation}$$

$$\boxed{\frac{H}{\pi f}\frac{d^2\delta}{dt^2} = (P_m - P_e);\ \text{in pu}}$$

where,
$M \rightarrow I\omega$ = Angular momentum
$P_m \rightarrow T_i\omega$ = Mechanical power input
$P_e \rightarrow T_e\omega$ = Electrical power output
$P_a \rightarrow P_m - P_e$ = Accelerating power
$T_i \rightarrow$ Mechanical input torque
$T_e \rightarrow$ Electromagnetic output torque
$I \rightarrow$ Moment of inertia

Stored Kinetic Energy of a Rotating Body

$$\boxed{K.E. = \frac{1}{2}I\omega^2 = \frac{1}{2}M\omega}$$

Inertia Constant

$$\boxed{H = \frac{\text{Kinetic energy stored}}{\text{Rating of machine}}}\ \text{MJ/MVA}$$

Angular Momentum

$$\boxed{M = \frac{GH}{\pi f}\ \frac{\text{MJ-sec}}{\text{Elec-rad}}}$$

$$\boxed{M = \frac{GH}{180°f}\ \frac{\text{MJ-sec}}{\text{Elec-degree}}}$$

For Multi-Machine System

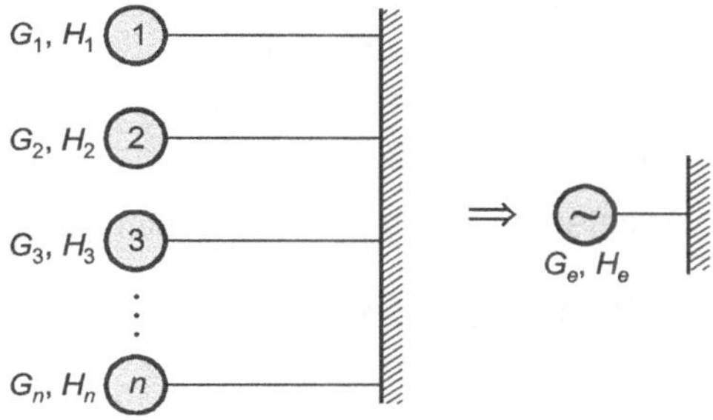

$$G_e H_e = G_1 H_1 + G_2 H_2 + G_3 H_3 + \cdots G_n H_n$$

If $\qquad G_e = G_{\text{base}}$

then, $\qquad H_{e(\text{p.u.})} = \frac{G_1 H_1}{G_{base}} + \frac{G_2 H_2}{G_{base}} + \cdots + \frac{G_n H_n}{G_{base}}$

- If n Machine Swing Together, $M_{eq} = M_1 + M_2 + M_3 + \cdots + M_n$
- If n Machine Do Not Swing Together, $M_{eq} = \frac{1}{M_1} + \frac{1}{M_2} + \frac{1}{M_3} + \cdots + \frac{1}{M_n}$

Steady State Stability Limit (P_{SSSL})

$$\boxed{P_{SSSL} = \frac{EV}{X_{eq}}}$$

Note:

☑ For better steady state stability margin a power system maintain $\delta = 30$ to $40°$.

Method to Improve Steady State Stability

- Operate system at high voltage. A 400 kV line has highest steady state power limit.
- Reduce the transfer reactance (X_{eq}) by
  - using parallel lines
  - using bundle conductor
  - using series capacitor

Transient Stability

- The ability of synchronous machine to deliver maximum power to the load, without losing synchronism for sudden and large disturbance, which is characterised as 3f S.C. fault for few cycles.

Transient Stability Evaluation Using Equal Area Criteria

- Equal area criteria is a graphical method.
- It gives absolute stability of the machine.
- It can applied only for single machine connected to infinite bus system.
- It can not use for multi machine system.

For stability : Accelerating area = Decellerating area

Method to Improve Transient Stability

- Higher inertia constant of machine
- By using high speed circuit breakers
- By using single pole switching
- By using autoreclosing units
- By using AVRs

Note:

A power system can not have steady state stability beyond $\delta = 90°$, however it can have transient stability beyond $\delta = 90°$ as long as the condition of equal area criteria is satisfy.

- **Critical Clearning Angle :**

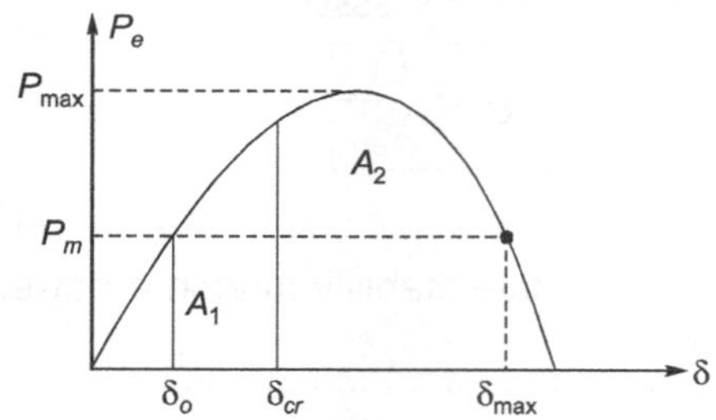

$$\delta_{cr} = \cos^{-1}[(\pi - 2\delta_o)\sin\delta_o - \cos\delta_o]$$

- **Critical Clearning Time :**

$$t_{cr} = \sqrt{\frac{2H(\delta_{cr} - \delta_o)}{\pi f P_a}}$$

where, P_a = Accelerating power

$P_a = P_m - P_e$

- **Natural Frequency of Oscillation of the Rotor**

$$\omega_n = \sqrt{\frac{(dP_e / d\delta)_o}{M}} \text{ rad/sec}$$

IX Economic Load Dispatch

Generator Constraints

- KVA loading on a generator should not exceed a prespecified value C_p because of the temperature rise condition.

$$P_p^2 + Q_p^2 \le C_p^2$$

- Maximum active power generator of a source is limited by thermal consideration and minimum power generation is limited by the flame instability of a boiler.

$$P_{pmin} \le P_p \le P_{pmax} \; ; \quad Q_{pmin} \le Q_p \le Q_{pmax}$$

where, $P_{p,min}, P_{p,max} \rightarrow$ Minimum and maximum active power generation of a source.

$Q_{p,min}, Q_{p,max} \rightarrow$ Minimum and maximum reactive power generation of a source.

Economic Dispatch

- **Optimum Load Dispatch by Neglecting Losses**

Economic dispatch problem is defined as : Min $F_T = \sum_{k=1}^{n} F_k$

Subject to $$P_D = \sum_{k=1}^{n} P_k$$

where, $F_T \rightarrow$ Total fuel input to the system
$F_K \rightarrow$ Fuel input to K^{th} unit
$P_K \rightarrow$ Generation of Kth unit
$P_D \rightarrow$ Sum of load demand at all buses

Auxiliary function, $F = F_T + \lambda\left(P_D - \sum_{k=1}^{n} P_k\right)$

Condition for optimum operation,

$$\boxed{\frac{dF_1}{dP_1} = \frac{dF_2}{dP_2} = \ldots\ldots = \frac{dF_n}{dP_n} = \lambda}$$

where, $\lambda \rightarrow$ Lagrangian multiplier

- **Optimum Load Dispatch Including Transmission Losses.**

Optimal load dispatch problem is defined as Min $F_T = \sum_{k=1}^{n} F_k$

Subject to $P_D + P_L - \sum_{k=1}^{n} P_k = 0$ where, $P_L \rightarrow$ Total loss

Auxiliary function, $F = F_T + \lambda(P_D + P_L - \Sigma P_n)$

Co-ordination equations

Condition for optimal load dispatch

$$\frac{dF_n}{dP_n} + \lambda\frac{\partial P_L}{\partial P_n} = \lambda$$

$$\boxed{\frac{dF_n / dP_n}{\left(1-\dfrac{\partial P_L}{\partial P_n}\right)} = \lambda}$$

where, $\frac{\partial P_L}{\partial P_n} \rightarrow$ Incremental transmission loss at plant n

$\lambda \rightarrow$ Incremental cost of received power

Loss Formulae

- Total Losses, $P_L = \sum_m \sum_n P_m B_{mn} P_n$

  where, $P_m, P_n \rightarrow$ Source loadings

  $B_{mn} \rightarrow$ Transmission loss coefficients

- Incremental Transmission Losses, $\frac{\partial P_L}{\partial P_n} = 2\sum_m B_{mn} P_m$

- Incremental Production Cost, $\frac{dF_n}{dP_n} = F_{mn} P_n + f_n$

$$P_n = \frac{1 - \dfrac{f_n}{\lambda} - 2\sum\limits_{m \neq n} B_{mn} P_m}{\dfrac{F_{mn}}{\lambda} + 2B_{nn}}$$

- Penalty Factor, $\boxed{L_n = \frac{1}{1-\dfrac{\partial P_L}{\partial P_n}}}$; where, $L_n \rightarrow$ Penalty factor of plant n

- For a two plant system, loss equation in terms of B coefficient.

$$P_{\text{Loss}} = B_{11}P_{G_1}^2 + B_{22}P_{G_2}^2 + 2B_{12}B_{21}P_{G_1}P_{G_2}$$

X Underground Cable

Classification of Cable

Based on Voltage

- Low voltage (LV) : 1 kV
- High voltage (HV) : 11 kV

- Super voltage (SV) : 22 kV-33 kV
- Extra high voltage (EHV) : 66 kV
- Extra super voltage (ESV) : 132 kV and above.

Based on Core

- Single core
- 3-core
- 3.5 core : it is used for secondary distribution purpose.

Note:

The most commonly used dielectrics in power cables are impregnated paper, butyl, rubber, PVC, polythene, cross linked polyethylene.

Parameters of Single Core Cables

Resistance

- **Insulation Resistance**

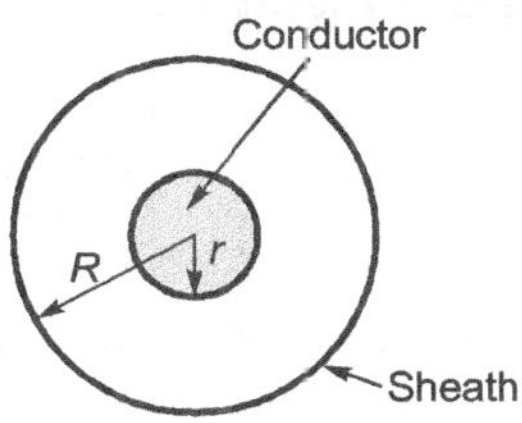

$$\boxed{R_{ins} = \frac{\rho}{2\pi} \ln\left(\frac{R}{r}\right)} \text{ ohms/metre}$$

where, $R_{ins} \rightarrow$ Insulation Resistance

$\rho \rightarrow$ Resistivity of the insulating material

$R \rightarrow$ Inside radius of sheath ; $r \rightarrow$ Conductor radius

- **Resistivity of insulating material at any temperature t**

$$\boxed{\rho_t = \rho_0 \, e^{-\alpha t}}$$

Note:

☑ In cable, insulation resistance is inversely proportion to length of the cable, $R_{ins} \propto \frac{1}{l}$

☑ As temperature increases R_{ins} decreases.

Capacitance

- **Capacitance between core and sheath**

$$\boxed{C = \frac{2\pi \epsilon_0 \epsilon_r \, r}{\ln\left(\frac{R}{r}\right)}} \text{ F/m}$$

where, $q \rightarrow$ Charge on the surface of the conductor per metre length of cable.

$\in_r \rightarrow$ Relative permittivity of dielectric

- **Dielectric power loss per phase**

$$P_d = VI\cos\phi = \omega CV^2 \tan\delta$$

where,

$C \rightarrow$ Capacitance of the cable

$V \rightarrow$ Line to neutral voltage

$P_d \rightarrow$ Dielectric loss

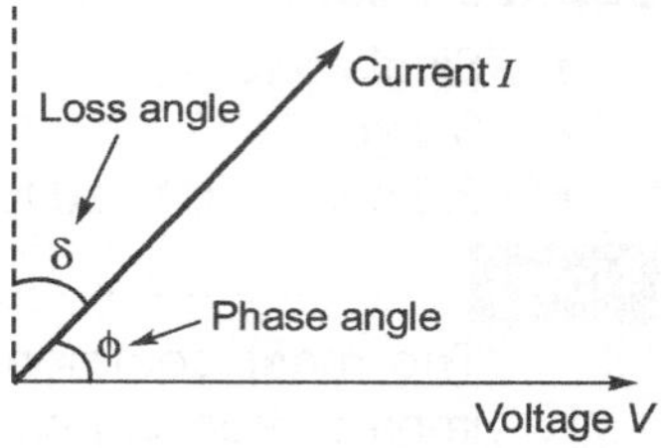

Electrostatic Stresses

Gradient at a Distance x from the Centre of the Conductor Within the Dielectric Material.

$$g = \frac{\lambda}{2\pi \in x} = E$$

where,

$\in \rightarrow$ Permittivity of the dielectric

$E \rightarrow$ Electric field intensity

$\lambda \rightarrow$ Charge per unit length

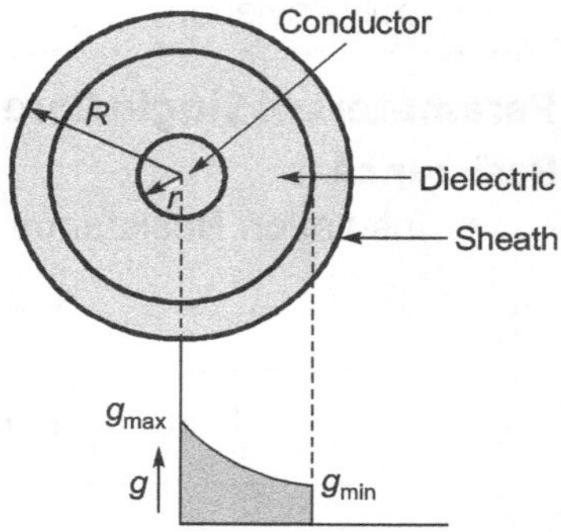

(Electric stress in a single core cable)

Potential

$$V = \frac{\lambda}{2\pi \in}\ln\left(\frac{R}{r}\right)$$

where, $r \rightarrow$ Radius of conductor ; $R \rightarrow$ Inner radius of the sheath

$V \rightarrow$ Potential of conductor with respect to the sheath

Gradient is Maximum at the Surface of the Conductor	Gradient is Minimum at the Inner Radius of the Conductor
$g_{max} = \dfrac{V}{r\ln\dfrac{R}{r}}$	$g_{min} = \dfrac{V}{R\ln\dfrac{R}{r}}$

Note: ....

☑ The most economical size of the cable is one in which $\frac{R}{r} = e = 2.718$. So that the stress at the surface of core is reduce which will increase the life of cable.

☑ Reason of failure of cable is formation of a void between any 2 layer of insulating material. Void near the surface of core will be more serious than that of the void in other location.

☑ Characteristic impedance of cable is 40 Ω.

☑ In overhead transmission line inductor is dominant but in cable capacitance is more dominant.

Grading of Cables

- The method of equilising the stress in the dielectric of the cable is known as grading of cable.

Dielectric Grading (or) Capacitance Grading

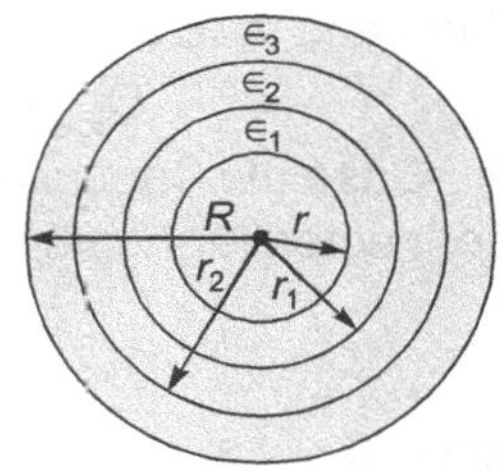

- In dielectric grading a homogeneous dielectric is replaced by different dielectrics such that $r < r_1 < r_2$. So, $\epsilon_1 > \epsilon_2 > \epsilon_3$.

The voltage between conductor and sheath

$$V = V_1 + V_2 + V_3$$

$$V = g_{1\max} r \ln\left(\frac{r_1}{r}\right) + g_{2\max} r_1 \ln\left(\frac{r_2}{r_1}\right) + g_{3\max} r_2 \ln\left(\frac{R}{r_2}\right)$$

For uniform dielectric stress

$$g_1 \max = g_2 \max = g_3 \max = g_{\max}$$

$$V = g_{\max}\left[r \ln\left(\frac{r_1}{r}\right) + r_1 \ln\left(\frac{r_2}{r_1}\right) + r_2 \ln\left(\frac{R}{r_2}\right)\right]$$

Capacitance of the cable

$$C = \frac{2\pi \epsilon_0}{\frac{1}{\epsilon_1}\ln\left(\frac{r_1}{r}\right) + \frac{1}{\epsilon_2}\ln\left(\frac{r_2}{r_1}\right) + \frac{1}{\epsilon_3}\ln\left(\frac{R}{r_2}\right)}$$

Intersheath grading

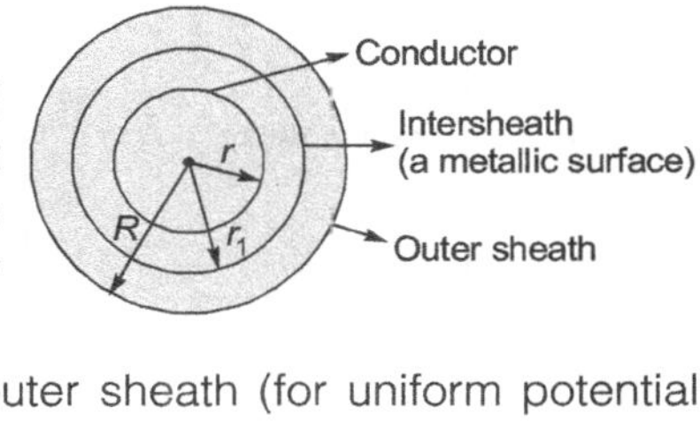

- In intersheath grading an identical dielectric material is utilised throughout the total thickness of the cable. It is divided into 2 or more layers by providing intersheath.
- Voltage between conductor and outer sheath (for uniform potential gradient, i.e., $g_{1\max} = g_{2(\max)} = g_{\max}$

$$V = g_{\max}\left[r \ln\left(\frac{r_1}{r}\right) + r_1 \ln\left(\frac{R}{r_1}\right)\right]$$

Capacitance of Three-Core Belted Type Cables

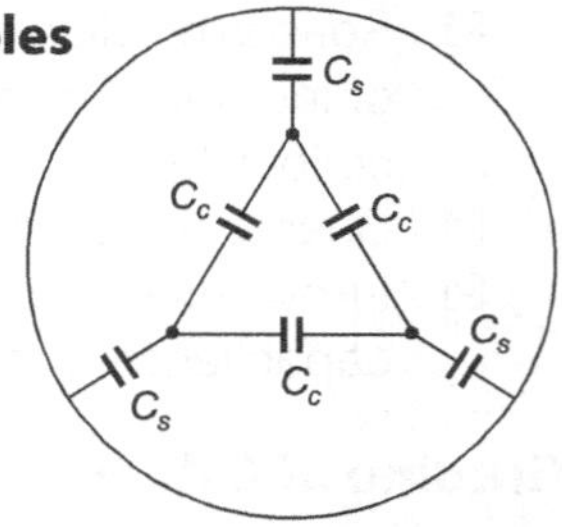

where C_s = Core-earth capacitance
C_c = Core-core capacitance

Case I : Three core are bunched together and capacitance is measured between bunched cores and the sheath.

$$C_{eq} = 3C_s$$

Case II : Two core are bunched with the sheath and capacitance between them and third one.

$$C_{eq} = 2C_c + C_s$$

Case III : Capacitance between the two cores or lines with the third core connected to sheath.

$$\boxed{C_{eq} = \frac{1}{2}(3C_s + C_s) = \frac{1}{2}C_{Ph}}$$

XI Protective Relays

- Relays are sensing device, which detect abnormal conditions in electric circuit like faults and send signal to operate circuit breaker to isolate faulty equipment from the system as quickly as possible.

Types of Relay

Based on Time of Operation

Instantaneous Relay :	Operating time ≤ 0.1 sec
Definite Minimum Time (DMT) Relay :	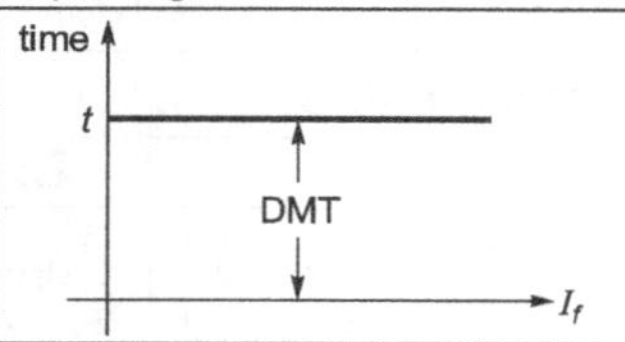
Inverse Relay :	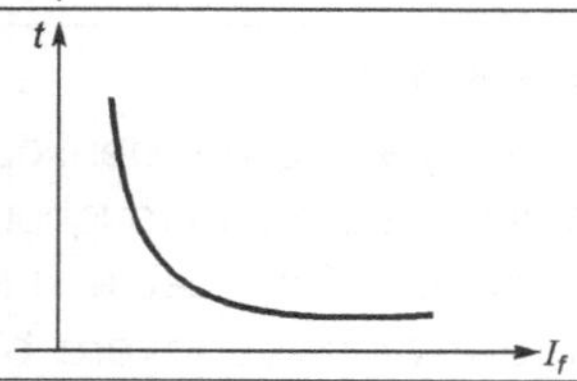
Inverse Definite Minimum Time (IDMT) Relay :	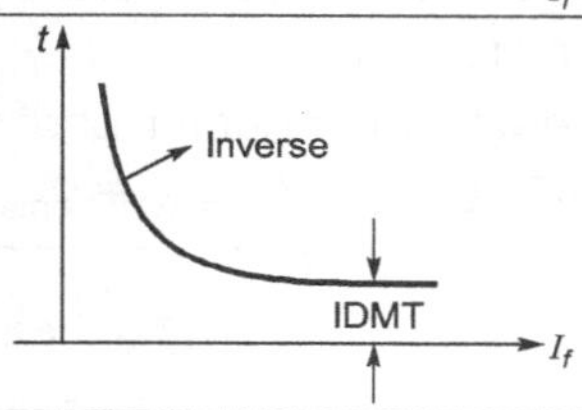

Based on Construction

- *Electromagnetic attraction type:*
  - Balanced beam type • Moving plunger type • Attracted armature type.
- *Electromagnetic induction type:*
  - Shaded pole type • Induction cup type • Wattmeter type
- *Gas operated relay:* Buchholz relay.
- *Thermal relay:* For over load protection.
- *Static/microprocessor based relay.*

Pickup Value

- It is the minimum value of operating quantity at which the relay is at the verge of operation.

Reset Value

- It is the maximum value of operating quantity at which the relay is at the verge of non-operation.

Note:

☑ For Well-design relay the ratio of reset to pickup value is unity.

☑ For induction type relay the ratio of reset to pickup value is 0.98.

Time Multiplier Setting (TMS)

- By setting different value of TMS, for the same operating current, we get different time of operation.

$$TMS_{(\text{required})} = \frac{t_{required}}{t_{(TMS=1)}}$$

where, $t_{\text{required}} \rightarrow$ Required time of operation

$t_{(\text{TMS}=1)} \rightarrow$ Time of operation when TMS = 1

$$\frac{TMS_2}{TMS_1} = \frac{t_2}{t_1}$$

Plug Setting Multiplier (PSM)

- PSM gives additional feature to the same relay, so that it can operate for different pickup current

$$\text{PSM} = \frac{\text{Fault current}}{\text{Current setting} \times \begin{pmatrix}\text{C.T. secondary} \\ \text{rated current}\end{pmatrix} \times C.T.\text{ ratio}}$$

Torque Equation

Universal Relay Torque Equation

$$T = K_1 I^2 + K_2 V^2 + K_3 VI\cos(\theta - \tau) + K$$

where, $I \rightarrow$ RMS value of current in current coil
$V \rightarrow$ RMS value of voltage fed to the voltage coil circuit
$\theta \rightarrow$ Angle between I and V
$\tau \rightarrow$ The maximum torque angle
$K \rightarrow$ Restraining torque including spring and friction.
$K_1, K_2, K_3 \rightarrow$ Relay constant

Torque Equation for Different Type of Relays

Over current relay	$T_{OC\ relay} = K_1 I^2 - K$
Directional relay	$T = K_3 VI\cos(\theta - \tau) - K$
Impedance relay	$T = K_1 I^2 - K_2 V^2$ For relay to operate, $Z = \frac{V}{I} < \sqrt{\frac{K_1}{K_2}}$
Reactance relay	$T = K_1 I^2 - K_3 VI\sin\theta;\ \ \tau = 90°$ For relay to operate, $\frac{V}{I}\sin\theta = X < \frac{K_1}{K_3}$
Mho relay	$T = K_3 VI\cos(\theta - \tau) - K_2 V^2$ For relay to operate, $\frac{V}{I} = Z < \frac{K_3}{K_2}\cos(\theta - \tau)$

Operating Characteristic of Relays

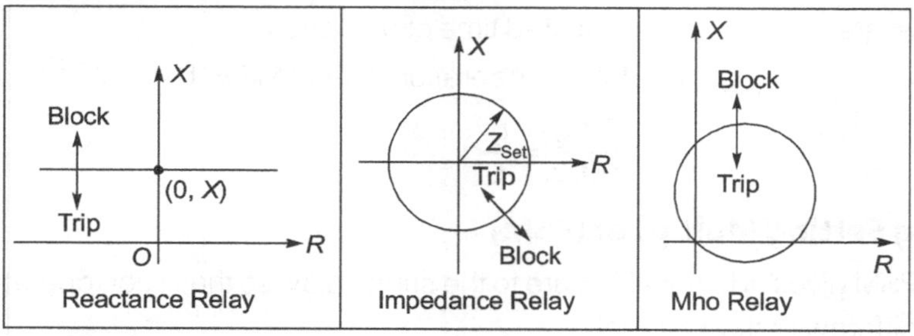

Note:

- Relay used for phase fault on
- Short line : Reactance relay, i.e., O.C. relay with directional restrained.
- Medium line : Impedance relay, i.e., voltage restrained O.C. relay.

- Long line : Mho relay, i.e., directional restraint O.C. relay.
- For earth fault generally reactance relay are used.

XII Circuit Breakers

- Restriking voltage

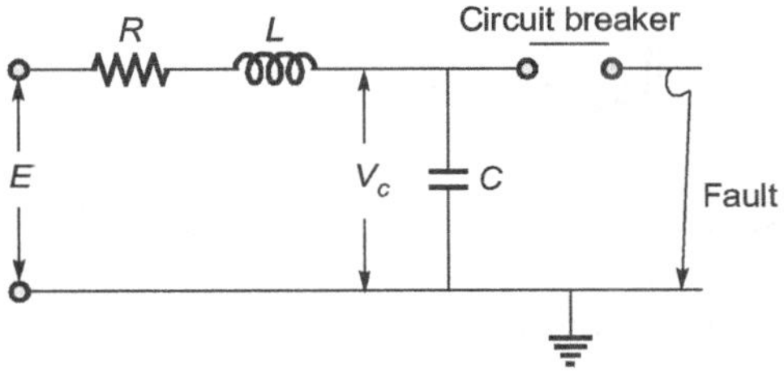

$$V = E\left(1 - \cos\frac{1}{\sqrt{LC}}t\right)$$

where, $L, C \rightarrow$ Inductance and capacitance per phase of the system upto the point of circuit breaker location.

$E \rightarrow$ System voltage at the instant of arc interruption

$V \rightarrow$ Restriking voltage

- **The maximum value of restriking voltage = 2 x E_{peak}**

  where, E_{peak} = Peak value of the system voltage

- **Natural frequency of oscillation,** $$f_n = \frac{1}{2\pi}\frac{1}{\sqrt{LC}}$$
- **Rate of rise of restriking voltage (RRRV)** $$RRRV = \omega_n E \sin\omega_n t$$
- **The maximum value of RRRV = $\omega_n E_{peak}$**

Resistance Switching

- Frequency of damped oscillation

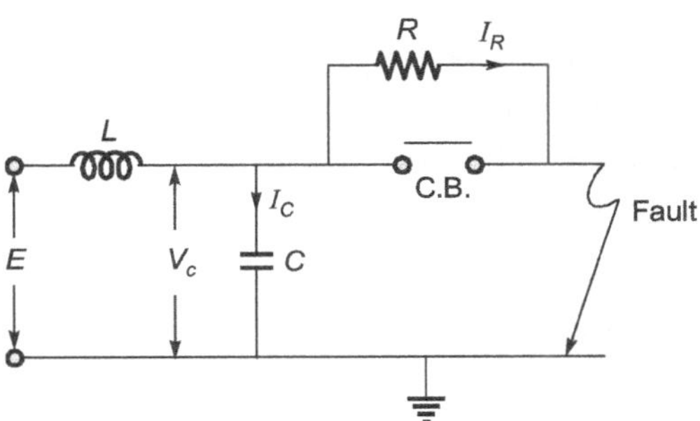

$$f = \frac{1}{2\pi}\sqrt{\frac{1}{LC} - \frac{1}{4R^2C^2}}$$

Note:

If value of the resistance connected across the contacts of circuit breaker, is equal to or less than $\frac{1}{2}\sqrt{\frac{L}{C}}$, then there will be no transient oscillation.

Transient Oscillation for Different Values of R

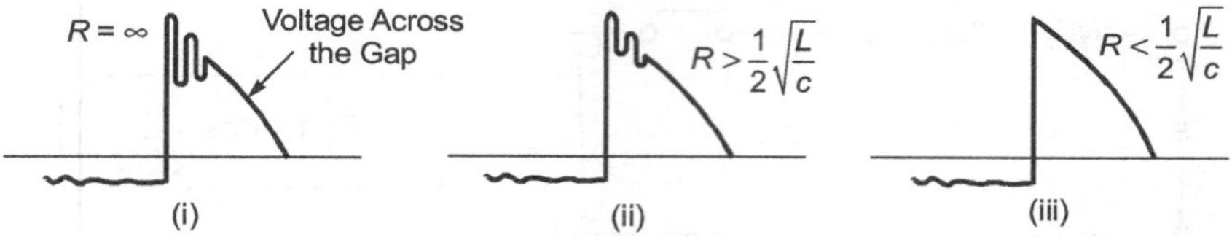

Transient oscilations for different values of *R*

- If $R > \frac{1}{2}\sqrt{\frac{L}{C}}$. There will be oscillation.
- $R = \frac{1}{2}\sqrt{\frac{L}{C}}$ is known as critical resistance. There will be no oscillation.
- 3ϕ circuit breaker capacity = $\sqrt{3} \times I_{\text{break}} \times V_{\text{rated}}$
- Making capacity of C.B. = 2.55 × (Symmetrical Breaking Capacity)
- Duty cycle of circuit breaker is given by :

  O – 30 sec – CO – 90 sec – CO – 3 min – CO

  (Here, O stands for open and C stands for closing of C.B.).

XIII Generating Power Stations

- **Water Power,** $P = \frac{0.736}{75} Q\eta H w \text{ kW}$

  where, $Q \rightarrow$ Discharge; m^3/sec; $H \rightarrow$ Water head; m
  $\eta \rightarrow$ Overall efficiency of turbine alternator set
  $w \rightarrow$ Specific weight of water in kg/m^3 = 1000 kg/m^3

- Specific speed of a turbine is the speed of a scale model of turbine which develops 1 metric h.p. under a head of 1 metre.

- **Specific Speed,** $N_s = \frac{N\sqrt{P_t}}{H^{1.25}}$

  where, $N_s \rightarrow$ Specific speed in metric units
  $N \rightarrow$ Speed of turbine in rpm ; $P_t \rightarrow$ Output in metric h.p.
  $H \rightarrow$ Effective head in metres

- Power output of Tidal scheme

$$P = Q\rho gH \text{ watts}$$

where, $Q \rightarrow$ Quantity of water flow, m^3/sec
$g \rightarrow$ Acceleration due to gravity = 9.81 m/sec^2
$H \rightarrow$ Water head, metre
$\rho \rightarrow$ Density of sea water = 1025 kg/m^3

Classification of Turbine

Based on Water Discharge	• High discharge : Kaplan • Medium discharge : Francis • Low discharge : Pelton
Based on Water Pressure	• Impulse : Pelton • Reaction : Kaplan, Francis, Propeller
Based on Direction of Water Flow	• Axial : Kaplan • Radial : Francis • Tangential : Pelton • Diagonal : Deriaz

Component of Hydroelectric Power Station

1. Reservoir 2. Dam 3. Trash rack 4. Spill-way 5. Gates
6. Intake gates 7. Forebay 8. Surge tank 9. Pen stock

Component of Nuclear Power Station

- **Reactor core:** Reactor core is made-up of stainless steel or zirconium.
- **Moderator:** It decrease speed of neutron. (i) Heavy water, (ii) Graphite and (iii) Beryllium.
- **Reflector:** Prevents escape of neutron from reactor core, and made-up of high grade graphite.
- **Control rod:** During earthquake it trips the generating station.
- **Cooler:** Na and Li are used as coolant material.
- **Shielding:** Shielding eliminates the effect of radiation.

Remember:

- The valve K indicates the chain reaction will continue at a steady state.

$$K = \frac{\text{No. of neutrons produced in one generation}}{\text{No. of neutrons produced in the proceeding generator}}$$

- If $K < 1$, the chain reaction stops.
- If $K > 1$, the chain reaction build up.

Component of Thermal Generating Stations

- Coal handling plant.
- **Boiler:** In boiler combustion takes place.
- **Super heater:** In super heater steam is converted into the super heated steam.
- **Airpreheater:** In airpreheater the atmospheric air absorb the heat of the flue gases and air at higher temperature send to boiler for effective combustion.
- **Economizer:** In economizer water absorbs the heat of flue gases and send to boiler.
- **Turbine:** Turbine runs by superheated steam and generate mechanical energy.
- **Condenser:** Here steam is converted into feed water.

XIV Loads and Load Curves

Demand Factor	$\frac{\text{Maximum demand}}{\text{Connected load}}$; Demand factor < 1
Load Factor (L.F.)	$\frac{\text{Average load}}{\text{Peak load}}$
Group Diversity Factor	$\frac{\text{Sum of individual maximum demand}}{\text{Maximum demand of the group}}$ Group diversity factor > 1
Peak Diversity Factor	$\frac{\text{Maximum demand of a consumer group}}{\text{Demand of the consumer group at the time of system peak demand}}$
Capacity Factor	$\frac{\text{Average Annual Load}}{\text{Rated Plant Capacity}}$
Capacity Factor	$\frac{\text{Maximum Load}}{\text{Plant Capacity}} \times \text{Load Factor}$
Capacity Factor	Load Factor × Utilisation Factor
Utilisation Factor	$\frac{\text{Maximum Load}}{\text{Rated Plant Capacity}}$
Reserve Capacity	Maximum Demand – Rated Plant Capacity

Remember:

☑ Plant capacity factor is also known as plant factor.

☑ A graph showing the variation of the system load during 24 hours of the day is known as the system chronological load curve.

- ☑ The area under a chronological load curve gives the energy consumed during the 24 hours.
- ☑ Load duration curve is a rearrangement of all the load elements of a chronological curve in a descending order
- ☑ Mass curve is plotted with energy as ordinate and time as abscissa.
- ☑ A mass curve is used in the study of variations between the rate of water flow and the electrical load and determination of the necessary storage.

■■■■

Electrical Machines

2

I Transformers

It is a static device which transfers electrical energy from one circuit to another circuit without change in frequency. It works on electromagnetic induction principle.

Induced EMF

Direction of induced emf can be found by **"Lenz's law"** and magnitude of induced emf can be given by **"Faraday's law of electromagnetic induction"**.

RMS value of induced Emf.

In primary winding

$E_1 = \sqrt{2}\,\pi N_1 f \phi_m$ Volts

In secondary winding

$E_2 = \sqrt{2}\,\pi N_2 f \phi_m$ Volts

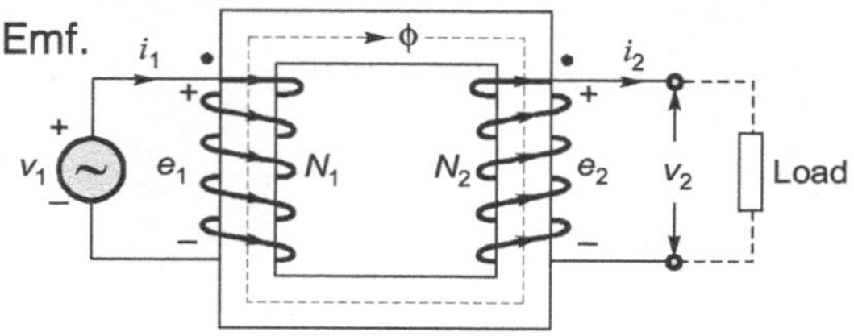

where, N_1 = Number of turns in primary winding;

N_2 = Number of turns in secondary winding;

ϕ_m = Maximum value of the magnetic flux (in webers) = $B_m.A$

B_m = Maximum flux density; A = Area of cross-section

f = Supply frequency, in Hz

Remember:

- ☑ Emf per turn in primary = Emf per turn in secondary
- ☑ Compensating primary mmf = Secondary mmf

  $I_1' N_1 = I_2 N_2$ (where, I_1' = Load component of primary current I_1)
- ☑ Primary volt-amperes = Secondary volt-amperes.
- ☑ Transformer is a constant frequency and constant power device.
- ☑ Step up transformer : $N_1 < N_2$.
- ☑ Step down transformer : $N_1 > N_2$.

Ideal Transformer

- The primary and secondary windings have zero resistance. So, there is no ohmic power loss and no resistive voltage drop.
- There is no magnetic leakage flux.

- The core loss considered to be zero.
- The core has infinite permeability.

Equivalent Circuit

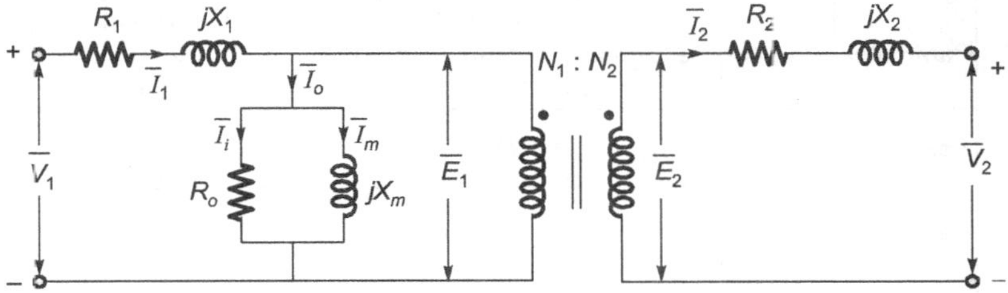

Exact equivalent circuit of transformer

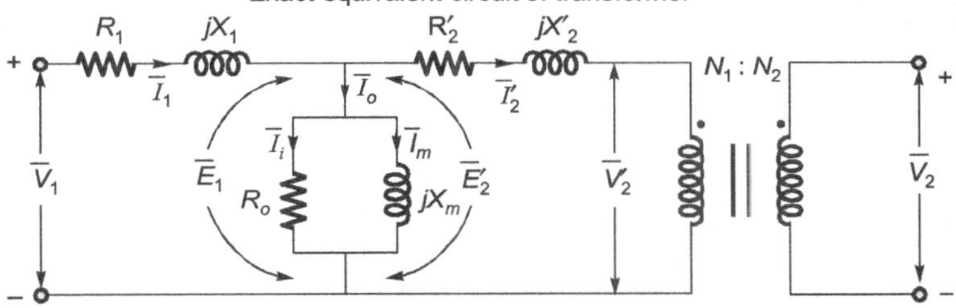

Equivalent circuit referred to primary side

where, $\bar{V}_1$ → Applied voltage to primary side.

$\bar{V}_2'$ → Secondary terminal voltage referred to primary side.

$\bar{E}_1, \bar{E}_2$ → Induced emf in primary and secondary side.

$\bar{E}_2'$ → Secondary induced emf referred to primary side.

$\bar{I}_o$ → No load current.

$\bar{I}_m, \bar{I}_i$ → Magnetizing and core loss component of exciting current.

R_1, R_2 → Primary and secondary winding resistances.

X_1, X_2 → Primary and secondary winding leakage reactances.

R_2', X_2' → Secondary resistance and leakage reactance referred to primary side.

R_o → Core loss equivalent resistance.

X_m → Magnetising reactance.

Note:

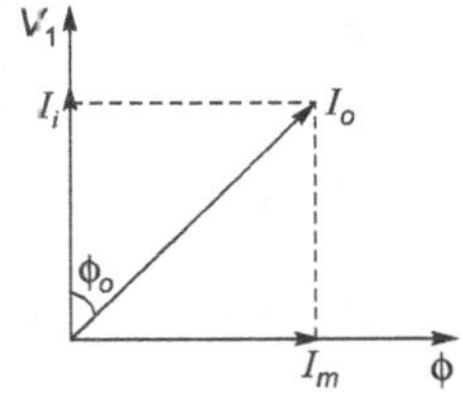

- At no load, current drawn from the supply is I_o

$\bar{I}_o = \bar{I}_m + \bar{I}_i = (2 \text{ to } 5\% \text{ of } I_{fl})$

$|\bar{I}_o| = \sqrt{I_m^2 + I_i^2}$

$\bar{I}_m = I_o \sin\phi_o$ = Magnetizing current

$\bar{I}_i = I_o \cos\phi_o$ = Core loss current

ϕ_0 = No load phase angle (80 to 85°)

- Core loss = $V_1 I_o \cos\phi_o = V_1 I_i$ Watt

Parameter	Primary	Secondary	Secondary side parameter referred to primary side	Primary side parameter referred to secondary side
Resistance	R_1	R_2	$R_2' = a^2 R_2$	$R_1' = \frac{R_1}{a^2}$
Reactance	X_1	X_2	$X_2' = a^2 \times X_2$	$X_1' = \frac{X_1}{a^2}$
Current	$\bar{I}_1$	$\bar{I}_2$	$\bar{I}_2' = \frac{\bar{I}_2}{a}$	$\bar{I}_1' = a\bar{I}_1$
emf	$\bar{E}_1$	$\bar{E}_2$	$\bar{E}_2' = a\bar{E}_2$	$\bar{E}_1' = \frac{\bar{E}_1}{a}$
Voltage	$\bar{V}_1$	$\bar{V}_2$	$\bar{V}_2' = a\bar{V}_2$	$\bar{V}_1' = \frac{\bar{V}_2}{a}$

where, $a = \frac{N_1}{N_2}$

Basic Equations for Transformer

$$\bar{E}_1 = a\bar{E}_2$$

$$\bar{I}_o = \bar{I}_m + \bar{I}_i$$

$$\bar{E}_2 = \bar{V}_2 + \bar{I}_2 R_2 + j\bar{I}_2 X_2$$

$$\bar{V}_1 = \bar{E}_1 + \bar{I}_1 R_1 + j\bar{I}_1 X_1$$

$$\bar{I}_1 = \frac{\bar{I}_2}{a} + \bar{I}_o$$

Phasor Diagram ⇒

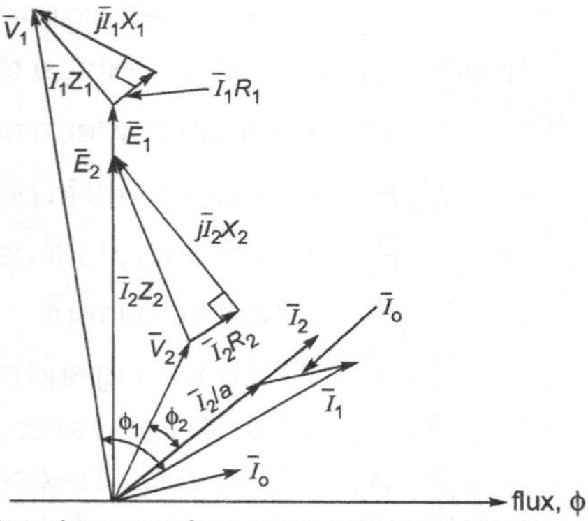

Step down transformer at lagging power factor

Note:

- ☑ $R_{01} = R_1 + R_2'$ = Equivalent resistance referred to primary side.
- ☑ $X_{01} = X_1 + X_2'$ = Equivalent reactance referred to primary side.
- ☑ $Z_{01} = \sqrt{R_{01}^2 + X_{01}^2}$ = Equivalent impedance referred to primary side.
- ☑ Core of transformer should have low reluctance and high permeability.
- ☑ It is made from CRGO (cold rolled grain oriented) material with 4% of silicon.
- ☑ It is laminated to reduce eddy current losses.

Stacking factor, $K_S = \dfrac{\text{Net cross section area}}{\text{Gross cross section area}} \Rightarrow K_S < 1$

Types of Transformer Construction

S.No.	Core Type	Shell Type
1.	Easier to construct	Complex construction
2.	Less mechanical support	More mechanical support
3.	More leakage flux	Less leakage flux
4.	Easy to repair	Difficult to repair
5.	Better cooling	Less cooling
6.	High voltage/power applications	Low voltage/power applications
7.	Less insulations required	More insulation required
8.	More copper required	Less copper required
9.	Winding surrounds core	Core surrounds windings
10.	Concentric winding	Sandwich winding

Losses in Transformer

- **Copper Loss (P_{cu}) :** The copper losses are I^2R losses in transformer windings due to winding resistance. This loss is variable loss.

  Total copper loss = Primary copper loss + Secondary copper loss

  $$P_{cu} = I_1^2 R_1 + I_2^2 R_2$$

  Resistance referred to any one side. $P_{cu} = I_1^2 R_{01} = I_2^2 R_{02}$
- **Iron Loss (P_i) :** This loss is constant loss. Iron loss = hysteresis loss + Eddy current loss
- **Hysteresis loss (P_h) :** It is due to reversal of magnetization of transformer core whenever it is subjected to alternating nature of magnetizing forces.
- **Eddy current loss (P_e) :** It is basically I^2R loss present in the core due to production of eddy currents in the core, because of its conducitivity.

Conditions	Hysteresis Loss $P_h = K_h f B_m^x$	Eddy Current Loss $P_e = K_e f^2 B_m^2 t^2$	Iron Loss $P_i = P_h + P_e$
$\frac{V}{f}$ = Constant	$P_h \propto f$	$P_e \propto f^2$	$P_i = Af + Bf^2$
$\frac{V}{f} \neq$ Constant	$P_h = K_h V^{1.6} f^{-0.6}$ (if x = 1.6, i.e., for Si steel)	$P_e \propto V^2$	$P_i = AV^{1.6} f^{-0.6} + BV^2$
Voltage constant & frequencies is decreased	Hysteresis losses increases	Eddy current loss constant	$P_i = P_h + P_e$ (As P_h increases, P_i also increases)
Frequency constant & V decreases	Hysteresis losses decreases	Eddy current loss decreases	Iron loss decreases

where, K_h = Proportionality constant depends upon volume, quality of the core material and units used.

B_m = Maximum flux density in the core.

f = Supply frequency

x = Steinmetz exponent (value vary from 1.5 to 2.5)

k_e = proportionality constant depends upon volume and resistivity of core material, thickness of laminations and unit employed

B_m = Maximum flux density

t = Thickness of lamination

Remember:

☑ p.u. resistance = p.u. full load copper loss

- **Separation of iron losses**

Plot $\left(\frac{P_i}{f}\right)$ vs frequency. The slope of the curve gives the constant B and the y-intercept gives the constant A.

$$m = \tan\theta$$
$$= \text{slope of straight line} = B$$
$$W_i = Af + Bf^2$$
$$\left(\frac{W_i}{f}\right) = A + Bf$$

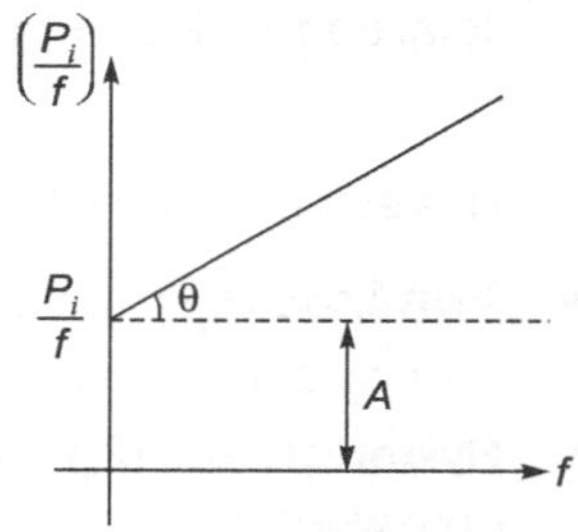

- **Stray load loss**

It largely results from leakage flux including eddy currents in the tank walls, conductors etc. To reduce stray load loss, outer shell of transformer can be made using Aluminium.

- **Dielectric loss**

These losses occur in inter-winding insulation and transformer oil. It also occurs due to leakage current through dielectric.

Remember:

☑ Rating of transformer is given in kVA because constant loss is proportional to voltage and variable loss is proportional to current so total loss in transformer is proportional to volt ampere.

☑ Voltage rating ∝ Insulation level

☑ Current rating ∝ Cross-sectional area of winding

☑ Power rating ∝ Cooling

Open Circuit Test and Short Circuit Test

- These helps to determine :
  - Parameters of equivalent circuit
  - Voltage regulation
  - Efficiency

Open Circuit (OC) Test

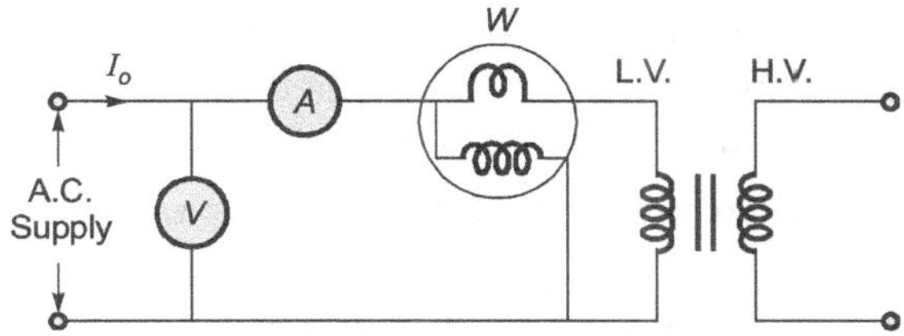

- OC test is carried out at rated frequency and rated voltage to determine the core losses
- OC test is always conducted on LV side.
- In OC test, LPF (Low power factor) wattmeter is generally used.

Sequence of equations used to find different parameters

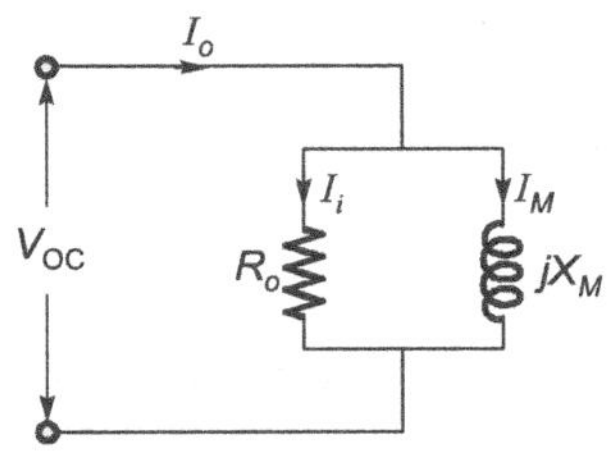

Wattmeter reading = P_i (core loss or iron loss)

Voltmeter reading = V_{oc} (Rated voltage)

Ammeter reading = I_o (No load current)

No load power factor = $\cos\phi_o$

$$P_i = V_{oc} I_o \cos\phi_o ; \quad \cos\phi_0 = \frac{P_i}{V_{oc} I_o}$$

$$I_i = I_o \cos\phi_o \text{ and } I_m = I_o \sin\phi_o$$

$$R_o = \frac{V_{oc}}{I_i}, \text{ and } X_m = \frac{V_{oc}}{I_m}$$

Short-Circuit (SC) Test

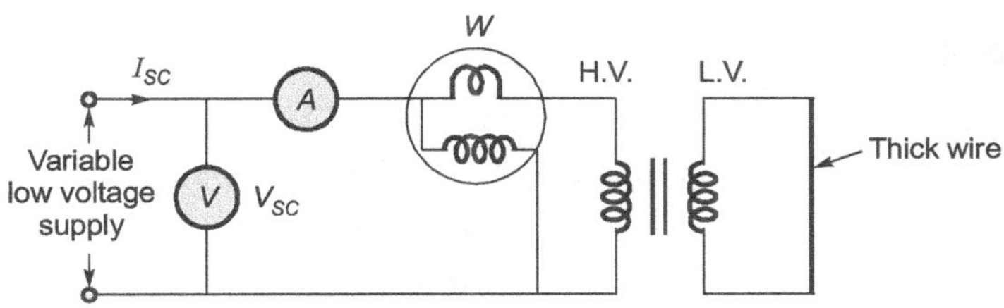

- SC test is carried out at rated current to determine the full load copper loss.
- SC test is conducted in HV side.

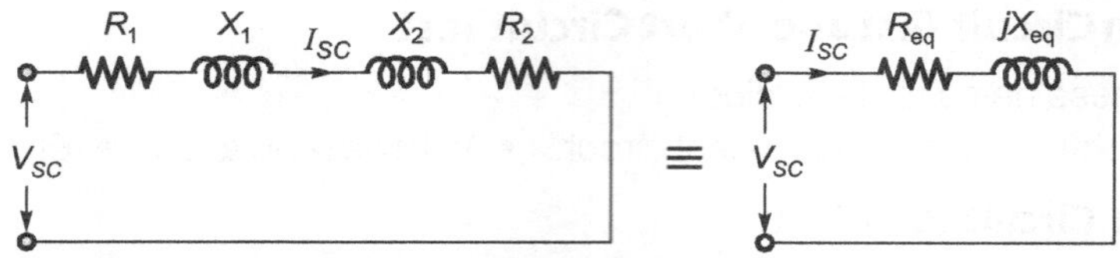

Wattmeter reading = P_{sc} (rated copper loss)
Voltmeter reading = V_{sc} (5 to 8% of rated voltage)
Ammeter reading = I_{sc} (rated current)

Sequence of equations used to find different parameters

$$R_{eq} = \frac{P_{SC}}{(I_{SC})^2} \quad ; \quad Z_{eq} = \frac{V_{SC}}{I_{SC}} \quad ; \quad X_{eq} = \sqrt{Z_{eq}^2 - R_{eq}^2}$$

Note:

☑ Short circuit test may not be conducted at rated current and hence we will transform copper loss to obtain full load copper loss.

$$P_{cuFL} = P_{cu}\left(\frac{I_{rated}}{I_{SC}}\right)^2$$

Sumpner's Test (Back to Back Test or Heat Run Test)

This test helps to determine :

- Maximum temperature rise • Losses • Efficiency

In Sumpner's test, due to unequal current in primary of both Transformers, the efficiencies of two transformers are different.

Polarity Test

- Polarity test is done to know the polarities of transformers.
- If transformers are connected in parallel without polarity test they may have opposite polarity which will lead to high circulating currents on secondary side.

Efficiency

$$\text{Efficiency } (\eta) = \frac{\text{Output power}}{\text{Input power}} = \frac{\text{Output}}{\text{Output + losses}}$$

Efficiency at any fraction 'x' of full load

$$(\eta)_{x \cdot \text{Full load}} = \frac{x(\text{kVA})_{fl}\cos\phi}{x(\text{kVA})_{fl}\cos\phi + P_i + x^2(P_{cu})_{fl}}$$

where, $P_i \rightarrow$ Core loss or iron loss; $(P_{cu})_{fl} \rightarrow$ Full load copper loss

For full load, $x = 1$

Condition for Maximum Efficiency, Copper loss = Iron loss

$$I_{\eta_{max}} = \sqrt{\frac{P_i}{R_{eq}}} \; ; \quad x = \sqrt{\frac{(P_i)}{(P_{Cu})_{fL}}}$$

KVA load for Maximum Efficiency, $S_{\eta_{max}} = S_{(\text{known load})} \cdot \sqrt{\frac{P_i}{P_{cu(\text{known load})}}}$

where, $S_{\eta_{max}}$ → KVA rating at maximum efficiency

$S_{(\text{known load})}$ → KVA rating at known load

$P_{cu(\text{known load})}$ → Copper loss at known load

For full load, $S_{\eta_{max}} = S_{\text{full load}} \cdot \sqrt{\frac{P_i}{P_{cu(\text{full load})}}}$

All day efficiency, $\eta_{\text{all day}} = \frac{\text{Output energy (kWh) in 24 hrs.}}{\text{Input energy (kWh) in 24 hrs.}}$

Remember:

☑ We compute efficiency for power transformer and all day efficiency for distribution transformer.

Voltage Regulation

$$\text{Voltage regulation} = \frac{\text{No load voltage} - \text{Full load voltage}}{\text{Full load voltage}} \times 100\%$$

where, full load voltage → Rated voltage

Remember:

☑ Regulation is always calculated on full load and should be as low as possible.

- **Voltage regulation**

  $V.R = x(R_{pu}\cos\phi + X_{pu}\sin\phi) \times 100\%$

  where, ϕ → angle between V_2' and I_1

  x → fraction of full load

  $R_{p.u.}$ → p.u. equivalent resistance

  $X_{p.u.}$ → p.u. equivalent reactance

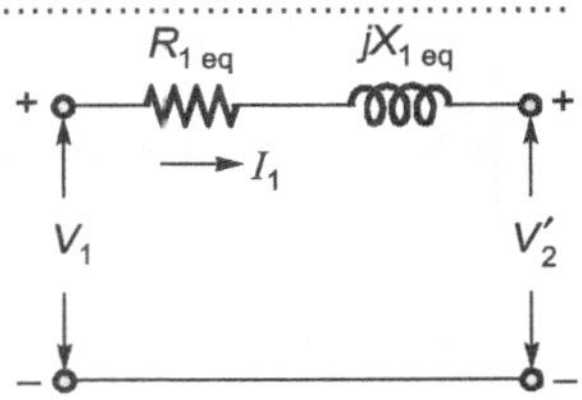

- o is positive for lagging power factor and negative for leading power factor.
- **Condition for maximum voltage regulation**

$$\cos\phi = \frac{|R_{p.u.}|}{|Z_{p.u.}|} = \frac{R_{1eq}}{Z_{1eq}} \quad \text{lagging}$$

- Maximum voltage regulation, $(V.R)_{max} = Z_{p.u.}$
- Condition for zero voltage regulation, $\cos\phi = \frac{X_{p.u.}}{Z_{p.u.}} = \frac{X_{1eq}}{Z_{1eq}}$ Leading

Regulation Curve

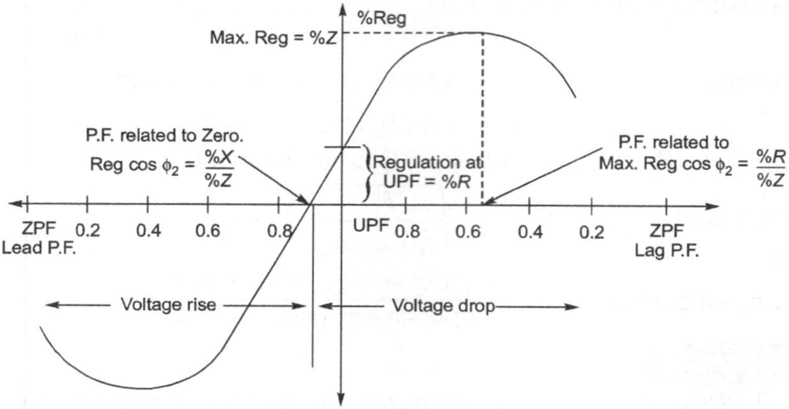

Regulation Curve of a Transformer

Remember:

☑ For leading p.f., greater than $\frac{X_{p.u.}}{Z_{p.u.}}$, the voltage regulation will be negative

☑ Regulation is always positive at all lagging p.f. loads and also at unity p.f. load.

☑ Regulation may be positive at large leading power factor, near to unity and regulation is negative at low leading power factor, away from unity.

S.No.	Power Transformer	Distribution Transformer
1.	Transmission network at higher voltages	Distribution network at lower voltage.
2.	Rating above 200 MVA	Rating less than 200 MVA
3.	Larger in size	Smaller in size
4.	Designed for η_{max} at 100% load	Designed for η_{max} at 50-70% load
5.	Copper & Iron losses occur entire day	Iron loss occurs entire day and copper loss depends on load cycle
6.	Designed to minimize copper loss	Designed to minimize core loss

Auto Transformer

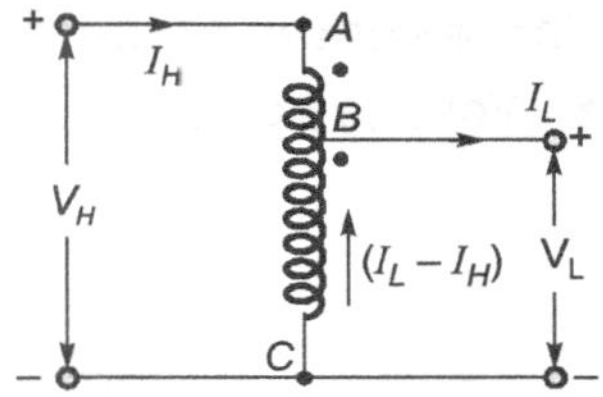

AB = Series winding

BC = Common winding

N_H = Number of turns in winding AC

N_L = Number of turns in winding BC

- **Ratio of transformation,** $a_{auto} = \frac{V_H}{V_L} = \frac{N_H}{N_L}$

Note:

- Always try to keep a_{auto} closer to 1 i.e. $a \approx 1$.
- Since there is no electrical isolation between primary winding and secondary winding, so when common winding is open circuit then full high voltage is appears across low voltage winding at the time of step down mode.

Comparison Between two Winding Transformer & Auto Transformer

- **Copper weight,** $(\text{Cu weight})_{auto} = \left(1 - \frac{1}{a_{auto}}\right) \times (\text{Cu weight})_{2\text{-wdg}}$
- **Copper saving,** $\%\text{Copper saving} = \frac{1}{a_{auto}} \times 100$
- $(\%\text{ Voltage drop})_{auto} = \left(1 - \frac{1}{a_{auto}}\right) \times (\%\text{ Voltage drop})_{2\text{-wdg}}$
- **kVA rating of autotransformer,** $\text{kVA}_{auto} = \left(\frac{a_{auto}}{a_{auto} - 1}\right) \text{kVA}_{2\text{-wdg}}$

where, $\text{kVA}_{auto} \rightarrow$ kVA rating of auto transformer

$\text{kVA}_{2\text{-wdg}} \rightarrow$ kVA rating of two winding-transformer

Inductive and Conductive Transfer

$$V_L I_L = \underbrace{V_L (I_L - I_H)}_{\text{inductive transfer}} + \underbrace{V_L I_H}_{\text{Conductive transfer}}$$

$$\frac{\text{Inductive transfer}}{\text{total transfer}} = \left(1 - \frac{1}{a_{(auto)}}\right) \; ; \quad \frac{\text{Conductive transfer}}{\text{Total transfer}} = \left(\frac{1}{a_{(auto)}}\right)$$

Remember:

☑ Auto transformer has higher efficiency, lower p.u. impedance and lower voltage regulation as compare to 2-wdg transformer.

☑ Losses remain same where 2-winding transformer is converted to auto transformer because amount of core & conductor material used is same.

Three Winding Transformer

- Voltage per turn

$$\frac{V_1}{N_1} = \frac{V_2}{N_2} = \frac{V_3}{N_3}$$

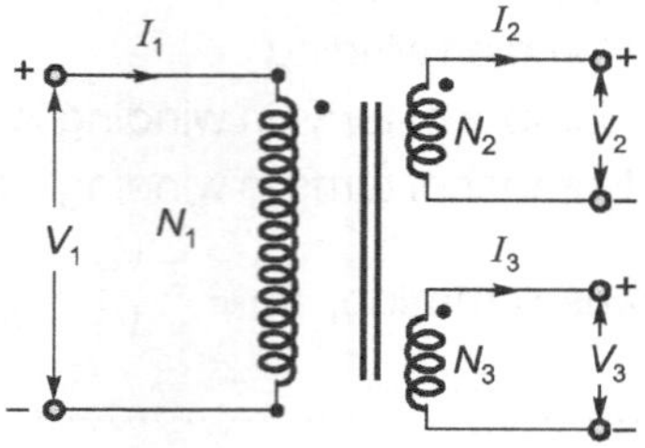

- MMF balance equation

$$N_1I_1 = N_2I_2 + N_3I_3 \text{ (neglect magnetizing current)}$$

$$I_1 = \frac{N_2}{N_1} \cdot I_2 + \frac{N_3}{N_1} I_3$$

- kVA balance equation, $V_1I_1 = V_2I_2 + V_3I_3$; $S_1 = S_2 + S_3$

Remember:

☑ Three winding transformer can be used to interconnect power systems at different voltage levels.

☑ Tertiary winding is used as a supporting winding in YY transformer to stabilize neutral and to avoid third harmonics.

Condition to be Satisfied for Parallel Operation

Must Condition

- Same polarity.
- Same voltage ratio and voltage rating.
- Same phase sequence.
- Zero phase difference, which means transformer belong to same phasor group can only operate in parallel.

Desirable Conditions

- Same X/R ratio for same power factor operation.
- Same p.u. impedance on their own kVA base for proportional load sharing.

Equivalent Circuit of Transformer connected to Parallel

$$I_A = I_L \times \frac{Z_B}{Z_A + Z_B}$$

$$I_B = I_L \times \frac{Z_A}{Z_A + Z_B}$$

$$S_A = S_L \times \frac{Z_B}{Z_A + Z_B}$$

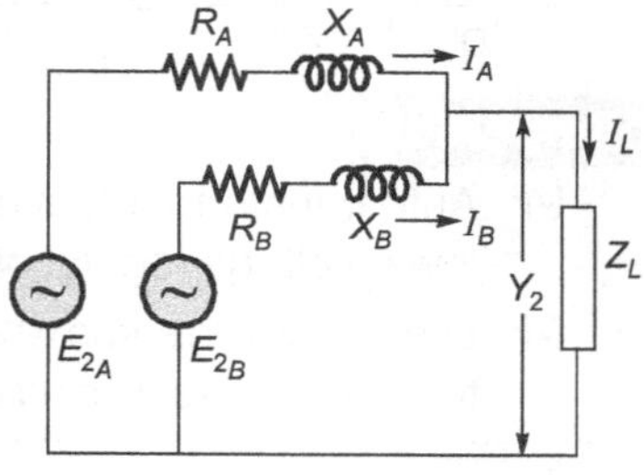

Similarly, $S_B = S_L \times \dfrac{Z_A}{Z_A + Z_B}$

where, I_A, I_B = Current in transformer A and B respectively

Z_A, Z_B = Impedance of transformer A and B respectively.

I_L = Load current

S_A, S_B = Apparent power of transformer A and B respectively.

S_L = Load Apparent power.

Three phase transformer connections

Group No. Symbol Phase P	Windings and Terminals	E.M.F. Vector Diagrams
Yy0 0°	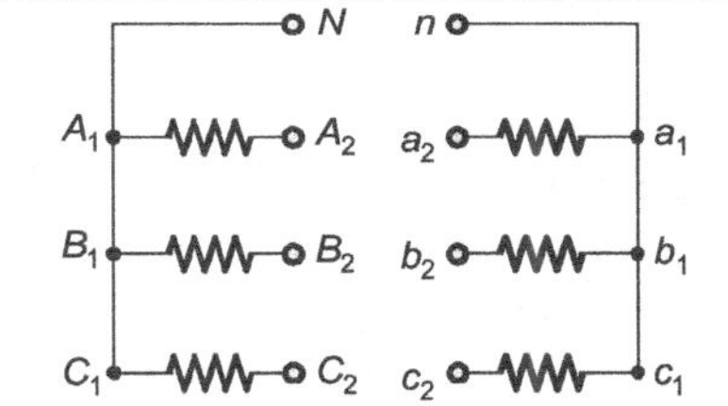	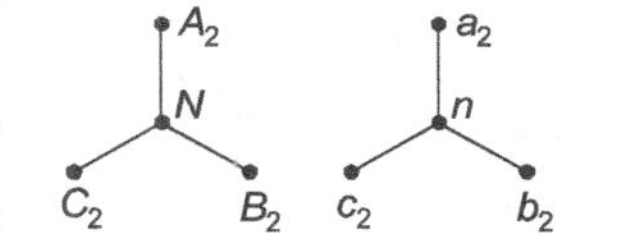
Dd0 0°	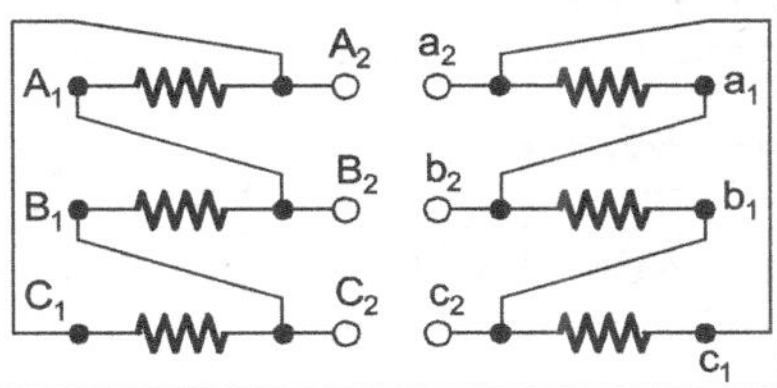	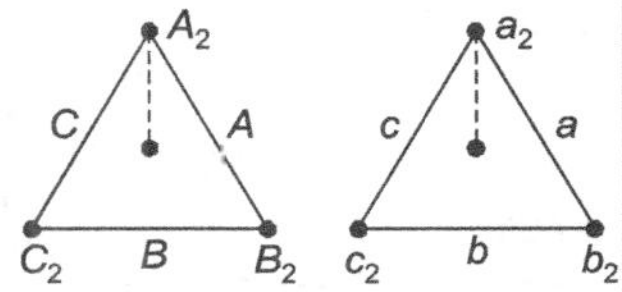
Yy6 180°	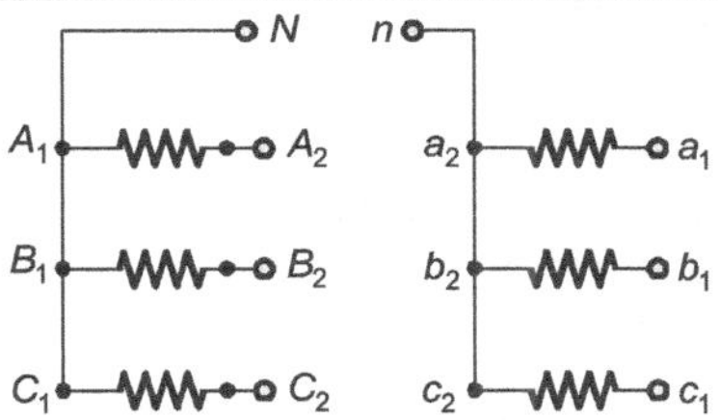	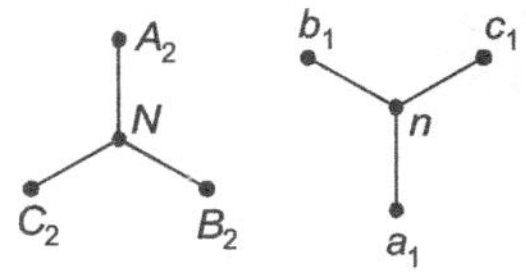
Dd6 180°	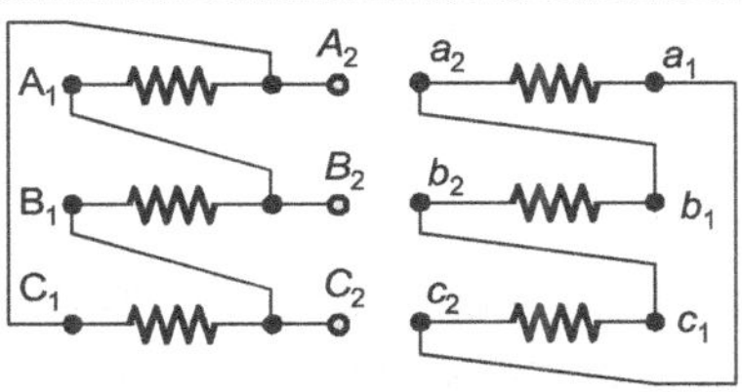	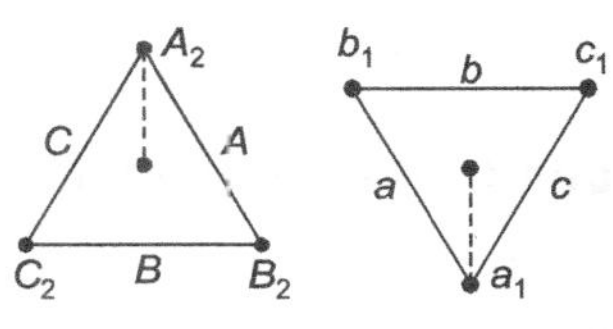

Dy1 –30°	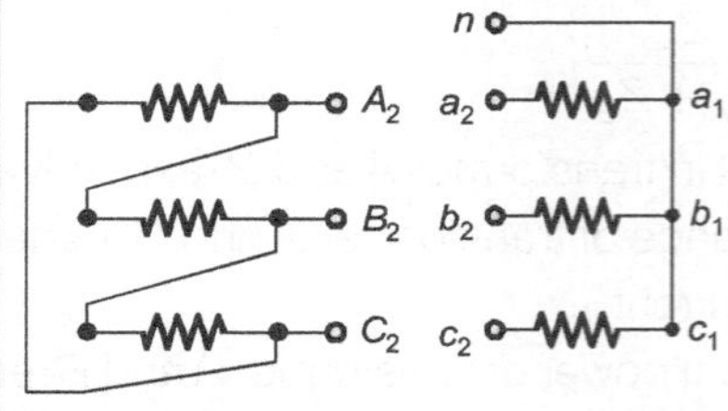	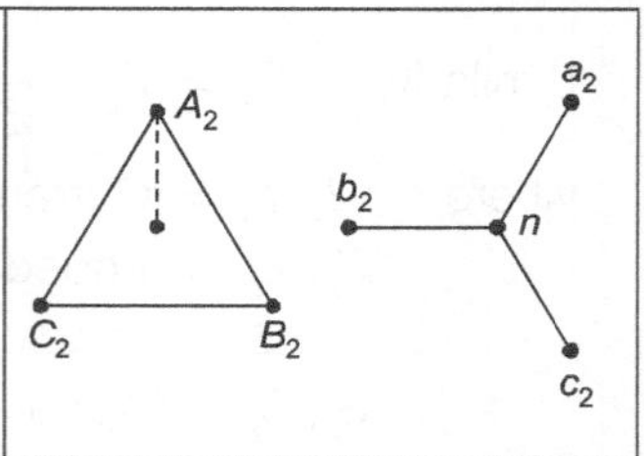
Yd1 –30°	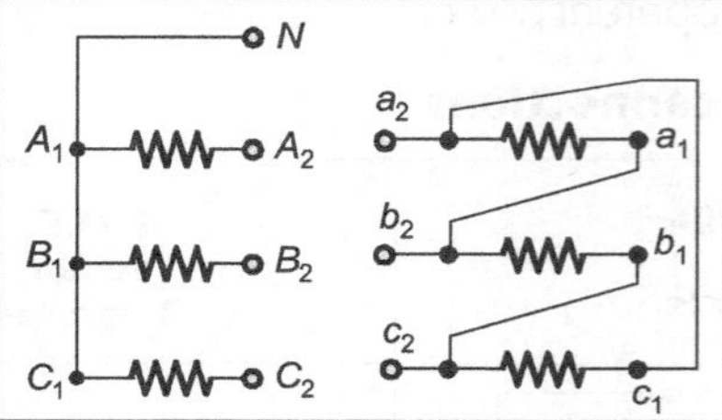	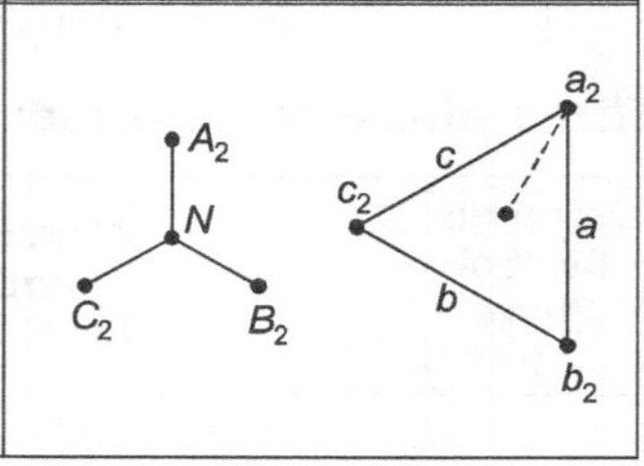
Dy11 +30°	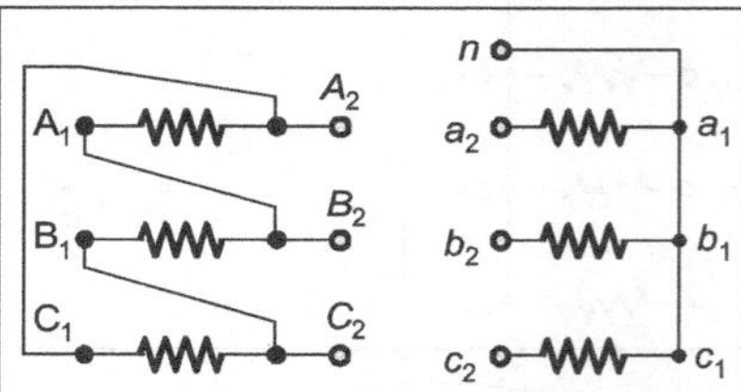	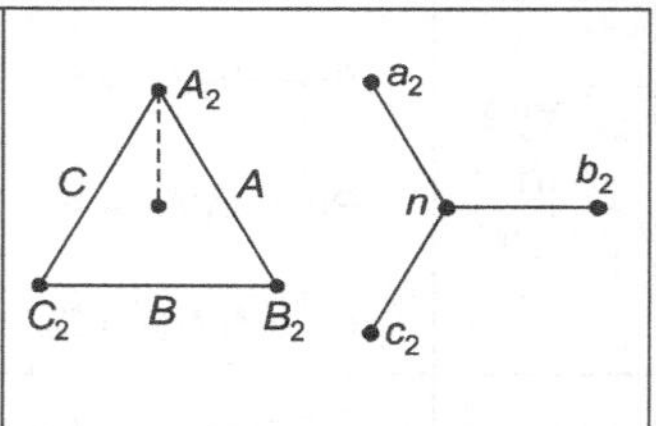
Yd11 +30°	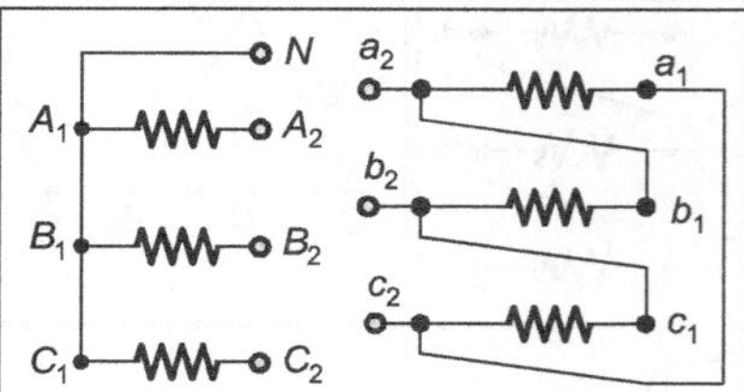	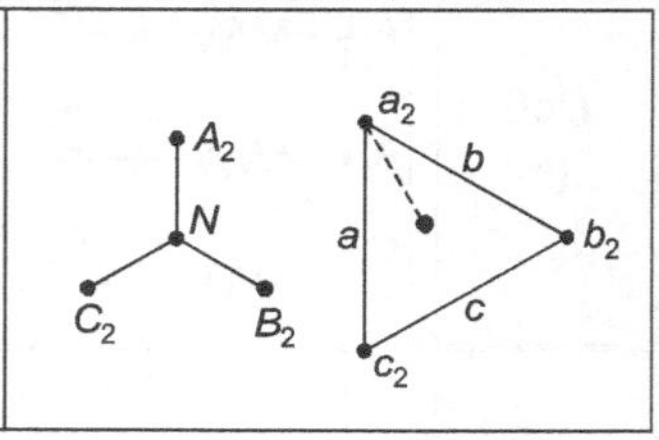

Note:

☑ In star-star transformer, there is no path for flow of 3rd harmonic currents and hence flux and emf one non-sinusoidal. It is not generally used without a tertiary delta.

☑ Star-delta and Delta star transformer can offer ± 30° and ± 150° phase shift whereas star-star and delta-delta transformer only offer 0° and 180° phase shift.

Open Delta Connection

If in three phase transformer bank, connected in delta-delta any one transformer if accidently damaged due to some reason it can be removed and three phase power can still be and supplied though at reduced level.

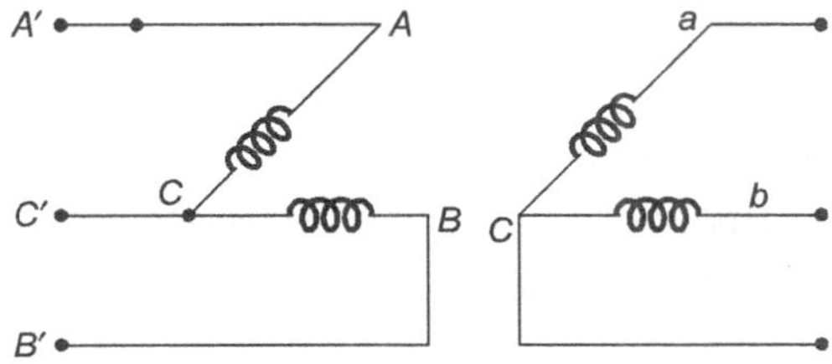

$$\frac{\text{Open delta KVA rating}}{\text{Closed delta KVA rating}} = \frac{\sqrt{3}V_{ph}I_{ph}}{3V_{ph}I_{ph}} = \frac{1}{\sqrt{3}} = 0.577$$

- The rating of open delta connection is 57.7% of delta-delta connection.

Transformer Utilisation Factor (TUF) :

$$\frac{\text{Actual KVA rating}}{\text{Available KVA rating}} = \frac{\sqrt{3}V_{ph}I_{ph}}{2\,V_{ph}I_{ph}} = \frac{\sqrt{3}}{2} = 0.866$$

So, we are only using 86.6% capacity of transformer and hence, capacity is not fully utilized.

II Electromagnetic System

Magneto motive force	MMF = No. of turns in the coil × current = NI
Magnetic flux	$\phi = \frac{\text{MMF}}{\text{Reluctance}}$
Reluctance	$R = \frac{l}{\mu A}$ where, $l \rightarrow$ Length of magnetic path, meter $A \rightarrow$ Area of cross-section normal to flux path, m^2 $\mu \rightarrow \mu_0\mu_r$ = Permeability of the magnetic material
Self inductance	$L = \frac{\Psi}{i}$
Magnetic flux density	$B = \frac{\text{Magnetic flux, }\phi}{\text{Cross-section area, }A}$ Tesla or Wb/m^2
Magnetic field intensity	$H = \frac{\text{MMF}}{\text{Mean length of magnetic circuits}}$ $H = \frac{\text{MMF}}{l}$ AT/m $B = \mu H$
Energy density in electric field	$W_d = \int_0^D E \cdot dl = \frac{1}{2}\frac{D^2}{\epsilon_0}$ where, $D \rightarrow$ Electric field flux density $E \rightarrow$ Electric field intensity or potential gradient = $\frac{D}{\epsilon_0}$

Remember:

☑ Electromagnetic torque developed in any physical systems acts in such a direction as to tend to :

- Decrease magnetic stored energy at constant ψ or ϕ.
- Increase both field energy stored or co-energy at constant current or mmf.
- Decrease reluctance.
- Increase permeance and inductance
- Decrease current i at constant flux linkages ψ.

- $$W_{fld} = W'_{fld} = \frac{1}{2}Li^2 = \frac{1}{2}\frac{\psi^2}{L}$$

  where, W_{fld} is the field energy stored; W'_{fld} is the co-energy.

- For **linear magnetizing curve**

  Force, $$f_e = -\frac{1}{2}\phi^2\frac{\partial R}{\partial x} = \frac{1}{2}i^2\frac{\partial L}{\partial x} = \frac{1}{2}\psi^2\frac{\partial}{\partial x}\left[\frac{1}{L}\right]$$

  Torque, $$T_e = -\frac{1}{2}\phi^2\frac{\partial R}{\partial \theta} = \frac{1}{2}i^2\frac{\partial L}{\partial \theta} = \frac{1}{2}\psi^2\frac{\partial}{\partial x}\left(\frac{1}{L}\right)$$

Electromagnetic Torque

$T_e \propto$ (stator field strength)(Rotor field strength) sin δ, where δ = Torque angle

III Basic Concepts in Rotating Electrical Machines

- **Relationship between Electrical and Mechanical degrees,**

  $$\theta_{elec} = \frac{P}{2}\theta_{mech}$$

  where, $\theta_{elec} \rightarrow$ angle in electrical degree

  $\theta_{mech} \rightarrow$ angle in mechanical degree ; $P \rightarrow$ number of poles

- The peripheral distance between adjacent poles is called pole pitch. It is always equal to 180 electrical degree.

- **Average value of flux density wave over one pole-pitch**

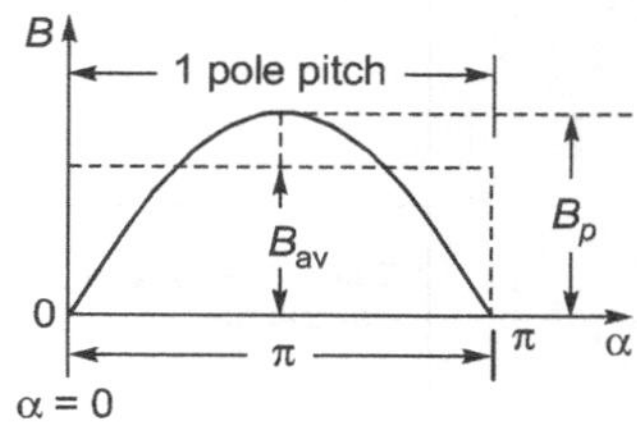

  $$B_{av} = \frac{2}{\pi}B_p$$ where, $B_p \rightarrow$ Max. flux density

- **Total flux per pole,** $\phi = \frac{4}{P}B_p rl$

  where, $l \rightarrow$ Axial length of the armature core

  $r \rightarrow$ Radius of armature core

- **Generated emf in a full-pitched coil,** $e = N\omega_r \phi \sin\omega_r t$

  where, $N \rightarrow$ Number of turns in single full pitched coil

  $\phi \rightarrow$ flux per pole

  $\omega_r \rightarrow$ relative velocity between field flux wave and armature coil

- **Rms value of generated emf in a full-pitched coil,** $E = 4.44 f_r N\phi$

  where, $f_r \rightarrow$ rotational or speed frequency

- **Rotational frequency,** $f_r = \dfrac{Pn_r}{2}$ Hz

  where, $n_r \rightarrow$ relative speed between armature coil and flux-density wave in rps.

- **Generated emf in a short-pitched coil,** $E = 4.44\, k_p f_r N\phi$

  where, $k_p \rightarrow$ coil pitch factor or coil-span factor or pitch factor

- **Pitch factor,** $k_p = \cos\dfrac{\varepsilon}{2}$

  where, $\varepsilon \rightarrow$ Angle of chording (or pitching) for fundamental flux wave.

- **Distribution factor,** $k_d = \dfrac{\sin\left(\dfrac{q\gamma}{2}\right)}{q\sin\left(\dfrac{\gamma}{2}\right)}$

  where, $q \rightarrow$ No. of slots per pole per phase; $\gamma \rightarrow$ Angular slot pitch

  $$\gamma = \frac{180\,P}{\text{Total no. of slots}} \quad \text{electrical degree}$$

Note:

☑ For n^{th} space field harmonic, chording angle becomes $n\varepsilon°$ electrical.

$$k_{pn} = \cos\frac{n\varepsilon}{2}$$

☑ Distribution factor for n^{th} harmonic, $k_{dn} = \dfrac{\sin\left(\dfrac{qn\gamma}{2}\right)}{q\sin\left(\dfrac{n\gamma}{2}\right)}$

- **Winding factor,** $k_w = k_p \times k_d$
- **Phase spread,** $\sigma = q\gamma$ ... in electrical space degrees
- **Synchronous speed**

  Speed at which rotating magnetic field revolves is called synchronous speed

  $$n_s = \frac{120f}{p} \text{ r.p.m.}$$

IV D.C. Machines

- D.C. machine is doubly excited machine.
- D.C. machine is suitable for drives requiring wide range of speed control.
- In DC machine, armature winding is kept on rotor whereas field winding is kept on stator.

Classification of D.C. Machines

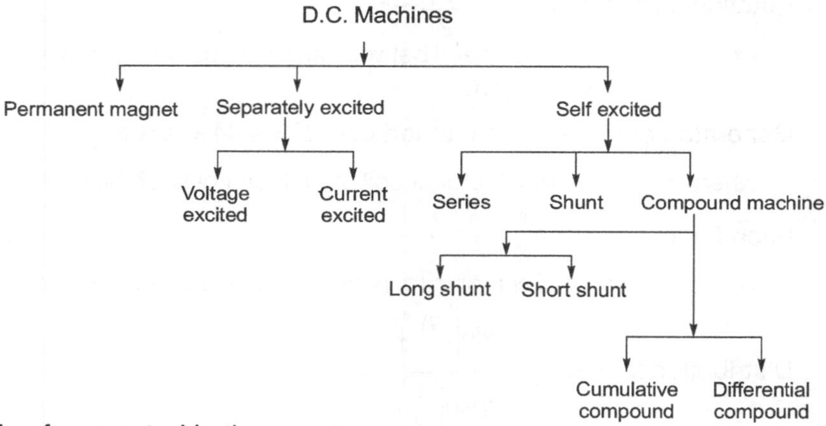

E.m.f generated in the armature

$$E_a = \frac{\phi ZN}{60} \times \left(\frac{P}{A}\right) = k\phi\omega_m$$

where, $\phi \rightarrow$ Flux per pole, wb ; $P \rightarrow$ Number of poles

$Z \rightarrow$ No. of conductor ;

$A \rightarrow$ No. of parallel paths = P for lap winding

= 2 for wave winding

$N \rightarrow$ Speed in rpm

$\omega_m \rightarrow \frac{2\pi N}{60}$ = Mech. angular velocity (mech. rad per sec.)

$k \rightarrow \frac{PZ}{2\pi A}$ = Constant of the machine

- **Torque equation,** $P_m = E_a I_a$; $T.\omega_m = E_a I_a$;

$$T = k\phi I_a = \frac{1}{\omega_m} E_a I_a$$

where, $T \rightarrow$ Torque developed, in N-m ; $I_a \rightarrow$ Armature current

Circuit Model

Separately Excited DC Machine

Field winding is energized by an external D.C. source.

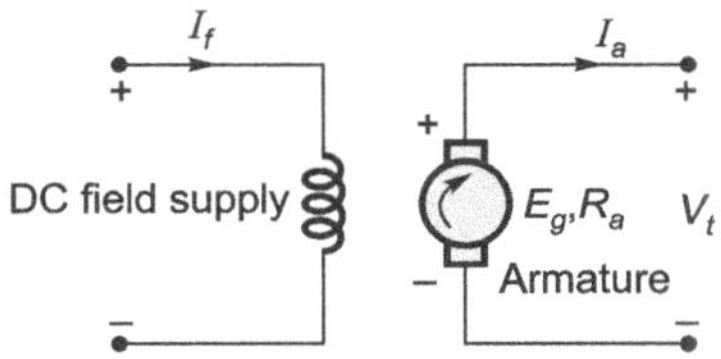

Separately Excited DC Generator

$$E_g = V_t + I_aR_a$$
$$P_g = V_tI_L + I_a^2R_a$$
$$I_a = I_L$$

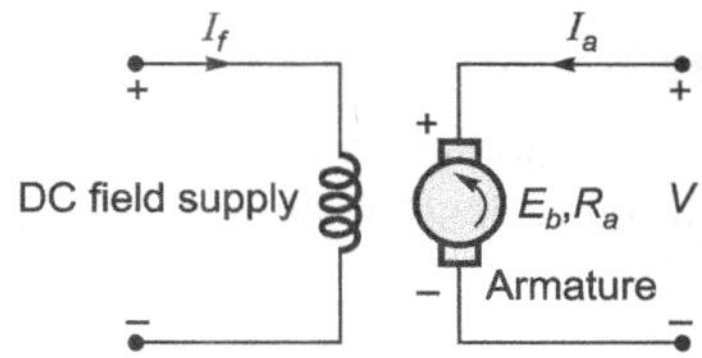

Separately Excited DC Motor

$$E_b = V - I_aR_a$$
$$P_a = VI_L - I_a^2R_a$$
$$I_a = I_L$$

Armature power = $P_a = E_aI_a$

Output = VI_L

Armature Cu-loss = $I_a^2R_a$

where, V_t = Terminal voltage;
E_g = Generated emf;
R_a = Armature resistance;
I_a = Armature current;
I_L = Load current;
V = Supply voltage to motor
E_b = Back emf
R_f = Field resistance
I_f = Field current
$E_g = E_b = E_a$ = Voltage developed in armature.

Shunt Wound DC Machine

A machine in which the field coils are connected in parallel with the armature is called a shunt machine.

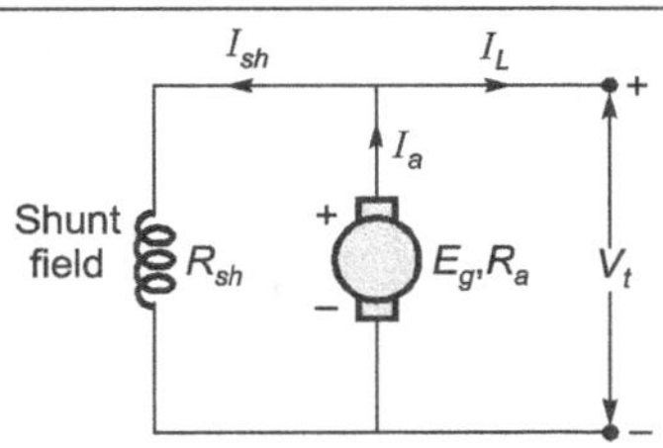

Shunt Wound DC Generator

$$I_a = I_L + I_{sh} \text{ and } I_{sh} = \frac{V}{R_{sh}}$$
$$E_g = V_t + I_aR_a$$
or $$E_gI_a = V_tI_a + I_a^2R_a$$
or $$P_a = VI_L + I_{sh}^2R_{sh} + I_a^2R_a$$

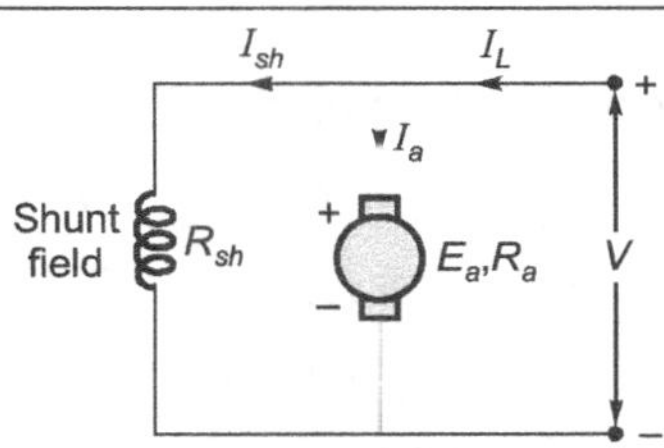

Shunt Wound DC Motor

$$I_a = I_L - I_{sh} \text{ and } I_{sh} = \frac{V}{R_{sh}}$$
$$E_b = V - I_aR_a$$
$$E_bI_a = VI_a - I_a^2R_a$$
or $$P_a = VI_L - I_{sh}^2R_{sh} - I_a^2R_a$$

where, $P_a = E_aI_a$ = Armature power (developed power)

$I_{sh}^2R_{sh}$ = Shunt field Cu loss

$I_{sh}^2R_{sh}$ = Armature cu loss

VI_L = Power delivered

Series Wound DC Machine

A DC machine in which the field coils are connected in series with the armature is called a series machine.

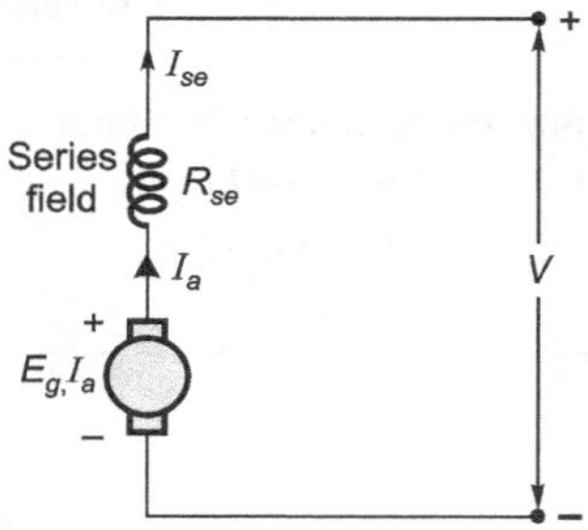

DC Series Generator

$$I_a = I_{se} = I_L$$
$$E_g = V + I_a(R_a + R_{se})$$
$$P_a = VI_L + I_{se}^2R_{se} + I_a^2R_a$$

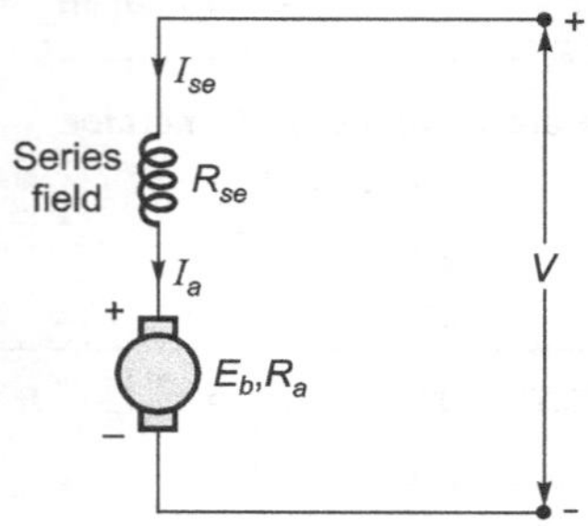

DC Series Motor

$$I_a = I_{se} = I_L$$
$$E_b = V - I_a(R_a + R_{se})$$
$$P_a = VI_L - I_{se}^2R_{se} - I_a^2R_a$$

where, $P_a = E_aI_a$ = armature power (developed power)

VI_L = Power delivered

$I_{se}^2R_{se}$ = Series field Cu loss

Compound Wound DC Machine

A DC machine having both shunt and series fields is called a compound machine.

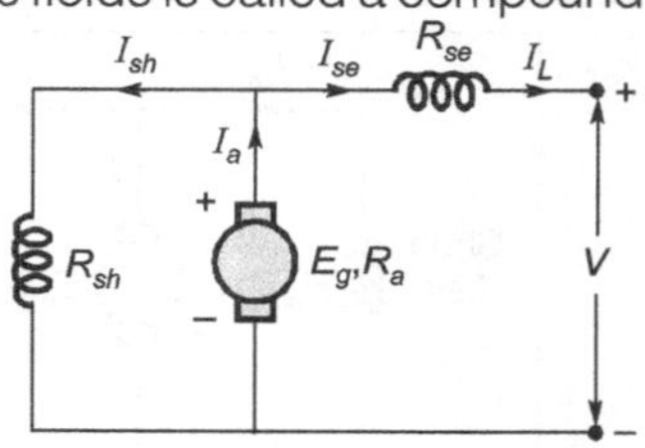

Short shunt compound DC generator :

$$I_a = I_{sh} + I_{se} = I_{sh} + I_L$$
$$E_g = V + I_{se}R_{se} + I_aR_a$$
$$P_a = VI_L + I_{sh}^2R_{sh} + I_{se}^2R_{se} + I_a^2R_a$$

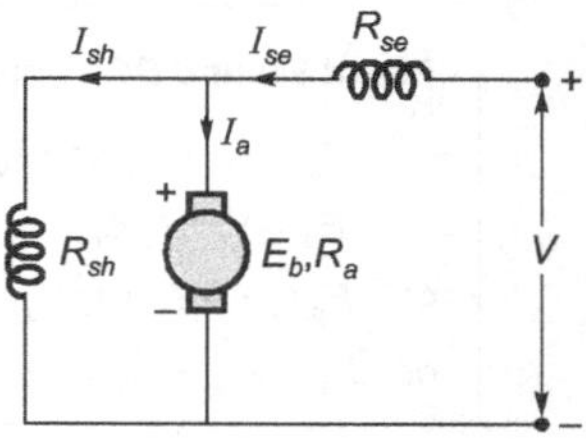

Short shunt compound DC motor :

$$I_a = I_L - I_{sh} \text{ and } I_{se} = I_L$$
$$E_b = V - I_{se}R_{se} - I_a = R_a$$
$$P_a = VI_L - I_{sh}^2R_{sh} - I_{se}^2R_{se} - I_a^2R_a$$

where, P_a = Power developed

VI_L = Power delivered

$I_{sh}^2 R_{sh}$ = Shunt field Cu loss

$I_{se}^2 R_{se}$ = Series field Cu loss

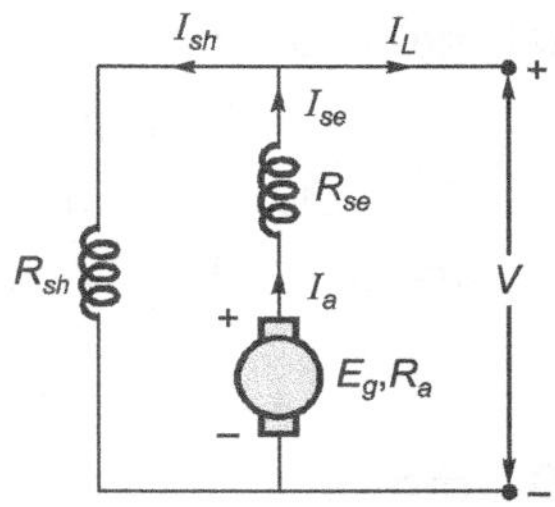

Long shunt DC generator :

$$I_a = I_{se} = I_L + I_{sh}$$
$$E_g = V + I_a(R_a + R_{se})$$
$$P_a = VI_L + I_{sh}^2 R_{sh} + I_{se}^2 R_{se} + I_a^2 R_a$$

Long shunt DC motor :

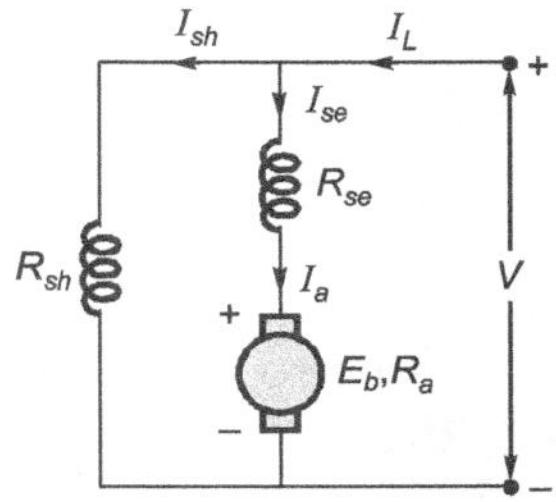

$$I_a = I_{se} = I_L - I_{sh}$$
$$E_b = V - I_{se}R_{se} - I_a R_a$$
$$P_a = VI_L - I_{sh}^2 R_{se} - I_{se}^2 R_{se} - I_a^2 R_a$$

Note:

☑ Generated emf in generator is called back emf in motor.

☑ For shunt motor flux, ϕ = Constant. $\therefore T \propto I_a$

☑ For series motor flux $\propto I_a$. $\therefore T \propto I_a^2$

Power Balance in D.C. Machine

Generator Action

Shaft power input —[− (Rotational losses)]→ Armature power Developed ($E_a I_a$) —[− (All copper losses including brush drop)]→ Electrical power output

Motor Action

Electrical power input —[− (All copper losses including brush drop)]→ Armature power Developed ($E_a I_a$) —[− (Rotational losses)]→ Shaft power output

- **Armature power developed,** $P_a = E_a I_a$
- **Rotational loss** = Mechanical loss + Magnetic loss

  Mechanical loss = Friction and windage loss

  Magnetic loss = Hysteresis and eddy current loss

Efficiency

$$\eta = 1 - \frac{\text{Losses}}{\text{Input power}} = \frac{\text{Output power}}{\text{Input power}} = \frac{\text{Output power}}{\text{Output power} + \text{Losses}}$$

DC generator

$$\eta_g = \frac{\text{Output power}}{\text{Output power + losses}}; \quad \eta_g = \frac{V_t I_L}{V_t I_L + (I^2R + V_{BD}I + P_K)}$$

where, $V_tI_L \rightarrow$ Output power ; $I^2R \rightarrow$ Total ohmic losses
$P_K \rightarrow$ Constant losses ; $V_{BD}I \rightarrow$ Brush contact losses
$V_t \rightarrow$ Output voltage ; $I_L \rightarrow$ Load current

DC motor

$$\eta_m = \frac{\text{Input power – Losses}}{\text{Input power}} \times 100\%;$$

$$\eta_m = \frac{VI - (I^2R + V_{BD}I + P_K)}{VI} \times 100\%$$

where, $VI \rightarrow$ Input electrical power

Note:

- ☑ For generator and motor maximum efficiency occurs when
  constant loss = variable loss, i.e., $P_k = I^2R$ (ohmic loss)
- ☑ Maximum power output by DC motor when
  $$E_a = \frac{V}{2} \text{ and } I_a = \frac{V}{2R_a}$$
- ☑ When motor operates at maximum power output, it gives only 50% efficiency.

Characteristics of D.C. Generator

- **No load (or) Magnetisation characteristic (or) Open circuit characteristics (O.C.C.) :** Magnetisation characteristic gives the variation of generated voltage and field current (E_g Vs I_f).
- **Internal characteristics (E_g Vs I_L) :** It is plot between the generated voltage and load current.
- **Load or external characteristics (V_t Vs I_L) :** Between terminal voltage Vs load current (I_L) at a constant speed.
- **Open circuit characteristics (OCC) (no load) for all types of generator**

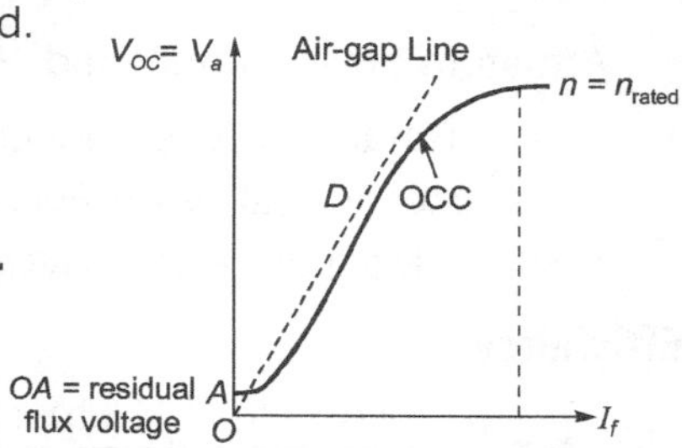

Open circuit characteristics of all types of generator

Note:

☑ OCC does not start from origin due to the presence of residual magnetism.

Characteristics for separately excited generator

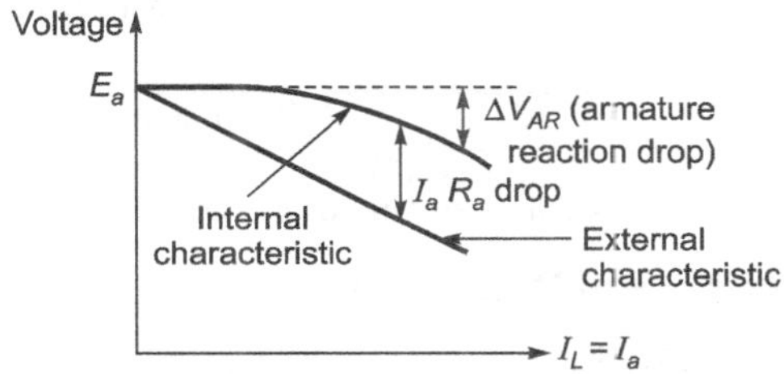

Characteristics of separately excited DC generator

Characteristics for shunt generator

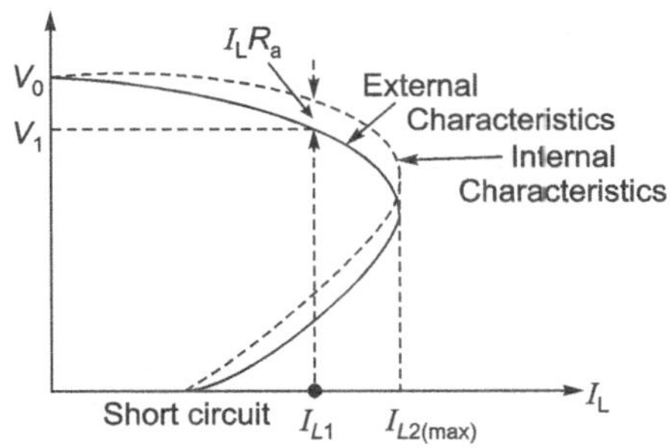

Characteristics for series generator

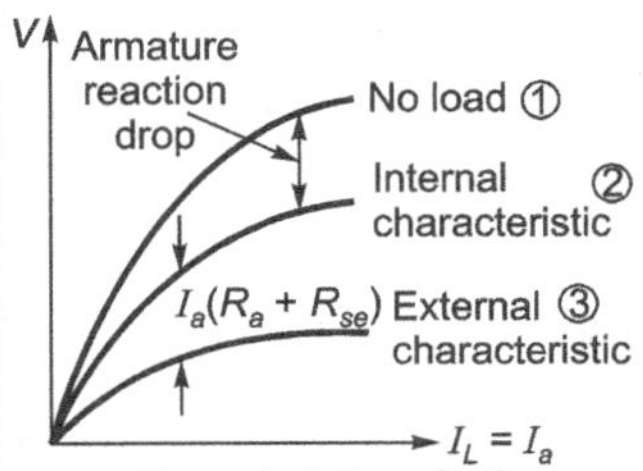

Characteristics of DC series generator

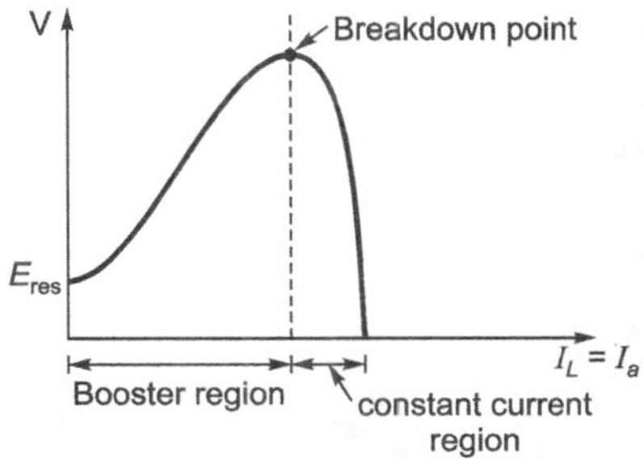

External characteristics of series generator

Note:

☑ Constant current region is suitable for welding.

☑ Series generator is used in booster region for line drop compensation.

- **Characteristics for compound generator**

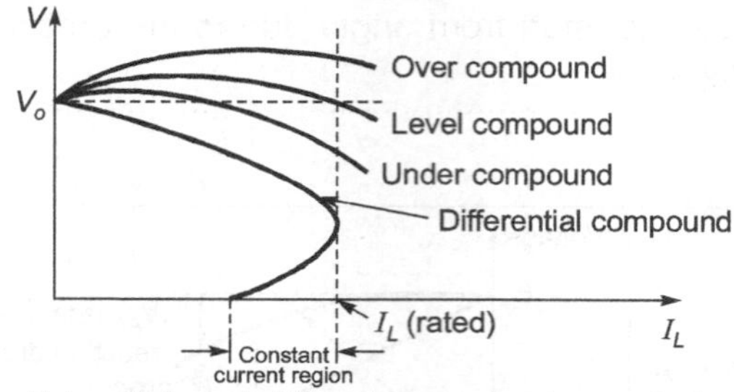

External characteristics of a compound generator

Characteristics of D.C. Motor

1. Speed Vs armature current (N Vs I_a).
2. Torque Vs armature current (T Vs I_a).
3. Speed Vs torque (N Vs T).

Characteristics for DC shunt motor

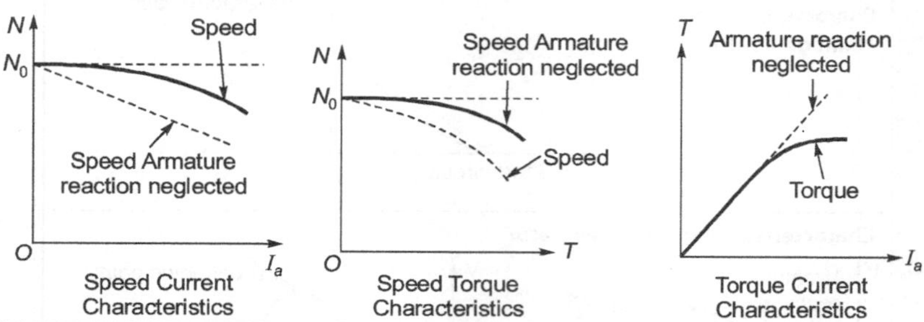

Characteristics for DC series motor

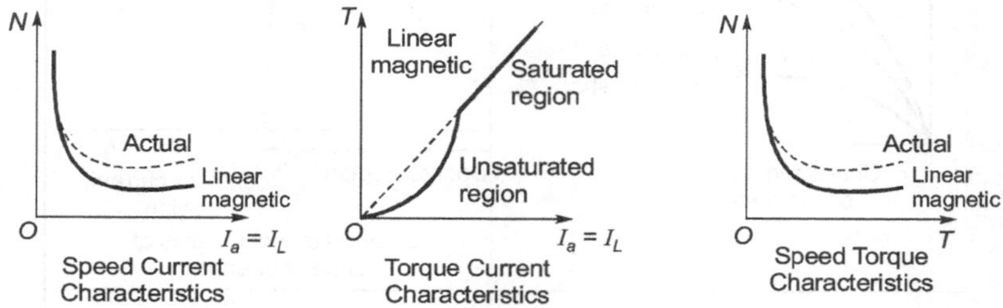

Note:

☑ In a traction system, we use series motor where high starting torque is required.

- **Characteristics of DC compound motor**

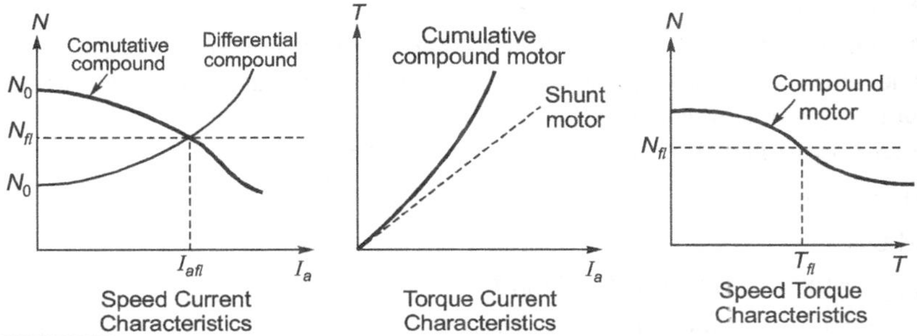

Note:

☑ Exciting or field windings produces the working flux or main flux.

☑ Armature winding, carries supply or load current and determines the rating of machine.

☑ For the development of electromagnetic torque in all rotating machine, the number of rotor poles should be equal to the number of stator poles.

☑ Order of air gap in different machine.

☑ Salient pole SM > DC machine > Non salient pole machine > Slip ring IM > Squirrel cage IM

Voltage Build up fails in generator if

- Residual flux absent.
- Field connection is reversed, I_f destroys ϕ_{Res}.
- Direction of rotation is wrong.
- Field resistance is more than critical field resistance.
- Speed less than critical speed.

Critical Speed : The minimum rotor speed at which voltage builds up occurs at particular field resistance.

Critical Resistance : The maximum resistance at which voltage build up occurs at particular speed.

Remember:

☑ Commutator serve as:
For D.C. generator: Mechanical rectifier (converts ac into dc)
For D.C. motor: Mechanical inverter (converts dc into ac)

☑ D.C. series motor never runs at no load.

Armature Reaction

- The effect of armature flux on the main field flux distribution in the air gap is called armature reaction.

- The armature mmf produces two undesirable effects on the main flux.
    - Net reduction in the main field.
    - Distortion of the main field flux wave along the air gap periphery.
- The effect of armature flux on the main field is cross-magnetizing as well as demagnetizing.
- Flux created by the armature mmf is called cross-flux,

$\overline{\phi}_R = \overline{\phi}_a + \overline{\phi}_f$

where,

$\phi_f \rightarrow$ Flux produced by field mmf

$\phi_a \rightarrow$ Flux produced by armature mmf

$\phi_R \rightarrow$ Resultant flux

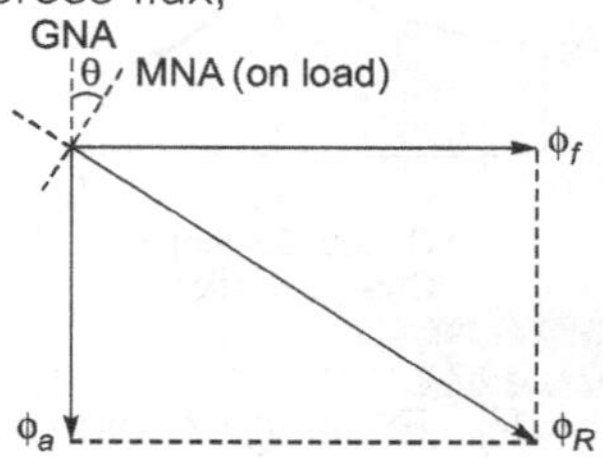

Remember:

☑ Magnetic neutral axis (MNA) is always perpendicular to the axis of resultant field flux.

☑ Geometric neutral axis (GNA) is along the quadrature axis of the d.c. machine.

Demagnetizing ampere turns

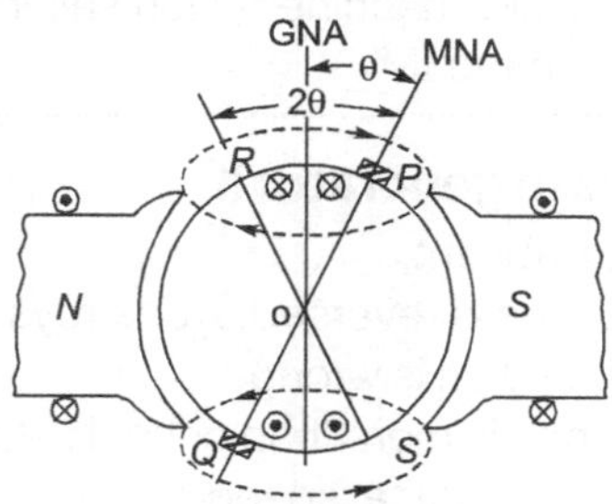

- Demagnetizing armature mmf per pole.

$$F_{ar(demag)} \text{ per pole} = \frac{(Z/2)}{P} \times \left(\frac{2\theta_{elc}}{180^\circ}\right) \times \frac{I_a}{A}$$

Cross-magnetizing ampere turns

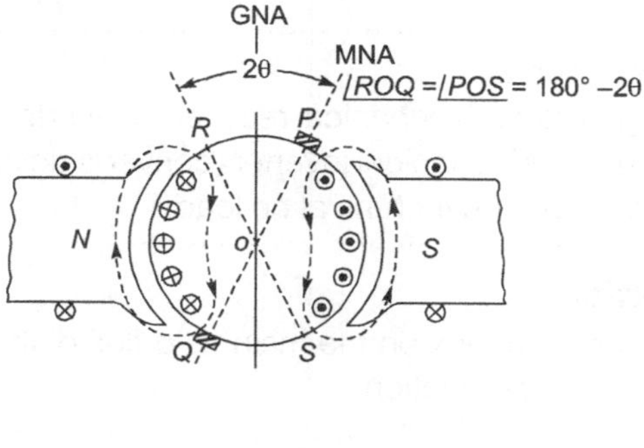

- **Cross magnetizing armature mmf per pole.**

$$F_{ar(cross)} \text{ per pole} = \frac{(Z/2)}{P} \times \left(1 - \frac{2\theta_{elc}}{180°}\right) \times \frac{I_a}{A}$$

- **Compensating winding mmf per pole (AT_c).**

$$AT_c = \frac{(Z/2)}{P} \cdot \frac{\text{Pole arc}}{\text{Pole pitch}} \times \frac{I_a}{A}$$

Note:

☑ Compensating winding mmf neutralizes the armature mmf only under the main pole face.

☑ Interpoles are use to neutrallised the armature reaction flux in the interpolar axis. It also produce some rotational voltage in the coil undergoing commutation and neutralize the reactance voltage and improve the commutation.

☑ Polarity of interpoles is same as the succeeding main pole in generator action and of the preceding main pole in motor action.

☑ The interpole winding and compensating winding carry armature current.

Methods of Speed Control in D.C. Machine

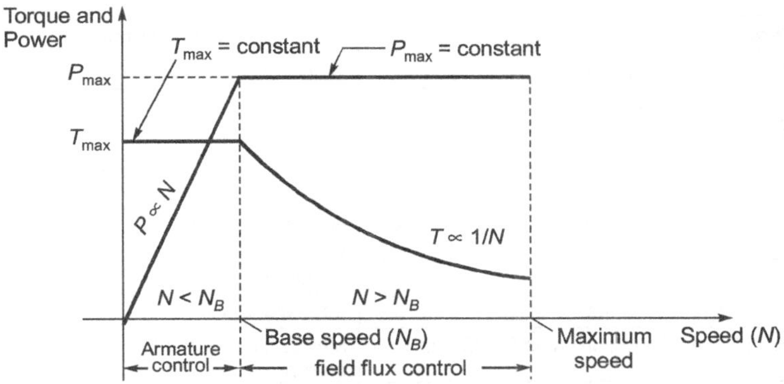

Torque and power characteristics is combined armature voltage and field control

- **Armature voltage control:**
  - Constant torque drive.
  - Speed control is possible only below base speed, i.e., $N < N_B$.
- **Armature resistance control:**
  - Constant torque drive.
  - Speed control is possible only below base speed, i.e., $N < N_B$.
  - Wide range of speed control is not possible.

- Field flux control:
  - It is constant power drive.
  - Speed control is possible above base speed i.e. $N > N_B$.

V Polyphase Induction Motors

Introduction

Induction Motor is singly-excited a.c. machine. Its stator winding is directly connected to a.c. source, whereas its rotor winding receives its energy from stator by means of induction.

Types of Induction Motor Rotors

Squirral - Cage rotor

- Robust construction and cheaper.
- Absence of brushes reduces the risk of sparking.
- Lesser maintenance.
- Higher efficiency and higher power factor.

Wound rotor or Slip-ring rotor

- Higher starting torque and low starting current.
- Additional resistance can be connected in the rotor circuit to control speed.

S.No.	Induction Machine	Transformer
1.	Singly excited machine	Singly excited machine
2.	Stator winding	Primary
3.	Rotor winding	Secondary
4.	Basic principle of operation is induction	Basic principle of operation is mutual induction
5.	Electro-mechanical energy conversion	No electro-mechanical energy conversion
6.	Air gap	No air gap
7.	High reluctance	Low reluctance
8.	Magnetizing current is 30-35% of rated current	Magnetizing current is 4-5% of rated current
9.	Variable frequency	Constant frequency

Slots

Parameter	Open	Semi-Open	Closed
Magnetizing current	Highest	Moderate	Least
No load Pf	Least	Moderate	Highest
Leakage reactance	Least	Moderate	Highest
Full load Pf	Least	Moderate	Highest
Harmonic Torque	Highest	Moderate	Least

- In 3-phase induction motors, semi open type of slots are generally adopted.
- In dc machines open slots are preferred, so that reactance voltage is reduced and commutation is improved.
- In synchronous machine, open slots are used to reduce leakage reactance so that maximum power is increased and stability improves.

Production of Rotating Magnetic Field in 3ϕ System

- If a 3ϕ balance winding is excited by 3ϕ balance current, a resultant mmf or flux is produced of constant magnitude of $\frac{3}{2}F_m$ or $\frac{3}{2}\phi_m$ in air gap of motor and rotate in space at synchronous speed (N_s).
- The direction of rotation of resultant flux in the air gap depends upon the phase sequence.

- **Synchronous speed,** $N_s = \frac{120f}{P}$

  where, $f \to$ Supply frequency ; $P \to$ Number of poles

- **Slip,** $s = \frac{N_s - N_r}{N_s}$

  where, $N_s \to$ Synchronous speed, rpm ; $N_r \to$ Rotor speed, rpm

- **Rotor frequency,** $f_2 = sf_1$ where, $f_1 \to$ Line frequency

Relative Speeds

1. Induction Motor

Speed of	With respect to			
	Stator	Rotor	Stator field	Rotor field
Stator	0	$-N_r$	$-N_S$	$-N_S$
Rotor	$N_r = N_S(1-S)$	0	$(N_r = N_S) = -SN_S$	$-SN_S$
Stator field	N_S	$(N_S - N_r) = SN_S$	0	0
Rotor field	N_S	SN_S	0	0

2. Inverted Induction Motor

Speed of	With respect to			
	Stator	Rotor	Stator field	Rotor field
Stator	0	N_r	$-SN_S$	$-SN_S$
Rotor	$-N_r$	0	$-N_S$	$-N_S$
Stator field	SN_S	N_S	0	0
Rotor field	SN_S	N_S	0	0

Per phase induced emf

- In stator winding, $E_1 = \sqrt{2}\pi N_1 f_1 \phi k_{\omega 1}$
- In rotor winding,

  At standstill $E_2 = \sqrt{2}\pi N_2 f_1 \phi k_{\omega 2}$

  At any slip 's' $E_2 = \sqrt{2}\pi N_2 (sf_1) \phi k_{\omega 2}$

  where, $k_{\omega 1}, k_{\omega 2}$ → Winding factors of stator and rotor windings respectively

  N_1, N_2 → Number of turns of stator and rotor winding respectively

Rotor Emf current and power

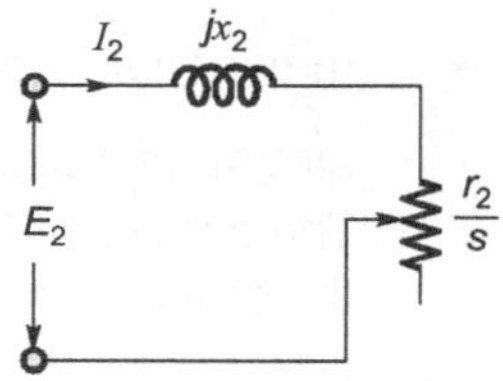

- Per phase rotor current at any slip 's'

$$I_2 = \frac{E_2}{\sqrt{\left(\frac{r_2}{s}\right)^2 + x_2^2}}$$

where, r_2 → Rotor resistance at standstill

x_2 → Rotor leakage reactance at standstill

E_2 → Per phase value of induced Emf at standstill

- Rotor power factor
  - Rotor current I_2 lags the rotor voltage E_2 by an angle θ_2.

$$\theta_2 = \tan^{-1}\left(\frac{sx_2}{r_2}\right)$$

  - Rotor power factor = $\cos\theta_2$

- **Per phase power input to rotor,** $P_g = E_2 I_2 \cos\theta_2 = I_2^2 \frac{r_2}{s}$

P_g is the power transferred from stator to rotor across the air gap also called air-gap power.

- **Rotor ohmic loss =** $I_2^2 r_2 = sP_g$
- **Internal mechanical power developed in rotor,** $P_m = (1-s)P_g$
- **Internal (or gross) torque developed per phase**

$$T_e = \frac{\text{Intermal mechanical power developed in rotor}}{\text{Rotor speed in mechanical rad. per sec.}}$$

$$T_e = \frac{P_g}{\omega_s} = \frac{P_m}{\omega_r} = \frac{\text{Rotor ohmic loss}}{\omega_s \times \text{slip}}$$

where, ω_s, ω_r → Synchronous speed and rotor speed in mechanical rad. per sec.

- Output or shaft power

$$P_{sh} = P_m - \text{Mechanical losses}$$

$$P_{sh} = P_g - \text{Rotor ohmic losses} - \text{Mechanical losses}$$

Note:

☑ Friction and windage losses are the mechanical losses.

- Output or shaft torque, $T_{sh} = \dfrac{P_{sh}}{\text{rotor speed}} = \dfrac{P_{sh}}{(1-s)\omega_s}$

- Air gap power

$$P_g = \text{Stator power input} - \text{Stator } I^2R \text{ loss} - \text{Stator core loss}$$

Power Flow Diagram

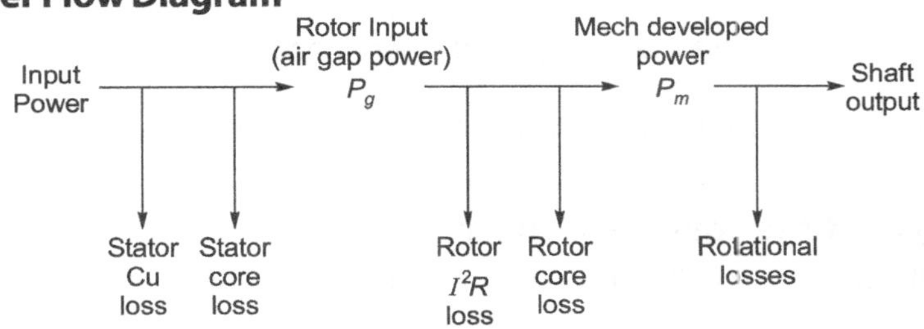

Efficiency

- Efficiency of a 3-phase induction motor, $\eta = \dfrac{P_{sh}}{P_{sh} + P_f + P_{oh}} \times 100$

where, $P_{sh} \rightarrow$ Shaft power

$P_f \rightarrow$ Fixed losses = core loss + friction and windage losses

$P_{oh} \rightarrow$ Stator and rotor ohmic losses

Equivalent Circuit

Exact equivalent circuit referred to stator

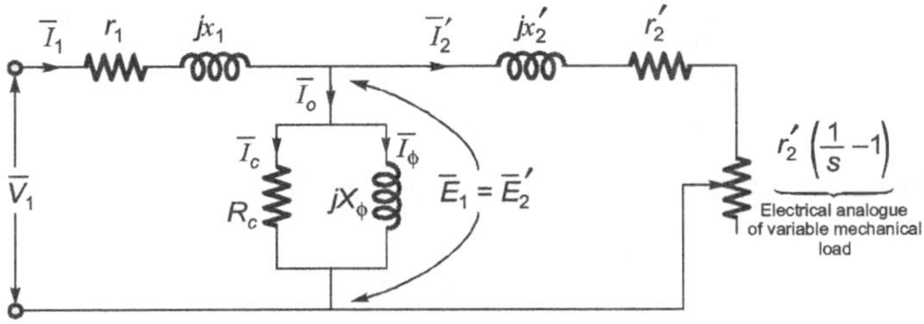

- Transformation ratio

$$a = \frac{E_1}{E_2} = \frac{(\text{Effective number of stator-series turns per phase } (N_1'))}{(\text{Effective number of rotor-series turns per phase } (N_2'))}$$

- Per phase generated Emf in rotor circuit referred to stator side

$$E_2' = aE_2$$

- Per phase rotor current referred to stator side, $I_2' = \dfrac{I_2}{a}$
- Per phase rotor resistance referred to stator side, $r_2' = a^2 r_2$
- Per phase rotor leakage reactance referred to stator side

$$x_2' = a^2 x_2$$

where, $r_1, r_2 \rightarrow$ Per phase stator and rotor resistances.
$x_1, x_2 \rightarrow$ Per phase stator and rotor leakage reactances at standstill.

Note:

☑ $\dfrac{R_2'}{s}$ may be divided into two components.

☑ R_s' represents rotor copper loss per phase.

☑ $R_2'\left(\dfrac{1-s}{s}\right)$ represents developed mechanical power.

Analysis of the equivalent circuit

All the rotor quantities have been referred to stator side.

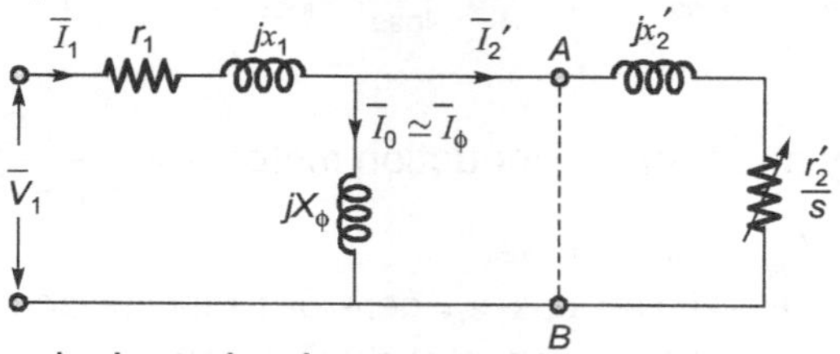

- Thevenin equivalent circuit

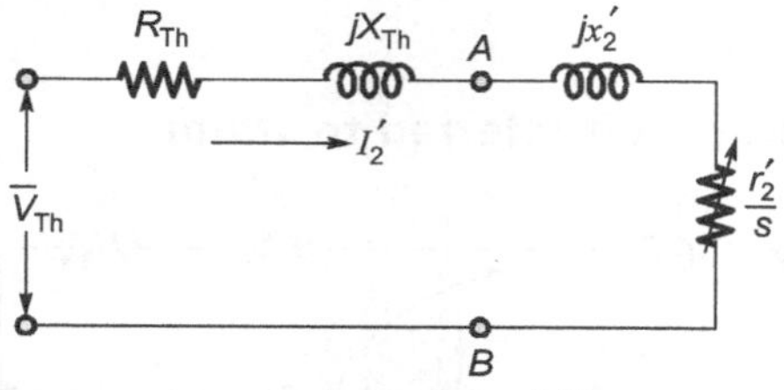

- Thevenin voltage across points A and B,

$$V_{Th} = \frac{V_1}{r_1 + j(x_1 + X_\phi)} \cdot jX_\phi$$

- Thevenin equivalent impedance across points A and B

$$Z_{Th} = \frac{(r_1 + jx_1) \times jX_\phi}{r_1 + j(x_1 + X_\phi)} = R_{Th} + jX_{Th}$$

Thevenin equivalent resistance and reactance across points A and B Neglecting r_1 as $r_1 << (x_1 + X_\phi)$

$$R_{Th} = \frac{r_1 X_\phi}{X} \quad \text{and} \quad X_{Th} = \frac{x_1 X_\phi}{X}$$

where, $X_\phi \rightarrow$ Magnetizing reactance

$V_1 \rightarrow$ Stator applied voltage; $X = x_1 + X_\phi$

- Total torque : $T_e = \dfrac{K_t}{\left(R_{Th} + \dfrac{r_2'}{s}\right)^2 + X_1^2} \cdot \dfrac{r_2'}{s}$, N-m

where, $K_t \rightarrow \dfrac{3V_{Th}^2}{\omega_s}$ and $X_1 = x_2' + X_{Th}$

- Maximum internal torque : $T_{em} = \dfrac{K_t}{2\left[R_{Th} + \sqrt{R_{Th}^2 + X_1^2}\right]}$, N-m

Note:

☑ Maximum internal torque is also referred as stalling torque, pull-out torque or breakdown torque.

Slip at maximum torque	$s_{mT} = \dfrac{r_2'}{\sqrt{R_{Th}^2 + X_1^2}}$
Starting torque	$T_{e,st} = \dfrac{K_t}{(R_{Th} + r_2')^2 + X_1^2} \cdot r_2'$
Motor torque in terms of maximum torque	$T_e = \dfrac{2T_{em}}{\dfrac{s_{mT}}{s} + \dfrac{s}{s_{mT}}}$
Total mechanical power developed	$P_m = \dfrac{3V_{Th}^2}{\left[R_{Th} + r_2' + \dfrac{r_2'(1-s)}{s}\right]^2 + X_1^2} \cdot r_2'\left(\dfrac{1-s}{s}\right)$
Slip at maximum power	$s_{mP} = \dfrac{r_2'}{\sqrt{(R_{Th} + r_2')^2 + X_1^2} + r_2'}$
Maximum power	$P_{m,\max} = \dfrac{3V_{Th}^2}{2\left[R_{Th} + r_2' + \sqrt{(R_{Th} + r_2')^2 + X_1^2}\right]}$

Remember:

☑ $P_{m,\max}$ depends on rotor resistance r_2' whereas T_{em} does not.

☑ At starting, internal mechanical power developed is zero.

- In order to get maximum power output from induction generator. The rotor speed (n_r)

$$n_r = n_s\left[1 + \frac{r_2'}{\sqrt{(R_{Th} + r_2')^2 + X_1^2} + r_2'}\right]$$

$$\left(\frac{I_{2,st}}{I_2}\right)^2 = \frac{1 + \left(\dfrac{s_{mT}}{s^2}\right)^2}{1 + s_{mT}^2} \Rightarrow \left(\frac{I_{2,mT}}{I_2}\right)^2 = \frac{1}{2}\left[1 + \left(\frac{s_{mT}}{s}\right)^2\right]$$

where, $I_{2,st} \rightarrow$ Stator load component of current at starting

$I_{2,mT} \rightarrow$ Stator load component of current at maximum torque

Approximate Analysis

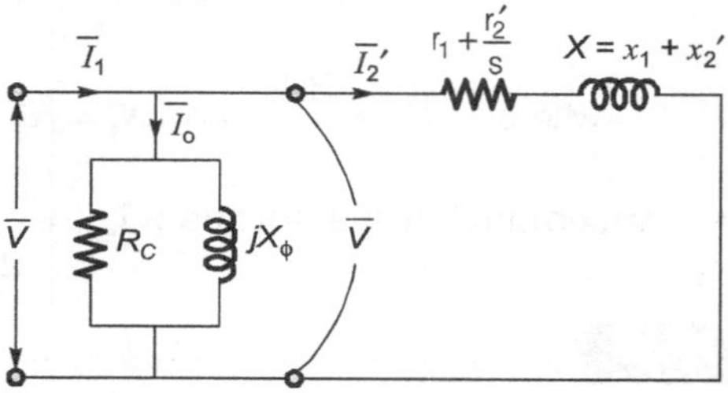

$$I_2' = \frac{V}{\sqrt{\left(r_1 + \dfrac{r_s'}{s}\right)^2 + X^2}}$$

$$T = \frac{3}{\omega_s}\frac{V^2 r_2'/s}{\left(r_1 + \dfrac{r_2'}{s}\right)^2 + X^2}$$

Starting torque	$T_{st} = \dfrac{3}{\omega_s}\cdot\dfrac{V^2 r_2'}{(r_1 + r_2')^2 + X^2}$
Maximum torque	$T_m = \dfrac{3}{\omega_s}\left(\dfrac{V^2}{2x_2'}\right)$
Slip at max. torque	$s_m = \dfrac{r_2'}{x_2'} = \dfrac{r_2}{x_2}$

Remember:

☑ At high slip (i.e., $s \approx 1$) : $T \propto \dfrac{1}{s}$ ☑ At low slip (i.e., $s \approx 0$) : $T \propto s$

Torque Slip Characteristics

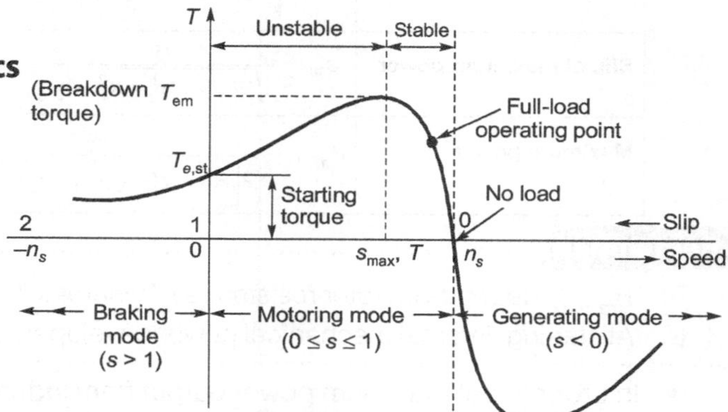

Torque-slip characteristics

Remember:

- ☑ $T_m \propto \frac{1}{f^2}$
- ☑ $T_{st} \propto \frac{1}{f^3}$
- ☑ Efficiency of IM, $\eta \propto \frac{1-s}{1+s}$; where, f = frecuency, s = slip

Determination of equivalent circuit parameters

No Load Test (O.C. Test):

Test performed at rated voltage to determine

- Iron loss at rated voltage, i.e., P_i.
- Shunt branch parameter, i.e., R_C and X_ϕ.

where,

I_o = No load current

V = Rated voltage applied to stator

P_i = Fixed losses read by watt meter

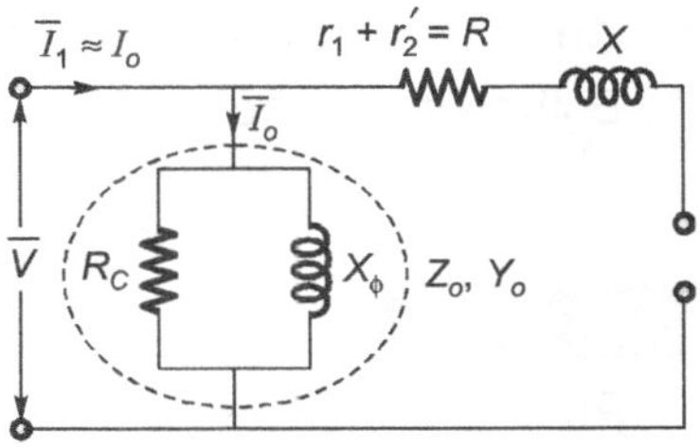

$$R_c = \frac{V^2}{P_i} \qquad G_C = \frac{1}{R_C}$$

$$Z_o = R_C \parallel jX_\phi \qquad Y_o = \frac{1}{Z_o} = G_C - jB_\phi$$

$$Y_o = \frac{I_o}{V} \qquad B_\phi = \sqrt{Y_0^2 - G_C^2}$$

$$X_\phi = \frac{1}{B_\phi}$$

Block Rotor Test

- Supply is given but the rotor is blocked to rotate mechanically.
- This test is also called short circuit test to determine copper loss at rated current. This test is performed at rated current condition.

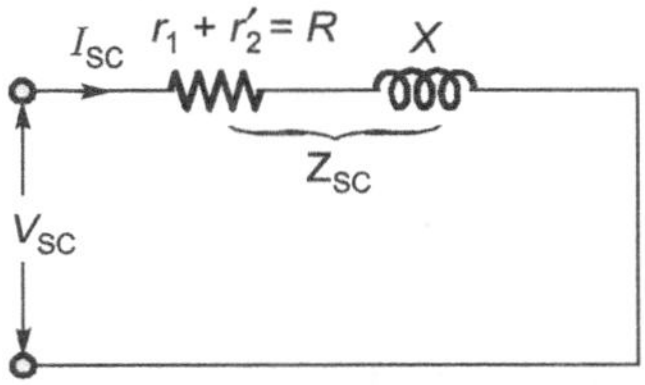

where, $I_{sc} \rightarrow$ Rated current

$V_{sc} \rightarrow$ Voltage that is required to circulate I_{sc} when rotor is blocked

$P_{sc} \rightarrow$ Total copper loss on full load at standstill, read by wattmeter

$$R \rightarrow \frac{P_{SC}}{I_{SC}^2}; \qquad Z_{SC} \rightarrow \frac{V_{SC}}{I_{SC}}; \qquad X \rightarrow \sqrt{Z_{sc}^2 - R^2}$$

Note:

☑ In block rotor condition, reduced voltage (up to 5% of rated) is required to flow the rated current.

Starting Method of 3ϕ Induction Motor

- **Direct Online Starting (D.O.L)**

  In direct online starting full supply voltage is applied across the winding.

- **Reduced Voltage Starting**

  To reduce the starting current if voltage is reduce by a fraction x, the starting current is also reduce by a fraction x but the starting torque gets reduce significantly by a factor x^2.

  $$I_{st} = xI_{sc}; \qquad T_{st} = x^2 T_{st}(\text{DOL})$$

- **Auto Transformer Starting**

  Both starting current from supply and torque are reduced by same factor x^2.

  $$I_{st} = x^2 I_{sc}; \qquad T_{st} = x^2 T_{st}(\text{DOL})$$

Remember:

☑ If $x = 1/\sqrt{3}$ results are same as Y-Δ starting.

- **Start-Delta Starting**

  In this, Δ connected induction motor started as Y connected, and as motor picks up the speed the connection are changed to Δ i.e. it runs as Δ connected.

  If delta connected I.M. is started as star connected both the starting current and torque are reduce by a factor of 1/3.

  $$I_{st} = \frac{1}{3} I_{sc}; \qquad T_{st} = \frac{1}{3} T_{st}(\text{DOL})$$

Remember:

☑ By increasing rotor resistance, starting torque increases but maximum torque remains unchanged.

Speed Control of Induction Motor

Rotor speed of induction motor is, $N_r = N_s(1-s)$

The rotor speed can be controlled by :
- slip control technique; • synchronous speed control technique.

- **Slip Control Technique**

$$T = \frac{sV^2}{R} \quad \text{where } T \text{ remains constant.}$$

  - **Voltage control technique:** Applicable for both I.M. i.e. slip ring induction motor (SRIM) and squirrel cage I.M.
  - **Rotor resistance method:** Only in SRIM.
  - **Rotor emf injection method:** Only in SRIM.

- **Synchronous Speed Control Technique**

$$N_s = \frac{120f}{P} \quad \text{and} \quad N_r = N_s(1-s)$$

  - **Frequency control technique:** Applicable in both types of induction motors.
  - **Pole changing technique:** Only in squirrel cage I.M.
  - **Cascading of 2 induction motor:** In this method one I.M. must be SRIM, while other can be either squirrel cage or SRIM.

Cogging

No. of Slot in stator = S_1; No. of slot in rotor = S_2

If either $S_1 = S_2$ or integral multiple to each other then the stator teeth and rotor teeth which are actually parallel, may develop very strong alignment force at the time of start and therefore the induction motor may fail to start, with rotor teeth blocked against the stator teeth. This phenomenon is known as cogging.

Crawling

The airgap flux in a 3ϕ induction motor contains harmonics of the order of $(6n \pm 1)$, then create parasitic torque. Due to these harmonic, motor continue to rotate stably at a speed N much less then full load speed N_{fl}. This phenomenon is called crawling.

Remember:

☑ Cogging and crawling phenomenon are usually not encountered in S.R.I.M. as their starting torque is high enough to accelerate it.

VI Polyphase Synchronous Machines

- A three-phase synchronous machine is a doubly excited ac machine.
- It rotates with speed of revolving field, i.e., synchronous speed.

Remember:

☑ In synchronous machine, windings are always connected in star.

☑ Armature winding is placed in stator and field winding is in rotor.

- **Synchronous speed,** $N_s = \dfrac{120f}{P}$ rpm

  where, f = Frequency of armature current ; P = Number of field poles

- **RMS induced emf per phase,** $E_a = 4.44 f N_e \phi_{av}$

  where, N_e = Effective turns per phase; ϕ_{av} = Average flux per pole

- **Effective number of turns,** $N_e = N_c k_p k_d$

  where, N_c = Coil turns, k_d = Distribution factor ; k_p = Pitch factor

- **Winding factor,** $k_\omega = k_p k_d$

- **Flux per pole,** $\phi_{av} = \dfrac{2 B_m D l}{P}$

  where, B_m = maximum flux density

  D = Diameter of stator; l = Axial length

- **Coil turns,** $N_C = \dfrac{Z}{2}$, where Z = Total number of conductors

- In **single layer winding**

  No. of slots = No. of coil sides; No. of coils = $\dfrac{\text{No. of slots}}{2}$

- In **double layer winding**

  No. of slots = 2 × No. of coil sides; No. of coils = No. of slots

- **Pitch factor,** $K_p = \cos\dfrac{\varepsilon}{2}$ where ε = chording angle (electrical angle)

- **Distribution factor,** $K_d = \dfrac{\sin\left(\dfrac{m\beta}{2}\right)}{m\sin\left(\dfrac{\beta}{2}\right)}$

  where, $m = \dfrac{\text{slots}}{\text{pole} \times \text{phase}}$

  $\beta = \dfrac{180^\circ}{\left(\dfrac{\text{Slots}}{\text{Pole}}\right)}$

Types of Rotor

S.No.	Salient Pole Rotor	Cylindrical Rotor
1.	Concentrated winding connected in series.	Field winding is distributed and all coils are connected in series.
2.	Air gap is minimum along *d*-axis and maximum along q-axis.	Air gap is uniform.
3.	Salient pole machine operates at low speed and hence is called as low speed alternator or hydro alternator.	It can be driven at high speed so it is also known as high speed alternator or turbo generator.
4.	Salient pole rotor has large diameter and small axial length.	Cylindrical rotor has smaller diameter and larger axial length.

Note:

☑ In alternator, main field mmf leads the air-gap mmf and air-gap mmf leads the armature mmf. Therefore, in alternator main field mmf is most leading.

☑ In motor, main field mmf lags the air-gap mmf and air-gap mmf lags the armature mmf. Therefore, in motor main field mmf is most lagging.

☑ For unidirection torque production, number of rotor poles must be equal to number of stator poles and they must be stationary with respect to each other.

Armature Reaction

The effect of armature flux (ϕ_a) on main field flux (ϕ_m) is known as armature reaction. Armature reaction depends upon the magnitude of load and load power factor.

P.F.	Alternator	Motor
1. Unity	Purely cross magnetizing	Purely cross magnetizing
2. ZPF lag	Purely demagnetizing	Purely magnetizing
3. ZPF lead	Purely magnetizing	Purely demagnetizing
4. 0.8 lag	Partly demagnetizing & Partly cross magnetizing	Partly magnetizing & Partly cross magnetizing
5. 0.8 lead	Partly magnetizing & Partly cross magnetizing	Partly demagnetizing & Partly cross magnetizing

Synchronous Generator Model:

$$\bar{E}_f = \bar{V}_a + \bar{I}_a R_a - \bar{E}_l - \bar{E}_{ar}$$

$$\bar{E}_f = \bar{V}_a + \bar{I}_a R_a + j\bar{I}_a X_l + j\bar{I}_a X_a$$

where, $\bar{E}_f \rightarrow$ Excitation voltage or internal voltage

$\bar{E}_{ar} \rightarrow$ Armature reaction voltage

$\bar{E}_l \rightarrow$ Armature leakage flux voltage

$R_a \rightarrow$ Armature resistance

$X_l \rightarrow$ Armature leakage reactance

$X_a \rightarrow$ Armature reaction reactance

$\bar{I}_a \rightarrow$ Armature current

$\bar{V}_a \rightarrow$ Terminal voltage

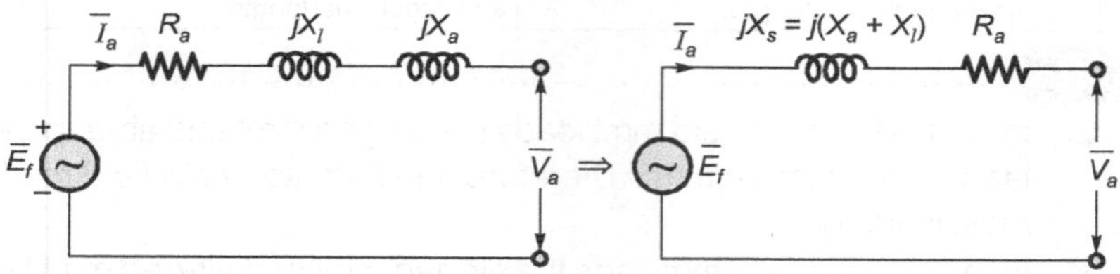

$$\bar{E}_f = \bar{V}_a + \bar{I}_a R_a + j\bar{I}_a X_s$$

$$|\vec{E}_f| = \sqrt{(V_a \cos\phi + I_a R_a)^2 + (V_a \sin\phi + I_a X_s)^2}$$

where, $X_s \rightarrow X_l + X_a$ = synchronous reactance

$$\bar{E}_f = \bar{V}_a + \bar{I}_a \bar{Z}_s$$

where, $Z_s \rightarrow (R_a + jX_s)$ = synchronous impedance

Remember:

☑ The terminal voltage of an a.c. generator depends upon the load and may be larger or smaller than the generated voltage.

Phasor diagram

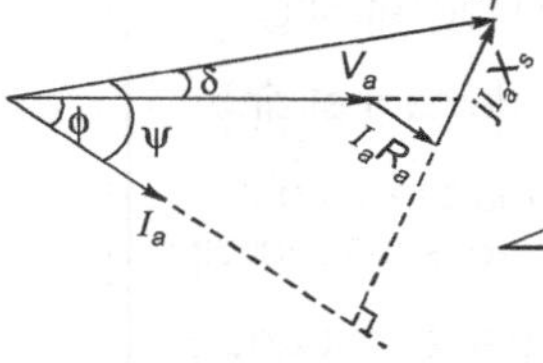

Lagging p.f.

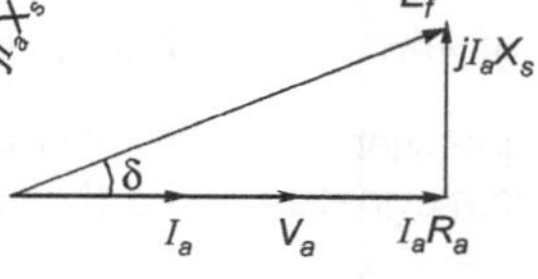

Unity p.f.

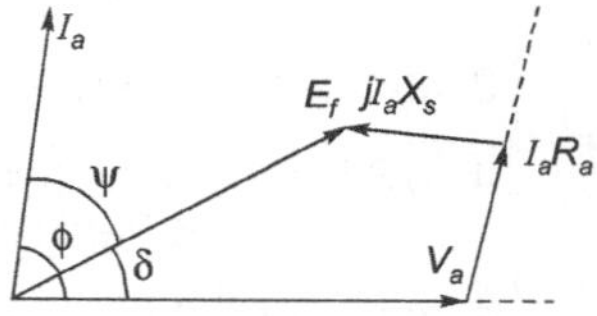

Leading p.f.

Note:

☑ When $E_f \cos\delta = V_a$ normal excitation (unity p.f.)

☑ When $E_f \cos\delta > V_a$ Over excited machine (lagging p.f.)

☑ When $E_f \cos\delta < V_a$ Under excited machine (leading p.f.)

Synchronous generator tests and characteristics

Resistance test

- Per phase resistance

  for star connected generator, $R_a = 0.5R_L$

  for delta connected generator, $R_a = 1.5R_L$

  where, $R_L \rightarrow$ measured value of resistance

Open circuit and Short circuit characteristics

- Find the value of the field current (I_{fr}) that gives the rated per-phase voltage (V_{anL}) from the OCC of the generator.
- Find the value of the short-circuit current (I_{sc}) from the SCC for the same value of the field current I_{fr}.

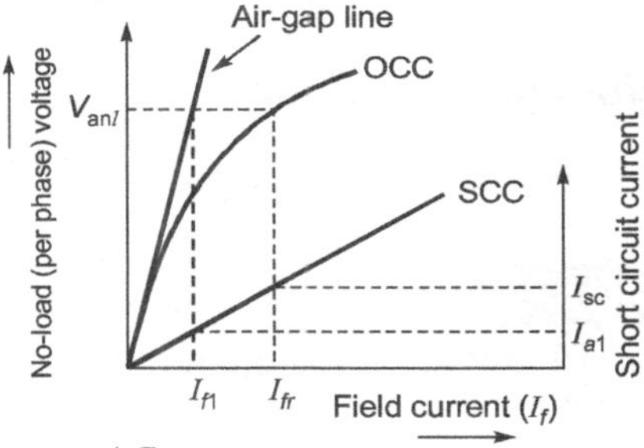

- Synchronous impedance

$$Z_s = \frac{\text{Open circuit terminal voltage for a certain field current}}{\text{Short-circuit current for the same field current}}$$

$$Z_s = \frac{V_{anL}}{I_{sc}}$$

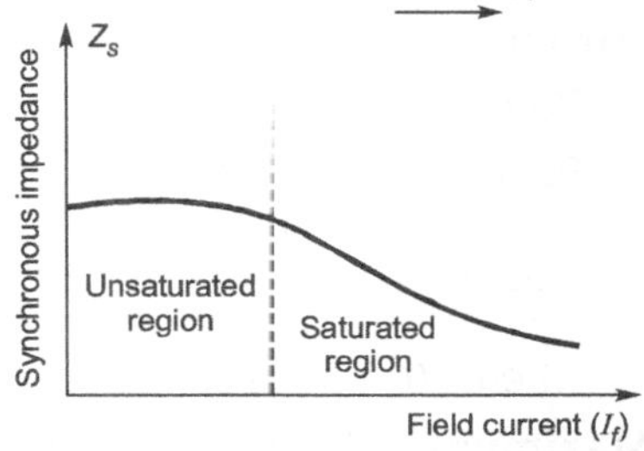

External characteristics

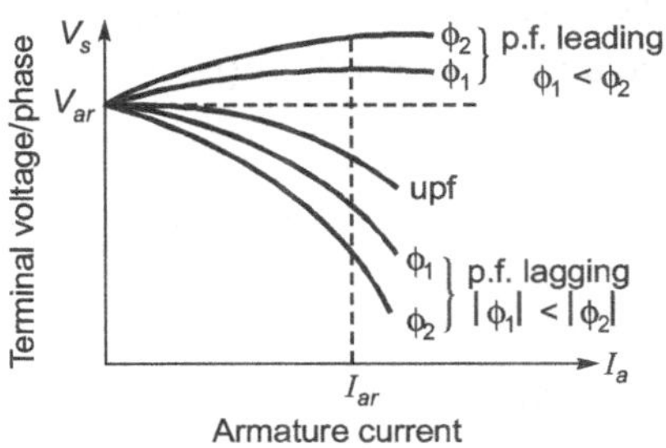

Short Circuit Ratio (SCR)

$$SCR = \frac{I_f \text{ for rated open circuit voltage}}{I_f \text{ for rated short circuit current}}$$

$$SCR = \frac{I_{SC}}{I_{rated}} = I_{SC(\text{p.u.})}$$

$$SCR = \frac{1}{X_{s(\text{saturated})\,\text{p.u.}}}$$

- $SCR \propto \dfrac{1}{\text{Armature Reaction}}$
- $SCR \propto \dfrac{1}{\text{Voltage Regulation}}$
- $SCR \propto$ Stability

Remember:

☑ For hydro generator SCR is high.
☑ For turbo generator SCR is low.

Voltage Regulation

$$\text{Regulation} = \frac{|E_f| - |V_a|}{|V_a|} \times 100\% ; \qquad \text{Regulation} \propto Z_s$$

Remember:

☑ Order of voltage regulation in alternator is 30 to 40%.

For minimum voltage regulation

Zero voltage regulation is possible only for leading p.f. load.

For zero regulation:

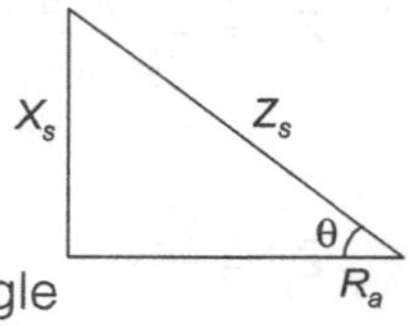

$$\cos(\theta + \phi) = \frac{-I_a Z_s}{2V} \quad \text{or} \quad \theta + \phi = 90^\circ$$

where, $\theta \rightarrow$ Impedance angle ; $\phi \rightarrow$ Power factor angle

Maximum voltage regulation

Maximum regulation is possible only for lagging pf load.

For maximum regulation : Load p.f. angle (ϕ) = Impedance angle (θ)

Remember:

☑ Voltage regulation is always positive for resistive load or U.P.F. load or lagging p.f. load.
☑ Voltage regulation could be positive (for high leading p.f.), zero and negative for leading p.f. load.

Voltage Regulation Method

- **Synchronous Impedance Method (EMF Method)**

Here, we assume drop due to armature reaction is considered drop due to leakage reactance. It gives regulation more then the actual value and hence it is called as pessimistic method.

- **M.M.F. Method (Amp-turn method)**

  Here, we assume drop due to leakage reactance is considered as drop due to armature reaction. This method gives regulation less than actual value therefore it is called as optimistic method.

- **Z.P.F. Method**

  Plot of V_a versus I_f corresponding to different field current. For maintaining rated armature current at zero p.f. lag called ZPF characteristic or potier triangle characteristic.

- **A.S.A. Method**

  Here effect of saturation is also considered. This method is combination of Z.P.F. and M.M.F. method.

Note:

☑ Voltage regulation EMF > ASA > ZPF > MMF.

Two Reaction Theory

This theory is applicable to salient pole machines only.

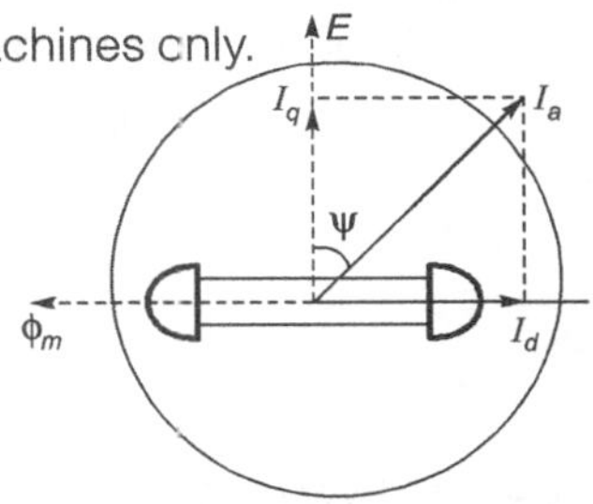

$$I_d = I_a \sin\psi$$

$$I_q = I_a \cos\psi$$

where,

I_d = Direct axis armature current

I_q = Quadrature axis armature current

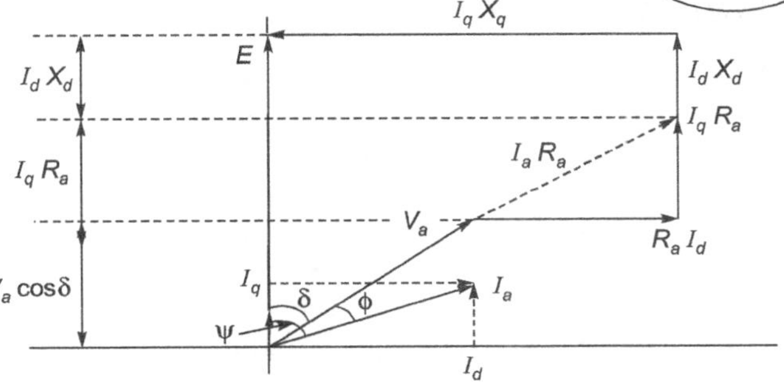

$$E = V_a \cos\delta + I_q R_a \pm I_d X_d$$

$$\tan\psi = \frac{V_a \sin\phi \pm I_a X_q}{V_a \cos\phi + I_a R_a} \quad \Rightarrow \quad \psi = \delta \pm \phi$$

Here, + sign for lagging p.f. and sign for leading p.f.

where, $\psi \rightarrow$ internal pf angle by which I_a lags E.

$\delta \rightarrow$ Load angle between E and V

$\phi \rightarrow$ Power factor angle between V_a and I_a

- If R_a neglected as $R_a << X_s$.

$$I_d = \frac{E - V_a \cos\delta}{X_d} \quad \text{and} \quad I_q = \frac{V_a \sin\delta}{X_q}$$

- Per Phase Power, $P = \underbrace{\frac{EV_a}{X_d}\sin\delta}_{\text{Electromagnetic power}} + \underbrace{\frac{V_a^2}{2}\left(\frac{1}{X_q} - \frac{1}{X_d}\right)\sin 2\delta}_{\text{Reluctance power } (P_{rel.})}$

Remember:

- ☑ In cylindrical rotor, P_{rel} is zero since it does not have saliency and $X_d = X_q = X_s$.
- ☑ Reluctance power is generated only when machine is connected to or operating on infinite bus bar.

Slip Test

From slip test, the direct and quadrature axis synchronous reactances, i.e., X_d and X_q can be determined.

$$X_d = \frac{\text{Maximum armature terminal voltage per phase}}{\text{Minimum armature current per phase}}$$

$$X_q = \frac{\text{Minimum armature terminal voltage per phase}}{\text{Maximum armature current per phase}}$$

Power Flow in Alternator

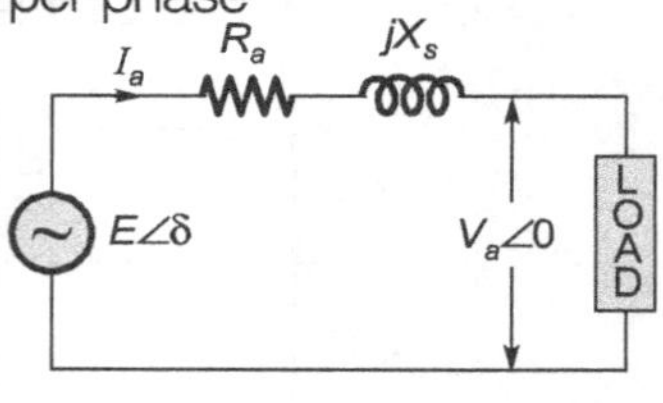

Complex power output

$$S = P + jQ = V_a I_a^*$$

$$P = \frac{EV_a}{Z_s}\cos(\theta - \delta) - \frac{V_a^2}{Z_s}\cos\theta$$

$$Q = \frac{EV_a}{Z_s}\sin(\theta - \delta) - \frac{V_a^2}{Z_s}\sin\theta$$

If, $R_a = 0$ then $Z_s = X_s$ and $\theta = 90°$

$$P = \frac{EV_a}{X_s}\sin\delta; \quad Q = \frac{V_a}{X_s}[E\cos\delta - V]$$

Condition for maximum power output, i.e., $\frac{dP}{d\delta} = 0$

impedance angle (θ) = load angle (δ)

Note:

☑ Always replace δ by θ.

$$P_{max} = \frac{EV_a}{Z_s} - \frac{V_a^2}{Z_s}\cos\theta$$

Power relationship

Mechanical power input to the generator	$P_{in,m} = T_s\omega_s$
D.C. power input to a wound rotor	$P_{in,e} = V_f I_f$
Total power input	$P_{in} = T_s\omega_s + V_f I_f$
Power output of a synchronous generator	$P_o = 3V_a I_a \cos\phi$
Copper loss in armature winding	$P_{cu} = 3I_a^2 R_a$

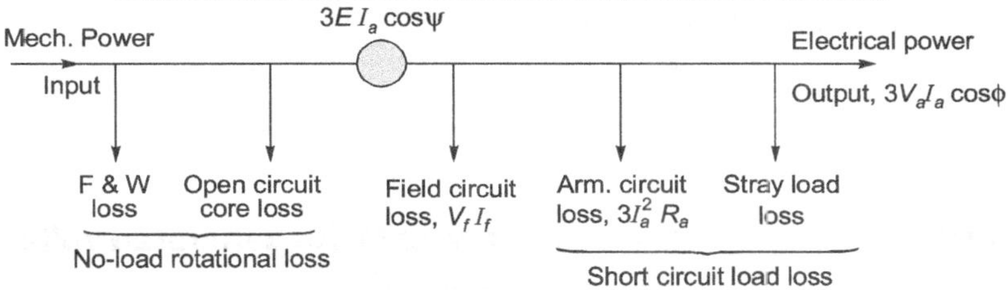

- Total power input to synchronous generator

$$P_{in} = 3V_a I_a\cos\phi + 3I_a^2 R_a + P_r + P_{st} + V_f I_f$$

where, $P_r \rightarrow$ Rotational losses of synchronous generator

$P_{st} \rightarrow$ Stray load losses

- Constant losses

Since the rotor revolves at a constant speed, the rotational loss is constant. The field-winding loss is constant. Assuming the stray-load loss to be a constant.

$$P_c = P_r + P_{st} + V_f I_f$$

Efficiency

- Efficiency of the generator, $\eta = \dfrac{3V_a I_a\cos\phi}{3V_a I_a\cos\phi + P_c + 3I_a^2 R_a} \times 100\%$

For the maximum efficiency, $3I_a^2 R_a = P_c$

Condition that must be satisfied for parallel operation

- Terminal voltage of incoming alternator should be equal to existing system and that can be done by varying excitation.
- Frequency of incoming alternator should be equal to existing system. Frequency is maintained same by adjusting primover speed.
- Phase sequences of incoming alternator must be same as that of existing system.

Droop Characteristics

Two generator are operating in parallel and we wish to determine load sharing.

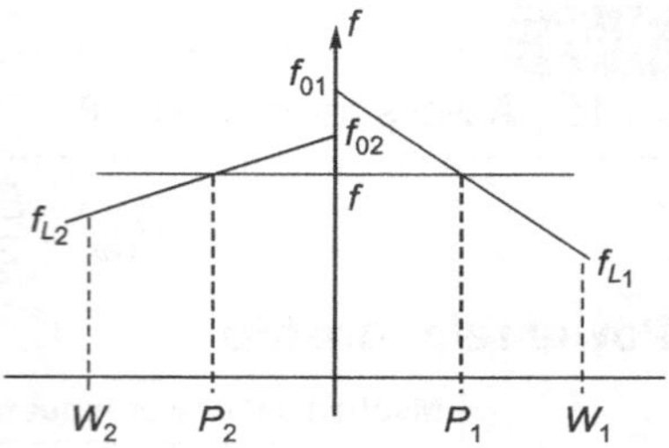

w_1 and w_2 : Rating of generator 1 and generator 2

f_{o1} and f_{o2} : No load frequency of generator 1 and generator 2

f_{L1} and f_{L2} : Full load frequency of generator 1 and generator 2.

f: Operating frequency

P_1 : Load and having by generator 1

P_2 : Load showing by generator 2

$$P_1 = \left(\frac{f_{o1} - f}{f_{o1} - f_{L1}}\right) w_1; \qquad P_2 = \left(\frac{f_{o2} - f}{f_{o2} - f_{L2}}\right) w_2$$

If load = P_L; $\quad \therefore P_1 + P_2 = P_L$

Constant power loci of armature current with generated voltage

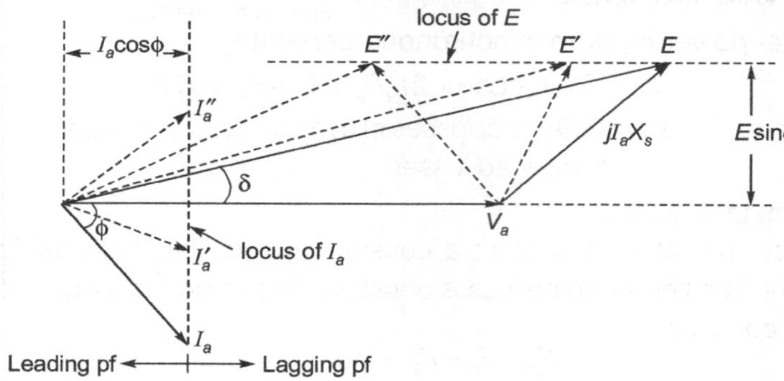

V curves

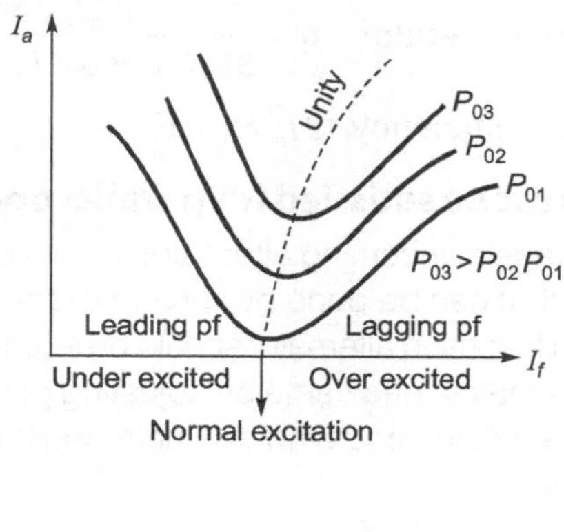

Round-Rotor synchronous motors

Equivalent circuit

$$\overline{V}_a = \overline{E}_a + \overline{I}_a R_a + j\overline{I}_a X_s$$

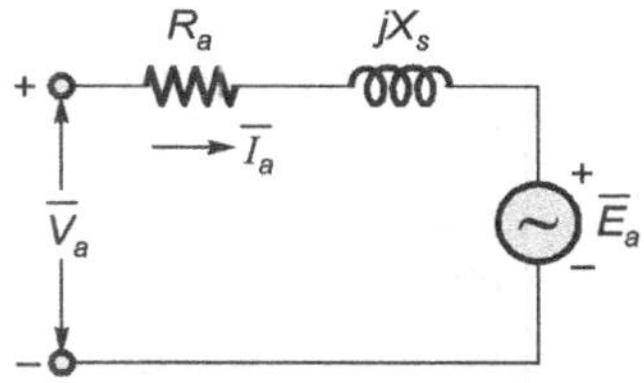

Phasor diagram

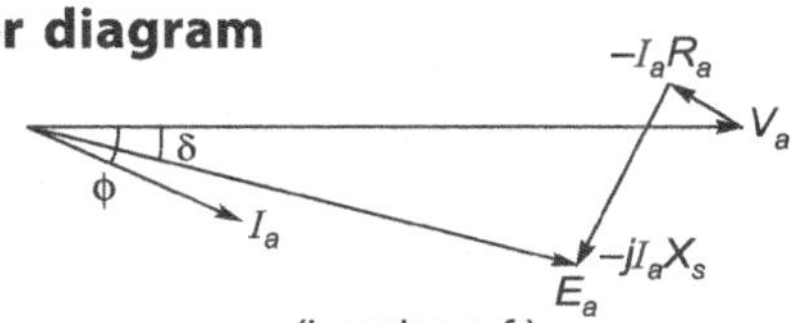

(Lagging p.f.)

Power flow diagram

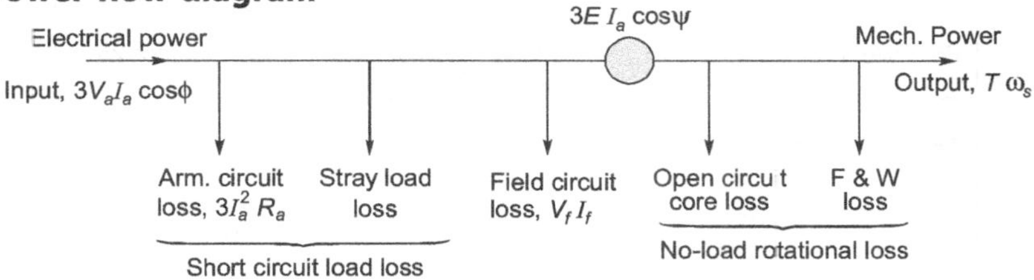

Average power input	$P_{in} = 3V_aI_a\cos\phi + V_fI_f$
Total copper loss	$P_{Cu} = 3I_a^2R_a$
Power developed	$P_d = 3V_aI_a\cos\phi - 3I_a^2R_a - V_fI_f$ – Stray load loss

Salient-pole synchronous motor

$$V_a = E_a + I_aR_a + jI_dX_d + jI_qX_q$$

or $$\overline{E}_a = \overline{V}_a - \overline{I}_aR_a - j\overline{I}_aX_q - j\overline{I}_d(X_d - X_q)$$

Phasor diagrams

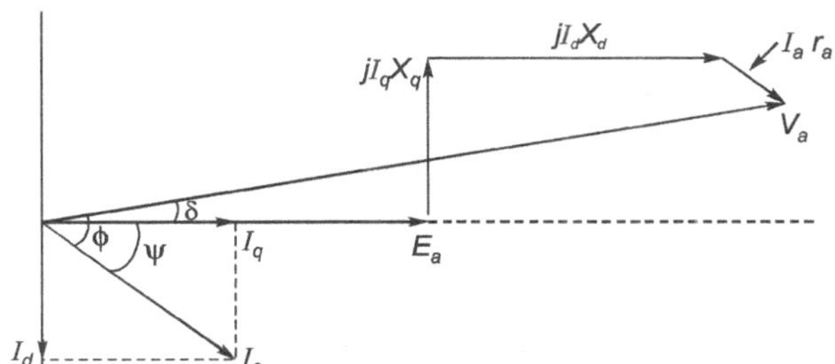

(Lagging pf)

Power developed	$P_d = \dfrac{3V_aE_a\sin\delta}{X_d} + \dfrac{3}{2}V_a^2\sin 2\delta\left[\dfrac{X_d - X_q}{X_dX_q}\right]$
Power developed owning to field excitation	$P_{df} = \dfrac{3V_aE_a}{X_d}\sin\delta$

Power developed due to the saliency of rotor	$P_{ds} = 3V_a^2\left(\frac{X_d - X_q}{2X_dX_q}\right)\sin 2\delta$
Condition for maximum power	$\tan\delta_m = \frac{X_s}{R_a}$
Per-phase maximum power developed	$P_{dma} = \frac{V_aE_a}{Z_s} - \frac{E_a^2R_a}{Z_s^2}$
Synchronizing power	$P_{sy} = \frac{\partial P}{\partial \delta}$
For cylindrical rotor :	$P_{sy} = \frac{E_aV_a}{X_s}\cos\delta$
For salient rotor :	$P_{sy} = \frac{E_aV_a}{X_s}\cos\delta + V_a^2\left(\frac{1}{X_q} - \frac{1}{X_d}\right)\cos 2\delta$
Synchronizing torque	$T_{sy} = \frac{P_{sy}}{\omega_{sm}}$ where, $\omega_{sm} = \frac{4\pi f}{P}$
For cylindrical rotor :	$T_{sy} = \frac{1}{\omega_{sm}}\left(\frac{E_aV_a}{X_s}\cos\delta\right)$
For salient rotor :	$T_{sy} = \frac{1}{\omega_{sm}}\left(\frac{E_aV_a}{X_s}\cos\delta + V_a^2\left(\frac{1}{X_q} - \frac{1}{X_d}\right)\cos 2\delta\right)$

Constant power loci of armature current and excitation voltage

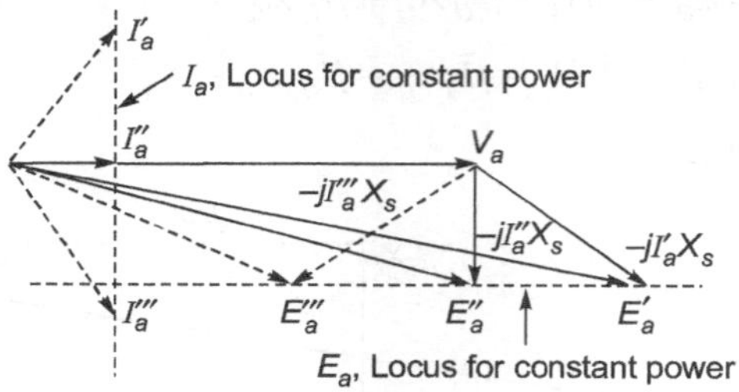

V curves

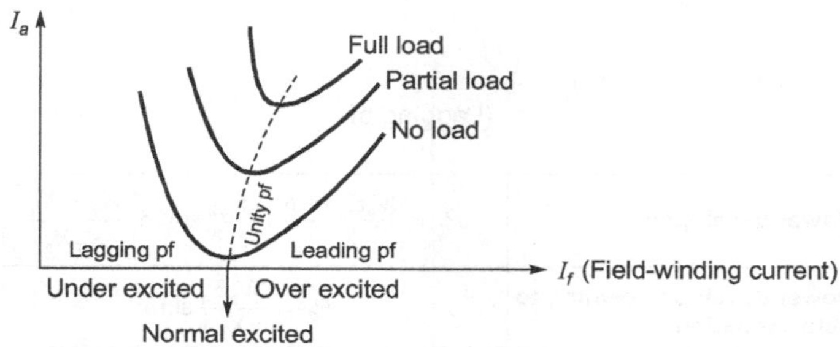

Starting of Synchronous Motors

A synchronous motor is not self-starting. It can be started by the following two methods :
- Starting with the help of an external prime mover.
- Starting with the help of damper windings.

Hunting or Phase Swinging

The phenomenon of oscillation of rotor about its final equilibrium position is called hunting. Since during rotor oscillations, the phase of the phasor E_f changes relation to phasor V hunting is known as phase swinging.

Hunting occurs not only in the synchronous motors but also in synchronous generators upon the abrupt change in loading.

Synchronous Condenser

It is a synchronous motor running without a mechanical load which can generate on absorb reactive VAr by varying the excitation of its field winding.

Remember:

- ☑ Since, a synchronous condenser behaves like variable inductor or a variable capacitor, it is used in power transmission systems to regulate line voltage.
- ☑ They are also used in constant speed applications.

VII Single Phase Induction Machine

Single phase induction machines are commonly used in domestic, commercial, agriculture, offices and small industries. Also called as Fractional H.P. or Fractional KW motors (upto 1 kW, speed upto 1500 rpm).

Remember:

- ☑ Main drawback of these motors is their non-starting nature.
- ☑ These are analyzed according to double revolving field theory.

Double Revolving Field Theory

This theory is based on the idea that pulsating field produced in a single phase motor can be resolved into two components of half its amplitude and rotating in opposite directions with synchronous speed.

$$F = \frac{1}{2}F_{max}\cos(\theta - \omega t) + \frac{1}{2}F_{max}\cos(\theta + \omega t)$$

where,

$$\frac{1}{2}F_{max}\cos(\theta - \omega t) = \text{Forward rotating field, } F_f$$

$$\frac{1}{2}F_{max}\cos(\theta + \omega t) = \text{Backward rotating field, } F_b$$

The slip of the rotor with respect to the forward rotating field, F_f,

$$S_f = \frac{N_s - N}{N_s} = S$$

The slip of the rotor with respect to the backward rotating field F_b,

$$S_b = \frac{N_s - (-N)}{N_s} = \frac{2N_s - (N_s - N)}{N_s} = 2 - S$$

Torque-Slip Characteristics

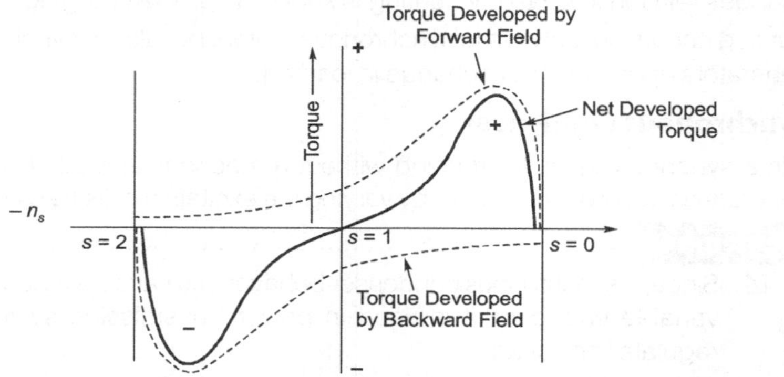

***Fig. :** Torque-Slip Curves for a Single Phase Induction Motor*

Equivalent Circuit of Single Phase Induction Motor

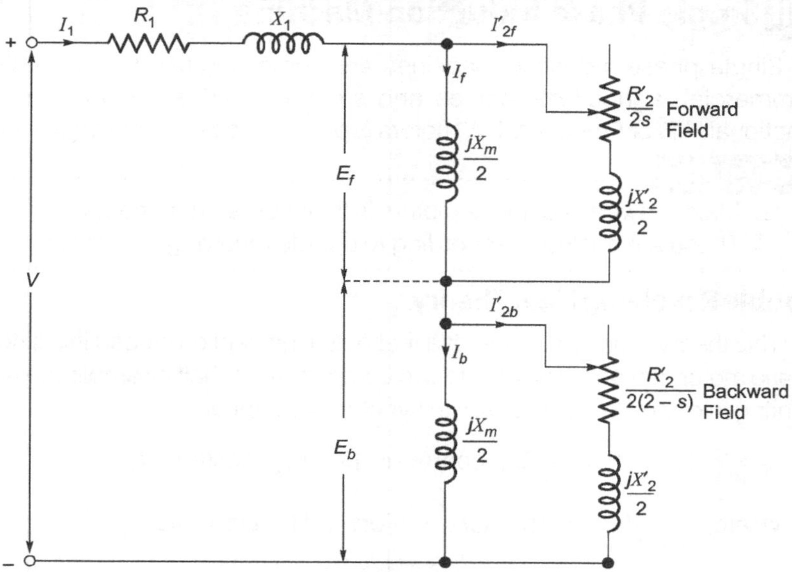

***Fig. :** Approximate Equivalent Circuit of a Single Phase Induction Motor Under Normal Operating Conditions*

where, R_1 = Resistance of the main stator winding

X_1 = Leakage reactance of the main stator winding

X_m = Magnetizing reactance

R'_2 = Standstill rotor resistance referred to the main stator winding

X'_2 = Standstill rotor leakage reactance referred to the main stator winding

S = Slip

Types of Single Phase Induction Motors

Single phase induction motors are usually classified according to the auxiliary means used to start the motor. They are classified as follows :

1. Split phase motors
2. Capacitor-start motors
3. Capacitor-start capacitor run motor (or two value capacitor motor)
4. Permanent-split capacitor (PSC) motor (or single value capacitor motor)
5. Shaded-pole motor

Split Phase Motor

It is also called resistance start motor. The motor employs an auxiliary winding with a higher $\frac{R}{X}$ ratio as compared to the main widing.

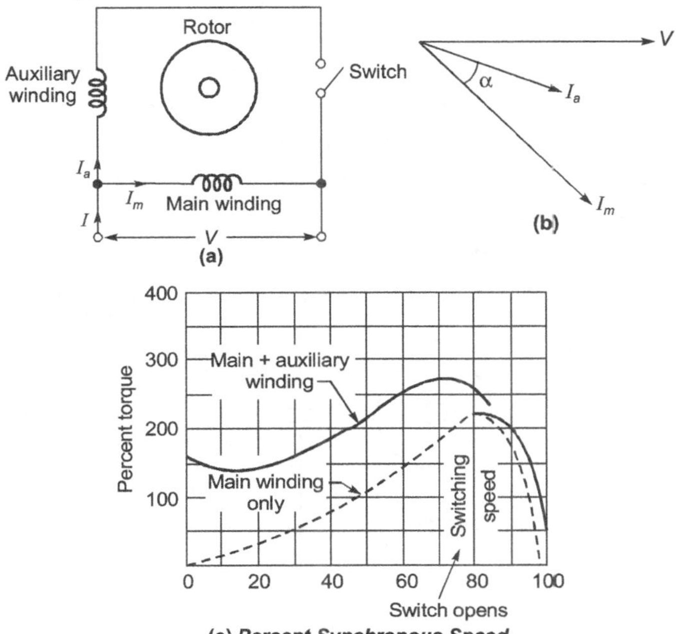

(c) ***Percent Synchronous Speed***

Note:

- ☑ The direction of rotation of the resistance-start inductions motor may be reversed by reversing the line connections of either the main winding or the starting winding.

Applications : Washing machine, air conditioning, fans, food mixers, grinders, floor polisher, blowers, centrifugal pumps, small drills, lathes, office machinery, dairy machinery etc.

Capacitor Start Motor

The motor is named so because it uses the capacitor for starting purpose only. The capacitor value is choosen to give $\alpha = 90°$.

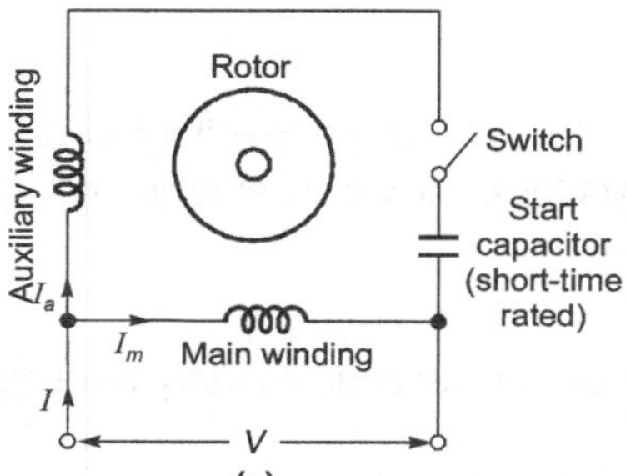

(a)

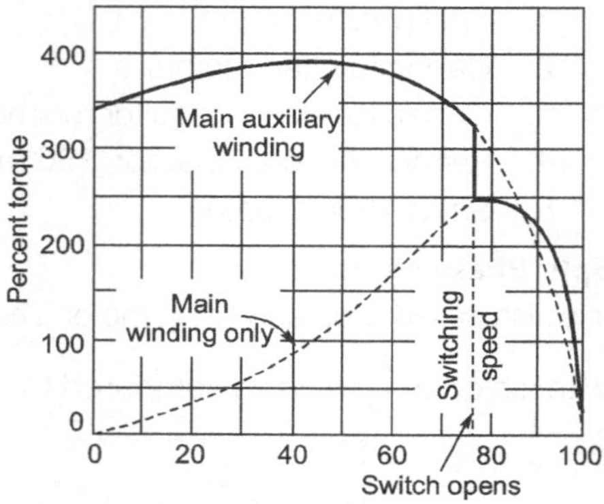

(b) : Percent Synchronous Speed

The capacitor-start motor may be reversed by reversing the connections of one of the windings. The motor is first bought to rest for this purpose.

Applications : Pumps and compressors, refrigerators, air-conditioner compressors, conveyors and machine tools etc.

Two-Value Capacitor Motor

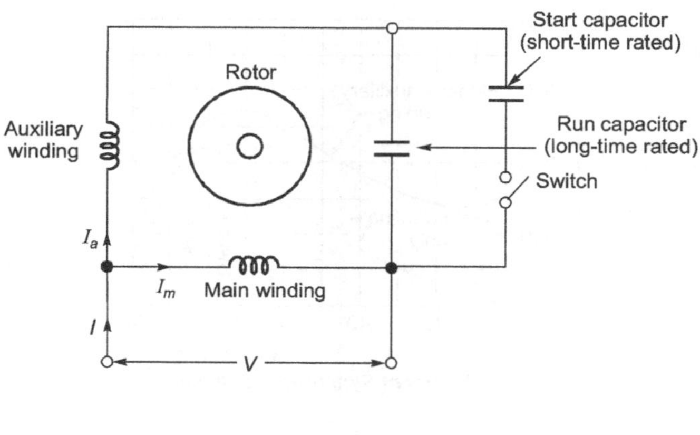

The two-value capacitor motor uses capacitors for starting as well as continuous running. The capacitor used permanently is called the run capacitor. Its use to improves the motor running performance.

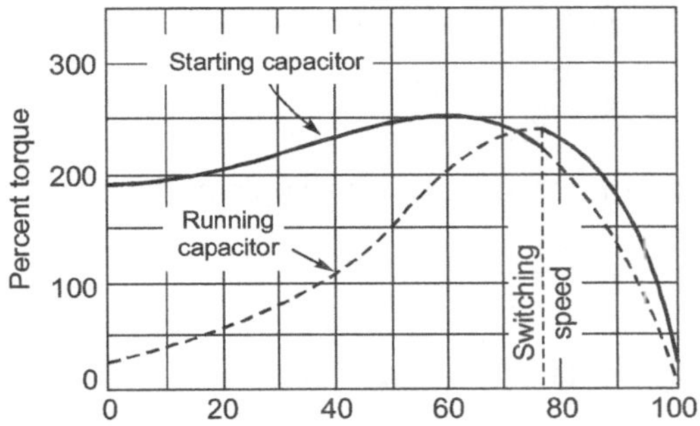

Fig. : Percent synchronous speed

Applications: Pumping equipment, refrigeration, air compressors etc.

Permanent Split Capacitor (PSC) Motor

It has only run capacitor.

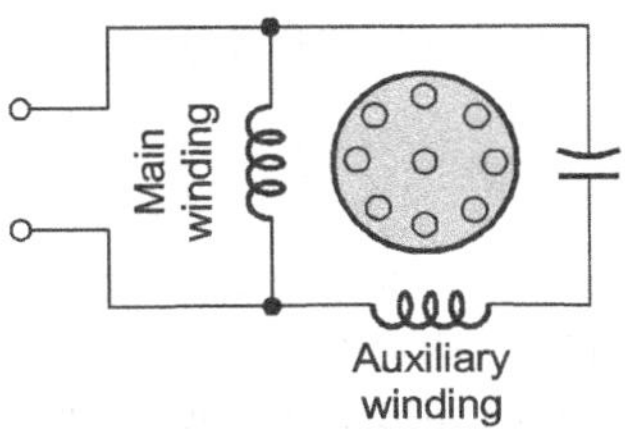

(a)

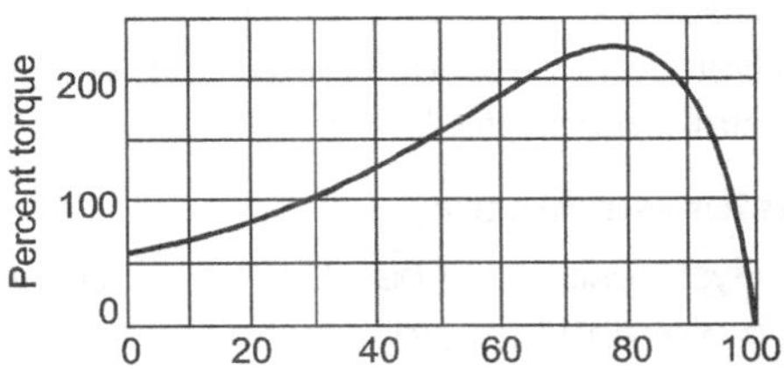

(b) *Percent synchronous speed*

Applications : Fans and blowers in heaters and air conditioners, refrigerator compressors, drive office machinery.

Shaded-Pole Motor

A shaded pole motor is a simple type of self starting single-phase induction motor.

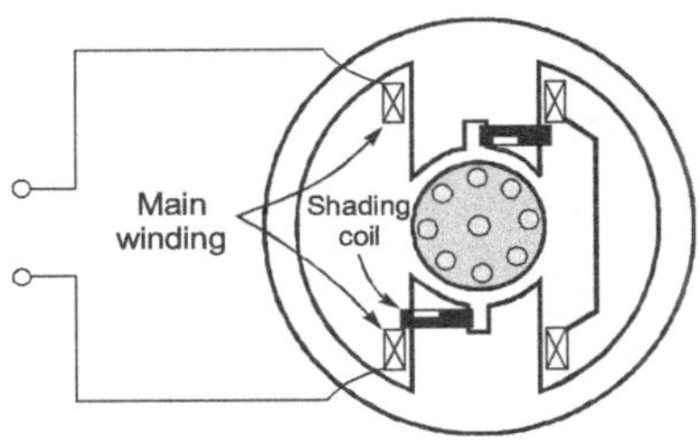

(a)

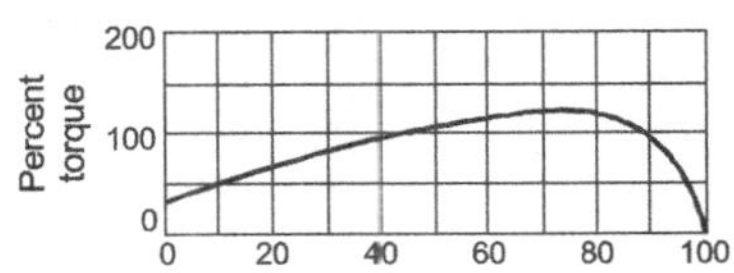

(b) *Percent synchronous speed*

Applications : Domestic fans, record players, tape recorders, humidifers, slide projectors, small business machines etc.

Note:

☑ For phase quadrature in single phase induction motor,

$$X_c = \frac{R_a R_m}{X_m} + X_a$$

☑ For maximum starting torque in single phase induction motor,

$$\theta_a = \frac{90 - \theta_m}{2}$$

where, $\theta_m = \tan^{-1}\frac{X_m}{R_m}; \ \theta_a = \tan^{-1}\left(\frac{X_c - X_a}{R_a}\right)$

VIII Special Machines

Reluctance Motor

- Depends on reluctance torque for its operation.
- Reluctance torque is maximum when the angle between the rotor and stator magnetic field is 45°.

Hysteresis Motor

- Hysteresis motor has its stator similar to that of permanent split capacitor motor and the rotor is solid cylinder of magnetically hard steel.
- Hysteresis torque remains constant with speed whereas eddy current torque linearly decreases with speed.

Stepper Motor

- It has rotor movement in discrete steps and angular rotation is determined by number of pulses fed into control circuit.
- Step angle, $\alpha = \frac{360°}{m_s N_r}; \quad \alpha = \frac{N_s - N_r}{N_s N_r} \times 360°$

  where, m_s = Number of stator stacks

  N_r = Number of rotor teeth

  N_s = Number of stator teeth

- $$\text{Speed} = \frac{\text{Step size} \times \text{Pulse rate}}{360°} \text{ rps}$$

■■■■

3 Power Electronics

I Power Electronics

- Power electronics is a subject that deals with the apparatus and equipment rated at high voltage, high current and high power (rather than signal level) working on the principle of electronics.

  Example: Thyristor, GTO, Power MOSFET, Power IGBT, TRIAC etc.

Power Device	Signal Device
1. Voltage and current rating is high	1. Voltage and current rating is low
2. Power handling capability is high	2. Power handling capability is low
3. Operate at power frequency	3. Operate at high frequency

Power Semiconductor Devices

- Semiconductor devices are used for high voltage, high current and high power applications.
- Power semiconductor devices have high efficiency due to low losses and high reliability.

  Power semiconductor devices can be classified based on their :
  - Turn-on and turn-off characteristics.
  - Gate signal requirements.

  (i) **Diodes** are uncontrolled rectifying devices and their ON state and OFF state are controlled by nature of power supply.

  (ii) **Thyristor** devices have controlled turned-on by a gate signal, these devices are also called as semicontrolled devices.

  (iii) **Controllable switches** turn-on and turn-off can be done by application of control signals.

Power Diodes

Power diode is a 2 layer, 2 terminal p-n junction semiconductor device. It has one p-n junction formed by alloying, diffusion or epitaxial growth.

Remember:

☑ When voltage rating is less than 400 V epitaxial process is used for diode fabrication.

☑ When voltage rating is greater than 400 V diffusion process is used for diode fabrication.

V-I Characteristics of Power Diodes :

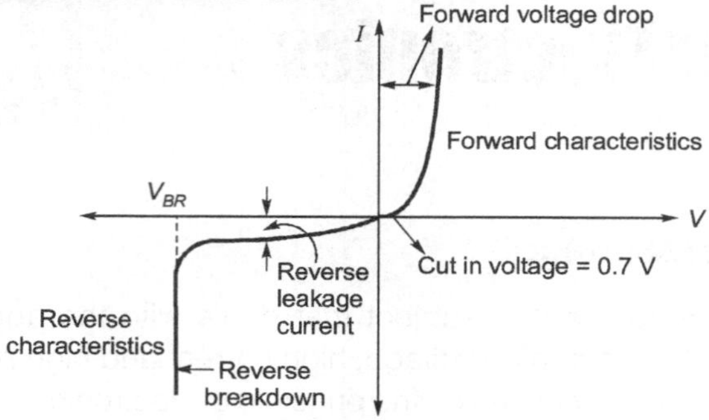

Characteristics of Semiconductor Diode

Note:

Peak inverse voltage (P.I.V.) specifies the maximum reverse voltage applied across diode by the source.

If P.I.V. < V_{BR}, diode remain in blocking state.

If P.I.V. > V_{BR}, breakdown occurs and diode starts conducting in reverse direction.

Reverse Recovery Characteristics of Power Diodes

- Due to the presence of excess stored charge carrier in the depletion region of diode, a reverse current immediately flow as soon as the forward diode current becomes zero.

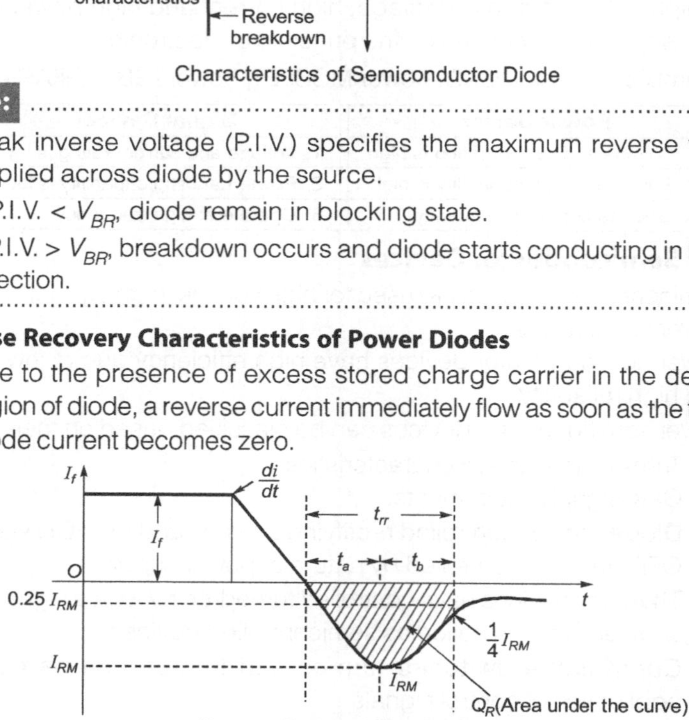

Reverse Recovery Characteristics

$$t_{rr} = t_a + t_b = t_{off}$$

where, $t_{rr} \rightarrow$ Reverse recovery time

$t_a \rightarrow$ Time between $I_A = 0$ to $I_A = I_{RM}$

$t_b \rightarrow$ Time between I_{RM} to 25% of I_{RM}

$I_{RM} \rightarrow$ Reverse peak current.

$$t_{rr} = \left[\frac{2Q_R}{di/dt}\right]^{1/2} \quad \text{and} \quad I_{RM} = \left[2Q_R\left(\frac{di}{dt}\right)\right]^{1/2}$$

where, Q_R gives the amount of excess charge stored.

- **Softness factor (S)** is measure of voltage transient appearing across the terminal of diode during recovery period.

$$S = \frac{t_b}{t_a}$$

when $S \to 1$; $t_a = t_b$; Soft recovery

$S << 1$; Voltage spikes will be present and it indicates fast recovery.

Types of Power Diode

Parameters	General Purpose Diode	Fast Recovery Diode	Schottky Diode
t_{rr} (reverse recovery time)	25 μsec.	5 μsec or less	In nano second
Voltage rating	50 V to 5 kV	50 V to 3 kV	Reverse voltage blocking capability limited to 100 Volts.
Current rating	1 A to 1000 A	1 A to 1000 A	1 A to 300A
Application	(i) Battery charging (ii) Electric traction (iii) UPS (iv) Welding	(i) Choppers (ii) Commutation circuits (iii) SMPS (iv) Induction heating	(i) High frequency Instrumentation (ii) Switching power supplies

Power Transistor

- Power BJT (Bipolar Junction Transistor).
- MOSFET (Metal-Oxide Semiconductor Field Effect Transistor)
- Power IGBT (Insulated Gate Bipolar Transistor)

Power BJT (Bipolar Junction Transistor)

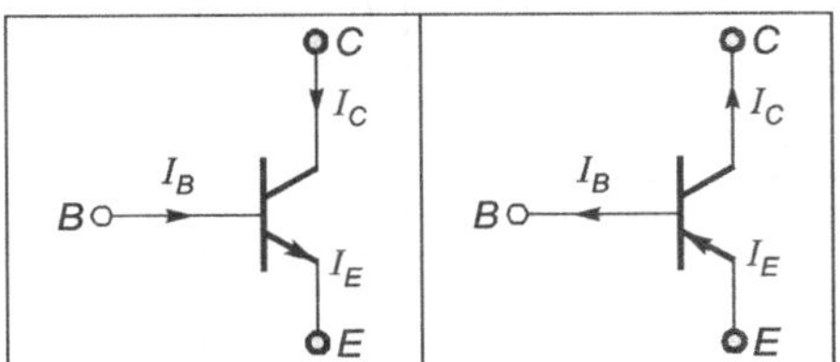

Symbol of (n-p-n type) | Symbol of (p-n-p type)

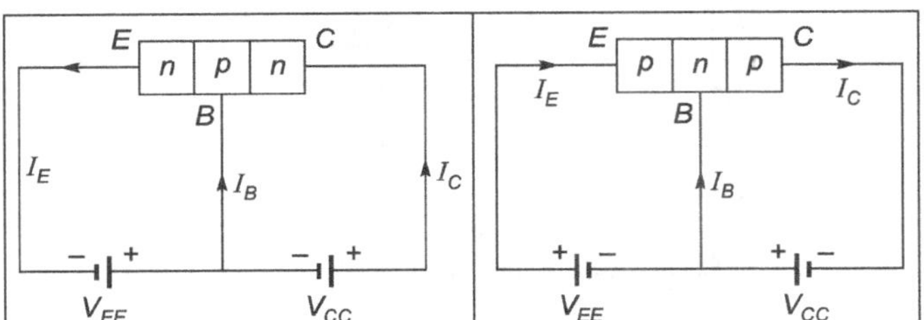

(n-p-n type) | (p-n-p type)

Note:

Current conduction in *pnp* transistor is carried out by hole whereas in *npn* transistor, it is due to electrons.

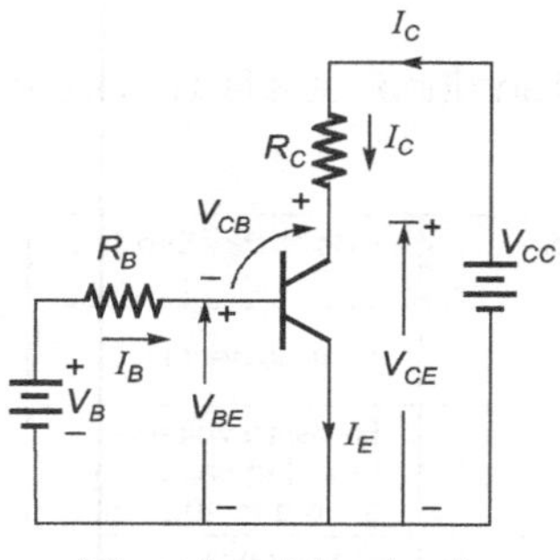

(a) npn transistor circuit characteristics

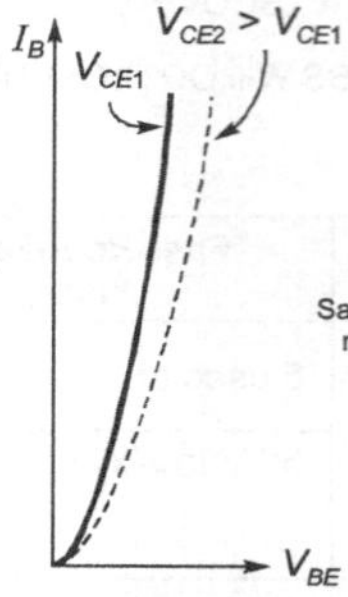

(b) Input characteristics

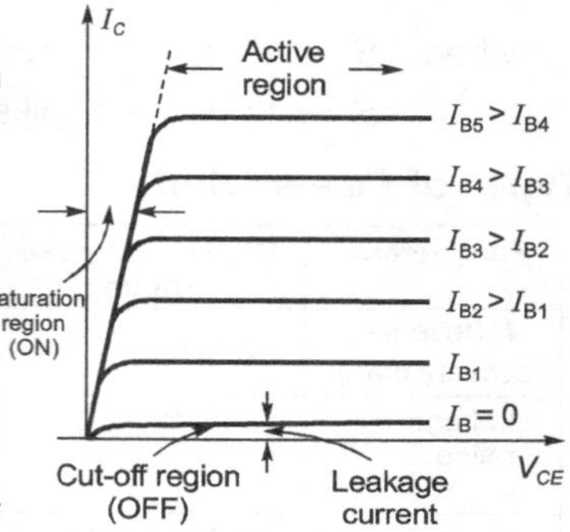

(c) Output characteristics

Remember:

Junctions		Operations
CB	**EB**	
F.B.	F.B.	Saturation (ON state)
F.B.	R.B.	Low gain amplifier
R.B.	R.B.	Cut-off
R.B.	F.B.	High gain amplifier

Switching Characteristics of n-p-n Transistor

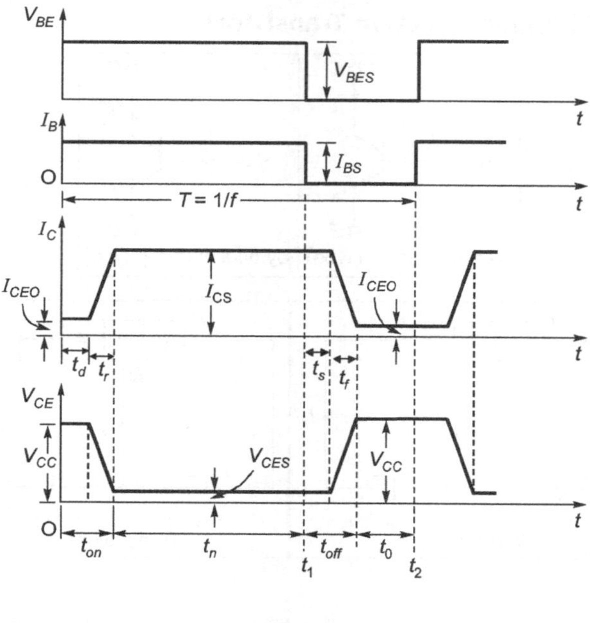

where, t_d (delay time) : t_d is defined as the time during which the collector current rises from 0 to $0.1I_{cs}$ and collector-emitter voltage falls from V_{cc} to $0.9V_{cc}$.

t_r (rise time) : t_r is defined as the time during which the collector current rises from $0.1I_{cs}$ to $0.9I_{cs}$ and collector-emitter voltage falls from $0.9V_{cc}$ to $0.1V_{cc}$.

Turn on time, $t_{on} = t_d + t_r$

t_n is conduction time period, t_o is OFF period.

t_s (storage time) : t_s is defined as the time during collector current falls from I_{cs} to $0.9I_{cs}$ and collector emitter voltage V_{CE} rises from V_{CES} to $0.1V_{CC}$.

t_f (fall time) : t_f is defined as the time during which collector current drops from $0.9I_{cs}$ to $0.1I_{cs}$ and collector-emitter voltage rises from $0.1V_{CC}$ to $0.9V_{CC}$.

$$t_{off} = t_s + t_r$$

V_{CES} : Small saturation voltage between collector and emitter

I_{CS} : Collector saturation current

$$T = t_{on} + t_n + t_{off} + t_o$$

Remember:

☑ t_{on} is in order of 30 to 300 ns.

Safe Operating Region for Power Transistor

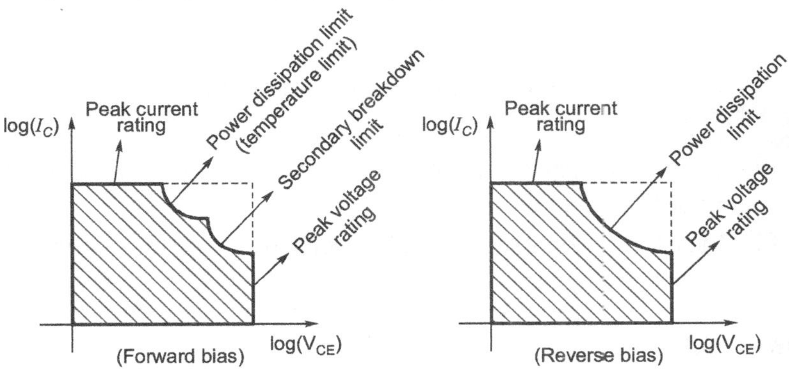

Power MOSFET

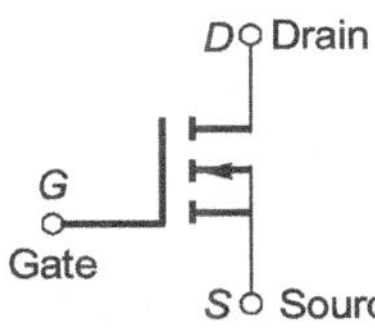

N-channel enhancement power MOSFET

where, D, $S \rightarrow$ Main terminal ; $G \rightarrow$ Control terminal

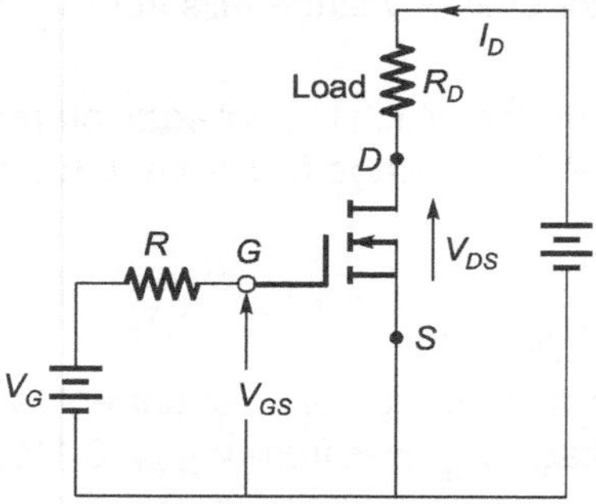

n-channel power MOSFET circuit diagram

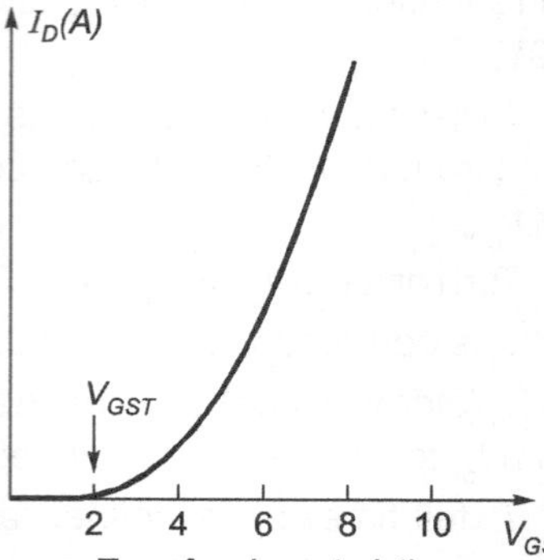

Transfer characteristic

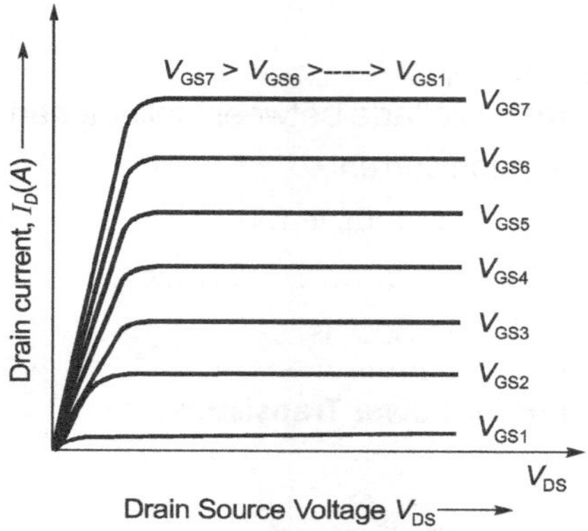

Output characteristics of a power MOSFET

Remember:

☑ n-channel enhancement MOSFET is more common because of higher mobility of electrons.

☑ The control signal (IB) in BJT is much larger than the control signal (Gate current) required in a MOSFET.

Power IGBT

This device combines into it the advantages of both MOSFET and BJT.

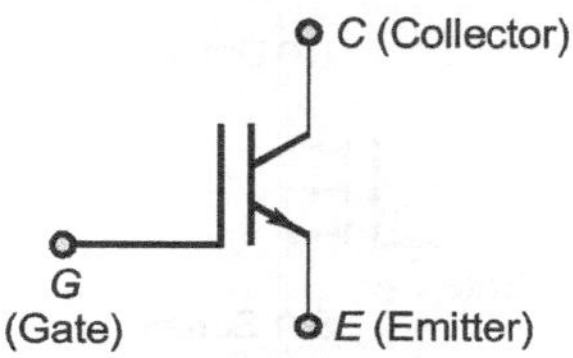

Circuit Symbol

IGBT Characteristics

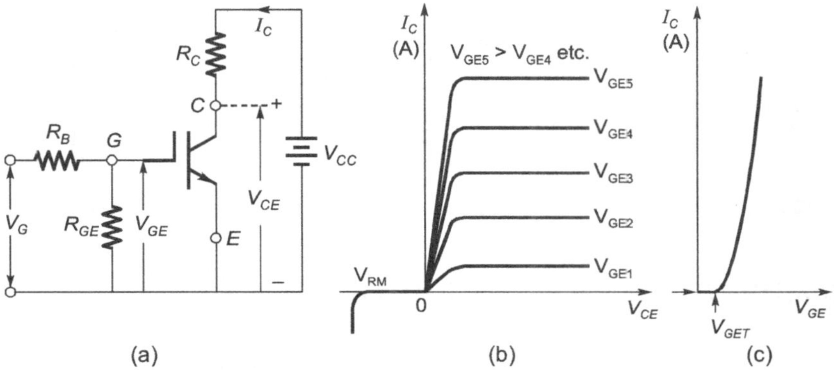

IGBT (a) Circuit diagram (b) Static V-I characteristics and (c) Transfer characteristics

Comparison

BJT	MOSFET	IGBT
1. Bipolar device	1. Unipolar device	1. Bipolar device
2. Low ON-state voltage drop	2. High ON-state voltage drop	2. Low ON-state voltage drop
3. Low ON state conduction power loss	3. High ON state conduction power loss	3. Low ON state conduction power loss
4. High switching power loss	4. Low switching power loss	4. Low switching power loss
5. Low input impedance	5. High input impedance	5. High input impedance
6. Current controlled device	6. Voltage controlled device	6. Voltage controlled device
7. Negative temperature coefficient for resistance	7. Positive temperature coefficient for resistance	7. Positive temperature coefficient for resistance
8. Secondary break down occur	8. Secondary break down will not occur	8. Secondary break down will not occur
9. Parallel operation is not advisable	9. Parallel operation is possible	9. Parallel operation is possible
10. 1200 V, 800 A, (10-20) kHz	10. 500 V, 140 A, 1 MHz	10. 1200 V, 500 A, 50 kHz
11. Application: UPS, charging batteries	11. SMPS	11. Inverters, choppers

Note:

- TRIAC only a device which operates on four modes all other devices not operates on all modes.
- For AC to AC convesion switch uses four mode of operation.
- For AC to DC conversion switch uses three mode of operation.
- In bipolar device conduction is due to both majority and minority carriers.

II Thyristor

Thyristor is a four layer, 3 junction, 3 terminal semicontrolled p-n-p-n semiconductor switching device.

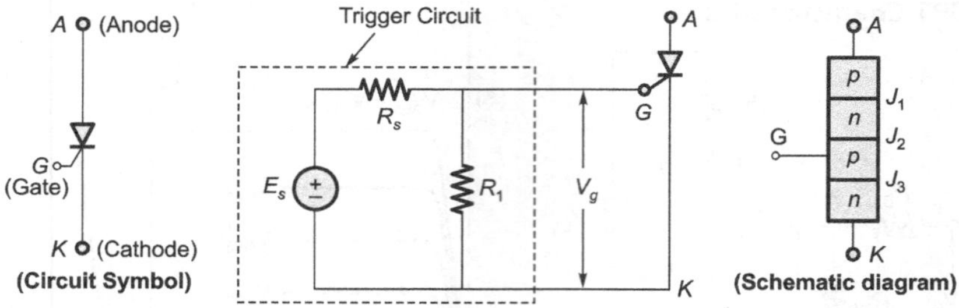

Static V-I Characteristics of Thyristor

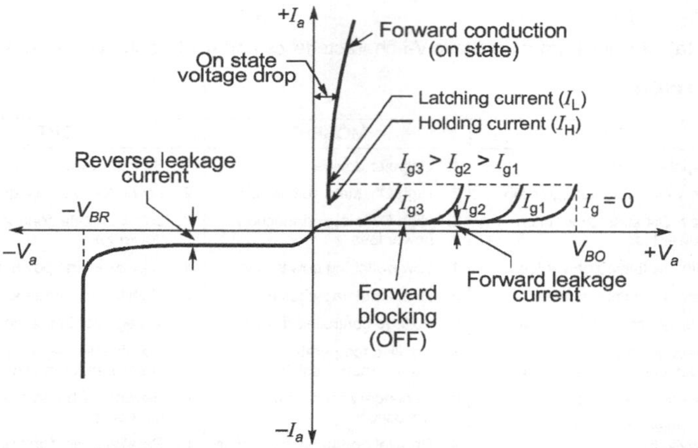

where, $V_{BO} \rightarrow$ Forward breakover voltage

$V_{BR} \rightarrow$ Reverse breakover voltage

$I_g \rightarrow$ Gate current

- **Latching Current (I_L)** : It is the minimum anode current to be attained above which the device continues to be in the ON state even after removal of the gate current.
- **Holding Current (I_H)** : It is the minimum anode current to be attained below which the device comes into the OFF state after applying a reverse voltage across it until it regains its blocking capability.
- $I_L > I_H$ or $I_L \approx I_H$.
- **Procedure to Turn-off the SCR** : Bring down the anode current below holding current. After that apply a reverse voltage across it till excess carrier are removed and it regains its blocking capability.

Thyristor Operates in Three-Region

1. Forward blocking mode :	Device is in OFF state. Anode is positive, cathode is negative and $V_a < V_{BO}$. $I_g = 0$ Junction : $J_1 \rightarrow$ Forward bias $J_2 \rightarrow$ Reverse bias $J_3 \rightarrow$ Forward bias Only forward leakage current flow.
2. Forward conduction mode :	Device is in ON state. Anode is positive, cathode is negative and $V_a > V_{BO}$. $I_g = 0$ Junction : $J_1 \rightarrow$ Forward bias $J_2 \rightarrow$ Breakdown occur $J_3 \rightarrow$ Forward bias
3. Reverse blocking mode :	Device is in OFF state. Anode is positive, cathode is positive. $I_g = 0$ Junction : $J_1 \rightarrow$ Reverse bias $J_2 \rightarrow$ Forward bias $J_3 \rightarrow$ Reverse bias

Remember:

- By increasing magnitude of Gate signal the breakdown voltage reduces.
- The Gate signal is needed until, the anode current is just above the latching value, so that it continues to be in ON state even after removing gate signal.
- Gate looses the control after SCR is turn-on, thats why it is called partially controlled device.
- By increasing the value of load inductance the minimum pulse width required to turn-on the SCR also increases.
- By increasing the value of resistance connected in series with inductive load, the minimum gate current width required is constant.
- By adding a resistance in parallel with load resistance the minimum pulse width is changed.

Switching Characteristics of Thyristor

Turn-on Characteristics	Turn-off Characteristics
Delay time (t_d) = 0.9 I_g to 0.1 I_a Rise time (t_r) = 0.1 I_a to 0.9 I_a Spread time (t_p) = 0.9 I_a to I_a	$t_q = t_{off} = t_{rr} + t_{gr}$ t_q = Device turn-off time t_{rr}= Reverse recovery time t_{gr} = Gate recovery time t_c= Circuit turn-off time

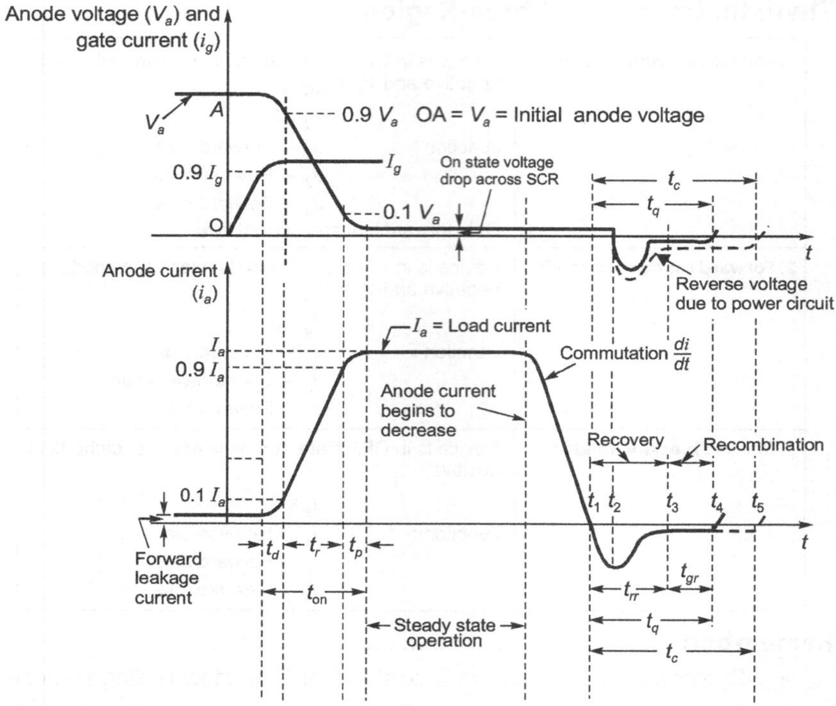

Remember:

For successful commutation, t_c (circuit turn-off time) > t_q (device turn-off time)

Triggering Methods of SCR

(i) Forward voltage triggering
(ii) Gate triggering
(iii) dv/dt triggering
(iv) Temperature triggering
(v) Light triggering

Remember:

☑ Light triggering thyristors are used in HVDC transmission system.

☑ The size of pulse transformer and average gate power dissipation can be reduced by using high frequency gating.

Series and Parallel Operation of Thyristors

String Efficiency

It is measure of utilisation of SCRs rating to its full capacity.

$$\text{String efficiency} = \frac{\text{Total string voltage/current rating}}{\text{n (individual voltage/current rating of one SCR)}}$$

where, $n \rightarrow$ Number of SCRs connected in series/parallel.

$$\text{Derating factor} = 1 - \text{String efficiency}$$

Series Connection of SCR

When system voltage is more than the voltage rating of a single thyristor, SCRs are connected in series in a string. On account of inherent variations in their V-I characteristics the voltage shared by each SCR is not the same and string efficiency is therefore less than one.

- A uniform voltage distribution in steady state can be achieved by connecting suitable resistance across each SCR such that each parallel combination has the same resistance.
- Likely that SCRs do not have identical dynamic characteristics. In such a case, series connected SCRs will have unequal voltage distribution during the transient conditions of turn-on, turn-off and high frequency operation.
- There is difference in forward blocking characteristics of series connected SCR. To overcome this problem, we use "static equalizing circuit".
- There is difference in the reverse recovery characteristic. To over come this problem, we use "dynamic equilising circuit".

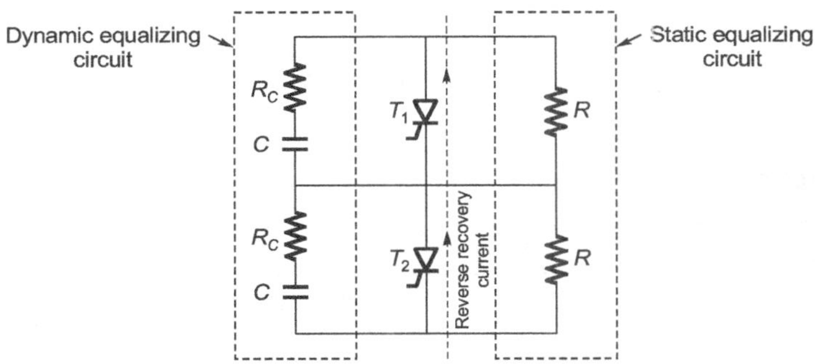

The value of static equalizing circuit resistance $R = \dfrac{nV_{bm} - V_s}{(n-1)\,\Delta I_b}$ and the value of dynamic equalizing capacitance, $C = \dfrac{(n-1)\Delta Q_R}{nV_{bm} - V_s}$

where, $V_{bm} \rightarrow$ Maximum permissible blocking voltage

Difference between maximum and minimum leakage current,

$\Delta I_b \rightarrow I_{b(\max)} - I_{b(\min)}$

$n \rightarrow$ Number of SCR connected in series

$V_s \rightarrow$ Total string voltage

$\Delta Q_R \rightarrow$ Difference in recovery charge

Parallel Connection of SCR

When available current rating of the SCR is not sufficient then we have to connect some of the SCRs in parallel.

Problem Related to Parallel Connected SCR

Unequal sharing of current :

(a) Due to difference in conduction characteristics of both SCR connected in parallel. To over come this problem connect current equilsing circuit.

(b) Due to temperature difference and to over come this problem we put all SCR in a common symmetrical heat sink.

Thermal Resistance

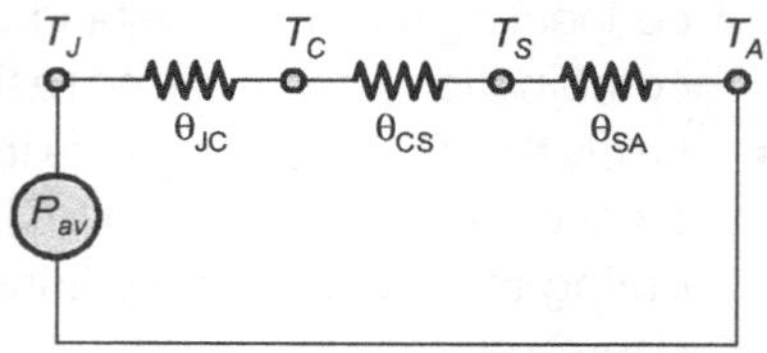

where, $P_{av} \rightarrow$ Average rate of heat generated

$T_J \rightarrow$ Junction temperature

$T_C \rightarrow$ Case temperature

$T_S \rightarrow$ Sink temperature; $T_A \rightarrow$ Ambient temperature

$\theta_{JC} \rightarrow$ Thermal resistance between junction and case

$\theta_{CS} \rightarrow$ Thermal resistance between case and sink

$\theta_{SA} \rightarrow$ Thermal resistance between sink and ambient

$$P_{av} = \frac{T_J - T_C}{\theta_{JC}} = \frac{T_C - T_S}{\theta_{CS}} = \frac{T_S - T_A}{\theta_{SA}} = \frac{T_J - T_A}{\theta_{JA}}$$

$$\theta_{JA} = \theta_{JC} + \theta_{CS} + \theta_{SA}$$

Protection of Thyristor

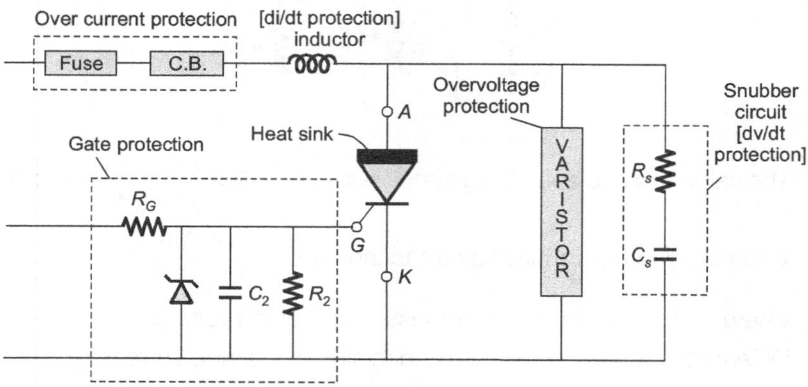

Circuit for protection of SCR

(i) **Over current protection:** Fuse or circuit breaker connected in series with SCR to limit over-current.
(ii) **Over voltage protection:** Varistor are connected across SCR.
(iii) **High dv/dt protection:** Snubber circuit is provided across SCR.
(iv) **High di/dt protection:** Connect a inductor in series with SCR.
(v) **Thermal protection:** Provide heat sink in SCR.

Gate Protection

(i) **Over current protection:** Connect a resistance in series with Gate.
(ii) **Over voltage protection:** Zener diode is connected across the gate and cathode junction.
(iii) **Protection against noise:** Connect a capacitor and a resistor across gate and cathode.

Remember:

- ☑ Gate current magnitudes are of the order of 20 to 200 mA.
- ☑ Triac is combination of antiparallel connection of two SCR.
- ☑ Diac is antiparallel connection of two SCR when $I_g = 0$.

Comparison between Different Types of Semiconductor Devices

Device	Circuit symbol	Voltage/current ratings
Diode		5000 V 5000 A
Transistors		
(a) BJT	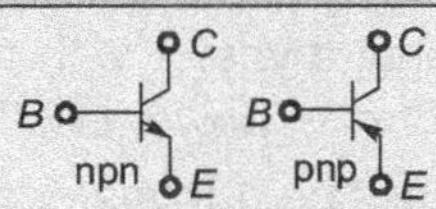	1400 V 400 A
(b) MOSFET (n-channel)	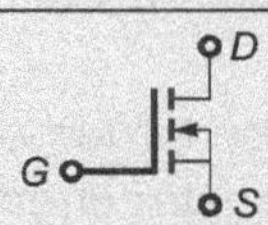	1000 V 50 A
(c) SIT	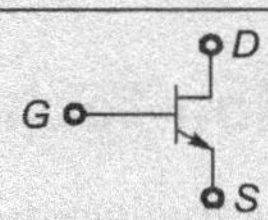	1200 V 300 A
(d) IGBT	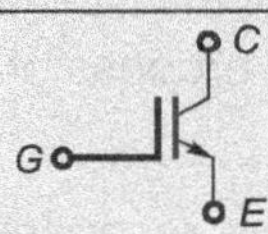	1200 V 500 A

Thyristors		
(a) SCR	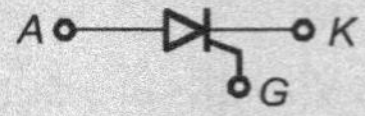	7000 V 5000 A
(b) LASCR	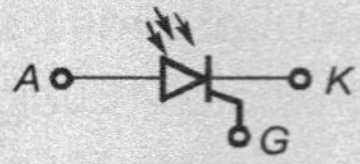	6000 V 3000 A
(c) ASCR/RCT	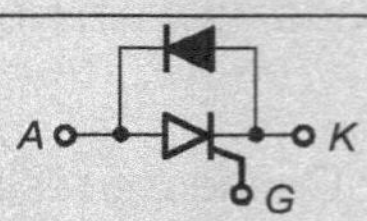	2500 V 400 A
(d) GTO	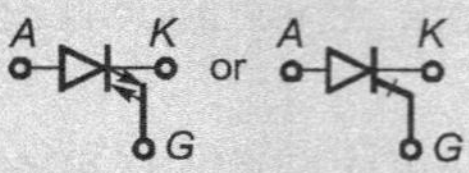	5000 V 3000 A
(e) SITH		2500 V 500 A
(f) MCT	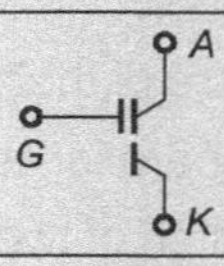	1200 V 40 A
(g) Triac	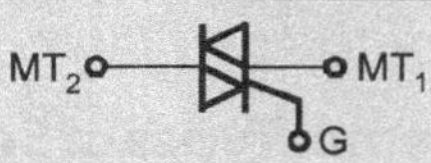	1200 V 1000 A

Remember:

- ☑ The term DIAC is obtained from capital letters Diode that can work on AC.
- ☑ A DIAC is sometimes called a gate less TRIAC.
- ☑ A TRIAC is a directional thyristor with three terminals. It is used extensively for the control of power in AC circuits.
- ☑ RCT may be considered as a thyristor with a built in anti-parallel diode.
- ☑ GTO is a device that can be turned-on by a positive gate current and turned-off by a negative gate current at its gate cathode terminals.

III Thyristor Commutation Techniques

Commutation means bringing the thyristor from forward conduction state to forward blocking state i.e. it is a process of turning off a thyristor.

Remember:

- ☑ Commutation technique use resonant LC or underdamped RLC circuits to force current or voltage of SCR to zero to turn-off device.

Types of Commutation

- **Natural Commutation (Line Commutation):** Here nature of supply take cares of commutation process.
  Example: Phase controlled rectifier and AC voltage controller.
- **Force Commutation:** Here external component bring anode current to zero forcefully.
  Example: Chopper and inverter.

Types of Force Commutation Circuit

Class A (Load Commutation):

Over all circuit must be underdamped.

i.e., $$R^2 < \frac{4L}{C}$$

- This type of commutation is possible in dc circuit not in ac circuit.
- Load commutation is also called resonant commutation or self commutation.

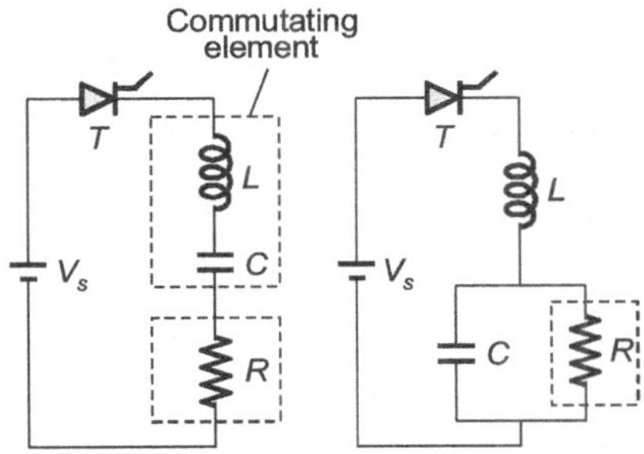

Class B Commutation:

- It is also called as **current commutation** or **resonant pulse commutation.**

  Assumption:

  1. Load current is assumed constant.
  2. The capacitor is initially charged with V_s volt with left plate positive and right plate negative.

  Here, T_A = Auxiliary thyristor
  T_M = Main thyristor

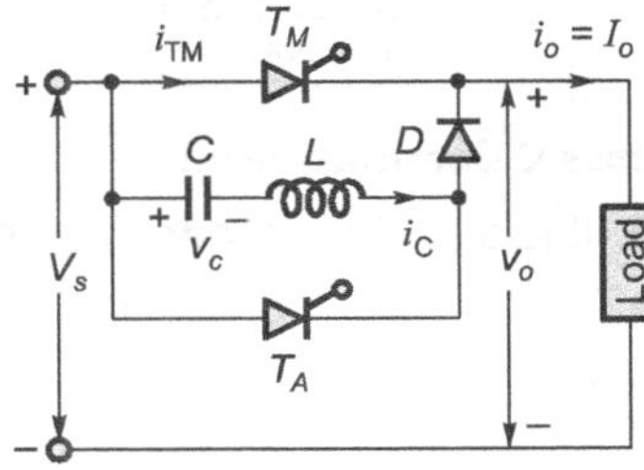

- Commutation process is initiated by switching ON the auxiliary thyristor.

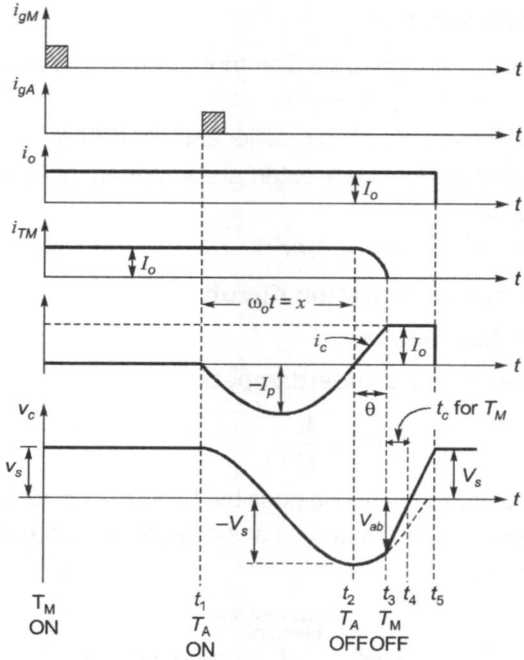

Parameters	Formula
Resonant current	$i_c = -V_s\sqrt{\frac{C}{L}}\sin\omega_o t = -I_p \sin\omega_o t$
Peak resonant current	$I_p = V_s\sqrt{\frac{C}{L}}$
Circuit turn-off time for main thyristor (T_m)	$t_c = C\frac{V_{ab}}{I_o}$
Reverse voltage across main thyristor	$V_{ab} = V_s \cos\left[\sin^{-1}\left(\frac{I_o}{I_p}\right)\right]$
Conduction time of auxiliary thyristor	$\pi\sqrt{LC}$

Class C Commutation:

It is also called, **complementary impulse commutation.**

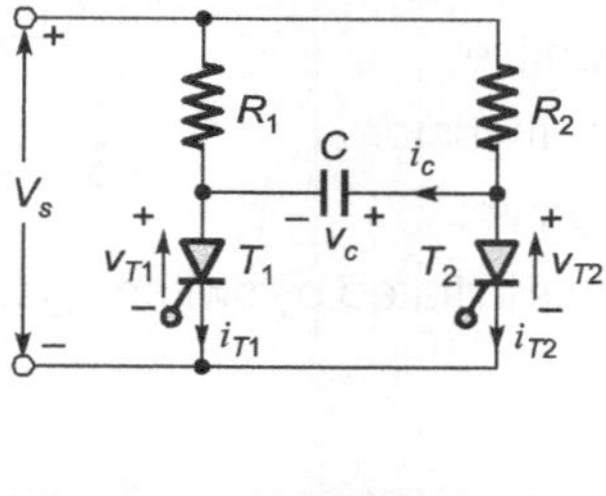

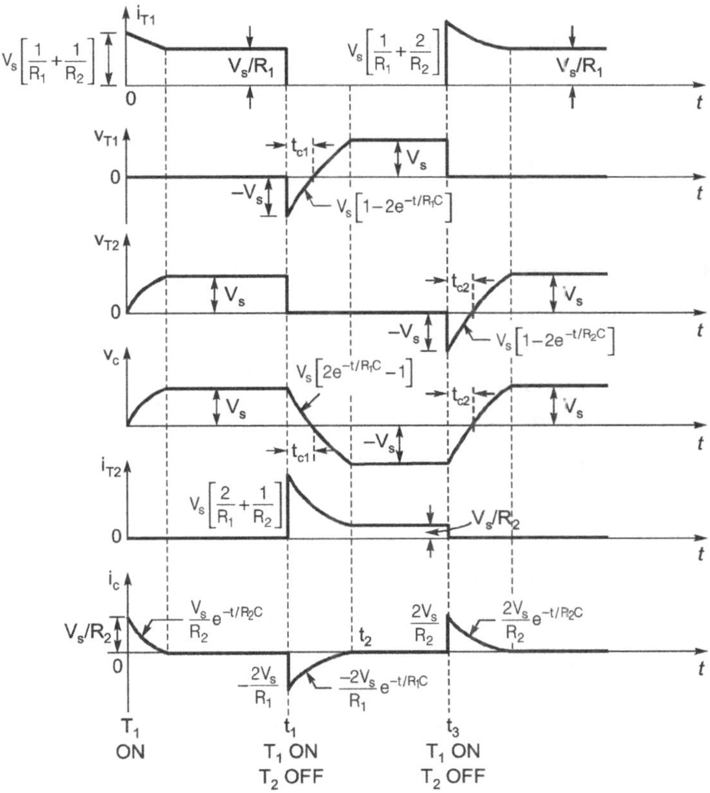

Case-1: When T_1 is turned ON at $t = 0$, T_2 is off

- The charging current (when capacitor is initially uncharged)

$$i_c = \frac{V_s}{R_2} \cdot e^{-t/R_2C}$$

- Voltage across capacitor C

$$v_c(t) = V_s\left(1 - e^{-t/R_2C}\right)$$

Case-2: When T_1 is to be turned OFF, T_2 is turned ON at t_1

- The charging current, $i_c(t) = -\dfrac{2V_s}{R_1}e^{-t/R_1C}$

- Voltage across capacitor C

$$v_c(t) = V_s \cdot \left[2e^{-t/R_1C} - 1\right]$$

Note:

☑ In last equation, time is measured from the instant t_1.

Peak current through T_1	$I_{T_1,P} = V_s\left[\frac{1}{R_1}+\frac{2}{R_2}\right]$
Peak current through T_2	$I_{T_2,P} = V_s\left[\frac{2}{R_1}+\frac{1}{R_2}\right]$
Circuit turn-off time for T_1	$t_{c1} = R_1 C \ln(2)$
Circuit turn-off time for T_2	$t_{c2} = R_2 C \ln(2)$

Class D Commutation:

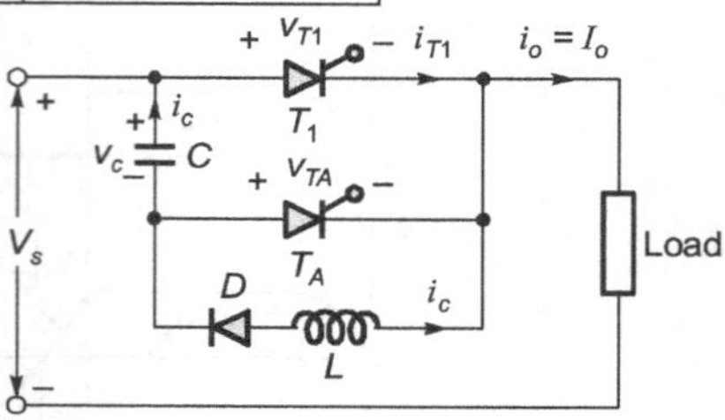

Also known as:

1. Impulse commutation
2. Voltage commutation
3. Auxiliary commutation
4. Parallel-capacitor commutation

where, $T_1 \rightarrow$ Main thyristor ; $T_A \rightarrow$ Auxiliary thyristor

Assumption:

1. Load current is assumed constant.
2. The capacitor is initially charged to V_s as shown in figure.

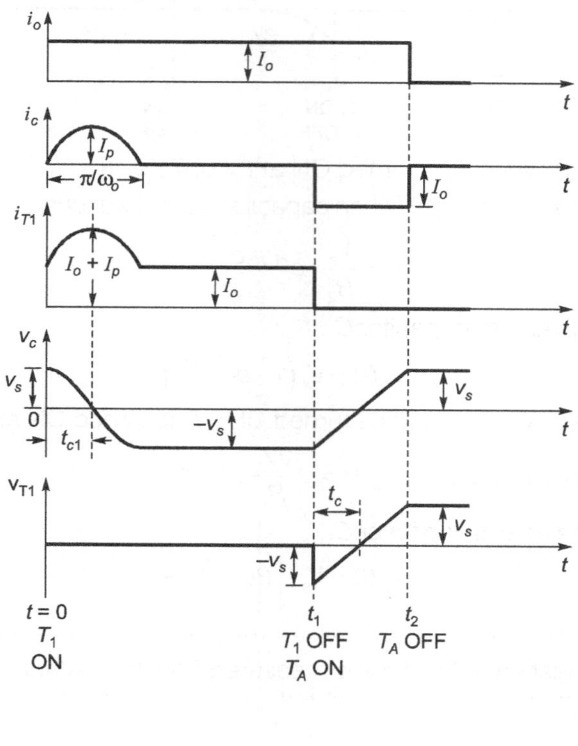

Capacitor current	$I_c = I_p \sin\omega_o t$ $I_c = V_s\sqrt{\frac{C}{L}}\sin\omega_o t$
Peak value of current through capacitor	$I_p = V_s\sqrt{\frac{C}{L}}$
Peak value of current through main thyristor (T_1)	$(I_p)_{T_1} = I_p + I_o$
Peak value of current through auxiliary thyristor (T_A)	I_o
Circuit turn-off time for main thyristor (T_1)	$t_c = C\frac{V_s}{I_o}$
Circuit turn-off time for auxiliary thyristor (T_A)	$t_{c_1} = \frac{\pi}{2\omega_o} = \frac{\pi}{2}\sqrt{LC}$

IV Diode Circuits and Rectifiers

Diode circuits with DC source

Resistive Load :

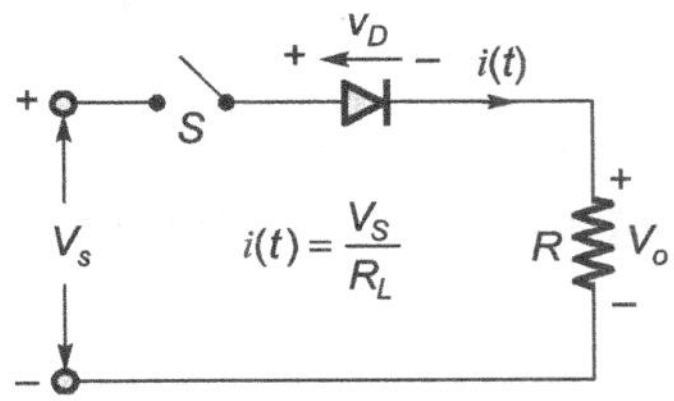

where, $V_s \to$ D.C. source voltage ; $R \to$ Load resistance

RC Load :

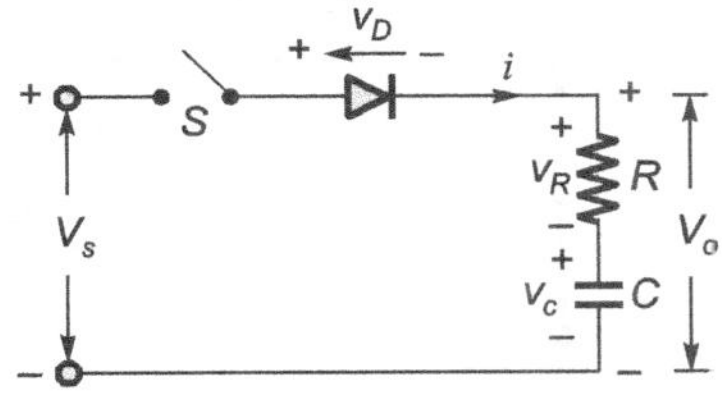

Load current	$i(t) = \frac{V_s}{R}\cdot e^{-t/RC}$
Voltage across capacitor	$v_c(t) = V_s(1 - e^{-t/RC})$ where, $v_c(t)$ = Voltage across capacitor at time t
Initial rate of rise of capacitor voltage	$\left(\frac{dV_c}{dt}\right)_{t=0} = \frac{V_S}{RC}$
Time constant	$\tau = RC = \frac{\text{Source voltage } (V_s)}{\left(\frac{dv_c}{dt}\right)_{t=0}}$

Remember:

At $t = 0$, capacitor acts as a conductor and at $t \to \infty$, it acts as an insulator.

RL load :

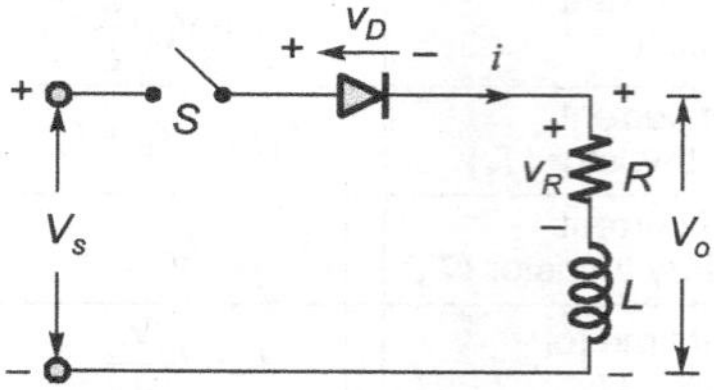

Load current	$i(t) = \frac{V_s}{R}\left(1 - e^{-\frac{R}{L}t}\right)$
Voltage across capacitor	$v_L(t) = V_s \cdot e^{-\frac{R}{L}t}$
Initial rate of rise of current	$\left(\frac{di}{dt}\right)_{t=0} = \frac{V_s}{L}$

Remember:

At $t = 0$, inductor acts as an insulator and at $t \to \infty$, it acts as a conductor.

LC load :

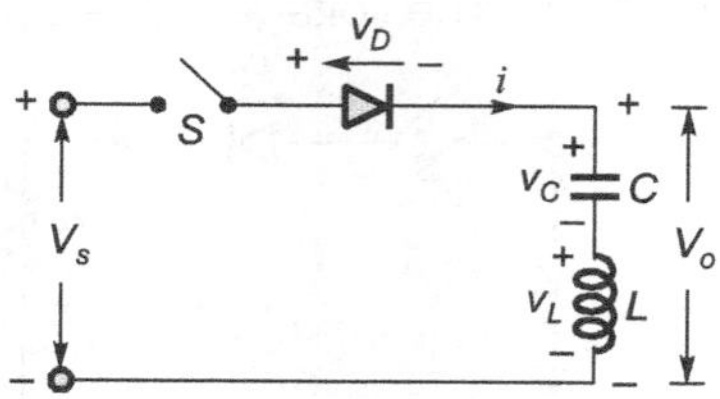

Load current :

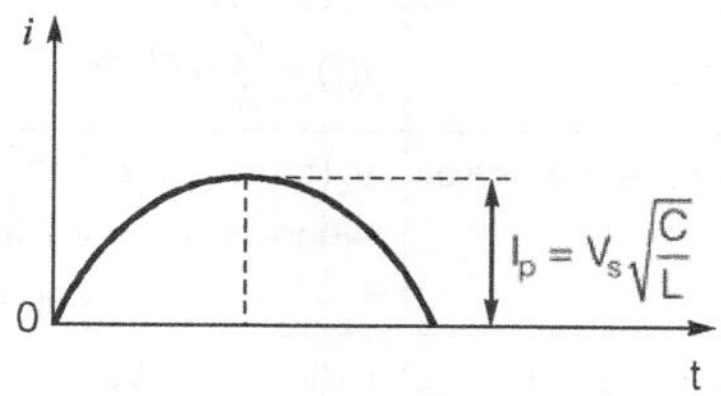

$i(t) = V_s\sqrt{\frac{C}{L}}\sin\omega_0 t$ where, $\omega_0 \to$ Resonant frequency of the circuit

$$\omega_0 = \frac{1}{\sqrt{LC}}$$

Voltage across capacitor	$v_C(t) = V_s(1 - \cos \omega_0 t)$ 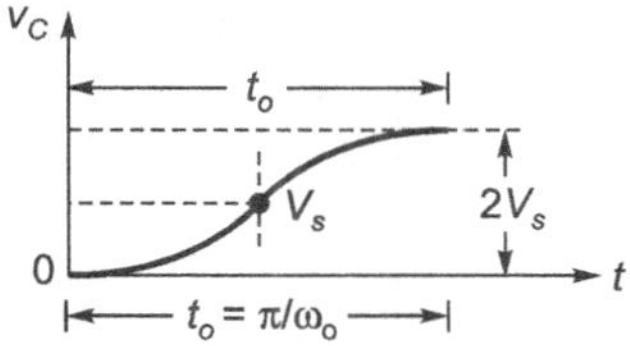
Voltage across inductor	$v_L(t) = V_s \cos \omega_0 t$ 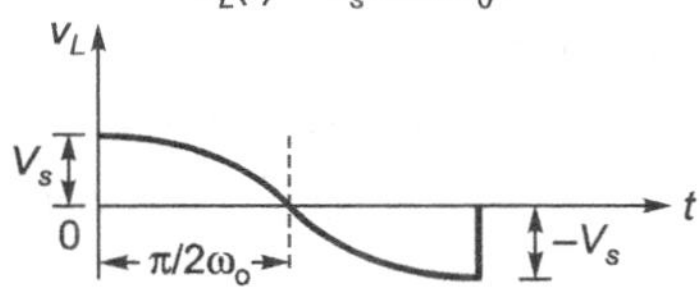
Conduction time	$t_0 = \dfrac{\pi}{\omega_0} = \pi\sqrt{LC}$

Remember:

- ☑ If we provide D.C. voltage to LC parameter, the current will be alternating in nature.
- ☑ In LC circuit when current follow sine curve, voltage follow cosine curve. This characteristic is used in commutation technique.

RLC load :

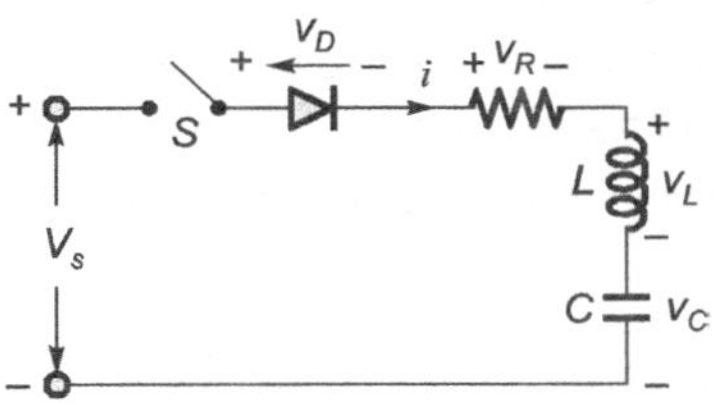

Characteristics equation	$s^2 + \dfrac{R}{L}s + \dfrac{1}{LC} = 0$
Damping factor	$\alpha = \dfrac{R}{2L} = \xi\omega_o$
Ringing frequency	$\omega_r = \sqrt{\omega_0^2 - \alpha^2}$ in rad/sec

Case 1. $\alpha < \omega_0$

- Roots of characteristic equation are complex.
- Circuit is under damped.

$$i(t) = \frac{V_s}{\omega_r L} \cdot e^{-\alpha t} \sin \omega_r t$$

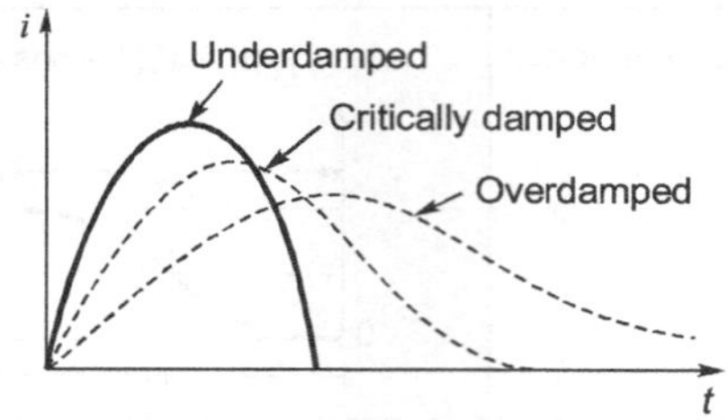

Case 2. $\alpha > \omega_0$

- Roots are real.
- Circuit is overdamped. $i(t) = \dfrac{V_s}{L\sqrt{\alpha^2 - \omega_0^2}} \cdot \sinh\left(\sqrt{\alpha^2 - \omega_0^2 t}\right)$

Case 3. $\alpha = \omega_0$

- Roots are equal.
- Circuit is critically damped. $i(t) = \dfrac{V_S}{L} \cdot t e^{-\alpha t}$

Performance Parameters

In order to evaluate the overall performance of rectifier load and the combinations certain performance parameters to their input and output must be known.

If V_S = RMS value of supply phase voltage

I_S = RMS value of supply phase current including fundamental and harmonics

I_{S1} = RMS value of fundamental component of supply current I_S

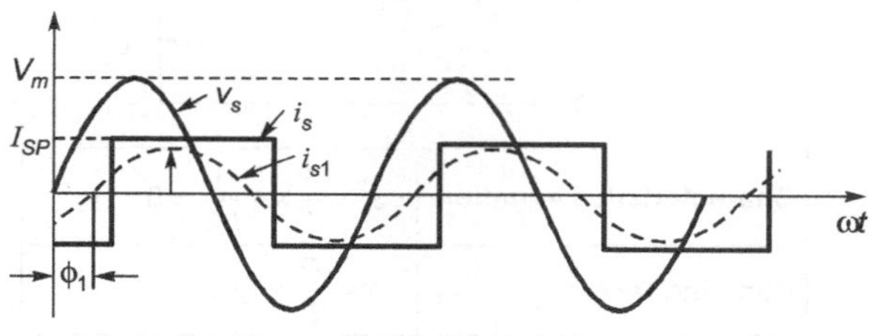

Parameter	Formula
Input power factor	P.F. = $\dfrac{\text{Mean ac input power}}{\text{Total rms input voltamperes}}$ P.F. = $\dfrac{I_{S1}}{I_S}\cos\phi_1$
RMS value of harmonic components	$I_h = \sqrt{I_S^2 - I_{S1}^2}$

Input current harmonic factor (HF)	$HF = \frac{I_h}{I_{S1}}$ or $HF = \sqrt{\frac{I_S^2 - I_{S_1}^2}{I_{S_1}^2}}$
Crest factor	$CF = \frac{I_{SP}}{I_S}$; where $I_{SP} \rightarrow$ Peak input current
Input current distortion factor	$CDF = \frac{I_{s1}}{I_s}$ $CDF = \frac{PF}{DF}$
Displacement factor	$DF = \cos\phi_1$
Form factor	$FF = \frac{V_{or}}{V_o}$ where $V_{or} \rightarrow$ RMS value of output voltage $V_o \rightarrow$ Average value of output voltage
Ripple voltage	$V_r = \sqrt{V_{or}^2 - V_o^2}$
Voltage ripple factor	$VRF = \frac{V_r}{V_o} = \left[\left(\frac{V_{or}}{V_o}\right)^2 - 1\right]^{1/2}$
Current ripple factor	$CRF = \frac{I_r}{I_o} = \left[\left(\frac{I_{or}}{I_o}\right)^2 - 1\right]^{1/2}$
Transformer utilization factor	$TUF = \frac{P_{dc}}{V_s I_s}$

where, $V_S \rightarrow$ RMS voltage of the secondary winding of transformer
$I_S \rightarrow$ RMS current of the secondary winding of transformer
$P_{dc} \rightarrow V_o I_o$ = DC output power
$V_o \rightarrow$ Average output voltage; $I_o \rightarrow$ Average output current

Diode Rectifiers

A rectifier employing diode is called an uncontrolled rectifier, because its average output voltage is a fixed D.C. voltage.

Single Phase Half-Wave Diode Rectifier

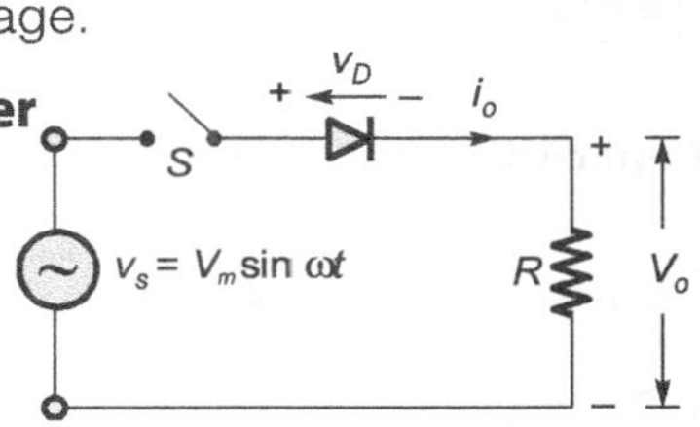

Resistive Load:

- Voltage rms value of output

$$V_{or} = \frac{V_m}{2}$$

where, V_m = Maximum value of source voltage (v_S)

- Average value of output voltage

$$V_o = \frac{V_m}{\pi}$$

- Power delivered

$$P = V_{or} I_{or}$$

$$P = I_{or}^2 R$$

where, I_{or} = RMS value of load current

and V_{or} = RMS value of load voltage

- Peak inverse voltage (PIV) = V_m
- Input power factor

$$\text{Input p.f.} = \frac{\text{Power delivered to load}}{\text{Input VA}}$$

$$\text{Input p.f.} = \frac{V_{or} \cdot I_{or}}{V_s \cdot I_{or}} = 0.707$$

Remember:

Peak inverse voltage (PIV) is the maximum voltage that appears across the device during its blocking state.

Inductive Load:

Output current	$i_o = \frac{V_m}{\omega L}(1 - \cos\omega t)$
Peak value of current	$I_{max} = \frac{2V_m}{\omega L}$
Average value of current	$I_o = \frac{1}{2} I_{max}$
RMS value of fundamental current	$I_{1r} = \frac{I_o}{\sqrt{2}}$
Output voltage	$v_o = V_m \sin\omega t = v_s$
Average value of output voltage	$V_o = 0$

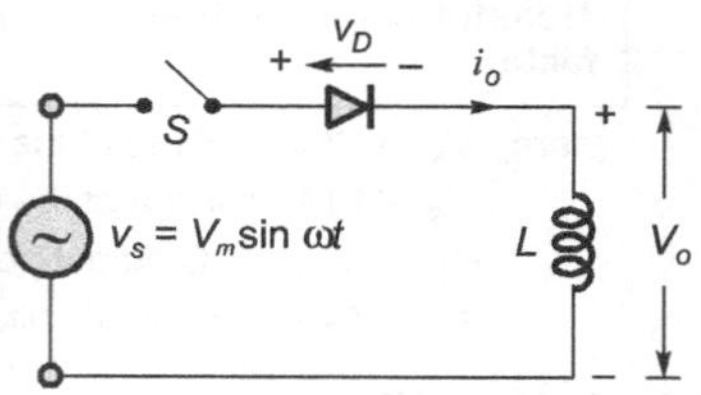

Capacitive Load

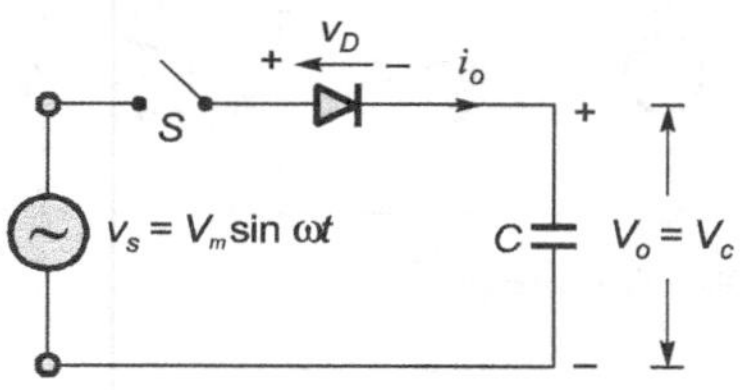

Output current	$i_o = \omega C V_m \cos\omega t$
Output voltage	$v_o = V_m \sin\omega t = v_s = v_c$
Diode voltage	$v_D = V_m(\sin\omega t - 1)$
Average value of Diode voltage	$V_D = V_m$
RMS value of diode voltage	$V_{rD} = 1.225 V_m$

RE Load

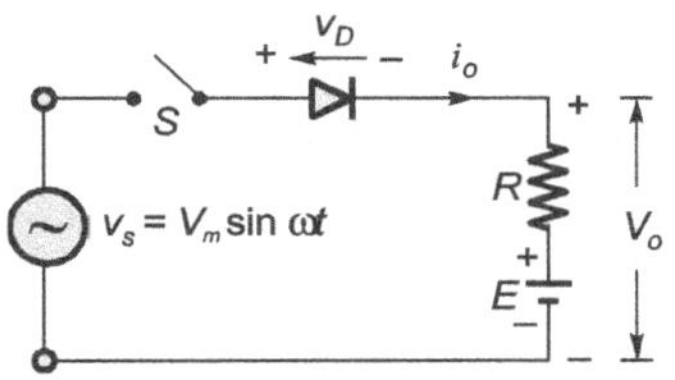

Turn-on angle	$\theta_1 = \sin^{-1}\left(\frac{E}{V_m}\right)$
Average value of output current	$I_o = \frac{1}{2\pi R}[2V_m \cos\theta_1 - E(\pi - 2\theta_1)]$

Note:

- ☑ Conduction angle for diode $= \pi - 2\theta_1$
- ☑ PIV for diode $= V_m + E$

RL Load

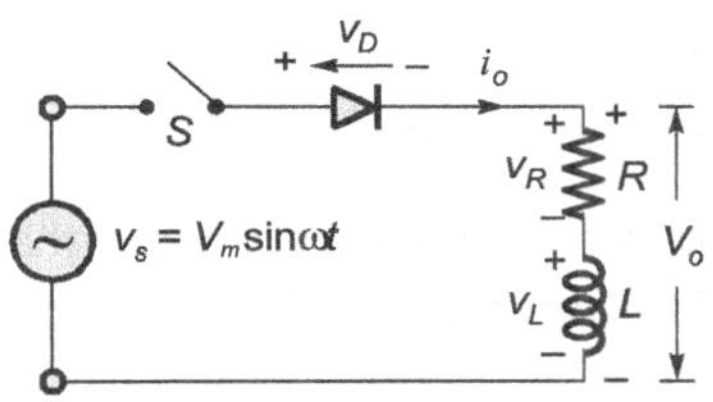

Average value of output voltage	$V_o = \frac{V_m}{2\pi}(1-\cos\beta)$ where, $\beta \rightarrow$ Extinction angle of the diode
Average value of output current	$I_o = \frac{V_m}{2\pi R}(1-\cos\beta)$ For $0 \le \omega t \le \beta$

Single phase full wave diode rectifier

Average output voltage	$V_o = \frac{2V_m}{\pi}$
RMS value of output voltage	$V_{or} = \frac{V_m}{\sqrt{2}}$

Comparison of various 1-ϕ diode rectifier

Parameters	Half-wave	Full-wave	
		Centre-tap (M-2)	Bridge (B-2)
DC output voltage, V_o	$\frac{V_m}{\pi}$	$\frac{2V_m}{\pi}$	$\frac{2V_m}{\pi}$
RMS output voltage, V_{or}	$\frac{V_m}{2}$	$\frac{V_m}{\sqrt{2}}$	$\frac{V_m}{\sqrt{2}}$
Ripple voltage, V_r	0.3856 V_m	0.3077 V_m	0.3077 V_m

Voltage ripple factor, VRF	1.211	0.482	0.482
Rectifier efficiency, η	40.53%	81.06%	81.06%
TUF	0.2865	0.672	0.8106
PIV	V_m	$2V_m$	V_m
Crest factor, CF	2	$\sqrt{2}$	$\sqrt{2}$
Number of diodes	1	2	4
Ripple frequency	f	$2f$	$2f$

Three-phase half-wave diode rectifier

R Load

- Average value of output voltage, $V_0 = \dfrac{3\sqrt{6}}{2\pi}V_{ph} = \dfrac{3}{2\pi}\cdot V_{ml}$

  where, $V_{mp} \rightarrow$ Maximum value of phase voltage, V_{ph}

  $V_{ml} \rightarrow$ Maximum value of line voltage, V_l

  $V_l \rightarrow \sqrt{3}\cdot V_{ph}$

Comparison of Various 3-ϕ Diode Rectifier

Parameters	3-Pulse Rectifier	6-Pulse Rectifier M-6 Type	6-Pulse Rectifier B-6 Type	12-Pulse Rectifier
DC output voltage, V_o	$\dfrac{3V_{ml}}{2\pi}$	$\dfrac{3V_{mp}}{\pi}$ or $\dfrac{\sqrt{3}V_{ml}}{\pi}$	$\dfrac{3\sqrt{3}V_{mp}}{\pi}$ or $\dfrac{3V_{ml}}{\pi}$	1.91 V_{ml}
RMS output voltage, V_{or}	0.4854 V_{ml}	0.55185 V_{ml}	0.9558 V_{ml}	1.9101 V_{ml}
Ripple voltage, V_r	0.0872 V_{ml}	0.02356 V_{ml}	0.0408 V_{ml}	0.019545 V_{ml}
Voltage ripple factor, VRF	0.1826 or 18.26%	0.043 or 4.3%	0.0427 or 4.27%	0.01023 or 1.023%
Rectifier efficiency, η	96.765%	99.82%	99.82%	–
TUF	0.6644	0.551	0.9541	–
PIV	V_{ml}	1.155 V_{ml}	V_{ml}	–
Form factor, FF	1.0165	1.0009	1.0009	1.00005

V Phase Controlled Rectifiers

Uncontrolled rectifier use diode and it convert fixed ac to fixed dc but controlled rectifier use SCR and it convert fixed ac to variable dc. By changing firing angle α, output also changes.

In all phase controlled rectifier natural commutation take place.

Single Phase Half Wave Rectifier

It is also called as one pulse converter because for one cycle of supply voltage we get one pulse at output voltage.

With R load:

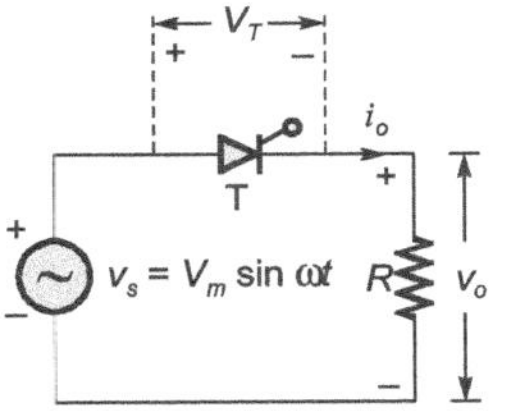

From: 0 to α $T \rightarrow$ OFF
α to π $T \rightarrow$ ON

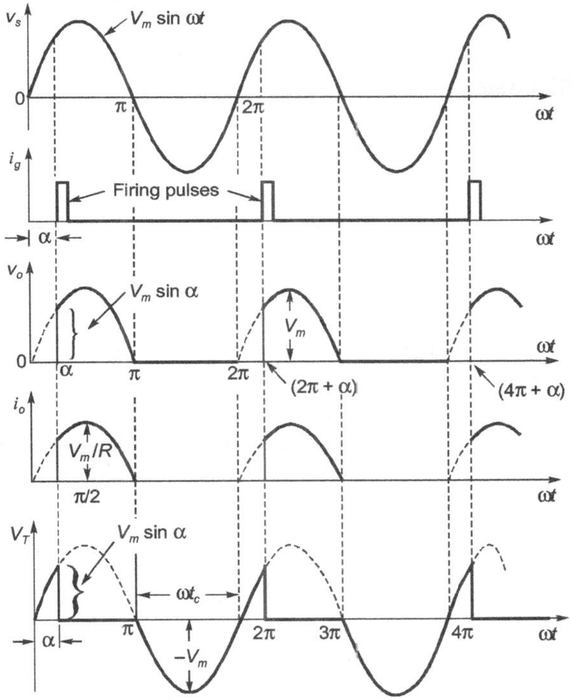

Average load voltage	$V_o = \frac{V_m}{2\pi}(1+\cos\alpha)$ where, $\alpha \rightarrow$ Firing angle; $V_m \rightarrow$ Maximum value of sinusoidal source voltage
RMS value of load voltage	$V_{or} = \frac{V_m}{2\sqrt{\pi}}\left[(\pi-\alpha)+\frac{1}{2}\sin 2\alpha\right]^{1/2}$
RMS current	$I_{or} = \frac{V_{or}}{R}$
Power dissipated in *R* load	$P = V_{or}I_{or} = I_{or}^2 R = \frac{V_{or}^2}{R}$
Input power factor	p.f. = $\frac{\text{Power delivered to load}}{\text{Input VA}}$ p.f. = $\frac{V_{or}}{V_s}$
Circuit turn-off time	$t_c = \frac{\pi}{\omega}$ sec.

Remember:

- ☑ Circuit turn-off time must be more than the SCR turn-off time for satisfactory commutation.
- ☑ The source current introduces DC component in the supply transformer and saturate the transformer. This is disadvantage of 1ϕ - half wave rectifier.

RL Load:

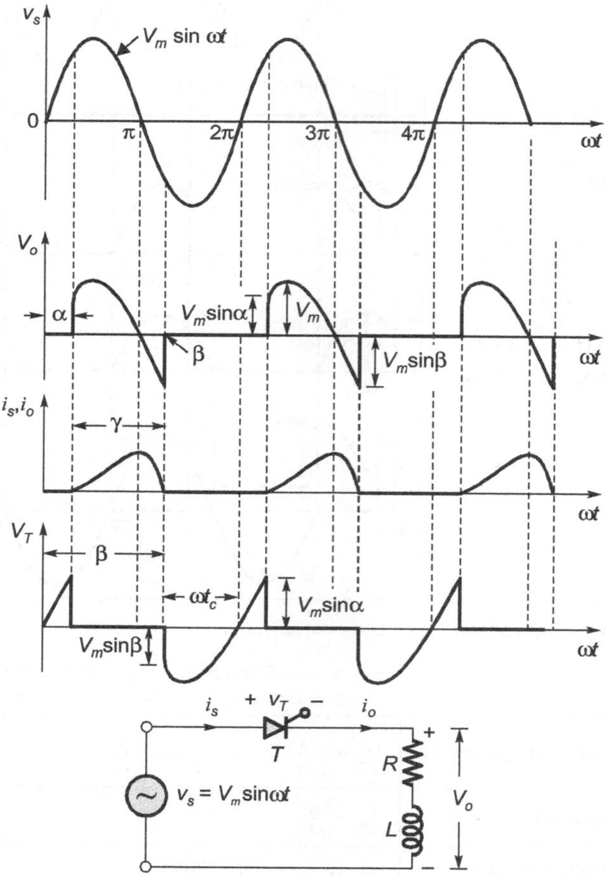

- Average load voltage, $V_o = \frac{V_m}{2\pi}(\cos\alpha - \cos\beta)$
- Rms load voltage, $V_{or} = \frac{V_m}{2\sqrt{\pi}}\left[(\beta-\alpha) - \frac{1}{2}\{\sin 2\beta - \sin 2\alpha\}\right]^{1/2}$

where, β = Extinction angle

- Conduction angle, $\gamma = \beta - \alpha$
- Circuit turn-off time, $t_c = \dfrac{2\pi - \beta}{\omega}$

Remember:

☑ From $\omega t = \pi$ to β, thyristor is forced to conduct in negative supply by the stored energy in the inductor, here energy is fed back to source.

☑ At $\omega t = \beta$, inductor fed back its whole energy to source.

RL Load with Freewheeling Diode:

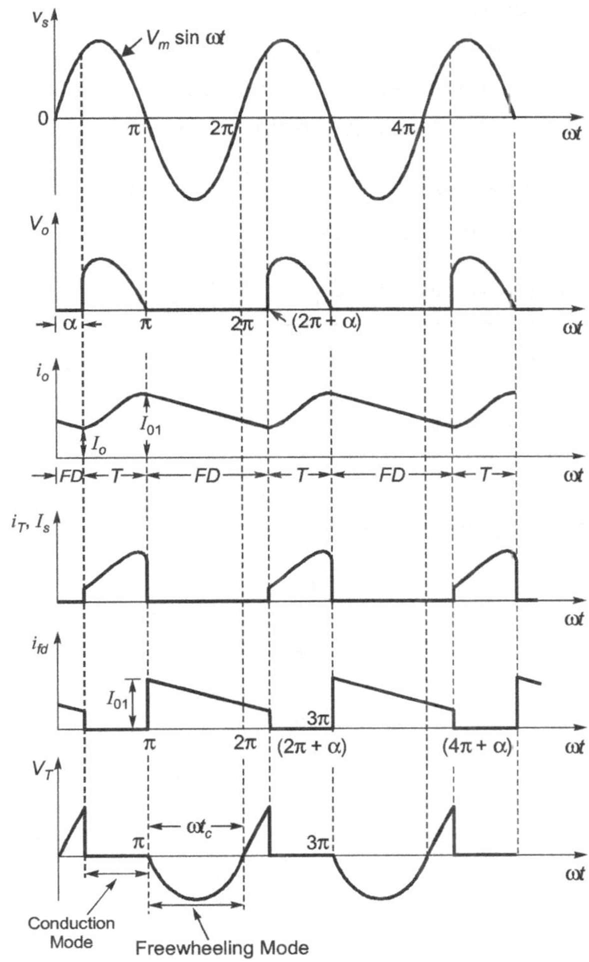

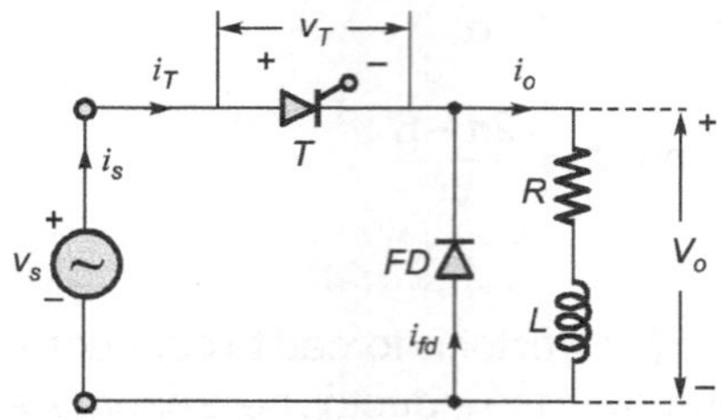

The performance of converter is improved by connecting a free wheeling diode across the load.

Advantage of freewheeling diode:

- Improves power factor.
- Negative spikes in output voltage are removed.
- The overall current waveform is improved.
- Overall converter efficiency improves.

Parameters	Formula
The average output voltage	$V_o = \frac{V_m}{2\pi}[1+\cos\alpha]$
The average output current	$I_o = \frac{V_m}{2\pi R}[1+\cos\alpha]$
The RMS output voltage	$V_{or} = \frac{V_m}{2\sqrt{\pi}}\left[(\pi-\alpha)+\frac{1}{2}\sin 2\alpha\right]^{1/2}$
Circuit turn-off time	$t_c = \frac{\pi}{\omega}$ sec

RE Load or Charging Battery

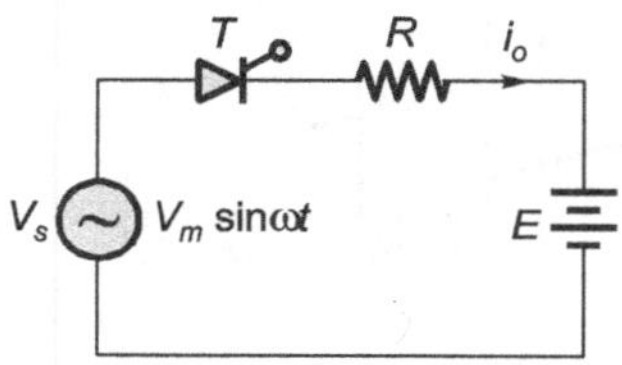

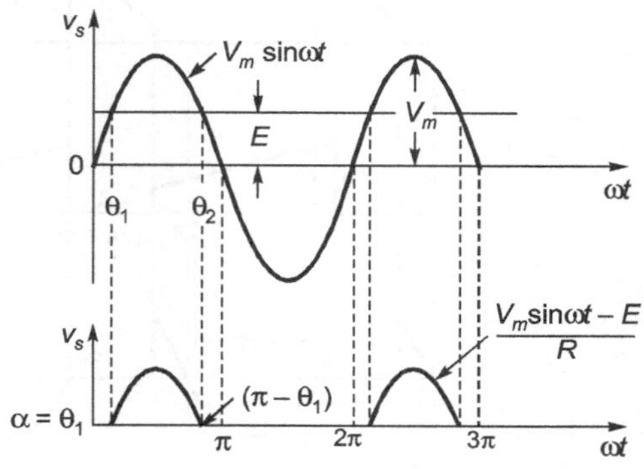

Parameters	Formula
Charging Current	$i_o = \frac{V_m \sin\omega t - E}{R}$
Minimum value of firing angle	$\theta_1 = \sin^{-1}\left(\frac{E}{V_m}\right)$
Maximum value of firing angle	$\theta_2 = \pi - \theta_1$
Range of firing angle	$\theta_1 \le \alpha \le \theta_2$
Conduction angle of thyristor	$\gamma = \theta_2 - \alpha$
Circuit turn-off time of thyristor	$t_c = \frac{2\pi + \theta_1 - \theta_2}{\omega} = \frac{\pi + 2\theta_1}{\omega}$
Average output voltage	$V_o = \frac{1}{2\pi}[V_m(\cos\alpha - \cos\theta_2) + E(2\pi + \alpha - \theta_2)]$
Average charging current of battery	$I_o = \frac{1}{2\pi R}[V_m(\cos\alpha - \cos\theta_2) - E(\theta_2 - \alpha)]$
Power supplied by battery	EI_o

Remember:

☑ When continuous gate signal is given to SCR, it behaves like diode.

RLE Load or DC Machine Load

Parameters	Formula
Minimum value of firing angle	$\theta_1 = \sin^{-1}\left(\frac{E}{V_m}\right)$
Maximum value of firing angle	$\theta_2 = (\pi - \theta_1)$
Average load current	$I_o = \frac{1}{2\pi R}[V_m(\cos\alpha - \cos\beta) - E(\beta - \alpha)]$
Average load voltage	$V_0 = E\left(1 - \frac{\gamma}{2\pi}\right) + \frac{V_m}{\pi}\sin\left(\alpha + \frac{\gamma}{2}\right)\cdot\sin\frac{\gamma}{2}$ or $V_0 = \frac{1}{2\pi}[V_m(\cos\alpha - \cos\beta) + E(2\pi + \alpha - \beta)]$
Maximum reverse voltage to thyristor	$(V_m + E)$
Circuit turn-off time	$t_c = \frac{2\pi - \beta}{\omega}$ sec
RMS load current	$I_{or} = \left[\frac{1}{2\pi R^2}\{(E^2 + V_s^2)(\pi - 2\alpha) + V_s^2 \sin 2\alpha - 4V_m E \cos\alpha\}\right]^{1/2}$
Power delivered to load	$P = I_{or}^2 R + I_o E$
Power factor	$PF = \frac{I_{or}^2 R + I_o E}{V_s I_{or}}$

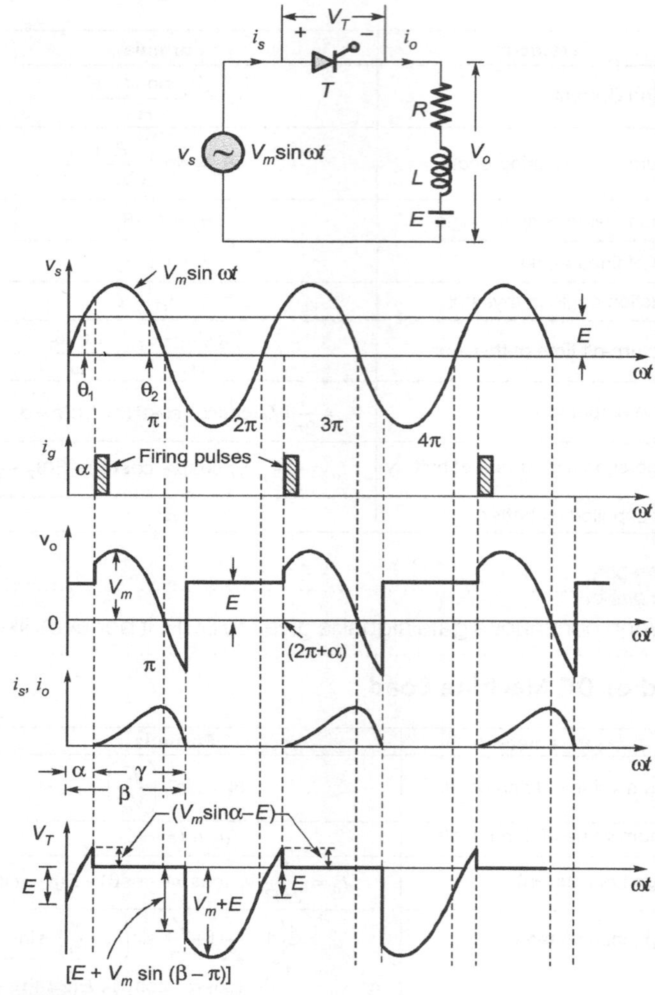

Remember:

☑ High value of inductance results in continuous conduction.

Single Phase Full-wave Converter

Full-wave rectifier is 2 pulse rectifier, for one cycle of voltage wave we get 2 pulse of current wave.

It has two configuration : Mid point converter and Bridge converter

- SCRs are subjected to a peak inverse voltage of 2 V_m in mid point converter and V_m in bridge converter.

- Mid point configuration is used in case, the terminals on D.C. side have to be grounded.
- Bridge configuration is preferred over midpoint configuration.

RL Load

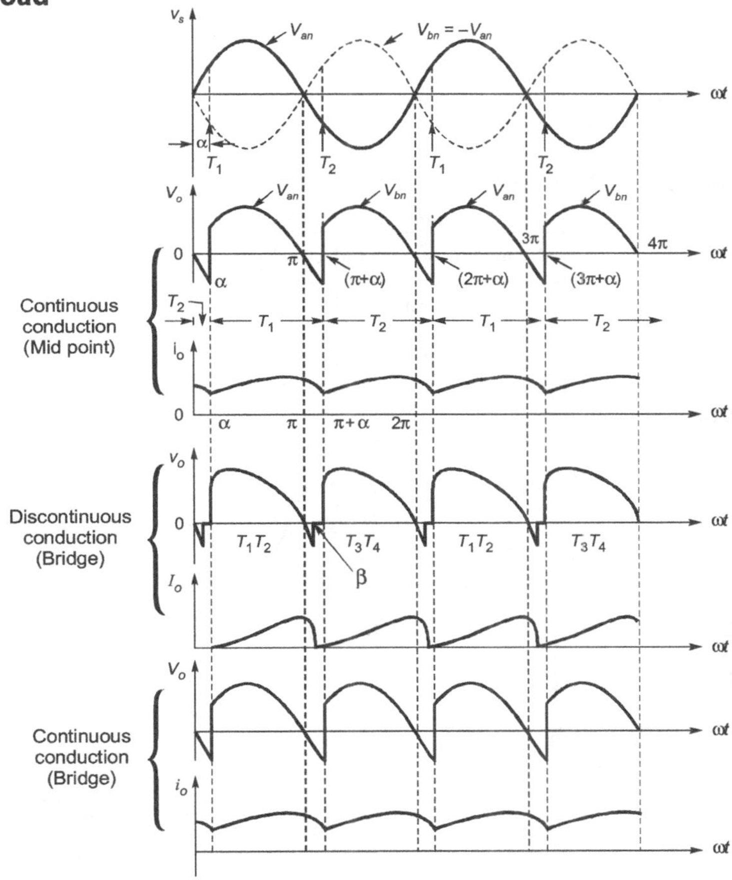

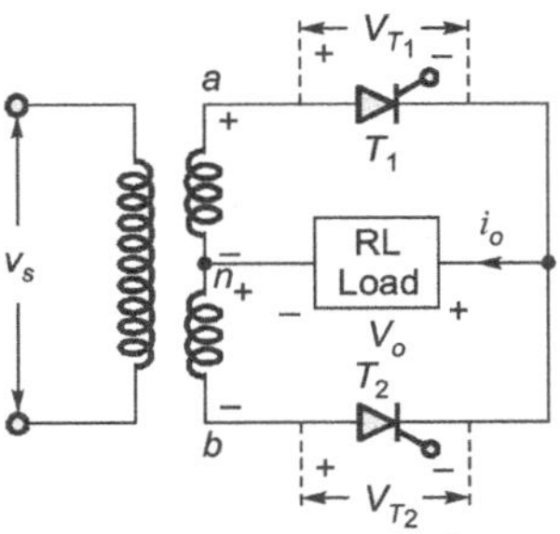

(Mid point connection)

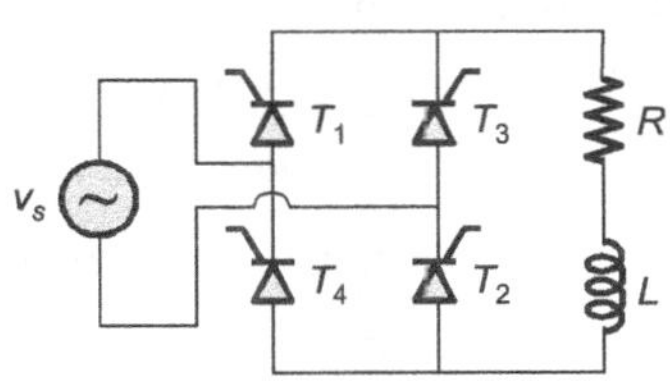

(Bridge connection)

Discontinuous conduction:

Parameters	Formula
Average output voltage	$V_o = \frac{V_m}{\pi}[\cos\alpha - \cos\beta]$
RMS output voltage	$V_{or} = \frac{V_m}{\sqrt{2\pi}}\left[(\beta-\alpha)+\frac{1}{2}(\sin 2\alpha - \sin 2\beta)\right]^{1/2}$
Power factor	$\cos\phi = \frac{V_{or}}{V_{sr}}$
Circuit turn-off time	$t_c = \frac{2\pi - \beta}{\omega}$

Continuous conduction:

Parameters	Formula
Average output voltage	$V_o = \frac{2V_m}{\pi}\cos\alpha$
RMS output voltage	$V_{or} = \frac{V_m}{\sqrt{2}}$
Circuit turn-off time	$t_c = \frac{\pi - \alpha}{\omega}$

RLE Load

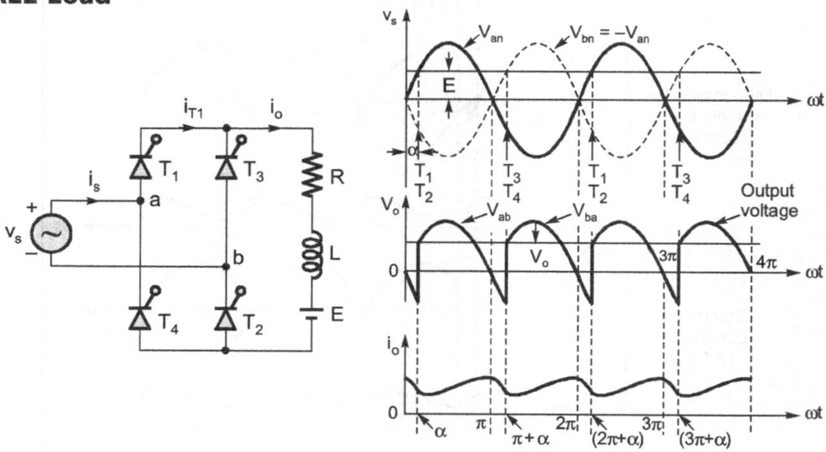

(Continuous conduction)

For continuous conduction

Parameters	Formula
Average output voltage	$V_o = \frac{2V_m}{\pi}\cos\alpha$
RMS output voltage	$V_{or} = V_s$
Circuit turn-off time	$t_c = \frac{\pi - \alpha}{\omega}$ sec

For discontinuous conduction

Parameters	Formula
Average output voltage	$V_o = \frac{1}{\pi}[V_m(\cos\alpha - \cos\beta) + E(\pi + \alpha - \beta)]$
RMS output voltage	$I_o = \frac{V_o - E}{R}$
Circuit turn-off time	$t_c = \frac{2\pi + \theta_1 - \beta}{\omega}$

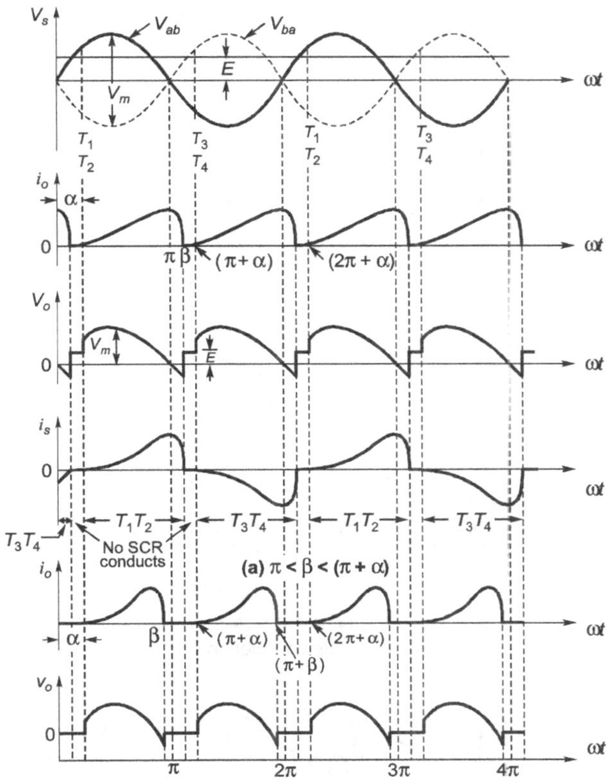

(b) β < π
(Discontinuous Conduction)

Single Phase Semiconverter

Semiconductor also known as half-controlled rectifier or 2 pulse converter.

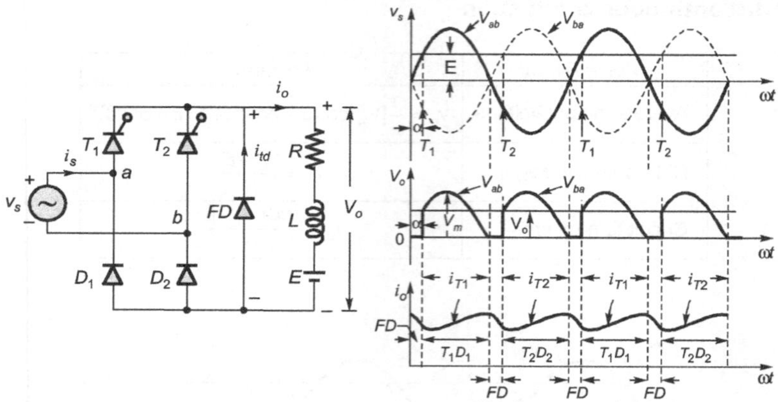

(Continuous conduction)

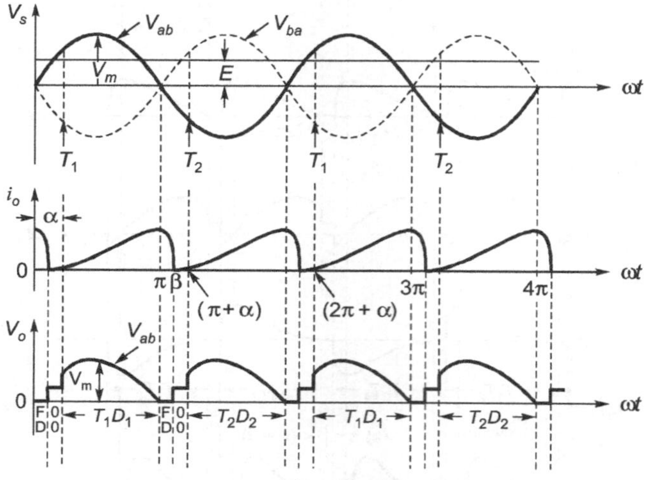

(a) π < β < (π + α)
Discontinuous conduction

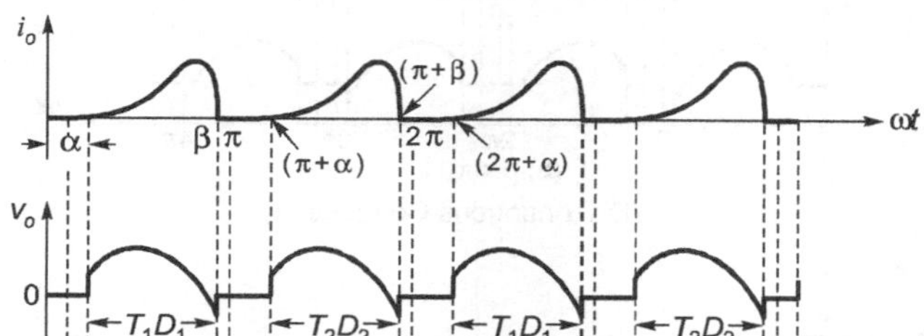

(b) $\beta < \pi$ and $V_m \sin\beta < E$ (Discontinuous conduction)

Continuous conduction

Parameter	Formula
Average output voltage	$V_o = \frac{V_m}{\pi}(1+\cos\alpha)$
RMS output voltage	$V_{or} = \frac{V_m}{\sqrt{2\pi}}\left\{(\pi-\alpha)+\frac{1}{2}\sin 2\alpha\right\}^{1/2}$
Circuit turn-off time	$t_c = \frac{\pi-\alpha}{\omega}$ (without freewheeling diode) $t_c = \frac{\pi}{\omega}$ (with freewheeling diode)

Discontinuous conduction

- Average output voltage, $V_o = \frac{1}{\pi}[V_m(1+\cos\alpha)+E(\pi+\alpha-\beta)]$

Note:

☑ Freewheeling diode prevents the occurrence of dead short circuit of source.

☑ For $\beta < \pi$, freewheeling diode does not conduct.

☑ Single phase full converter is two quadrant converter and single phase semiconverter is one quadrant converter.

3ϕ Half-Wave Controlled Converter

- It is a 3-pulse converter

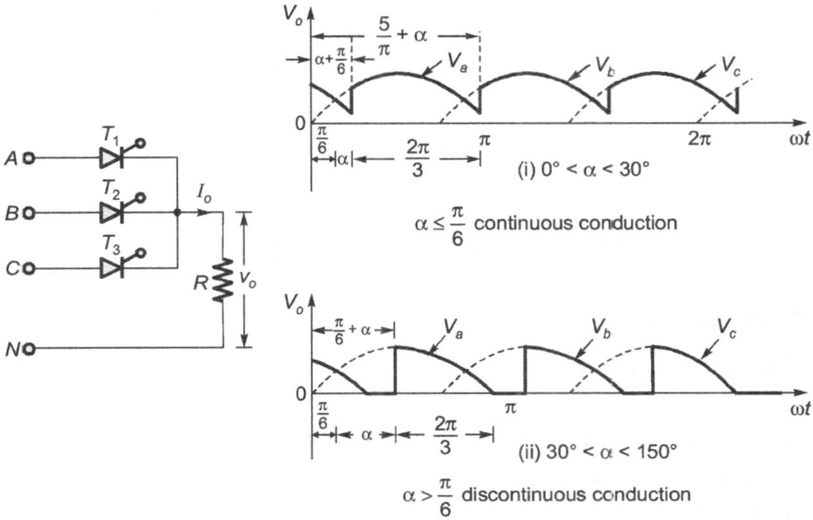

R Load

Case-1: when $\alpha < 30°$

- Average output voltage, $V_0 = \dfrac{3\sqrt{3}}{2\pi} V_{mp} \cos\alpha = \dfrac{3V_{ml}}{2\pi} \cos\alpha$
- RMS value of output voltage, $V_{or} = V_{ml} \left[\dfrac{1}{6} + \dfrac{\sqrt{3}}{8\pi} \cos 2\alpha \right]^{1/2}$

Both formula valid for (i) R load when $\alpha < 30°$. (ii) RL load for any value of α.

where V_{mp} = Maximum value of phase voltage

$V_{ml} = \sqrt{3}\, V_{mp}$ = Maximum value of line voltage

Case-2: when $\alpha > 30°$

- Average output voltage, $V_0 = \dfrac{3V_{mp}}{2\pi} \left[1 + \cos\left(\alpha + \dfrac{\pi}{6} \right) \right]$
- RMS value of output voltage, $V_{or} = \dfrac{V_{ml}}{2\sqrt{\pi}} \left[\left(\dfrac{5\pi}{6} - \alpha \right) + \dfrac{1}{2} \sin\left(2\alpha + \dfrac{\pi}{3} \right) \right]^{1/2}$

Both formula valid for

(i) R load when $\alpha > 30°$.

(ii) RL load with freewheeling diode when $\alpha > \pi/6$.

(iii) RLE load with F.D. when $\alpha > \pi/6$.

3ϕ-Full converter

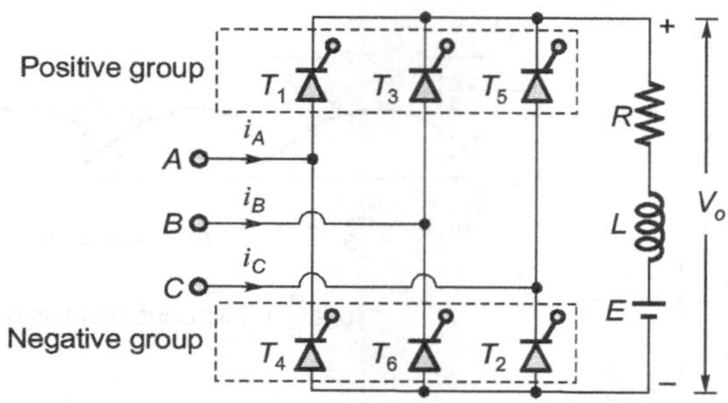

6 Pulse Converter

- Positive group of SCR fired at an interval of 120°, similarly negative group of SCR are fired with an interval of 120° amongst them.
- But SCR from both the group are fired at an interval of 60°.
- At any time two SCRs, one from positive group and other from negative group must conduct together.

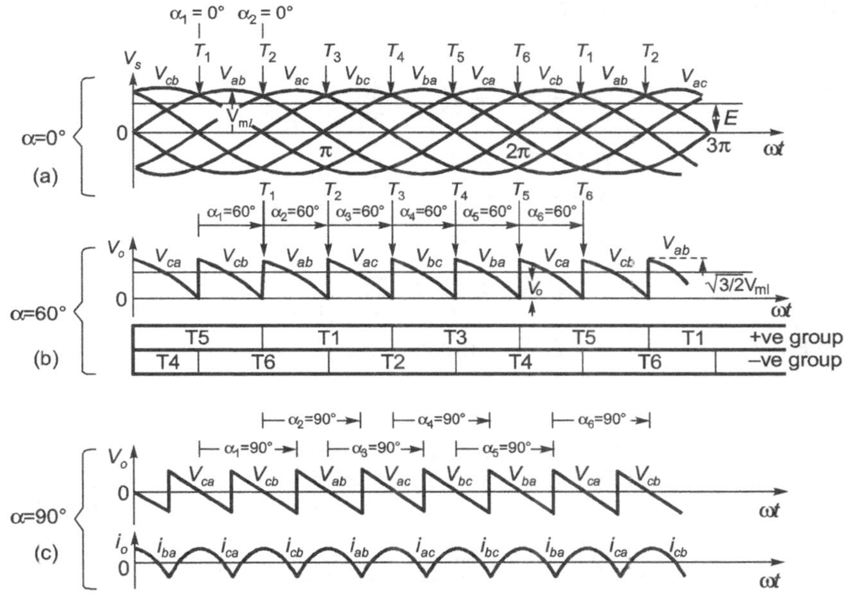

For R Load

Case 1: when $\alpha \le 60°$

- Average output voltage, $V_o = \dfrac{3V_{ml}}{\pi}\cos\alpha$

- RMS output voltage, $V_{or} = \sqrt{\dfrac{3}{2\pi}}\, V_{ml}\left[\dfrac{\pi}{3}+\dfrac{\sqrt{3}}{2}\cos 2\alpha\right]^{1/2}$

Above formulae valid for

(i) R load when $\alpha \le 60°$.

(ii) RL and RLE load any value of α.

Case 2: when $\alpha > 60°$

Average output voltage, $V_o = \dfrac{3\,V_{ml}}{\pi}\left[1+\cos\left(\alpha+\dfrac{\pi}{3}\right)\right]$

Note: Formula valid for R load $\alpha > 60°$ and for RL, RLE load with FD.

3φ Half Controlled Rectifier (or) 3φ Semiconverter

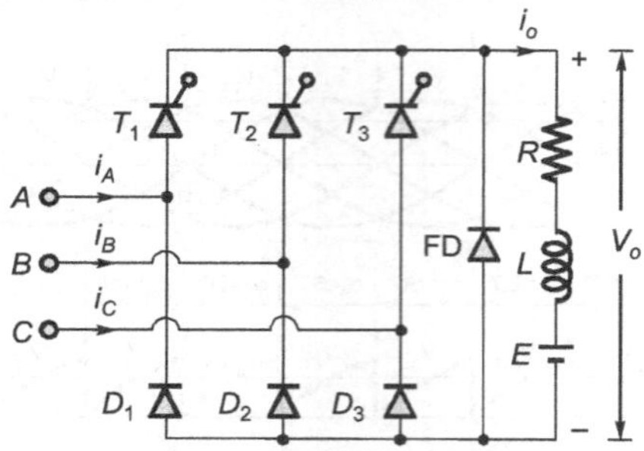

For $\alpha < 60° = 6$ pulse ; $\alpha \geq 60° = 3$ pulse

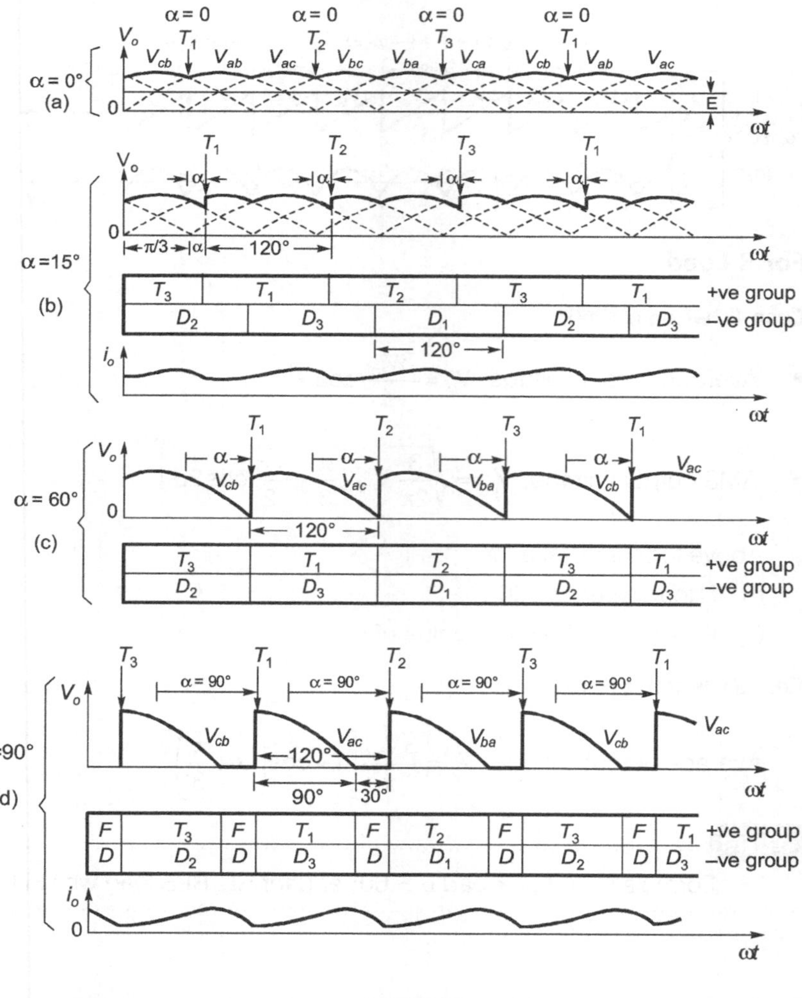

Case-1 : when $\alpha < 60°$	
Average output voltage	$V_0 = \frac{3V_{ml}}{2\pi}(1+\cos\alpha)$
RMS value of output voltage	$V_{or} = \frac{V_{ml}}{2}\sqrt{\frac{3}{\pi}}\left[\frac{2\pi}{3}+\frac{\sqrt{3}}{2}(1+\cos 2\alpha)\right]^{1/2}$
Case-2 : when $\alpha \geq 60°$	
Average output voltage	$V_0 = \frac{3V_{ml}}{2\pi}(1+\cos\alpha)$
RMS value of output voltage	$V_{or} = \frac{V_{ml}}{2}\sqrt{\frac{3}{\pi}}\left[(\pi-\alpha)+\frac{1}{2}\sin 2\alpha\right]^{1/2}$

Remember:

☑ For $\alpha \leq 60°$, freewheeling diode does not conduct.
☑ 3ϕ-semiconverter is only one quadrant converter.
☑ In 3ϕ-semiconverter, breaking mode is not possible.

VI Choppers

- A chopper is a high speed on/off semiconductor switch.
- Chopper is a static device that converters fixed dc input voltage to a variable dc output voltage directly.
- For medium power application we use IGBT and GTO, in chopper.

Step Down Chopper

Average output voltage V_o < input voltage V_s.

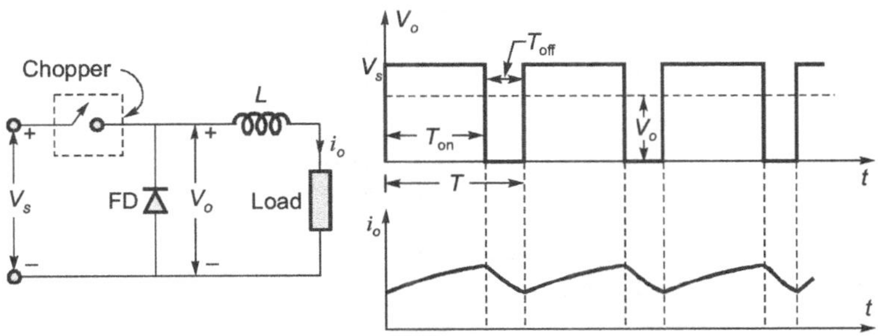

Parameter	Formula
Duty cycle	$\alpha = \frac{T_{ON}}{T} = \frac{T_{ON}}{T_{ON}+T_{OFF}}; \alpha < 1$
Average load voltage	$V_o = \alpha V_s$
RMS load voltage	$V_{RMS} = \sqrt{\alpha}V_s$
Average load current	$I_o = \frac{V_o}{R} = \frac{\alpha V_s}{R}$
RMS load current	$I_{RMS} = \frac{\sqrt{\alpha}V_s}{R}$

Note:

☑ Above formulae valid only for continuous conduction.

☑ Inductor connected in series with load is used to reduce the ripple content in output current.

Step Up Chopper

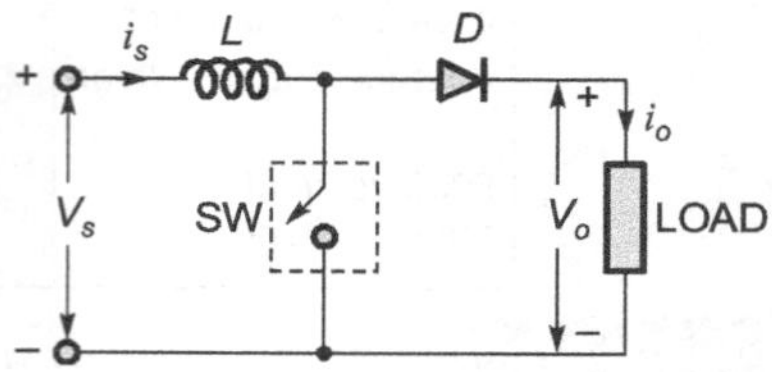

Average output voltage V_o is more than input DC voltage V_s.

- Average load voltage

$$V_o = \frac{1}{1-\alpha} V_s$$

For a basic DC to DC converter, the critical inductance of the filter circuit is given by

$$L = \frac{V_o^2 (V_s - V_o)}{2f V_s P_o}$$ where, $V_o \rightarrow$ Load voltage ; $V_s \rightarrow$ Source voltage

$P_o \rightarrow$ Load power ; $f \rightarrow$ Chopping frequency

Steady State Time-Domain Analysis of Type-A Chopper

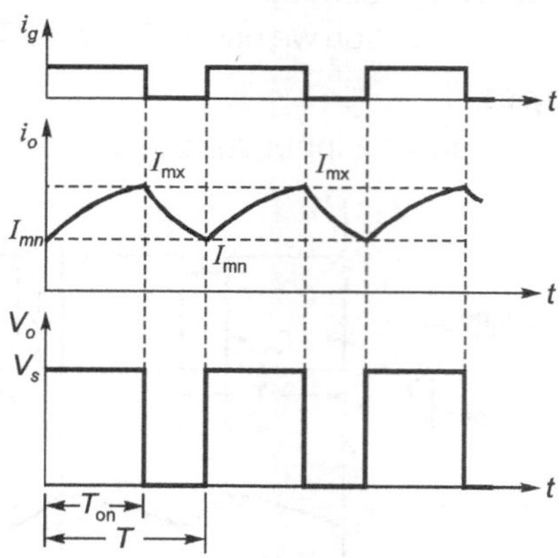

Maximum value of current	$I_{mx} = \frac{V_s}{R}\left[\frac{1-e^{-T_{on}/T_a}}{1-e^{-T/T_a}}\right] - \frac{E}{R}$
Minimum value of current	$I_{mn} = \frac{V_s}{R}\left[\frac{e^{T_{on}/T_a}-1}{e^{T/T_a}-1}\right] - \frac{E}{R}$ where, $T_a = \frac{L}{R}$
Per unit ripple current	p.u. ripple current $= \frac{(I_{mx} - I_{mn})}{V_s/R}$ $= \frac{(1-e^{-\alpha T/T_a})(1-e^{-(1-\alpha)T/T_a})}{(1-e^{-T/T_a})}$

Note:

- ☑ The peak to peak ripple current has maximum value ΔI_{mx} when duty cycle $\alpha = 0.5$.
- ☑ For $\alpha = 0.5$, $\Delta I_{mx} = \frac{V_s}{R}\tanh\frac{R}{4fL}$
- ☑ If $4fL >> R$ then $\tanh\frac{R}{4fL} \approx \frac{R}{4fL}$

$$\Delta I_{mx} = \frac{V_s}{4fL}$$

- ☑ Higher the inductance, ripple is minimum.

- **The value of duty cycle at the limit of continuous conduction**

$$\alpha' \geq \frac{T_a}{T}\ln\left[1+\frac{E}{V_s}\left(e^{T/Ta}-1\right)\right]$$

where, α' measures the limit of continuous conduction.

Load Current Discontinuous

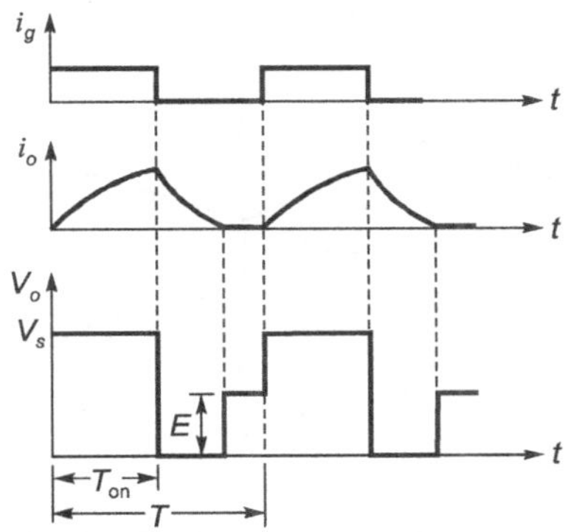

Parameter	Formula
Minimum value of current	$I_{mn} = 0$
Maximum value of current	$I_{mx} = \frac{V_s - E}{R}(1-e^{-T_{on}/T_a})$
Extinction time	$t_x = T_{on} + T_a\ln\left[1+\frac{V_s-E}{E}(1-e^{-T_{on}/T_a})\right]$
Average output voltage	$V_o = \alpha V_s + \left(1-\frac{t_x}{T}\right)E$

Buck Converter

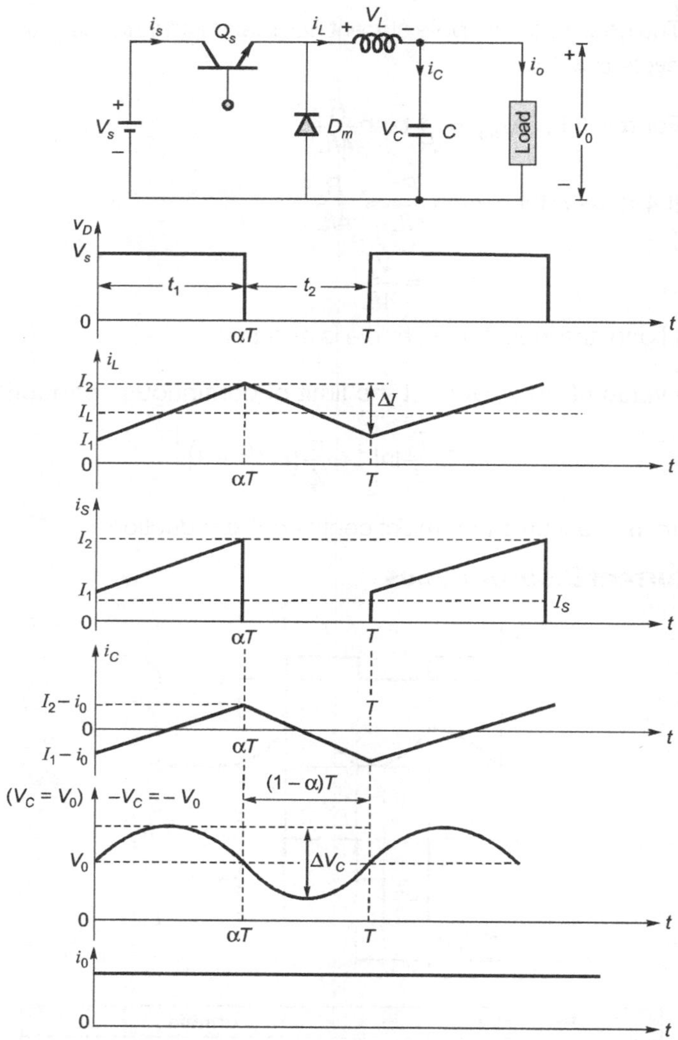

At boundary condition

$$i_{L(avg)} = \frac{\Delta I_L}{2}$$

$$I_{max} = I_0 + \frac{\Delta I_L}{2}$$

$$I_{min} = I_0 - \frac{\Delta I_L}{2}$$

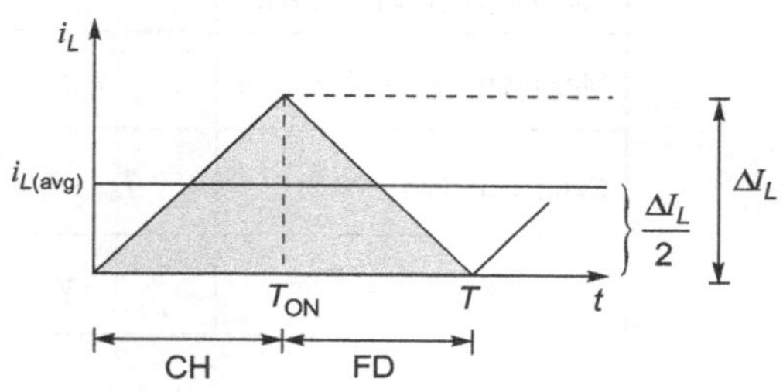

$$\Delta I_L = \frac{\alpha(1-\alpha)V_s}{fL} \quad ; \quad \Delta V_c = \frac{\alpha(1-\alpha)V_s}{8f^2LC}$$

Critical Values of Filters

$$L_c = \frac{(1-\alpha)R}{2f} \quad ; \qquad C_c = \frac{1-\alpha}{16f^2L_c}$$

Boost Converter

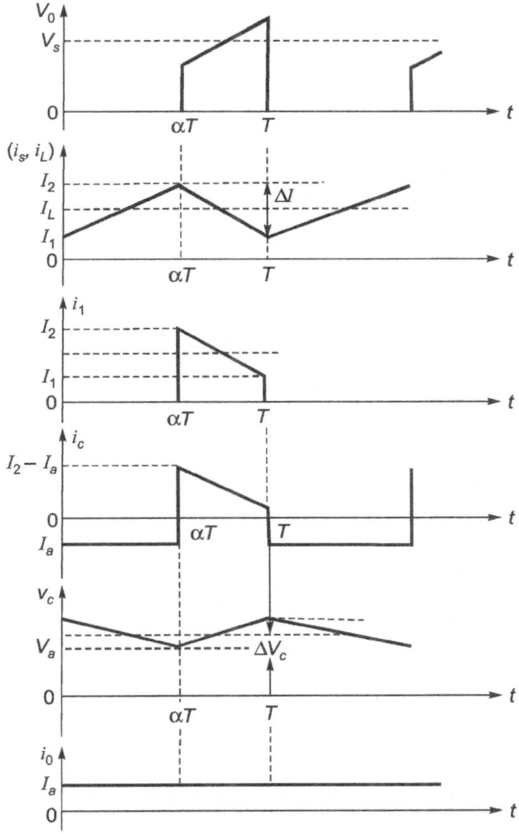

$$V_0 = \frac{V_s}{1-\alpha}$$

$$I_s = \frac{I_0}{1-\alpha}$$

$$\Delta I_L = \frac{\alpha V_s}{fL}$$

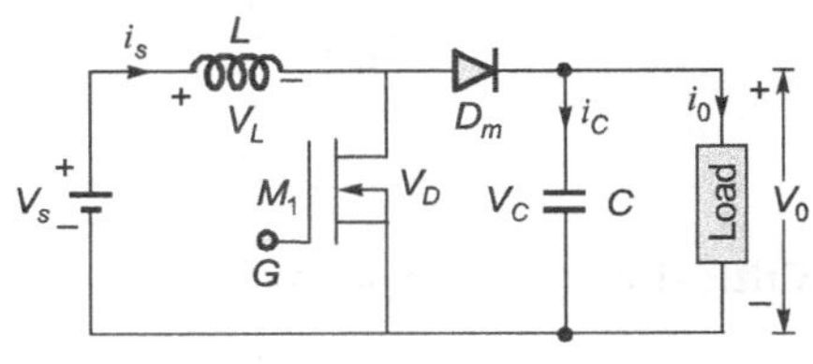

Critical Value of Filter

$$C_C = C = \frac{\alpha}{2fR}$$

Buck Boost Converter

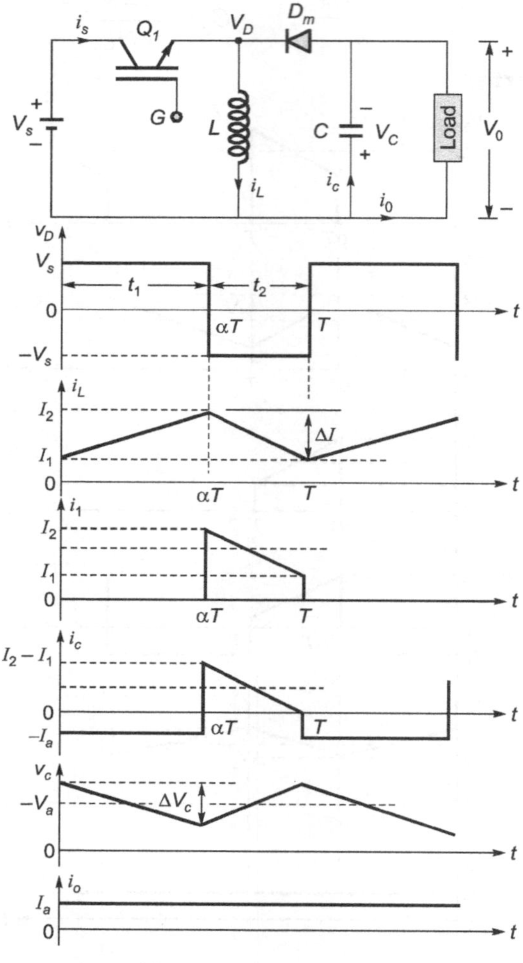

$$V_0 = \frac{\alpha V_s}{1-\alpha}; \qquad I_s = \frac{\alpha I_0}{1-\alpha}$$

$$\Delta V_0 = \Delta V_c = \frac{\alpha I_0}{fC} \qquad \Delta I_L = \frac{\alpha V_s}{fL}$$

Critical Value of Capacitance

$$C_C = C = \frac{\alpha}{2fR}$$

VII Inverters

- An inverter is a circuit which converts a dc power into an ac power at desired output voltage and frequency.
- For low and medium power output, we use BJTs, MOSFET, IGBT, GTO.
- For high power output, thyristor should be used.

Types of Inverter

Voltage Source Inverter (VSI)

- VSI is one in which dc source has small impedance. Because of a low internal impedance, the terminal voltage of a voltage source inverter remains substantially constant with variation in load.

Current Source Inverter (CSI)

Current source inverter is supplied with a controlled current from a dc source of high impedance. Typically, a phase controlled thyristor rectifier feeds the inverter with a regulated current through a large series inductor. Thus, load current is controlled.

Remember:

☑ CSI does not require any feedback diodes whereas, these are required in a VSI.

☑ To make input current almost ripple free ,L-filter is used before CSI.

☑ CSI may be load or force commutated. Load commutation is possible when load p.f. is leading. For lagging p.f. loads, force commutation is essential.

Single Phase Bridge Inverter

Single-Phase Half Bridge Inverter

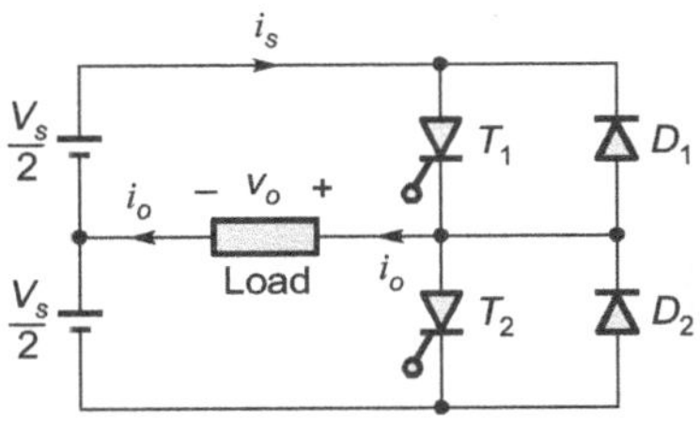

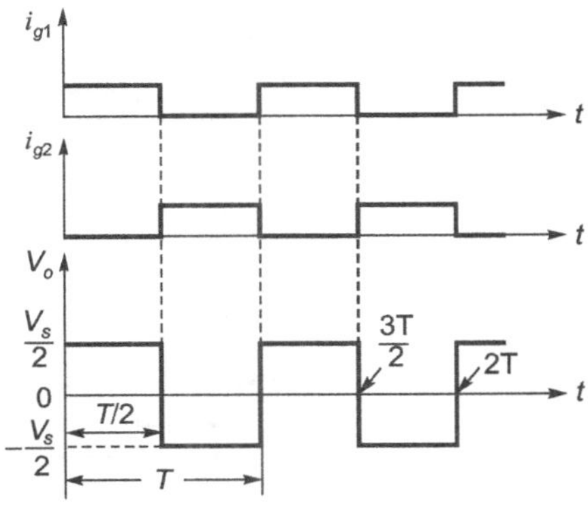

For R load

Parameters	Formula
RMS output voltage	$V_{or} = \frac{V_s}{2}$
Voltage waveform resolved into Fourier series,	$V_o = \sum_{n=1,3,5,...}^{\infty} \frac{2V_s}{n\pi} \sin n\omega t$ V
n^{th} harmonic output voltage	$V_{on} = \frac{2V_s}{n\pi} \sin n\omega t$
n^{th} harmonic current,	$I_{on} = \frac{2V_s}{n\pi R} \sin n\omega t$
Fundamental output voltage	$V_{01} = \frac{2V_s}{\pi} \sin \omega t$
RMS fundamental output voltage	$(V_{01})_{RMS} = \frac{\sqrt{2}V_s}{\pi}$
Fundamental distortion factor (FDF)	$g = \frac{(V_{01})_{RMS}}{V_{or}} = \frac{2\sqrt{2}}{\pi}$
Total harmonic distortion	$THD = \left(\frac{1}{g^2} - 1\right)^{1/2} = 48.34\%$

Single Phase-Full Bridge Inverter

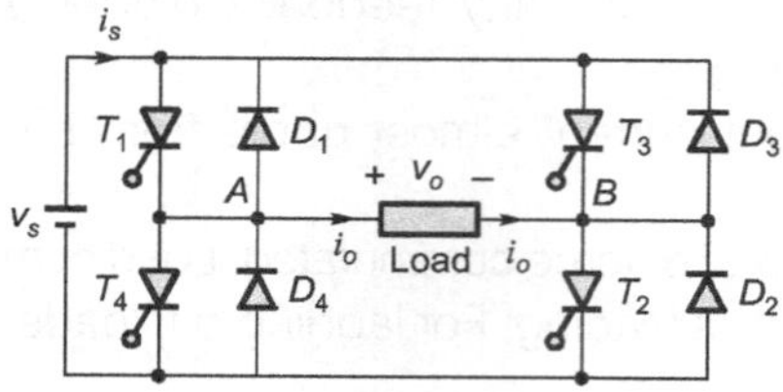

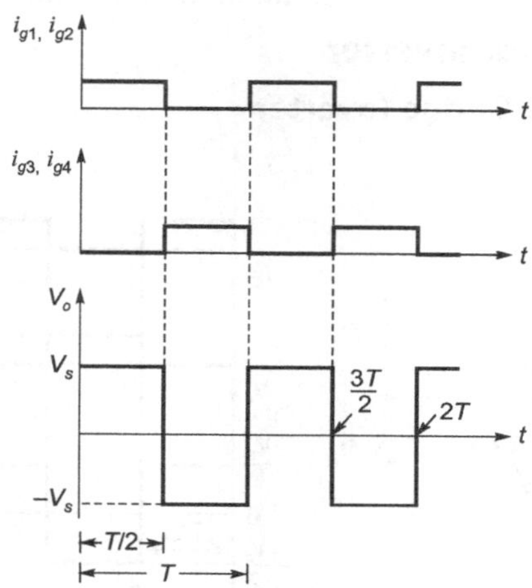

For R load

Parameters	Formula
RMS output voltage	$V_{or} = V_s$
Voltage waveform resolved into Fourier series,	$V_o = \sum_{n=1,3,5,...}^{\infty} \frac{4V_s}{n\pi} \sin n\omega t$ V
n^{th} harmonic output voltage	$V_{on} = \frac{4V_s}{n\pi} \sin n\omega t$
Fundamental output voltage	$V_{01} = \frac{4V_s}{\pi} \sin \omega t$
RMS fundamental output voltage	$(V_{01})_{RMS} = \frac{2\sqrt{2}}{\pi} V_s$
Fundamental distortion factor (FDF)	$g = \frac{2\sqrt{2}}{\pi}$
Total harmonic distortion (THD)	THD = 48.34%

Single Phase VSI for Different Loads

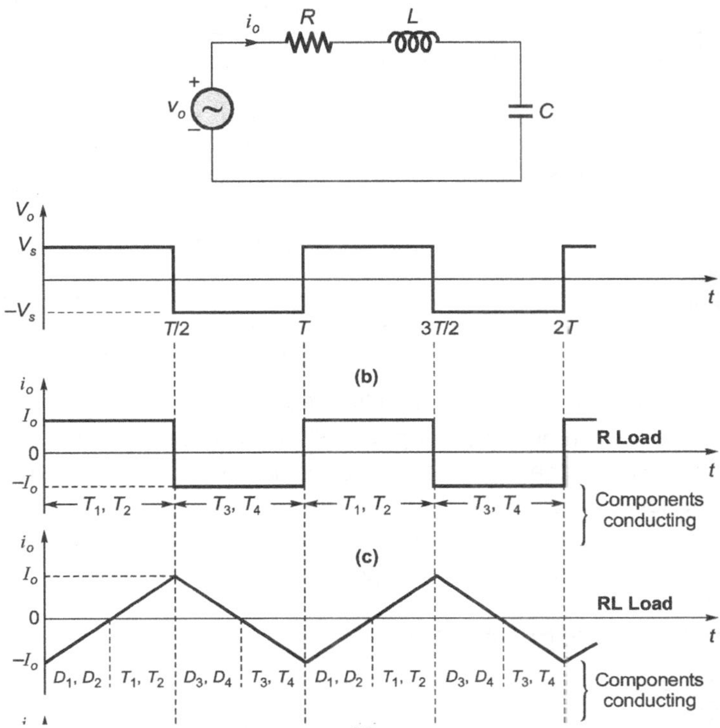

(d)

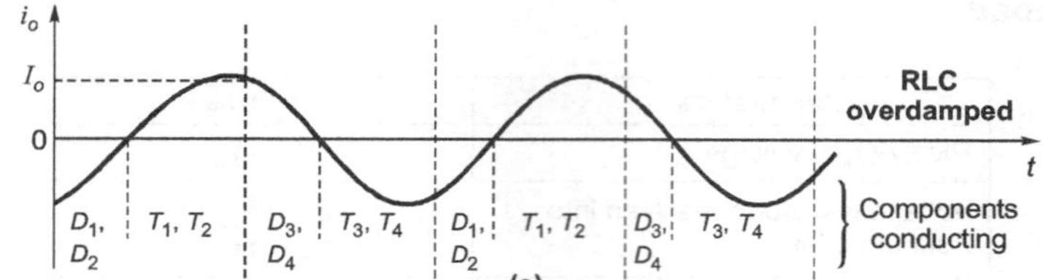

(e)

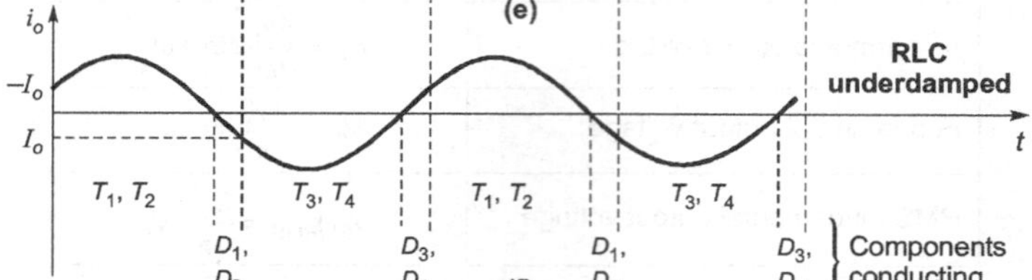

(f)

Parameters	Formula
Load impedance at frequency (nf)	$Z_n = \left[R^2 + \left(n\omega L - \frac{1}{n\omega C}\right)^2\right]^{1/2}$
Phase angle	$\phi_n = \tan^{-1}\frac{\left[n\omega L - \frac{1}{n\omega C}\right]}{R}$ rad
Output current at the instant of commutation	$I_o = \left.\frac{V_o}{Z_n}\right\rvert_{\omega t=\pi}$
Fundamental load power	$P_{o1} = I_{o1}^2 R = V_{o1} I_{o1} \cos\theta_1$ where I_{o1} = RMS value of fundamental output current V_{o1} = RMS value of fundamental output voltage

Note:

- ☑ If circuit turn-off time (t_c) is more than device turn-off time (t_q) then it does not require force commutation.
- ☑ If $t_q > t_c$ then force commutation required.
- ☑ The fundamental output power P_{o1} does the useful work and the output power associated with harmonic current is dissipated as heat, leading to rise in load temperature.
- ☑ Feedback diode (D) conducts only when there is presence of energy storing element L and C in load.

Three Phase Bridge Inverter

For providing adjustable frequency power to industrial application, 3ϕ bridge inverter are used. A large capacitor is connected at the input terminals tend to make the input DC voltage constant.

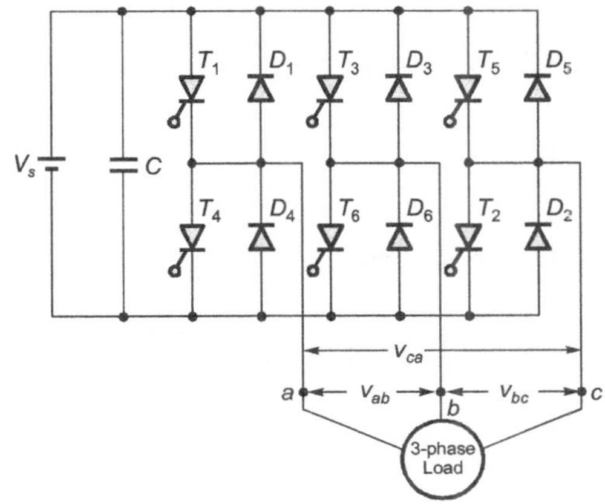

3φ 180° Mode VSI

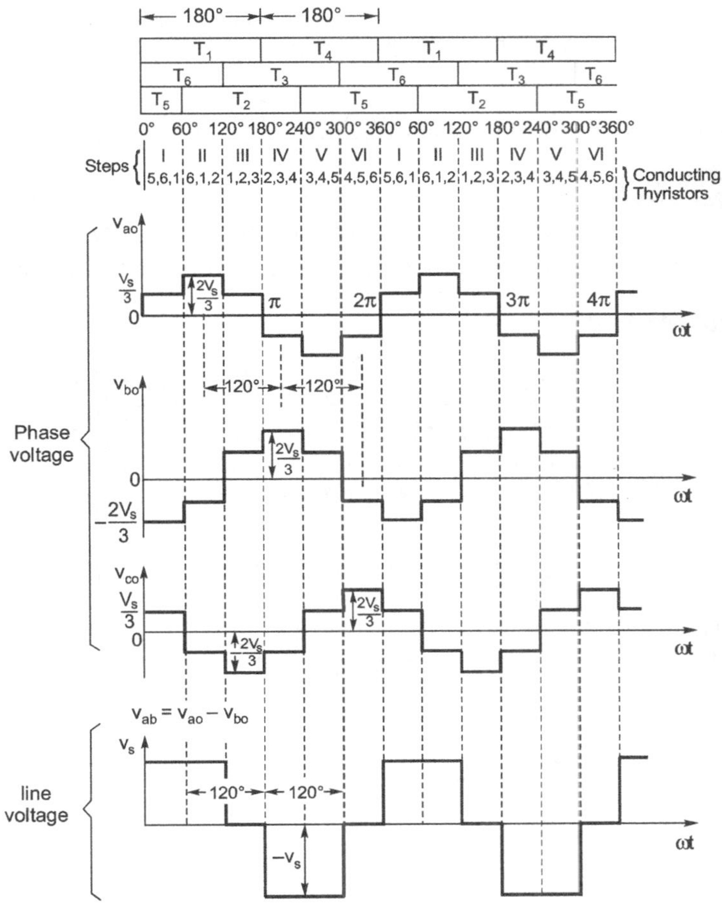

- Phase RMS voltage, $(V_{ph})_{RMS} = \frac{\sqrt{2}}{3}V_s$
- Line RMS voltage, $(V_L)_{RMS} = \sqrt{\frac{2}{3}}V_s$
- n^{th} harmonic output voltage for phase "a", $V_{an} = \frac{2V_s}{n\pi}\sin n\omega t$
- Fundamental phase RMS voltage, $(V_{a1}) = \frac{\sqrt{2}V_s}{\pi}$
- $FDF\ (g) = \frac{3}{\pi}$
- $THD = 31\%$
- n^{th} harmonic line output voltage, $(V_{ab})_n = \frac{4V_s}{n\pi}\cos\frac{n\pi}{6}\sin n\left(\omega t + \frac{\pi}{6}\right)$
- Fundamental RMS line value, $(V_{ab})_{RMS} = \frac{2\sqrt{2}}{\pi}\cos\frac{\pi}{6} = \frac{\sqrt{6}}{\pi}V_S$

Remember:

☑ The phase as well as line voltage are out of phase by 120°.

☑ For $n = 3$, all triple harmonics are absent from the line voltage.

3ϕ 120° Mode VSI

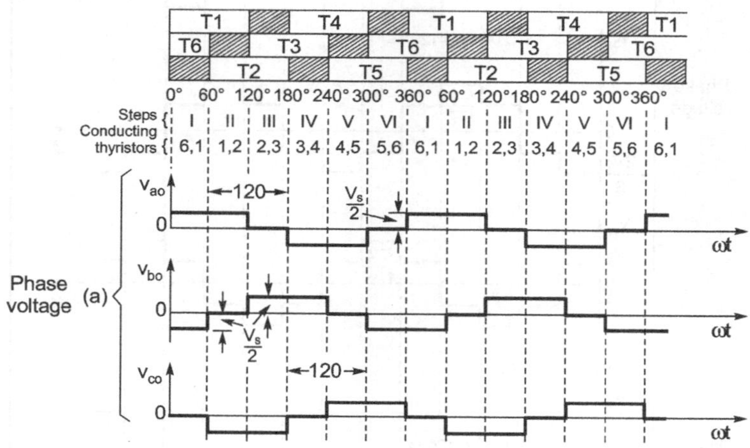

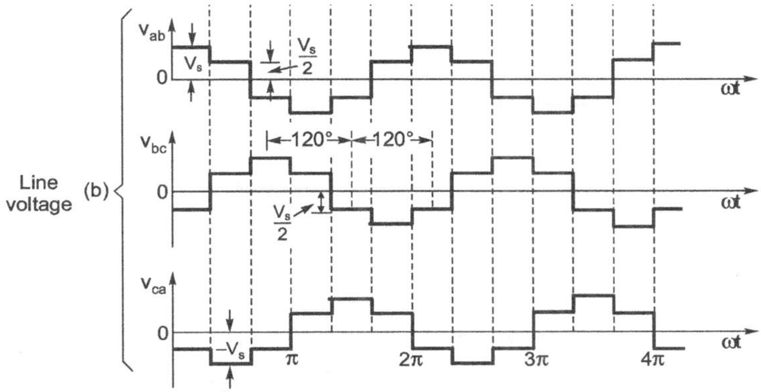

Parameters	Formula
Line voltage	$V_{ab} = \sum_{n=6k\pm1}^{\infty} \frac{3V_s}{n\pi} \sin n\left(\omega t + \frac{\pi}{3}\right)$
Phase voltage	$V_{a0} = \sum_{n=1,3,5,...}^{\infty} \frac{2V_s}{n\pi} \cos\frac{n\pi}{6} \sin n\left(\omega t + \frac{\pi}{6}\right)$
RMS value of phase voltage	$V_P = \frac{V_s}{\sqrt{6}} = 0.4082V_s$
RMS value of line voltage	$V_L = \frac{V_s}{\sqrt{2}} = 0.7071V_s$

where $k = 0, 1, 2, ...$

For both line and phase voltage:

- $\text{FDF} = \frac{3}{\pi}$ and $\text{THD} = 31\%$

Remember:

- Generally we used IGBT in 1ϕ and 3ϕ inverter.

Pulse Width Modulated Inverters

Single-pulse Modulation

Parameters	Formula
Output voltage	$V_0 = \sum_{n=1,3,5}^{\infty} \frac{4V_s}{n\pi} \cdot \sin\frac{n\pi}{2} \cdot \sin nd \cdot \sin n\omega t$
Pulse width	$2d$
Peak value of nth harmonic	$V_{o,nm} = \frac{4V_s}{n} \sin nd$
RMS value of output voltage	$V_{or} = V_s \cdot \left[\frac{2d}{\pi}\right]^{1/2}$
FDF (g)	$\frac{2\sqrt{2}\sin d}{\sqrt{\pi \cdot 2d}}$
THD	$\sqrt{\frac{1}{g^2} - 1}$

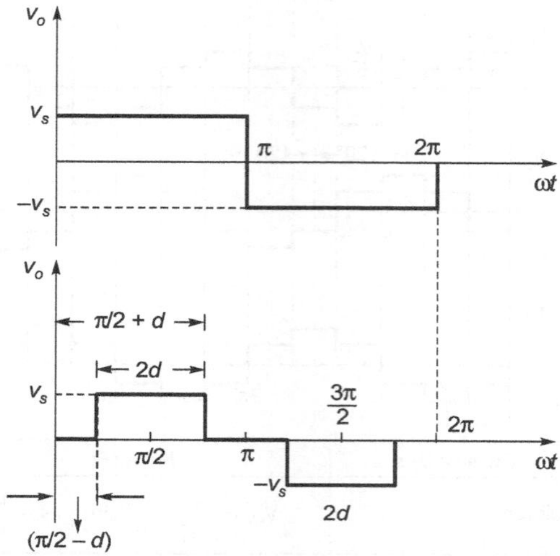

Remember:

☑ When pulse width $2d = \dfrac{2\pi}{n}$, then nth harmonic eliminated.

Multiple Pulse Width Modulation

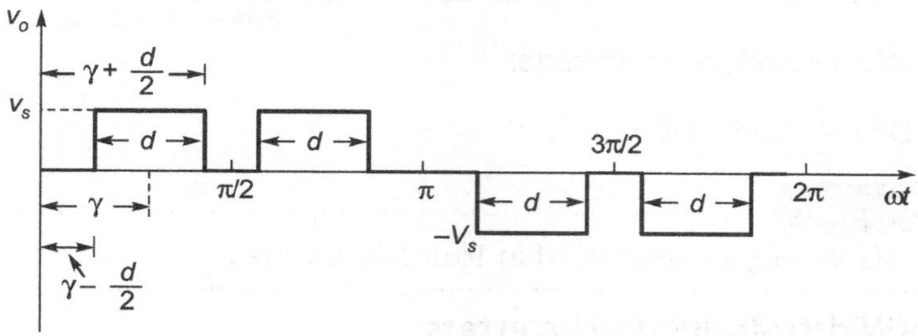

- Output voltage

$$V_o = \sum_{n=1,3,5...}^{\infty} \frac{8V_s}{n\pi} \cdot \sin n\gamma \cdot \sin\frac{nd}{2} \cdot \sin n\omega t$$

$$\gamma = \frac{\pi - 2d}{N+1} + \frac{d}{N}$$

where, N = Number of pulses per half cycle.

- Amplitude of the nth harmonic of the two pulse wave form

$$V_n = \frac{8V_s}{n\pi}\sin n\gamma \cdot \sin\frac{nd}{2}$$

- Condition to eliminate n^{th} harmonic

(i) $\gamma = \frac{\pi}{n}$ (ii) $2d = \frac{4\pi}{n}$

Remember:

☑ By using PWM technique, we can get variable voltage and frequency within inverter itself by eliminating some of the lower order harmonics.

☑ In PWM inverter, force commutation is essential.

VIII AC Voltage Controllers

A.C. voltage controllers are thyristor based devices which convert fixed alternating voltage directly to variable alternating voltage without, change in the frequency.

Single phase half wave ac voltage controller

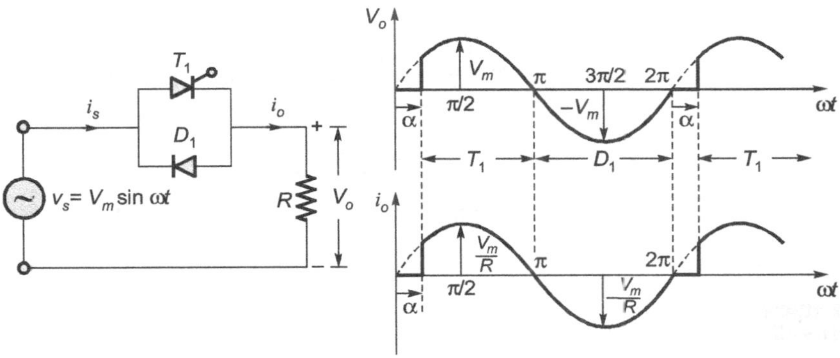

- RMS value of output voltage

$$V_{or} = \frac{V_m}{2}\left[\frac{1}{\pi}\left\{(2\pi - \alpha) + \frac{\sin 2\alpha}{2}\right\}\right]^{1/2}$$

- Average value of output voltage

$$V_o = \frac{V_m}{2\pi}(\cos\alpha - 1)$$

Single phase full wave ac voltage controller

- RMS value of output voltage for R load

$$V_{or} = \frac{V_m}{\sqrt{2}}\left[\frac{1}{\pi}\left\{(\pi-\alpha)+\frac{1}{2}\sin 2\alpha\right\}\right]^{1/2}$$

- RMS value of output voltage for RL load

$$V_{or} = \frac{V_m}{\sqrt{2\pi}}\left[(\beta-\alpha)+\frac{1}{2}(\sin 2\alpha - \sin 2\beta)\right]^{1/2}$$

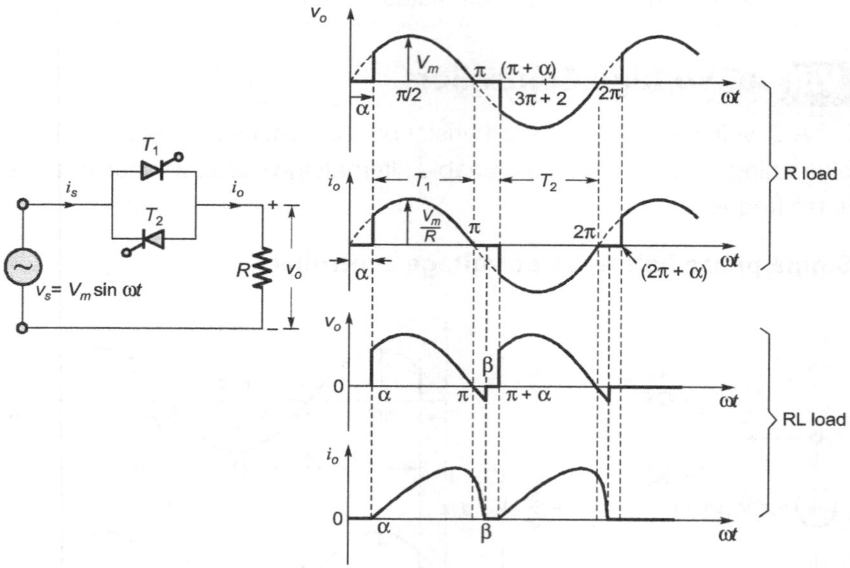

Remember:

☑ For RL load, output voltage is controllable only when $\alpha > \phi$ where $\phi = \tan^{-1}\frac{\omega L}{R}$.

☑ Range of a for getting controllable output voltage

$$\phi \le \alpha \le 180°$$

☑ Average value of output voltage would be zero.

Integral Cycle Control

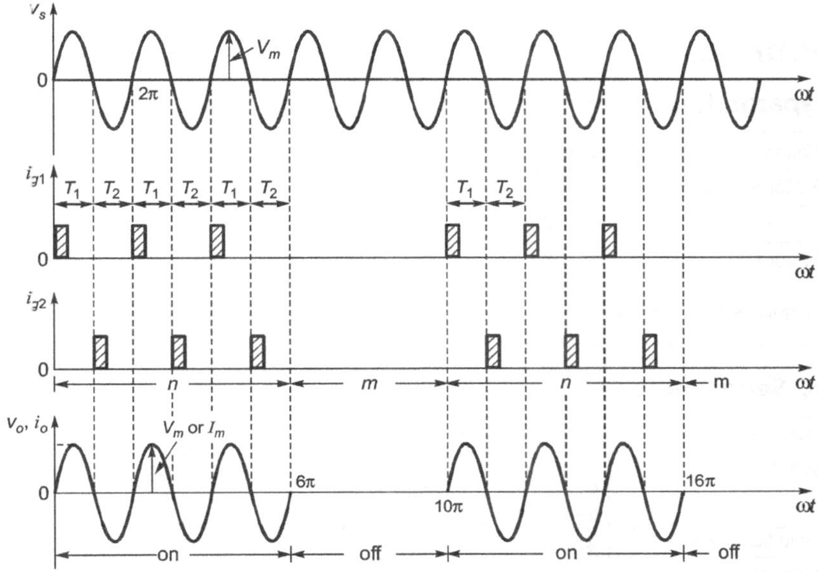

Integral cycle control for *n* on-cycles and *m* off cycles

- Rms value of output voltage, $V_{or} = \dfrac{V_m}{\sqrt{2}}\sqrt{\dfrac{n}{n+m}} = V_s\sqrt{k}$

  where, $k \rightarrow \dfrac{n}{n+m}$ = duty cycle of ac voltage controller

  $V_{or} \rightarrow$ RMS value of output voltage

  $V_s \rightarrow$ RMS value of source voltage

- Power delivered to load, $P = \dfrac{kV_s^2}{R}$

- Input power factor, $p.f. = \sqrt{\dfrac{n}{n+m}} = \sqrt{k}$

Note:

- Integral cycle control method is used only for those load which have high time constant.
- AC voltage controller is used for domestic and industrial heating, speed control of 1ϕ and 3ϕ AC drives, starting of induction motors.
- Disadvantage of AC voltage controller is introduction of objectionable harmonics in the supply current and load voltage waveforms.

IX Electric Drives

DC Drives

Separately-excited DC motor

Voltage across field winding	$V_f = I_f \cdot r_f$
Motor terminal voltage	$V_t = E_a + I_a r_a$
Motor back emf	$E_a = k_a \phi \omega_m = k_m \omega_m$
Motor torque	$T_e = k_m I_a$
Angular speed of motor	$\omega_m = \dfrac{V_t - I_a r_a}{k_m}$

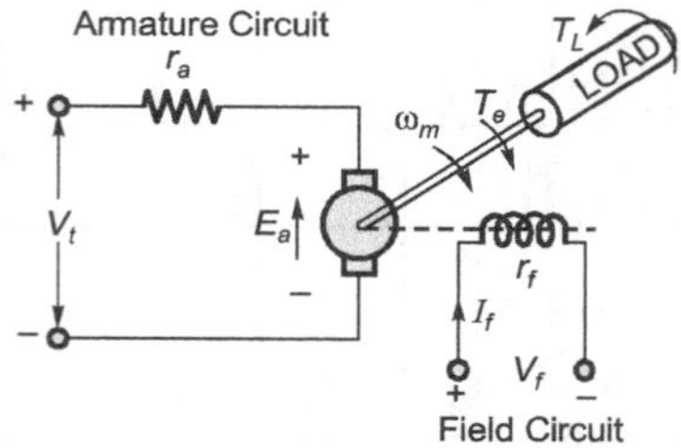

DC Series motor

Motor terminal voltage	$V_t = E_a + I_a(r_a + r_s)$
Motor torque	$T_e = k_a \phi I_a$ for no saturation in the magnetic circuit
Field flux per pole	$\phi = kI_a$
Motor torque	$T_e = KI_a^2$
Motor back emf	$E_a = KI_a \omega_m$
Angular speed of motor	$\omega_m = \dfrac{V_t}{KI_a} - \dfrac{(r_a + r_s)}{K}$ where, $r_s \rightarrow$ Series-field resistance, Ω

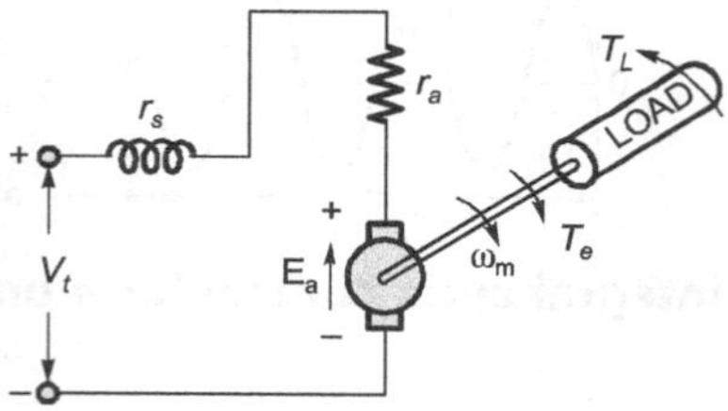

where $K = kk_a$

Single Phase DC Drives

Single Phase Half Wave Converter Drives

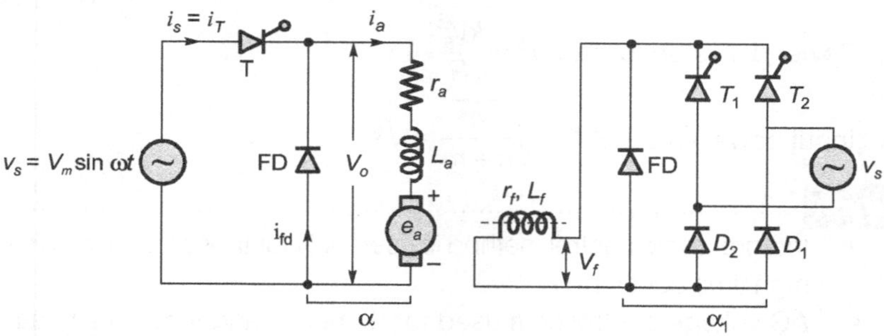

- Average output voltage of converter

  For $0 < \alpha < \pi$

$$V_o = V_t = \frac{V_m}{2\pi}(1 + \cos\alpha)$$

where $V_m \rightarrow$ Maximum value of source voltage, V

$V_0 \rightarrow V_t$ = Armature terminal voltage, V

For single-phase semiconverter in the field circuit

Average output voltage	For $0 < \alpha_1 < \pi$ $V_f = \frac{V_m}{\pi}(1+\cos\alpha_1)$
RMS value of source current	$I_{sr} = I_a\sqrt{\frac{\pi-\alpha}{2\pi}}$
Apparent input power	$P_i = V_s I_{sr}$
Power delivered to motor	$P = V_t I_a$

Single phase semiconverter Drives

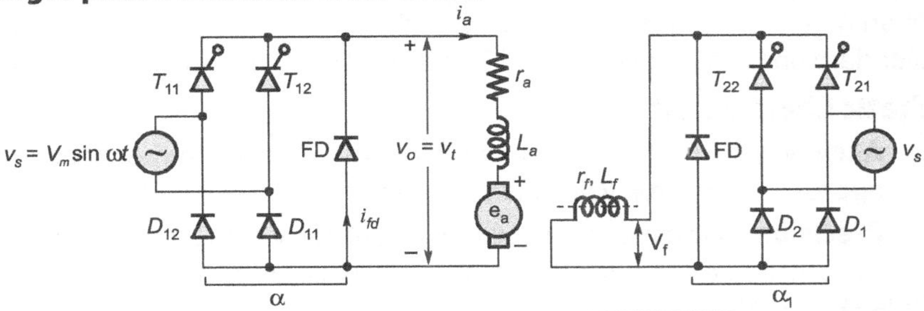

Average output voltage	$V_o = V_t = \frac{V_m}{\pi}(1+\cos\alpha)$
For field circuit	$V_f = \frac{V_m}{\pi}(1+\cos\alpha_1)$
RMS value of source current	$I_{sr} = I_a\sqrt{\frac{\pi-\alpha}{\pi}}$

Single Phase Full Converter Drives

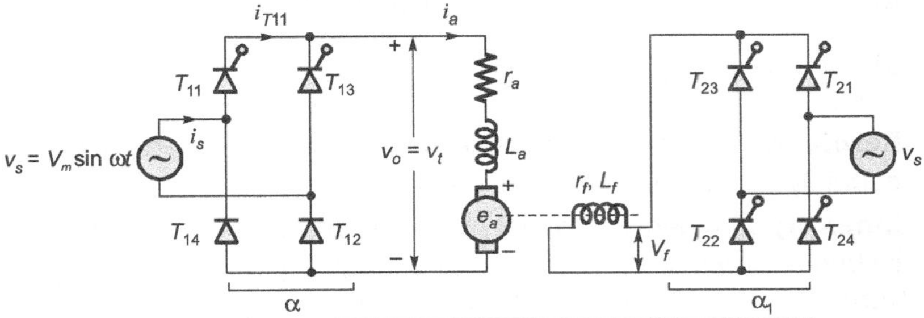

For armature converter-1	$V_0 = V_t = \frac{2V_m}{\pi}\cos\alpha$ .... for $0 < \alpha < \pi$
For field converter-2	$V_f = \frac{2V_m}{\pi}\cdot\cos\alpha_1$ .... for $0 < \alpha_1 < \pi$

- RMS value of source current, $I_{sr} = I_a$

Measurements and Instrumentation

4

I Characteristics of Instruments and Measurement Systems

Measurements

Measurement is a process by which one can convert physical parameters to meaningful numbers. The measuring process is one in which the property of an object or system under consideration is compared to an accepted standard unit, a standard defined for that particular property.

Static Characteristics

Accuracy : It is the closeness with which an instrument reading approaches the true value of the quantity being measured.

Precision : It is a measure of the reproducibility of the measurements. It is a measure of degree of agreement within a group of measurements.

Remember:

- ☑ Precision is not the guarantee of accuracy.
- ☑ An instrument with more significant figure has more precision.

Sensitivity : It is the ratio of the magnitude of output signal to the magnitude of input signal applied to the instrument.

Note:

- ☑ An instrument requires high degree of sensitivity.
- ☑ Sensitivity $\propto \dfrac{1}{\text{Deflection factor}}$

Resolution : The smallest change in input which can be detected with certainty by an instrument is its resolution.

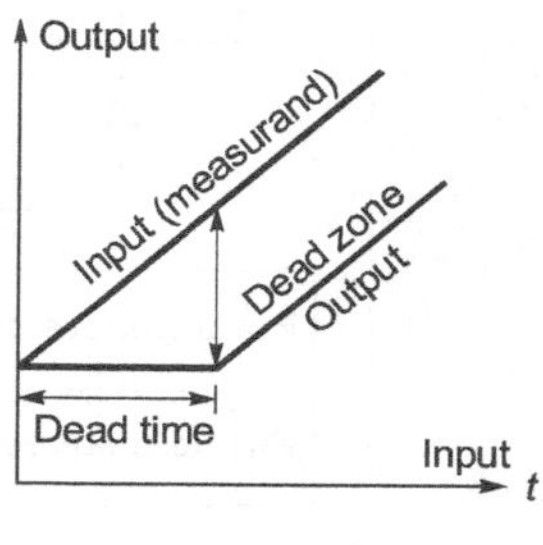

Linearity : Linearity is a behaviour of an instrument in which the output signal strength varies in direct proportion to input signal strength.

Dead Zone : It is the largest change of input quantity for which there is no output of the instrument.

Dead Time : Time required by an instrument to begin to respond to the change in a measurand.

Range and Span : The difference between the maximum and minimum values of the scale is called range. The maximum value of the scale is called span.

Errors : Measurement error is the difference between the measured value and true value.

$$\text{Error} = \text{Measured value} - \text{True value} = -\text{Accuracy}$$

Error Analysis

- Static error, $\delta A = A_m - A_t$

  where, $A_m \rightarrow$ Measured value of quantity or Actual value

  $A_t \rightarrow$ True value of quantity or Nominal value

- Relative static error, $\epsilon_r = \dfrac{\delta A}{A_t}$
- Static correction, $\delta C = A_t - A_m = -\delta A$
- Static sensitivity $= \dfrac{\Delta q_0}{\Delta q_i}$

  where, $\Delta q_o \rightarrow$ Infinitesimal change in output

  $\Delta q_i \rightarrow$ Infinitesimal change in input

- Non-linearity (N.L.)

$$N.L. = \frac{\text{(Max. deviation of output from the idealized straight line)}}{\text{Full scale deflection}} \times 100$$

$$\text{Error at desired value} = \frac{\text{Full scale value} \times \text{Error at full scale}}{\text{Desired value}}$$

Combination of Quantities with Limiting Errors

Sum or Difference of Two or more than Two Quantities

Let $X = \pm x_1 \pm x_2 \pm x_3 \pm x_4$

$$\frac{\delta X}{X} = \pm\left(\frac{x_1}{X}\frac{\delta x_1}{x_1} + \frac{x_2}{X}\frac{\delta x_2}{x_2} + \frac{x_3}{X}\frac{\delta x_3}{x_3} + \frac{x_4}{X}\frac{\delta x_4}{x_4}\right)$$

where $\pm\delta x_i \rightarrow$ Relative increment in quantity x_i

$\pm\delta X \rightarrow$ Relative increment in X

$\dfrac{\delta x_i}{x_i} \rightarrow$ Relative limiting error in quantity x_i

$\dfrac{\delta X}{X} \rightarrow$ Relative limiting error in X

Product or Quotient of Two or more than Two Quantities

Let $X = x_1 x_2 x_3$ or $X = \dfrac{x_1}{x_2 x_3}$ or $X = \dfrac{1}{x_1 x_2 x_3}$

$$\frac{\delta X}{X} = \pm\left(\frac{\delta x_1}{x_1} + \frac{\delta x_2}{x_2} + \frac{\delta x_3}{x_3}\right)$$

Composite Factors

Let $X = x_1^n \cdot x_2^m$, $\quad \dfrac{\delta X}{X} = \pm\left(n\dfrac{\delta x_1}{x_1} + m\dfrac{\delta x_2}{x_2}\right)$

Mean, Deviation and Variance

- Arithmetic Mean, $\bar{X} = \dfrac{\Sigma x}{n} = \dfrac{x_1 + x_2 + ... + x_n}{n}$

  where, $x_1, x_2, ..., x_n \rightarrow$ Readings or samples ; $n \rightarrow$ Number of readings

$$d_n = x_n - \bar{X}$$

Note:

Algebraic sum of deviation is zero.

- Average deviation, $\bar{D} = \dfrac{\Sigma|d|}{n} = \dfrac{|d_1| + |d_2| + \cdots |d_n|}{n}$
- Standard deviation

  For $n > 20$, S.D. $= \sigma = \sqrt{\dfrac{\Sigma d^2}{n}}$ and for $n < 20$, S.D. $= s = \sqrt{\dfrac{\Sigma d^2}{n-1}}$
- Variance

  For $n > 20$, $V = \sigma^2 = \dfrac{\Sigma d^2}{n}$ and for $n < 20$, $V = s^2 = \dfrac{\Sigma d^2}{n-1}$

Normal or Gaussian Curve of Errors

For Infinite Numbers of Reading

$$y = \frac{1}{\sigma\sqrt{2\pi}} \exp(-x^2/2\sigma^2)$$

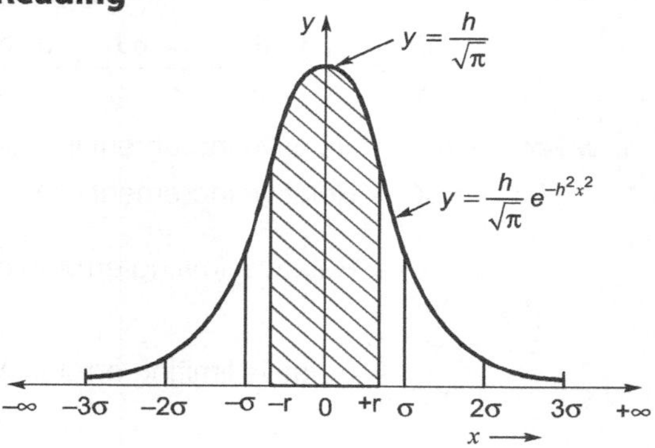

where, $x \rightarrow$ Magnitude of deviation from mean ; $y \rightarrow$ Number of readings at any deviation x, (the probability of occurrence of deviation x)
$\sigma \rightarrow$ Standard deviation

Precision Index	$h = \frac{1}{\sigma\sqrt{2}}$
Probable error (P.E.)	$r = \frac{0.4769}{h}$
Average deviation	$\bar{D} = \frac{r}{0.8453} = \frac{1}{\pi h^2}$
Standard deviation	$\sigma = \frac{r}{0.6745} = \frac{1}{h\sqrt{2}}$ P.E. $= r = 0.8453\bar{D} = 0.6745\sigma$

For Finite Numbers of Reading

- For $n > 20$, $\quad$ P.E. $= r = 0.6745\sqrt{\frac{\sum |d|^2}{n}}$
- For $n < 20$, $\quad$ P.E. $= r = 0.6745\sqrt{\frac{\sum |d|^2}{n-1}}$
- Standard deviation of mean, $\sigma_m = \frac{\sigma}{\sqrt{n}}$
- Standard deviation of standard deviation, $\sigma_\sigma = \frac{\sigma_m}{\sqrt{2}}$
- Variance of combination of components

  Let $\quad X = f(x_1, x_2, \cdots, n)$

$$V_x = \left(\frac{\partial X}{\partial x_1}\right)^2 V_{x_1} + \left(\frac{\partial X}{\partial x_2}\right)^2 V_{x_2} + \cdots + \left(\frac{\partial X}{\partial x_n}\right)^2 V_{x_n}$$

  where, $V_{x1}, V_{x2}, \cdots, V_{xn} \rightarrow$ Variance of $x_1, x_2, \cdots, x_n$

- **Standard Deviation of combination of components**

  Let $\quad X = f(x_1, x_2, , \ldots\ldots, x_n)$

$$\sigma_x = \sqrt{\left(\frac{\partial X}{\partial x_1}\right)^2 \sigma_{x_1}^2 + \left(\frac{\partial X}{\partial x_2}\right)^2 \sigma_{x_2}^2 + \cdots + \left(\frac{\partial X}{\partial x_n}\right)^2 \sigma_{x_n}^2}$$

  where, $\sigma_{x1}, \sigma_{x2}, \cdots, \sigma_{xn} \rightarrow$ Standard deviation of $x_1, x_2, \cdots, x_n$

- **Probable Error of combination of components**

  Let $\quad X = f(x_1, x_2, \ldots\ldots, x_n)$

$$r_x = \sqrt{\left(\frac{\partial X}{\partial x_1}\right)^2 r_{x_1}^2 + \left(\frac{\partial X}{\partial x_2}\right)^2 r_{x_2}^2 + \cdots + \left(\frac{\partial X}{\partial x_n}\right)^2 r_{x_n}^2}$$

where, $r_{x_1}, r_{x_2}, \cdots, r_{x_n} \rightarrow$ Probable error of $x_1, x_2, \cdots, x_n$

- **Uncertainty of combination of components**
  Uncertainty is the deviation of measured value from a unique value from a unique sample of data specified in term of degree of confidence of user.
  Let $X = f(x_1, x_2, \ldots\ldots, x_n)$

$$w_x = \sqrt{\left(\frac{\partial X}{\partial x_1}\right)^2 w_{x_1}^2 + \left(\frac{\partial X}{\partial x_2}\right)^2 w_{x_2}^2 + \cdots + \left(\frac{\partial X}{\partial x_n}\right)^2 w_{x_n}^2}$$

where, $w_{x_1}, w_{x_2}, \cdots, w_{x_n} \rightarrow$ Uncertainties of $x_1, x_2, \cdots, x_n$

Order of Instrument

Zero Order System

As input changes, output also changes immediately called zero order system. Example : Resistor.

First Order System

As input changes, output also changes but not immediately, it takes some delay but without oscillation. Example : heater.

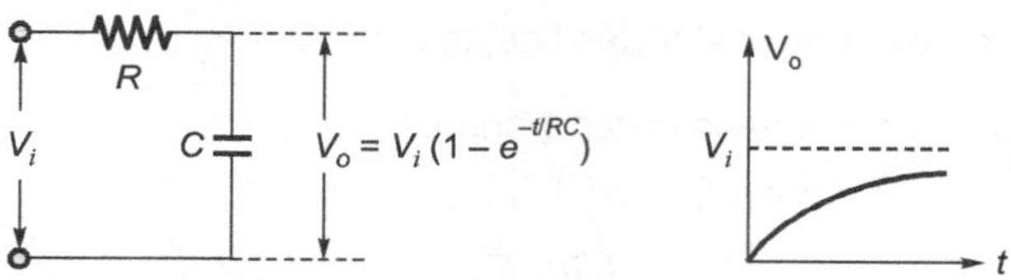

Second Order System

As input changes, output also changes, with some delay and oscillation.

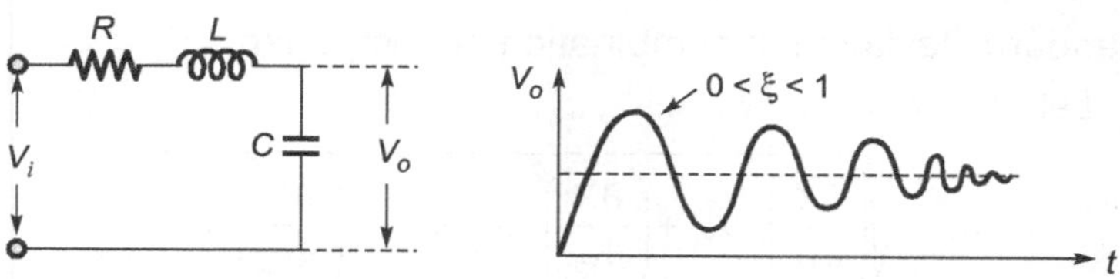

Remember:

The analog instruments are of second order which have damping factor (ξ) between 0.6 to 0.8.

Standards

Quantity	Unit	Definition
Length	Metre	The length of path travelled by light in an interval of $\frac{1}{299792458}$ sec.
Time	second	9.192631770×10^9 cycles of radiation from vapourised Caesium-133 atom.
Temp.	Kelvin	The temperature difference between the absolute and the triple point of water is defined as 273.16°K.
Voltage	Volt	Standard cell voltage of weston cell i.e. 1.0186 V.
Current	Ampere	One ampere is the current flowing through two infinite long parallel conductor of negligible cross section placed 1 meter apart producing a force of 2×10^{-7} N/m.

II Circuit Components (Resistors, Inductors, Capacitors)

Frequency Errors in Resistors

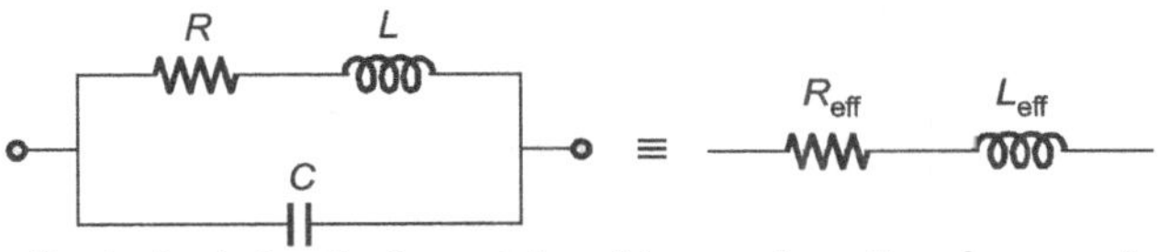

Equivalent circuit of a resistor at low and medium frequencies

Effective resistance	$R_{eff} = \frac{R}{1+\omega^2 C(CR^2-2L)}$
Effective inductance or residual inductance	$L_{eff} = \frac{L-CR^2}{1+\omega^2 C(CR^2-2L)}$
Time constant	$\tau = \frac{L_{eff}}{R_{eff}} = \frac{L-CR^2}{R} = \frac{L}{R} - CR$
Condition for resistance to remain independent of frequency	$CR^2 = 2L$
Condition for resistance to show no inductive effect	$CR^2 = L$
Effective resistance for zero effective inductance	$R_{eff} = \frac{R}{1-\omega^2 LC}$
Quality factor	$Q = \frac{\omega L}{R}$

$$\tan\phi = \frac{X_{eff}}{R_{eff}} = \frac{\omega L_{eff}}{R_{eff}} = \frac{\omega(L-CR^2)}{R} = \omega\left(\frac{L}{R} - CR\right)$$

where, $\phi \rightarrow$ Phase deflection angle

Frequency Errors in Inductors

- Effective resistance, $R_{eff} = \dfrac{R}{(1-\omega^2 LC)^2}$
- Effective inductance, $L_{eff} = L(1+\omega^2 LC)$

Capacitor

Parallel Representation

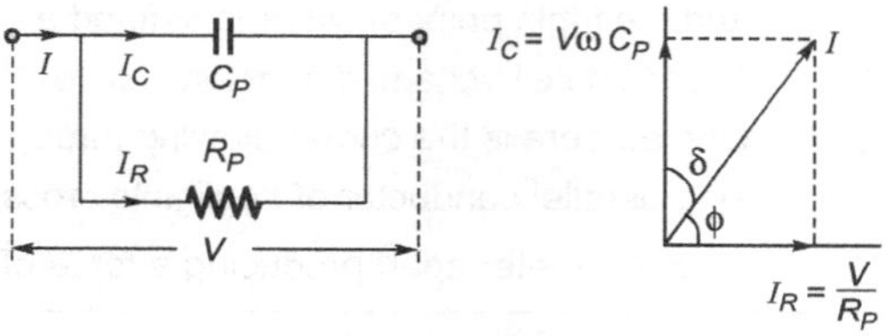

Dielectric loss	$P_L = \omega C_P V^2 \tan\delta$
Dissipation factor	$D = \tan\delta = \dfrac{1}{\omega C_P R_P}$ where $\delta \rightarrow$ loss angle of the capacitor

Series Representation

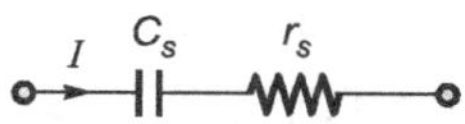

Dielectric loss	$P_L = \dfrac{I^2}{\omega C_s}\tan\delta$
Dissipation factor	$D = \tan\delta = \omega C_s r_s$

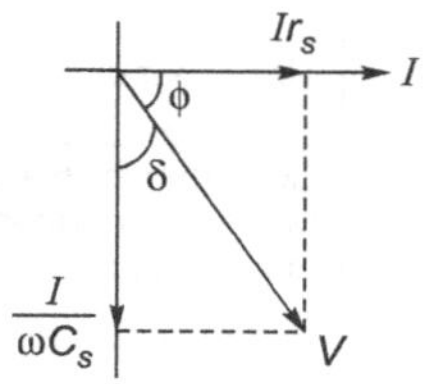

Frequency Errors in Capacitors

- Effective capacitance

$$C_{\text{eff}} = \frac{C}{1-\omega^2 LC}$$

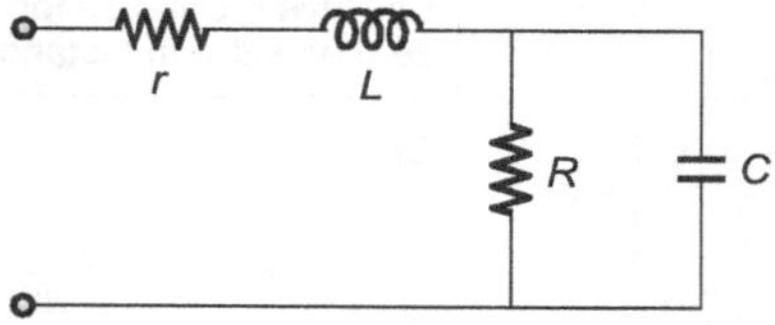

Equivalent Circuit of a Capacitor

For Medium Frequency

Effective capacitance	$C_{eff} = C(1+\omega^2 LC)$
Effective series resistance	$R_{eff} = r + \dfrac{R}{1+\omega^2 R^2 C^2}$
Loss angle	$\tan\delta = \dfrac{1-\omega^2 LC}{\omega r + \dfrac{1}{\omega CR}}$

For Low Frequency

Effective capacitance	$C_{eff} = C + \dfrac{1}{\omega^2 CR^2}$
Effective series resistance	$R_{eff} = \dfrac{R}{1+\omega^2 C^2 R^2}$
Loss angle	$\tan\delta = \omega Cr + \dfrac{1}{\omega CR}$

III Galvanometers

D' Arsonval Galvanometer

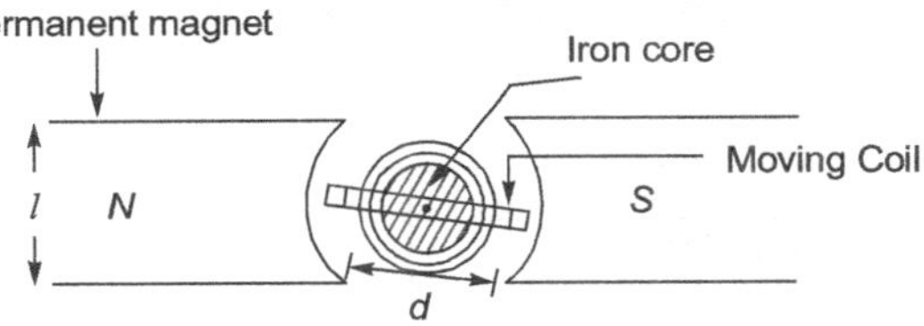

- Deflecting torque, $T_d = BiNA = Gi$

  where, $B \rightarrow$ Flux density in air gap; Wb/m^2

  $i \rightarrow$ Current through moving coil; Amperes

  $N \rightarrow$ Number of turns in coil ;

  $A \rightarrow ld$ = Area of coil; m^2

  l, $d \rightarrow$ Length of vertical and horizontal side (width) of coil respectively; m

  $G \rightarrow$ Displacement constant of galvanometer

- Controlling torque, $T_c = K\theta_f$

  where, K = Spring constant of suspension; Nm/rad

  θ_f = Final steady deflection of moving coil; rad

- Final steady deflection, $\theta_f = \left(\dfrac{NBA}{K}\right)i = \left(\dfrac{G}{K}\right)i$

Dynamic behaviour of Galvanometers

Torques in Galvanometers

Inertia torque	$T_j = J\frac{d^2\theta}{dt^2}$ where, $\theta \rightarrow$ deflection at any time t; rad $J \rightarrow$ moment of inertia of moving system about the axis of rotation; kg-m^2
Damping torque	$T_D = D\frac{d\theta}{dt}$ where, $D \rightarrow$ Damping constant
Controlling torque	$T_C = K\theta$ where, $K \rightarrow$ Control constant
Deflecting torque	$T_d = Gi$
Equation of motion	$T_j + T_D + T_C = T_d$ $J\frac{d^2\theta}{dt^2} + \frac{Dd\theta}{dt} + K\theta = Gi$

Note:

- ☑ If $D^2 < 4$ kJ, galvanometer is underdamped.
- ☑ If $D^2 = 4$ kJ, galvanometer is critically damped.
- ☑ If $D^2 > 4$ kJ, galvanometer is overdamped.

Total resistance of galvanometer circuit for critical damping	$R = \frac{G^2}{2\sqrt{KJ}}$
External series resistance required for critical damping	$R_e = R - R_g = \frac{G^2}{2\sqrt{KJ}} - R_g$ where R_g = Resistance of galvanometer

Sensitivity

Current sensitivity	$S_i = \frac{\theta_f}{i} = \frac{G}{K}$ rad/A $S_i = \frac{d}{i \times 10^6}$ scale divisions/μA $S_i = \frac{2000G}{K \times 10^6}$ mm/μA
Voltage sensitivity	$S_v = \frac{d}{iR_g \times 10^6}$ scale division/μV
Megohm sensitivity	$S_o = \frac{d}{i \times 10^{-6}}$ MΩ/scale division

Remember:

Sensitive galvanometer is one which produces a large deflection for a small current.

IV Analog Meters

Classification of Analog Meters

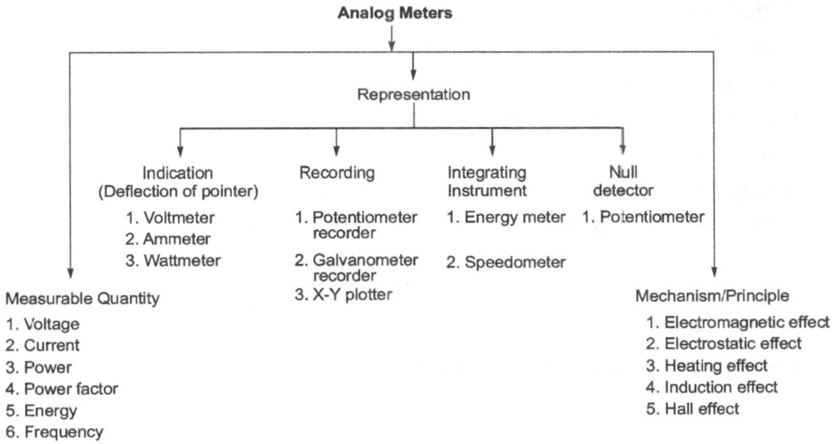

Torque in Analog Meter

Deflecting Torque (T_D) : Deflecting torque is proportional to quantity under measurement. This torque deflect the pointer away from initial or zero position.

$$T_D \propto \text{Measurable quantity}$$

Controlling Torque (T_C) : The controlling torque is opposite to deflecting torque. When, deflecting torque equals to controlling torque, pointer comes to final steady state position.

At equilibrium, $T_C = T_D$

Note:

- ☑ Control torque is also used to bring the pointer in zero initial position, if there is no deflecting torque.
- ☑ Except in PMMC, in all other instruments if the control spring is broken then pointer moves to the maximum position of scale.
- ☑ Control torque is provided by : (i) Spring control, (ii) Gravity control

Damping Torque

It is used to damp out oscillation at final steady state position. The time response of the instrument depends on damping torque.

Damping torque is provided by :

(i) **Air friction damping:** Used where low magnetic fields are produced

(ii) **Fluid friction damping:** Used where deflecting torque is minimum.

(iii) **Eddy current damping:** Used where permanent magnet produces the required deflecting torque.

Error in Analog Meters

Frictional Error : To reduce the frictional error, the torque to weight ratio of the instrument should be high.

Temperature Error : Due to change in temperature, change in resistance of meters and shunts and series multiplier occurs. To reduce this effect, resistances are made up of manganin material.

Frequency Error : Due to change in frequency, error is produced in instrument because change in frequency cause change in reactance. To reduce this error, a capacitance is used in case of voltmeter and for ammeter, the time constant and shunt impedances are maintained at same value.

Permanent Magnet Moving Coil (PMMC)

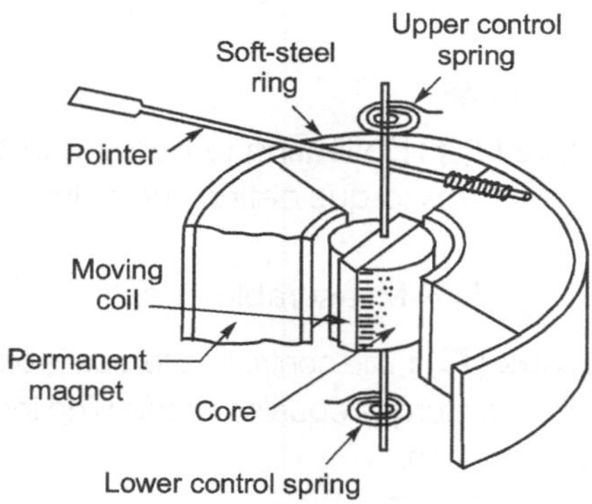

- Deflection torque : $T_D = nBAI$; $T_D = GI$

  where, $n \rightarrow$ Number of turns ; $B \rightarrow$ Flux density ; $A \rightarrow$ Area of core
  $I \rightarrow$ Current to be measured

- Final steady state deflection, $\theta = \left(\frac{G}{K}\right)I$ where, $K \rightarrow$ Spring constant

Note:

☑ PMMC instrument measures only DC or average values.

☑ Scale is linear.

☑ Spring is used for controlling torque.

☑ Damping torque provided by eddy current damping.

☑ It has more torque to weight ratio so accuracy and sensitivity is higher compared to other instrument.

☑ In direct measurement, the PMMC measures upto a current of 50 mA or a voltage of 100 mV, without any external device.

Enhancement of Scale of Ammeters and Voltmeters

Ammeter Shunts

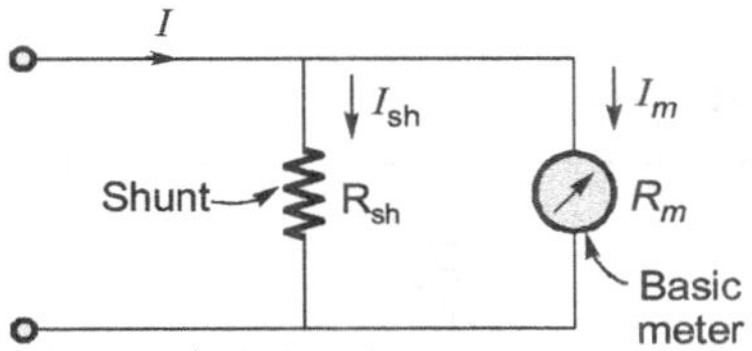

Basic ammeter circuit

$$I_{sh} R_{sh} = I_m R_m \quad ; \quad I = \left(1 + \frac{R_m}{R_{sh}}\right) I_m$$

where, $I \rightarrow$ Current to be measured ;

$I_m = I_{fs} \rightarrow$ Full scale deflection current ; A

$R_m \rightarrow$ Internal resistance of meter ; Ω

$R_{sh} \rightarrow$ Resistance of the shunt ; Ω

- Shunt resistance, $R_{sh} = \dfrac{R_m}{m-1}$

where, $m = \dfrac{I}{I_m} = 1 + \dfrac{R_m}{R_{sh}}$; $m \rightarrow$ Multiplying factor for shunt

Note:

To reduce the temperature effect, swamp resistance made up of manganin is added in series with ammeter.

Universal or Ayrton Shunt

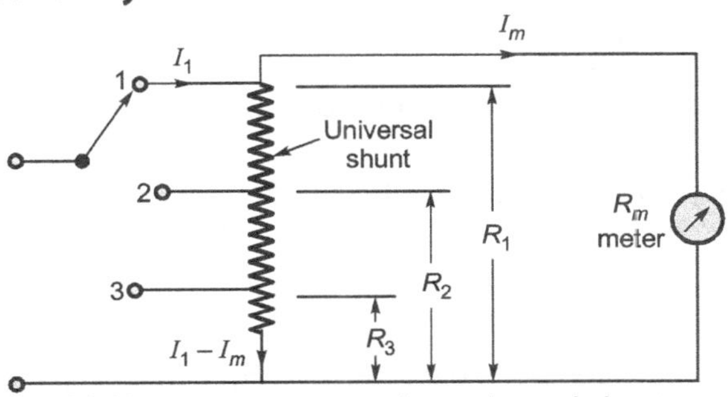

Multi-range ammeter using universal shunt

For switch at a position 1	$R_1 = \dfrac{R_m}{(m-1)}$
For switch at a position 2	$R_2 = \dfrac{(R_1 + R_m)}{m_2}$
For switch at a position 3	$R_3 = \dfrac{(R_1 + R_m)}{m_3}$

where, $m_1 = \dfrac{I_1}{I_m}$, $m_2 = \dfrac{I_2}{I_m}$, $m_3 = \dfrac{I_3}{I_m}$

Voltmeter Multipliers

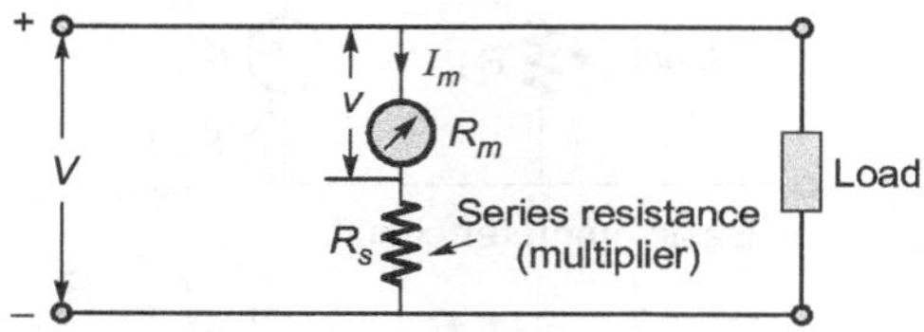

- Multiplying factor for multiplier, $m = \dfrac{V}{v} = 1 + \dfrac{R_s}{R_m}$
- Resistance of multiplier, $R_s = (m-1)R_m$

  where, $R_s \rightarrow$ Multiplier resistance; $R_m \rightarrow$ Internal resistance of meter

4. Potential Divider Arrangement

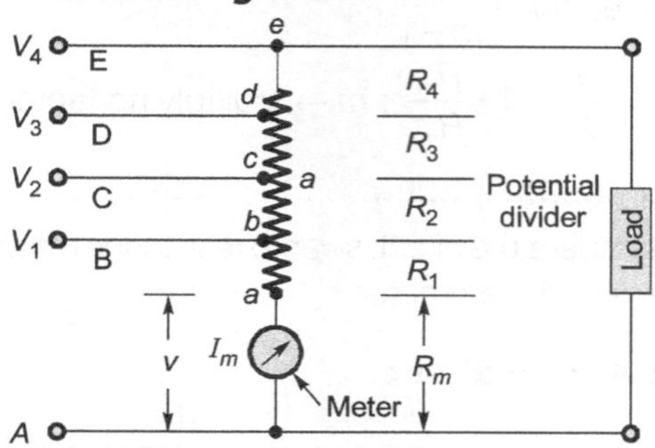

where, Resistance between point *a* & *b* $R_1 = (m_1 - 1)\,R_m$; $m_1 = \dfrac{V_1}{v}$

Resistance between point *b* and *c* $R_2 = (m_2 - m_1)\,R_m$; $m_2 = \dfrac{V_2}{v}$

Resistance between point *c* and *d* $R_3 = (m_3 - m_2)\,R_m$; $m_3 = \dfrac{V_3}{v}$

Resistance between point *d* and *e* $R_4 = (m_4 - m_3)\,R_m$; $m_4 = \dfrac{V_4}{v}$

- **Voltmeter Sensitivity (S_v)**

  It is the smallest value of input that the instrument can sense.

$$S_V = \frac{1}{I_{fs}} = \frac{R_s + R_m}{V}\ \Omega/\text{V} \quad \text{where, } I_{fs} \rightarrow \text{Full scale current}$$

Remember:

To reduce loading effect, a voltmeter with higher value of sensitivity is preferred.

Moving Iron Instruments

The moving iron instruments are based on the principle of magnetic effect of electric current.

Deflecting torque	$T_d = \frac{1}{2} I^2 \frac{dL}{d\theta}$
Deflection	$\theta = \frac{1}{2} \frac{I^2}{K} \frac{dL}{d\theta}$
For linear scale	$\theta \cdot \frac{dL}{d\theta} = \text{constant}$

Scale is cramped at lower and higher end.

where, $L \rightarrow$ Inductance of moving iron instrument

Note:

☑ Moving iron instrument measure both A.C. and D.C. quantities.
☑ In case of A.C., it measure RMS value.
☑ Scale is non linear.
☑ Controlling torque is provided by spring and air friction damping is used.
☑ Curve between $\frac{dL}{d\theta}$ and θ is rectangular hyperbola.

Shunts for Moving Iron Instruments

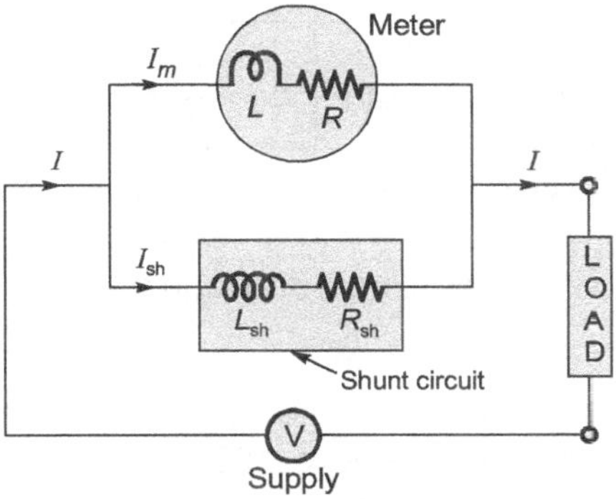

$$\frac{I_{sh}}{I_m} = \frac{R}{R_{sh}} \frac{\sqrt{1+(\omega L/R)^2}}{\sqrt{1+(\omega L_{sh}/R_{sh})^2}}$$

For the instrument to measure true value at all frequencies

$$\frac{L_{sh}}{R_{sh}} = \frac{L}{R}$$

Multipliers for Moving Iron Instruments

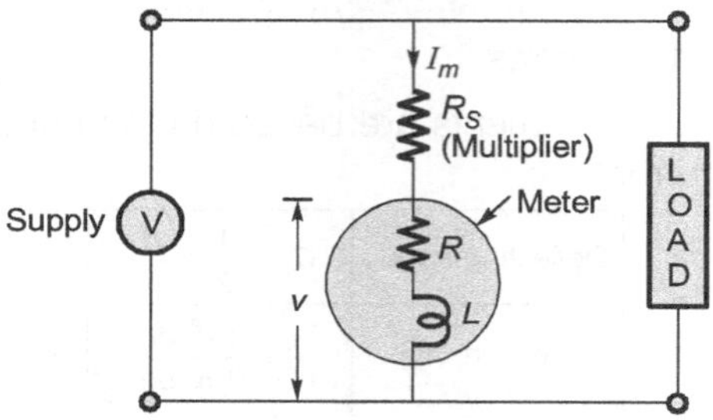

- Voltage multiplying factor, $m = \frac{V}{v} = \frac{\sqrt{(R+R_s)^2 + \omega^2 L^2}}{\sqrt{R^2 + \omega^2 L^2}}$

Errors in Moving Iron Instruments

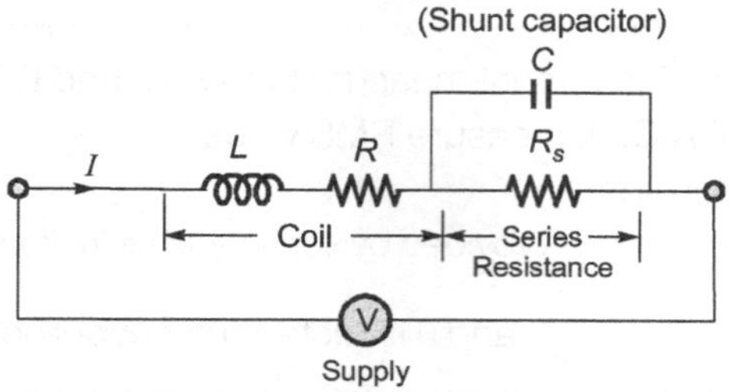

Shunt capacitance	$C = 0.41\frac{L}{R_s^2}$
Eddy currents When ω is small :	$I_e' = \frac{\omega^2 M L_e I}{R_e^2}$
When ω is large :	$I_e' = \frac{MI}{L_e}$ = Constant

where, R_e, $L_e \rightarrow$ Resistance and inductance of eddy current path

Note: ....

☑ Moving iron instrument is not suitable for measurement of current or voltage for frequency above 125 Hz because eddy current is constant at higher frequency.

☑ The voltmeter is made independent of input supply frequency by connecting a capacitor in parallel to the series multiplier resistance R_s.
☑ To reduce hysteresis error, the iron part of moving iron is made up of Nickel iron alloy.
☑ To reduce the external stray magnetic field, the instrument is kept inside the iron case or iron shielding is done.

Electrodynamometer

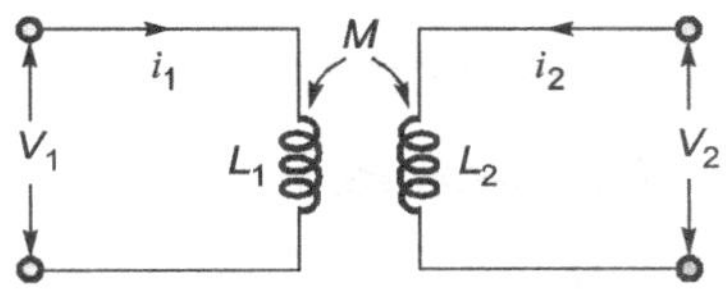

(a) If i_1 and i_2 are D.C. current, i.e., $i_1 = i_2 = I$

$$T_d = I^2 \frac{dM}{d\theta} \quad \text{(Measure average value)}$$

(b) If i_1 and i_2 are A.C. current and no phase shift

$$i_1 = i_2 = I$$

$$T_d = I^2 \frac{dM}{d\theta} \quad \text{(Measure RMS value)}$$

(c) If $i_1 = I_{m_1} \sin\omega t$ and $i_2 = I_{m_2} \sin(\omega t - \phi)$

$$T_d = I_1 I_2 \cos\phi \frac{dM}{d\theta} \quad \text{(Measure RMS value)}$$

Where, $I_1 = \frac{I_{m_1}}{\sqrt{2}}$ and $I_2 = \frac{I_{m_2}}{\sqrt{2}}$

Note:

☑ Electrodynamometer instrument is a transfer instrument.
☑ It measures both A.C. and D.C.
☑ Scale is nonlinear.
☑ Its sensitivity is lesser than PMMC and M.I. type instruments.

Ratiometer

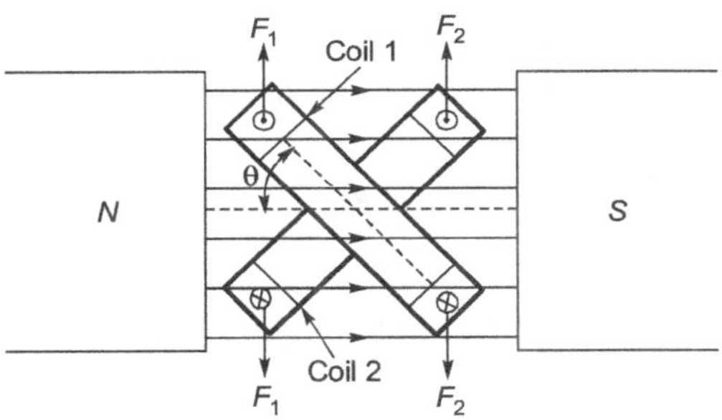

- Deflecting torque acting on coil 1, $T_{d1} = N_1 B l_1 d_1 I_1 \cos\theta$
- Deflecting torque acting on coil 2, $T_{d2} = N_2 B l_2 d_2 I_2 \cos\theta$

  where, $I_1, I_2 \rightarrow$ Current in coil 1 and 2 ; $N_1, N_2 \rightarrow$ Number of turns in coil 1 and 2 ; $l_1, l_2 \rightarrow$ Length of coil 1 and 2 ; $d_1, d_2 \rightarrow$ Width of coil 1 and 2 ; $B \rightarrow$ Flux density of magnetic field

- Deflection at equilibrium, $\theta = k\left(\frac{I_1}{I_2}\right)$

V Instrument Transformers

Ratios of Instrument Transformers

Transformation Ratio (*R*)

It is the ratio of magnitude of the primary phasor to the secondary phasor.

$$R = \frac{|\text{primary phasor}|}{|\text{secondary phasor}|}$$

- **For current transformer (C.T.) :** $R = \frac{\text{primary winding current}}{\text{secondary winding current}}$
- **For potential transformer (P.T.) :** $R = \frac{\text{primary winding voltage}}{\text{secondary winding voltage}}$

Nominal Ratio (K_n)

It is the ratio of rated primary winding current (or voltage) to the rated secondary winding current (or voltage).

- For C.T. : $K_n = \frac{\text{rated primary winding current}}{\text{rated secondary winding current}}$
- For P.T. : $K_n = \frac{\text{rated primary winding voltage}}{\text{rated secondary winding voltage}}$

Turns Ratio (*n*)

- For C.T. : $n = \frac{\text{number of turns of secondary winding}}{\text{number of turns of primary winding}}$
- For P.T. : $n = \frac{\text{number of turns of primary winding}}{\text{number of turns of secondary winding}}$

Ratio Correction Factor : $RCF = \frac{R}{K_n}$

Remember:

☑ The ratio marked on the transformers is their nominal ratio.

Current Transformer

Equivalent Circuit

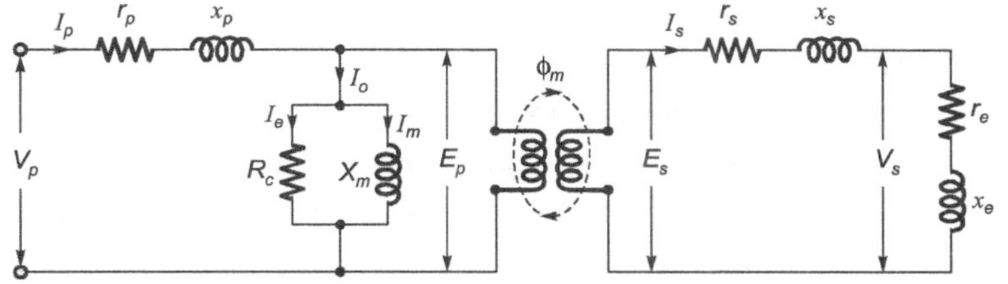

where, r_s, x_s → resistance, reactance of secondary winding
r_e, x_e → resistance, reactance of external burden
E_p, E_s → primary and secondary winding induced voltage
N_p, N_s → number of primary and secondary winding turns
I_p, I_s → primary and secondary winding current
ϕ → working flux of transformer
θ → phase angle of transformer
δ → angle between secondary winding induced voltage and secondary winding current
Δ → phase angle of secondary winding load circuit
I_0 → exciting current
I_m → magnetizing component of exciting current
I_e → loss component of exciting current
α → angle between exciting current and flux

Phasor Diagram

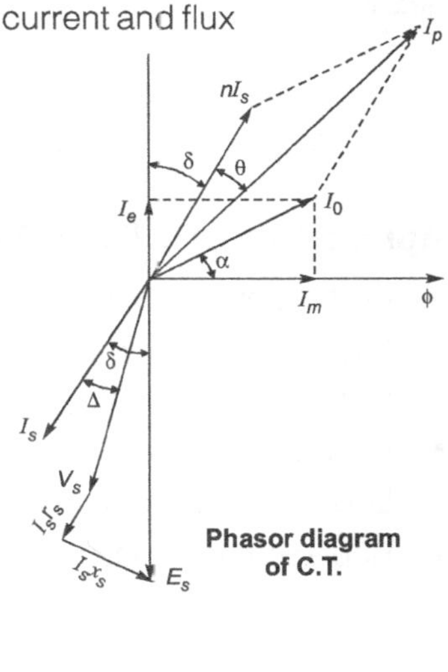

Phasor diagram of C.T.

- Transformation ratio,

$$R \simeq n + \frac{(I_m \sin\delta + I_e \cos\delta)}{I_s}$$

$$R \simeq n + \frac{I_e}{I_s} \simeq n\left(1 + \frac{I_e}{I_p}\right)$$

where, $I_m = I_o \cos\alpha$; $I_e = I_o \sin\alpha$

- Phase angle,

$$\theta \simeq \frac{180}{\pi}\left(\frac{I_m \cos\delta - I_e \sin\delta}{nI_s}\right) \text{ degree}$$

$$\theta \simeq \frac{180}{\pi}\frac{I_m}{I_p} \text{ degree}$$

- Ratio error $= \dfrac{\text{nominal ratio}(K_n) - \text{actual ratio}(R)}{\text{actual ratio}(R)}$

Remember:

- ☑ The primary current of C.T. is depending on load connected to system but it is not depending secondary winding burden.
- ☑ Primary winding is single turn or bar winding and secondary has more number of turns to reduce the current at secondary.
- ☑ If primary current is very high, it causes reduction in ratio error and phase angle error. So to increase value of primary current the primary is maintain with single turn.
- ☑ The secondary number of turns are reduce by 1 or 2 turns then the ratio error reduces.

Potential transformer

- Actual transformation (voltage) ratio,

$$R = n + \frac{\frac{I_s}{n}\left[R_p \cos\Delta + X_p \sin\Delta\right] + I_e r_p + I_m x_p}{V_s}$$

- Phase angle,

$$\theta = \frac{I_s}{V_s}(X_s \cos\Delta - R_s \sin\Delta) + \frac{I_e x_p - I_m r_p}{nV_s} \text{ rad}$$

Note:

- ☑ C.T. never operates with secondary winding open but P.T. can be operated with secondary winding open.
- ☑ Strip wound core is used to reduce ratio error and phase angle error.

Application of C.T. and P.T.

- Multiple operation with a single device.
- Higher current and higher voltages are step down to lower current and lower voltage so that metering is easier.
- Measuring circuit is isolated from the power circuit.
- Low power consumption.
- Replacement is easier.

VI Measurement of Power and Wattmeters

Measurement of Power

D.C. Circuits

Ammeter connected between load and voltmeter

Ammeter connected between load and voltmeter	Power consumed by load, $P = VI - I^2R_a$	Percentage error in reading, $\frac{I^2R_a}{VI} \times 100$	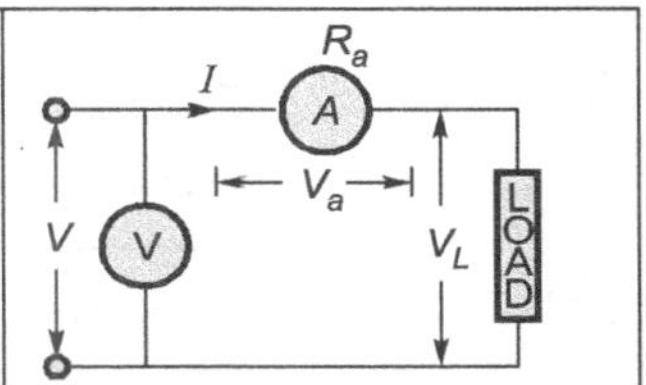
Voltmeter connected between load and ammeter	Power consumed by load, $P = VI - \frac{V^2}{R_V}$	Percentage error in reading, $\frac{V^2}{R_V \times VI} \times 100$	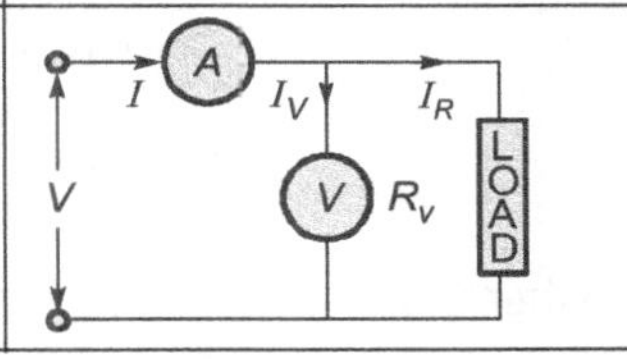

where, $V \rightarrow$ Voltage across voltmeter; $I \rightarrow$ Current through ammeter
$R_a \rightarrow$ Resistance of ammeter; $R_V \rightarrow$ Resistance of voltmeter

A.C. Circuits

- Instantaneous power, $p = vi = V_m I_m \sin\omega t \cdot \sin(\omega t - \phi)$

  where, $v = V_m \sin\omega t$; $i = I_m \sin(\omega t - \phi)$

- Average power, $P = VI\cos\phi = \frac{V_m I_m}{2}\cos\phi$

  where, V, I = RMS values of voltage and current ;
  $\cos\phi$ = Power factor of the load

- Let, $v = V_o + \sum_{n=1}^{m} V_n \sin(n\omega t + \theta_n)$

  and $i = I_o + \sum_{n=1}^{m} I_n \sin(n\omega t + \phi_n)$

  then $P_{avg.} = V_o I_o + \frac{1}{2}\sum_{n=1}^{m} V_n I_n \cos[\theta_n - \phi_n]$

Remember:

Wattmeter reads average active power.

Electrodynamometer Wattmeter

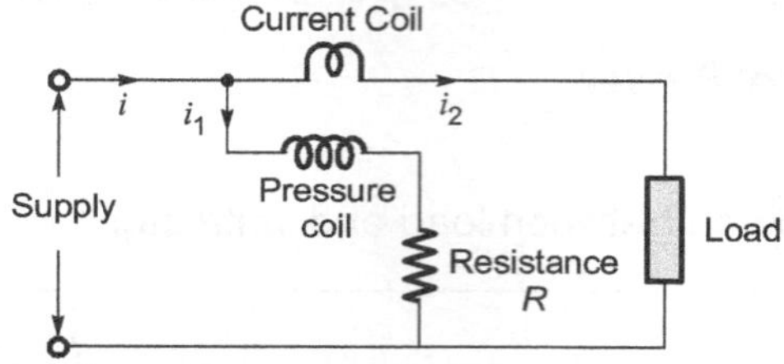

Instantaneous torque	$T_i = i_1 i_2 \left(\frac{dM}{d\theta}\right)$
Deflecting torque	$T_d = \frac{VI}{R_p}\cos\phi \cdot \frac{dM}{d\theta}$
Controlling torque	$T_c = K\theta$
Deflection	$\theta = \left(K_1 \frac{dM}{d\theta}\right)P$

where, i_1, i_2 → Instantaneous value of current in pressure and current coils

M → Mutual inductance

R_p → Resistance of pressure coil circuit

K → Spring constant ; θ → Final steady deflection

Power being measured, $P = VI\cos\phi$; $K_1 \rightarrow \frac{1}{R_p K}$

Note:

Scale is linear in terms of power as $\theta \propto P$.

Errors in Electrodynamometer Wattmeters

Correction Factor (*K*)

The correction factor is a factor by which the actual wattmeter reading is multiplied to get the true power.

(i) For lagging power factor, $K = \frac{\cos\phi}{\cos\beta\cos(\phi-\beta)}$

(ii) For leading power factor, $K = \frac{\cos\phi}{\cos\beta\cos(\phi+\beta)}$

where, ϕ → Angle between current in the current coil and voltage of pressure coil

β → Angle between current and voltage of pressure coil

True power = Correction factor × Actual wattmeter reading

For very small β :

$$\text{Actual wattmeter reading} = \text{True power} \times (1 + \tan\phi \tan\beta)$$

$$\text{Error} = \tan\phi \tan\beta \times \text{True power} = VI \sin\phi \tan\beta$$

$$\%\text{error} = \tan\phi \tan\beta \times 100$$

$$\text{True power} = VI \cos\phi$$

where, $V \to$ Voltage applied to pressure coil; $I \to$ Current in current coil

Power in Poly-Phase Systems

Blondel's Theorem

If a network is supplied through n conductors, the total power is measured by summing the reading of n wattmeters so arranged that a current element of a wattmeter is in each line and the corresponding voltage element is connected between that line and a common point, if the common point is located on one of the lines, then the power may be measured by $(n-1)$ wattmeters.

Two wattmeter method

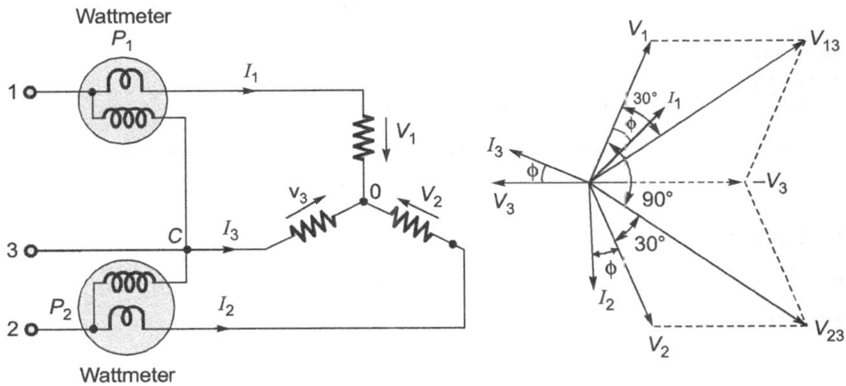

Reading of P_1 wattmeter	$P_1 = \sqrt{3}VI\cos(30° - \phi)$
Reading of P_2 wattmeter	$P_2 = \sqrt{3}VI\cos(30° + \phi)$
Total power consumed by load	$P = P_1 + P_2$
Power factor	$\cos\phi = \cos\left[\tan^{-1}\left(\sqrt{3}\frac{P_1 - P_2}{P_1 + P_2}\right)\right]$ $\tan\phi = \sqrt{3}\frac{P_1 - P_2}{P_1 + P_2}$

where, $V \to$ Phase voltage ; $I \to$ Phase current ;
$\phi \to$ Angle between phase current and phase voltage

Reading of Wattmeter at Different Power Factor

S.No.	ϕ	$\cos\phi$	P_1	P_2	$P = P_1 + P_2$	Comment
1.	0	1	$\frac{\sqrt{3}}{2}V_L I_L$	$\frac{\sqrt{3}}{2}V_L I_L$	$\sqrt{3}V_L I_L$	$P_1 = P_2$ (Equal reading)
2.	30°	0.866	$V_L I_L$	$\frac{V_L I_L}{2}$	$1.5 V_L I_L$	$P_1 = 2P_2$
3.	60°	0.5	$\frac{\sqrt{3}}{2}V_L I_L$	0	$\frac{\sqrt{3}}{2}V_L I_L$	$P_2 = 0, P_2 = P$
4.	90°	0	$+\frac{V_L I_L}{2}$	$-\frac{V_L I_L}{2}$	0	$P_1 = -P_2$

Note:

When wattmeter reading comes into negative, reverse either P.C. or C.C. terminal and then take the reading of negative wattmeter.

Measurement of Energy

Energy meter is an integrating instrument which adds the energy cumulatively over a period of time.

$$\text{Energy} = \text{Power} \times \text{time}\ ; \quad \text{Energy} = \int_0^t P \cdot dt \ \text{kWhr}$$

Note:

- ☑ Energy meter works on principle of induction motor.
- ☑ The meter which measure A.C. energy is called watt hour meter.
- ☑ The meter which measure D.C. energy is called amp-hour meter.

- Deflection torque, $T_D \propto P$
- Breaking torque, $T_B \propto N$ where, $N \rightarrow$ Speed of disc in rps
- At balance,

$$T_d = T_B$$

$$\int P \cdot dt = K \int N \cdot dt$$

$$\text{Energy} \propto \int N \cdot dt$$

- Energy meter constant (EMC)

$$\text{EMC} = \frac{\text{Number of revolution made by disc}}{\text{Energy recorded in kWhr.}}$$

$$K = \frac{N}{P \times t}$$

where, $P \rightarrow$ Power in kW ; $t \rightarrow$ Time in hrs.

- $$\% \text{ Creeping Error} = \frac{\text{Revolution of disc due to creeping per hour}}{\text{Revolution of disc due to total load per hour}} \times 100$$

Remember:

Potential coil of energy meter should be highly inductive so that it measures true energy.

Compensation in Energy Meter

- **Lag compensation** : Through lag coil or shading coil.
- **Low load or friction adjustment** : By using shading loop.
- **Over friction or creeping** : By providing holes or slots on rotating disc.
- **Over load compensation** : By keeping saturable shunt magnet in series magnet or current coil.
- **Over voltage compensation** : By keeping saturable shunt magnet in shunt magnet.
- **Temperature compensation** : By making permanent magnet of "mutemp" material.
- **Speed adjustment** : By adjusting position of break magnet.

Remember:

☑ Creeping error is always positive.

☑ If either potential coil or current coil is wrongly connected then the disc rotates in opposite direction.

VII Measurement of Resistance

Classification of Resistance

- **Low resistance:** All resistance of the order of 1 Ω and below.
  Example: Winding coils of electrical motors, generators and transformers.
- **Medium resistance:** Resistances from 1 Ω upwards to about 0.1 MΩ.
  Example: Resistance of heaters, potentiometers.
- **High resistance:** All resistances of the order of 0.1 MΩ and above.
  Example: Insulation of electrical cable and windings, insulation of motors, generators and transformers.

Measurement of Medium Resistance

The different methods employed are:

(i) Ammeter-voltmeter method

(ii) Wheatstone bridge method

(iii) Ohmmeter method

(iv) Substitution method

Ammeter Voltmeter Method

$$R_m = \frac{\text{Voltmeter reading}}{\text{Ammeter reading}} = \frac{V}{I}$$

where, $R_m \rightarrow$ Measured value of resistance

- **Circuit for higher resistance**

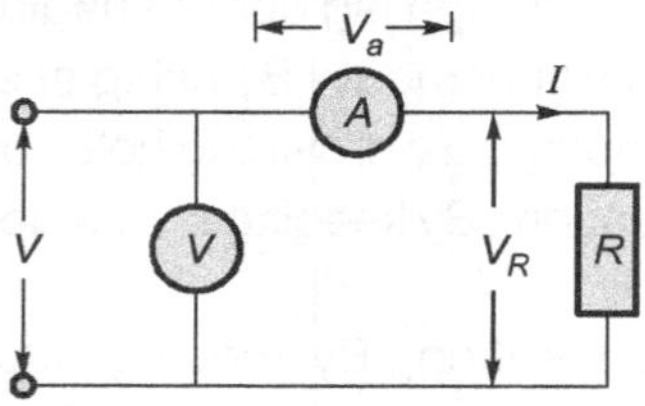

- True value of resistance

$$R = R_{m_1} - R_a \quad ; \quad R_m = R_{m1}\left(1 - \frac{R_a}{R_{m1}}\right)$$

where, $R_{m_1} \rightarrow$ Measured value of resistance

$R_a \rightarrow$ Resistance of ammeter

- Relative error

$$\epsilon_r = \frac{R_{m1} - R}{R} = \frac{R_a}{R}$$

To get minimum error, the test resistance should be more than the ammeter resistance so that this adjustment is suitable for measurement of high resistance.

- **Circuit for lower resistance**

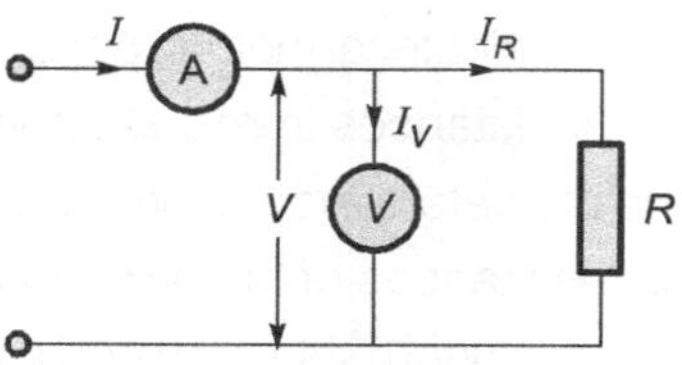

- True value of resistance

$$R = \frac{R_{m_2} R_v}{R_v - R_{m_2}}$$

where, $R_{m_2} \rightarrow$ Measured value of resistance

$R_v \rightarrow$ Resistance of voltmeter

For $R_v >> R_{m_2}$ $\qquad R = R_{m_2}\left(1 + \frac{R_{m_2}}{R_v}\right)$

- Relative error

$$\epsilon_r = \frac{R_{m2} - R}{R} = \frac{-R^2_{m2}}{R_V R} \qquad \left[\text{For } R_V >> R_{m_2}\right]$$

- Approximate relative error

$$\epsilon = -\frac{R}{R_V} \qquad \left[\text{For } R_{m_2} \approx R\right]$$

This circuit is suitable for measurement of low resistance under medium scale range to get minimum error.

Note: ....

Relative errors for above two cases are equal when true value of resistance, $R = \sqrt{R_a R_V}$.

....

Wheatstone Bridge

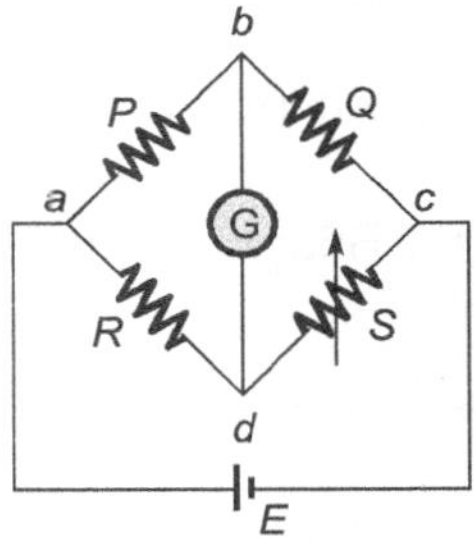

- At balance, $R = S\dfrac{P}{Q}$
- Sensitivity of Wheatstone bridge

$$S_B = \frac{\theta}{\Delta R / R} = \frac{S_V ESR}{(R+S)^2}; \text{mm}; \quad S_B = \frac{S_V E}{\frac{P}{Q} + 2 + \frac{Q}{P}}$$

where, $S_V \rightarrow$ Voltage sensitivity of galvanometer, mm/volt
$E \rightarrow$ Bridge voltage ; P, $Q \rightarrow$ Branch resistances
$\theta \rightarrow$ Deflection of galvanometer, mm

- For a bridge with equal arms, $S_B = \dfrac{S_V E}{4}$

Note: ....

☑ For maximum bridge sensitivity : $\dfrac{P}{Q} = \dfrac{R}{S} = 1$

☑ Sensitivity of bridge is most important parameter as compared to accuracy, precision and resolution.

....

Equivalent circuit of Wheatstone bridge

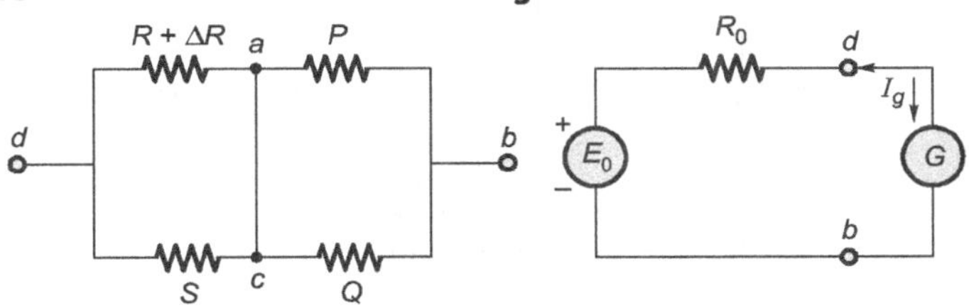

- Galvanometer current

$$I_g = \frac{E_0}{R_0 + G}$$

where $E_0 \rightarrow$ Thevenin's or open circuit voltage appearing between terminals b and d with galvanometer circuit open circuited.

$G \rightarrow$ Resistance of the galvanometer circuit

$$E_0 = E\left[\frac{R+\Delta R}{2R+\Delta R} - \frac{1}{2}\right] \approx E\left(\frac{\Delta R}{4R}\right) \text{ as } \Delta R << R$$

$\Delta R \rightarrow$ Change in resistance R

- Thevenin equivalent resistance of bridge : $R_0 = \dfrac{RS}{R+S} + \dfrac{PQ}{P+Q}$

- Galvanometer deflection : $\theta = \dfrac{S_v \, ES \, \Delta R}{(R+S)^2} = \dfrac{S_i \, ES\Delta R}{(R_0 + G)\,(R+S)^2}$

where, $S_i \rightarrow$ Current sensitivity of galvanometer

- Bridge sensitivity, $S_B = \dfrac{\theta}{\Delta R / R} = \dfrac{S_i \, E \, SR}{(R_0 + G)\,(R+S)^2}$

- Current Sensitivity, $S_i = \dfrac{\theta}{I_g}$; mm/μA

where, $\theta \rightarrow$ Deflection in galvanometer ; $I_g \rightarrow$ Current in galvanometer

- Voltage Sensitivity, $S_v = \dfrac{\theta}{V_{Th}}$; mm/V

where, $V_{Th} \rightarrow$ Voltage across galvanometer

Note: In Wheatstone bridge method, the effect of lead resistance is not eliminated hence it is not suitable for measurement of low resistance.

Ohmmeters

- **Series Type Ohmmeter**

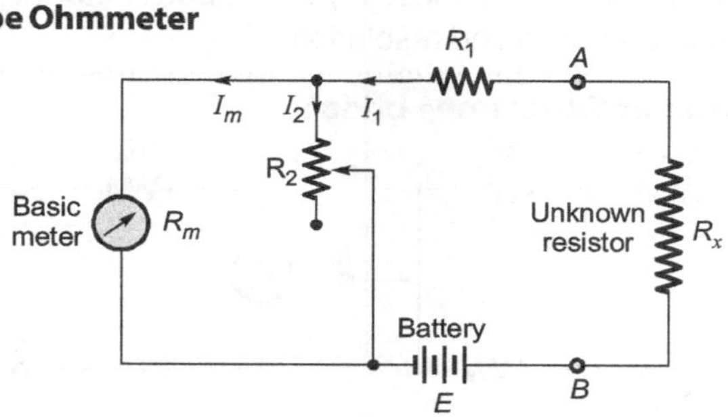

Half scale resistance	$R_h = R_1 + \frac{R_2 R_m}{R_2 + R_m}$
Meter current	$I_m = \frac{ER_2}{(R_h + R_x)(R_2 + R_m)}$
Full scale deflection current	$I_{fs} = \frac{ER_2}{R_h(R_2 + R_m)}$
Fraction of full scale reading	$s = \frac{I_m}{I_{fs}} = \frac{R_h}{R_x + R_h}$

Shunt Type Ohmmeter

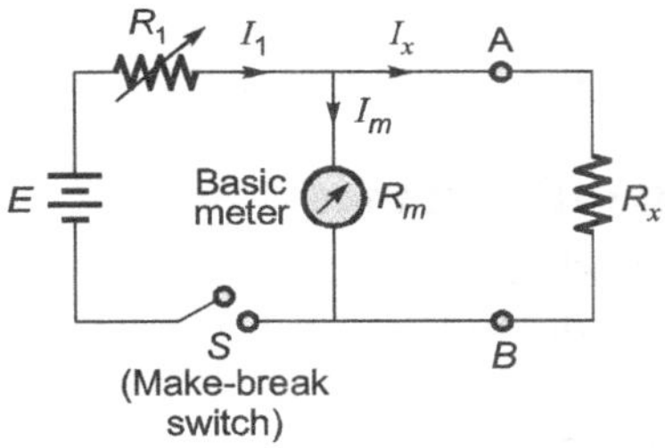

Half scale reading of unknown resistance R_x is	$R_h = \frac{R_1 R_m}{R_1 + R_m}$
Half scale reading of the meter	$I_h = 0.5 I_{fs} = \frac{ER_h}{R_1 R_m + R_h(R_1 + R_m)}$

where, $R_m \rightarrow$ Internal resistance of meter
$R_1 \rightarrow$ Adjustable resistor ; $E \rightarrow$ Supply voltage

Measurement of Low Resistance

The different methods employed are : (i) Kelvin's double bridge method; (ii) Ammeter voltmeter method and (iii) Potentiometer method

Kelvin's Double Bridge Method

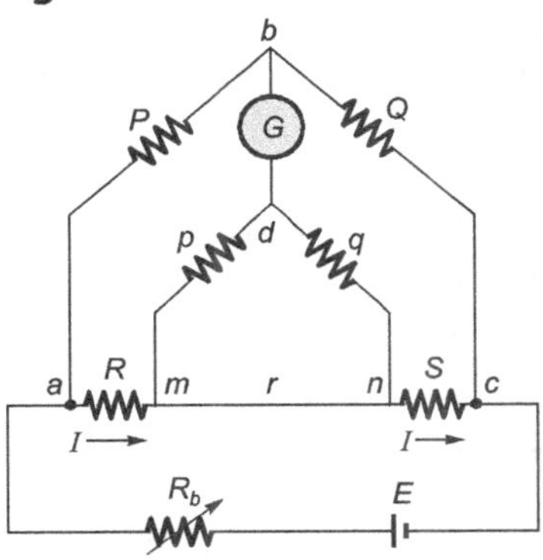

- For zero galvanometer deflection, $E_{ab} = E_{amd}$

$$R = \frac{P}{Q} \cdot S + \frac{qr}{p+q+r}\left[\frac{P}{Q} - \frac{p}{q}\right]$$

- If $\frac{P}{Q} = \frac{p}{q}$ then $R = \frac{P}{Q} \cdot S$

Note:

Accuracy of Kelvin double bridge method

(i) From 1000 μΩ to 1.0 Ω : 0.05%

(ii) From 100 μΩ to 1000 μΩ : 0.5% to 0.05%

(iii) From 10 μΩ to 100 μΩ : 0.5% to 0.2%

Measurement of High Resistance

The difference methods employed are:

(i) Loss of charge method

(ii) Megger

(iii) Direct deflection method

(iv) Megohm bridge

Loss of Charge Method

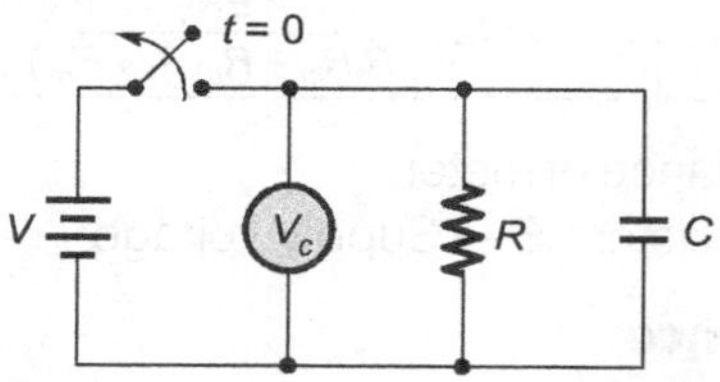

$$R = \frac{0.4343t}{C\log_{10}\left(\frac{V}{V_c}\right)}$$

Megger

- Megger works on the principle of electrodynamometer.
- Megger is used to measure the insulation resistance of cable, motor and generator, etc.
- Deflecting torque angle is proportional to the resistance of the insulator, which is under test.
- It is also used to check the continuity of cable.
- No external control torque provided.
- Air friction damping is used.
- No need of external power supply.

Note:

High resistance have a guard terminal which is used to avoid leakage current.

VIII A.C. Bridges

Introduction

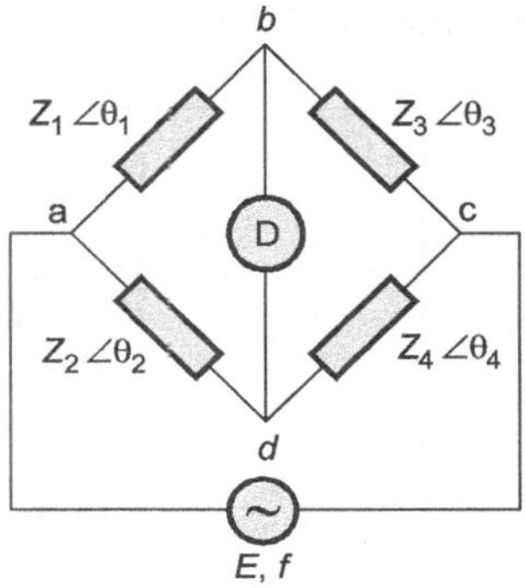

They are used to measure self inductance, Mutual inductance, capacitance, and frequency.

- **General equation for bridge balance**

$$\bar{Z}_1 \bar{Z}_4 = \bar{Z}_2 \bar{Z}_3$$

- **Magnitude condition**

$$|Z_1||Z_4| = |Z_2||Z_3|$$

- **Angle condition**

$$\angle\theta_1 + \angle\theta_4 = \angle\theta_2 + \angle\theta_3$$

Note:

☑ Magnitude condition and angle condition both must be satisfied for the bridge to be balanced.

☑ Depending upon the frequency, different null detectors are used:

Vibration galvanometer – 5 Hz to 1 kHz
Head phones – 250 Hz to 4 kHz
Tuned amplifier detector – 10 Hz to 100 kHz
D'Arsonval Galvanometer – DC frequency = 0 Hz

☑ Depending upon phase angle θ, elements are

θ	Elements
0°	R
90°	L_1
–90°	C_1
$0° < \theta_1 < 90°$	R_1, L_1
$-90 < \theta_1 < 0°$	R_1, C_1

- **Convergence to balance point**

If the variables are in the same arm of bridge then minimum time is required for balancing of bridge. This is called convergence to balance point.

- **Quality factor (Q.F.)** : $Q.F. = \dfrac{\text{Energy stored}}{\text{Energy dissipated}}$

Measurement of Self Inductance

- **Maxwell's Inductance Bridge**

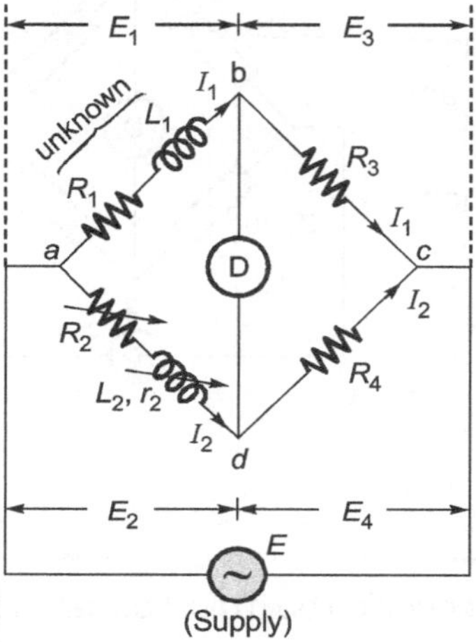

$$R_1 = \frac{R_3}{R_4}(R_2 + r_2)\,; \quad L_1 = \frac{R_3}{R_4} \cdot L_2$$

where, L_1 → Unknown inductance of resistance R_1

L_2 → Variable inductance of fixed resistance r_2

R_2 → Variable resistance connected in series with L_2

R_3, R_4 → Known non-inductive resistances

- **Maxwell's Inductance-Capacitance Bridge**

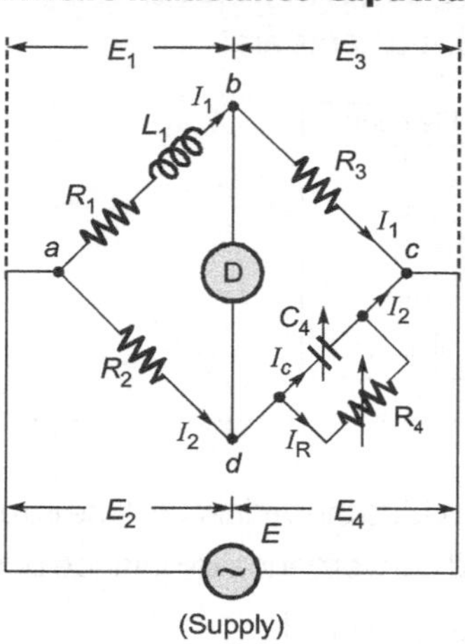

$$R_1 = \frac{R_2 R_3}{R_4} \text{ and } L_1 = R_2 R_3 C_4$$

- ***Q*-factor of the coil**

$$Q = \frac{\omega L_1}{R_1} = \omega C_4 R_4$$

Note:

- ☑ Not suitable for measurement of high Q coil because phase angle criteria does not satisfy.
- ☑ Not suitable for measurement of low Q-coil because of sliding balance problem.
- ☑ Suitable for measurement of medium Q coil i.e. $(1 < Q < 10)$.

- **Hay's Bridge**

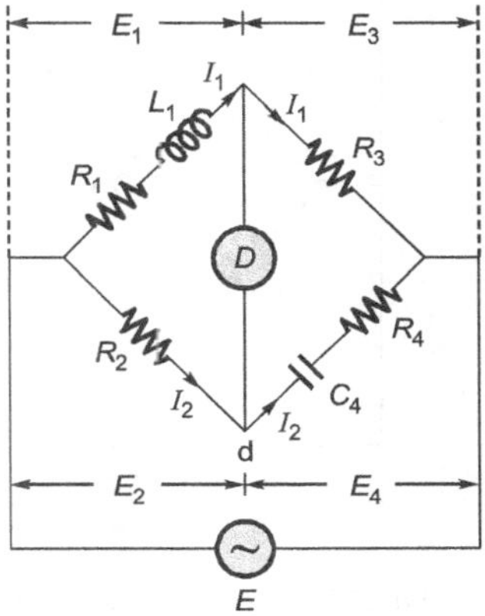

$$R_1 = \frac{\omega^2 R_2 R_3 R_4 C_4^2}{1+\omega^2 C_4^2 R_4^2} = \frac{R_2 R_3}{R_4}\left(\frac{1}{1+Q^2}\right)$$

$$L_1 = \frac{R_2 R_3 C_4}{1+\omega^2 C_4^2 R_4^2} = \frac{R_2 R_3 C_4}{1+\left(\frac{1}{Q}\right)^2}$$

- For $Q > 10$, $\quad L_1 = R_2 R_3 C_4$

$$Q = \frac{\omega L_1}{R_1} = \frac{1}{\omega C_4 R_4}$$

where, $L_1 \rightarrow$ Unknown-inductance having a resistance R_1

$R_2, R_3, R_4 \rightarrow$ known non-inductive resistance

$C_4 \rightarrow$ Standard capacitor

Note:

The Hay's bridge is suited for the measurement of the high Q-inductors.

- **Anderson's Bridge**

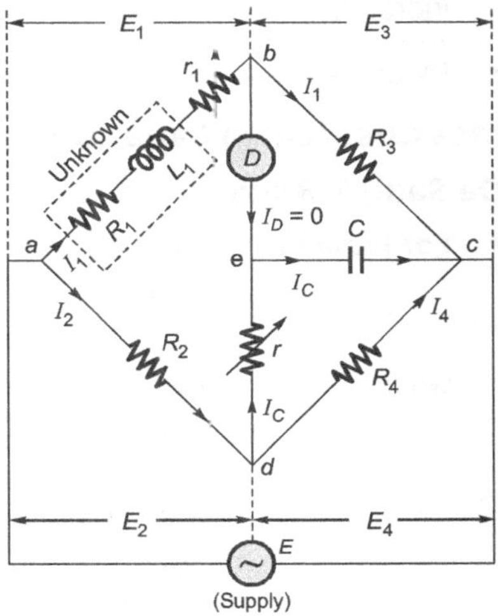

$$R_1 = \frac{R_2 R_3}{R_4} - r_1$$

$$L_1 = \frac{CR_3}{R_4}\left[r(R_4 + R_2) + R_2 R_4\right]$$

where, $L_1 \rightarrow$ Self-inductance to be measured

$R_1 \rightarrow$ Resistance of self-inductor

$r_1 \rightarrow$ Resistance connected in series with self-inductor

$r, R_2, R_3, R_4 \rightarrow$ Known non-inductive resistances

$C \rightarrow$ Fixed standard capacitor

Note:

- ☑ Measure low $Q(Q < 1)$.
- ☑ Difficult to obtain balance condition.

- **Owen's Bridge**

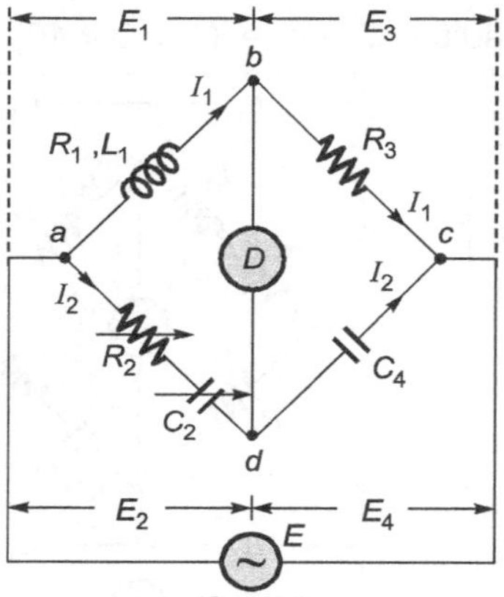

$$L_1 = R_2 R_3 C_4 ; \quad R_1 = R_3 \frac{C_4}{C_2}$$

where, $L_1 \rightarrow$ Unknown self inductance of resistance R_1
$R_2 \rightarrow$ Variable non-inductive resistance
$R_3 \rightarrow$ Fixed non-inductive resistance
$C_2 \rightarrow$ Variable standard capacitor
$C_4 \rightarrow$ Fixed standard capacitor

Note:

Owen's bridge is used for measurement of unknown inductance and incremental inductance and incremental permeability (μ).

- Measure low $Q(Q < 1)$.

Measurement of Capacitance

De Sauty's Bridge

- **For lossless capacitor**

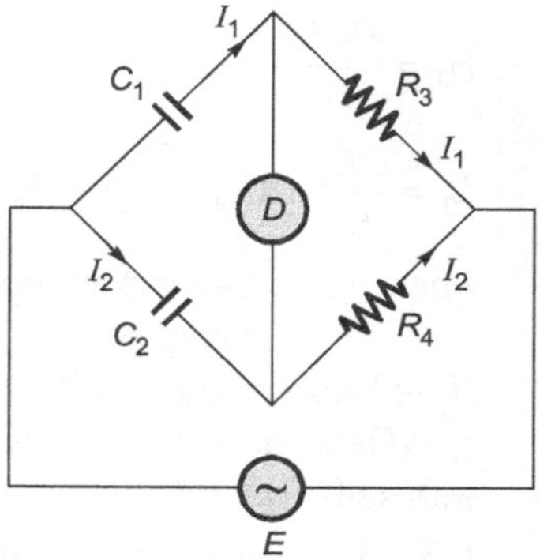

$$C_1 = \frac{C_2 R_4}{R_3}$$

where, $C_1 \rightarrow$ Capacitor whose capacitance to be measured
$C_2 \rightarrow$ A standard capacitor; $R_3, R_4 \rightarrow$ Non-inductive resistors

- **For imperfect capacitor having dielectric loss**

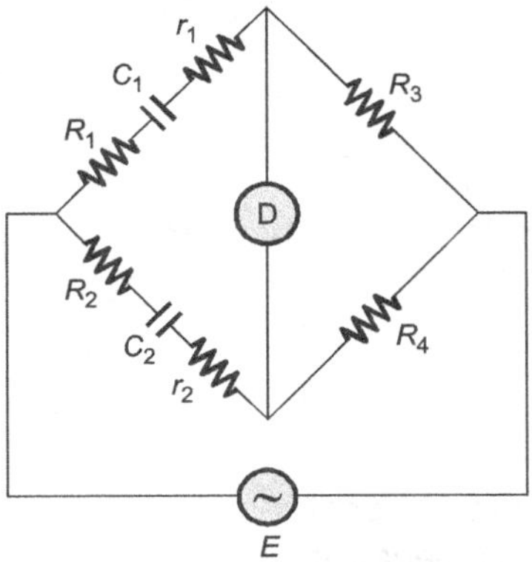

$$\frac{C_1}{C_2} = \frac{R_4}{R_3} = \frac{R_2 + r_2}{R_1 + r_1}$$

where, $r_1, r_2 \rightarrow$ Resistance representing the loss component of the two capacitor.

Dissipation factor : $D = \tan\delta = \omega C_1 r_1 = \omega C_2 r_2$

Schering Bridge

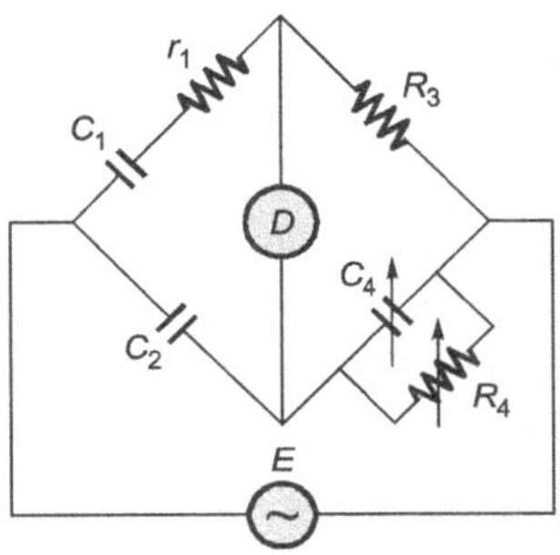

$$r_1 = \frac{R_3 C_4}{C_2};$$

$$C_1 = C_2\left(\frac{R_4}{R_3}\right)$$

Dissipation factor : $D = \omega C_1 r_1 = \omega C_4 R_4$

where, $C_1 \rightarrow$ Capacitor whose capacitance is to be determined
$r_1 \rightarrow$ Series resistance representing the loss in the capacitor C_1
$C_2 \rightarrow$ Standard loss-free capacitor
$R_3 \rightarrow$ Non-inductive resistance
$C_4 \rightarrow$ Variable capacitor
$R_4 \rightarrow$ Variable non-inductive resistance in parallel with variable capacitor C_4

Note:

Schering bridge is shielded with metal screen to reduce the stray capacitance which exists between the arms and arms to the earth.

Measurement of Frequency

Wein's Bridge

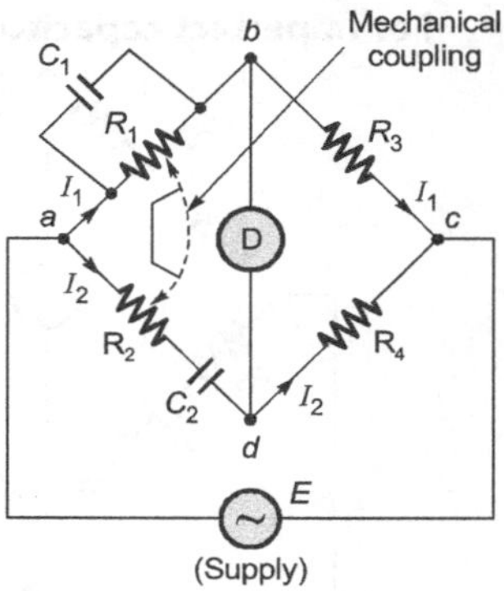

- Frequency

$$\frac{R_4}{R_3} = \frac{R_2}{R_1} + \frac{C_1}{C_2} \;;\; f = \frac{1}{2\pi\sqrt{R_1 R_2 C_1 C_2}} \text{ Hz}$$

- For $R_1 = R_2 = R$ and $C_1 = C_2 = C$.

$$f = \frac{1}{2\pi RC} \;;\; \frac{R_4}{R_3} = 2$$

Limitation of Wein's Bridge

If the input signal is not a sinusoidal or signal containing harmonics then balancing of bridge is not possible because null detector is sensitive to the frequencies.

IX Magnetic Measurements

Flux Density

$$B = \frac{\phi}{A_s} = \frac{RK_q\theta_1}{2NA_s}$$

where, $\phi \rightarrow$ Flux linking search coil
$A_s \rightarrow$ Cross-sectional area of specimen
$R \rightarrow$ Resistance of the ballistic galvanometer circuit
$K_q\theta_1 \rightarrow$ Charge indicated by ballistic galvanometer
$N \rightarrow$ Number of turns in the search coil

Hysteresis Loss Per Unit Volume

$$p_h = \eta f B_m^k$$

where, $\eta \rightarrow$ Hysteresis coefficient
$f \rightarrow$ Frequency; Hz
$B_m \rightarrow$ Maximum flux density; Wb/m^2
$k \rightarrow$ Steinmetz coefficient

Note:

☑ The value of k varies from 1.6 to 2.

Eddy Current Loss Per Unit Volume for Laminations

$$p_e = \frac{4k_f^2 f^2 B_m^2 t^2}{3\rho}$$

where, $k_f \rightarrow$ Form factor ; $t \rightarrow$ Thickness of laminations; m
$\rho \rightarrow$ Resistivity of material; Ω-m

Total Iron Loss Per Unit Volume

$$p_i = p_h + p_e = \eta f B_m^k + \frac{4k_f^2 f^2 B_m^2 t^2}{3\rho}$$

Maximum Flux Density

$$B_m = \frac{E_2}{4k_f f A_s N_2}$$

For Sinusoidal Supply

$$B_m = \frac{E_2}{4.44 f A_s N_2}$$

where, $E_2 \rightarrow$ Voltage induced in secondary winding ; $E_2 = 4k_f f \phi_m N_2$
$k_f \rightarrow$ Form factor (= 1.11 for sinusoidal supply) ;
$f \rightarrow$ Frequency
$A_s \rightarrow$ Cross-sectional area of specimen
$N_2 \rightarrow$ Number of turns in secondary winding
$\phi_m \rightarrow$ Maximum flux linkage

X Electronic Instruments

Average current through diode vacuum tube voltmeters

$$I_{av} = \frac{E_{av}}{2R} = \frac{E_{rms}}{2\times 1.11\times R} = 0.45\frac{E_{rms}}{R}$$

where, E_{rms} = RMS value of applied voltage
E_{av} = Average value of applied voltage
R = Load resistance

Difference amplifier type of electronic voltmeter

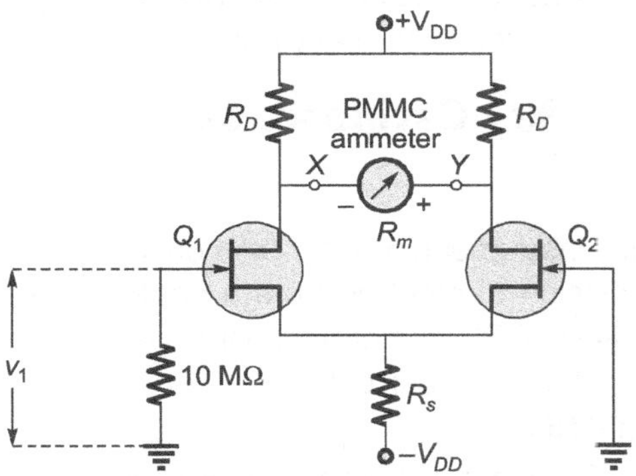

Thevenin's voltage across terminal X-Y

$$V_{Th} = g_m\left(\frac{r_d R_D}{r_d + R_D}\right)v_1$$

Thevenin's resistance looking into terminals X-Y

$$R_{\text{Th}} = \frac{2r_d R_D}{r_d + R_D}$$

where, $R_d \rightarrow$ A.C. drain ressitance in Ω

$g_m \rightarrow$ Transconductance in mho

Current through ammeter

$$i = \frac{V_{Th}}{R_{Th} + R_m} = \frac{g_m r_d R_D / (r_d + R_D)}{2r_d R_D / (r_d + R_D) + R_m} \cdot v_1$$

when $R_D << r_d$

$$i = \frac{g_m R_D}{2R_D + R_m} \cdot v_1$$

Digital Meters

Basic measurable quantity in digital meter is D.C.

Resolution (*R*) of Digital Meter

The smallest change in the input which a digital meter can be able to detect is called resolution.

$$R = \frac{1}{10^n}$$ where, $n \rightarrow$ Number of full digit.

Sensitivity (S)

The smallest change in input that can be displayed within given range.

$$S = \text{Resolution} \times \text{Range of meter}$$

Over ranging

Switch on the extra half (1/2) is called over ranging. Due to this over ranging the range of the instrument increases.

XI Cathode Ray Oscilloscope

Introduction

CRO is a digital instrument, which works on the principle of thermionic emission, i.e., emission of electron from a heated surface. It is a linear device. With the use of CRO one can measure peak to peak, rms, peak or average value of voltage and current.

- **Calibration of CRO**

  Calibration of CRO is done by applying a known quality of square signal having a frequency of 1 kHz and peak to peak magnitude of 1 mV.

- **The rise time (t_r), of signal applied to CRO and bandwidth of CRO are related as**

$$t_r \times B.W. = 0.35$$

If this condition fails then the signal is distorted at the output of CRO.

- **Electrostatic Deflection**

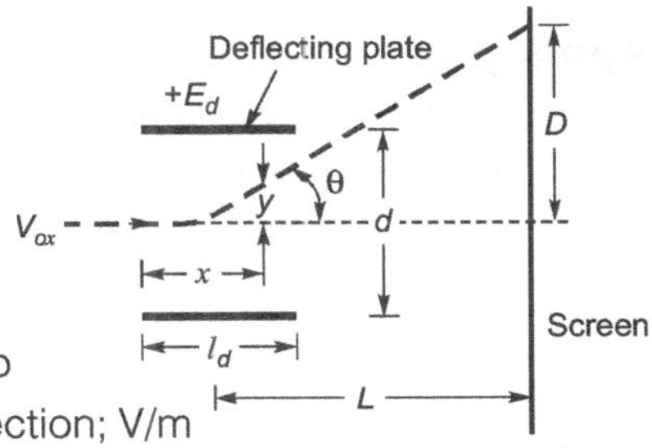

$$y = \frac{1}{2}\frac{eE_y}{mv_{ox}^2}x^2$$

where, $y \rightarrow$ Displacement in y-direction; m

$e \rightarrow$ Charge of an electron; Coulomb

$E_y \rightarrow$ Electric field intensity in Y-direction; V/m

$m \rightarrow$ Mass of electron; kg

$v_{ox} \rightarrow$ Velocity of electron when entering the fields of deflecting plates; m/s

$x \rightarrow$ Displacement in x-direction ; m

- **Deflection :** $D = \frac{L l_d E_d}{2d E_a}$

where, $L \rightarrow$ Distance between screen and the centre of deflecting plates

$l_d \rightarrow$ Length of deflecting plates

$E_d \rightarrow$ Potential between deflecting plates

$d \rightarrow$ Distance between deflecting plates

$E_a \rightarrow$ Voltage of accelerating anode

- **Deflection sensitivity :** $S = \frac{D}{E_d} = \frac{L l_d}{2d E_a}$ m/V

- **Deflection factor :** $G = \frac{1}{S} = \frac{2d E_a}{L l_d}$ V/m

Sawtooth Generator

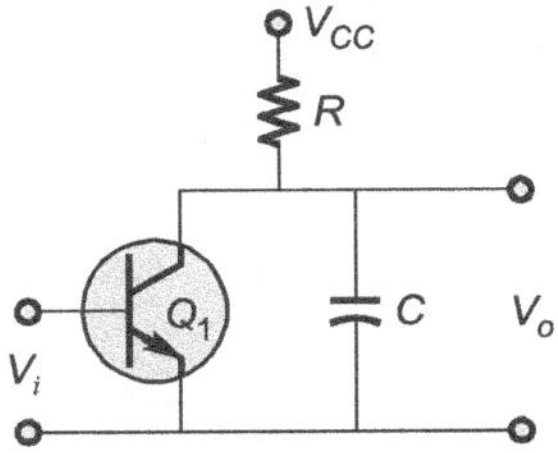

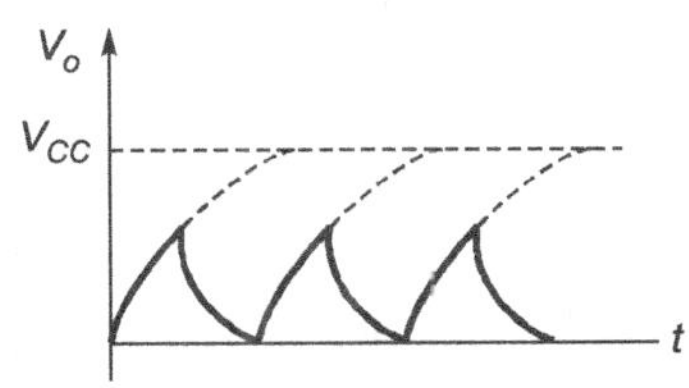

$$V_0 = V_{cc}[1 - \exp(-t / RC)]$$

where, $V_0 \rightarrow$ Instantaneous voltage across the capacitor at time t

$V_{CC} \rightarrow$ Supply voltage

Lissajous patterns

If horizontal and vertical deflecting plate are applied with sinusoidal signal, the waveform pattern appearing on the screen is called Lissajous pattern.

Application

- Used for finding the phase angle difference between the signal applied to vertical and horizontal plate.
- Used for finding the frequency ratio between vertical and horizontal plates voltage.

Phase angle (ϕ) between V_x and V_y	Lissajous pattern
0° or 360°	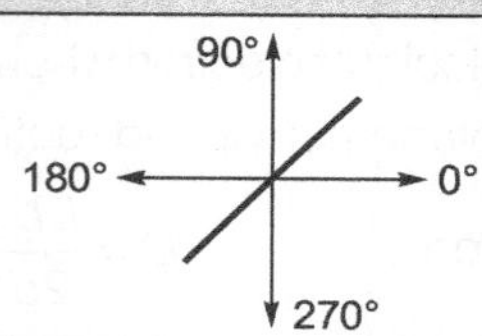
0° < ϕ < 90° (or) 270° < ϕ < 360°	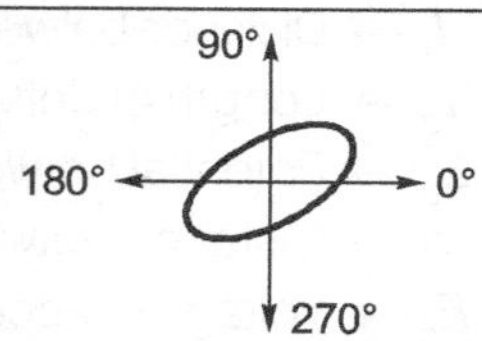
ϕ = 90° or 270°	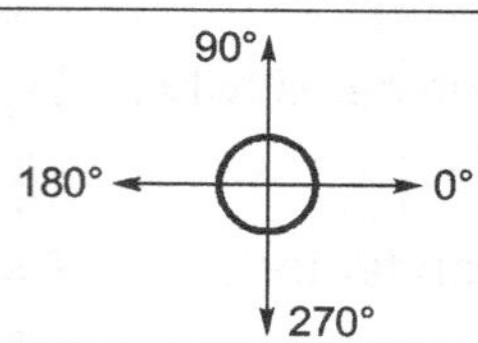
90° < ϕ < 180° (or) 180° < ϕ < 270°	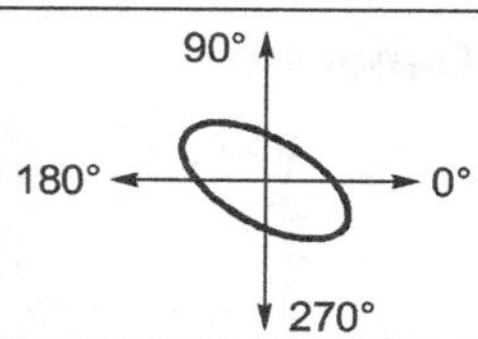
ϕ = 180°	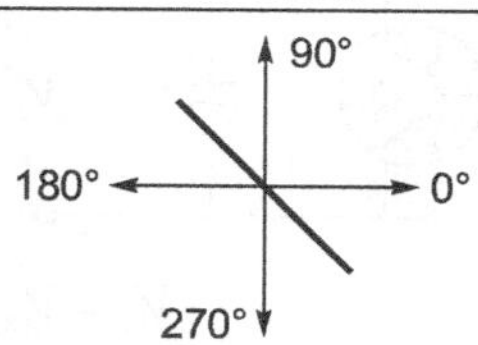

Finding the phase angle ϕ from given Lissajous pattern

When Lissajous pattern is in first and third quadrant

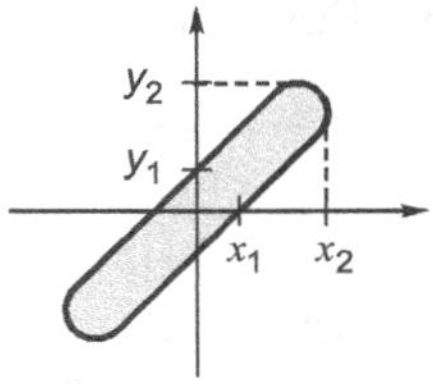

- First possibility : $\phi = \sin^{-1}\left(\frac{x_1}{x_2}\right) = \sin^{-1}\left(\frac{y_1}{y_2}\right)$
- Second possibility = $360° - \phi$

When Lissajous pattern is in second and fourth quadrant

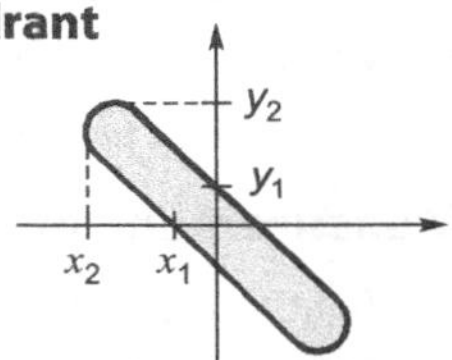

- First possibility : $\phi = 180° - \sin^{-1}\left(\frac{x_1}{x_2}\right)$
- Second possibility = $360° - \phi$

Measurement of Frequency Using Lissajous Pattern (for open shape)

$$\frac{f_y}{f_x} = \frac{\text{(number of intersections of the horizontal line with the curve)}}{\text{(number of intersections of the vertical line with the curve)}}$$

where, $f_y \rightarrow$ Frequency of signal applied to Y plates
$f_x \rightarrow$ Frequency of signal applied to X plates

XII High Frequency Measurements

Q-meter

It works on the principle of series resonance.

Measurement of the Storage Factor-Q

Resonant frequency of Q-Meter	$f_0 = \frac{1}{2\pi\sqrt{LC}}$
Measured value of Q	$Q_m = \frac{\omega_0 L}{R + R_{sh}}$
True value of Q	$Q_t = \frac{\omega_0 L}{R}$ $Q_t = Q_m\left(1+\frac{R_{sh}}{R}\right) = Q_m\left(1+\frac{C_d}{C}\right)$
Measurement of inductance	$L = \frac{1}{4\pi^2 f_o^2 C}$
Measurement of effective resistance	$R = \frac{\omega_o L}{Q_t}$

where, $R \rightarrow$ Resistance of coil; $L \rightarrow$ Inductance of coil;
$R_{sh} \rightarrow$ Shunt resistance; $C \rightarrow$ Tuning capacitance
$C_d \rightarrow$ Distributed or self-capacitance

Measurement of Distributed or self-capacitance

- Resonance frequency

$$f_1 = \frac{1}{2\pi\sqrt{L(C_1 + C_d)}} \quad ; \quad f_2 = \frac{1}{2\pi\sqrt{L(C_2 + C_d)}}$$

when, $f_2 = nf_1$ then, $C_d = \dfrac{C_1 - n^2 C_2}{n^2 - 1}$

where, $C_1 \rightarrow$ Tuning capacitance at frequency f_1

$C_2 \rightarrow$ Tuning capacitance at frequency f_2

Measurement of Unknown Capacitance C_x

Adjust capacitor $C = C_1$ to get resonance frequency f_1 with unknown capacitance C_x in parallel.

$$f_1 = \frac{1}{2\pi\sqrt{L(C_x + C_1)}} \quad \text{...(i)}$$

Now remove C_x and again adjust $C = C_2$ to get same resonance frequency f_1

$$f_1 = \frac{1}{2\pi\sqrt{LC_2}} \quad \text{...(ii)}$$

By equating equation (i) and (ii), $C_x = C_2 - C_1$

XIII Transducers

Transducer Classifications

Transducer is a device which converts one form of energy to another. They are classified as :

- **Primary and Secondary**

  Primary transducers are those devices which actually sense the parameters under measurement. Ex: Diaphragm, bellow etc.

  Secondary transducers are those which take the output of primary transducer and convert it into analogous signal. Ex: LVDT.

- **Active and Passive**

  Active Transducers are those which do not require any external source of supply. Ex. Thermocouple, quartz, photo voltaic cell.

  Passive transducers are those which require an external source of electrical supply. Ex: LVDT, strain gauges.

Strain Gauge

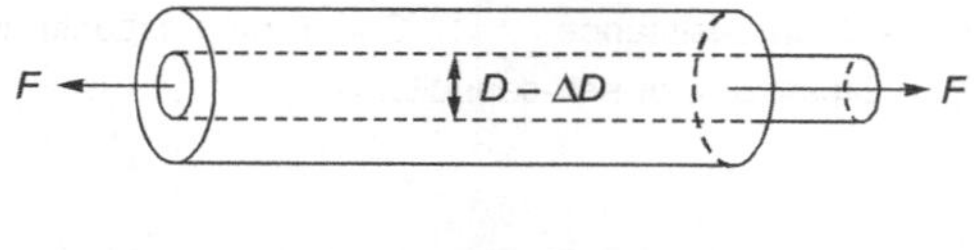

- **Gauge factor of strain gauge**

$$G_f = \frac{\Delta R / R}{\Delta L / L} = 1 + 2\nu + \frac{\Delta\rho / \rho}{\in}$$

where, $\frac{\Delta\rho}{\rho} \rightarrow$ Per unit change in resistivity

$\nu \rightarrow$ Poisson's ratio ; $\in \rightarrow$ Strain

Note: ....

For metal ; $\frac{\Delta\rho}{\rho} \rightarrow 0$; $\quad G_f \approx 1 + 2\nu$

....

- **Poisson's ratio**

$$\text{Poisson's ratio}(\nu) = \frac{\text{Lateral strain}}{\text{Longitudinal strain}} = \frac{-\Delta D / D}{\Delta L / L}$$

- **Strain**

$$\text{Strain } (\in) = \frac{\Delta L}{L} \quad \text{where, } \frac{\Delta L}{L} \rightarrow \text{Per unit change in length}$$

Thermistor

- Resistance of thermistor : $R_{T_1} = R_{T_2} \exp\left[\beta\left(\frac{1}{T_1} - \frac{1}{T_2}\right)\right]$

  where, $R_{T_1} \rightarrow$ Resistance of thermistor at absolute temperature T_1 ; °K

  $R_{T_2} \rightarrow$ Resistance of thermistor at absolute temperature T_2 ; °K

  $\beta \rightarrow$ A constant depending upon the material of thermistor

- Thermistor resistance : $R_T = aR_o \exp(\beta / T)$

  where, $R_T, R_o \rightarrow$ Resistance of thermistor at temperature T °K and ice point respectively

Thermocouple

E.M.F. produced in a thermocouple

$$E = a(\Delta\theta) + (b / T)^2$$

where $\Delta\theta \rightarrow$ Difference in temperature between the hot thermocouple junction and the reference junction of thermocouple; °C

$a, b \rightarrow$ Constants

LVDT

- Output voltage, $\quad \Delta e_o = \left| e_{s_2} - e_{s_1} \right|$

- Sensitivity of LVDT, $\quad \text{Sensitivity} = \frac{\text{Output voltage}}{\text{Displacement}}$

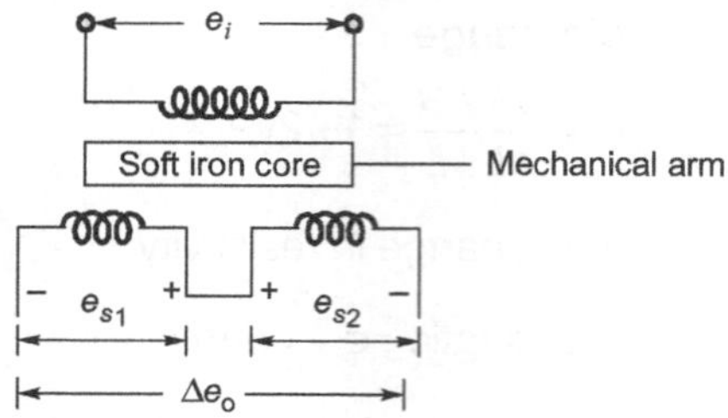

Capacitive Transducers

Capacitance

- Capacitance of parallel plate capacitor

$$C = \frac{\in A}{d} = \frac{\in xw}{d}$$

where, $A \rightarrow$ Overlapping area of plates
$x \rightarrow$ Length of overlapping part of plates
$w \rightarrow$ Width of overlapping part of plates
$d \rightarrow$ Distance between two plates
$\in \rightarrow$ Permittivity of medium

Piezo-Electric Transducer

Voltage sensitivity of crystal	$g = \frac{\text{Electric field}}{\text{Stress}} = \frac{E}{P}$ Vm/N
Charge sensitivity	$d = \in_r \in_0 g$ C/N
Output voltage	$e_o = gtP$

where, $P \rightarrow$ Pressure or stress; N/m^2
$E \rightarrow$ Electric field;
$t \rightarrow$ Thickness of crystal; m

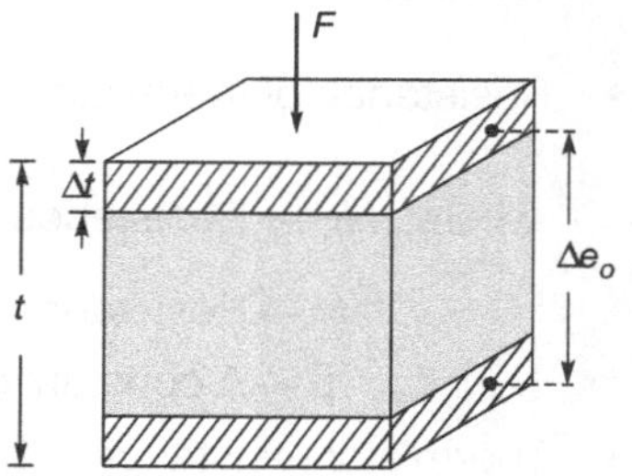

Note:

It is not used under static condition.

XIV Digital Instrumentation

Introduction

The advantages of digital instruments are:

(i) The output is in *BCD* format that can be directly fed to computer for processing.
(ii) Observational errors are minimized.
(iii) They have a very high accuracy and resolution.
(iv) Operational power consumption is zero.
(v) Loading effects are eliminated.
(vi) High speed of response.

Digital Voltmeters

- **Ramp Type Digital Voltmeter**

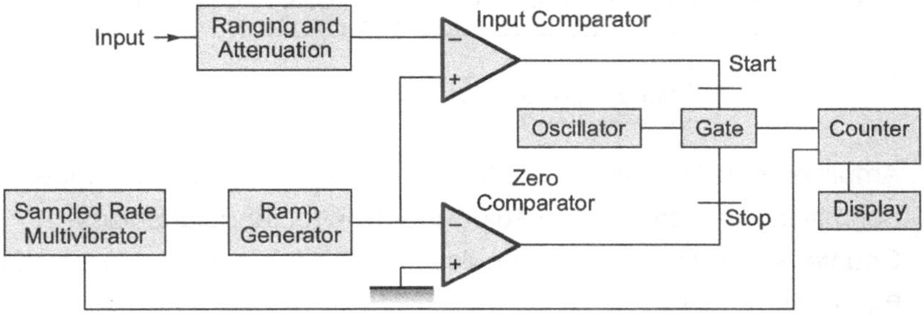

- Voltage to time conversion is the basic principle.
- High level of accuracy.
- **Integrating-dual slope digital voltmeter**

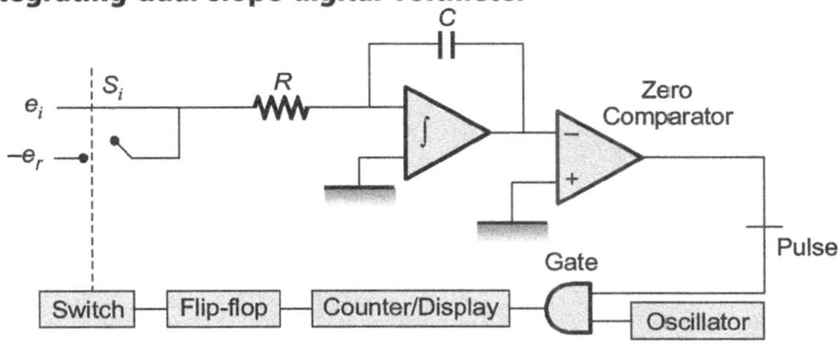

- Voltage to time conversion is followed.
- The count stored in the counter represents the magnitude of input.
- The input voltage is given by, $e_i = e_r \times \frac{t_2}{t_1}$

  where, $e_r \rightarrow$ Reference voltage ; $t_1 \rightarrow$ Time taken by counter to reach full count ; $t_2 \rightarrow$ Discharging time

Resolution of a Digital Voltmeter

Resolution is minimum value of input that can be clearly indicated at the output. The output of digital instrument is specified in term of $N\frac{1}{2}$ digit display.

$$\text{Resolution of meter} = \frac{1}{10^N} \times \text{Full scale reading}$$

Digital Frequency Meter

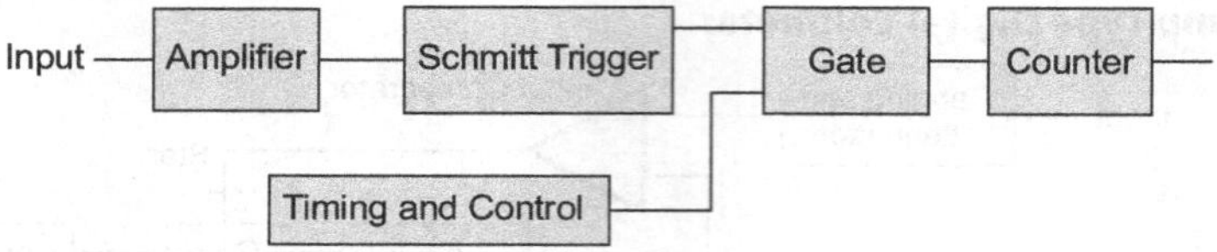

- Amplifier amplifies the input to the level required by Schmitt trigger.
- Schmitt trigger converts sinusoidal input the rectangular pulses.
- Counter stores the number of pulses reaching it in 1 second which is equal to frequency.

■■■■

5 Network Theory

I Basic Definitions and Circuits Element

Definitions

1. **Charge:** It is an electrical property of an atomic particle of which the matter consists of and its unit is Coulombic (C).

   It can be positive or negative : Charge of $1e^- = 1.6 \times 10^{-19}$ C

2. **Current:** The flow of electron or the time rate of change of charge is called current.

$$\boxed{i = \frac{dq}{dt}} \text{ Ampere}$$

3. **Voltage:** It is an external force required to move the electron from one point to another in a particular direction in an electrical circuit. This force is provided by the electromotive force (emf).

$$\boxed{V = \frac{dw}{dq}} \text{ J/C or Voltage}$$

4. **Power:** It is the time rate of change of energy.

$$\boxed{P = \frac{dw}{dt} = v(t) \cdot i(t)}$$

   Its unit is watts.

5. **Law of Conservation of Charge:** It states that the charge can neither be created nor be destroyed. It can only be transferred from one body to another body.

6. **Law of Conservation of Energy:** The sum of total instantaneous power in a circuit must be equal to zero.

$$\Sigma P = 0$$

Electrical Network Components

Resistor

It is the property of a substance due to which it opposes the flow of current (i.e., electrons) through it.

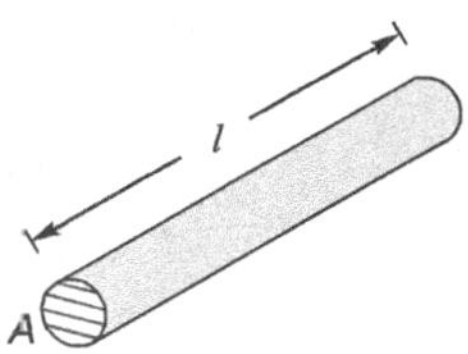

Resistance, $R = \rho\frac{l}{A}$, Ohm (Ω)

where, $l \rightarrow$ Length of conductor, metre (m)
$A \rightarrow$ Area of cross-section, m^2
$\rho \rightarrow$ Resistivity of the material, Ω-m

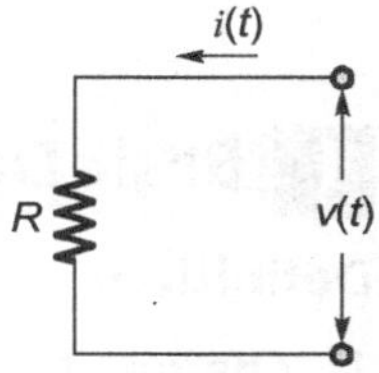

$v(t) = Ri(t)$; $i(t) = \frac{1}{R}v(t) \cdots$ in time domain

$V(s) = RI(s)$; $I(s) = \frac{1}{R}V(s) \cdots$ in s-domain

- Power loss in resistor, $p(t) = v(t)\, i(t) = i^2(t) = \frac{v^2(t)}{R}$
- Energy dissipated in resistor, $E_R = \int_{t_1}^{t_2} p(t)\,dt$

Combination of Resistors

Resistors in series

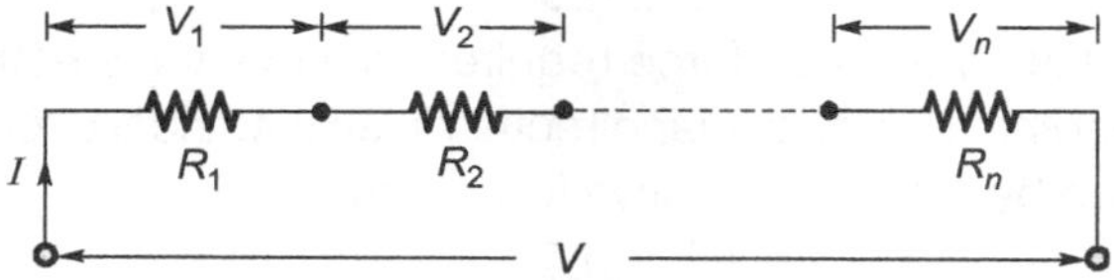

$V = I R_{eq}$; $R_{eq} = R_1 + R_2 + ... + R_n$

Resistors in parallel

$$I = \frac{V}{R_{eq}}$$

$$\frac{1}{R_{eq}} = \frac{1}{R_1} + \frac{1}{R_2} + ...\frac{1}{R_n}$$

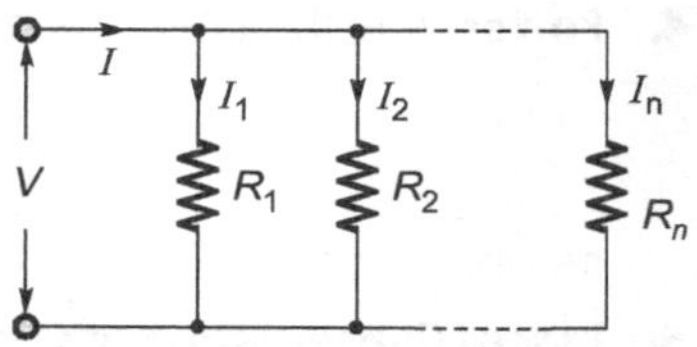

Capacitor

The circuit element that stores energy in the form of electric field is a capacitor or capacitance.

Capacitance $C = \frac{\epsilon_0 \epsilon_r A}{d}$ Farad(F)

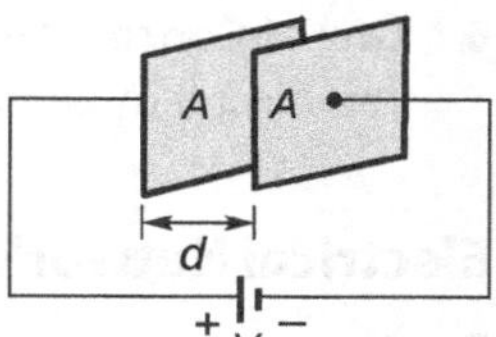

where $\epsilon_0 \rightarrow$ Permittivity of free space, F/m ,
$\epsilon_r \rightarrow$ Relative permittivity of the dielectric ; $A \rightarrow$ Cross-sectional area of parallel plates, m^2
$d \rightarrow$ Separation of plates, m

Note: ....

When the capacitance is removed from source, the capacitor retains the charge and the electric field until a discharge path is provided.

....

$$v(t) = \frac{1}{C}\int_0^t i(t)\,dt \;;\; i(t) = C\frac{dv(t)}{dt} \cdots \text{in time domain}$$

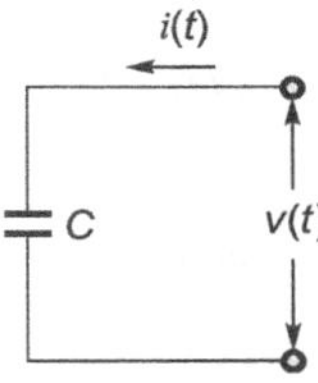

Power in capacitance, $p(t) = v(t)\,i(t) = Cv(t)\,\dfrac{dv(t)}{dt}$

Energy stored in capacitance, $E_c = \dfrac{1}{2}Cv^2$

Combination of Capacitors

- **Capacitors in series**

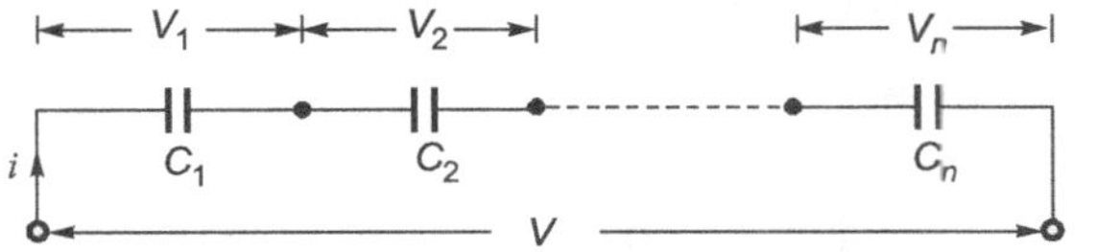

$$V = \frac{1}{C_{eq}}\int_0^t i\,dt \quad ; \quad \frac{1}{C_{eq}} = \frac{1}{C_1} + \frac{1}{C_2} + \frac{1}{C_3} + \ldots.. + \frac{1}{C_n}$$

- **Capacitors in parallel**

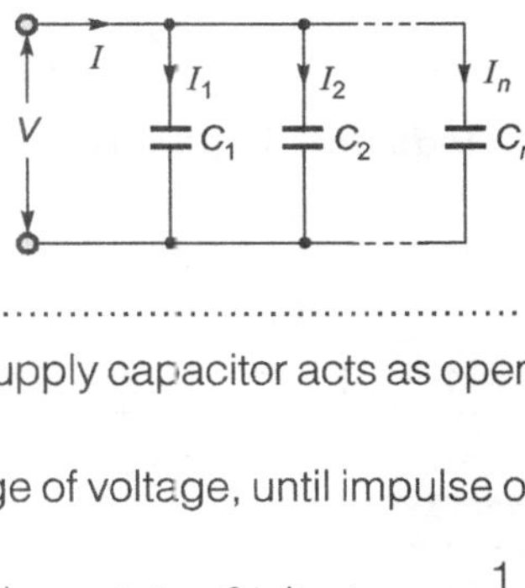

$$I = C_{eq}\frac{dv}{dt}$$

$$C_{eq} = C_1 + C_2 + C_3 + \ldots\ldots.. + C_n$$

Note: ....

☑ Under steady state condition for D.C. supply capacitor acts as open circuit.

☑ Capacitor doesn't allow sudden change of voltage, until impulse of current is applied.

☑ If unit impulse of current is applied then at $t = 0^+$ it stores $\dfrac{1}{2C}$ Joules energy.

☑ Power dissipation in ideal capacitor is zero.

....

Inductor

The circuit element that stores energy in the form of a magnetic field is an inductor or inductance.

Inductance $\qquad L = \dfrac{\mu_0 N^2 A}{l}$ Henry (H)

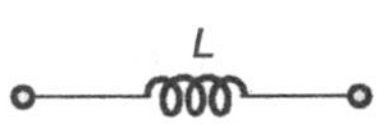

where $\mu_0 \rightarrow$ Permeability of free space, H/m ; $N \rightarrow$ Total number of turns in coil ; $A \rightarrow$ Cross-sectional area of coil, m^2 ; $l \rightarrow$ Length of coil, m

Note:

When the inductance is removed from the source, the magnetic field will collapse, i.e., no energy is stored without a connected sources.

$$v(t) = L\frac{di(t)}{dt} \; ; \; i(t) = \frac{1}{L}\int_0^{\infty} v(t)\,dt \quad \text{..... in time domain}$$

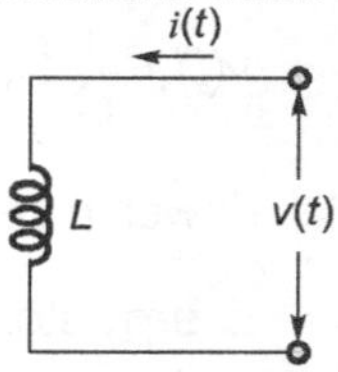

- Power in inductance, $p(t) = v(t)\,i(t) = Li(t)\dfrac{di(t)}{dt(t)}$
- Energy stored in an inductance, $E_L = \dfrac{1}{2}Li^2(t)$

Combination of Inductor

- **Inductors in series**

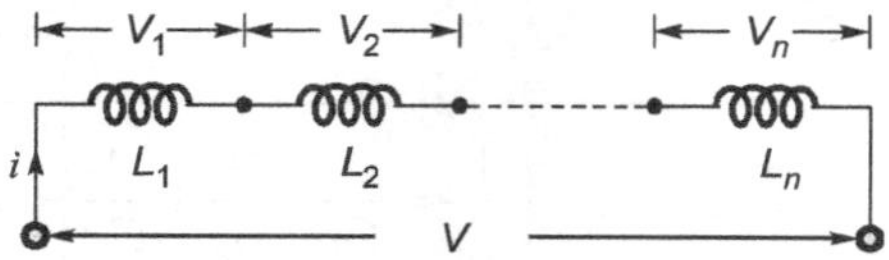

$$V = L_{eq}\frac{di}{dt} \quad ; \quad L_{eq} = L_1 + L_2 + L_3 + \ldots\ldots + L_n$$

- **Inductors in parallel**

$$I = \frac{1}{L_{eq}}\int_0^t v\,dt$$

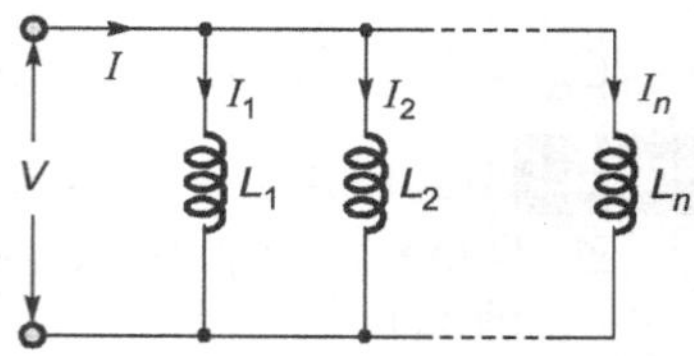

$$\frac{1}{L_{eq}} = \frac{1}{L_1} + \frac{1}{L_2} + \frac{1}{L_3} + \ldots\ldots \frac{1}{L_n}$$

Note:

☑ Under steady state condition for D.C. supply, inductor acts as short circuit.

☑ Inductor doesn't allow sudden change of voltage, until impulse of voltage is applied.

☑ If unit impulse of voltage is applied across it then at t = 0^+ it stores energy = $\dfrac{1}{2L}$ Joules.

☑ Power dissipation in ideal inductor is zero.

Ideal Transformer

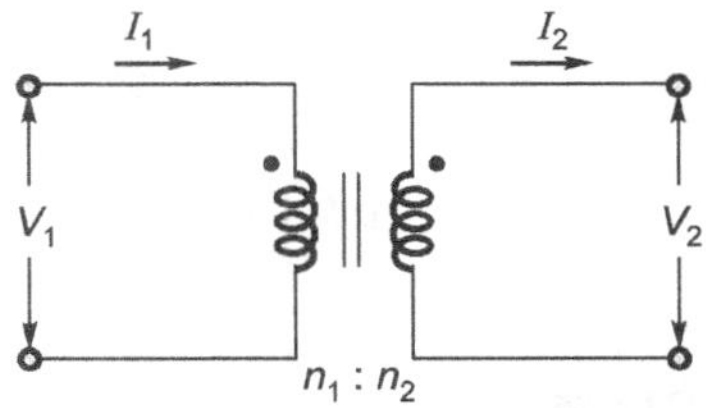

$$\frac{V_2}{V_1} = \frac{n_2}{n_1} = \frac{I_1}{I_2}$$

If $n_1 < n_2$, step up transformer

and $n_1 > n_2$, step down transformer

Gyrator

Gyrator shows an impedance inversion.

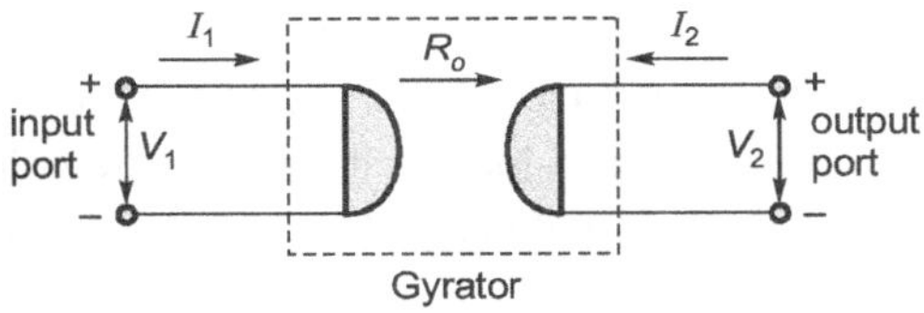

$$V_1 = +R_o I_2$$
$$V_2 = -R_o I_1$$

where, $R_o \rightarrow$ Coefficient of gyrator and it depends upon

(i) Op-amp parameter (ii) Externally connected R and C values

Remember:

- ☑ Linear network is one which holds the principle of superposition and principle of homogeneity both.
- ☑ Element which conducts in both directions is called bilateral element.
- ☑ A network containing circuit elements without any energy source is called passive network. e.g. R, L, C.
- ☑ A network containing energy source together with other circuit elements is called active network. e.g. Op-Amp, transistors.

Ideal Sources

Ideal Voltage Source

Voltage always remains constant for any value of current passing through it.

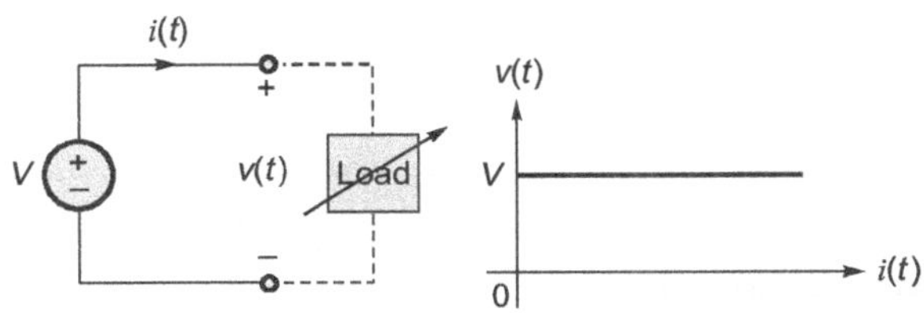

Practical Voltage Source

It has small internal resistance. Voltage across the element varies with respect to current.

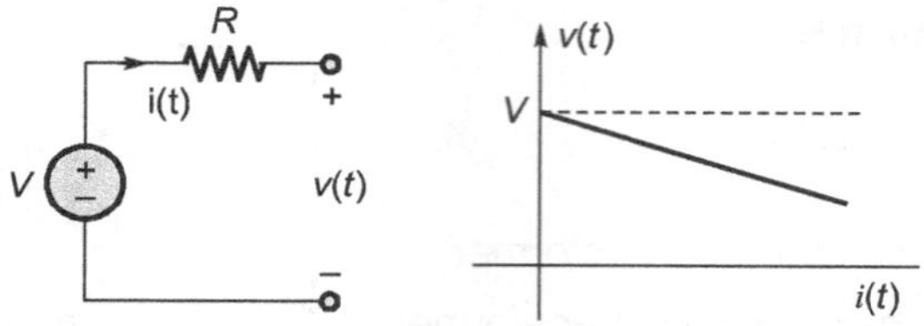

Note:

☑ The current through any voltage source is purely arbitrary. It will depend upon
   (a) The current source which is connected in series with it.
   (b) The load resistance which is connected in parallel with it.

☑ We cannot write KCL equation at a load with which any voltage source is connected as the current through this voltage source is purely arbitrary.

Ideal current source

Current always remains constant for any value of voltage across it

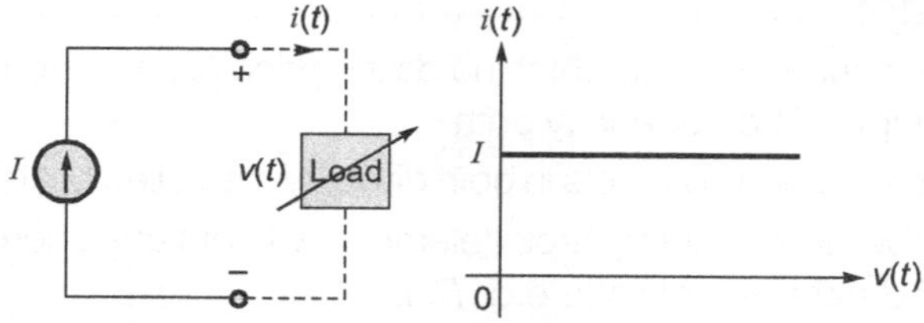

Practical current source

It has large value of internal resistance. Current varies with respect to the voltage across the element.

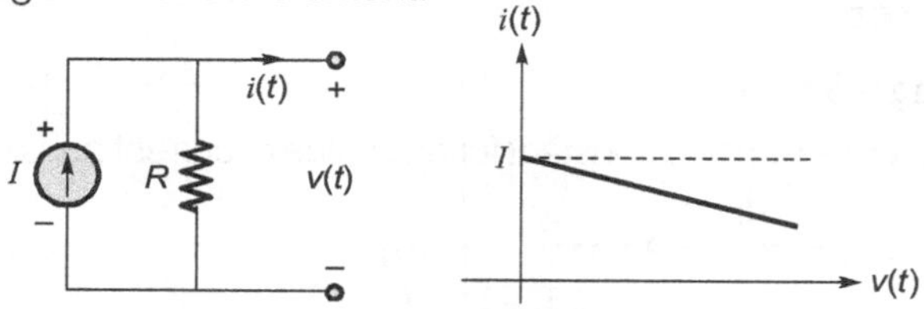

Note:

☑ The voltage across current source is purely arbitrary. It will depend upon
   (a) Voltage source which is connected in parallel with it.
   (b) The load resistance which is connected in series with it.

☑ We cannot write any KVL equation in a loop in which a current source is present as the voltage across the current source is purely arbitrary.

Dependent Sources

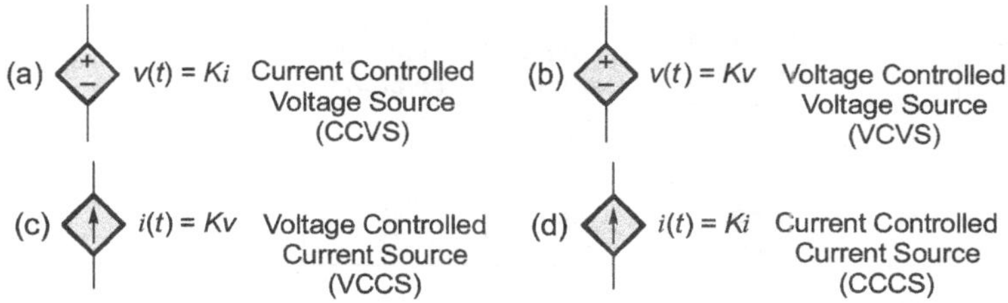

where, K is a constant.

Classification of Circuit Elements

Linear and non linear: Elements which obey Ohm's law are linear elements. Ex: R, L, C.

Elements which do not follow Ohm's law are non-linear.
Ex. Diode, transistor.

Active and passive: Elements which supply finite power for infinite time are active. Ex: ideal voltage and current sources.

Elements which either absorbs or supplies finite power only for finite time. Ex. R, L, C.

Unilateral and bilateral: Elements which allow flow of current only in one direction are unilateral. Ex. diode.

Elements which allow current flow in both the directions. Ex. R, L, C.

Properties of Periodic Signal

For periodic signal $y(t)$:

- Average value, $Y_{av} = \frac{1}{T}\int_0^T y(t)dt$

  where T is the time period of periodic function $y(t)$

- RMS or effective value, $Y_{rms} = \sqrt{\frac{1}{T}\int_0^T [y(t)]^2 dt}$

  If $y(t) = a_0 + (a_1 \cos\omega t + a_2 \cos 2\omega t + ...) + (b_1 \sin \omega t + b_2 \sin 2\omega t + ...)$

$$Y_{rms} = \sqrt{a_0^2 + \frac{1}{2}(a_1^2 + a_2^2 + ...) + \frac{1}{2}(b_1^2 + b_2^2 + ...)}$$

- Form factor, $FF = \frac{Y_{rms}}{Y_{av}} = \frac{\sqrt{\frac{1}{T}\int_0^T [y(t)]^2 dt}}{\frac{1}{T}\int_0^T y(t)dt}$

- Peak factor, $PF = \frac{Y_{\max}}{Y_{\text{rms}}}$

Remember:

-

	Pure sine wave	Half-wave rectified waveform	Full-wave rectified waveform
Y_{av}	0	$\frac{Y_{max}}{\pi}$	$\frac{2Y_{max}}{\pi}$
Y_{rms}	$\frac{Y_{max}}{\sqrt{2}}$	$\frac{Y_{max}}{\sqrt{2}}$	$\frac{Y_{max}}{\sqrt{2}}$
FF	$\frac{\pi}{2\sqrt{2}}$	$\frac{\pi}{2}$	$\frac{\pi}{2\sqrt{2}}$
PF	$\sqrt{2}$	2	$\sqrt{2}$

- $\sin \omega t = \dfrac{e^{j\omega t} - e^{-j\omega t}}{2j}$ and $\cos \omega t = \dfrac{e^{j\omega t} + e^{-j\omega t}}{2}$

  $\sinh \omega t = \dfrac{e^{\omega t} - e^{-\omega t}}{2}$ and $\cosh \omega t = \dfrac{e^{\omega t} + e^{-\omega t}}{2}$

Miscellaneous

Average value, $Y_{av} = \dfrac{1}{T}\int_0^T y(t)dt$

where T is the time period of periodic function $y(t)$

RMS or effective value, $Y_{rms} = \sqrt{\dfrac{1}{T}\int_0^T [y(t)]^2 dt}$

- If $y(t) = a_0 + (a_1 \cos\omega t + a_2 \cos 2\omega t + \ldots) + (b_1 \sin\omega t + b_2 \sin 2\omega t + \ldots)$

$$Y_{rms} = \sqrt{a_0^2 + \frac{1}{2}(a_1^2 + a_2^2 + \ldots) + \frac{1}{2}(b_1^2 + b_2^2 + \ldots)}$$

Form factor : $\mathrm{FF} = \dfrac{Y_{rms}}{Y_{av}} = \dfrac{\sqrt{\dfrac{1}{T}\int_0^T [y(t)]^2 dt}}{\dfrac{1}{T}\int_0^T y(t)dt}$

Remember:

☑ RMS value of $\sin\omega t$ and $\cos\omega t$ is $\dfrac{1}{\sqrt{2}}$.

☑ $\sin \omega t = \dfrac{e^{j\omega t} - e^{-j\omega t}}{2j}$ and $\cos \omega t = \dfrac{e^{j\omega t} + e^{-j\omega t}}{2}$

$\sinh \omega t = \dfrac{e^{\omega t} - e^{-\omega t}}{2}$ and $\cosh \omega t = \dfrac{e^{\omega t} + e^{-\omega t}}{2}$

Representation of Shifted Unit Step Function

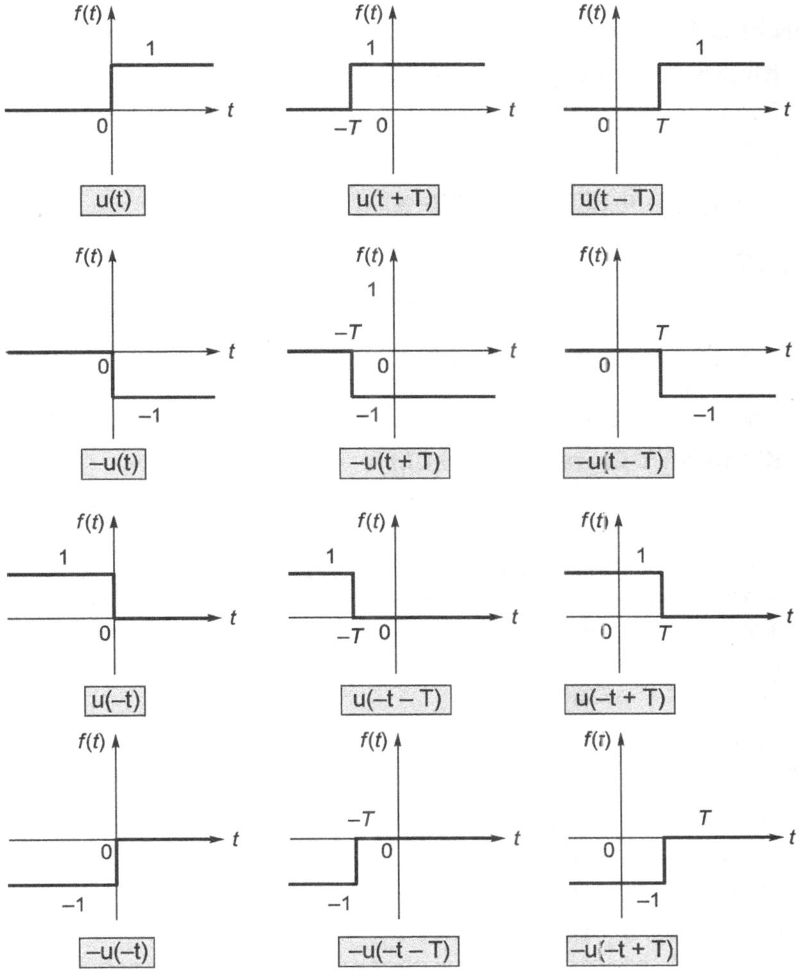

II Network Laws and Theorems

Ohm's Law

The ratio of potential difference (V) between any two points on a conductor to the current (I) flowing through them is constant, provided the temperature of the conductor does not change.

$$\frac{V}{I} = \text{constant} \quad \text{or} \quad \frac{V}{I} = R$$

where, R is the resistance of the conductor between the two points considered.

Kirchhoff's Laws

Kirchhoff's Voltage Law (KVL)

For any closed path in a network, the algebraic sum of the voltages is zero.

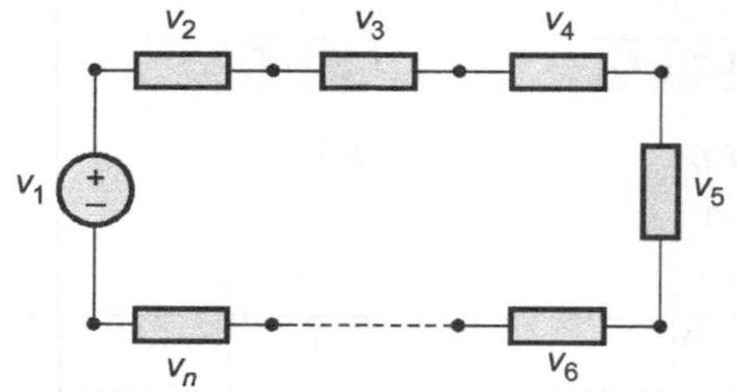

$$\sum_{k=1}^{n} v_k(t) = 0$$ .... in a closed loop

where, v_k is the voltage drop or voltage gain across k^{th} element

Kirchhoff's Current Law (KCL)

The algebraic sum of the currents at a node is zero. Alternatively the sum of the currents entering a node is equal to the sum of the currents leaving that node.

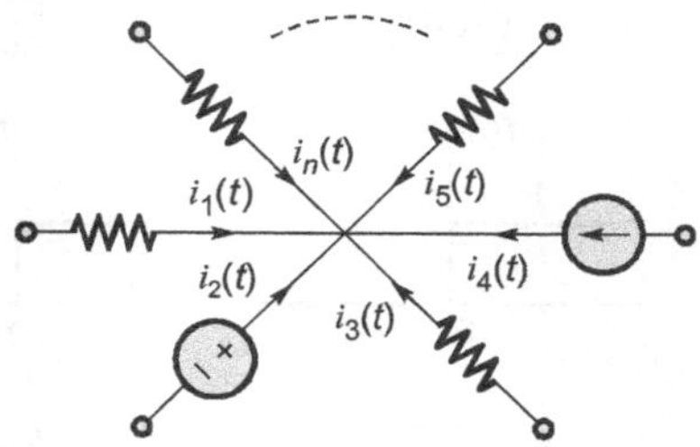

$$\sum_{k=1}^{n} i_k(t) = 0$$ ..... at any node

where $i_k(t)$ is the current through k^{th} branch

Note:

- ☑ A network is an interconnection of elements or devices, whereas a circuit is a network providing one or more closed paths.
- ☑ Number of KVL equations = $b - (n - 1)$
- ☑ Number of KCL equations = $(n - 1)$
  where, b is number of branches and n is number of nodes.
- ☑ KVL is based on the law of conservation of energy.
- ☑ KCL is based on the law of conservation of charge.
- ☑ KVL and KCL are applicable to lumped networks.

Voltage Division Equations

$$V_1 = \frac{R_1}{R_1 + R_2} V \quad ; \quad V_2 = \frac{R_2}{R_1 + R_2} V$$

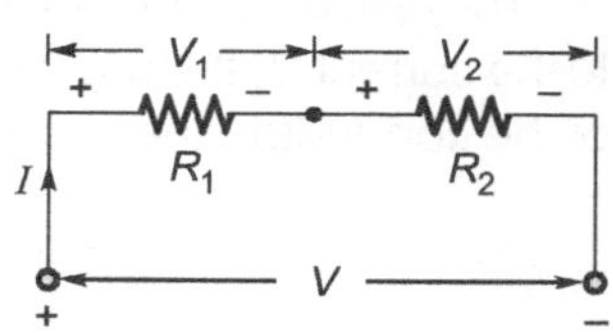

Current Division Equations

$$I_1 = \frac{R_2\, I}{R_1 + R_2}$$

$$I_2 = \frac{R_1\, I}{R_1 + R_2}$$

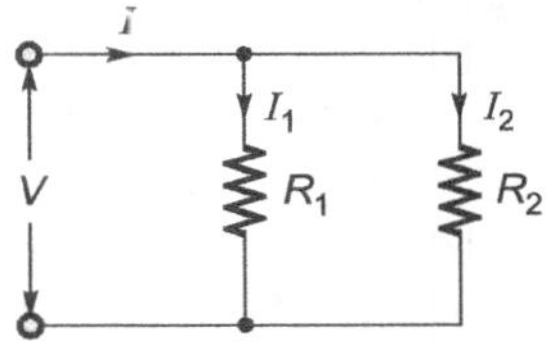

Star to Delta Transformation

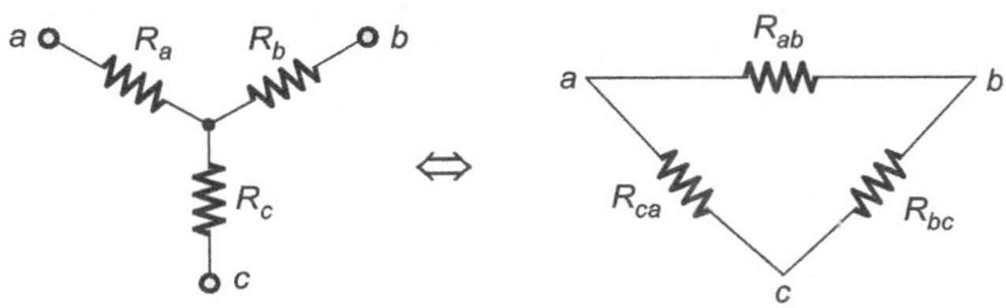

$$R_{ab} = \frac{\Delta}{R_c},\quad R_{bc} = \frac{\Delta}{R_a},\quad R_{ca} = \frac{\Delta}{R_b}$$ where, $\Delta \rightarrow (R_aR_b + R_bR_c + R_cR_a)$

Delta to Star Transformation

$$R_a = \frac{R_{ca}R_{ab}}{R_{ab} + R_{bc} + R_{ca}}\ ;\ R_b = \frac{R_{ab}R_{bc}}{R_{ab} + R_{bc} + R_{ca}}\ ;\ R_c = \frac{R_{bc}R_{ca}}{R_{ab} + R_{bc} + R_{ca}}$$

Source Transformation

Transformation of a resistive voltage source to a resistive current source or vice-versa.

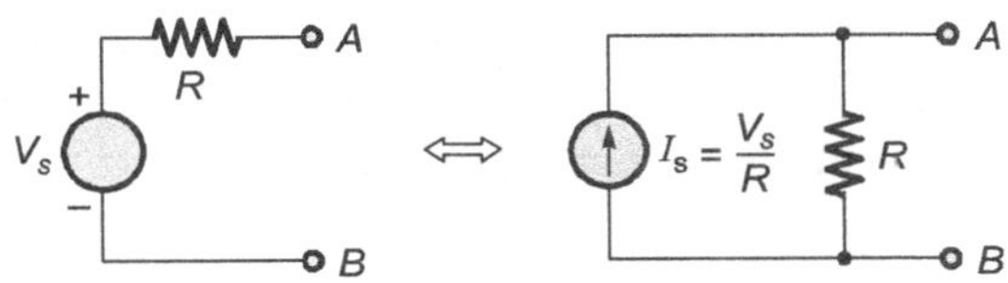

Network Theorems

Superposition Theorem

The response in any element of a linear, bilateral RLC network containing more than one independent voltage or current source is the algebraic sum of responses produced by the independent source when each of them is acting alone with

(a) All other independent voltage sources are short circuited (S.C.).

(b) All other independent current sources are open circuited (O.C.).

(c) All dependent voltage and current sources remain as they are and therefore, they are neither S.C. nor O.C.

Note:

- ☑ The theorem is not applicable to the network containing
  - (a) Non linear elements.
  - (b) Unilateral elements such as diode or BJT.
- ☑ The theorem is not applicable to power since it is a non linear parameter.
- ☑ The theorem is also applicable for circuit having initial condition.

Thevenin's Theorem

A linear active RLC network which contains one or more independent or dependent voltage or current sources can be replaced by a single voltage source V_{OC} in series with equivalent impedance Z_{eq}.

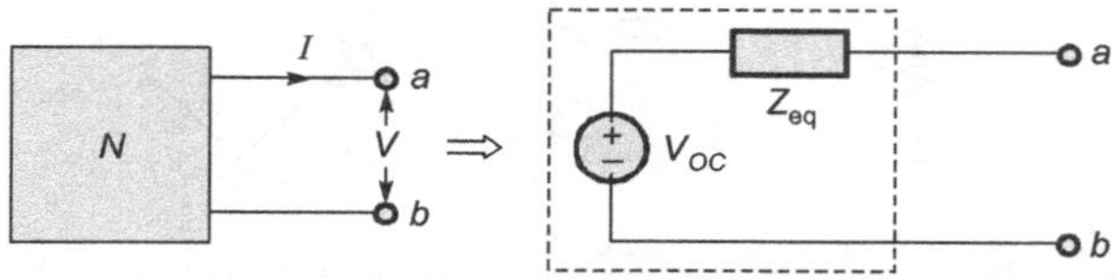

where, V_{OC} → Open circuit voltage between a and b (when $I = 0$).
Z_{eq} → Equivalent impedance between a and b, when

(a) All independent sources are replaced by their internal impedances.
(b) All dependent voltage and current sources remain as they are.

Note:

Theorem is not applicable to the network containing :

- Non linear element.
- Unilateral element.

Norton's Theorem

A linear, active RLC network which contains one or more independent or dependent voltage or current sources can be replaced by a single current source I_{SC} in shunt with equivalent impedance Z_{eq}.

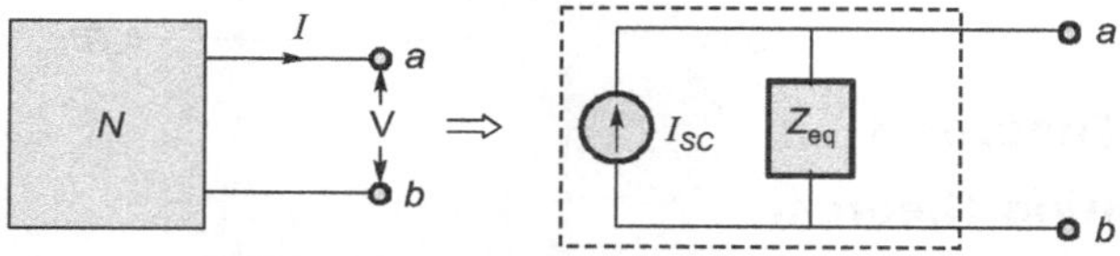

where, I_{SC} → Short circuit current between a and b (when $V = 0$)
Z_{eq} → Same as that of Thevenin's theorem

Maximum Power Transfer Theorem

In any linear bilateral network, the power transferred to the load will be maximum when the load resistance is equal to Thevenin's equivalent resistance.

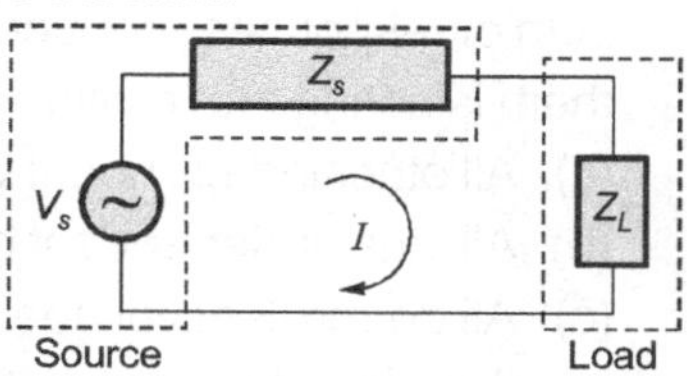

$Z_L = Z_S^*$ ......for maximum power transfer

Case 1: If $Z_s = R_s + jX_s$ and $Z_L = R_L + jX_L$

then $R_L = R_s$ and $X_L = -X_s$

Case 2: If $Z_s = R_s + jX_s$ and $Z_L = R_L$ then $R_L = \sqrt{R_s^2 + X_s^2}$

Case 3: If $Z_L = R_L$ and $Z_s = R_s$ then $R_L = R_s$

Note:

☑ Efficiency at maximum power transfer is 50%.

☑ If Z_s(source resistance) is varied then maximum power is transferred to the load if $Z_s = 0$.

Tellegen's Theorem

- In any network, the sum of instantaneous power consumed by various elements of the branches is always equal to zero.
- Total power supplied by different voltage sources is equal to total power consumed by various passive elements in various branches of the network.

$$\sum_{k=1}^{b} v_k \cdot i_k = 0 \qquad \text{where, } b \rightarrow \text{Number of branches}$$

Note:

☑ The theorem is valid for any type of network as long as KVL and KCL equations are valid.

Millman's Theorem

When number of voltage sources ($E_1, E_2 \cdots E_n$) are in parallel with internal resistances $\left(\frac{1}{Y_1}, \frac{1}{Y_2} \cdots \frac{1}{Y_n}\right)$, the arrangement can be replaced by a single equivalent voltage source E_{eq} with an equivalent series resistance $\left(\frac{1}{Y_{eq}}\right)$.

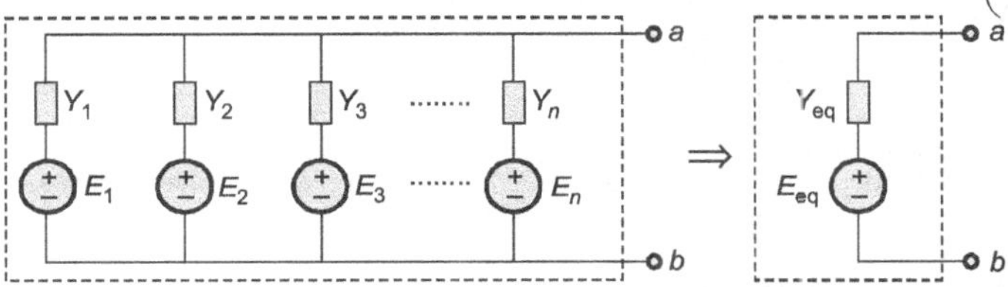

$$E_{eq} = \frac{\sum_{i=1}^{n} E_i Y_i}{\sum_{i=1}^{n} Y_i} \quad \text{and} \quad Y_{eq} = \sum_{i=1}^{n} Y_i$$

Reciprocity Theorem

In a linear bilateral single source network, the ratio of excitation to the response is constant even when the position of excitation and response are interchanged.

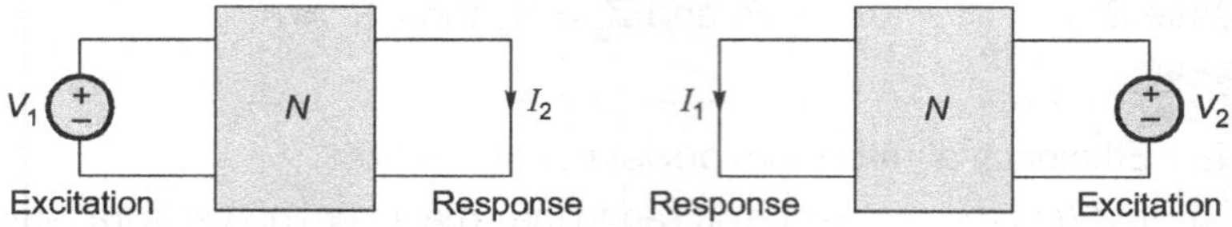

$$\frac{V_1}{I_2} = \frac{V_2}{I_1} \; ; \; Z_{12} = Z_{21}$$

Note:

- ☑ The basis of the theorem is the symmetry of impedance or admittance of matrix.
- ☑ The theorem is valid for network in which linear and bilateral elements are present.
- ☑ The theorem is valid only when single independent voltage or current source is present.
- ☑ The initial conditions are assumed to be zero in reciprocity theorem.

Compensation Theorem

If impedance 'Z' of any branch of a network is changed by 'δZ', then the incremental current 'δI' in such branch is that which will be produced by a compensating voltage source $V_c = I\,\delta Z$ introduced in the same branch with polarity opposing the original direction of current I.

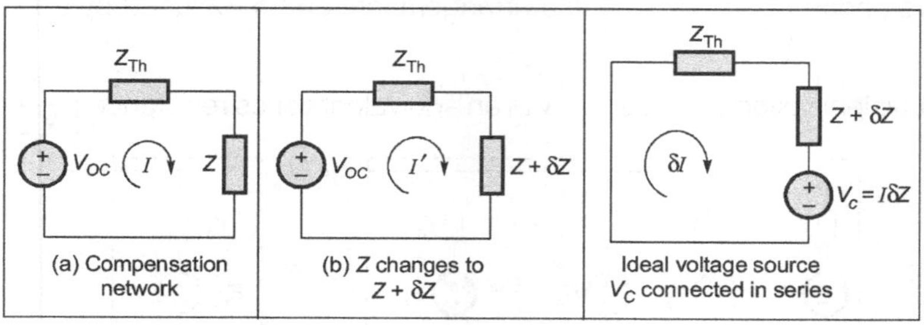

Substitution Theorem

Any branch of a bilateral network may be substituted by a different branch without disturbing the voltages and current in the entire network, provided the new branch has the same set of terminal voltage and current as the original branch.

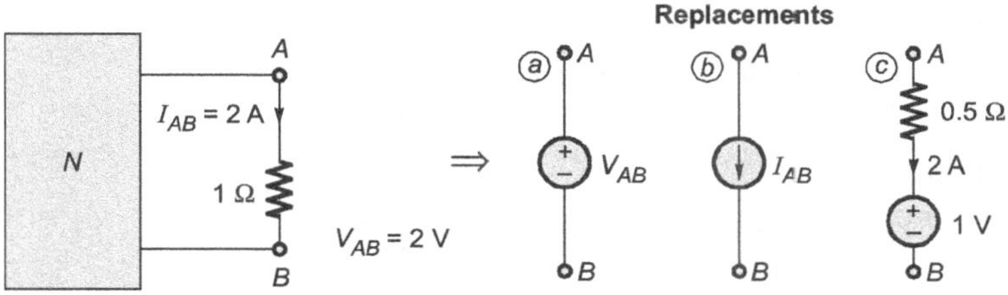

III Graph Theory

Network Topology

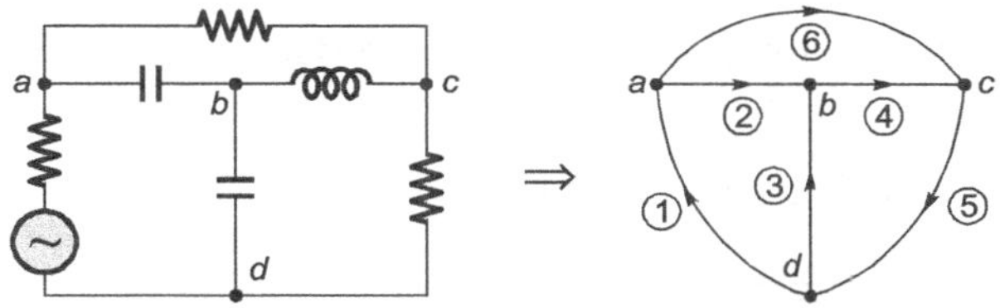

Circuit diagram of linear graph   Oriented graph

For the graph shown in the above figure

Node or vertices : [a b c d] ; Branch or edge: [1, 2, 3, 4, 5, 6]

- In fully connected graph, each node is connected to all other nodes of the graph.
- A closed loop or a closed circuit may not contain all the nodes of the graph.
- ***Degree*** of any node represents the number of branches which are connected to it.
- For a fully connected graph, the degree of each node is equal to the rank of the graph.
- Tree of the graph
  (a) It contain all the nodes of the graph.
  (b) If graph contains n nodes, its tree will contain $n-1$ branches.
  (c) There is no closed path and hence, a tree is circuitless.
  (d) The tree of a graph is not unique.
- The branches of the ***tree*** are represented by the tree branches or the ***twigs***, whereas the branches of the ***co-tree*** are represented by ***link*** or the ***chords***. A co-tree is a part of the directed graph that is not covered by the tree.

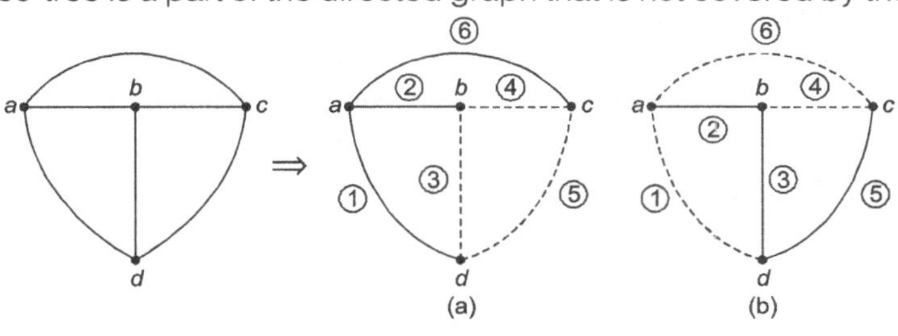

For figure (a) :

Tree: $\underbrace{[1, 2, 6]}_{\text{twig}}$ Co-tree $[\underbrace{3, 4, 5}_{\text{link or chords}}]$

For figure (b) :

Tree: $\underbrace{[2, 3, 5]}_{\text{twig}}$ Co-tree $[\underbrace{1, 4, 6}_{\text{link or chords}}]$

- (a) Number of trees equals to
  - (i) $n^{(n-2)}$ (for fully connected graph)
  - (ii) det $\left\{[A][A]^T\right\}$ (general expression)

  where, $n \rightarrow$ Number of nodes ; $A \rightarrow$ Reduced incident matrix (RIM)
  $[A]^T \rightarrow$ Transpose of RIM

  (b) Number of branches in a fully connected graph = $\dfrac{n(n-1)}{2}$

  (c) Number of tree branches
  = $n - 1$ = Number of KCL equations
  = Degree of each node of fully connected graph
  = Rank of graph
  = Number of fundamental cut-sets.

  (d) Number of link/chords
  = $b - (n - 1)$ = Number of KVL equations
  = Number of tie-sets

  where, $b \rightarrow$ Number of edges
- A tree can be used to solve the electrical network using:
  (a) tie-set schedule
  (b) cut-set schedule

Incidence Matrix

(i) This matrix translates all the geometrical feature of the graph into an algebraic expression.

(ii) Every graph has incidence matrix and vice-versa.

(iii) Each row of matrix contains +1, –1, 0 depending upon the orientation of the branches with the node.

•——➤—— = +1 (if branch is pointing away from node)
——➤——• = –1 (if branch is pointing towards a node)
• = 0 (if branch is not connected to the node)

(iv) Each column of matrix contain only one entry of +1 and only one entry of –1 so that the sum of the element of each column is always equal to zero.

(v) The determinant of the incidence matrix of a closed loop is equal to zero.
(vi) Order of matrix is $[n \times b]$ where n = node and b = branch
(vii) Two graphs having same incidence matrix are called isomorphic graph.

Reduced Incidence Matrix

(i) A particular node is taken as a reference node and the row corresponding to that node is deleted, resulting in the reduced incidence matrix.
(ii) The order of matrix is $[(n-1) \times b]$.
(iii) This matrix can be utilised to find number of trees in a graph whether that graph is fully connected or not.
(iv) Even if one of the row of the incidence matrix is deleted, the information of the oriented graph is still preserved.

Cut set Matrix

It is a group of branches containing only one twig and a number of links.

Fundamental Cut Set Matrix

(i) Fundamental cut set is a group of branch containing only one twig and the minimum number of links.
(ii) Fundamental cut set matrix can be used to write KCL equations for the given network.

Procedure to construct Cut-Set Matrix

1. Select a tree by selecting the number of links.
2. If the graph contains 'n' number of nodes, then there will be $n-1$ cut sets. The number of cut sets to the assigned equals the number of twigs.
3. The cut-set contains any number of links but should cover only one twig.
4. The cut-set is oriented in the same direction as the twig it covers.
5. Assign +1 in the matrix if the direction of cut-sets and branches are same, otherwise-1.

Note:

☑ Number of fundamental cut sets of a graph
= Number of twigs = Number of KCL equations
= $(n-1)$
= Number of node pair voltage

☑ Number of rows of the matrix = Number of fundamental cut sets

☑ Number of columns = Number of branches.

☑ Number of KCL equation = Number of fundamental cut set.

Tie set Matrix

It is a group of branches containing only one link and a number of twigs.

Fundamental Tie Set Matrix

(i) It is a mathematical representation of fundamental tie sets of a graph in form of matrix.

(ii) Fundamental Tie set is a group of branches containing only one link and minimum number of twigs.

Procedure to construct Tie Set Matrix

1. Select a tree from the oriented graph by selecting a number of links.
2. Number of tie-sets = number of links.
3. Assign a tie-set current such that it flows through only one link in the path.
4. The direction of tie-set current is same as the direction of link.
5. Assign '+1' if tie-set current direction and the branch current are same, otherwise '–1' is assigned. Assign '0' if the tie-set current does not flow through the branch.

Remember:

Number of fundamental tie sets for a graph.

= Number of links

= Number of KVL(mesh equation) = $b - n + 1$

IV Laplace Transform Analysis and Circuit Transients

Laplace Transform

One sided Laplace transform of a function $f(t)$: $\mathcal{L}[f(t)] = \int_0^\infty f(t)e^{-st}dt = F(s)$

Note:

☑ For the two sided Laplace transformation, the limits on the integration are from $-\infty$ to ∞.

☑ In practical circuits like *RL*, *RC*, one sided Laplace transform is used for analysis purpose where initial conditions are used.

Representation of Circuit Elements in s-domain

Resistor

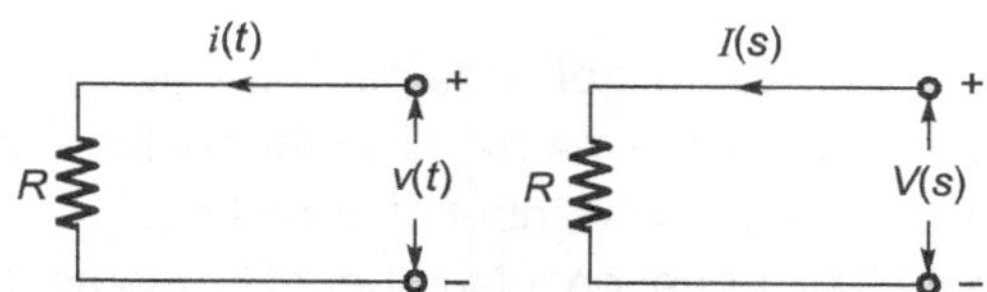

Time domain s-domain

- In time domain, $v(t) = R \cdot i(t)$
- In s-domain, $V(s) = R \cdot I(s)$ where, $s \to \sigma + j\omega$
- For sinusoidal excitation, $\sigma = 0$ and $s = j\omega$

Inductor

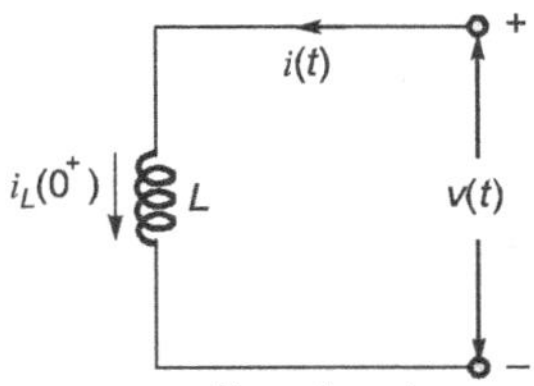

Time domain

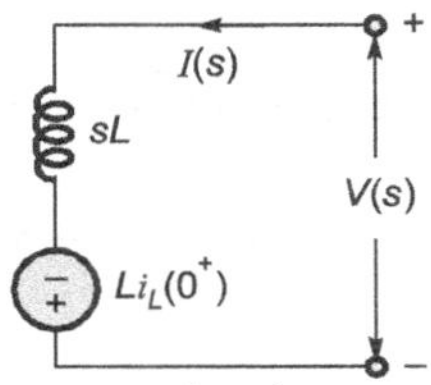

s-domain

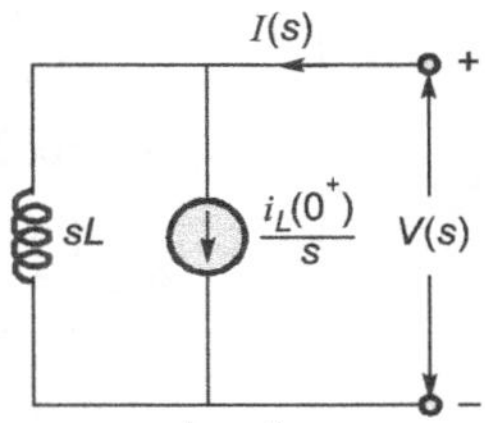

s-domain

- In time domain : $v(t) = L\dfrac{di(t)}{dt}$; $i(t) = \dfrac{1}{L}\displaystyle\int_{-\infty}^{t} v(t)dt$
- In s-domain : $V(s) = sL \cdot I(s) - Li_L(0^+)$; $I(s) = \left[\dfrac{V(s)}{sL} + \dfrac{i_L(0^+)}{s}\right]$

Capacitor

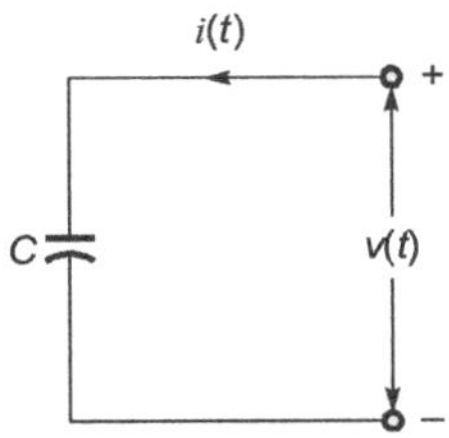

Time domain

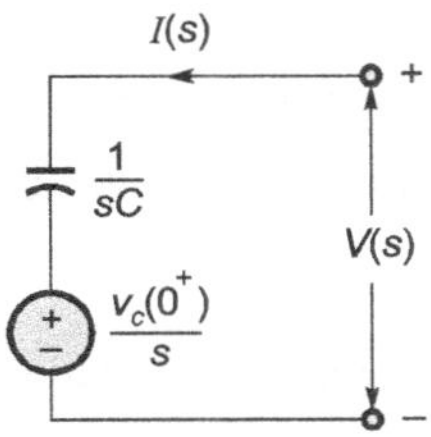

s-domain

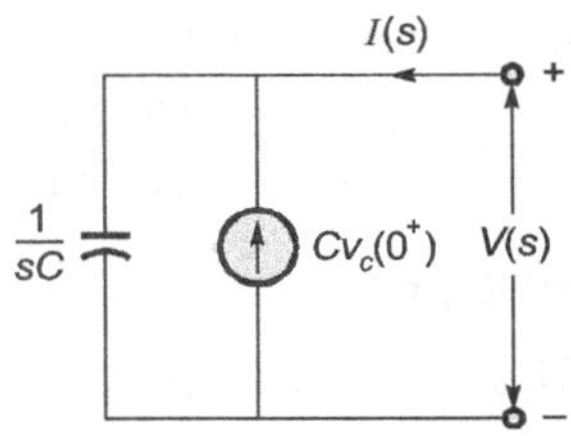

s-domain

- In time domain : $i(t) = C\dfrac{dv(t)}{dt}$;

$$v(t) = \frac{1}{C}\int_{-\infty}^{t} i(t)dt$$

- In s-domain : $I(s) = sCV(s) - Cv_C(0^+)$;

$$V(s) = \frac{I(s)}{sC} + \frac{v_C(0^+)}{s}$$

Transient Response

A system is in transient state if its output changes with time.

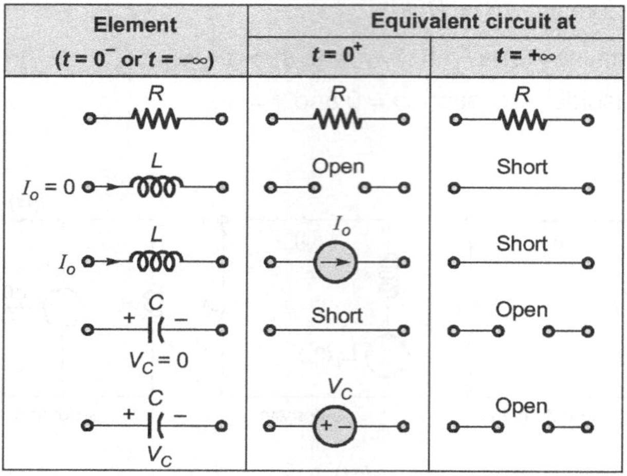

Disturbance takes place at $t = 0$.

RL Circuit

The transient current through the inductor L at any time t

$$i_L(t) = i_\infty - [i_\infty - i_{0^+}]e^{-R_{eq}t/L}$$

where, $i_\infty \rightarrow$ Current through L as $t \rightarrow \infty$

i.e., steady-state current through L

$i_{0^+} \rightarrow$ Current through L as $t \rightarrow 0^+$

$R_{eq} \rightarrow$ Thevenin's equivalent resistance seen across L for $t > 0$.

RC Circuit

The transient voltage across capacitor C at any time t

$$v_C(t) = v_\infty - [v_\infty - v_{0^+}]e^{-t/R_{eq}C}$$

where, $v_\infty \rightarrow$ Voltage across capacitor as $t \rightarrow \infty$,

i.e., steady state voltage across C

$v_{0^+} \rightarrow$ Voltage across C as $t \rightarrow 0^+$

$R_{eq} \rightarrow$ Thevenin's equivalent resistance seen across C for $t > 0$

Transient in Simple Circuits

The transients depends upon the driving source, whether it is d.c. source or an a.c. source.

DC Source

- **Resistance only**

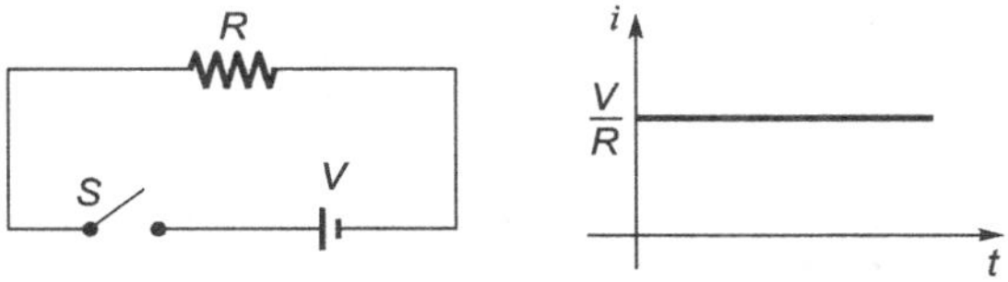

Transient current : $I = \frac{V}{R}$

- **Inductance only**

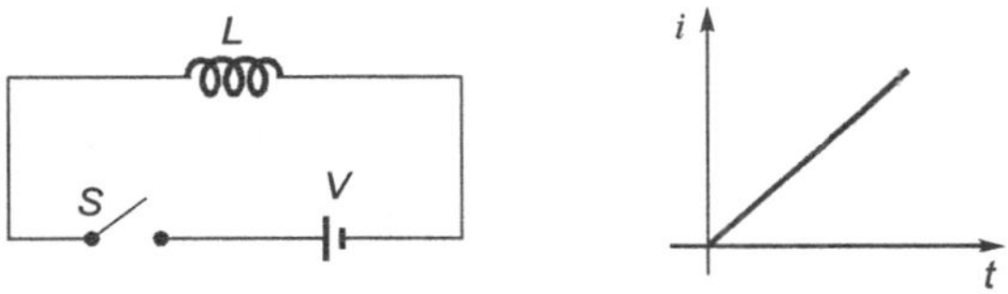

Transient current: $i(t) = \frac{V}{L}t$

- **Capacitance only**

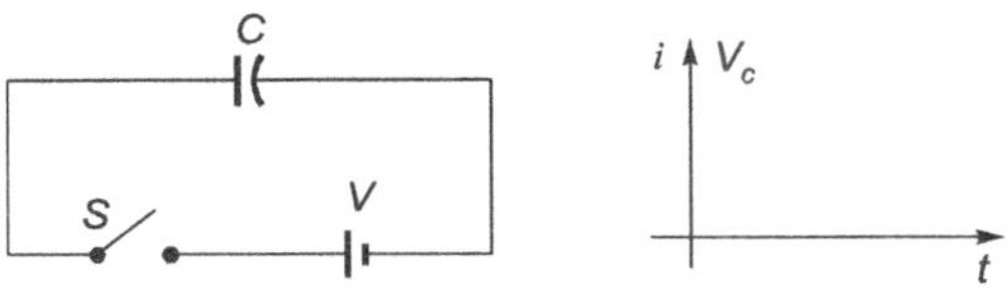

Transient current: $I(s) = \frac{V(s)}{Z(s)} = \frac{V}{s} \cdot sC = VC$

which is an impulse of magnitude VC.

- **R-L circuit**

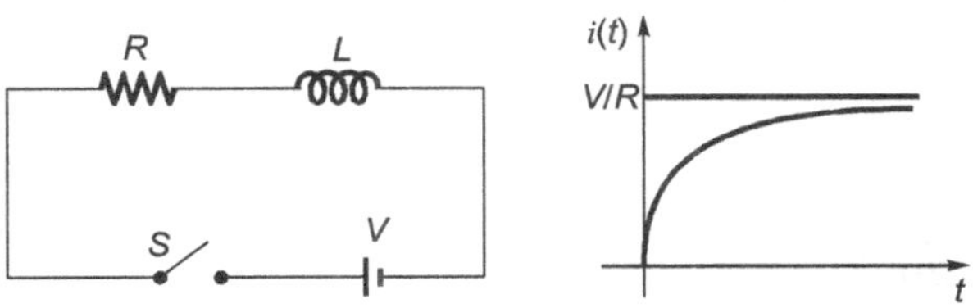

Transient current: $i(t) = \frac{V}{R}\left[1 - \exp\left(-\frac{R}{L}t\right)\right]$

- **R-C circuit**

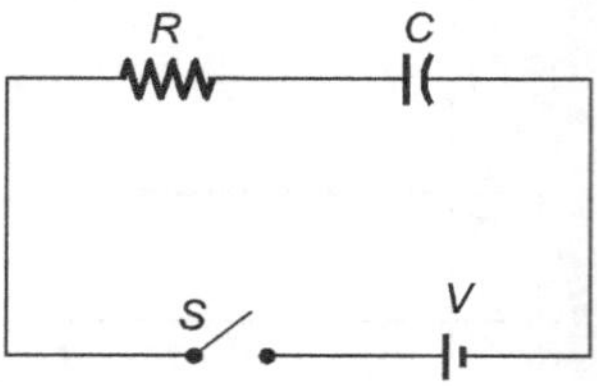

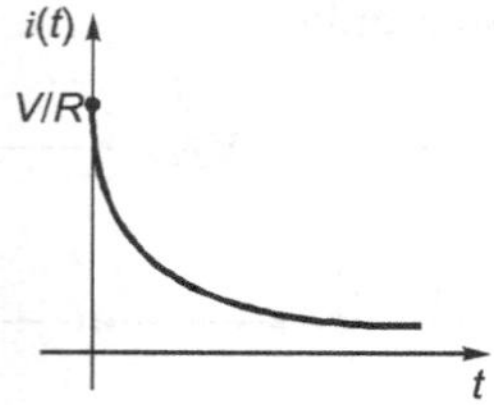

Transient current: $i(t) = \frac{V}{R}\exp\left(-\frac{t}{RC}\right)$

- **R-L-C circuit**

Transient current

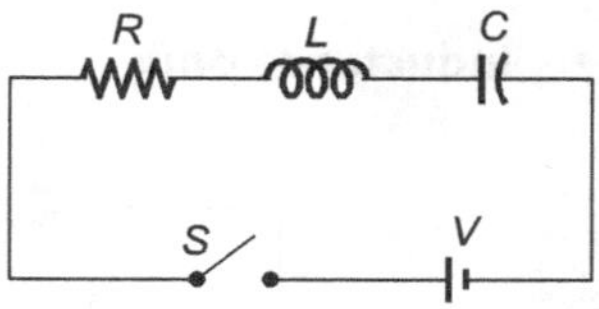

$$i(t) = \frac{V}{2bL}\left[e^{-(a-b)t} - e^{-(a+b)t}\right]$$

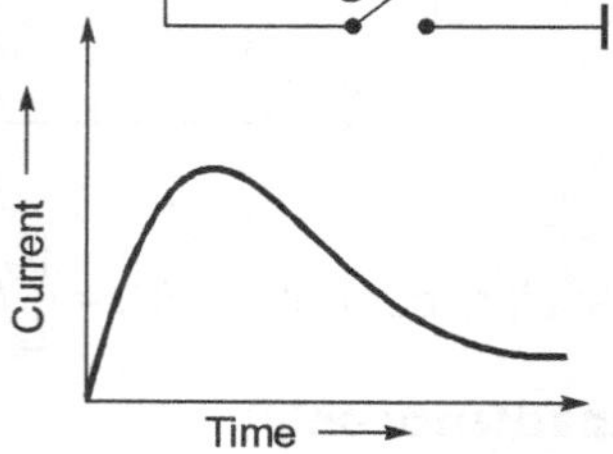

where, $a = \frac{R}{2L}$ and $b = \sqrt{\frac{R^2}{4L^2} - \frac{1}{LC}}$

- **Case-1**

When b is real

then $\frac{R^2}{4L^2} > \frac{1}{LC}$

and $i(t) = \frac{V}{2bL}\left[e^{-(a-b)t} - e^{-(a+b)t}\right]$

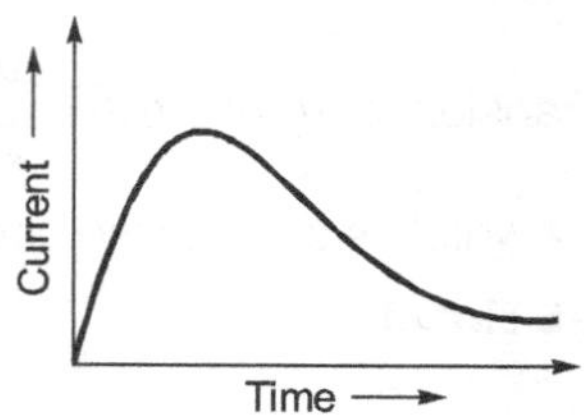

- **Case-2**

When $b = 0$

$$i(t) = \frac{V}{2bL}\left(e^{-at} - e^{-at}\right)$$

which is indeterminate form or $\left(\frac{0}{0}\right)$ form

$$i(t) = \frac{V}{L}te^{-at}$$

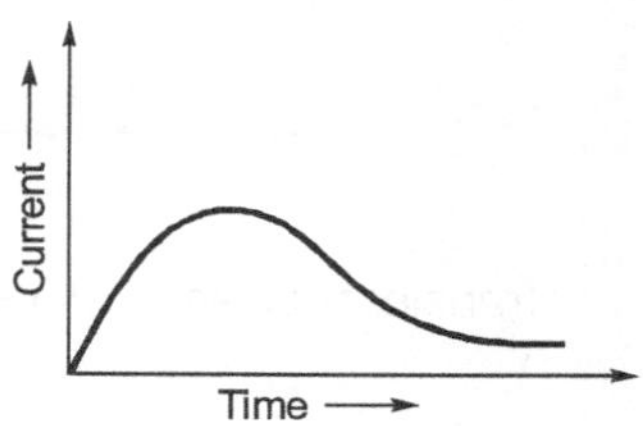

- **Case-3 :** When b is imaginary

then $\frac{1}{LC} > \frac{R^2}{4L^2}$

and $i(t) = \frac{V}{2bL} e^{-at} \cdot 2\sin kt$

where , $k = \sqrt{\frac{1}{LC} - \frac{R^2}{4L^2}}$

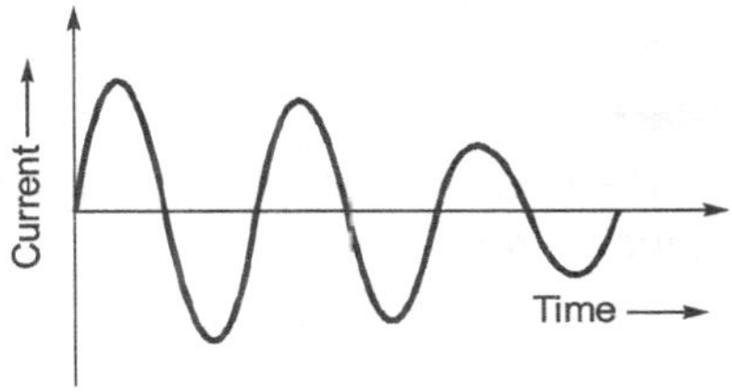

V Resonance

- Resonance is the condition when the voltage across a circuit becomes in phase with the current supplied to the circuit.
- At resonance, the circuit behaves like a resistive circuit.
- Power factor of the circuit at resonance is unity.

Series Resonance

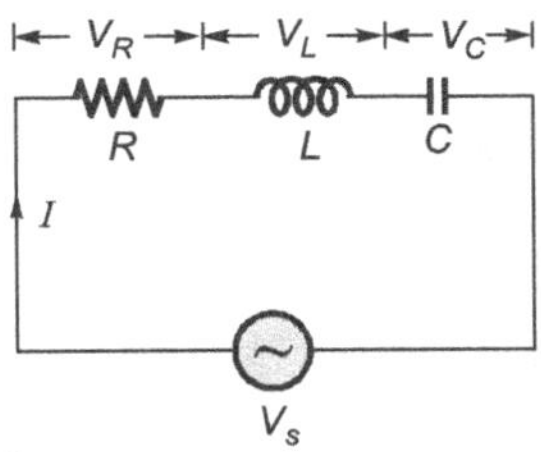

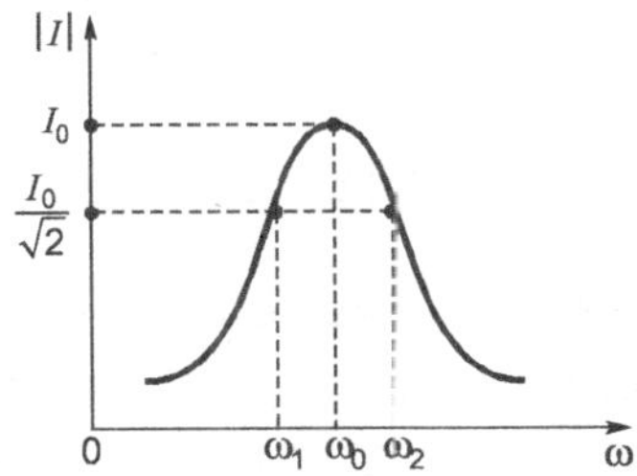

At Resonance

- $|V_L| = |V_C|$ and these are 180° out of phase
- Imaginary part of input impedance = 0

$Z_{in}|_{\omega=\omega_0} = R$ ......minimum ; $I|_{\omega=\omega_0} = \frac{V_s}{R}$ ......maximum

where, $\omega_0 = \frac{1}{\sqrt{LC}}$ → Resonant frequency in rad/sec

Remember:

☑ For $\omega < \omega_0$, series *RLC* circuit behaves like an *RC* circuit.

☑ For $\omega > \omega_0$, series *RLC* circuit behaves like an *RL* circuit.

☑ For $\omega = \omega_0$, series *RLC* circuit behaves like resistive circuit.

Bandwidth

The bandwidth of the network represent the range of the frequencies at which the current drawn by network becomes 0.707 of its maximum value.

$$\Delta\omega = (\omega_2 - \omega_1) = \frac{R}{L} \; ; \; \omega_1 = \omega_0 - \frac{1}{2}\Delta\omega \text{ and } \omega_2 = \omega_0 + \frac{1}{2}\Delta\omega$$

$\omega_0 = \sqrt{\omega_1 \cdot \omega_2}$ where, $\omega_1, \omega_2 \rightarrow$ Cut-off frequency or half power frequency

***Q*-factor**

The quality factor of the network is a measure of sharpness of the curve for the current drawn by circuit.

$$Q\text{-factor} = 2\pi\left(\frac{\text{Maximum energy stored}}{\text{Energy dissipated per cycle}}\right)$$

$$Q_0 = \frac{\omega_0}{\Delta\omega} = \frac{1}{\omega_0 RC} = \frac{\omega_0 L}{R} = \frac{1}{R}\sqrt{\frac{L}{C}}$$

- Frequency at which voltage across capacitor is maximum

$$f_C = \frac{1}{2\pi}\sqrt{\frac{1}{LC} - \frac{R^2}{2L^2}}$$

- Frequency at which voltage across inductor is maximum

$$f_L = \frac{1}{2\pi\sqrt{LC - \frac{C^2R^2}{2}}}$$

- Selectivity = $\dfrac{\text{Resonance frequency}}{\text{Bandwidth}} = \dfrac{f_0}{f_2 - f_1}$

Features of Series Resonance Circuit

- The resonant frequency is a geometric mean of the two half power frequency.
- The resonant frequency represent the rate at which the electrical energy stored in the capacitor is transformed to the magnetic energy stored in the inductor and vice versa.
- The quality factor represents the voltage amplification factor and must have high value for any tuned circuit.
- The quality factor of the resonant curve depends upon the numerical value of RLC components.
- For ***tuned network***, ***quality factor*** must be ***high***, ***bandwidth*** should be ***small*** and therefore the resistance used in the network should be small.
- Any ***series RLC*** network represents a ***band pass filter***.

Parallel Resonance

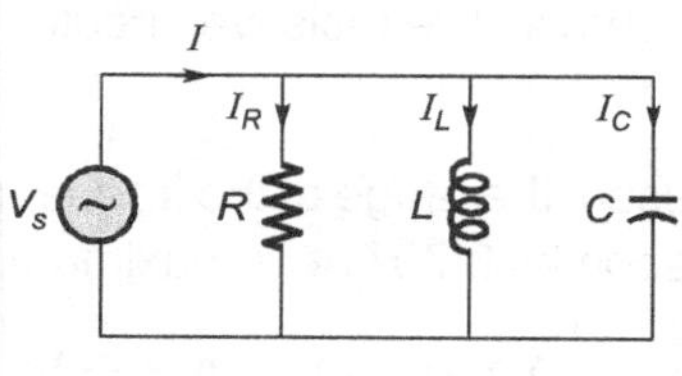

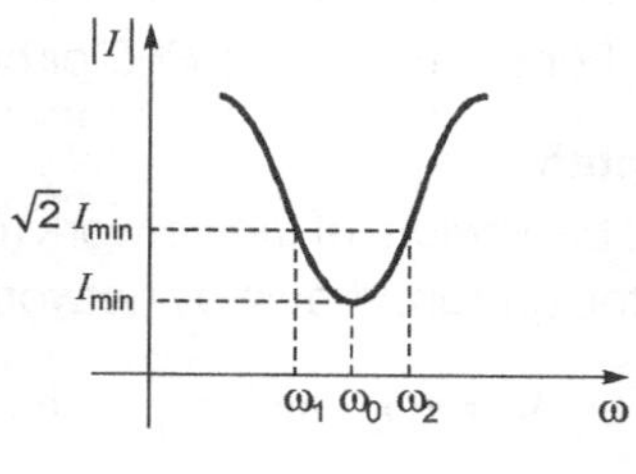

At Resonance

- $|I_L| = |I_C|$ and these are 180° out of phase
- Imaginary part of input impedance = 0

$Z_{in}|_{\omega=\omega_0} = R$ ...... maximum ; $I|_{\omega=\omega_0} = \dfrac{V_s}{R}$ ...... minimum

where, $\omega_0 \rightarrow \dfrac{1}{\sqrt{LC}}$ = resonant frequency in rad/sec

Remember:

- For $\omega < \omega_0$ the circuit behaves like an *RL* circuit.
- For $\omega > \omega_0$ the circuit behaves like an *RC* circuit.
- For $\omega = \omega_0$ the circuit behaves like a resistive circuit.

- Bandwidth, $\Delta\omega = (\omega_2 - \omega_1) = \dfrac{1}{RC}$
- *Q*-Factor, $Q_0 = \dfrac{\omega_0}{\Delta\omega} = \omega_0 RC = \dfrac{R}{\omega_0 L} = R\sqrt{\dfrac{C}{L}}$

Features of Parallel RLC Circuit

- The ***quality factor*** of the network represents the ***current amplification*** factor and must have a high value for any tuned circuit.
- For a ***tuned network***, ***quality factor*** should be ***high*** and therefore should have ***low bandwidth*** therefore the ***capacitance*** value is ***kept high***.
- A ***parallel RLC*** network behaves as a ***band stop filter*** or ***band reject filter***.

Note:

☑ Two important circuits with their resonance frequency is given as

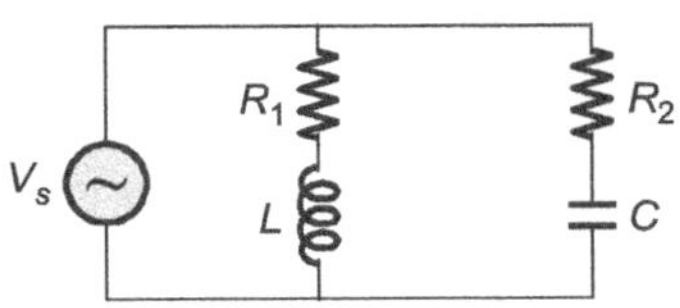

$$f_o = \frac{1}{2\pi\sqrt{LC}}\sqrt{\frac{R_1^2 - L/C}{R_2^2 - L/C}}$$

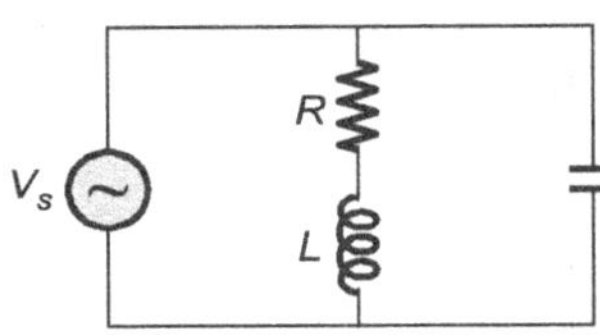

$$f_o = \frac{1}{2\pi}\sqrt{\frac{1}{LC} - \frac{R^2}{L^2}}$$

VI Magnetically Coupled Circuit

Magnetic Coupling

- When the two inductor are placed physically close to each other, then due to current flow in the first inductor there will be some magnetic flux, part of which will link with the second inductor. Due to rate of change of magnetic flux and by using Faraday's law of electromagnetic induction some induced voltage is produced between the two terminals of the second inductor.
- The polarity of the induced voltage depends upon the relative sense of windings of the two inductors and therefore, the induced voltage may be additive or substractive in nature.

Case: (A)

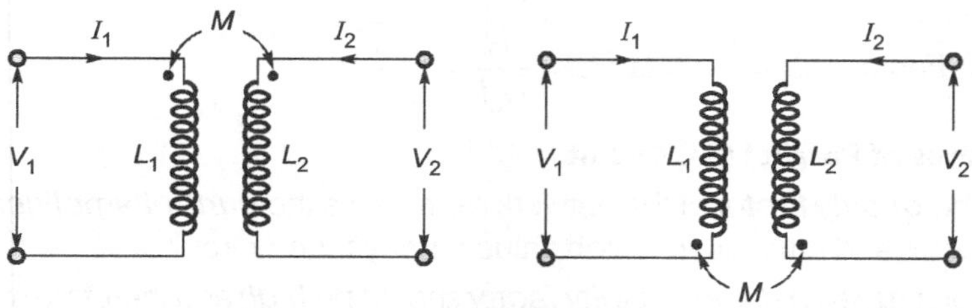

(Case-1) (Case-2)

For above two cases, the effect of mutual inductance is positive. Also induced emf (e_{ind}) is additive in nature:

KVL in time domain:

$$V_1 = L_1\frac{dI_1}{dt} + M\frac{dI_2}{dt} \text{ and } V_2 = M\frac{dI_1}{dt} + L_2\frac{dI_2}{dt}$$

KVL in s-Domain:

$$V_1(s) = L_1 s I_1(s) + Ms I_2(s)$$

and

$$V_2(s) = Ms I_1(s) + L_2 s I_2(s)$$

$$\begin{bmatrix} V_1(s) \\ V_2(s) \end{bmatrix} = \begin{bmatrix} sL_1 & sM \\ sM & sL_2 \end{bmatrix} \begin{bmatrix} I_1(s) \\ I_2(s) \end{bmatrix}$$

For sinusoidal excitation:

$$\begin{bmatrix} V_1 \\ V_2 \end{bmatrix} = \begin{bmatrix} j\omega L_1 & j\omega M \\ j\omega M & j\omega L_2 \end{bmatrix} \begin{bmatrix} I_1 \\ I_2 \end{bmatrix}$$

Case: (B)

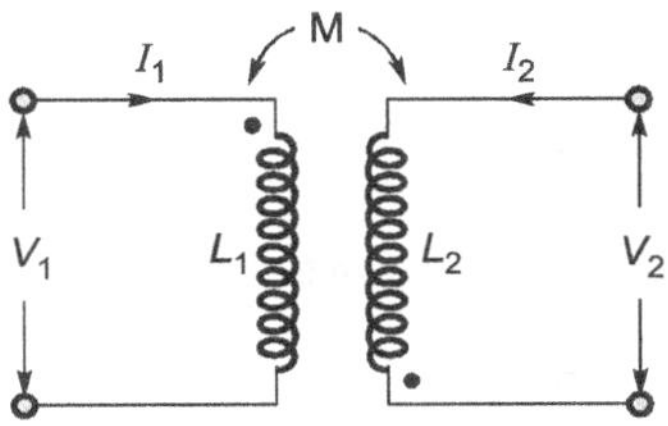

(Case-3)

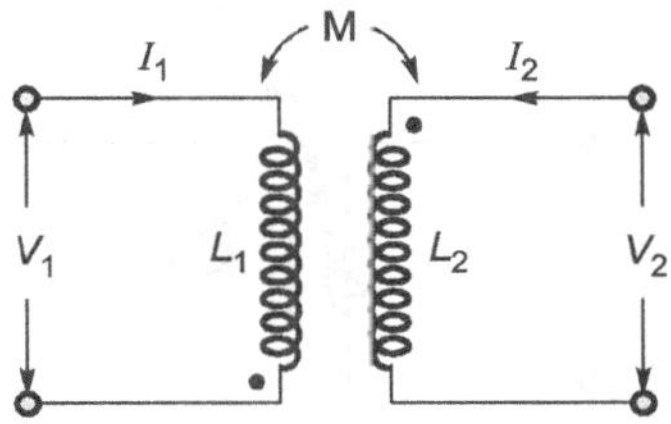

(Case-4)

For above two cases, effect of the mutual coupling is negative. So e_{ind} is substractive in nature.

KVL in time domain	$V_1 = L_1\frac{dI_1}{dt} - M\frac{dI_2}{dt}$ and $V_2 = -M\frac{dI_1}{dt} + L_2\frac{dI_2}{dt}$
KVL in s-domain	$\begin{bmatrix} V_1(s) \\ V_2(s) \end{bmatrix} = \begin{bmatrix} sL_1 & -sM \\ -sM & sL_2 \end{bmatrix}\begin{bmatrix} I_1(s) \\ I_2(s) \end{bmatrix}$
For sinusoidal excitation	$\begin{bmatrix} V_1 \\ V_2 \end{bmatrix} = \begin{bmatrix} j\omega L_1 & -j\omega M \\ -j\omega M & j\omega L_2 \end{bmatrix}\begin{bmatrix} I_1 \\ I_2 \end{bmatrix}$

Representation of Mutual Inductance

Case-1: In terms of M.

Case-2: In terms of 'K', i.e., coefficient of coupling.

$$K = \frac{M}{\sqrt{L_1 L_2}}$$

(a) If $K < 1$; loose coupling (under coupling circuit).

(b) If $K = 1$ i.e. $M = \sqrt{L_1 L_2}$; (critical coupling)

(c) $K > 1$; tight coupling (over coupled circuit).

Case-3: In terms of turns ratio

$$\frac{V_1}{V_2} = \frac{n_1}{n_2} = \frac{I_2}{I_1}$$

and $$\frac{n_1}{n_2} \cong \frac{L_1}{M} \cong \frac{M}{L_2} = \sqrt{\frac{L_1}{L_2}}$$

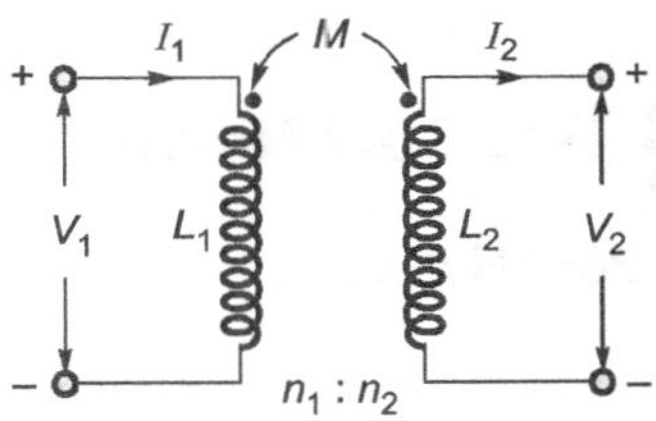

Note:

☑ Equivalent Circuits

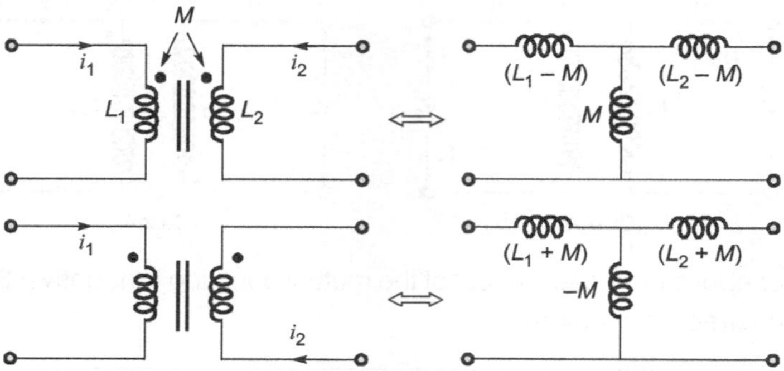

☑ Equivalent Inductance

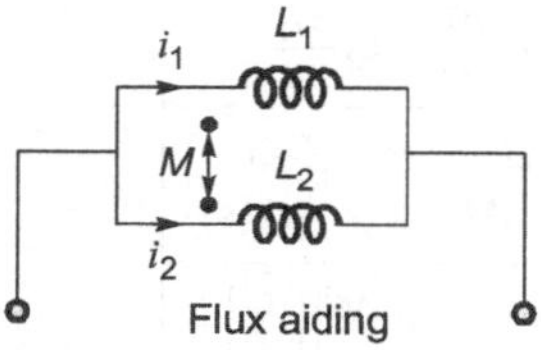

$$L_{eq} = \frac{L_1 L_2 - M^2}{L_1 + L_2 - 2M}$$

☑ Equivalent Inductance

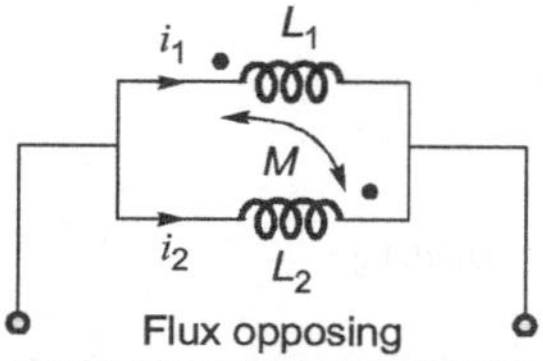

$$L_{eq} = \frac{L_1 L_2 - M^2}{L_1 + L_2 + 2M}$$

☑ Energy stored in coupled circuit

$$\frac{1}{2} L_1 i_1^2 + \frac{1}{2} L_2 i_2^2 \pm M_1 M_2$$

(+ : Flux aiding; − : Flux opposing)

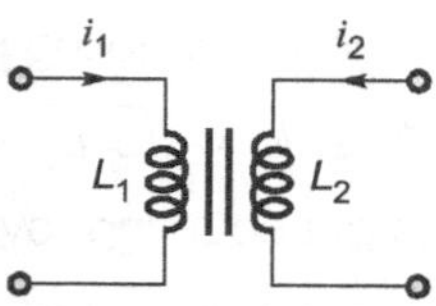

VII Two Port Network

Two Port Network

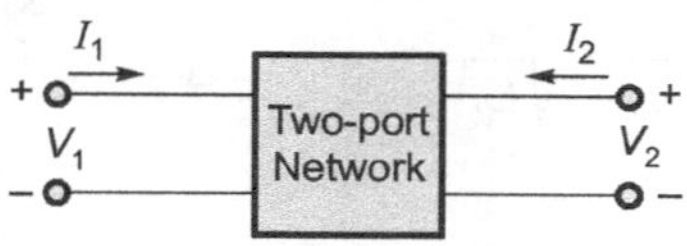

Driving Point Impedance and Admittance Functions	
• Driving point impedance function	$Z_{11}(s)=\frac{V_1(s)}{I_1(s)}$ and $Z_{22}(s)=\frac{V_2(s)}{I_2(s)}$
• Driving point admittance function	$Y_{11}(s)=\frac{I_1(s)}{V_1(s)}$ and $Y_{22}(s)=\frac{I_2(s)}{V_2(s)}$
Impedance and Admittance Transfer Function	
• Impedance Transfer Function	$Z_{21}(s)=\frac{V_2(s)}{I_1(s)}$ and $Z_{12}(s)=\frac{V_1(s)}{I_2(s)}$
• Admittance Transfer Function	$Y_{21}(s)=\frac{I_2(s)}{V_1(s)}$ and $Y_{12}(s)=\frac{I_1(s)}{V_2(s)}$
Voltage and Current Transfer Function	
• Voltage Transfer Function	$G_{12}(s)=\frac{V_1(s)}{V_2(s)}$ and $G_{21}(s)=\frac{V_2(s)}{V_1(s)}$
• Current Transfer Function	$\alpha_{12}=\frac{I_1(s)}{I_2(s)}$ and $\alpha_{21}=\frac{I_2(s)}{I_1(s)}$

Parameters

Impedance or *Z*-parameters (Open circuit parameters)

$$V_1 = Z_{11}I_1 + Z_{12}I_2 \text{ and } V_2 = Z_{21}I_1 + Z_{22}I_2$$

Admittance or *Y*-parameters (Short circuit parameters)

$$I_1 = Y_{11}V_1 + Y_{12}V_2 \text{ and } I_2 = Y_{21}V_1 + Y_{22}V_2$$

h-parameters (Hybrid parameters)

$$V_1 = h_{11}I_1 + h_{12}V_2 \text{ and } I_2 = h_{21}I_1 + h_{22}V_2$$

g-parameters (Inverse hybrid parameters)

$$I_1 = g_{11}V_1 + g_{12}I_2 \text{ and } V_2 = g_{21}V_1 + g_{22}I_2$$

ABCD parameters (Transmission parameters)

$$V_1 = AV_2 - BI_2 \text{ and } I_1 = CV_2 - DI_2$$

Conditions for a network to be symmetrical and reciprocal

Reciprocal	Symmetrical
$Z_{12}=Z_{21}$	$Z_{11}=Z_{22}$
$Y_{12}=Y_{21}$	$Y_{11}=Y_{22}$
$\begin{vmatrix} A & B \\ C & D \end{vmatrix}=1$	$A=D$
$h_{12}=-h_{21}$	$\begin{vmatrix} h_{11} & h_{12} \\ h_{21} & h_{22} \end{vmatrix}=1$
$g_{12}=-g_{21}$	$\begin{vmatrix} g_{11} & g_{12} \\ g_{21} & g_{22} \end{vmatrix}=1$

Interconnection Between two-port Network

Network	Circuit Diagram	Parameters
Cascaded Network	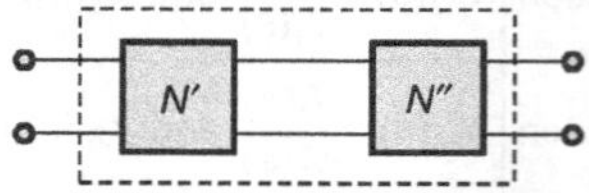	$\begin{bmatrix} A & B \\ C & D \end{bmatrix} = \begin{bmatrix} A' & B' \\ C' & D' \end{bmatrix} \begin{bmatrix} A'' & B'' \\ C'' & D'' \end{bmatrix}$
Series Network	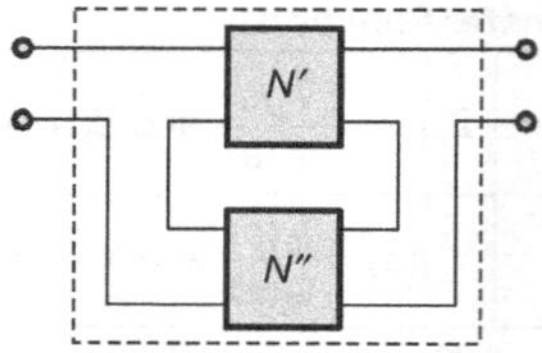	$\begin{bmatrix} Z_{11} & Z_{12} \\ Z_{21} & Z_{22} \end{bmatrix} = \begin{bmatrix} Z'_{11}+Z''_{11} & Z'_{12}+Z''_{12} \\ Z'_{21}+Z''_{21} & Z'_{22}+Z''_{22} \end{bmatrix}$
Parallel Network	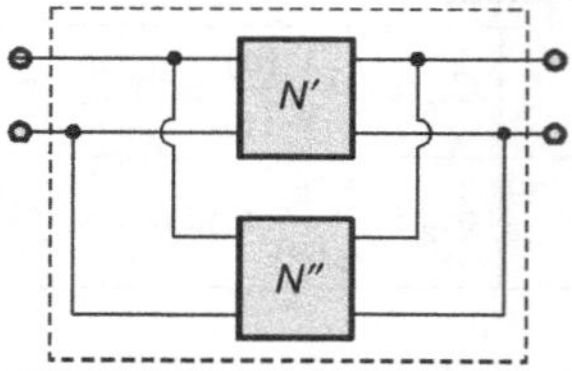	$\begin{bmatrix} Y_{11} & Y_{12} \\ Y_{21} & Y_{22} \end{bmatrix} = \begin{bmatrix} Y'_{11}+Y''_{11} & Y'_{12}+Y''_{12} \\ Y'_{21}+Y''_{21} & Y'_{22}+Y''_{22} \end{bmatrix}$
Series-Parallel Network	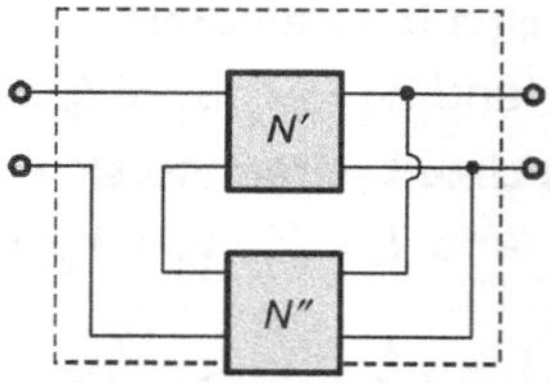	$\begin{bmatrix} h_{11} & h_{12} \\ h_{21} & h_{22} \end{bmatrix} = \begin{bmatrix} h'_{11}+h''_{11} & h'_{12}+h''_{12} \\ h'_{21}+h''_{21} & h'_{22}+h''_{22} \end{bmatrix}$
Parallel-Series Network	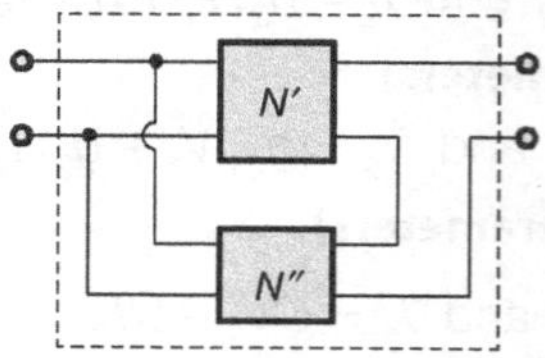	$\begin{bmatrix} g_{11} & g_{12} \\ g_{21} & g_{22} \end{bmatrix} = \begin{bmatrix} g'_{11}+g''_{11} & g'_{12}+g''_{12} \\ g'_{21}+g''_{21} & g'_{22}+g''_{22} \end{bmatrix}$

Lattice Network

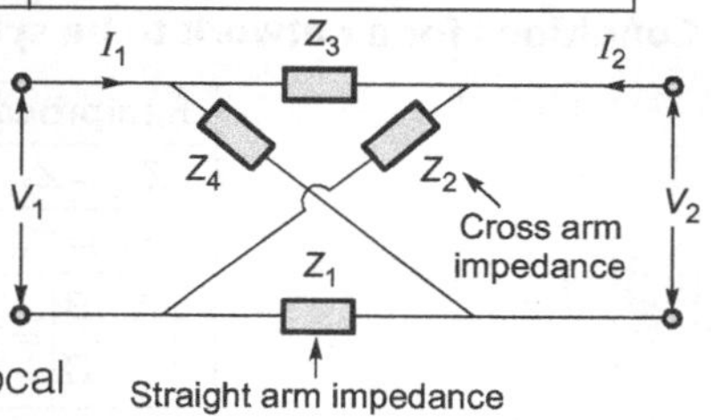

- A lattice network is said to be symmetrical if the two straight arm impedances are equal, i.e. $Z_1 = Z_3$. Therefore, $Z_{11} = Z_{22}$
- A lattice network is said to be reciprocal if the two cross arm impedances are equal, i.e., $Z_2 = Z_4$. Therefore, $Z_{12} = Z_{21}$
- For a symmetrical and reciprocal lattice network = $\begin{bmatrix} \dfrac{Z_1+Z_2}{2} & \dfrac{Z_2-Z_1}{2} \\ \dfrac{Z_2-Z_1}{2} & \dfrac{Z_1+Z_2}{2} \end{bmatrix}$

Remember:

☑ Z-parameters of T-model are given as

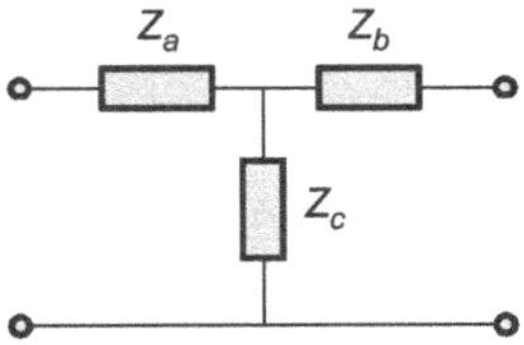

$$[Z] = \begin{bmatrix} Z_a + Z_c & Z_c \\ Z_c & Z_b + Z_c \end{bmatrix}$$

☑ Y-parameter of π-model are given as

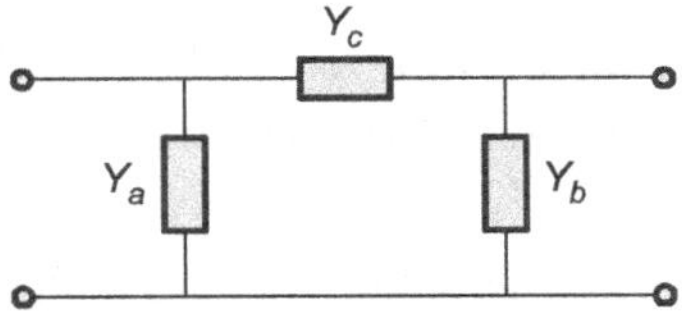

$$[Y] = \begin{bmatrix} Y_a - Y_c & -Y_c \\ -Y_c & Y_b + Y_c \end{bmatrix}$$

☑ For an ideal transformer, $ABCD$ parameters are given as

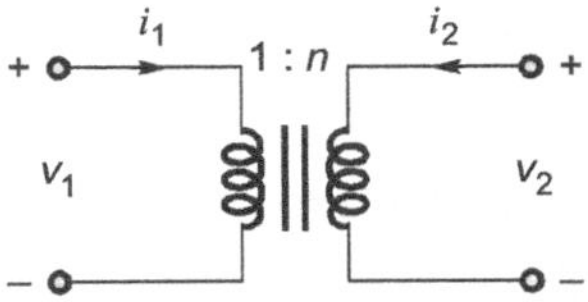

$$[T] = \begin{bmatrix} \frac{1}{n} & 0 \\ 0 & n \end{bmatrix}$$

VIII Miscellaneous

Dual Networks

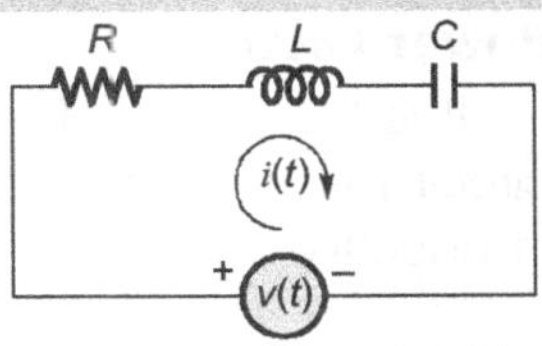

Series RLC circuit (N_1)

By applying KVL,

$$v(t) = Ri(t) + L\frac{di}{dt} + \frac{1}{C}\int_0^t i(t)\,dt$$

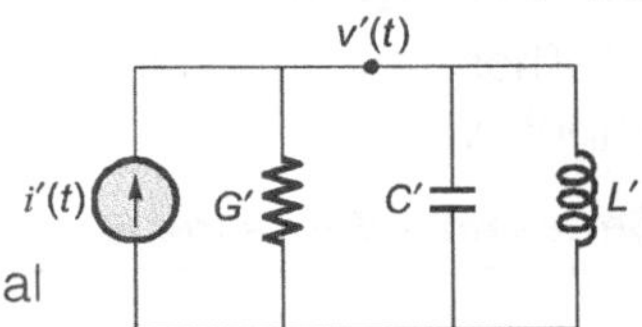

Parallel RLC circuit (N_2)

By applying KCL,

$$i'(t) = v'(t)\,G' + C'\frac{dv'(t)}{dt} + \frac{1}{L'}\int_0^t v'(t)\,dt$$

- The networks N_1 and N_2 are called dual networks but these networks are not equivalent networks and therefore, the network N_1 cannot be replaced by N_2 network.
- The KVL equation for network N_1 has similar format as for the KCL equation for network N_2. Therefore, if the response of the first network is known is terms of current $i(t)$ then the response of its dual network for

voltage $v'(t)$ can be directly written without actually solving it by using necessary transformation by various elements as shown.

Power Relations in AC Circuits

Instantaneous Power

If v is the voltage instantaneous voltage and i is the instantaneous current, then instantaneous power P is given as : $P = vi$

Real power

Average power, P is given as : $P = \frac{V_m}{\sqrt{2}} \frac{I_m}{\sqrt{2}} \cos\phi \quad VI\cos\phi$ Watts

where, $V_m, I_m \rightarrow$ Maximum values of v and i ; $\phi \rightarrow$ Phase difference between voltage and current ; $V, I \rightarrow$ RMS value of v and i

Reactive power

The product of rms values of voltage and current with the sine of angle between then is called reactive power. It is expressed in (VAR)-volt ampere reactive.

$$Q = VI\sin\phi \text{ VAR}$$

Apparent power

The product VI^* is called apparent power. It is expressed in (VA).

$$S = VI^* \; VA \quad S = P + j(Q_L - Q_C)$$

where, $Q_L \rightarrow$ is used for inductive reactive power; $Q_C \rightarrow$ is used for capacitive reactive power

Power Factor

It is the cosine of angle between the voltage and current in any AC circuit. It is unity in resistance circuit, leading in capacitive circuit and lagging in inductive circuit.

$$\text{Power factor} = \cos\phi$$

Locus Diagrams

These are important to understand the variations in current for a fixed supply voltage.

Series impedance locus

(i)	$Z = R + jX_L$ 	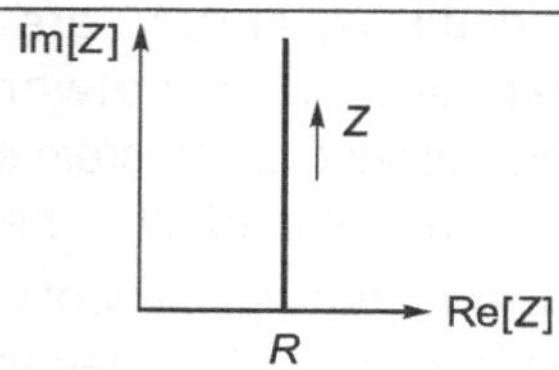

(ii)	$Z = R + jX_L$ 	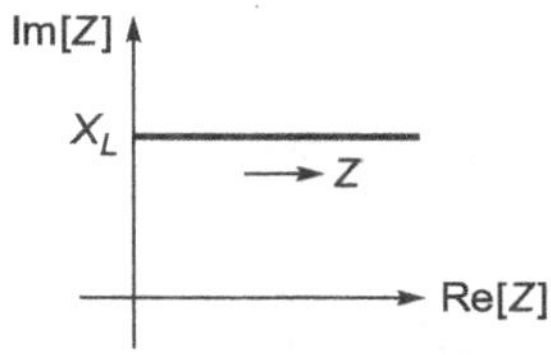
(iii)	$Z = R - jX_C$ 	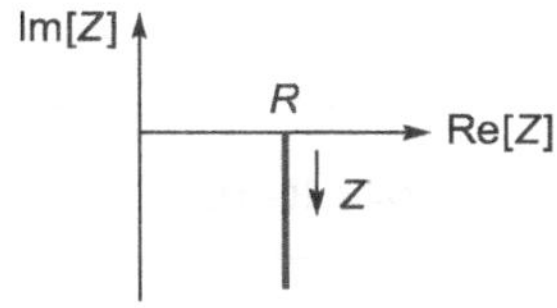
(iv)	$Z = R - jX_C$ 	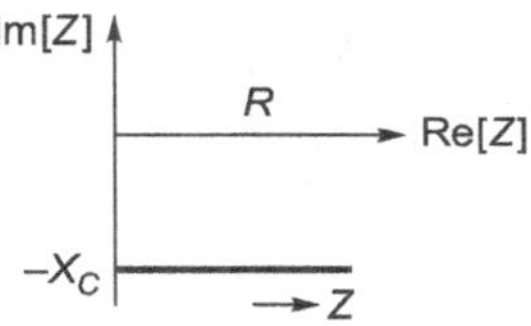

Series admittance locus

(i)	$Z = R + jX_L;\ Y = \frac{1}{Z}$ 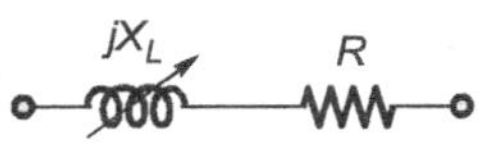	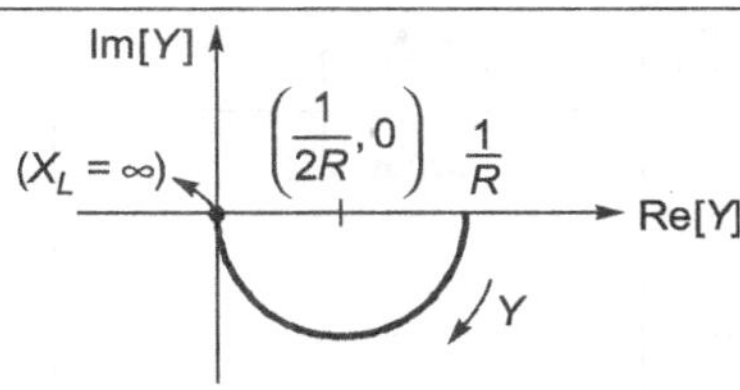
(ii)	$Z = R + jX_L;\ Y = \frac{1}{Z}$ 	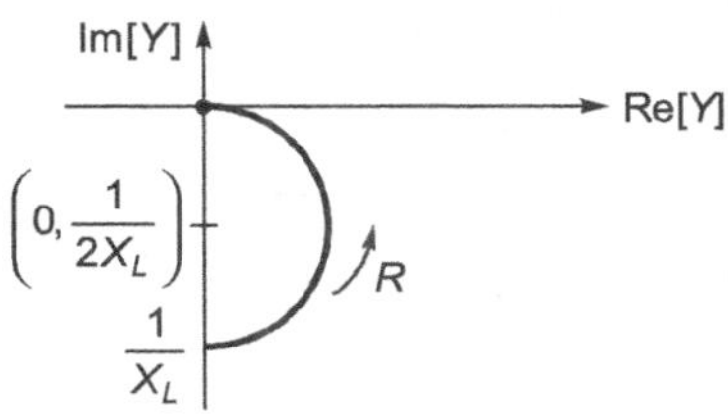
(iii)	$Z = R - jX_C\ ;\ Y = \frac{1}{Z}$ 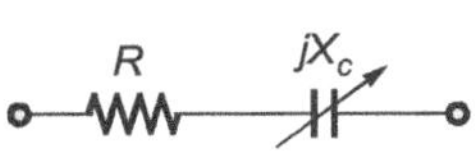	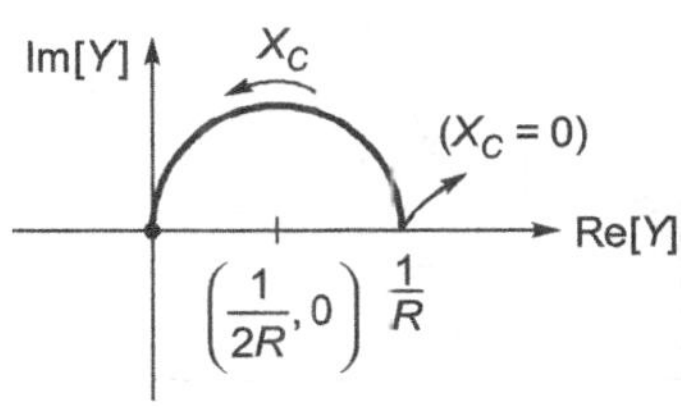

(iv)	$Z = R - jX_C\,;\ Y = \frac{1}{Z}$ 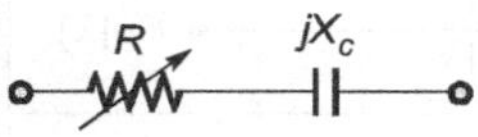	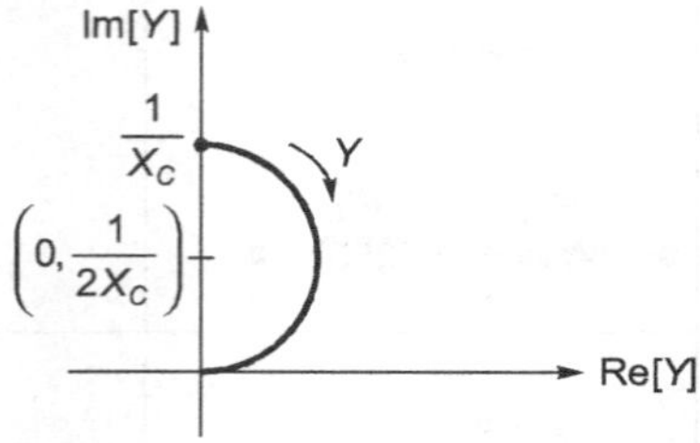

Parallel admittance locus

(i)	$Y = G - jB_L$ 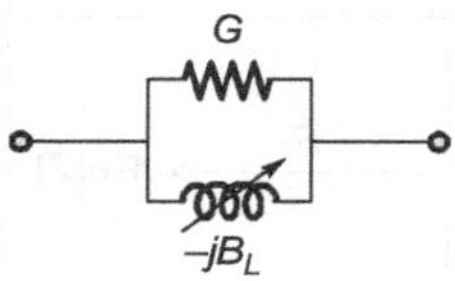	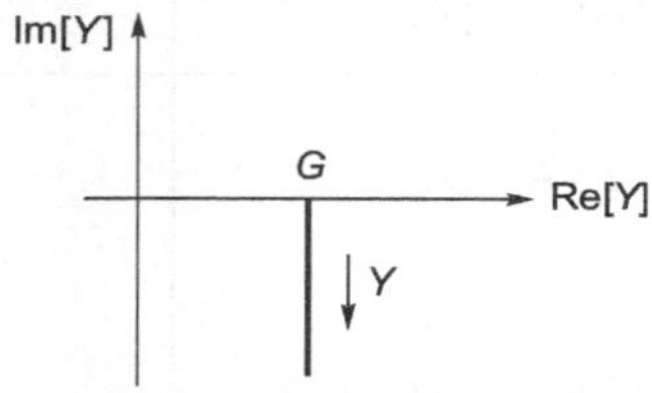
(ii)	$Y = G - jB_L$ 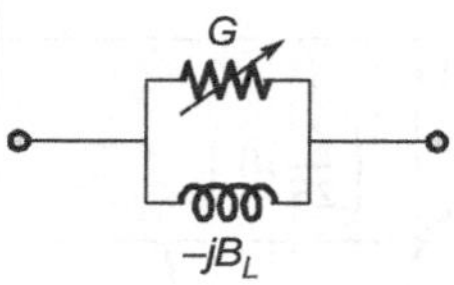	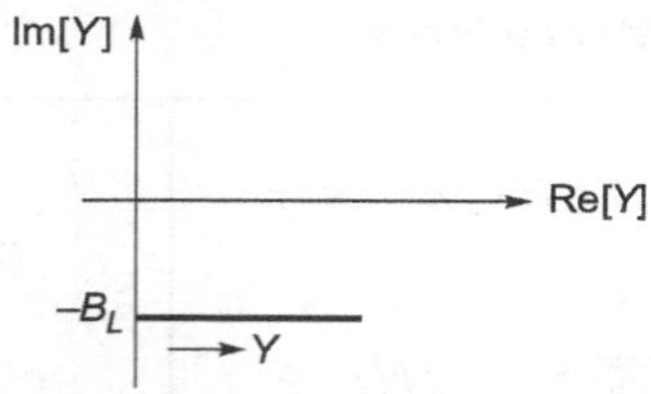
(iii)	$Y = G + jB_C$ 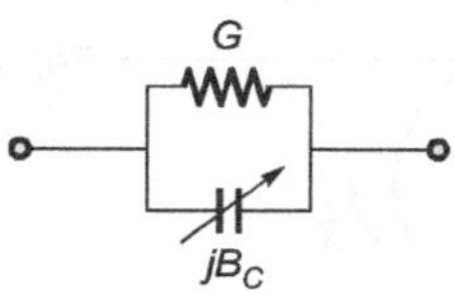	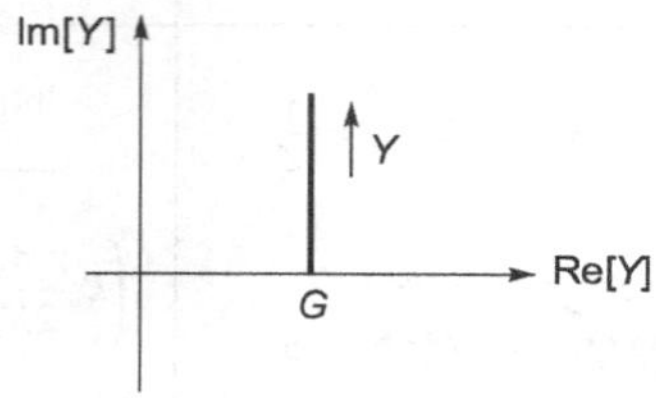
(iv)	$Y = G + jB_C$ 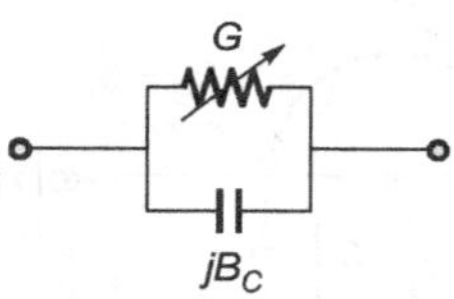	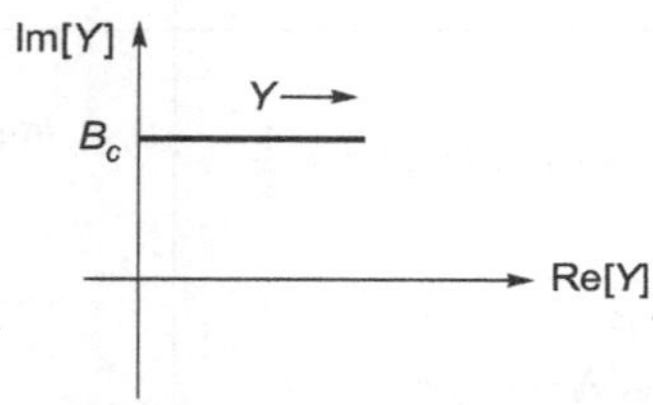

Parallel impedance locus

	Circuit	Locus
(i)	$Y = G - jB_L$ $Z = \frac{1}{Y}$ 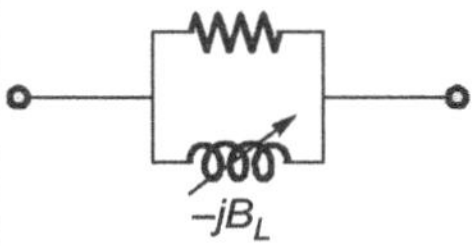	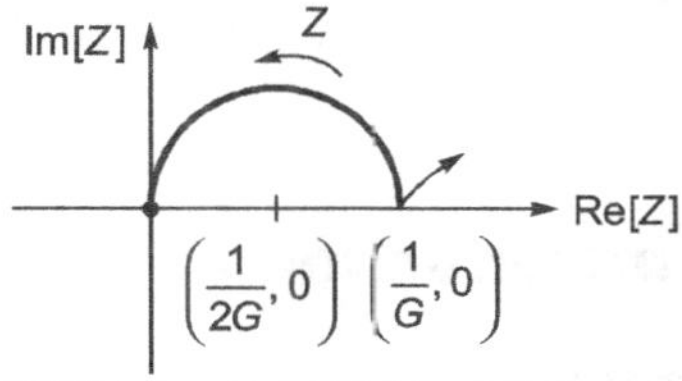
(ii)	$Y = G - jB_L$ $Z = \frac{1}{Y}$ 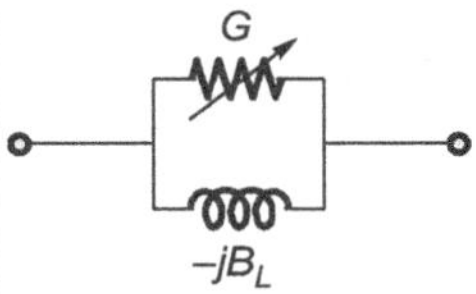	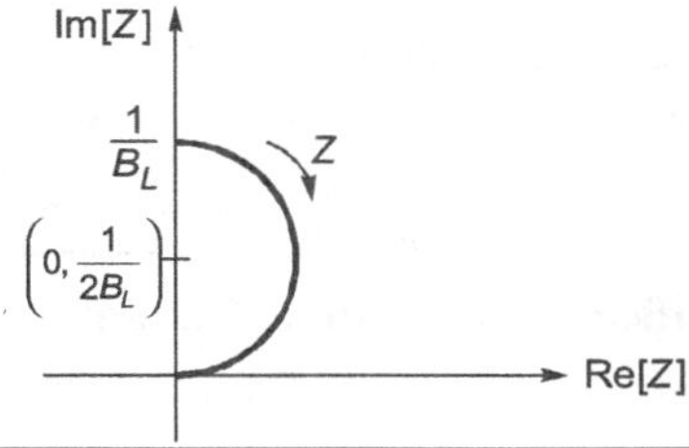
(iii)	$Y = G + jB_C$ $Z = \frac{1}{Y}$ 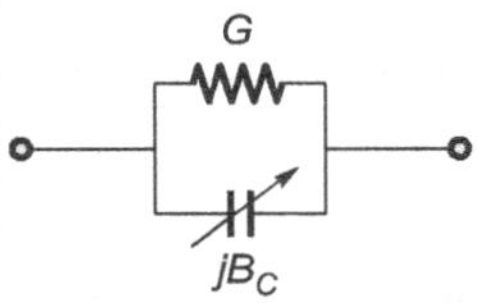	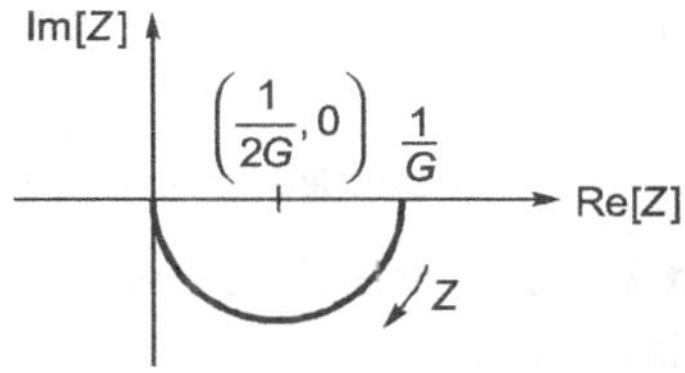
(iv)	$Y = G + jB_C$ $Z = \frac{1}{Y}$ 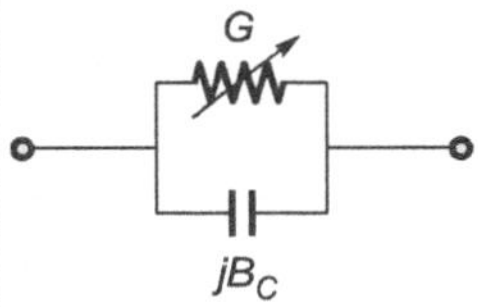	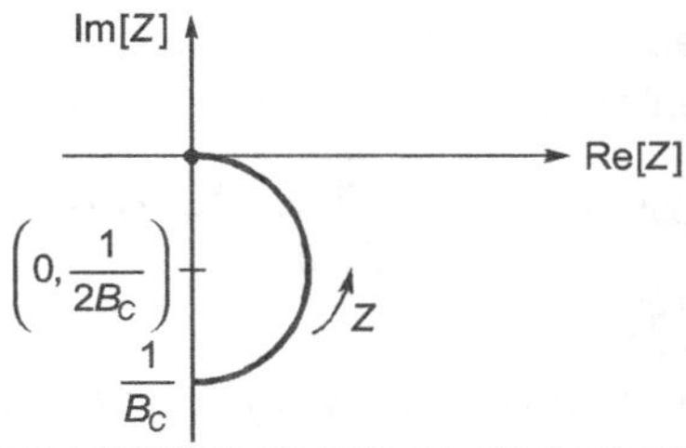

R-L-C Series Admittance locus

$Z = R + j(X_L - X_C)$

$Y = \frac{1}{Z}$

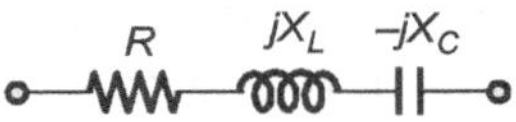

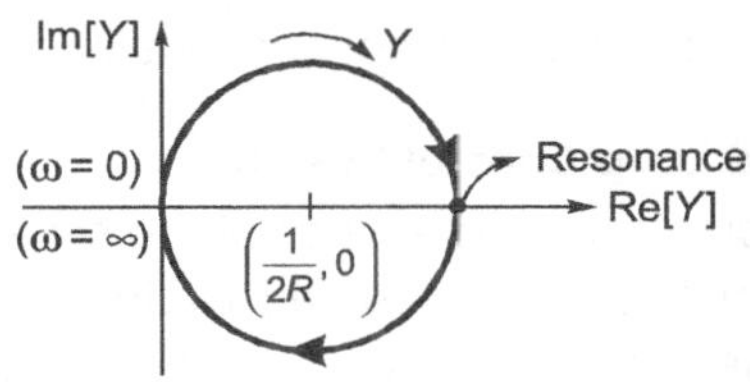

6

Control Systems

I Introduction

Control System

A control system is a combination of elements arranged in a planned manner wherein each element causes an effect to produce a desired output.

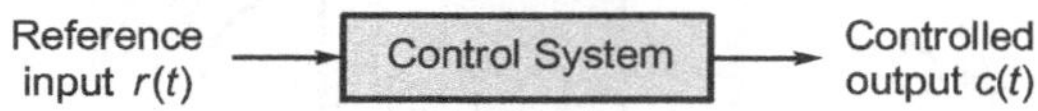

Classification of Control System

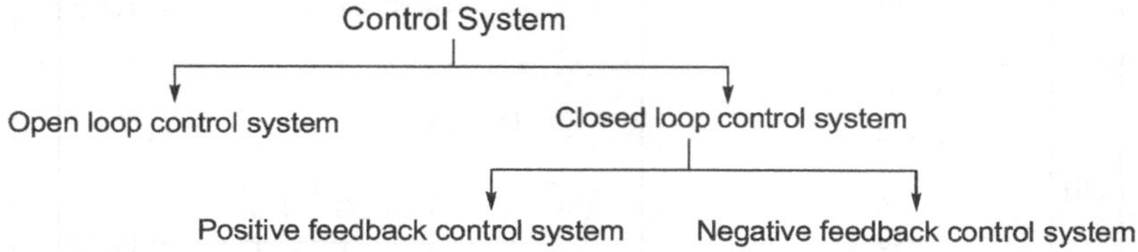

Open Loop Control System

It is a conditional control system, formulated under the basic condition that, the system is not subjected to any type of the disturbances. Control characteristics of such systems are independent of output of the system. The output is neither measured nor feedback for comparison with the input.

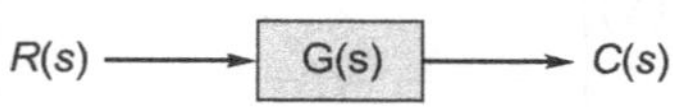

Note:

- ☑ Performance analysis is not applicable to open loop system because they are highly stable system.
- ☑ Faithfulness of an open loop control system depends upon the accuracy of input calibration.

Closed Loop Control System

Control characteristics of the system depends upon the output of the system. It is also termed as feedback control system. The control action is actuated by an error signal 'e', which is the difference between input signal and output signal. The purpose of feedback is to reduce the error between the reference input and the system output.

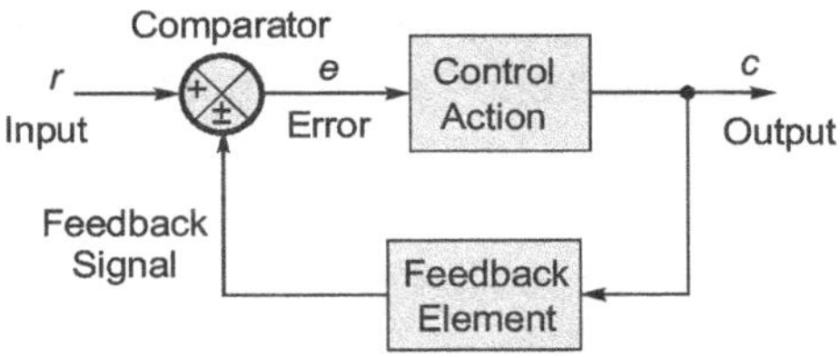

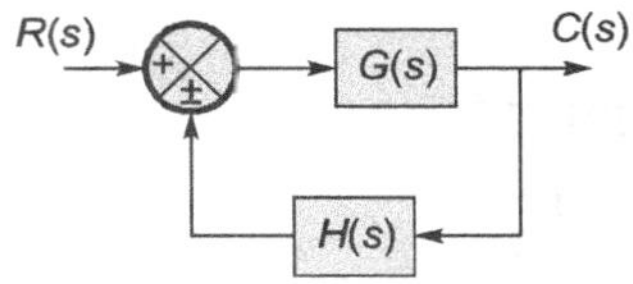

Effect of Negative Feedback

1. Effect of parameter variation reduces.
2. The gain of system reduces by a factor $(1 + GH)$.
3. The bandwidth of the system increases.
4. Effect of internal disturbance reduces.

Comparison of Open Loop and Closed Loop Control System

Open Loop System	Close Loop System
1. As long as the calibration is good, open-loop system will be accurate	1. Due to feedback, the closed-loop system is more accurate
2. Organization is simple and easy to construct	2. Complicated and difficult
3. Generally stable in operation	3. Stability depends on system components
4. If non-linearity is present, system operation degrades	4. Comparatively, the performance is better than open-loop system if non-linearity is present

II Mathematical Modelling

Mechanical Systems

All mechanical systems are classified in two types.

(i) Mechanical translational system; **(ii)** Mechanical rotational system.

Mechanical Translational System

Input : Force; **Output** : Linear displacement (x) OR Linear velocity (v)

Inertia force

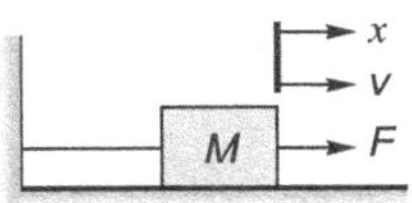

$$F = M\frac{d^2x}{dt^2} = M\frac{dv}{dt}$$

where, $F \rightarrow$ Force on block M ; $x \rightarrow$ Displacement of block M

$v \rightarrow$ Velocity of block M ; $M \rightarrow$ Mass of block M

Damping force

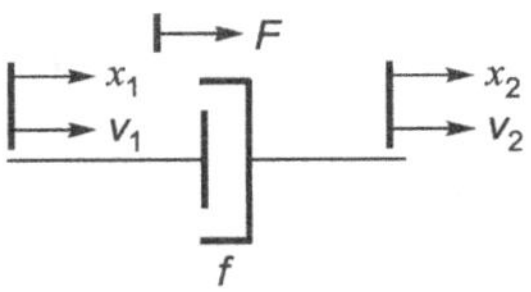

$$F = f\frac{d(x_1 - x_2)}{dt} = f(v_1 - v_2)$$

where, $F \rightarrow$ Damper force ; $x_1, x_2 \rightarrow$ Displacement at side 1 and 2 of damper ; $v_1, v_2 \rightarrow$ Velocity at side 1 and 2 ; $f \rightarrow$ Damper constant

Spring force

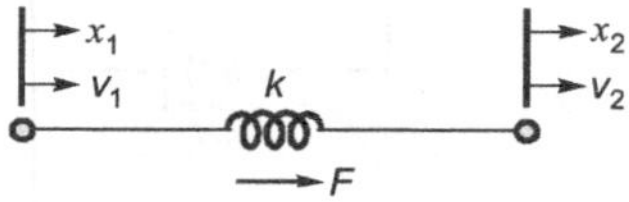

$$F = K(x_1 - x_2) = k\int (v_1 - v_2)dt$$

where $F \rightarrow$ Spring force ; $x_1, x_2 \rightarrow$ Displacement at side 1 and 2 of spring ; $v_1, v_2 \rightarrow$ Velocity at side 1 and 2 ; $K \rightarrow$ Spring constant

Mechanical Rotational System

Input : Torque (T)

Output : Angular displacement (θ) OR Angular velocity (ω)

Inertia torque

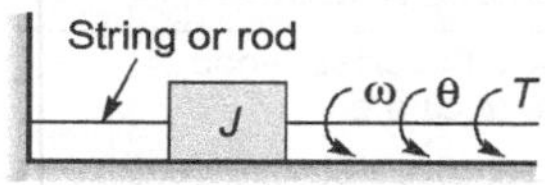

$$T = J\frac{d^2\theta}{dt^2} = J\frac{d\omega}{dt}$$

Damper torque

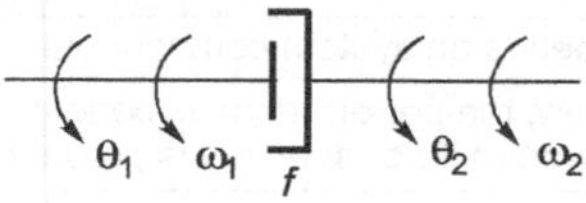

$$T = f\frac{d\theta}{dt} = f(\omega_1 - \omega_2)$$

Spring torque

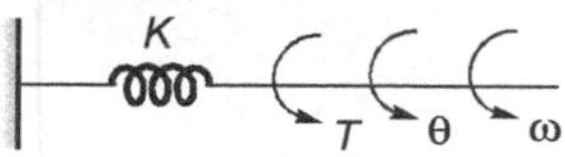

$$T = K\theta = K\int \omega dt$$

Analogous System

Electrical equivalent of mechanical system is called analogous system.

Mechanical Translation of Systems

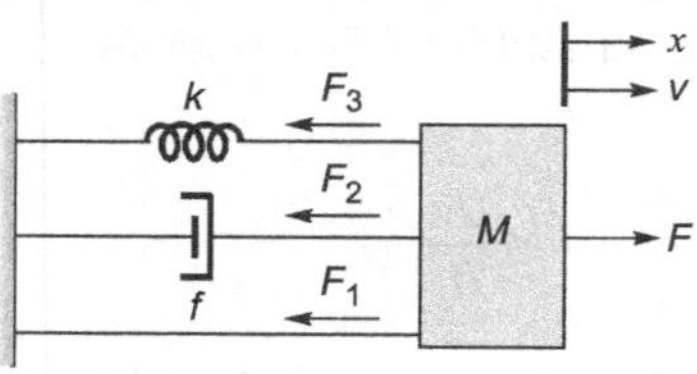

$$F = F_1 + F_2 + F_3$$

$$F = M\frac{dv}{dt} + fv + k\int vdt$$

$$F = M\frac{d^2x}{dt^2} + f\frac{dx}{dt} + kx$$

Mechanical Rotational Systems

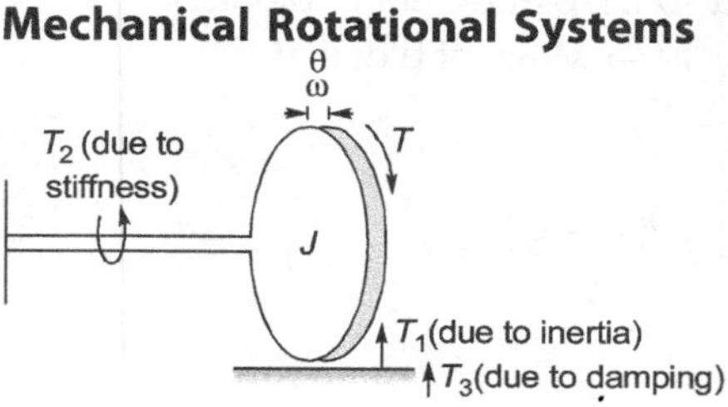

$$T = T_1 + T_2 + T_3$$

$$T = J\frac{d\omega}{dt} + f\omega + K\int \omega dt$$

$$T = J\frac{d^2\theta}{dt^2} + f\frac{d\theta}{dt} + K\theta$$

Electrical Systems

A. Series RLC Circuit

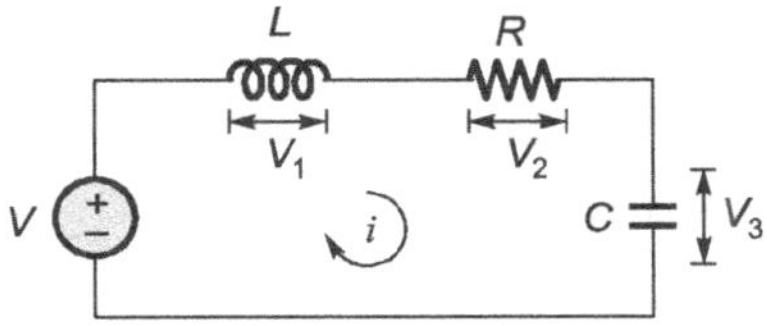

$$V = V_1 + V_2 + V_3$$

$$V = L\frac{di}{dt} + iR + \frac{1}{C}\int i dt$$

$$V = L\frac{d^2q}{dt^2} + R\frac{dq}{dt} + \frac{q}{C}$$

B. Parallel RLC Circuit

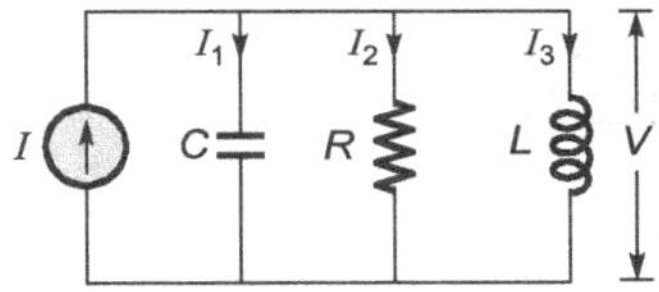

$$I = I_1 + I_2 + I_3$$

$$I = C\frac{dV}{dt} + \frac{V}{R} + \frac{1}{L}\int V dt$$

$$I = C\frac{d^2\phi}{dt} + \frac{1}{R}\frac{d\phi}{dt} + \frac{\phi}{L}$$

Force Voltage Analogy & Force Current Analogy

Series RLC	Parallel RLC	Mechanical Translation System	Mechanical Rotational System
V	I	F	T
q	ϕ	x	θ
R	$1/R$	f	f
I	V	Linear velocity (v)	Angular velocity (ω)
$1/C$	$1/L$	k	K
L	C	M	J

Nodal Method

- Number of nodes = Number of displacement.
- Take an additional node which is reference node.
- Connect the mass or inertia element between principal node and reference node only.
- Connect the spring and damping elements either between the principal nodes or between the principal node and reference node, depending on their position.
- Obtain the nodal diagram and write the differential equation at each node.

Servomotors

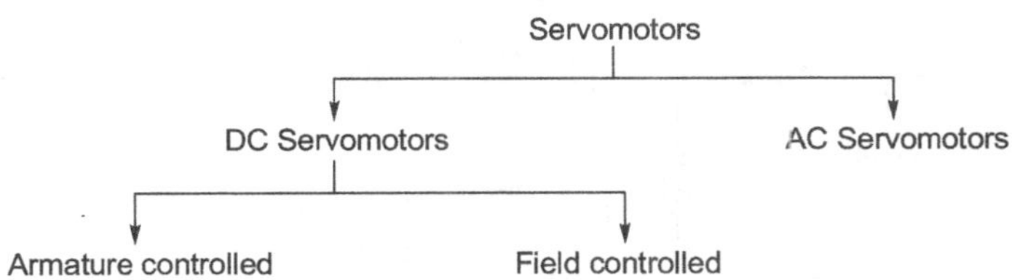

Armature Controlled DC Servo Motors

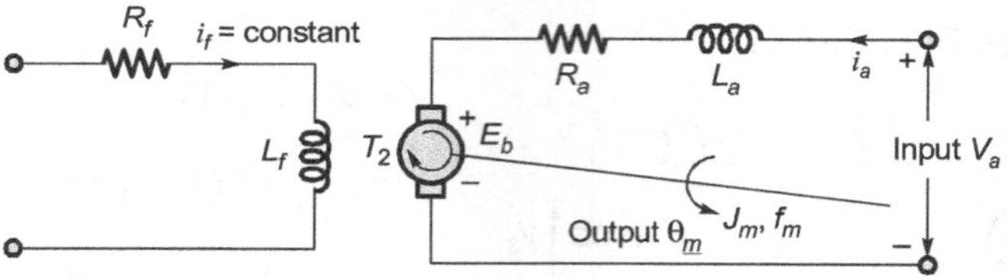

Block diagrams

(a) Relating $\theta_m(s)$ and $V_a(s)$

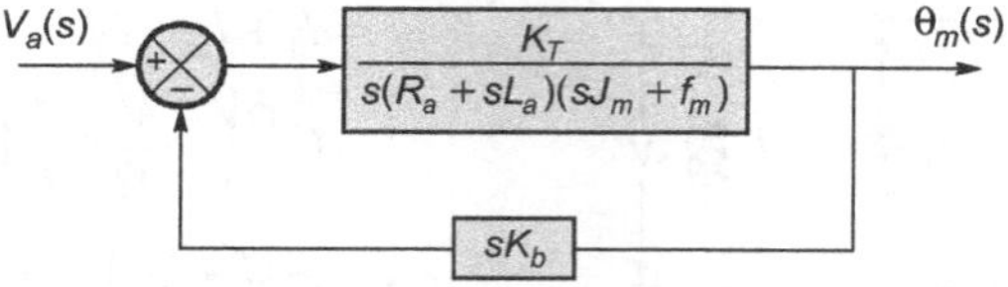

(b) Relating $\omega_m(s)$ and $V_a(s)$

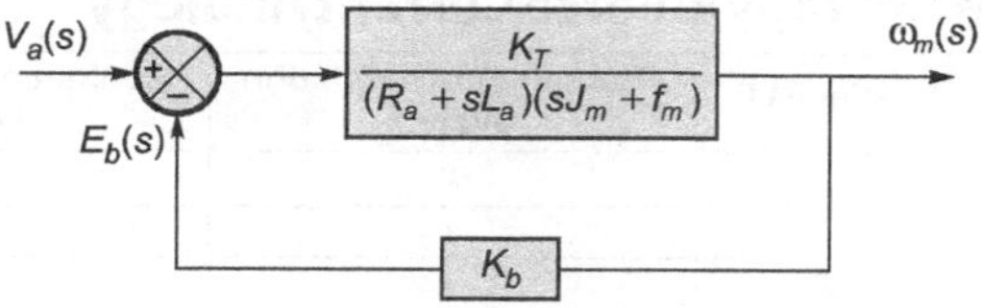

Dynamic equations

Dynamic equations in time domain	Corresponding equation in s-domain
$V_a - e_b = R_a i_a + L_a \dfrac{di_a}{dt}$	$V_a(s) - E_b(s) = R_a I_a(s) + sL_a I_a(s)$
$e_b = K_b \dfrac{d\theta_m}{dt} = K_b\omega_m$	$E_b(s) = sK_b\theta_m(s) = K_b\omega_m(s)$
$T_M = K_T i_a$	$T_M(s) = K_T I_a(s)$
$T_M = J_m \dfrac{d^2\theta_m}{dt^2} + f_m \dfrac{d\theta_m}{dt}$	$T_M(s) = s^2 J_m\theta_m(s) + sf_m\theta_m(s)$
$T_M = J_m \dfrac{d\omega_m}{dt} + f_m\omega_m$	$T_M(s) = sJ_m\omega_m(s) + f_m\omega_m(s)$

where, $e_b \rightarrow$ Back emf ; $T_m \rightarrow$ Torque

- **Field Controlled DC Servomotor**

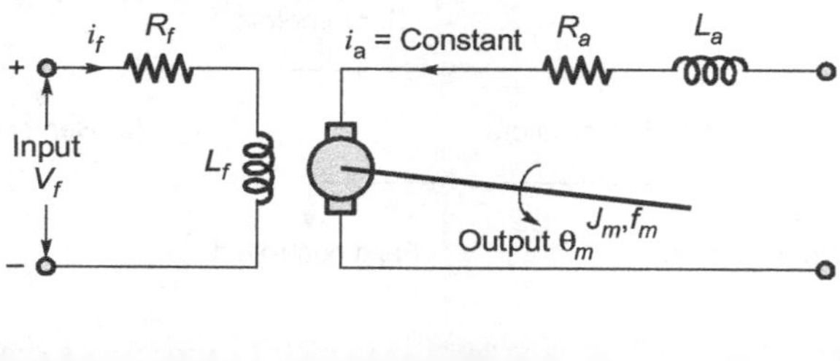

Block diagrams

(a) Relating $\theta_m(s)$ and $V_f(s)$

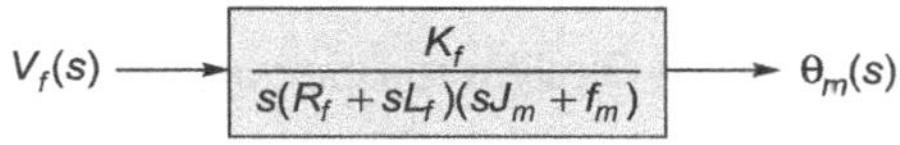

(b) Relating $\omega_m(s)$ and $V_f(s)$

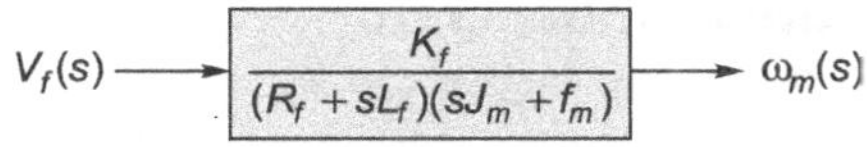

Dynamic equations

Dynamic equations in time domain	Corresponding equation in s-domain
$V_f = R_f i_f + L_f \frac{di_f}{dt}$	$V_f(s) = R_f I_f(s) + sL_f I_f(s)$
$T_M = K_f i_f$	$T_M(s) = K_f I_f(s)$
$T_M = J_m \frac{d^2\theta_m}{dt^2} + f_m \frac{d\theta_m}{dt}$	$T_M(s) = s^2 J_m \theta_m(s) + sf_m \theta_m(s)$
$T_M = J_m \frac{d\omega_m}{dt} + f_m \omega_m$	$T_M(s) = sJ_m \omega_m(s) + f_m \omega_m(s)$

- **Two Phase AC Servomotor**

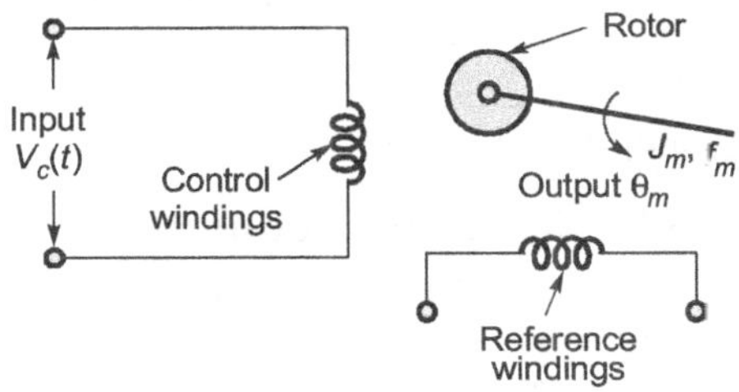

Block diagrams

(a) Relating $\theta_m(s)$ and $V_c(s)$:

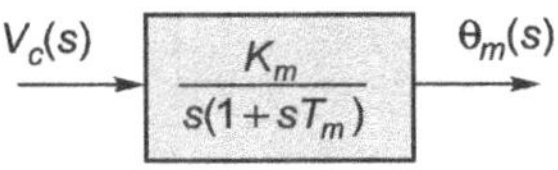

(b) Relating $\omega_m(s)$ and $V_c(s)$:

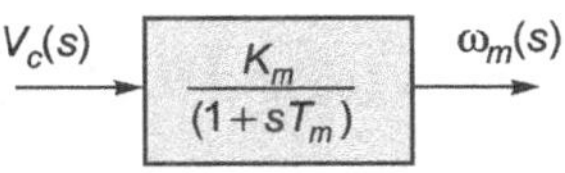

Dynamic equations

Dynamic equations in time domain	Corresponding equations in s-domain
$T_M = m\frac{d\theta_m}{dt} + KV_c$	$T_M(s) = sm\theta_m(s) + KV_c(s)$
$T_M = J_m \frac{d^2\theta_m}{dt^2} + f_m \frac{d\theta_m}{dt}$	$T_M(s) = s^2 J_m \theta_m(s) + sf_m \theta_m(s)$

where, $m = \frac{T_o}{\omega_o}$ (slope of torque speed characteristic)

$$K = \frac{T_o}{V_c}\ ;\ T_o \rightarrow \text{Stalling torque}\ ;\ \omega_0 \rightarrow \text{No load speed}$$

Generators

- **Separately Excited DC Generator**

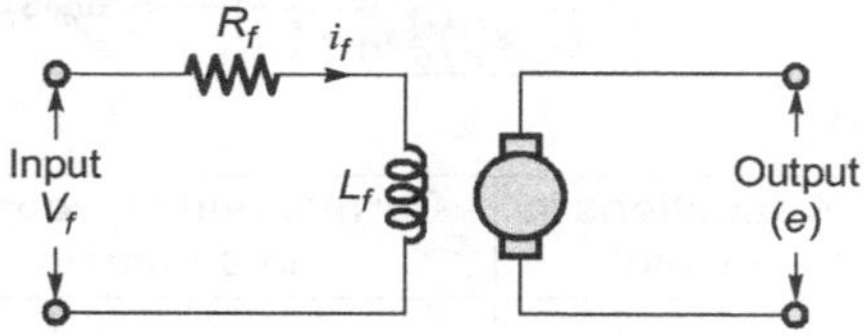

Block diagram

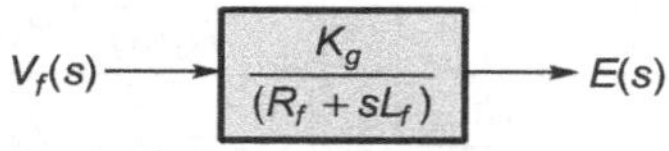

Dynamic equations

Related equations in time domain	Corresponding equations in s-domain
$V_f = R_f i_f + L_f \frac{di_f}{dt}$	$V_f(s) = R_f I_f(s) + sL_f I_f(s)$
$e = K_g i_f$	$E(s) = K_g I_f(s)$

where, $K_g \rightarrow$ Generator constant

Separately Excited DC Generator Connected to Load

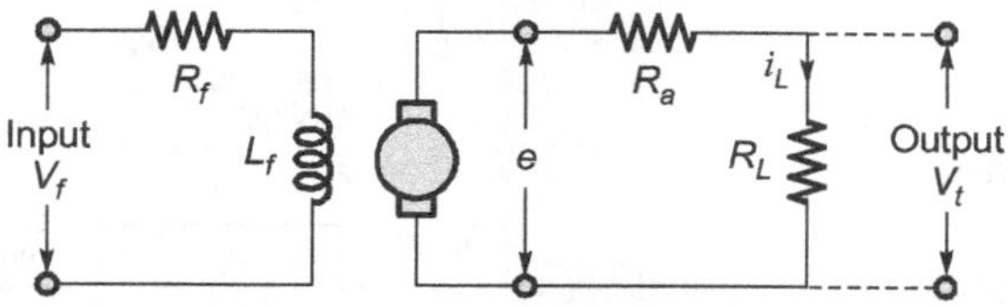

Block diagram

Relating $V_t(s)$ and $V_f(s)$:

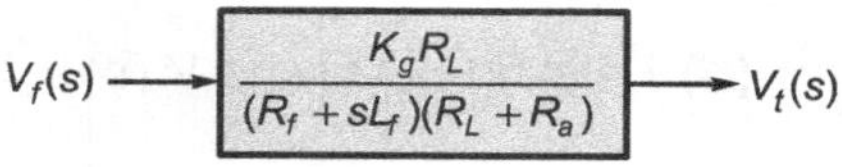

Dynamic equations

Related equations in time domain	Corresponding equations in s-domain
$V_t = e - R_a i_L$	$V_t(s) = E(s) - R_a I_L(s)$
$I_L = \frac{V_t}{R_L}$	$I_L(s) = \frac{V_t(s)}{R_L}$

Note:

- ☑ Servomechanism is electromechanical system whose input is electrical ***voltage*** and output is mechanical ***position*** or its time derivative.
- ☑ Field control DC servo motor is used for low power applications while armature control DC servomotor is used for high power applications.
- ☑ AC servomotor is generally used for low power application.
- ☑ Torque speed characteristic is linear for AC servomotor whereas, it is nonlinear for general purpose induction motor
- ☑ For the fast response of the servomotor system, the inertia of the rotor should be kept to minimum. This is achieved by reducing the diameter and increasing the length of the rotor.

Tachometer

Tachometer is a speed transducer, used as feedback element in control system.

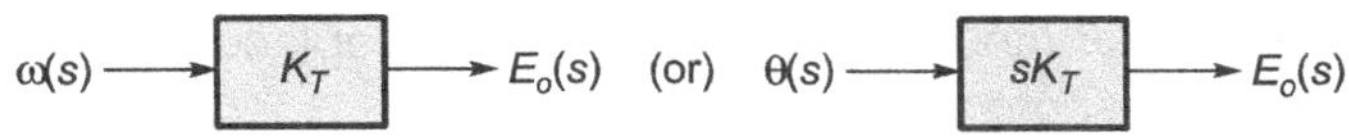

where, $sK_T \rightarrow$ Tachometer constant

Potentiometer

It is variable resistive displacement transducer. A pair of potentiometer act as error detectors in control system application.

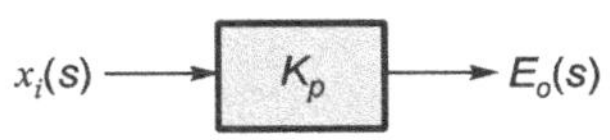

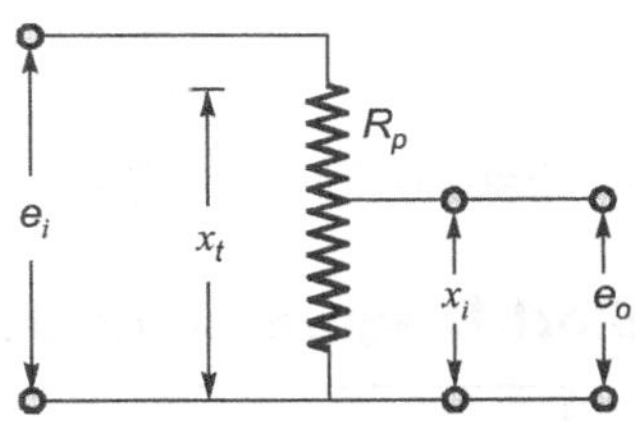

Input = Wiper displacement (x_i)

Output = Voltage (e_o)

$$K_P = \frac{e_i}{x_t} = \text{Potentiometer gain}$$

III Transfer Function

The transfer function of a linear, time-invariant, differential equation system is defined as the ratio of the Laplace transform of the output to the Laplace transform of the input, under the assumption that all initial conditions are zero.

$$G(s) = \frac{L[y(t)]}{L[u(t)]} = \frac{Y(s)}{U(s)}$$

Note:

☑ A system is said to be linear if principle of superposition and principle of homogeneity applies.

☑ A differential equation is linear if the coefficients are constants or function only of independent variable.

Open Loop Transfer Function

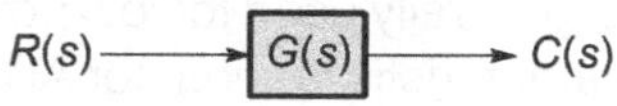

- Transfer function, $\text{T.F.} = \dfrac{C(s)}{R(s)} = G(s)$

  where, $C(s) \rightarrow$ Laplace transform of output ; $R(s) \rightarrow$ Laplace transform of input

- Impulse response of a system, $g(t) = c(t)$

  where, $g(t) \rightarrow$ Impulse response ; $c(t) \rightarrow$ Output of the system

Note:

☑ The transfer function of a system is the Laplace transform of its impulse response.

☑ Transfer function is always calculated with zero initial conditions.

Closed Loop Transfer Function

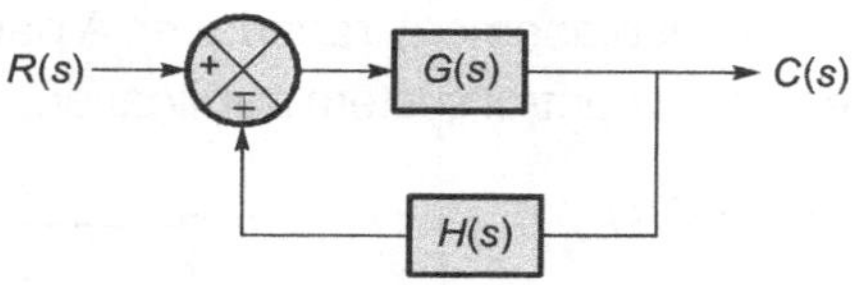

- Transfer function, $\text{T.F.} = \dfrac{G(s)}{1 \pm G(s)H(s)}$

Block Diagram Reduction

Original Block Diagram	Equivalent Block Diagram
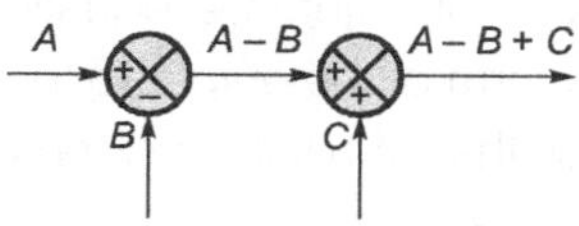	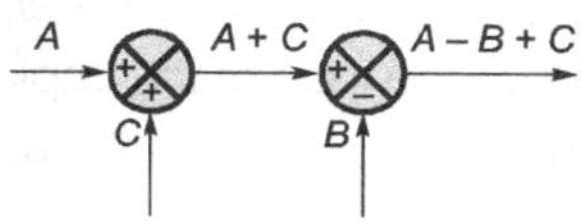

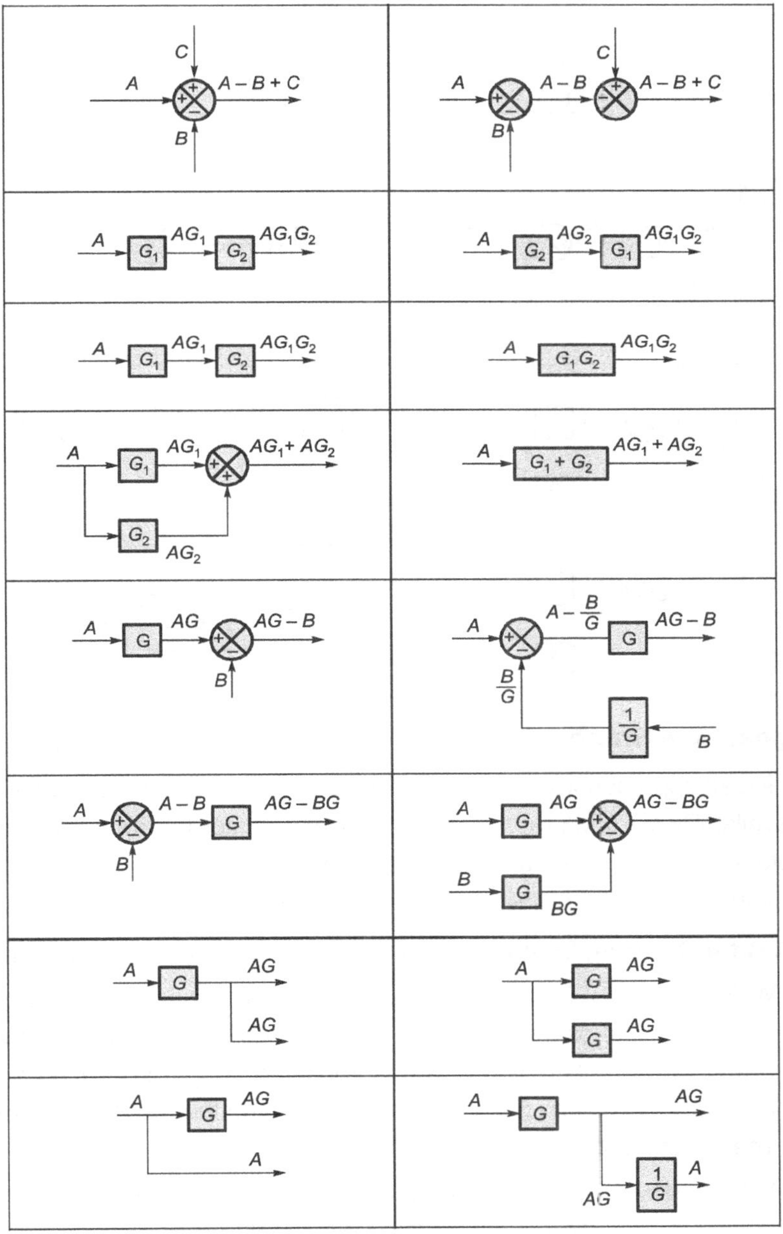
C
A
A – B + C
B
C
A
A – B
A – B + C
B
A
AG1
AG1G2
G1
G2
A
AG2
AG1G2
G2
G1
A
AG1
AG1G2
G1
G2
A
G1G2
AG1G2
A
AG1
AG1+ AG2
G1
G2
AG2
A
G1 + G2
AG1 + AG2
A
G
AG
AG – B
B
A – B/G
A
G
AG – B
B/G
1/G
B
A
A – B
G
AG – BG
B
A
G
AG
AG – BG
B
G
BG
A
G
AG
AG
A
G
AG
G
AG
A
G
AG
A
A
G
AG
1/G
A
AG

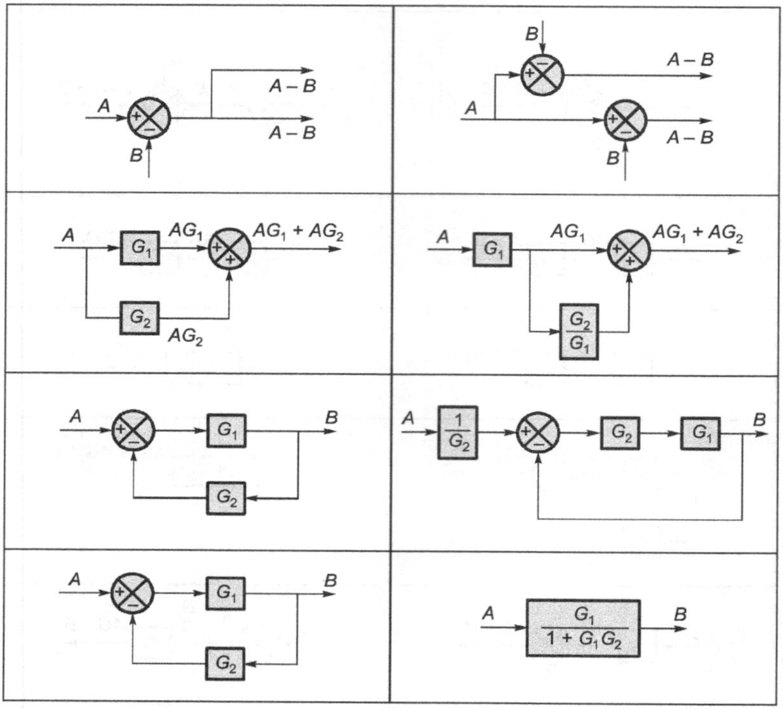

Signal Flow Graph

It is an approach which does not require any reduction process because of availability of a flow graph gain formula which relates the input and output system variables. It is a graphical representation of the relationships between the variables of a set of linear algebraic equations.

Rules for Drawing Signal Flow Graph From a given Block Diagram

1. While drawing SFG from a given block diagram, the adjacent summing points and take-off points (but not a take-off point preceding a summing point in the direction of SFG) are represented by a node and the block transfer function is represented by a line joining the respective nodes. The direction of signal flow is indicated by an arrow on the line.
2. If in the direction of signal flow, a take-off point precedes a summing point then such points are represented by two separate nodes with a transmittance of unity between them.

Original Block Diagram	Equivalent SFG
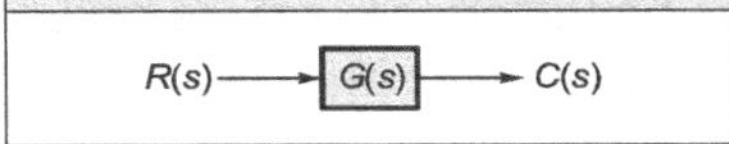	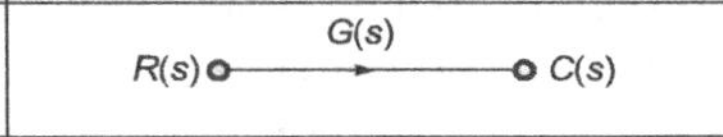
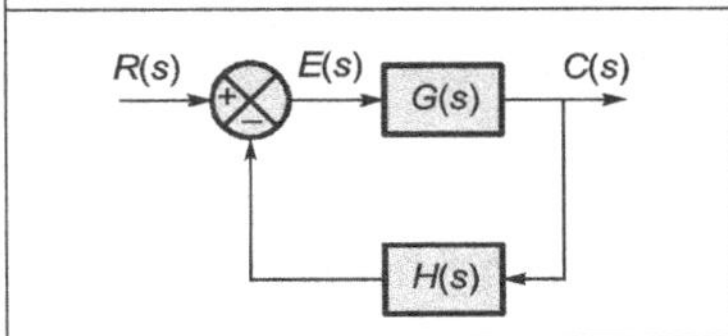	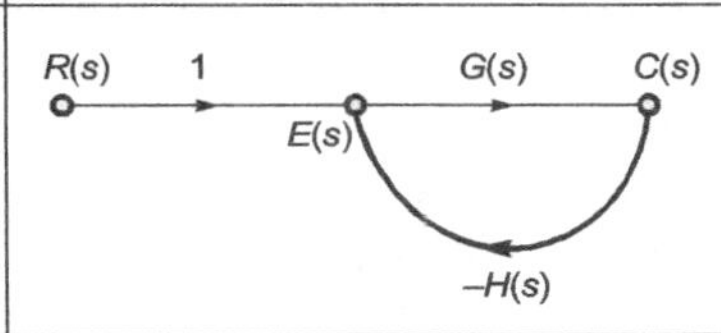
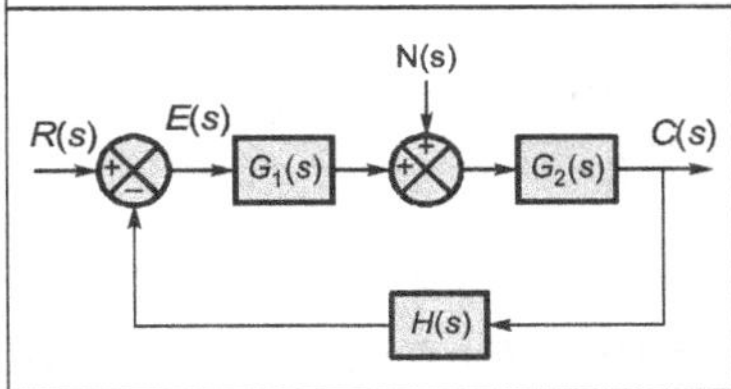	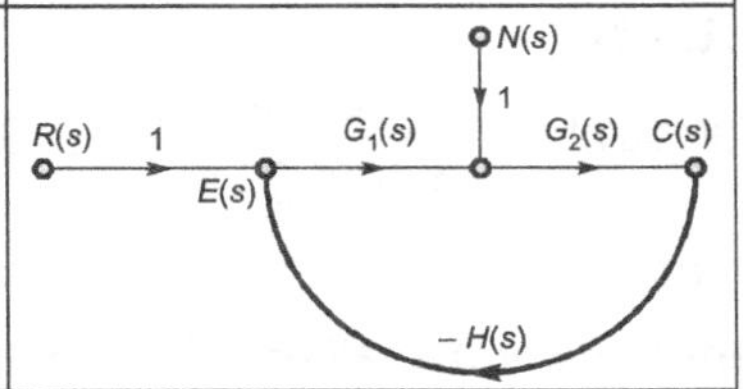
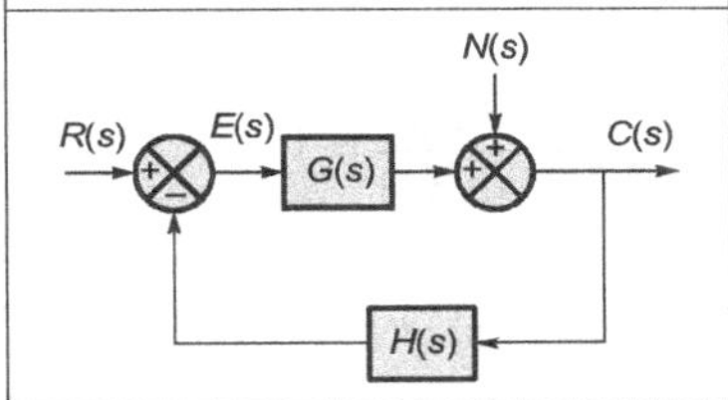	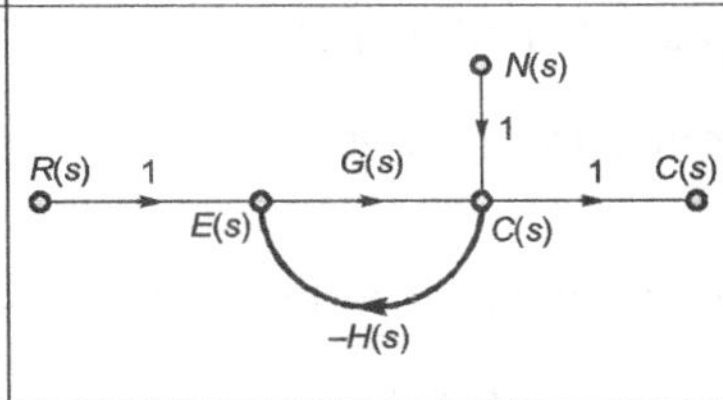
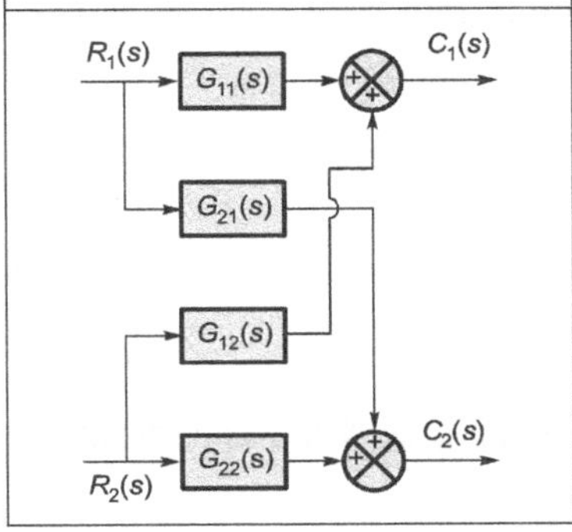	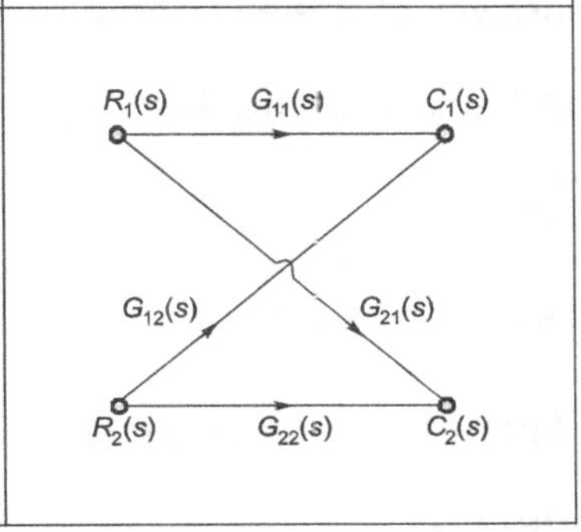

- Mason's gain formula, $P = \frac{1}{\Delta}\sum_{k} P_k \Delta_k$

where, $P \rightarrow$ Overall gain or transfer function

$P_k \rightarrow$ Path gain of kth forward path

$\Delta \rightarrow$ 1 – (sum of loop gains of all individual loops) + (sum of gain products of all possible combinations of two non-touching loops) – (sum of gain products of all possible combinations of three non-touching loops) + ...

$\Delta_k \rightarrow$ Value of Δ obtained by removing all the loops touching k^{th} forward path

Note:

☑ Forward path is a path that can be traced through the graph from the input node to the output node without touching a node twice.
☑ Path gain is the product of all branch gain in a path.
☑ The individual loop is a closed path starting and ending at the same node.
☑ Loop gain is the product of all branch gains in a loop.
☑ Non touching loops are loops which do not have a common node.

Sensitivity

$$\text{Sensitivity} = \frac{\%\text{ change in } A}{\%\text{ change in } K} \; ; \; S_K^A = \frac{\partial A / A}{\partial K / K} = \frac{\partial A}{\partial K} \times \frac{K}{A}$$

where, $S_K^A \rightarrow$ Sensitivity of variable A with respect to parameter K

- **Sensitivity of Overall Transfer Function $M(s)$ With Respect to Forward Path Transfer Function $G(s)$**
- For open loop control system, $S_G^M = \dfrac{G(s)}{M(s)} \cdot \dfrac{\partial M(s)}{\partial G(s)} = 1$
- For closed loop control system, $S_G^M = \dfrac{1}{1 + G(s)H(s)}$
- **Sensitivity of Overall Transfer Function $M(s)$ With Respect to Feedback Path Transfer Function $H(s)$:** $S_H^M = -\dfrac{G(s)H(s)}{1 + G(s)H(s)}$

Note:

Closed loop system is less sensitive to parameter variation. Hence, closed loop system is better.

IV Time Response Analysis of Control System

Introduction

The time response of a control system means as to how a system behaves in accordance with time when a specified input test signal is applied.

The time response of a control system is divided in two parts:

(i) Transient response (ii) Steady state response

Transient part of time response reveals the nature of response and its speed, where as steady state part of time response reveals the accuracy of a control system.

Standard Input Test Signals

Step function	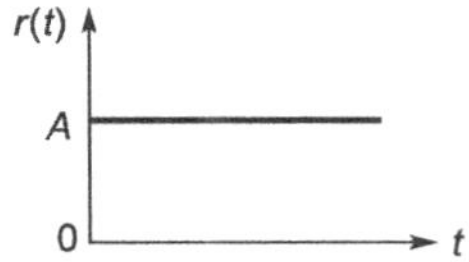	$r(t) = Au(t)$ where, $u(t) = \begin{cases} 1 \; ; & t \geq 0 \\ 0 \; ; & t < 0 \end{cases}$
Ramp function	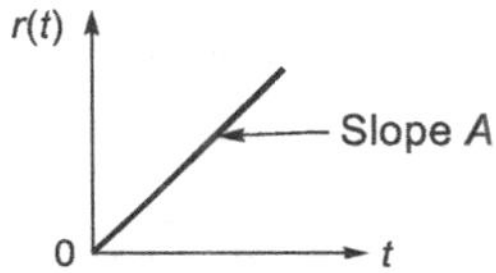	$r(t) = \begin{cases} At \; ; & t \geq 0 \\ 0 \; ; & t < 0 \end{cases}$
Parabolic function	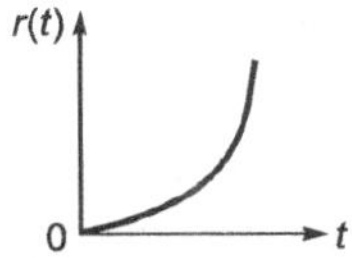	$r(t) = \begin{cases} \frac{At^2}{2} \; ; & t \geq 0 \\ 0 \; ; & t < 0 \end{cases}$
Impulse Function	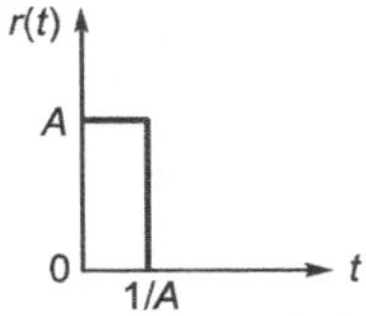	$r(t) = A\delta(t) = \begin{cases} A \; ; & t = 0 \\ 0 \; ; & t \neq 0 \end{cases}$ $\lim_{\alpha \to 0} \int_{-\alpha}^{\alpha} \delta(t)dt = 1$

Time response of a first order control system

Transfer Function, $\dfrac{C(s)}{R(s)} = \dfrac{1}{sT+1}$

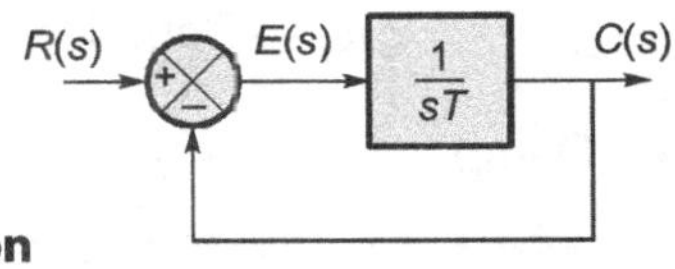

(i) Response to Unit Impulse Input Function

Time response expression, $c(t) = \dfrac{1}{T} e^{-t/T} \; u(t)$

(ii) Response to Unit Step Input Function

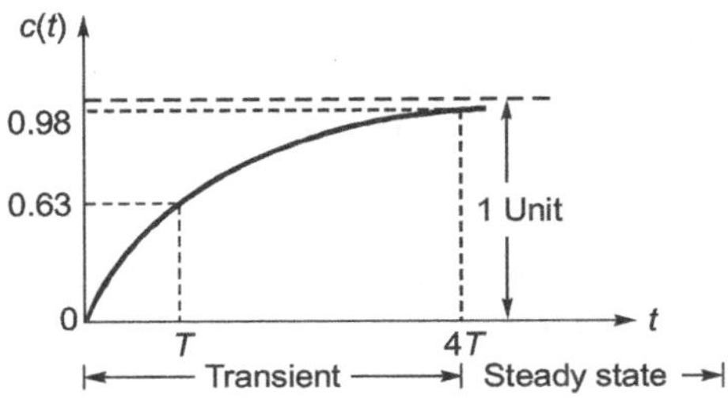

$$C(s) = \frac{1}{s(sT+1)}$$

- Time response expression, $c(t) = \mathcal{L}^{-1}\{C(s)\} = (1 - e^{-t/T})\, u(t)$

- Error, $e(t) = r(t) - c(t) = e^{-t/T}u(t)$
- Steady state error, $e_{ss} = \lim_{t \to \infty} e^{-t/T} = 0$

Note:

Lower the time constant, faster is the time response of a control system.

(iii) Response to Unit Ramp Input Function

- Time response expression, $c(t) = (t - T + Te^{-t/T})u(t)$
- Error, $e(t) = r(t) - c(t) = (T - Te^{-t/T})u(t) = (1 - e^{-t/T})\,u(t)$
- Steady State Error, $e_{ss} = \lim_{t \to \infty}(T - Te^{-t/T}) = T$

Time Response of Second Order Control System

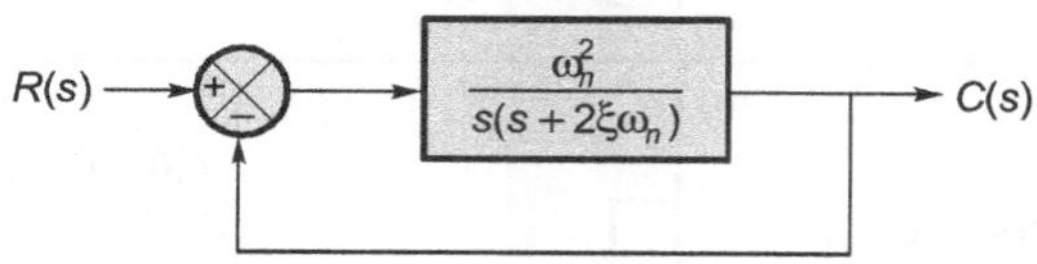

- Transfer function, $\dfrac{C(s)}{R(s)} = \dfrac{\omega_n^2}{s^2 + 2\xi\omega_n s + \omega_n^2}$

  where, $\omega_n \rightarrow$ Natural frequency of oscillations ; $\xi \rightarrow$ Damping ratio
  $\xi\,\omega_n \rightarrow$ Damping factor

- Damped frequency, $\omega_d = \omega_n\sqrt{1-\xi^2}$

  where, $\omega_d \rightarrow$ Damped frequency of oscillations

Response to Unit Step Input Function

- Time response expression

  Case-1: $\xi < 1$, i.e., Underdamped oscillations

$$c(t) = \left[1 - \frac{e^{-\xi\omega_n t}}{\sqrt{1-\xi^2}}\sin(\omega_d t + \phi)\right]u(t); \quad \phi = \tan^{-1}\left(\frac{\sqrt{1-\xi^2}}{\xi}\right)$$

Note:

Response settles within 2% of the desired value (1 unit) after damping out the oscillations in a time 4T (or $4/\xi\omega_n$).

Case-2: $\xi = 0$ i.e. Sustained (undamped) Oscillations

$$c(t) = (1 - \cos\omega_n t)u(t)$$

Case-3: $\xi = 1$, i.e., Critically Damped Oscillations

$$c(t) = [1 - e^{-\omega_n t}(1 + \omega_n t)]$$

Case-4: $\xi > 1$, i.e., Overdamped Oscillations

$$c(t) = 1 - \frac{e^{-\left(\xi - \sqrt{\xi^2 - 1}\right)\omega_n t}}{\left(2\sqrt{\xi^2 - 1}\right)\left(\xi - \sqrt{\xi^2 - 1}\right)}$$

Effect of ξ on Time Response

S.No.	Value of x	Type of closed poles	Nature of response	System classification
1.	$\xi = 0$	Purely imaginary	Oscillations with constant frequency and amplitude	Undamped
2.	$0 < \xi < 1$	Complex conjugate	Damped oscillation	Underdamped
3.	$\xi = 1$	Real, equal and negative	Critical and pure exponential	Critically damped
4.	$1 < \xi < \infty$	Real, unequal and negative	Purely exponential slow and sluggish	Overdamped

Pole Locations for Different Cases, for a Second Order System

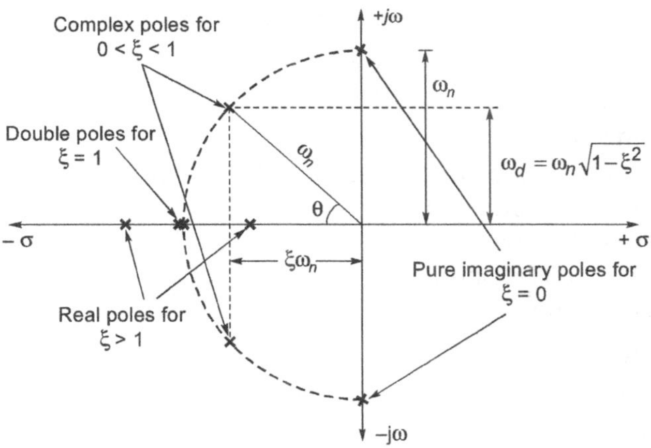

Note:

☑ As ξ increases from 0 to 1 then the frequency of oscillation reduces.

☑ For all positive value of ξ, system are stable.

☑ Pole away from origin are called insignificant pole and does not affect the stability of system.

☑ $\theta = \cos^{-1}\xi$.

Some Practical Example of Second Order System

Series RLC Circuit	Parallel RLC Circuit	Translatory System	Rotational System
Characteristic equation	Characteristic equation	Characteristic equation	Characteristic equation
$s^2+\frac{R}{L}s+\frac{1}{LC}=0$	$s^2+\frac{1}{RC}s+\frac{1}{LC}=0$	$s^2+\frac{f}{M}s+\frac{K}{M}=0$	$s^2+\frac{f}{J}s+\frac{K}{J}=0$
$\omega_n=\frac{1}{\sqrt{LC}}$	$\omega_n=\frac{1}{\sqrt{LC}}$	$\omega_n=\sqrt{\frac{K}{M}}$	$\omega_n=\sqrt{\frac{K}{J}}$
$\xi=\frac{R}{2}\sqrt{\frac{C}{L}}$	$\xi=\frac{1}{2R}\sqrt{\frac{L}{C}}$	$\xi=\frac{f}{2\sqrt{KM}}$	$\xi=\frac{f}{2\sqrt{KJ}}$

Transient Response Specifications of Second Order Control System

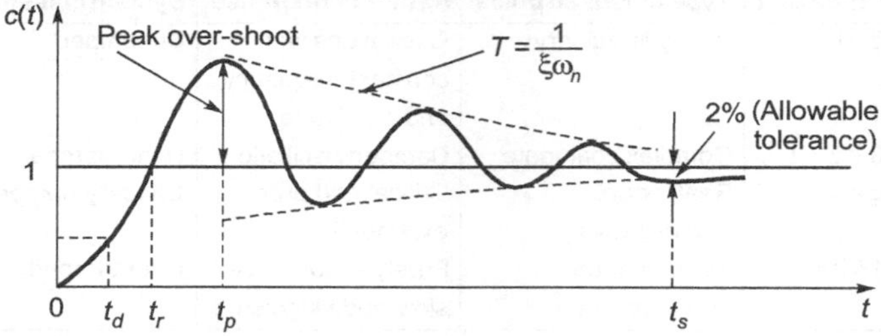

Delay Time (t_d)

It is the time required for the response to rise from 0 to 50% of the final value.

$$t_d = \frac{1+0.7\xi}{\omega_n}$$

Rise Time (t_r)

It is the time required for the response to rise from
0 to 100% (underdamped)
5 to 95% (critical damped)
10 to 90% (over damped)
of the final value

$$t_r = \frac{\pi-\tan^{-1}\left(\frac{\sqrt{1-\xi^2}}{\xi}\right)}{\omega_d} = \frac{\pi-\cos^{-1}\xi}{\omega_d}$$

Peak Time (t_p)

It is the time required for the response to rise from zero to peak of the time response.

$$t_p = \frac{n\pi}{\omega_d}$$ where, $n \to 1, 2, 3, \ldots$
for $n \to 1$, first overshoot; $n \to 2$, first undershoot

Settling Time (t_s)

It is the time required for response to rise and reach to the tolerance band

For 2% tolerance band, $t_s = 4\tau = \frac{4}{\xi\omega_n}$

For 5% tolerance band, $t_s = 3\tau = \frac{3}{\xi\omega_n}$

Peak Over Shoot (M_p)

It gives the normalised difference between the steady state output to first peak of the time response.

$$\%M_p = e^{-\left(\frac{\pi\xi}{\sqrt{1-\xi^2}}\right)} \times 100\%$$

Note:

☑ The time period of the oscillation before reaching the steady state.

$$T_{\text{oscillation}} = \frac{2\pi}{\omega_d}$$

☑ Number of cycles before reaching steady state, $N = \dfrac{T_{\text{settling}}}{T_{\text{oscillation}}}$

Type and Order of System

- Every Transfer function representing the control system has certain type and order.
- The steady state analysis depends on type of the system.
- The type of the system is obtained from open loop transfer function i.e. $G(s)\,H(s)$.
- The number of open loop poles occurring at origin determines the type of the system.

Let $\quad G(s)\,H(s) = \dfrac{K(1+sT_a)}{s^p(1+sT_1)}$

For
$p = 0$ for Type-0 system
$p = 1$ for Type-1 system
$p = n$ for Type-n system

- Transient state analysis depends on order of the system.
- The order of system is obtained from closed loop transfer function, i.e., $\dfrac{G(s)}{1+G(s)H(s)}$.
- The highest power of characteristic equation, i.e., $1 + G(s)\,H(s) = 0$ determines the order of control system.

Steady State Error

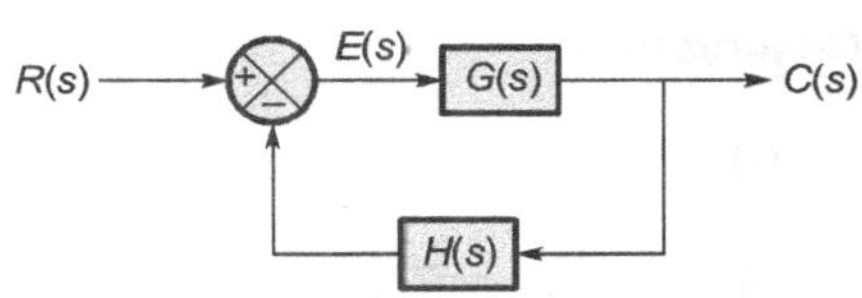

$$E(s) = \frac{R(s)}{1+G(s)H(s)}$$

Steady state error, $\quad e_{ss} = \lim_{t\to\infty} e(t)$ (in time-domain)

$$e_{ss} = \lim_{s \to 0} sE(s) \text{ (in } s\text{-domain)}$$

$$e_{ss} = \lim_{s \to 0} s \frac{R(s)}{1+G(s)H(s)}$$

Note:

Steady state error depends on two factor :

(i) Type of input applied i.e. $R(s)$.

(ii) Type of system, i.e., $G(s)\ H(s)$.

Steady State Error for Different Types of Input

(i) Step input

$$R(s) = \frac{A}{s} \quad ; \quad e_{ss} = \frac{A}{1+K_p} \quad ; \quad K_p = \lim_{s \to 0} G(s)H(s)$$

where, $K_p \rightarrow$ Static position error constant

(ii) Ramp input

$$R(s) = \frac{A}{s^2} \quad ; \quad e_{ss} = \frac{A}{K_v} \quad ; \quad K_v = \lim_{s \to 0} sG(s)H(s)$$

where, $K_v \rightarrow$ Static velocity error constant

(iii) Parabolic input

$$R(s) = \frac{A}{s^3} \quad ; \quad e_{ss} = \frac{A}{K_a} \quad ; \quad K_a = \lim_{s \to 0} s^2 G(s)H(s)$$

where, $K_a \rightarrow$ Static acceleration error constant

Steady State Error for Different Types of System

Type	Step input	Ramp input	Parabolic input
Type-0	$\frac{A}{1+K}$	∞	∞
Type-1	0	$\frac{A}{K}$	∞
Type-2	0	0	$\frac{A}{K}$

Where, K is the system gain.

Observation

(i) $e_{ss} \propto \frac{1}{K}$, so, as system gain increases, steady state error decreases.

(ii) For Linear Time Invariant (LTI) system, the maximum type number is two. Beyond type-2, the system exhibits non-linear characteristics.

Note:

- ☑ Steady state error is valid only for closed loop stable system.
- ☑ Steady state error are calculated for the closed loop system by using open loop transfer function.

V Stability in Time-Domain

Stability

A Linear Time Invariant (LTI) system is said to be stable if the following conditions are satisfied:

(i) If the system is excited by a bounded input, the output must be bounded.

(ii) If input to the system is zero, the output must be zero, irrespective of all the initial condition.

Marginal Stable System

A LTI system is said to be marginal stable if for the bounded input, the output maintains the constant amplitude and frequency.

Absolute Stable System

System is stable for all the value of system gain (K) from 0 to ∞.

Conditional Stable System

System is stable for a certain range of K.

Routh Hurwitz (R-H) Criterion

Purpose

To find:

1. Closed loop system stability.
2. Number of closed loop pole in right side or left side of the s-plane.
3. Range of K for conditionally stable system.
4. Value of K required for marginal stability and thereby, determine the frequency of oscillations.

Note:

To find a closed loop system stability by using R-H criteria we need characteristic equation i.e. $1 + G(s)H(s) = 0$.

Routh Array

- General n^{th} order characteristics equation is
  $a_0s^n + a_1s^{n-1} + a_2s^{n-2} + \cdots + a_{n-1}s^1 + a_n = 0$

- Routh array

$$
\begin{array}{c|cccc}
s^n & a_0 & a_2 & a_4 & \cdots \\
s^{n-1} & a_1 & a_3 & a_5 & \cdots \\
s^{n-2} & \dfrac{a_1 a_2 - a_0 a_3}{a_1} & \dfrac{a_1 a_4 - a_0 a_5}{a_1} & & \cdots \\
\vdots & \vdots & & & \\
s^0 & a_n & & &
\end{array}
$$

1. For closed loop system to be stable, all the coefficient in the first column must be positive and there should not be any missing coefficient in the first column.
2. If there is any sign change in the first column of the Routh array then system is unstable.
3. Number of sign changes is equal to number of roots lying in the right side of the s-plane.

Note:

☑ Magnitude and sign of every term in each row and column depend on first column element, therefore we always see first column term.

☑ The presence of a zero in the first column of the Routh's tabulation leads to following conclusions
   (i) Equal real roots with opposite signs.
   (ii) Pair of conjugate roots on imaginary axis.

Difficulties and Limitations of Routh Array

1. The Routh array is applicable to LTI system only.
2. It determines poles in LHS or RHS of s-plane but not their exact location.
3. Whenever any one coefficient is zero in first column then replace zero by smallest positive constant ($\in$) and continue the Routh array, finally substitute $\in = 0$ and check for the sign change.
4. When Routh array ends abruptly, construct an auxiliary equation and differentiate it to get new coefficient to complete the Routh array.
5. When the system is marginally stable, find the frequency of sustained oscillations from auxiliary equation. The auxiliary equation should be an even polynomial of order two only.

Root Locus

Root locus is defined as the locus of closed loop poles obtained when system gain K is varied from 0 to ∞.

Angle Condition

Angle condition is used for checking whether certain points lie on root locus or not and hence, the validity of root locus for closed loop poles.

For a point to lie on root locus, the angle evaluated at that point must be an odd multiple of ±180°, i.e., $\pm(2q + 1)180°$.

Magnitude Condition

The magnitude condition is used for finding the system gain K at any point on root locus. $|G(s)\, H(s)| = 1$

Construction Rules of Root Locus

- Root locus is symmetrical about real axis.
- Let P = Number of open loop poles
  Z = Number of open loop zeros

  If $P > Z$

  then (a) The number of branches of root locus = P
  (b) The number of branches terminating at zero = Z
  (c) The number of branches terminating at infinity = $P - Z$
- A point on real axis is said to be on root locus if, to the right side of this point, the sum of number of open loop poles and zeros is an odd number.
- **Angle of Asymptotes**

  Number of asymptotes = $P - Z$

  $$\text{Angle of asymptotes} = \frac{(2q+1)180°}{P-Z} \qquad (\text{where, } q = 0, 1, 2, \cdots)$$
- **Centroid**

  Centroid is the point of intersection of asymptote on the real axis.

  $$\text{Centroid} = \frac{\Sigma\text{Real part of open loop poles} - \Sigma\text{Real part of open loop zeros}}{P-Z}$$
- **Break away point**

  They are the points where multiple roots of characteristic equation occurs.

  Procedure to find out break away point :

  (a) Construct characteristic equation, i.e., $1 + G(s)\, H(s) = 0$.

  (b) Write K in terms of s.

  (c) Find $\frac{dK}{ds}$.

  (d) The root of $\frac{dK}{ds} = 0$ give break away points.
- **Intersection of root locus with imaginary axis**

  Roots of auxiliary equation at $K = K_{(\text{marginal})}$ from Routh array gives intersection of root locus with imaginary axis.
- **Angle of departure and arrival**

  The angle of departure is tangent to the root locus at the complex poles.

$$\phi_D = 180° + \phi$$

The angle of arrival is tangent to the root locus at the complex zero.

$$\phi_A = 180° - \phi \; ; \; \phi = \phi_Z - \phi_P$$

where, $\phi_z \rightarrow$ Sum of all the angles subtended by remaining zeros

$\phi_p \rightarrow$ Sum of all the angle subtended by poles

Note:

- ☑ Whenever the system transfer function consist of the poles at origin then the root locus is nothing but angle of asymptotes.
- ☑ Value of system gain K on root locus can be found out by

$$K = \frac{\Pi(\text{Phasor length from open loop poles})}{\Pi(\text{Phasor length from open loop zeros})}$$

Effect of Addition of Pole

1. Operating **range of K** decreases for system to be stable.
2. **Relative stability** reduces.
3. System becomes more **oscillatory**.
4. The **break away** point shift towards the imaginary axis.
5. **Damping factor** decreases and system becomes unstable.
6. **Settling time** (T_s) and bandwidth increases.
7. **Rise time** (T_r) decreases.

Effect of Addition of Zero

1. System become stable.
2. Relative stability improves.
3. Range of K for stability increases.
4. System becomes less oscillatory.
5. Break away point shift towards the left side of the s-plane.
6. Damping factor increases.
7. Bandwidth decreases.

VI Industrial Controller

Controller

It is a device that is used usually in series with a plant to drive the plant towards decreased output.

Types of Industrial Controller

Proportional Controller

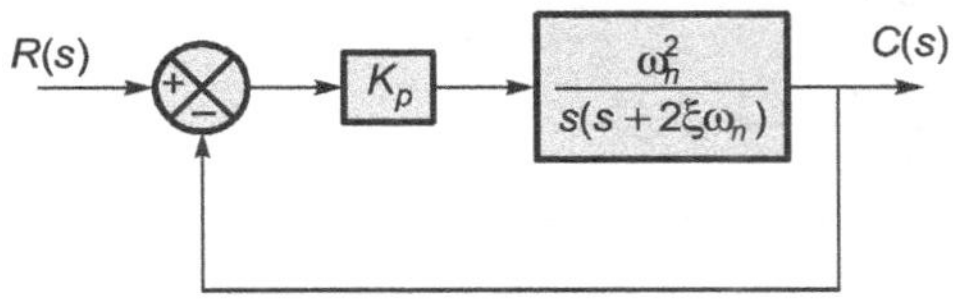

$$\omega_n' = \omega_n\sqrt{K_p} \text{ and } \xi' = \frac{\xi}{\sqrt{K_p}}$$

where, $K_p \rightarrow$ Proportional gain

Effect

(i) Natural frequency of oscillation (ω_n) increases by $\sqrt{K_p}$.

(ii) Damping ratio (ξ) decreases by $\sqrt{K_p}$.

(iii) Peak overshoot (M_p) increases.

(iv) Steady state error reduces.

Derivative Controller

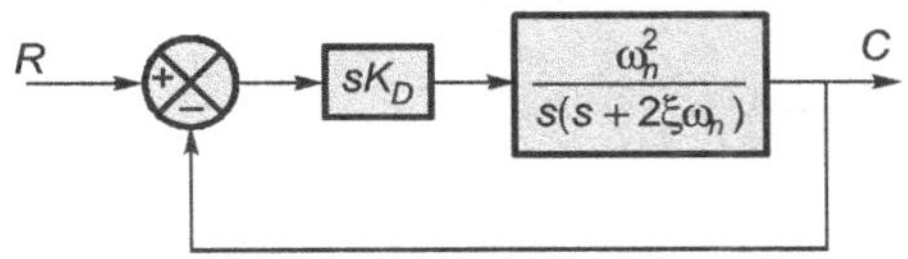

where, $K_D \rightarrow$ Rate constant

Effect

(i) Type and order of the system reduces by one.

(ii) Oscillations dies out, hence transient response improves.

(iii) Not used in isolation.

Integral Controller

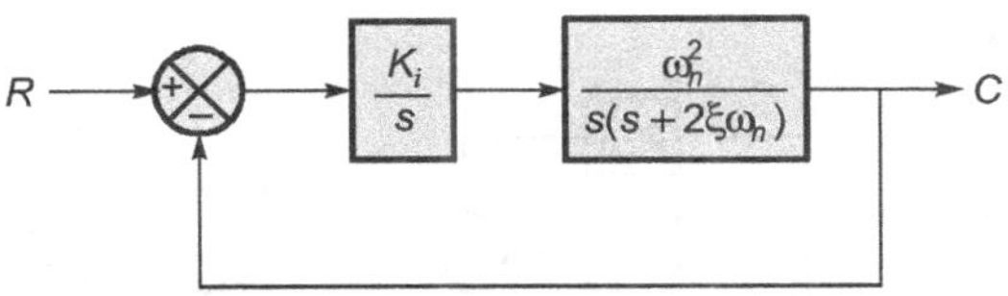

where, $K_i \rightarrow$ Integral scaling

Integral controller is a memory based controller.

Effect

(i) It increases type and order by one.

(ii) Makes the system less stable.

(iii) Steady state error reduces.

(iv) It improves the steady state response.

P-D Controller

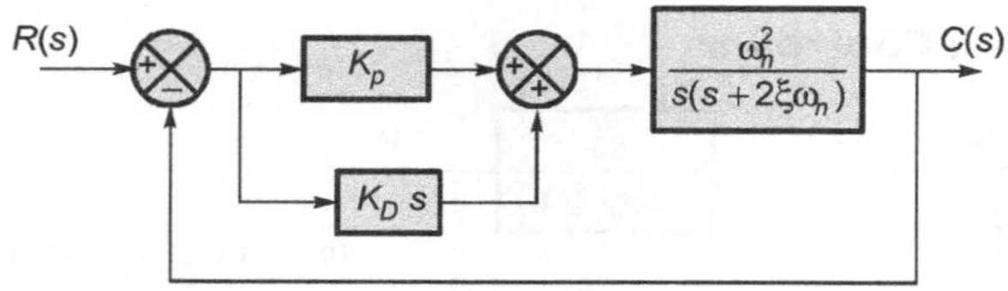

Effect:

(i) Transient response improves.
(ii) Damping ratio increases.
(iii) Peak overshoot decreases.
(iv) Bandwidth increases.
(v) Noise level increases.
(vi) Improves gain margin, phase margin and resonant peak.

P-I Controller

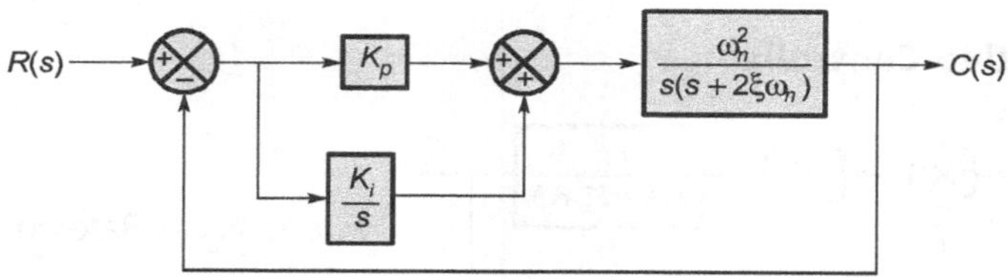

Effect

(i) Improves steady state response.
(ii) Improves type and order of system by one.
(iii) Noise level reduces.
(iv) Increases error constant.
(v) Bandwidth reduces.

PID Controller

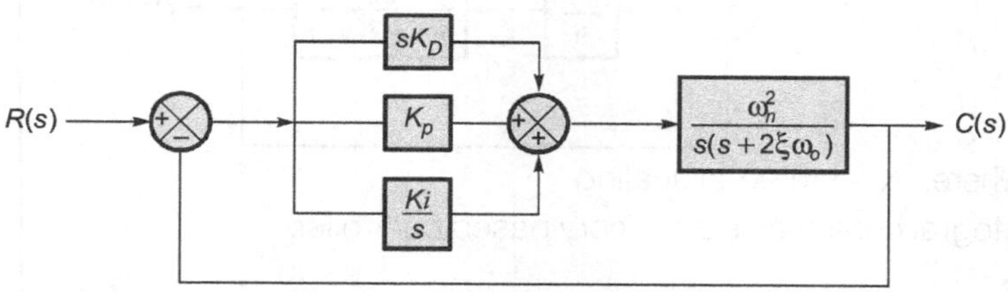

It is similar to lead-lag compensator and band reject filter.

Effect

(i) It improves both steady state as well as transient response.
(ii) It reduces rise-time.

(iii) It increases bandwidth.
(iv) It increases stability.
(v) If eliminate steady state error between input and output.
(vi) It increases type and order of system by one.

VII Compensator

Introduction

Compensator in control system are used for improving the performance specifications, i.e., the transient and steady state response characteristics.

Lead Compensator

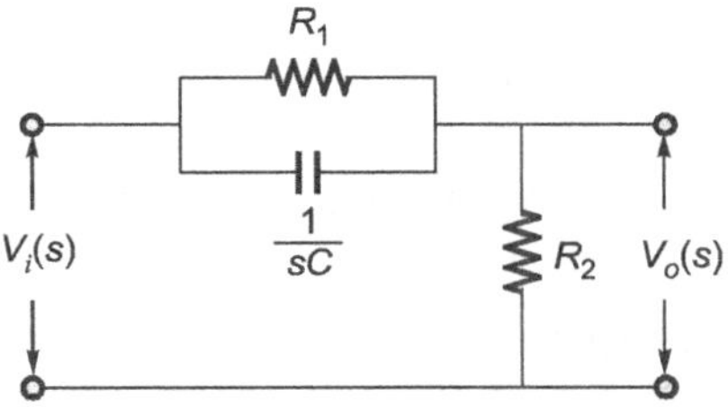

$$\frac{V_o(s)}{V_i(s)} = \frac{\alpha(1+Ts)}{(1+\alpha Ts)}$$

where, $T = R_1 C$ and $\alpha = \frac{R_2}{R_1 + R_2}$ $(\alpha < 1)$

Note:

☑ Zero is closer to origin than the pole.
☑ It is similar to the PD controller.

Effect of Lead Compensator

1. It improves the transient response.
2. It increases the margin of stability.
3. It increases the bandwidth.
4. $\left(\frac{\text{Signal}}{\text{Noise}}\right)_{\text{output}} < \left(\frac{\text{Signal}}{\text{Noise}}\right)_{\text{input}}$
5. It helps to increase error constant upto some extent.
6. It allows to pass high frequencies and attenuates low frequencies.
7. Increase the phase shift.

Lag Compensator

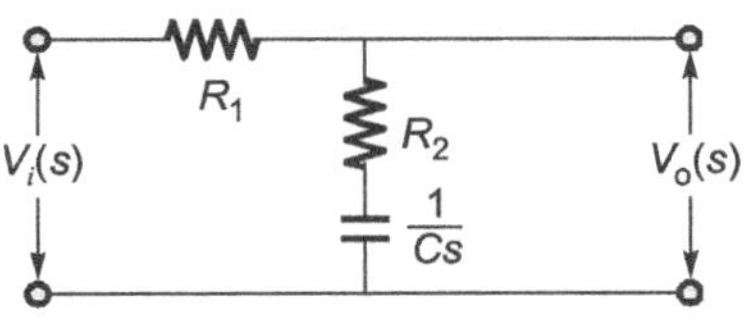

$$\frac{V_o(s)}{V_i(s)} = \frac{1+Ts}{1+\beta Ts}$$

where, $T = R_2 C$ and $\beta = \frac{1}{\alpha} = \frac{R_1 + R_2}{R_2}$ $(\beta > 1)$

Note:

☑ Pole is closer to origin than the zero.
☑ It is similar to the PI controller.

Effect of Lag Compensator

1. It improves the steady state response.
2. It increases the error constant.
3. It decreases the bandwidth.
4. It reduces the effect of noise.
5. It reduces the stability margin.
6. It does not affect the transient response.
7. System become lesser stable.
8. It allows to pass low frequencies and attenuates the high frequencies.
9. Decreases the phase shift.

Lead-Lag/Lag-Lead Compensator

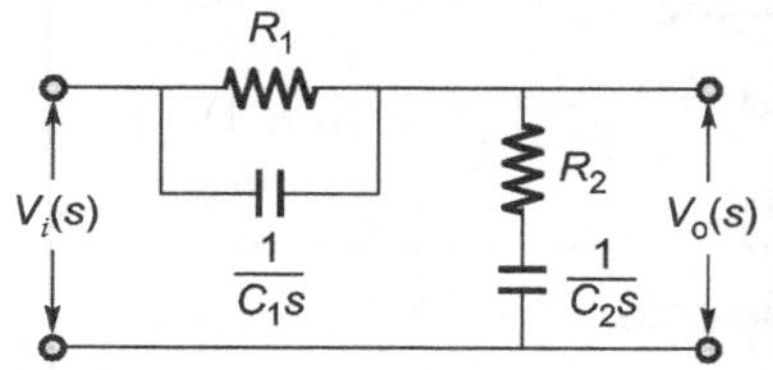

$$\frac{V_o(s)}{V_i(s)} = \frac{\alpha(1+T_1 s)(1+T_2 s)}{(1+\alpha T_1 s)(1+\beta T_2 s)}$$

where, $T_1 = R_1 C_1$ and $T_2 = R_2 C_2$

$$\alpha = \frac{R_2}{R_1 + R_2} \text{ and } \beta = \frac{R_1 + R_2}{R_2}$$

Effect of lead-lag/lag-lead compensator

Lead-Lag network improves both steady state and transient response of the system.

Note:

☑ Compensator in the form $G(s) = \frac{s+a}{s+b}$ will work as lead compensator if $a < b$, and as lag compensator if $a > b$.
☑ Frequency at which maximum phase occurs is $\omega = \sqrt{ab}$ rad/s.
☑ Maximum phase (ϕ_m) is given as, $\sin\phi_m = \frac{1-\alpha}{1+\alpha}$, where $\alpha = \frac{a}{b}$.

VIII Frequency Response Analysis

Frequency Response Analysis for Second Order System

Transfer Function

$$\frac{C(s)}{R(s)} = \frac{\omega_n^2}{s^2 + 2\xi\omega_n s + \omega_n^2}$$

Resonant Frequency

It is defined as, the frequency at which the magnitude has maximum value.

$$\omega_r = \omega_n\sqrt{1-2\xi^2}$$

Resonant Peak

It is maximum value of magnitude occurring at resonant frequency ω_r.

$$|M_r| = \frac{1}{2\xi\sqrt{1-\xi^2}}$$

Phase Angle at Resonant Frequency

$$\phi_r = -\tan^{-1}\left[\frac{\sqrt{(1-2\xi^2)}}{\xi}\right]$$

Remember:

☑ As ξ approaches zero, ω_r approaches ω_n and M_r approaches to infinity.

☑ For $0 < \xi \le 1/\sqrt{2}$, the resonant frequency always has a value less than ω_n and the resonant peak has a value greater than 1.

Cut-off Frequency and Bandwidth

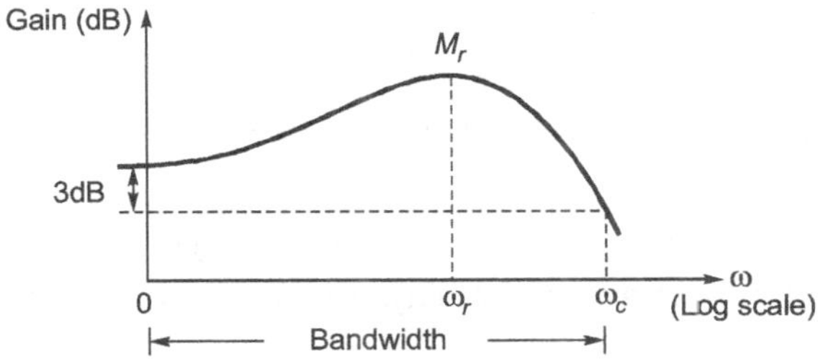

- Cut-off frequency, $\omega_c = \omega_n\left[1-2\xi^2 + \sqrt{4\xi^4 - 4\xi^2 + 2}\right]^{1/2}$
- Bandwidth $= \omega_c$

Stability from Frequency Response Plot

- Phase Margin, P.M. $= 180° + \angle G(j\omega)\,H(j\omega)$

....at gain cross over frequency

- Gain Margin,

$$G.M. = \frac{1}{|G(j\omega)\,H(j\omega)|_{\omega=\omega_{pc}}} = \frac{1}{X} \quad \cdots \text{at phase crossover frequency}$$

$$\text{G.M. (dB)} = 20\log\left(\frac{1}{X}\right)$$

Remember:

☑ **For stable systems**
(i) Gain cross over frequency < phase cross-over frequency
(ii) G.M. and P.M. both are positive

☑ **For unstable systems**
(i) Gain cross over frequency > phase cross-over frequency
(ii) G.M. and P.M. both are negative

☑ **For marginally stable systems**
(i) Gain cross over frequency = Phase cross-over frequency
(ii) G.M. and P.M. both are zero

G.M. and P.M. for Second Order System

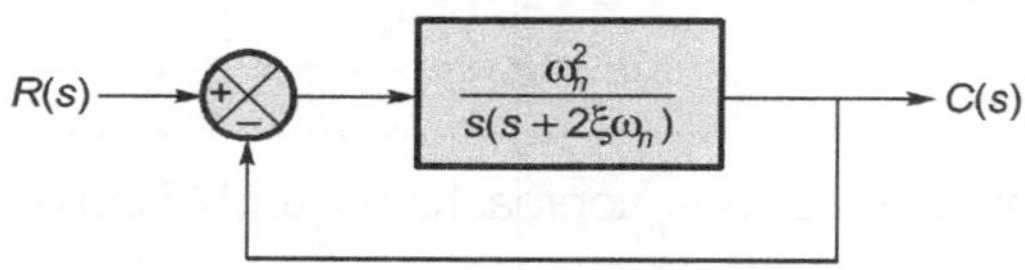

- Gain cross-over frequency, $\omega_{gc} = \omega_n\sqrt{-2\xi^2 + \sqrt{4\xi^4 + 1}}$

- Phase margin, $\text{P.M.} = \tan^{-1}\left[\frac{2\xi}{\sqrt{-2\xi^2 + \sqrt{4\xi^4 + 1}}}\right] \approx 100\xi$

  Gain margin of first and second order system is ∞.

Polar Plot

It is a plot of absolute values of magnitude and phase angle in degree of open loop transfer function $G(j\omega)\,H(j\omega)$ versus ω drawn on polar coordinates.

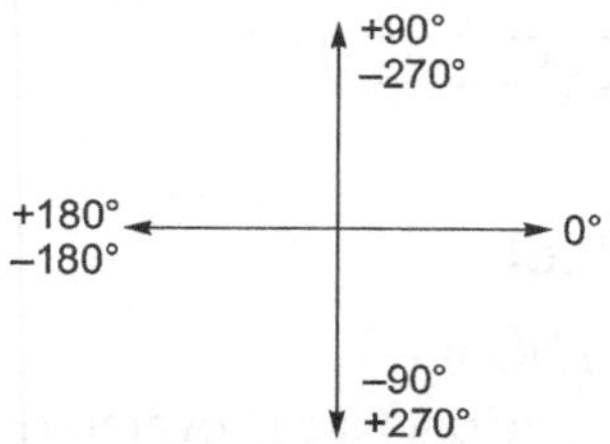

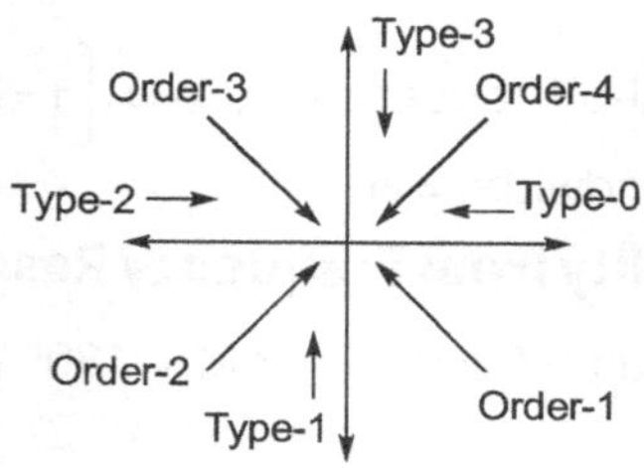

General Shapes of Polar Plot

Type/Order	Transfer Function	Polar Plot
Type-0/Order-1	$G(s)=\dfrac{1}{1+Ts}$	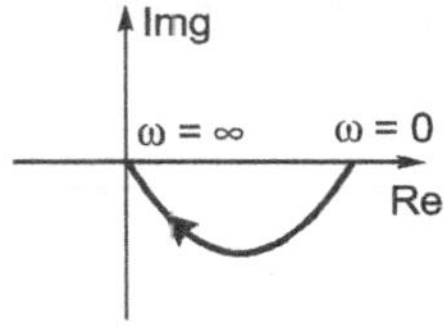
Type-0/Order-2	$G(s)=\dfrac{1}{(1+T_1s)(1+T_2s)}$	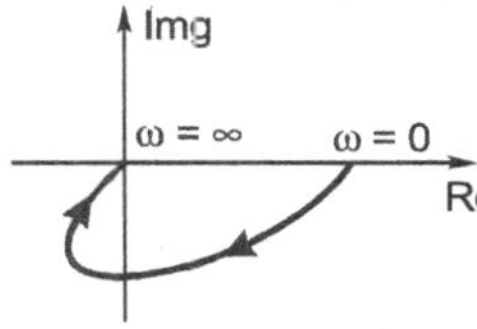
Type-0/Order-3	$G(s)=\dfrac{1}{(1+T_1s)(1+T_2s)(1+T_3s)}$	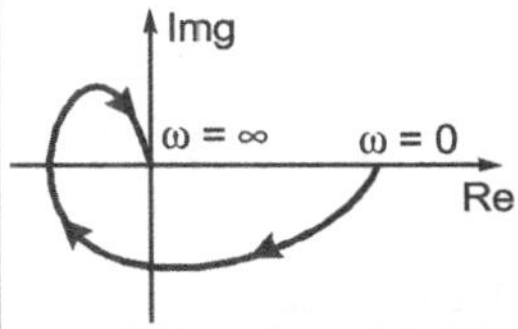
Type-0/Order-4	$G(s)=$ $\dfrac{1}{(1+T_1s)(1+T_2s)(1+T_3s)(1+T_4s)}$	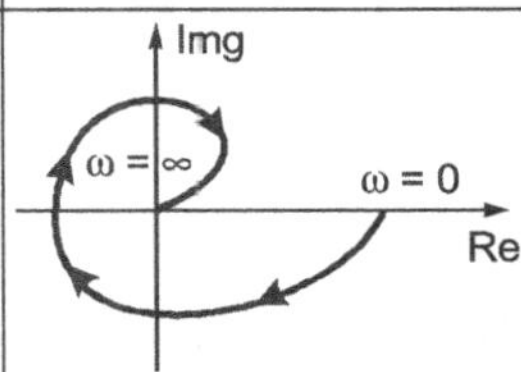
Type-1/Order-1	$G(s)=\dfrac{1}{s}$	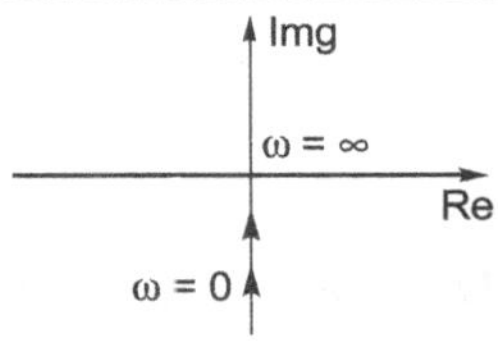
Type-1/Order-2	$G(s)=\dfrac{1}{s(1+Ts)}$	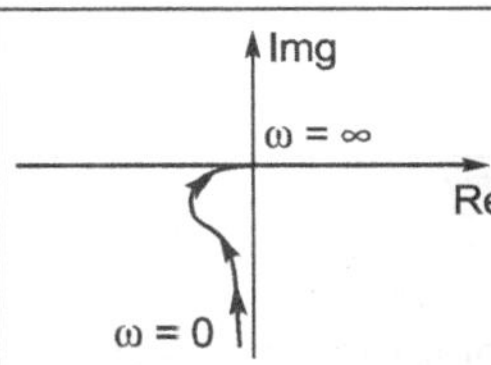
Type-1/Order-3	$G(s)=\dfrac{1}{s(1+T_1s)(1+T_2s)}$	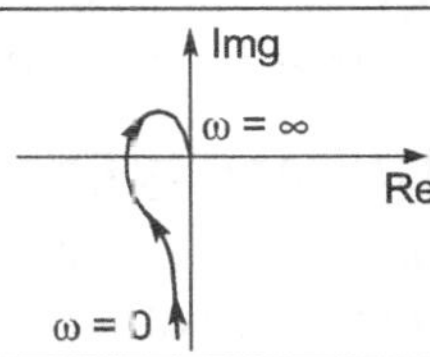

Type-1/Order-4	$G(s)=\dfrac{1}{s(1+T_1s)(1+T_2s)(1+T_3s)}$	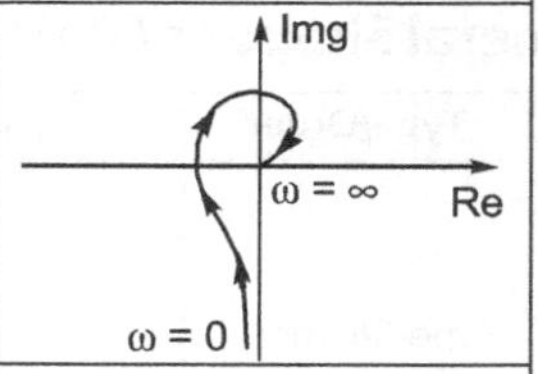
Type-2/Order-2	$G(s)=\dfrac{1}{s^2}$	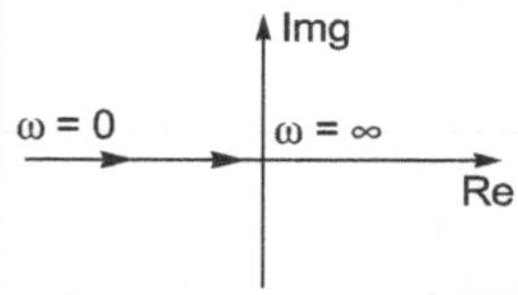
Type-2/Order-3	$G(s)=\dfrac{1}{s^2(1+Ts)}$	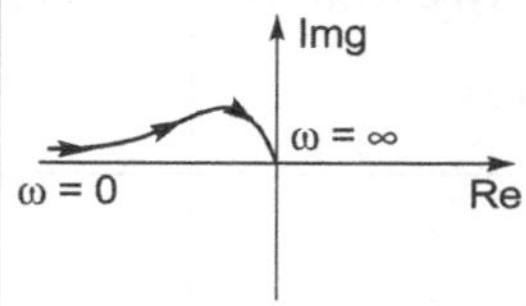
Type-2/Order-4	$G(s)=\dfrac{1}{s^2(1+T_1s)(1+T_2s)}$	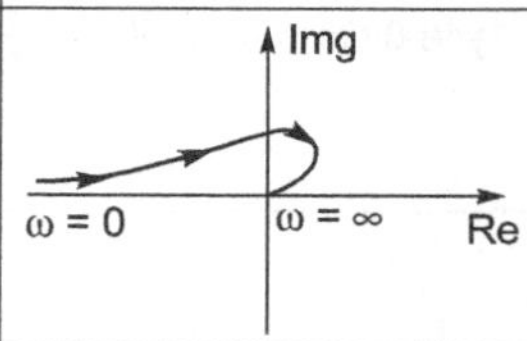
Type-3/Order-3	$G(s)=\dfrac{1}{s^3}$	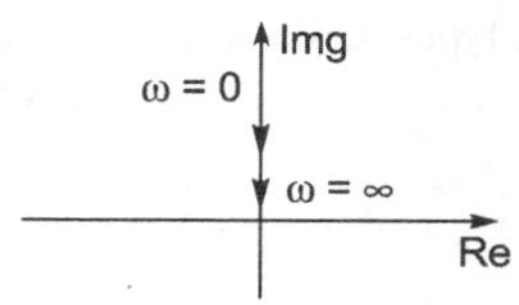
Type-3/Order-4	$G(s)=\dfrac{1}{s^3(1+Ts)}$	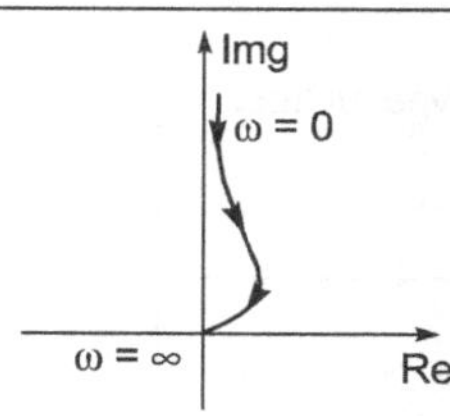

Note:

For standard transfer function of type-2 and type-3 systems, the polar plot intersects negative real axis as many times as there are zeros in open loop transfer function.

Nyquist Stability Criteria

Nyquist stability criteria states that the number of encirclements about the critical point (–1 + j0) should be equal to poles of open loop transfer function $G(s)\,H(s)$, which are in the right half of s-plane.

i.e., $N = P - Z$

For stability (i) $Z = 0$ and (ii) $N = P$

where, P = Number of open loop poles in RHS of s-plane

Z = Number of closed loop poles in RHS of s-plane

N = Number of encirclements about of the point $(-1 + j0)$ by $G(s)H(s)$ plot, the positive direction of encirclements being anti-clockwise.

Bode Plot

The variation of magnitude of sinusoidal transfer function expressed in decibel and the corresponding phase angle in degrees being plotted w.r.t. frequency on a logarithmic scale in rectangular axis, the plot thus obtained is known as Bode plot.

Procedure to Draw Bode Plot

- s is replaced by $j\omega$ to convert into frequency domain.
- Find magnitude in terms of ω and write in terms of dB.

  Magnitude in dB $M_{dB} = 20\log|G(j\omega)\, H(j\omega)|$
- Find the phase angle $\angle\phi$

$$\angle\phi = \tan^{-1}\left(\frac{\text{imaginary part}}{\text{real part}}\right)$$

- With required approximation and by varying the frequency from minimum to maximum, draw the magnitude and phase plot.
- If $G(s)\,H(s) = K$, then $|M_{dB}| = 20\log K$

  System gain K shift the magnitude plot either in upward or downward by "20 logK".
- **Slope :** $\text{Slope} = \dfrac{d|M|_{dB}}{d\log\omega}$

Note:

☑ The magnitude plot should start at the frequency of 0.1 with opposite sign of the slope and it should pass through 0 dB line and intersect at $\omega = 1$, when $K = 1$.

- n poles at origin gives Slope = –20n dB/dec; $\angle\phi = -90n°$
- n zeros at origin gives Slope = +20n dB/dec; $\angle\phi = +90n°$

Remember:

The initial slope of the magnitude plot is given by the poles or zero located at origin.

Corner frequency

The frequency at which slope changes from one level to another level.

- If $G(s)H(s) = \dfrac{1}{(1+sT)^n}$

Frequency	Slope	Phase (ϕ)
Below corner frequency	0	0
Above corner frequency	$-20n$ dB/dec	$-90n°$

- If $G(s)H(s) = (1 + sT)^n$

Frequency	Slope	Phase (ϕ)
Below corner frequency	0	0
Above corner frequency	$20n$ dB/dec	$90n°$

Remember:

☑ The error is nothing but difference between actual value and asymptotic value.

☑ Error is maximum at corner frequency.

☑ Error is even function w.r.t. corner frequency.

☑ 6 dB/oct = 20 dB/dec.

Classification of system

- **Minimum Phase System (MPS)**

  A system in which all finite poles and finite zeros are located in left half of s-plane then it is called as minimum phase system.

- **Non Minimum Phase System (NMPS)**

  A system in which one or more zeros lies in the right half of *s*-plane, remaining all the poles and zeros lies on the left half of *s*-plane.

- **All Pass System (APS)**

  A system in which zeros lies on the right half of *s*-plane, poles lies on the left half of *s*-plane and which are symmetrical about imaginary axis. For all pass system magnitude should be 1 and phase angle is –180°.

Note:

☑ $\text{Gain}_{NMPS} = \text{Gain}_{MPS} \times \text{Gain}_{APS}$

☑ $\angle\phi_{NMPS} = \angle\phi_{MPS} + \angle\phi_{APS}$

Constant Magnitude Loci (*M*-circles)

M-circles represents magnitude of closed loop frequency response plotted against frequency.

- Radius of *M*-circles, $\text{Radius} = \frac{M}{1-M^2}$
- Centre of *M*-circles, $\text{Centre} = \left(\frac{M^2}{1-M^2}, 0\right)$

Constant Phase Angles Loci (*N*-circles)

N-circles represent phase of closed loop frequency response plotted against frequency.

- Radius of *N*-circles : $\text{Radius} = \sqrt{\frac{1}{4} + \frac{1}{4N^2}}$
- Centre of *N*-circles : $\text{Centre} = \left(-\frac{1}{2}, \frac{1}{2N}\right)$ where $N = \tan\phi$

IX State Space Analysis

State

It describes the future behaviour of a system based on present and past data. The knowledge of present is given by input and knowledge of past is given by state variables.

State Variables

These are the minimum set of variables such that knowledge of these variables alongwith the input can completely describe behaviour of the system.

State Space

The set of possible values that a state vector can assume is called state space.

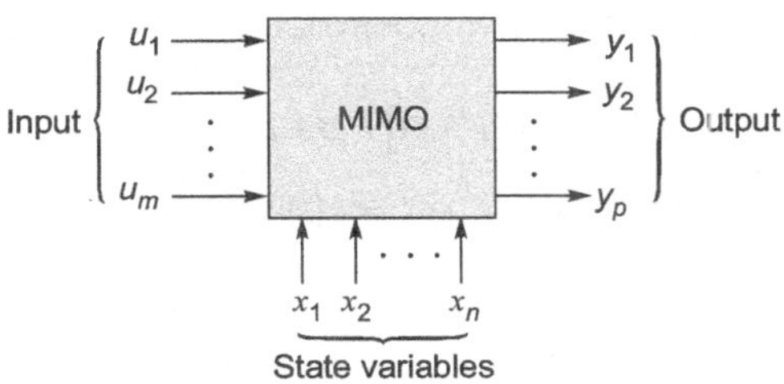

General Representation of State Model

- State equation, $\dot{x} = Ax + Bu$

- Output equation, $y = Cx + Du$

  where, $\dot{x} \rightarrow$ Velocity vector $(n \times 1)$; $x \rightarrow$ State vector $(n \times 1)$;
  $u \rightarrow$ Input vector $(m \times 1)$; $y \rightarrow$ Output vector $(p \times 1)$;
  $A \rightarrow$ State matrix $(n \times n)$; $B \rightarrow$ Input matrix $(n \times m)$;
  $C \rightarrow$ Output matrix $(p \times n)$; $D \rightarrow$ Transmission matrix $(p \times m)$
  $n \rightarrow$ Number of state variables ; $p \rightarrow$ Number of outputs ;
  $m \rightarrow$ Number of inputs

Controllability

A system is said to be controllable if it is possible to transform the initial state of the system to any desired state with application of input within a finite interval of time.

$Q_c = \left[B : AB : A^2B : ...... : A^{n-1}B\right]$ where, $Q_c \rightarrow$ Controllability test matrix

Condition for State Controllability

$$|Q_c| \neq 0 \text{ .... (Matrix be non singular)}$$

Observability

A system is said to be the observable if it is possible to measure all state variables by measuring the output.

$$Q_o = \left[C^T : A^TC^T : (A^T)^2C^T : ..... : (A^T)^{n-1}C^T\right]$$

where, $Q_o \rightarrow$ Observability test matrix

Condition for State Observability

$$|Q_o| \neq 0 \text{ .... (Matrix be non singular)}$$

Note:

☑ If AB is controllable, A^TB^T is observable.

☑ If AC is observable, A^TC^T is controllable.

☑ If input-output transfer function of a linear time invariant system does not have pole-zero cancellation, the system can always be represented by completely controllable and observable state model.

- Transfer function, T.F. $= C[sI - A]^{-1}B + D$
- State transition matrix (STM), $e^{At} = \phi(t) = L^{-1}\left[(sI - A)^{-1}\right]$

Properties of STM :

(i) $\phi(0) = I$(identity matrix)

(ii) $\phi^{-1}(t) = \phi(-t)$

(iii) $[\phi(t)]^K = \phi(Kt)$

(iv) $\phi(t_1 + t_2) = \phi(t_1)\,\phi(t_2)$

(v) $\phi(t_2 - t_1)\,\phi(t_1 - t_0) = \phi(t_2 - t_0)$

Signals and Systems

7

I Introduction to Signals

Signals

A signal can be defined as a function of one or more independent variable, which conveys information about the behaviour or nature of some phenomenon.

Elementary Signals

Unit step function

For continuous time	$u(t)=\begin{cases}1; & t>0\\0; & t<0\end{cases}$	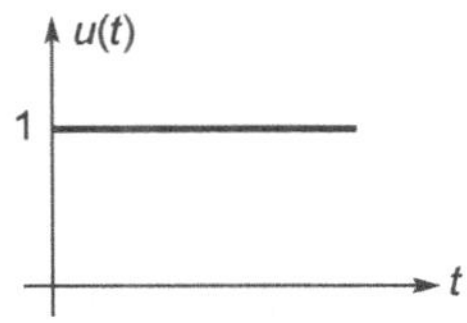
For discrete time	$u(n)=\begin{cases}1; & n\geq 0\\0; & n<0\end{cases}$	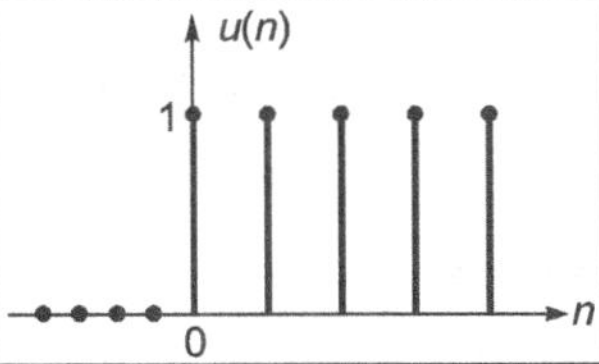

Remember:

☑ Unit step function is not continuous and differentiable at $t = 0$.

☑ Mathematically $u(0)=\frac{1}{2}$; average value

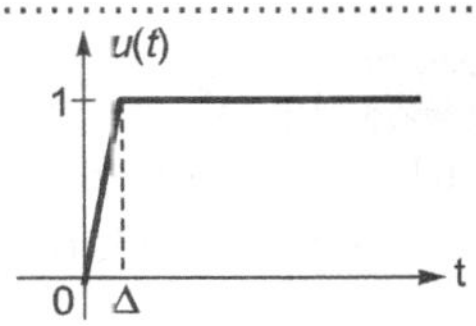

Unit impulse function

For continuous time

$$\delta(t) = \begin{cases}\infty & ; \quad t = 0\\ 0 & ; \quad t \neq 0\end{cases}$$

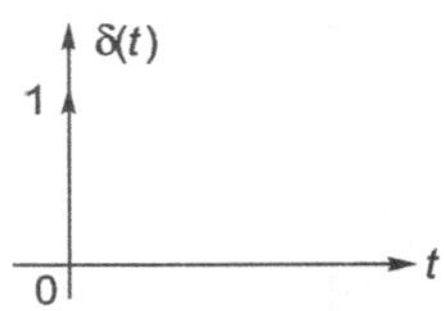

Properties of the impulse function

(i) Impulse function is a continuous function and the area under this function is equal to one

$$\int_{-\infty}^{\infty} \delta(t)dt = 1$$

(ii) Time reversal, $\delta(-t) = \delta(t)$

(iii) Scaling, $\delta(at) = \frac{1}{|a|}\delta(t)$; for $a \neq 0$

(iv) Product, $x(t)\,\delta(t-t_0) = x(t_0)\,\delta(t-t_0)$; where, t_0 is time shift

(v) Sampling property, $\int_{t_1}^{t_2} x(t)\,\delta(t-t_0) = x(t_0)$; $t_1 < t_0 < t_2$

(vi) Delta function of a function of x

$$\delta[g(x)] = \sum_i \frac{\delta(x-x_i)}{|g'(x_i)|}$$

where x_i are the roots of $g(x)$.

Ex: $$\delta(x^2 - a^2) = \frac{1}{2|a|}[\delta(x+a) + \delta(x-a)]$$

Remember:

Relationship between unit impulse and unit step function :

$$u(t) = \int_{-\infty}^{t} \delta(t)\,dt \quad \text{and} \quad \delta(t) = \frac{du(t)}{dt}$$

For discrete-time

$$\delta[n] = \begin{cases} 1 & ; \quad n = 0 \\ 0 & ; \quad n \neq 0 \end{cases}$$

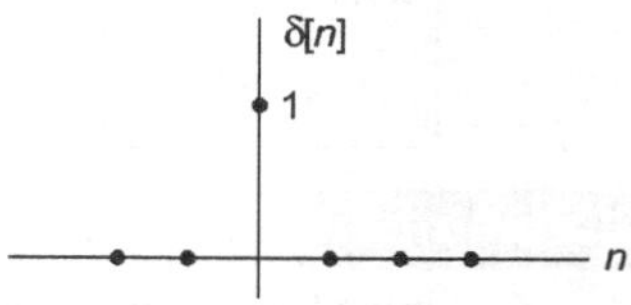

Remember:

☑ The discrete-time unit impulse is the first difference of the discrete-time unit step

$$\delta[n] = u[n] - u[n-1]$$

☑ Relationship between unit impulse and unit step

$$u[n] = \sum_{k=0}^{\infty} \delta[n-k] = \sum_{k=-\infty}^{n} \delta[k]$$

☑ Sampling property, $x[n]\,\delta[n-n_0] = x[n_0]\,\delta[n-n_0]$

Triangular Function	$\text{Tri}\left(\frac{t}{T}\right)=\begin{cases}\frac{1}{T}t+1 & ; -T\le t\le 0\\ -\frac{1}{T}t+1 & ; 0\le t\le T\\ 0 & ; \text{otherwise}\end{cases}$	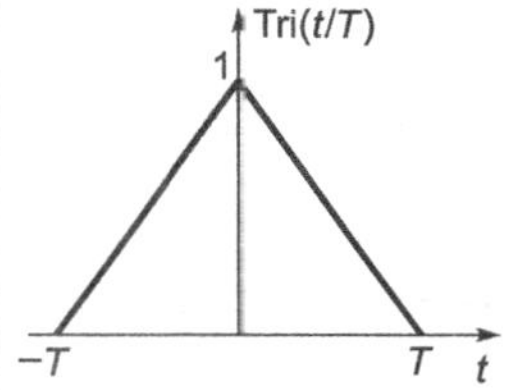
Rectangular or Gate Function	$\text{rect}\left(\frac{t}{T}\right)=\begin{cases}1 & ; -\frac{T}{2}<t<\frac{T}{2}\\ 0 & ; \text{otherwise}\end{cases}$	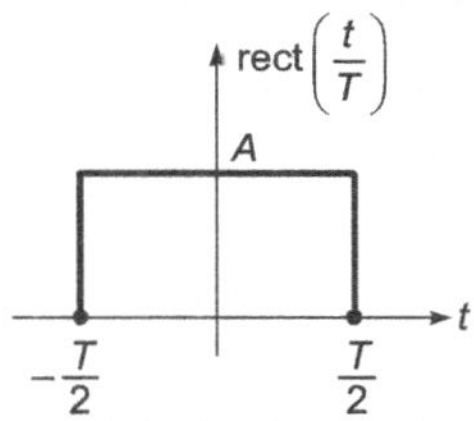
Unit Ramp Function	$r(t)=\begin{cases}t & ; t>0\\ 0 & ; t<0\end{cases}$ $\frac{d}{dt}r(t)=u(t)$	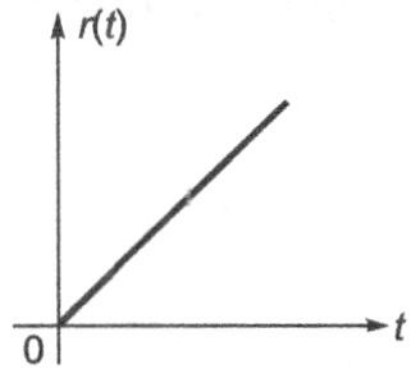
Signum or sgn Function	sgn(t) = 2u(t) – 1	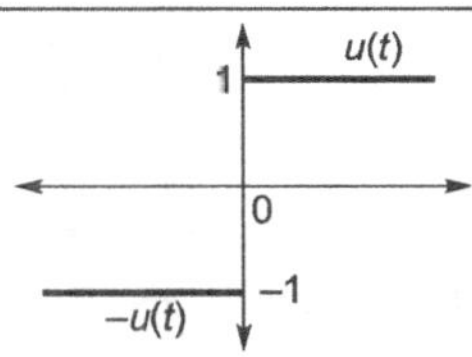

Note:

☑ $u(-t) = 1 - u(t)$

☑ sgn function is not defined at $t = 0$ and is chosen as 0 at $t = 0$

sinc and sinc² function

- The sinc and sinc² functions are defined in terms of an independent variable λ.

$$\text{sinc}(\lambda)=\frac{\sin(\pi\lambda)}{\pi\lambda}$$

- $\text{sinc}(\lambda)$ is equal to zero for $\lambda = \pm n\ (n \ne 0)$, n an integer.

Sine integral function

- Sine integral function is an odd function

$$\text{Si}(y)=\int_0^y \frac{\sin(\alpha)}{\alpha}d\alpha \;;\quad \text{Si}(y)=\frac{y}{(1)!1}-\frac{y^3}{(3)3!}+\frac{y^5}{(5)5!}-\frac{y^7}{(7)7!}+\cdots$$

Remember:

☑ $Si(0) = 0, Si(\pi) \simeq 1.8519, Si(\infty) = \left(\frac{\pi}{2}\right)$

☑ Si function converges fast and only a few terms in above equation are needed for a good approximation

Signal Operations

Time scaling for analog signals : Let $x(t)$ be an arbitrary signal, a time scaled version of $x(t)$ is obtained by replacing 't' by 'at' where 'a' is scaling factor.

$$\varphi(t) = x(at)$$

- $a > 1$ shows compression of $x(t)$.
- $0 < a < 1$ shows expansion of $x(t)$.

Time scaling for discrete signals : For a discrete time sequence $x[n]$, compression of a signal by factor M is given by $y[n] = x[Mn]$; M and n both are integers.

Time shifting for analog signals : Shifting in time may results in time delay or time advancement. For a continuous-time signals $x(t)$, time shifting is given as

$\varphi(t) = x(t - t_0)$ ··· delay or shift right by 't_0'.

$\varphi(t) = x(t + t_0)$ ··· advance or shift left by 't_0'.

Time shifting for discrete signals : For a discrete time sequence $x[n]$ time shifting is given as

$\varphi[n] = x[n - n_0]$ ··· delay or shift right by n_0 samples.

$\varphi[n] = x[n + n_0]$ ··· advance or shift left by n_0 samples.

Time reversal for analog signals : Time reversal $x(t)$ is achieved by rotation of signal 180° about vertical axis. This operation is also called as folding or reflection about vertical axis.

For a continuous-time signals $x(t)$, time reversal is given as

$$\varphi(t) = x(-t)$$

Time reversal for discrete signals : For discrete time sequence $x[n]$, time reversal is given as

$$\varphi[n] = x[-n]$$

Note:

☑ In priority order: Shifting > Scaling > Reversal

☑ There is no effect of scaling on unit step signal.

☑ The time scaling on ramp signal will result into magnitude scaling as

$$r(at) \longrightarrow ar(t)$$

Classification of Signals

Continuous Time and Discrete Time Signals : $x(t)$ is a continuous-time signal if 't' is a continuous variable. But if 't' is a discrete variable that is $x(t)$ is defined at discrete times, then $x(t)$ is a discrete-time signal.

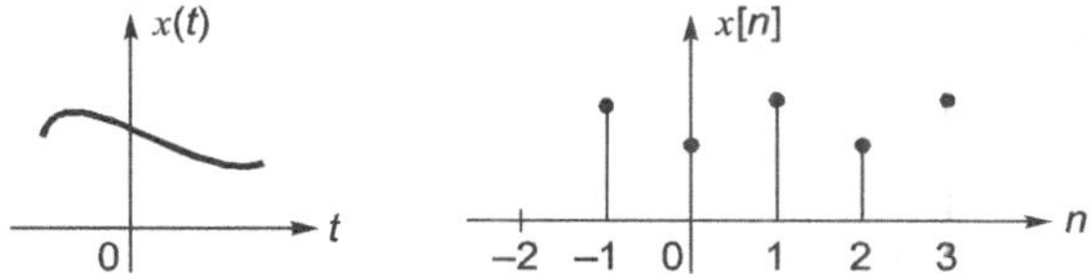

(a) Continuous-time (b) Discrete-time

Note:

A discrete-time signal $x[n]$ may be obtained by sampling a continuous-time signal $x(t)$.

Analog and Digital Signals : If a continuous-time signal $x(t)$ can take on any value in the continuous interval (a, b) where 'a' may be $-\infty$ and 'b' may be $+\infty$, then the continuous time signal $x(t)$ is called an analog signal. If a discrete-time signal $x[n]$ can take on only a finite number of distinct value, then this signal is called a digital signal.

Real and Complex Signal : A signal $x(t)$ is a real signal if its value is a real number and a signal $x(t)$ is a complex signal if its value is a complex number.

Deterministic and Random Signals : If $x(t)$ can be perfectly known for any time 't' then it is called deterministic signal. If $x(t)$ cannot be exactly determined at any given time then it is called Random signal.

Even and Odd Signals : A signal $x(t)$ or $x[n]$ is an even signal if $x(-t) = x(t)$ and $x[-n] = x[n]$. A signal $x(t)$ or $x[n]$ is referred to as an odd signal if $x(-t) = -x(t)$ and $x[-n] = -x[n]$

Note:

☑ The product of two even signals or of two odd signals is an even signal.

☑ The product of an even signal and an odd signal is an odd signal.

For complex valued signal $x(t)$,

If $x(t) = x^*(-t)$; Even conjugate or Conjugate symmetric

If $x(t) = -x^*(-t)$; Odd conjugate or Conjugate Anti-symmetric

		Continuous Time	Discrete Time
Real valued	Even part	$\frac{x(t)+x(-t)}{2}$	$\frac{x[n]+x[-n]}{2}$
	Odd part	$\frac{x(t)-x(-t)}{2}$	$\frac{x[n]-x[-n]}{2}$
Complex Valued	Even conjugate	$\frac{x(t)+x^*(-t)}{2}$	$\frac{x[n]+x^*[-n]}{2}$
	Odd conjugate	$\frac{x(t)-x^*(-t)}{2}$	$\frac{x[n]-x^*[-n]}{2}$

Energy and Power Signals

Energy of the signal : $E = \int_{-\infty}^{\infty} |x(t)|^2 dt$; $E = \sum_{n=-\infty}^{\infty} |x[n]|^2$

Average power of a signal : $P = \lim_{T\to\infty} \frac{1}{T} \int_{-T/2}^{T/2} |x(t)|^2 dt = \lim_{N\to\infty} \frac{1}{2N+1} \sum_{n=-N}^{N} |x[n]|^2$

Note:

- ☑ An energy signal has zero average power always.
- ☑ A power signal has infinite energy
- ☑ Signal having infinite duration with their value tending to a non-zero constant as $t \to \infty$ and signals which are periodic are in general power signal.
- ☑ As $t \to \pm\infty$, if amplitude tends to ∞, it is neither energy nor power signal.
- ☑ All finite duration and bounded signals are energy signal.
- ☑ Energy of a signal is only affected by scaling operation as

$$E(at) \longrightarrow \frac{E(t)}{a}$$

- ☑ Power is not affected by time scaling, reversal or shifting.

Periodic and Non Periodic signals

A continuous time signal $x(t)$ is said to be periodic with period T if there is a positive non zero value of T for which

$$x(t + T) = x(t)\ ; \quad \forall t$$

In discrete-time signal, a sequence $x[n]$ is periodic with period N, if there is a positive integer N for which

$$x[n + N] = x[n]\ ; \quad \forall n$$

Note:

- ☑ The fundamental period for analog signal is : $T_0 = \frac{2\pi}{\omega_0}$
- ☑ For discrete signal, $\frac{N}{K} = \frac{2\pi}{\omega_0}$; where $K = 0, 1, 2, 3 \cdots$
- ☑ The fundamental period T_0 of $x(t)$ and N of $x[n]$ is the smallest positive integer for which above equation holds good.
- ☑ Sum of two continuous-time periodic signals may not be always periodic but the sum of two discrete periodic sequences is always periodic.
- ☑ If two continuous time signals have period T_1, T_2 then for their sum to be

periodic, $\frac{T_1}{T_2}$ should be rational and the period of the sum is given by $T = \text{LCM}(T_1, T_2)$.

☑ If two discrete time signals have period N_1, N_2 then their sum is always periodic with period $N = \text{LCM}(N_1, N_2)$.

For periodic signal $x(t)$

- Average value of the signal, $x_{avg} = \frac{1}{T}\int x_T(t)dt$
- Average signal power, $P_x = \frac{1}{T}\int_T \left| x_T(t)^2 \right| dt$
- Effective or rms value of the signal, $x_{rms} = \sqrt{P_x}$

  where, $x_T(t) \rightarrow$ One period of $x(t)$

Note:

☑ RMS value of $A\cos(\omega t + \phi)$ is $\frac{A}{\sqrt{2}}$.

☑ With DC offset, RMS value of $A_0 + A\cos(\omega t + \phi)$ is $\sqrt{A_0^2 + \frac{A^2}{2}}$.

Symmetry of Periodic Function

- Odd half-wave symmetry, $x(t) = -x\left(t \pm \frac{T}{2}\right)$
- Even half-wave symmetry, $x(t) = x\left(t \pm \frac{T}{2}\right)$ where, $T \rightarrow$ Period of $x(t)$

II Linear Time Invariant Systems

System

A system is a quantity which maps a set of input signal to a set of output signals. Linear Time Invariant (LTI) systems are used to represent signals as linear combinations of basic signals.

Continuous-Time and Discrete-Time Systems

A continuous time system (CTS) is one in which continuous time input signals are transformed into continuous time output signals.

$$y(t) = T\{x(t)\}$$

A discrete time system (DTS) is one which transform discrete time input signal into discrete time output signal.

$$y[n] = T\{x[n]\}$$

Moreover, a continuous time signal can be processed by a discrete time system. On the other hand processing of discrete time signal by a continuous time system is also possible because discrete time systems have several significant advantages over continuous time systems.

Classification of Systems

Linear Systems and Non Linear Systems : When system satisfy principle of superposition and homogeneity, it is called linear system, else non linear system.

(i) Superposition (Additivity)

if $x_1(t) \rightarrow y_1(t)$ and $x_2(t) \rightarrow y_2(t)$

then $x_1(t) + x_2(t) \rightarrow y_1(t) + y_2(t)$

(ii) Homogeneity (Scaling)

$$ax_1(t) + bx_2(t) \rightarrow ay_1(t) + by_2(t)$$

Ex. of linear systems: $y(t) = 2x(t)$, $y(t) = \int_{-\infty}^{t} x(t)dt$

Time-Invariant and Time Varying Systems : A system is called time-variant if a time shift in the input signal causes the same time-shift in the output signal.

If $x(t) \rightarrow y(t)$, then $x(t-t_0) \rightarrow y(t-t_0)$

Ex. of time invariant system: $y(t) = x^2(t)$

Causal and Non Causal systems : The output of the causal system at the present time depends on only the present and/or past value of the input, not on its future values. A system is called non causal if it is not causal.

Ex. of causal system $\rightarrow y(t) = x(\sqrt{t})$.

Note:

- ☑ All memory-less systems are causal, but not vice-versa.
- ☑ Causal system are referred as non-anticipative as the system output does not anticipate future values of input.

Static and Dynamic Systems : System is said to be static or memory less if the output at any instant depends on only the input at that instant otherwise, the system is a dynamic system with memory.

Ex. of static system: $y(t) = 2x(t)$.

Stable and Unstable Systems : A system is bounded input bounded output, i.e., BIBO stable if for any bounded input $x(t)$ the corresponding output $y(t)$ is also bounded.

Ex. of stable system: $y(t) = e^{x(t)}$.

Note:

A consequence of the homogeneity of linear system is that a zero input yield a zero output.

Invertible and Non-Invertible System : If it is possible to define an inverse system of a given system, then the system is invertible, otherwise it is called non-invertible system.

Ex. $y(t) = \int_{-\infty}^{t} x(t)\,dt \leftarrow$ Invertible ; $y(t) = \frac{d}{dt}x(t) \leftarrow$ Inverse system

Convolution

The convolution integral or the superposition integral, represent a continuous-time linear time invariant (LTI) system in terms of its response to a unit impulse.

Continuous-time LTI system

Convolution of two functions, $x_1(t)$ and $x_2(t)$

$$y(t) = x_1(t) * x_2(t) = \int_{-\infty}^{\infty} x_1(\tau)x_2(t-\tau)d\tau = \int_{-\infty}^{\infty} x_1(t-\tau)\,x_2(\tau)\,d\tau$$

Note:

If two analog signals get convolved:

(i) The resultant of convolution of two signals will have a width equal to the sum of the individual width of the two signals being convolved

(ii) The lower extent of resultant signal will be equal to sum of individual lower extents $l = l_1 + l_2$, similarly upper extent is sum of individual upper extents $u = u_1 + u_2$.

(iii) The area of resultant convolution is equal to the product of the area of the signals being convolved.

Properties of the Convolution Integral

- Commutative property, $y(t) = x_1(t) * x_2(t) = x_2(t) * x_1(t)$
- Distributive property, $x_1(t) * [x_2(t) + x_3(t)] = x_1(t) * x_2(t) + x_1(t) * x_3(t)$
- Associative property, $x_1(t) * (x_2(t) * x_3(t)) = (x_1(t) * x_2(t)) * x_3(t)$
- Derivative of the convolution, $\frac{dy(t)}{dt} = \frac{dx_1(t)}{dt} * x_2(t) = x_1(t) * \frac{dx_2}{dt}(t)$
- Convolution of two delayed functions

  If $y(t) = x_1(t) * x_2(t)$

$$x_1(t-t_1) * x_2(t-t_2) = y((t-(t_1+t_2))$$

- **Time scaling property:**

  If $\quad y(t) = x_1(t) * x_2(t)$

$$x_1(\alpha t) * x_2(\alpha t) = \frac{1}{|\alpha|} y(\alpha t) \;;\alpha \neq 0$$

  Special Cases:

$$\phi(t) * \delta(t) = \int_{-\infty}^{\infty} \phi(\tau)\, \delta(t-\tau)\, d\tau = \phi(t) \;; \quad \delta(t) * \delta(t) = \int_{-\infty}^{\infty} \delta(\tau)\, \delta(t-\tau) d\tau = \delta(t)$$

Discrete-time LTI system

$$y[n] = x_1[n] * x_2[n] = \sum_{k=-\infty}^{\infty} x_1[k]\, x_2[n-k] = \sum_{K=-\infty}^{\infty} x_1[n-k]\, x_2[k]$$

Note:

If two discrete signals get convolved :

(i) The resultant of convolution of two signals will have a length equal to the sum of the individual length of the two signals being convolved minus 1, i.e., $(L_1 + L_2 - 1)$.

(ii) Resultant of convolution has extent equal to the sum of the individual extents of the signals being convolved.

(iii) The sum of sampled values in resultant convolution is equal to the product of sum of the individual sampled values of the signals being convolved.

$$\sum_{n=-\infty}^{\infty} y[n] = \left(\sum_{n=-\infty}^{\infty} x_1[n] \right) \left(\sum_{n=-\infty}^{\infty} x_2[n] \right)$$

III Fourier Series

Introduction

It is an approximated process by which any non-standard signal is converted into a standard signal. The approximation of a given function by Fourier series gives a smooth function even when the function being approximated has discontinuities.

Eigen Function

- If the response of a system for an input signal is same as the input signal except for a scalar multiple, then this input signal is called Eigen function of the system.
- A phasor signal $e^{j\omega_0 t}$ can be defined as the Eigen function of any general LTI system.

- If the input of a system is in terms of phasors, then the response evaluation can be simplified from a more difficult convolution to simpler multiplication. This is the basic principle of Fourier series.

Convergence of Fourier Series (Dirichlet Condition)

Periodic signal $x(t)$ has a Fourier series representation if it satisfies the following Dirichlet condition.

- $x(t)$ is absolutely integrable over any period, i.e., $\int_{T_0} |x(t)|dt < \infty$
- $x(t)$ has a finite number of maxima and minima within any finite interval of t.
- $x(t)$ has a finite number of discontinuities within any finite interval of t, and each of these discontinuities is finite.

Fourier Series Representation of Continuous Time Periodic Signals

By using Fourier series, a non-sinusoidal periodic function can be expressed as an infinite sum of sinusoidal function.

Trigonometric Fourier series : Any practical periodic function of frequency ω_0 can be expressed as an infinite sum of sine (or) cosine functions that are integral multiples of ω_0

$$f(t) = a_0 + \sum_{n=1}^{\infty}(a_n \cos \omega_0 t + b_n \sin \omega_0 t)$$

where, ω_0 = Fundamental frequency ; a_0, a_n, b_n = Tr gonometric Fourier series coefficient ; $n\omega_0$ = n^{th} harmonic of ω_0

$$a_0 = \frac{1}{T}\int_0^T f(t)dt \ ; \quad a_n = \frac{2}{T}\int_0^T f(t)\cos n\omega_0 t \, dt \ ; \quad b_n = \frac{2}{T}\int_0^T f(t)\sin n\omega_0 t \, dt$$

Note:

☑ Here a_0 is the value of constant component of the signal $f(t)$.

☑ The Fourier coefficient a_n and b_n are maximum amplitude of nth harmonic component.

Polar form of Trigonometric Fourier Series

$$f(t) = c_0 + \sum_{n=1}^{\infty} c_n \cos(n\omega_0 t - \phi_n)$$

where, $c_0 = a_0$

$|c_n| = \sqrt{a_n^2 + b_n^2}$ ... Magnitude spectrum

$\phi_n = \tan^{-1}\left(\frac{b_n}{a_n}\right)$ ... Phase spectrum

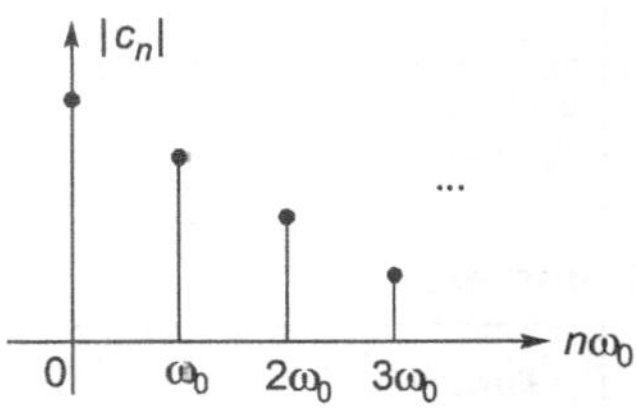

Magnitude spectrum

Exponential Fourier Series

$$f(t) = \sum_{n=-\infty}^{\infty} c_n e^{jn\omega_0 t} \qquad \text{where, } c_0 = a_0 = \frac{1}{T}\int_0^T f(t)dt$$

$$c_n = \frac{1}{T}\int_0^T f(t)e^{-jn\omega_0 t}dt$$

Note:

☑ Exponential Fourier series is compact form of Fourier series

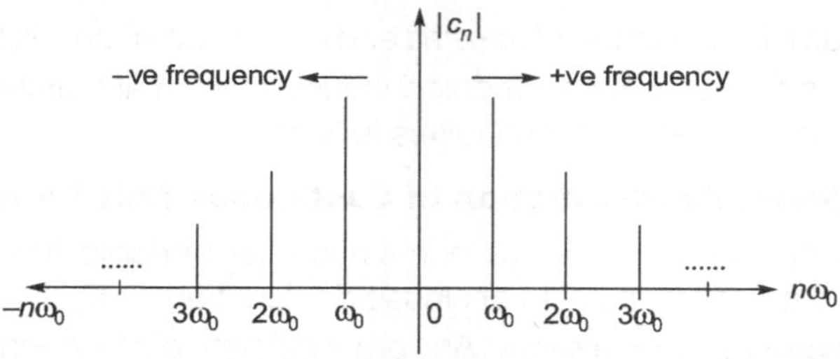

☑ Positive and negative frequency indicate different phase of rotation, they maintain same magnitude but different phase.

Relation between Exponential and Trigonometric Fourier Series

$$a_n = (c_n + c_{-n}) \quad ; \quad c_n = \frac{a_n}{2} - j\frac{b_n}{2}$$

$$b_n = j(c_n - c_{-n}) \quad ; \quad c_{-n} = \frac{a_n}{2} + j\frac{b_n}{2}$$

Effect of Symmetry of Fourier Coefficients

Odd symmetry	Even symmetry	Half-wave symmetry
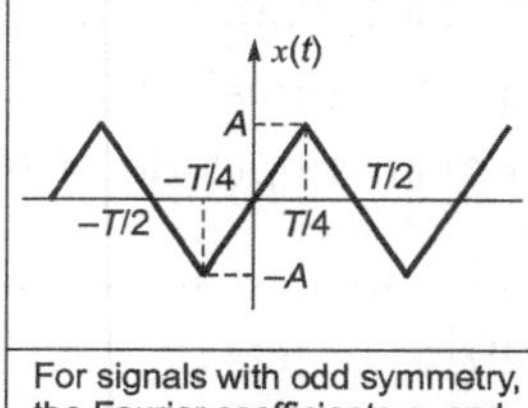	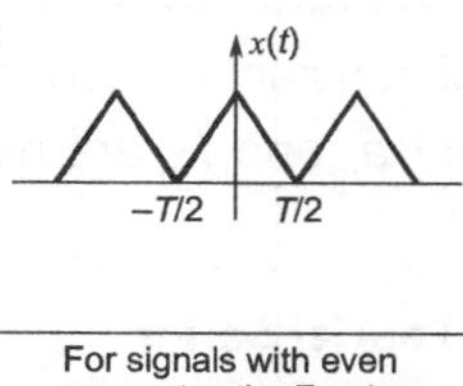	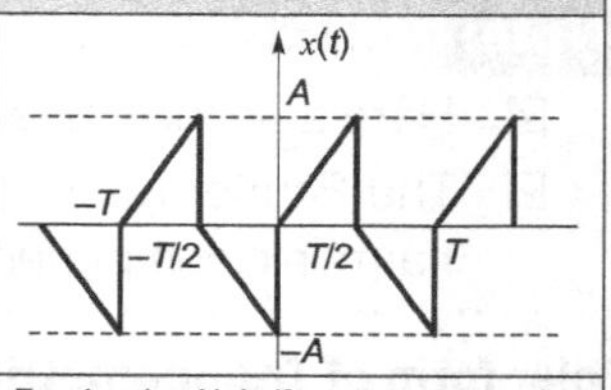
For signals with odd symmetry, the Fourier coefficients a_0 and a_n are zero. $x(t) = -x(-t)$	For signals with even symmetry, the Fourier coefficient b_n is zero. $x(t) = x(-t)$	For signals with half-wave symmetry, the Fourier series will consist of odd harmonic terms of sine and cosine signals. $x(t \pm T/2) = -x(t)$

Summary

Function	C_n	Fourier Coefficient	Trigonometric Fourier Series
Real (Neither even nor odd)	Generally Complex $C_n = C_{-n}^*$	$a_0 \neq 0$ $a_n = b_n \neq 0$	DC term, sine terms and cosine terms are present.

Even	Real (Even in nature)	$a_0 \neq 0$ $a_n \neq 0$ $b_n = 0$	DC term and cosine terms are present.
Odd	Imaginary (odd in nature)	$a_0 = 0$ $a_n = 0$ $b_n \neq 0$	Only sine terms are present.
Half wave symmetry	$C_n = 0$; For n = even	$\left.\begin{matrix} a_n = 0 \\ b_n = 0 \end{matrix}\right\}$ $a_0 = 0$; n = even	Odd sine and odd cosine terms are present.
Even and Half wave symmetry	C_n = Real & even $C_n = 0$; n = even	$\left.\begin{matrix} a_n = 0 \\ b_n = 0 \end{matrix}\right\}$ $a_0 = 0$; n = even	Only odd cosine terms are present.
Odd and Half wave symmetry	C_n = Imaginary & odd $C_n = 0$; n = even	$\left.\begin{matrix} a_n = 0 \\ b_n = 0 \end{matrix}\right\}$; n = even	Only odd sine terms are present.

Properties of Exponential Form of Fourier Series Coefficients

Property	Continuous Time Periodic Signal	Fourier Series Coefficients
Linearity	$Ax(t) + By(t)$	$Ac_n + Bd_n$
Time shifting	$x(t \pm t_0)$	$c_n e^{\pm jm\omega_0 t_0}$
Frequency shifting	$e^{\pm jm\omega_0 t} x(t)$	$c_{n \mp m}$
Conjugation	$x^*(t)$	c^*_{-n}
Time reversal	$x(-t)$	c_{-n}
Time scaling	$x(\alpha t)$; $\alpha > 0$ [$x(t)$ is periodic with period T/α]	c_n (No change in Fourier coefficient
Multiplication	$x(t)\, y(t)$	$\sum_{m=-\infty}^{+\infty} c_m d_{n-m}$
Differentiation	$\frac{d}{dt}x(t)$	$jn\omega_0 c_n$
Integration	$\int_{-\infty}^{t} x(t)\, dt$ (Finite valued and periodic only if $a_0 = 0$)	$\frac{1}{jn\omega_0} c_n$
Periodic Convolution	$\int_T x(\tau)\, y(t-\tau) d\tau$	$Tc_n d_n$
Symmetry of real signals	$x(t)$ is real	$c_n = c^*_{-n}$ $\lvert c_n\rvert = \lvert c_{-n}\rvert$; $\angle c_n = -\angle c_{-n}$ Re $\{c_n\}$ = Re $\{c_{-n}\}$ Im $\{c_n\}$ = –Im $\{c_{-n}\}$
Real and even	$x(t)$ real and even	c_n are real and even
Real and odd	$x(t)$ real and odd	c_n are imaginary and odd

where, $c_n \rightarrow$ Fourier series coefficients of $x(t)$

$d_n \rightarrow$ Fourier series coefficients of $y(t)$

Fourier Series Representation of Discrete Time Periodic Signals

Fourier series representation of $x[n]$, with period 'N' is given as

$$x[n] = \sum_{k=0}^{N-1} c_k\, e^{jk\left(\frac{2\pi}{N}\right)n}$$

where, $c_k = \dfrac{1}{N}\sum_{n=0}^{N-1} x[n]\, e^{-jk\left(\frac{2\pi}{N}\right)n}$

c_k is periodic with period 'N', $\quad c_k = c_{k+N}$

Properties of Discrete Time Fourier Series

Property	Continuous Time Periodic Signal	Fourier Series Coefficients
Linearity	$Ax[n] + By[n]$	$Ac_k + Bd_k$
Time shifting	$x[n \pm n_0]$	$c_k e^{\pm jk\left(\frac{2\pi}{N}\right)n_0}$
Frequency shifting	$e^{\pm jm\left(\frac{2\pi}{N}\right)n} x[n]$	$c_{k\,\mu m}$
Conjugation	$x^*[n]$	c^*_{-k}
Time reversal	$x[-n]$	c_{-k}
Time scaling	$x_{(m)}[n] = \begin{cases} x\left[\dfrac{n}{m}\right]; & \text{if } n \text{ is a multiple of } M \\ 0 \;; & \text{otherwise} \end{cases}$	$\dfrac{1}{m}c_k$ (Period mN)
Multiplication	$x[n]\, y[n]$	$\sum_{l=0}^{N-1} c_l\, b_{k-l}$
First Difference	$x[n] - x[n-1]$	$\left(1 - e^{-j\left(\frac{2\pi}{N}\right)k}\right)c_k$
Running sum	$\sum_{k=-\infty}^{n} x[k]$	$\left(\dfrac{1}{1 - e^{-jk\left(\frac{2\pi}{N}\right)}}\right)c_k$
Convolution	$\sum_{r=0}^{N-1} x[r]\, y[n-r]$	$Nc_k d_k$

Parseval's Power Theorem

Parseval's power theorem is given by

For continuous time signals: $P = \dfrac{1}{T}\int_0^T |x(t)|^2\, dt = \sum_{n=-\infty}^{\infty} |c_n|^2 = a_0^2 + \dfrac{1}{2}\sum_{n=1}^{\infty}(a_n^2 + b_n^2)$

For discrete signals: $P = \dfrac{1}{N}\sum_{n=0}^{N-1} |x[n]|^2 = \sum_{k=0}^{N-1} |c_k|^2$

IV Fourier Transform

Fourier Transform of Continuous Time Signals

Fourier transform provides a frequency domain description of time domain signals and is extension of Fourier series to non-periodic signals.

- Fourier transform, $F.T\left[x(t)\right] = X(\omega) = \int_{-\infty}^{\infty} x(t)\, e^{-j\omega t} dt$
- Inverse Fourier transform, $I.F.T\left[X(\omega)\right] = x(t) = \frac{1}{2\pi} \int_{-\infty}^{\infty} X(\omega)\, e^{j\omega t} d\omega$

Dirichlet's Conditions of Fourier Transformation

For existence of Fourier transform :

- Fourier transform is defined for all stable signals, $\int_{-\infty}^{\infty} \left|x(t)\right| dt < \infty$
- Periodic signals, which are neither absolutely integrable nor square integral over an infinite interval, can be considered to have Fourier transform if impulse functions are permitted in the transform.
- $f(t)$ have a finite number of discontinuities and finite number of maxima and minima within any finite interval.

The Properties of Fourier Transform

Properties	$X(f)$ - form	$X(\omega)$ - form
Linearity : $ax_1(t) + bx_2(t)$	$aX_1(f) + bX_2(f)$	$aX_1(\omega) + bX_2(\omega)$
Time-scaling : $x(at)$	$\frac{1}{\lvert a\rvert} X\left(\frac{f}{a}\right)$	$\frac{1}{\lvert a\rvert} X\left(\frac{\omega}{a}\right)$
Time-reversal : $x(-t)$	$X(-f)$	$X(-\omega)$
Time-shift : $x(t \pm t_0)$	$e^{\pm j2\pi f t_0} X(f)$	$e^{\pm j\omega t_0} X(\omega)$
Frequency shift : $x(t)^{\pm j\omega_0 t}$	$X(f \mu f_0)$	$X(\omega \mu \omega_0)$
Differentiation in time	$\frac{d}{dt}x(t) \leftrightarrow j2\pi f X(f)$	$\frac{d}{dt}x(t) \leftrightarrow j\pi X(\omega)$
Frequency Differentiation	$-j2\pi t x(t) \leftrightarrow \frac{d}{df}X(f)$	$-jtx(t) \leftrightarrow \frac{d}{d\omega}X(\omega)$
Convolution in time : $x(t) * h(t)$	$X(f)H(f)$	$X(\omega)H(\omega)$
Frequency Convolution : $x_1(t) . x_2(t)$	$X_1(f) * X_2(f)$	$\frac{1}{2\pi}[X_1(\omega) * X_2(\omega)]$
Integration $\int_{-\infty}^{t} x(\tau)d\tau$	$\frac{X(f)}{j2\pi f} + 0.5X(0)\delta(f)$	$\frac{X(\omega)}{j\omega} + \pi X(0)\delta(\omega)$
Parseval's theorem $\int_{-\infty}^{+\infty} x^2(t)dt$	$\int_{-\infty}^{+\infty} \lvert X(f)\rvert^2 df$	$\frac{1}{2\pi}\int_{-\infty}^{+\infty} \lvert X(\omega)\rvert^2 d\omega$
Conjugation : $x^*(t)$	$X^*(-j2\pi f)$	$X^*(-j\omega)$
Duality	$x(t) \leftrightarrow X(f)$ $X(t) \leftrightarrow -x(f)$	$x(t) \leftrightarrow X(\omega)$ $X(t) \leftrightarrow 2\pi x(-\omega)$

where, $x_1(t) \leftrightarrow X_1(f)$ or $X_1(\omega)$; $x_2(t) \leftrightarrow X_2(f)$ or $X_2(\omega)$

Fourier Transform of Useful Signals

Signal, $x(t)$	$F(f)$ form	$F(\omega)$ form
$e^{-at}u(t)$, a>0	$\frac{1}{a+j2\pi f}$	$\frac{1}{a+j\omega}$
$e^{at}u(-t), a>0$	$\frac{1}{a-j2\pi f}$	$\frac{1}{a-j\omega}$
$\delta(t)$	1	1
A, Constant	$A\delta(f)$	$2\pi A\delta(\omega)$
A rect $\cdot (t/T)$	$AT \operatorname{sinc}(fT)$	$AT \operatorname{sinc}(\omega T/2)$
$e^{-a\|t\|}$	$\frac{2a}{a^2+4\pi^2 f^2}$	$\frac{2a}{a^2+\omega^2}$
e^{-at^2}	$\sqrt{\pi/a}\ e^{-\pi^2 f^2/a}$	$\sqrt{\pi/a}\ e^{-\omega^2/4a}$
$Sgn(t)$	$\frac{1}{j\pi f}$	$\frac{2}{j\omega}$
$u(t)$	$\frac{1}{j2\pi f}+0.5\delta(f)$	$\frac{1}{j\omega}+\pi\delta(\omega)$
$\cos\omega_0 t$	$\frac{\delta(f-f_0)+\delta(f+f_0)}{2}$	$\pi[\delta(\omega-\omega_0)+\delta(\omega+\omega_0)]$
$\sin\omega_0 t$	$\frac{\delta(f-f_0)-\delta(f+f_0)}{2j}$	$\frac{\pi}{j}[\delta(\omega-\omega_0)-\delta(\omega+\omega_0)]$
$t^n e^{-at}u(t)$	$\frac{n!}{(a+j2\pi f)^{n+1}}$	$\frac{n!}{(a+j\omega)^{n+1}}$

Hilbert Transform

$$H\{x(t)\} = x(t) * \frac{1}{\pi t}$$

- Hilbert transform provides a phase shift of –90° to all signals.
- Signals and its Hilbert transform have same Energy Spectral Density or Power Spectral Density and Auto-correlation.
- Impulse response: $h(t) = \frac{1}{\pi t}$

$$H(j\omega) = -j\operatorname{sgn}(\omega) = \begin{cases} -j & ; \quad \omega > 0 \\ j & ; \quad \omega < 0 \end{cases}$$

Remember:

☑ Fourier transform of a real signal is always even conjugate in nature.

☑ Fourier transform of real and even signal is purely real and even.

☑ Fourier transfer of real and odd signal is purely imaginary and odd.

☑ Shifting in time domain only changes the phase spectrum of signal.

☑ Compression in time domain leads to extension in frequency domain and vice-versa.

☑ Time differentiation property is only valid if $f(t) \to 0$ as $|t| \to \infty$.

☑ $|F(\omega)|^2$ is Energy Spectral Density of signal for energy signal.

Fourier Transfer of Discrete Time Signals

DTFFT of an aperiodic signal $x[n]$ is given by

$$X(e^{j\omega}) = \sum_{n=-\infty}^{\infty} x[n]e^{-j\omega n}$$

Inverse DTFT of $X(e^{j\omega})$ is given by

$$x[n] = \frac{1}{2\pi}\int_0^{2\pi} X(e^{j\omega})e^{j\omega n}d\omega$$

DTFT is always periodic with period '2π'.

Properties of DTFT

Property	Aperiodic signal	Fourier Transform
Linearity	$ax[n] + by[n]$	$aX(e^{j\omega}) + bY(e^{j\omega})$
Time shifting	$x[n \mp n_0]$	$e^{\mp j\omega n_o} X(e^{j\omega})$
Frequency shifting	$e^{\pm j\omega_0 n}x[n]$	$X(e^{j(\omega \mp \omega_0)})$
Conjugation	$x^*[n]$	$X^*(e^{-j\omega})$
Time reversal	$x[-n]$	$X(e^{-j\omega})$
Time expansion	$x_{(k)}[n] = \begin{cases} x\left(\frac{n}{k}\right) & ; \text{ if } n \text{ is multiple of k} \\ 0 & ; \text{ Otherwise} \end{cases}$	$X(e^{jk\omega})$
Convolution	$x[n] * y[n]$	$X(e^{j\omega})\,Y(e^{j\omega})$
Multiplication	$x[n]\,y[n]$	$\frac{1}{2\pi}\left[X(e^{j\omega}) * Y(e^{j\omega})\right]$
Differencing	$x[n] - x[n-1]$	$(1-e^{-j\omega})\,(X(e^{j\omega}))$
Accumulation	$\sum_{k=-\infty}^{n} x[k]$	$\frac{1}{1-e^{-j\omega}} X(e^{(j\omega)})$
Differentiation in frequency	$nx[n]$	$j\frac{d}{d\omega} X(e^{j\omega})$
Parseval's relation	$\sum_{n=-\infty}^{\infty} \lvert x[n]\rvert^2$	$\frac{1}{2\pi}\int_0^{2\pi} \lvert X(e^{j\omega})\rvert^2 d\omega$

where, $x[n] \leftrightarrow X(e^{j\omega})$; $y[n] \leftrightarrow Y(e^{j\omega})$

Discrete Fourier Transform

DFT of a periodic $x[n]$ with period N is given by, $X(k) = \sum_{n=0}^{N-1} x[n]\, e^{-jk\left(\frac{2\pi}{N}\right)n}$

'k' varies from $k = 0$ to $N - 1$.

Inverse DFT of periodic $X(k)$ with period N is given by

$$x[n] = \frac{1}{N}\sum_{k=0}^{N-1} X(k)\, e^{jk\left(\frac{2\pi}{N}\right)n}$$

DFT Pairs

- $x[n] = \delta[n]\ ; 0 \le n \le N-1$; $X(k) = 1\ ; 0 \le k \le N-1$
- $x[n] = 1\ ; 0 \le n \le N-1$; $X(k) = N\delta(k)\ ;\ 0 \le k \le N-1$
- $x[n] = e^{j2\pi k_0 n}\ ; 0 \le n \le N-1$; $X(k) = N\delta(k - k_o)\ ; 0 \le k \le N-1$
- $x[n] = \cos(2\pi\, k_0\, n/N)\ ;\ 0 \le n \le N-1$

  $X(k) = \frac{N}{2}\left[\delta(k - k_o) + \delta\left(k - (N - k_o)\right)\right];\ 0 \le k \le N-1$

Properties of DFT

Property	DFT
Periodicity	$f[n], F[k] = F^*[N-k]$
Time reversal of sequence	$x[[-n]]_N = x[N-n]$
Circular time shift	$x[[n-l]]_N \longleftrightarrow X(k)e^{-j2\pi k l/N}$
Circular frequency shift	$x[n]\, e^{j2\pi ln/N} \longleftrightarrow X[k-l]_N$
Complex conjugate property	$x^*[n] \longleftrightarrow X^*[[-k]]_N$
Circular Convolution	(i) $x_2[n] \otimes_N x_2[n] \longleftrightarrow X_1[k] \cdot X_2[k]$ (ii) $f[n] \otimes_N h[n] = \sum_{k=0}^{N-1} f[k]\, h[(n-k)]_N$
Multiplication of two sequences	$x_1[n]\, x_2[n] \longleftrightarrow \frac{1}{N} X_1[k] \otimes_N X_2[k]$
Parseval's Theorem	$\sum_{n=0}^{N-1} \lvert x[n]\rvert^2 = \frac{1}{N}\sum_{k=0}^{N-1} \lvert X[k]\rvert^2$

Remember:

☑ Circular convolution can be found using linear convolution. Method is divide the result of linear convolution into two parts such that first part has N samples and then add corresponding part in order and write as circular convolution. Here N is no. of samples in each signal.

☑ Given two signals $f[n]$ and $h[n]$ each consisting of N-samples, if we zero pad each of the signal with N-zeros and carry out circular convolution of order 2 N this will be equal to linear convolution of two signals $f[n]$ and $h[n]$.

Fast Fourier Transform

The Fast Fourier Transform is a discrete Fourier transform algorithm which is reduces the number of the computations needed to find transform. For N-point DFT

$$\text{Number of multiplications} = \frac{N}{2}\log_2 N \text{ ; Number of additions} = N\log_2 N$$

V Laplace Transform

Introduction

The main drawback of (continuous) the Fourier transform is that the Fourier transform (F.T) can be defined only for stable systems where as Laplace transform (L.T) can be defined for both stable and unstable systems.

Laplace Transform

Laplace transform of a general signal $f(t)$

- Bilateral (or two sided) Laplace transform, $F(s) = \int_{-\infty}^{\infty} f(t)e^{-st}dt$
- Unilateral (or one-sided) Laplace transform,

$$F(s) = \int_{0}^{\infty} f(t)e^{-st}dt$$

- Inverse Laplace Transform, $f(t) = \int_{-\infty}^{\infty} F(s)e^{st}ds$
- Relation Between Laplace Transform and Fourier Transform

$$F(\sigma + j\omega) = \int_{-\infty}^{\infty} f(t)e^{-\sigma t}e^{-j\omega t}dt = e^{-\sigma t}F(\omega), \quad \text{where, } s \rightarrow \sigma + j\omega$$

Note:

☑ At $\sigma = 0$, $s = j\omega$, the Laplace transform become equal to Fourier transform (or) L.T. calculated on $j\omega$ axis is F.T.

☑ Fourier transform is used for signal analysis and Laplace transform is used for designing the system.

Region of Convergence (ROC)

The range of values of 's' for which $F(s)$, i.e., $\int_{-\infty}^{\infty} \left| f(t)e^{-\sigma t} \right| < \infty$ is define its known as ROC.

Properties of ROC :

(a) ROC does not contain any poles, it is bounded by the poles.

(b) ROC of $F(s)$ contains strips parallel to imaginary axis, i.e., $j\omega$ axis.

(c) If $f(t)$ is the finite duration and the absolutely integrable than the ROC is entire s-plane.

(d) If $f(t)$ is a right sided and if the line $\text{Re}\{s\} = \sigma_0$ is in the ROC then all values of s for which $\text{Re}\{s\} > \sigma_0$ will also be in ROC.

(e) If $f(t)$ is a left sided and if the line $\text{Re}\{s\} = \sigma_0$ is in the ROC then all values of s for which $\text{Re}\{s\} < \sigma_0$ will also be in ROC.

(f) For two sided signal, ROC is the combination of two region.

Remember:

☑ For stability ROC must include imaginary axis (i.e., $j\omega$ axis)

☑ Solution of L.T. is unique only when the ROC is given.

Properties of Laplace Transform

Property	Signal	Laplace transform	ROC
Linearity	$ax_1(t) + bx_2(t)$	$aX_1(s) + bX_2(s)$	$R_1 \cap R_2$
Time shifting	$x(t - t_0)$	$e^{-st_o} X(s)$	R
Shifting in s-domain	$e^{s_o t} x(t)$	$X(s - s_o)$	Shifted version of R
Time scaling	$x(at)$	$\frac{1}{\lvert a \rvert} X\left(\frac{s}{a}\right)$	Scaled ROC
Conjugation	$x^*(t)$	$X^*(s^*)$	R
Differentiation in time domain	$\frac{d}{dt} x(t)$	$sX(s)$	At least R
Differentiation in s-domain	$-tx(t)$	$\frac{dX(s)}{ds}$	R
Convolution	$x_1(t) * x_2(t)$	$X_1(s) X_2(s)$	$R_1 \cap R_2$
Integration	$\int_{-\infty}^{t} x(\tau) d\tau$	$\frac{1}{s} X(s)$	$R_1 \cap \{\text{Re}(s) > 0\}$

where, $x(t) \rightarrow X(s)$ with ROC R; $x_1(t) \rightarrow X_1(s)$ with ROC R_1
$x_2(t) \rightarrow X_2(s)$ with ROC R_2

- Initial value theorem, $f(0) = \lim_{s \to \infty} sF(s)$
- Final value theorem, $f(\infty) = \lim_{s \to 0} sF(s)$

Note:

☑ Initial value theorem is applicable only for proper Laplace i.e. denominator power of function is more than numerator power.

☑ Final value theorem is not applicable if poles are conjugate on imaginary axis or poles lie in right side of s-plane.

Laplace transform of elementary function

Transform pair	Signal	Transform	ROC
1	$\delta(t)$	1	All s
2	$u(t)$	$\frac{1}{s}$	$\text{Re}\{s\} > 0$
3	$-u(-t)$	$\frac{1}{s}$	$\text{Re}\{s\} < 0$
4	$\frac{t^{n-1}}{(n-1)!}u(t)$	$\frac{1}{s^n}$	$\text{Re}\{s\} > 0$
5	$-\frac{t^{n-1}}{(n-1)!}u(-t)$	$\frac{1}{s^n}$	$\text{Re}\{s\} < 0$
6	$e^{-\alpha t}u(t)$	$\frac{1}{s+\alpha}$	$\text{Re}\{s\} > -\alpha$
7	$-e^{-\alpha t}u(-t)$	$\frac{1}{s+\alpha}$	$\text{Re}\{s\} < -\alpha$
8	$\frac{t^{n-1}}{(n-1)!}e^{-\alpha t}u(t)$	$\frac{1}{(s+\alpha)^n}$	$\text{Re}\{s\} > -\alpha$
9	$-\frac{t^{n-1}}{(n-1)!}e^{-\alpha t}u(-t)$	$\frac{1}{(s+\alpha)^n}$	$\text{Re}\{s\} < -\alpha$
10	$\delta(t-T)$	e^{-sT}	All s
11	$[\cos\omega_0 t]u(t)$	$\frac{s}{s^2+\omega_0^2}$	$\text{Re}\{s\} > 0$
12	$[\sin\omega_0 t]u(t)$	$\frac{\omega_0}{s^2+\omega_0^2}$	$\text{Re}(s) > 0$
13	$\left[e^{-\alpha t}\cos\omega_0 t\right]u(t)$	$\frac{s+\alpha}{(s+\alpha)^2+\omega_0^2}$	$\text{Re}\{s\} > -\alpha$
14	$\left[e^{-\alpha t}\sin\omega_0 t\right]u(t)$	$\frac{\omega_0}{(s+\alpha)^2+\omega_0^2}$	$\text{Re}\{s\} > -\alpha$
15	$u_n(t) = \frac{d^n\delta(t)}{dt^n}$	s^n	All s

Characterization of LTI Systems Using Laplace Transform

Causality :

- ROC associated with the system function for causal system is a right-half plane.

- For a system with rational transfer function, causality of the system is equivalent to ROC being the right-half plane to right of the rightmost pole.
- A system is anticausal if its impulse response $h(t) \neq 0$ for $t < 0$.

Stability :

- An LTI system is stable if and only if the ROC of its system function $H(s)$ includes the $j\omega$-axis.
- A causal system with rational system functions $H(s)$ is stable if and only if all of the poles of $H(s)$ lie in the left-half of the s-plane-i.e. all of the poles have negative real parts.

VI Discrete Time Fourier Transform

The Definition: DTFT

Consider a discrete-time signal $x[n]$. Its discrete-time Fourier transform (DTFT) is defined as

$$\text{DTFT}\{x(n)\} = X(\Omega) = \sum_{n=-\infty}^{+\infty} x[n]e^{-j\Omega n}$$

Where 'Ω' is discrete time frequency in 'rad/second'. The transform $X(\Omega)$ is frequency domain description of $x[n]$. Thus $x[n]$ and $X(\Omega)$ makes Fourier transform pair as

$$\underbrace{x(n)}_{\substack{\text{Discrete}\\ \text{Aperiodic}}} \xleftrightarrow{\text{DTFT}} \underbrace{X(\Omega)}_{\substack{\text{Periodic}\\ \text{continuous}}}$$

and the inverse DTFT which maps the frequency domain description $X(\Omega)$ back into time is given as

$$IDTFT\{X(\Omega)\} = x[n] = \frac{1}{2\pi}\int_{2\pi} X(\Omega)e^{j\Omega n}d\Omega$$

Properties of DTFT

Sl. No.	$x[n]$	$X(\Omega)$
1.	Linearity	$\alpha x_1[n]+\beta x_2[n] \xleftrightarrow{DTFT} \alpha X_1(\Omega)+\beta X_2(\Omega)$
2.	Time-shifting	$x[n-n_o] \xleftrightarrow{DTFT} e^{-j\Omega n_0} X(\Omega)$
3.	Frequency shifting	$x[n]\, e^{j\Omega_0 n} \xleftrightarrow{DTFT} X(\Omega-\Omega_0)$
4.	Modulation property	$x_1[n]\cos \Omega_0 n \xleftrightarrow{DTFT} \frac{1}{2}[X(\Omega-\Omega_0)+X(\Omega+\Omega_0)]$
5.	Time reversal	$x[-n] \xleftrightarrow{DTFT} X(-\Omega)$

6.	Conjugate symmetry	$x^*[n] \xleftrightarrow{DTFT} X^*(-\Omega)$
7.	Frequency differentiation	$nx[n] \xleftrightarrow{DTFT} j\dfrac{dX(\Omega)}{d\Omega}$
8.	Convolution n-domain	$x_1[n] * x_2[n] \xleftrightarrow{DTFT} X_1(\Omega)\cdot X_2(\Omega)$
9.	Multiplication property	$x_1[n]\, x_2[n] \xleftrightarrow{DTFT} \dfrac{1}{2\pi}\left[X_1(\Omega) * X_2(\Omega)\right]$
10.	First difference	$x[n]-x[n-1] \xleftrightarrow{DTFT} (1-e^{-j\Omega})\, X(\Omega)$
11.	Accumulation property	$\sum_{k=-\infty}^{m} x[k] \xleftrightarrow{DTFT} \pi X(1)\sum_{K=-\infty}^{K=\infty}\delta(\Omega-2\pi k)+\dfrac{e^{j\Omega}}{e^{j\Omega}-1}X(\Omega)$
12.	Parseval's relation	$\sum_{n=-\infty}^{+\infty} \lvert x[n]\rvert^2 = \dfrac{1}{2\pi}\int_{2\pi} \lvert X(\Omega)\rvert^2\, d\Omega$

VII Z -Transform

- Z-transform is a discrete-time counterpart of Laplace transform.
- For a discrete-time LTI system with the impulse response $h[n]$, the response $y[n]$ of the system to a complex exponential input of the form z^n is $y[n] = z^n H(z)$

  where, $H(z) \rightarrow$ Transform function of the system
- Z-transform of a general discrete time signal, $X(z) = \sum_{n=-\infty}^{+\infty} x[n]z^{-n}$

Note:

☑ The range of values of 'z' for which above equation is defined give ROC of Z-transform.

☑ Z-transform are used for digital filter design, sampled data control system.

Region of Convergence (ROC)

ROC is the region of range of values for which the summation

$$\sum_{n=-\infty}^{+\infty} x[n]z^{-n} \text{ converges.}$$

Properties of ROC :

- The ROC of $X(z)$ consists of a circle in the z-plane centered about the origin.
- ROC does not contain any poles, it is bounded by the poles.
- If $x[n]$ is of finite duration then ROC is entire z-plane except possibly $z = 0$ and/or $z \rightarrow \infty$.

- If $x[n]$ is a right sided sequence and if circle $|z| = a$ is in the ROC then all finite values of z for which $|z| > a$ will also be in ROC.
- If $x[n]$ is a left sided sequence and if circle $|z| = a$ is the ROC then all finite values of z for which $|z| < a$ will also be in ROC.
- If $x[n]$ is two sided and if the circle $|z| = a$ is in the ROC, then the ROC will consist of a ring in the z-plane that includes the circle $|z| = a$.

Unilateral Z-transform

$$X(z) = \sum_{n=0}^{\infty} x[n]z^{-n}$$

- Left shift, $x[n+1] \leftrightarrow zX[z] - zx[0]$
- Right shift,

$$x[n-1] \xrightarrow{\text{Z.T.}} z^{-1}X(z) + x(-1)$$
$$x[n-2] \xrightarrow{\text{Z.T.}} z^{-2}X(z) + z^{-1}x(-1) + x(-2)$$

- First difference, $x[n] - x[n-1] = (1 - z^{-1})X(z)$
- Conjugation, $x^*[n] = X^*(z^*)$

Properties of Z-transform

Property	Signal	z - Transform	ROC
Linearity	$ax_1[n] + bx_2[n]$	$aX_1(z) + bX_2(z)$	$R_1 \cap R_2$
Time shifting	$x[n - n_0]$	$z^{-n_0}X(z)$	R
Scaling	$a^n x[n]$	$X\left(\frac{z}{a}\right)$	Scaled version of R
Time reversal	$x[-n]$	$x(z^{-1})$	Invert R
Time expansion	$x_{(k)}[n] = \begin{cases} x[r], & n = rk \\ 0, & \text{otherwise} \end{cases}$	$X(z^k)$	$R^{1/k}$
Conjugation	$x^*[n]$	$X^*(z^*)$	R
Convolution	$x_1[n] * x_2[n]$	$X_1(z)\, X_2(z)$	$R_1 \cap R_2$
First difference	$x[n] - x[n-1]$	$(1 - z^{-1})X(z)$	$R_1 \cap \{\lvert z\rvert > 0\}$
Accumulation	$\sum_{k=-\infty}^{n} x[k]$	$\frac{1}{1 - z^{-1}} X(z)$	$R \cap \{\lvert z\rvert > 1\}$
Differentiation	$nx[n]$	$-\frac{zdX(z)}{dz}$	R

where, $x[n] \to X(z)$ with ROC R; $x_1[n] \to X_1(z)$ with ROC R_1
$x_2[n] \to X_2(z)$ with ROC R_2

Initial value theorem

If $x[n] = 0$ for $n < 0$, $x[0] = \lim_{Z \to \infty} X(z)$

Final value theorem

$$x[\infty] = \lim_{z\to 1}(1 - z^{-1})X(z) \quad \text{or} \quad x[\infty] = \lim_{z\to 1}(z-1)X(z)$$

Characterization of LTI Systems Using Z-Transform

Causality :

- A discrete-time LTI system is causal if and only if the ROC of its system function is the exterior of a circle, including infinity.
- A discrete-time LTI system with rational system function $H(z)$ is causal if and only if. Denominator degree is greater than numerator degree.

Stability :

- An LTI system is stable if and only if the ROC of its system function $H(z)$ includes the unit circle, $|z| = 1$.
- A causal LTI system with rational system function $H(z)$ is stable if and only if all of the poles of $H(z)$ lie inside the unit circle i.e., they must all have magnitude smaller than 1.

Inverse Z-transform

As in the case of Laplace transform there is a formal expression for the inverse z-transform in terms of an integration in the z-plane.

i.e.,
$$x[n] = \frac{1}{2\pi j}\oint_C X(z)\cdot z^{n-1}dz$$

where, c is a counter clockwise contour of integration enclosing the origin.

Some Common Z-transform Pairs

Sl.No.	Signal	Transform	ROC
1.	$\delta[n]$	1	All z
2.	$u[n]$	$\frac{1}{1-z^{-1}}$	$\|z\| > 1$
3.	$-u[-n-1]$	$\frac{1}{1-z^{-1}}$	$\|z\| < 1$
4.	$\delta[n-m]$	z^{-m}	All z, except 0 (if $m > 0$) or ∞(if $m < 0$)
5.	$\alpha^n u[n]$	$\frac{1}{1-\alpha z^{-1}}$	$\|z\| > \|\alpha\|$
6.	$-\alpha^n u[-n-1]$	$\frac{1}{1-\alpha z^{-1}}$	$\|z\| < \|\alpha\|$
7.	$n\alpha^n u[n]$	$\frac{\alpha z^{-1}}{(1-\alpha z^{-1})^2}$	$\|z\| > \|\alpha\|$

8.	$-n\alpha^n u[-n-1]$	$\dfrac{\alpha z^{-1}}{(1-\alpha z^{-1})^2}$	$\lvert z\rvert < \lvert\alpha\rvert$
9.	$[\cos\omega_0 n]u[n]$	$\dfrac{1-[\cos\omega_0]z^{-1}}{1-[2\cos\omega_0]z^{-1}+z^{-2}}$	$\lvert z\rvert > 1$
10.	$[\sin\omega_0 n]u[n]$	$\dfrac{[\sin\omega_0]z^{-1}}{1-[2\cos\omega_0]z^{-1}+z^{-2}}$	$\lvert z\rvert > 1$
11.	$[a^n\cos\omega_0 n]u[n]$	$\dfrac{1-[a\cos\omega_0]z^{-1}}{1-[2a\cos\omega_0]z^{-1}+a^2z^{-2}}$	$\lvert z\rvert > a$
12.	$[a^n\sin\omega_0 n]u[n]$	$\dfrac{[a\sin\omega_0]z^{-1}}{1-[2a\cos\omega_0]z^{-1}+a^2z^{-2}}$	$\lvert z\rvert > a$

VIII Discrete Fourier Transform

The Discrete Fourier Transform (DFT) is basically Fourier transform of "sampled signal repeated periodically". The number of points in DFT is known as N-point DFT.

The Definition

Consider a sampled and periodic signal $x[n]$. Its DFT is given as

$$\text{DFT}\{x[n]\} = X[k] = \sum_{n=0}^{N-1} x[n]e^{-j\frac{2\pi k}{N}n}$$

The signal $x[n]$ and $X[k]$ makes DFT pair as

$$\underbrace{x[n]}_{\text{Discrete periodic}} \xleftrightarrow{\text{DFT}} \underbrace{X[k]}_{\text{Discrete periodic}}$$

and the inverse DFT can be obtained as

$$x[n] = \frac{1}{N}\sum_{k=0}^{N-1} X[k]e^{j\frac{2\pi k}{N}\cdot n}; \qquad \Omega = \frac{2\pi}{N}$$

Where 'N' is period of the signal $x[n]$.

Properties of DFT

If $x[n] \xleftrightarrow{\text{DFT}} X[k]$, then

Property	DFT
Linearity	$\alpha x_1[n]+\beta x_2[n] \xleftrightarrow{\text{DFT}} \alpha X_1[k]+\beta X_2[k]$
Time Shifting	$x[n-n_0] \xleftrightarrow{\text{DFT}} e^{-j\frac{2\pi}{N}kn_0}X[k]$

Shift in k-domain	$x[n] \cdot e^{j\frac{2\pi}{N}k_0 n} \xleftrightarrow{\text{DFT}} X[k-k_0]$
Conjugate Symmetry	$x^*[n] \xleftrightarrow{\text{DFT}} X^*[-k]$
Time Reversal	$x[-n] \xleftrightarrow{\text{DFT}} X[-k]$
Linear Convolution	$x_1[n] * x_2[n] \xleftrightarrow{\text{DFT}} X_1[k]\, X_2[k]$ with period $N_1 + N_2 - 1$
Periodic Convolution	$x_1[n]\ (N)\ x_2[n] \xleftrightarrow{\text{DFT}} \frac{1}{N} X_1[k] * X_2[k]$
Parseval's Relation	$E = \sum_{n=0}^{N-1} \lvert x[n] \rvert^2 = \frac{1}{N} \sum_{k=0}^{N-1} \lvert X[k] \rvert^2$

FFT (Fast Fourier Transform)

The N-point DFT is given as

$$X[k] = \sum_{n=0}^{N-1} x[n] \left(e^{-j\frac{2\pi}{N}} \right)^{kn}$$

- To compute one point (one sample) DFT, it is required 'N' multiplications and $N-1$ additions.
- To compute N-point DFT, it is required 'N^2' multiplications and $N(N-1)$ additions.
- In general we can say as N increases these computation become time consuming even for high-speed computer.
- Number of addition = $N \log_2 N$.
- Number of multiplication = $\frac{N}{2} \log_2 N$.

IX Digital Filters

Digital filter is a system that performs mathematical operations on a sampled, discrete time signal to reduce or enhance certain parameters of that signal. There are two types:

Finite Impulse Response filter

The impulse response is of finite duration. The difference equation is given by

$$y[n] = \sum_{i=0}^{N} b_i\, x[n-i]$$

where, $x[n]$, $y[n] \rightarrow$ Input and output respectively ; $N \rightarrow$ Filter order
$b_i \rightarrow$ Value of impulse response at i^{th} instant

Properties :

(i) No feedback is required.

(ii) They are inherently stable.

(iii) They can be easily designed to be linear phase by making the coefficient sequence symmetric.

(iv) Computationally complex.

Ex: Moving average filter, $y[n] = b_0 x[n] + b_1 x[n-1] + b_2 x[n-2]$.

Infinite Impulse Response filter

The impulse response of IIR filters continues indefinitely. The difference equation is given as

$$y[n] = \frac{1}{a_0}\left(\sum_{i=0}^{P} b_i x[n-i] - \sum_{j=1}^{Q} a_j y[n-j]\right)$$

where, $P \to$ Feedforward filter order ; $Q \to$ Feedback filter order ; $b_i \to$ Feedforward filter coefficients ; $a_i \to$ Feedback filter coefficients

Properties:

(i) Feedback is required.

(ii) Stability depends on the filter coefficients.

(iii) They are preferred over FIR filters for their efficiency in implementation.

Ex: $y[n] = y[n-1] + x[n]$.

Basics Structures for IIR Systems

Causal IIR systems are characterized by the constant coefficient difference equation or equivalently, by the real rational transfer function of equation. From these equations, it can be seen that the realization of infinite duration impulse response (IIR) systems involves a recursive computational algorithm. In this section, the most important filter structures namely direct Forms-I and II, cascade and parallel realizations for IIR systems are discussed.

$$H(z) = \frac{Y(z)}{X(z)} = \frac{\sum_{k=0}^{M} b_k z^{-k}}{\sum_{k=1}^{N} a_k z^{-k}}$$

Basic Structures for FIR Systems

The system function of the FIR system whose realization take the form of a non-recursive computational algorithm may be written as

$$H(z) = \sum_{n=0}^{M-1} h(n) z^{-n}$$

That is, if the impulse response is M samples in duration, the $H(z)$ is a polynomial in z^{-1} of degree $M-1$. Thus $H(z)$ has $(M-1)$ poles at $z=0$ and $(M-1)$ zeros that can be anywhere in the finite z-plane.

FIR filters are often preferred in many applications, since they provide an exact linear phase over the whole frequency range and they are always BIBO stable independent of the filter coefficients. Two simple realization methods for FIR filters, viz direct form and cascade form.

IIR Filter	FIR Filter
• Length of $h[n]$ is infinite.	• Length of $h[n]$ is finite.
• Design of IIR filter may become unstable.	• It is always stable.
• It's design involves less number of (i) delay elements (ii) adders (iii) multipliers i.e. design is simple.	• Design involves more number of (i) delay elements (ii) adders (iii) multipliers i.e. design is complicated.
• Less memory required.	• More number of memory required.
• It cannot be designed with linear phase.	• It can be designed with linear phase.
• It will have delay or phase distortion in the output.	• Linear, phase FIR filter will NOT have delay or phase distortion.
• Faster response.	• Slower response.
• Recursive.	• Non-recursive.

Realisation of Linear Phase FIR Systems

If the impulse response is symmetric about its origin, linear phase results. If all zero filter has a linear phase response, a special non-recursive structure that reduces the number of multiplications by approximately one-half can be implemented. The impulse response for a causal filter begins at zero and ends at $M-1$.

As discussed, a linear phase FIR filter to length M is characterized by

$$h(n) = h(M-1-n)$$

The symmetry property of a linear phase FIR filter is used to reduce the multipliers required in these realisation.

Comparison of Designing methods

Three methods of FIR filter design, i.e. windows, frequency sampling and Chebyshev approximation.

Advantages of window design method:

(i) Simple method for design.

(ii) Various window functions can be used depending upon the application.

Disadvantages of window design method:

(i) Lack of precise control of the critical frequencies ω_p and ω_s.

(ii) ω_p and ω_s depend upon type of window and filter length M.

Advantages of frequency sampling method:

(i) Simple method of design.

(ii) Simple realization in frequency domain.

(iii) Better control over ω_p and ω_s.

Disadvantages of frequency sampling method : This method is suitable only for frequency domain realization.

X Miscellaneous

Unit Doublet Function

The derivative of a unit impulse function is called a unit doublet $(\delta'(t))$.

Properties :

(i) $\delta'(t) = \dfrac{d\delta(t)}{dt}$ (ii) $\delta'(at) = \dfrac{1}{a|a|}\delta'(t)$

All even derivatives of $\delta(t)$ are even functions of time while all odd derivatives are odd functions.

(iii) If $f(t)$ is any function, then

$$f(t)\,\delta'(t-a) = f(a)\,\delta'(t-a) - f'(a)\,\delta(t-a)$$

(iv) $\int_{-\infty}^{\infty} f(t)\delta'(t-a) = -\dfrac{d}{dt}f(t)\Big|_{t=a}$

Response of LTI System to a Sinusoidal Input

Any LTI system does not effect the frequency of input signal, only amplitude and phase change takes place.

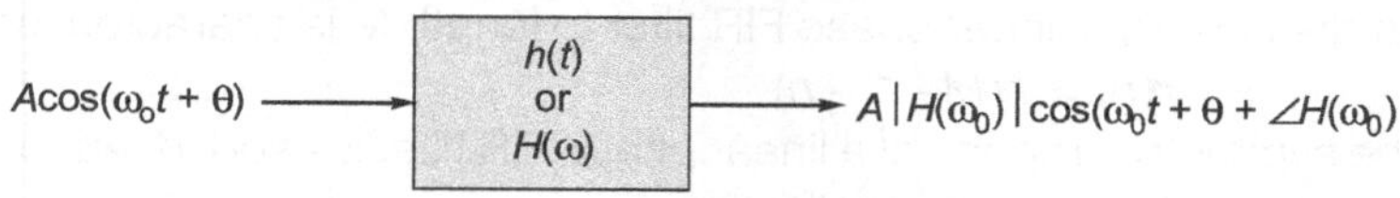

where, $|H(\omega_o)| \rightarrow$ Magnitude response at $\omega = \omega_0$

$\angle H(\omega_0) \rightarrow$ Phase response at $\omega = \omega_0$

Phase delay is the delay offered to the phase of input consisting of single frequency.

$$\text{Phase delay} = -\frac{\phi(\omega)}{\omega}$$

Group delay is the delay offered to phase of input consisting of multiple frequencies.

$$\text{Group delay} = -\frac{d\phi(\omega)}{d\omega}$$

For a distortionless LTI system, both group delay and phase delay are constant and are equal.

Sampling

Any bandlimited signal $x(t)$ is uniquely determined by its samples if the sampling frequency (ω_s) is greater than twice the maximum frequency (ω_m) present in the signal. This is sampling theorem.

$$\omega_s > 2\omega_m$$

There are two types of sampling :

(a) Ideal/Instantaneous sampling (zero pulse width).

(b) Practical sampling (non-zero pulse width) :

(i) Natural sampling (ii) Flat top sampling

Remember:

☑ The reconstruction of a continuous time signal from the samples can be done using a low pass filter.

☑ For a bandpass signal with the spectral components from ω_L to ω_H, the sampling frequency (ω_s) is given by $\omega_s \geq \frac{2\omega_H}{k}$

where, $k = \frac{\omega_H}{\omega_H - \omega_L}$ [k must be integer]

For any integer n satisfying

$$1 \leq n \leq \left\lfloor \frac{\omega_H}{\omega_H - \omega_L} \right\rfloor \quad (\lfloor\ \rfloor \rightarrow \text{floor function})$$

Correlation

Correlation is a measure of relation between two signals. For two signals $f(t)$ and $g(t)$, correlation is defined as

$$R_{fg}(\tau) = \int_{-\infty}^{\infty} f(t)\, g(t-\tau)dt = [f(t) \otimes g(-t)]_{t=\tau}$$

Autocorrelation function is a special case where both the signals are same.

For energy signals, $R_{ff}(\tau) = \int_{-\infty}^{\infty} f(t)\, f(t-\tau)dt = [f(t) \otimes f(-t)]_{t=\tau}$

For power signals, $R_{ff}(\tau) = \lim_{x \to \infty} \frac{1}{T} \int_{-\infty}^{\infty} f(t)\, f(t-\tau)d\tau$

Properties of auto-correlation function:

(i) $R_{ff}(\tau)$ is even ; $R_{ff}(\tau) = R_{ff}(-\tau)$.

(ii) $R_{ff}(0) > R_{ff}(\tau)$

(iii) For energy signals, energy = $R_{ff}(0)$. For power signals, power = $R_{ff}(0)$

(iv) For energy signals, Fourier transform of $R_{ff}(\tau)$ gives Energy Spectral Density, i.e., $R_{ff}(\tau) \leftrightarrow F(\omega)\,F^*(\omega)$ [ESD_f]

For power signals, Fourier transform of $R_{ff}(\tau)$ gives power spectral density, i.e., $R_{ff}(\tau) \leftrightarrow \lim\limits_{T\to\infty} \dfrac{|F_T(\omega)|^2}{T}$ [PSD_f]

Remember:

☑ The autocorrelation function of a binary rectangular random pulse train $x(t) = \sum\limits_{-\infty}^{\infty} a_n p(t - nT + t_o)$ where, a and n are independent variables with values ±1 (equal probability), t_o (time offset, uniformly distributed between ± $T/2$) and $p(t)$ is pulse with amplitude 'A' and width 'T' is given by $R_{xx}(\tau) = A^2\left[1 - \dfrac{|\tau|}{T}\right]; |\tau| \le T$.

Pre-envelope and Complex Envelope

Let $x(t)$ be a real bandpass signal with Fourier transform as $X(\omega)$, then $x_p(t)$ is its pre-envelope and $x_c(t)$ is its complex envelope.

Pre-envelope

$x_p(t) = x(t) + jx_h(t)$ and $X_p(\omega) = 2X(\omega)\,u(\omega)$

where, $x_h(t) \rightarrow$ Hilbert transform of $x(t)$; $X_p(\omega) \rightarrow$ Fourier transform of $x_p(t)$; $u(\omega) \rightarrow$ Unit step signal

Complex Envelope

$$x_c(t) = x_p(t)e^{-j\omega_c t} = x_i(t) + jx_Q(t)$$

$$x(t) = \text{Re}\{x_p(t)\} = x_i(t)\cos\omega_c t - x_Q(t)\sin\omega_c t$$

where, $x_i(t), x_Q(t) \rightarrow$ In phase and quadrature components respectively

$\omega_c \rightarrow$ Centre frequency of carrier

Remember:

☑ Complex envelope is a low pass signal.

☑ Pre-envelope is a band pass signal.

☑ Complex envelope and pre-envelope are frequency shifted version of each other.

■■■■

Analog Electronics

8

I Semiconductor Physics

Energy Band

1. **Valence band:** The band occupied by the valence electrons or a band having highest occupied band energy is called valance band.
2. **Conduction band:** The band occupied by the conduction electrons or the band having the lowest unfilled energy is called conduction band.
3. **Forbidden energy gap:** The separation between conduction band and valence band is known as forbidden energy gap.

Remember:

- ☑ Valence band can never be empty.
- ☑ When a substance has empty conduction band the current conduction is not possible.
- ☑ In order to push an electron from valence band to conduction band, external energy is required which is equal to the forbidden energy gap.

Energy Band Diagram

Where,

- E_V is highest energy for the electron when it is inside the atom. It is the minimum energy to be given externally to make that electron free from atom.
- E_C is the lowest energy of the electron participating in the current conduction.
- $E_g = E_C - E_V$, is the amount of energy require to break the covalent bond and make the electron participate in the current conduction.

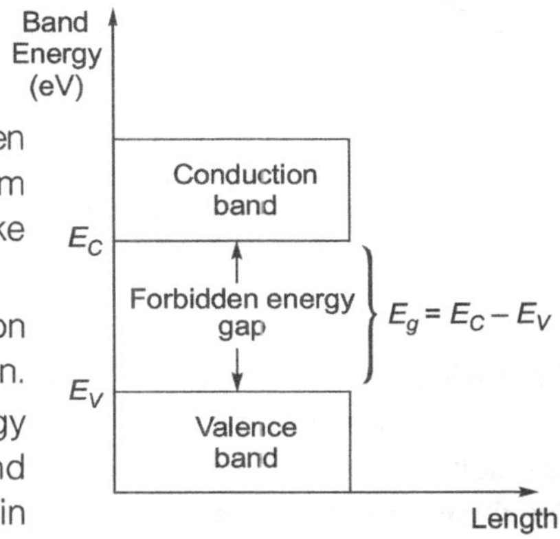

Note:

- ☑ Electron volt (eV) is smallest unit of energy.
- ☑ 1 eV is defined as the amount of energy accepted or released by a single electron when it is moving in the potential difference of 1 V.

$$1 \text{ eV} = 1.6 \times 10^{-19} \text{ Joule}$$

☑ With increase in the temperature energy gap decreases as

$$(E_g)_{T°K} = (E_g)_{0°K} - \beta T$$

where, β = Weiss constant

= 2.36×10^{-4} for Ge = 3.6×10^{-4} for Si

☑

	Ge	Si	GaAs
$(E_g)_{0°K}$	0.782 eV	1.21 eV	1.56 eV
$(E_g)_{300°K}$	0.72 eV	1.12 eV	1.43 eV

Insulators, Conductors and Semiconductors

Insulator : In insulators, the forbidden energy band is very wide, thus electrons cannot jump from valence band to conduction band. The resistivity of insulators is of the order of 10^7 ohm-metre.

Semiconductor : In semiconductor, the forbidden band is very small. A semiconductor material is one whose electrical properties lies between insulators and good conductors. In semiconductors, the conductivities are of order of 10^2 mho/m.

Conductor : In conductors, there is no forbidden band and the valence band and the conduction band overlap each other. The total current in conductor is simply due to the flow of electrons.

Note:

In germanium the forbidden band is of the order of 0.7 eV while in case of silicon, the forbidden band is of the order of 1.1 eV.

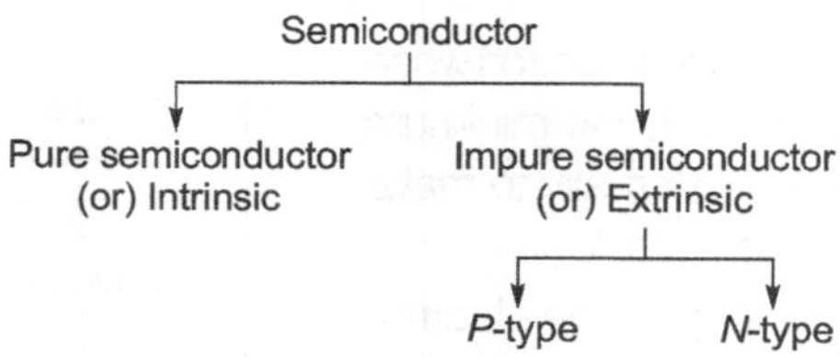

Intrinsic semiconductor : A semiconductor in an pure form is known as intrinsic semiconductor. In this type of semiconductor electrons and holes are solely created by thermal excitation.

Extrinsic semiconductor : In extrinsic semiconductor, a small percentage of trivalent or pentavalent atoms impurity are added to the pure semiconductor. The process of adding impurities to pure semiconductor is called doping.

***P*-type extrinsic semiconductor :** When a small amount of trivalent impurity (Boron, Gallium or Indium) is added to a pure crystal during the crystal growth, the resulting crystal is called *P*-type extrinsic semiconductor.

Note:

☑ In P-type semiconductor materials, the majority carriers are holes while the minority carriers are electrons.

N-type extrinsic semiconductor : When a small amount of pentavalent impurity (Antimony, Phosphorous or Arsenic) is added to a pure semiconductor crystal during the crystal growth, the resulting crystal is called as *N*-type extrinsic semiconductor.

Note:

☑ In *N*-type semiconductor, the electrons are the majority carriers while holes are minority carriers.

Mass Action Law

It states that at thermal equilibrium the product of concentration of free electrons and holes is equal to the square of intrinsic concentration at that temperature, i.e.,

$$n_o\, p_o = n_i^2$$

where, $n_o \rightarrow$ Concentration of electron in conduction band

$p_o \rightarrow$ Concentration of holes in valence band

$n_i \rightarrow$ Intrinsic concentration at given temperature

Intrinsic concentration

$$n_i^2 = A_o T^3\, e^{-E_g/kT} \qquad \text{where, } A_o \text{ is a constant}$$

Note:

Intrinsic concentration depends on temperature. As temperature increases the intrinsic concentration increases as $T^{3/2}$.

Mobility

- It is drift velocity per unit electric field.
- It defines how fast the charge travels from one place to other and is given by

$$\mu = \frac{v_d}{E}$$

where, v_d = drift velocity ; E = electric field

Note:

Electron's mobility is always greater than hole mobility in a given material. Hence electron can travel faster so contribute more current for same electric field than hole as explained by quantum mechanical physics.

- Mobility of charge carriers decreases with temperature and varies as :

$$\mu \propto T^{-m}$$

where, m is a constant

$m = 1.66$ for e^- and 2.33 for hole in Ge

$m = 2.5$ for e^- and 2.7 for hole in Si

- Mobility also varies with electric field applied as

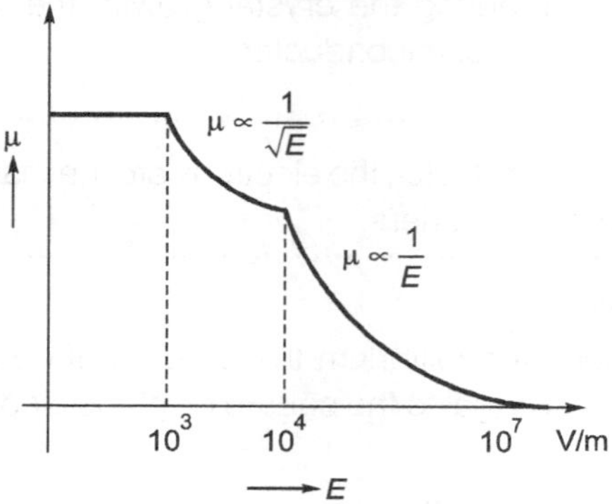

Note:

☑ At smaller electric field mobility is constant.

☑ At very high electric field product of mobility and electric field becomes constant and is equal to saturation value of drift velocity.

Remember:

☑

Mobility (cm^2/V-sec)	Ge	Si	GaAs
μ_n	3800	1300	4600
μ_h	1800	500	300

At $T = 300°K$

Drift current and conductivity

Drift current I_D is defined as the charge flowing per second across any normal plane of the conductor.

$$I_D = nAqv_d = nAq\mu E$$

Drift current density (J) $J = \frac{I_D}{A} = nq\mu E$

But $J = \sigma E$ ... According to Ohm's law

So, $\sigma = nq\mu$ or, $\sigma = qn\,\mu_n + qn\,\mu_p$

In intrinsic semiconductor, $n = p = n_i$

$$\sigma_{total} = \sigma_n + \sigma_p$$

$$\sigma = n_i q(\mu_n + \mu_p)$$

Conductivity of extrinsic semiconductor

1. **For *n*-type :** $\sigma \cong qN_D \mu_n$ where, N_D = Donor concentration
2. **For *p*-type :** $\sigma \cong qN_A \mu_p$ where, N_A = Accepter concentration

Remember:

- ☑ Conductivity of pure semiconductor increases with temperature.
- ☑ Conductivity of pure semiconductor at 0°K is zero.
- ☑ Conductivity of extrinsic semiconductor decrease with increase in temperature above normal temperature.
- ☑ Conductivity of extrinsic semiconductor initially increases when temperature rises from 0°K.
- ☑ Conductivity increases with increase in doping temperature.
- ☑ Conductivity of extrinsic semiconductor at 0°K is zero.
- ☑ At Curie temperature conductivity becomes equal to intrinsic conductivity.

Concentration of Charge Carriers

Number of electrons in the conduction band

$$n_o = N_C \, e^{-(E_C - E_F)/kT}$$

where, E_F = Fermi energy level; E_C = Energy level of lowest

k = Boltzman's constant; T = Absolute temperature

N_C = Effective density of states in conduction band

$$N_C = 2\left[\frac{2\pi \, m_e \, kT}{h^2}\right]^{3/2}$$

where, m_e = Mass of electron = 9.1×10^{-3}/kg

h = Plank's constant = 6.636×10^{-31} J-s

Number of holes in the valence band

$$p_o = N_V \, e^{-(E_F - E_V)/kT}$$

where, N_V = Effective density of states in valence band

E_V = Highest energy level of valence band

$$N_V = 2\left[\frac{2\pi m_h \, kT}{h^2}\right]^{3/2}$$

where, m_h = Mass of hole = 9.1×10^{-31} kg

Fermi Level

- Fermi level is energy state having probability 1/2 of being occupied of an electron if there is no forbidden band exists.
- Energy of fastest moving electron at 0 K is called fermi energy level.

- *Fermi dirac function* $f(E)$ gives the probability that an available energy state E will be occupied by an electron at absolute temperature T, under conditions of thermal equilibrium.

Fermi-Dirac function

$$f(E) = \frac{1}{1 + e^{(E - E_F)/kT}}$$ where, $E_F \rightarrow$ Energy of Fermi level in eV.

Fermi Level in Intrinsic Semiconductor

$$E_F = \frac{E_C + E_V}{2} - \frac{kT}{2} \ln \frac{N_C}{N_V}$$

where, $E_C \rightarrow$ Maximum energy of conduction band

$E_V \rightarrow$ Maximum energy of valence band

For $m_e = m_h$ at $T = 0$ K

$$E_F = \frac{E_C + E_V}{2}$$

- In intrinsic semiconductor Fermi level stays at the centre of energy gap. Such that Fermi level is independent of temperature.
- In intrinsic semiconductor the probability of occupancy of charge carrier is 50%.

Fermi level in *N*-type extrinsic semiconductor

$$E_F = E_C - kT \log_e \left(\frac{N_C}{N_D} \right)$$

where, $N_D \rightarrow$ Concentration of donor ion

$N_C \rightarrow$ Material constant and can be considered as a function of temperature

Note:

- ☑ In *N*-type semiconductor Fermi-level depends on both temperature and donor concentration.
- ☑ At 0°K Fermi level coincides with E_C.
- ☑ As temperature increases Fermi-level moves towards the middle of band gap.
- ☑ As donor concentration increases Fermi level moves towards the E_C.
- ☑ Normally Fermi level lies close to E_C.

Fermi level in *p*-type extrinsic semiconductor

$$E_F = E_V + kT \log_e \left(\frac{N_V}{N_A} \right)$$

where, $N_V \rightarrow$ Density of states in valence band

$N_A \rightarrow$ Concentration of acceptors.

Note:

- ☑ In p-type semiconductor Fermi level E_F at 0°K coincides with E_V.
- ☑ In p-type semiconductor Fermi level lies close to valence band.
- ☑ In p-type semiconductor Fermi level moves away from valence band as temperature increases.
- ☑ In p-type semiconductor fermi level moves towards valence band as N_A increases.

Diffusion Current

- The migration of charge carrier from higher concentration to lower concentration or from higher density to lower density is called diffusion. The current due to this diffusion process is called diffusion current.
- Diffusion current flows only in semiconductor.
- Diffusion is associated with random motion of charge carriers due to thermal vibrations.

Diffusion current density due to electrons and holes

$$J_n = qD_n\left(\frac{dn}{dx}\right) \quad ; \quad J_p = -q\,D_p\left(\frac{dp}{dx}\right)$$

where, D_n = Electron diffusion coefficient

D_p = Hole diffusion coefficient

$\frac{dn}{dx}$ = Electron concentration gradient

$\frac{dp}{dx}$ = Hole concentration gradient

Length of diffusion

Average distance travelled by charge carriers before recombination.

$$L = \sqrt{D\tau}$$

where, $L \rightarrow$ Diffusion length

$D \rightarrow$ Diffusion constant of charge carrier

$\tau \rightarrow$ Carrier life time

We know, $D = \mu V_T$ and $V_T = \frac{kT}{q}$

So, $L = \sqrt{\mu V_T \tau}$ where, $\mu \rightarrow$ Mobility of charge

$V_T \rightarrow$ Thermal voltage

- Diffusion length of electron, $L_n = \sqrt{D_n \tau_n}$
- Diffusion length of hole, $L_p = \sqrt{D_p \tau_p}$

Remember:

Diffusion length depends upon:
- Diffusion constant (D)
- Temperature (T)
- Carrier life time (τ)
- Mobility of charge carrier (μ)

Einstein Relation

This gives the relation between diffusion constant D and mobility μ.

$$\frac{D_n}{\mu_n} = \frac{D_p}{\mu_p} = V_T \quad \text{where, } V_T \rightarrow \text{Volt equivalent of temperature}$$

$$V_T = \frac{kT}{q} = \frac{T}{11600} \quad \text{where, } k \rightarrow \text{Boltzman constant}$$

Note:

At room temperature, i.e., 300°K, $V_T = 26$ mV.

Overall Current Density

The conduction of current in semiconductor materials may occur due to
- Drift of charged particles under the effect of applied electric field.
- Diffusion of carriers from a region of high charge density to region of lower charge density.

and it is possible that a potential gradient and a concentration gradient may exist within a semiconductor.

Total hole current density : $$J_p = q\mu_p Ep - qD_p \frac{dp}{dx}$$

Total electron current density : $$J_e = q\mu_n En - qD_n \frac{dn}{dx}$$

Remember:

☑ Conductors have positive temperature coefficient of resistivity whereas semiconductors have negative temperature coefficient of resistivity.

☑ Highly doped semiconductors exhibits metallic property.

☑ For intrinsic semiconductors
Carrier concentration $\propto$ Temperature
Conductivity $\propto$ (Temp.)$^{3/2}$ and
Conductivity $\propto$ Doping.

☑ For extrinsic semiconductors.
Majority carrier concentration $\propto$ doping
Minority carrier concentration $\propto$ 1/Doping
Minority carrier concentration $\propto$ Temperature
Majority carrier concentration is almost independent of temperature.

II Junction Diode Characteristics

Junction Diode

- A diode is the second generation semiconductor family.
- It is a single layer, two region, single junction, two terminal unilateral active semiconductor device.
- A *p-n* junction is formed from a piece of semiconductor by diffusing *p*-type material to one half side and *n*-type material to other half side.
- The plane dividing the two zones is known as a junction.
- A region near the junction is without any free charge particles called depletion region or charged free region or space charge region because there is no charge available for conduction.

Note:

- Depletion layer consists of immobile charged particles i.e. ions only.
- Depletion layer consists of negative ion (Acceptor ions) on the *p*-side and positive ion (Donor ions) on the *n*-side.
- Depletion layer opposes majority carriers in crossing the junction.

Contact Potential or Build in Potential (V_o)

$$V_o = \frac{kT}{q} \ln\left[\frac{N_A N_D}{n_i^2}\right]$$

where, $V_o \to$ Contact potential; $k \to$ Boltzman's constant$(1.38 \times 10^{-23}$ J/K)
$q \to$ Electron charge $(1.6 \times 10^{-19}$ Coulomb)
$T \to$ Temperature in Kelvin
$N_A \to$ Concentration of accepters on *p*-side
$N_D \to$ Concentration of donor on *n*-side
$n_i \to$ Intrinsic concentration at given temperature

Note:

☑ In case of Ge diode typical value of V_o is 0.2 V & for Si diode it is 0.7 V.

☑ In unbiased *p-n* junction diode electric field is maximum at the junction and decreases on either side of junction and zero outside of the depletion layer.

Contact potential in the terms of maximum electric field

$$V_o = -\frac{1}{2} E_o w$$

where, $E_o \to$ Maximum electric field at junction; $w \to$ Width of depletion region

Remember:

Depletion width increases with reverse and decrease with forward biased.

Depletion Width (*w*)

- Width of depletion region in unbiased condition is given by

$$w = \left[\frac{2\in V_0}{q}\left(\frac{1}{N_a}+\frac{1}{N_d}\right)\right]^{1/2}$$

where, $\in \rightarrow$ permitivity of material used for formation p-n junction diode ; $V_0 \rightarrow$ contact potential

$q \rightarrow$ electron charge (1.6×10^{-19} Coulomb)

$N_a \rightarrow$ concentration of acceptors (/cm^3) on p-side

$N_d \rightarrow$ concentration of donors (/cm^3) on n-side

Note:

☑ If we reverse bias the diode by voltage V then in formulae of depletion width V_0 is replaced by $|V_0 + V|$.

☑ It is clear that depletion width increases with reserve and decreases with forward biased.

☑ $W \propto \dfrac{1}{\sqrt{\text{concentration}}}$

Current Equation of Diode

Boltzman's equation of diode current

$$I = I_o\left[\exp\left(\frac{V}{\eta V_T}\right)-1\right]$$

where, $I \rightarrow$ Diode current ; $I_o \rightarrow$ Diode reverse saturation current

$V \rightarrow$ Diode voltage (Positive for forward bias and negative for reverse bias)

$\eta \rightarrow$ 1 for germanium and 2 for silicon ; $V_T \rightarrow$ Thermal voltage

$T \rightarrow$ Temperature in Kelvin

$$V_T \rightarrow \frac{KT}{q} = \frac{T(K)}{11600} \approx 26\text{ mV} \quad (T = 25°\text{C})$$

In forward bias : $I_F = I_o \exp\left(\dfrac{V_F}{\eta V_T}\right)$

In reverse bias : $I_R \cong -I_o$

Note:

Reverse current I_0 is temperature dependent. It gets double for every 10°C rise.

$$I_0(T_2) = I_0(T_1) \times 2^{\left[\frac{T_2 - T_1}{10}\right]}$$

Resistance of Diode

Resistance of diode can be calculated only when the diode is in forward bias.

- **Static resistance (DC resistance):** It is ratio of voltage and current at any point.

$$R_{DC} = \frac{V}{I}$$

- **Dynamic resistance (AC resistance):** It is define as reciprocal of slope, which is the smallest linear region in the entire non-linear curve.

$$R_{AC} = \frac{1}{\text{Slope}} \quad ; \quad R_{AC} = \frac{dV}{dI}$$

$$R_{AC} = \frac{\eta V_T}{I} = \frac{\eta kT}{Iq}$$

Capacitance of Diode

Diffusion capacitance or storage capacitance

When a diode is forward biased, a capacitance called diffusion capacitance or storage capacitance (C_D) is formed due to junction.

$$C_D = \frac{q_F}{\eta V_T} = \frac{\tau I}{\eta V_T}$$

where, $C_D \rightarrow$ Diffusion capacitance

$\tau \rightarrow$ Mean life time of minority carriers on either side or time constant of diode

$\eta \rightarrow$ Recombination of factor, 1 for Ge, 2 for Si

$V_T \rightarrow$ Thermal voltage ; $I \rightarrow$ Forward diode current

Note:

☑ Diffusion capacitance is proportional to the forward current and forward current depends on forward voltage.

$$C_D \propto \sqrt{\text{Doping}}$$

Transition capacitance or space charge capacitance

It is also known as depletion capacitance. It is the capacitance when diode is in reverse bias.

$$C_T = \frac{\epsilon_o \epsilon_r A}{\omega_d}$$

where, $C_T \rightarrow$ Transition capacitance ; $\omega_d \rightarrow$ Width of depletion layer

Also $\omega_d \propto V_r$ (Reverse bias voltage)

So, $$C_T \propto \frac{1}{V_r^n}$$

where, $n \rightarrow \frac{1}{2}$ = Step graded; $\rightarrow \frac{1}{3}$ = Linearly graded

The transition capacitance depends upon the magnitude of the reverse voltage applied.

Note:

- ☑ Transition capacitance can be used as variable capacitance for tunning the frequencies of television and radio channel.
- ☑ Maximum value of C_T is 40 pF.
- ☑ Maximum value of C_D is 0.02 μF.
- ☑ Diffusion capacitance is always greater than transition capacitance, i.e., $C_D > C_T$.

Zener Diode

- It is a heavily-doped *p-n* junction diode which is operated under reverse bias condition in the breakdown region.
- Zener current is independent of the supply voltage and only depends on external circuit resistance. Therefore Zener diode is known as constant voltage or voltage reference device.
- At reverse voltages less than 6 V, Zener breakdown predominates while at about 8 V, Avalanche breakdown predominates.
- Zener breakdown diodes are negative temperature coefficient of voltage and Avalanche breakdown diodes are positive temperature coefficient of voltage.
- These are used for voltage regulators.

Zener Voltage Regulator

$$I_{z\min} = I_{s\min} - I_{L\max}$$
$$I_{z\max} = I_{s\max} - I_{L\min}$$

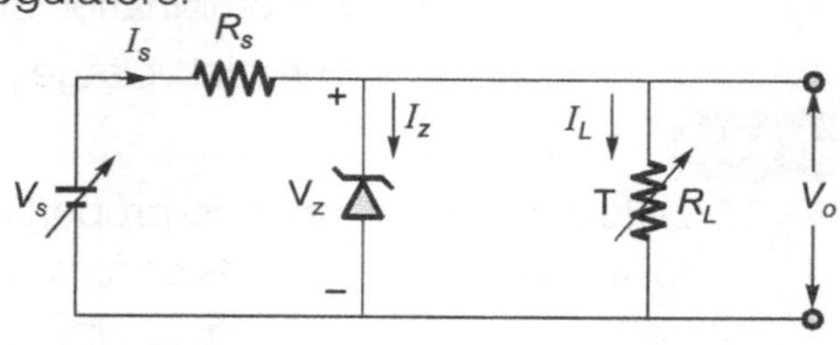

Schottky Diode

- It is a metal-semiconductor junction diode without depletion layer.
- In Schottky diode no depletion layer is formed and there will be no holes in this diode.
- Cut-in voltage of this diode is 0.3 V.
- This diode is used at high frequencies.

Photo Diode

- Principle of operation is ***photo conductive effect***.
- Photo sensitive material used are CdS, Se, ZnS.
- It is also called ***light operated switch***.
- Ge-photo diode respond to visible light while Si-photo diode respond to infrared light.

- Photo sensitive coating is provided at junction only.
- Compare to normal diode photo diode has larger depletion width obtained from lower level of doping.
- It is always operated under ***reverse biased condition***.
- Compare to normal diode it is 10 times faster, 100 times higher sensitive but power handling capacity is low.
- Magnitude of photo current increases with increase in intensity of light falling at junction.
- Current in photo diode is given by, $I = I_S + I_0\left[1 - e^{V/\eta V_T}\right]$

  where, $I_s \rightarrow$ short circuit current of photo diode
  $I_0 \rightarrow$ reverse saturation current
  $V \rightarrow$ voltage applied ; $V_T \rightarrow$ thermal voltage
- Photo current flows from *n* to *p*.
- Photo current is a minority carrier current.
- It does not provide gain.
- Photo current is a ***diffusion current***.
- It is used in remote control sensor, in designing of optocouplers and to read audio track recorded on motion picture film.
- When photo diode is forward biased it behaves as a normal diode and effect of light on current is zero.

Tunnel Diode

- It is fastest switch.
- Its response time is of the order of peco second.
- It is a $p^+ n^+$ diode having doping level of 1 : 10^3.
- Worked on the principle of ***tunneling effect***.
- It has very narrow depletion layer 100 Å to 200 Å.
- It is used as linear device as well as negative resistance device.
- Best material is GaAs having highest swing.
- It is used in designing microwave oscillators, as a relaxation oscillator, in designing of pulse and switching circuits, and as parametric amplifier.

Power Supply

- Power supply is an electronic system which converts AC power into DC power.
- A regulated power supply is interconnection of rectifier, filter, regulator.

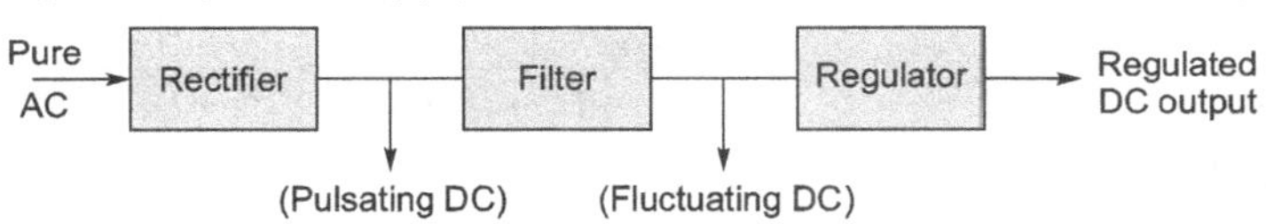

- AC to DC conversion is required because majority of the electronic systems require DC power.

Rectifiers

An electronic circuit which converts pure AC signal into pulsating DC or which converts a bidirectional waveform into unidirectional waveform.

Ripple factor (γ)

The amount of AC present in the output of rectifier is called as ripple. The amount of AC from the signal which contains AC and DC, is calculated by ripple factor.

$$\gamma = \frac{\text{rms value of AC component}}{\text{DC value}}; \quad \gamma = \sqrt{\left(\frac{V_{rms}}{V_{DC}}\right)^2 - 1}; \quad \gamma = \sqrt{F^2 - 1}$$

Form Factor (*F*)	$F = \frac{\text{rms value}}{\text{DC value}} = \frac{V_{rms}}{V_{DC}}$
Crest Factor (*C*)	$C = \frac{\text{Peak value}}{\text{rms value}}$
Efficiency (η)	$\eta = \frac{\text{Output DC power}}{\text{AC input power}}$

Half Wave Rectifier

Average value of current and voltage

Ideal case : $I_{DC} = \frac{I_m}{\pi}$; $I_m = \frac{V_m}{R_L}$ and $V_{DC} = \frac{V_m}{\pi}$

where, $I_{DC} \rightarrow$ Average value of current ; $V_{DC} \rightarrow$ Average value of voltage; $I_m \rightarrow$ Maximum value of current; $V_m \rightarrow$ Maximum value of voltage; $R_L \rightarrow$ Load resistance

Practical case : $I_{DC} = \frac{I'_m}{\pi}$; $V_{DC} = \frac{V'_m}{\pi}$

$$I'_m = \frac{V_m}{R_s + R_f + R_L} ; \quad V'_m = I'_m R_L$$

where, $R_s \rightarrow$ Coil resistance ; $R_f \rightarrow$ Diode forward resistance ; $R_L \rightarrow$ Load resistance

RMS value of current and voltage

$$I_{rms} = \frac{I_m}{2} ; \quad V_{rms} = \frac{V_m}{2}$$

Efficiency of half-wave rectifier

$$\eta = \frac{I_{DC}^2 R_L}{I_{rms}^2 [R_s + R_f + R_L]} = \frac{4}{\pi^2}\left[\frac{R_L}{R_s + R_f + R_L}\right]$$

- Ripple factor of half wave rectifier : $\gamma = 1.21$
- Crest factor of half-wave rectifier : $C = 2$
- Form factor of half-wave rectifier : $F = 1.58$

Remember:

☑ Maximum efficiency (when $R_S = R_f = 0$) ideal case

$$\eta = \frac{4}{\pi^2} = 40.6\%$$

☑ Peak inverse voltage for half wave rectifier PIV = V_m.

☑ Transformer utilization factor (TUF) for half wave rectifier TUF = 0.286.

☑ Output frequency = Input frequency for half wave rectifier.

☑ Conduction angle for half wave rectifier: $\theta = \pi$ for ideal case

$$\theta = \pi - 2\sin^{-1}\frac{V_\gamma}{V_m} \quad \text{for practical case}$$

$V_\gamma \rightarrow$ cut-in voltage of diode ; $V_m \rightarrow$ maximum voltage

☑ Ripple frequency (f_r) for half wave rectifier

$$f_r = f; \quad f \rightarrow \text{input frequency}$$

Full Wave Centre Tap Rectifier

Average value of current and voltage

$$I_{DC} = \frac{2I_m}{\pi}, \text{ ideal case} \qquad V_{dc} = \frac{2V_m}{\pi}, \text{ ideal case}$$

RMS value of current and voltage

$$I_{rms} = \frac{I_m}{\sqrt{2}}, \text{ ideal case} \qquad V_{rms} = \frac{V_m}{\sqrt{2}}, \text{ ideal case}$$

For practical case I_m is replaced by I'_m.

where, $I'_m = \dfrac{V_m}{\dfrac{R_s}{2} + R_f + R_L}$ and $V'_m = I'_m R_L$

Efficiency

$$\eta = \frac{8}{\pi^2} \frac{R_L}{\dfrac{R_s}{2} + R_f + R_L}$$

- Ripple factor (γ) = 0.48 for full wave rectifier.
- Form factor (F) = 1.11 for full wave rectifier.
- Crest factor (C) = $\sqrt{2}$ for full wave rectifier.

Remember:

☑ η is maximum when $R_S = R_f = 0 \rightarrow$ ideal case and $\eta_{max} = \dfrac{8}{\pi^2} = 81.2\%$

☑ Peak inverse voltage for full wave rectifier (centre tap) is PIV = $2V_m$.

☑ Transformer utilization factor (TUF): TUF = 0.57 for full wave centre tap.

Full Wave Bridge Rectifier

For full wave rectifier (bridge type) in all the formulas $\dfrac{R_S}{2}$ is replaced by R_S and R_f is replaced by $2R_f$.

Remember:

- ☑ For bridge type PIV = V_m.
- ☑ TUF = 0.812 for bridge type.
- ☑ Conduction angle: $\theta = 2\pi$ ideal case

$$\theta = 2\pi - 4\sin^{-1}\frac{V_\gamma}{V_m} \qquad \theta = 2\pi - 4\sin^{-1}\frac{2V_\gamma}{V_m}$$

for center tap $\qquad V_\gamma \rightarrow$ cut in voltage of diode

Remaining parameters are same as full wave centre tap rectifier.

Remember:

Parameter	Half-wave rectifier	Center tap based full wave rectifier	Bridge circuit based full wave rectifier
Type of transformer	Normal step-down	Center tapped	Normal step down
V_{DC}	$\frac{V_m}{\pi}$	$\frac{2V_m}{\pi}$	$\frac{2V_m}{\pi}$
I_{DC}	$\frac{I_m}{\pi} = \frac{V_{DC}}{R_s + R_f + R_L}$	$\frac{2I_m}{\pi} = \frac{V_{DC}}{R_s/2 + R_f + R_L}$	$\frac{2I_m}{\pi} = \frac{V_{DC}}{R_s + 2R_f + R_L}$
V_{rms}	$\frac{V_m}{2}$	$\frac{V_m}{\sqrt{2}}$	$\frac{V_m}{\sqrt{2}}$
Ripple factor	1.21	0.48	0.48
Efficiency, η%	$40.5 \times \frac{R_L}{R_L + R_s + R_f}\%$	$81.2 \times \frac{R_L}{R_L + R_s/2 + R_f}$	$81.2 \times \frac{R_L}{R_s + 2R_f + R_L}$
PIV	V_m	$2V_m$	V_m
Regulation (%)	$\frac{R_s + R_f}{R_L} \times 100$	$\frac{R_s/2 + R_f}{R_L} \times 100$	$\frac{R_s + 2R_f}{R_L} \times 100$
TUF%	$\frac{28.7 \times R_L}{R_L + R_s + R_f}$	$\frac{69.3 \times R_L}{R_L + \frac{R_s}{2} + R_f}$	$\frac{81.2 \times R_L}{2R_f + R_L + R_s}$
Output frequency	f	$2f$	$2f$
Numbers of diodes	1	2	4
Form factor	$\frac{\pi}{2}$	1.11	1.11

Filters

A filter is a frequency selective network which can be used to pass a desired range of frequencies and eliminate unwanted frequencies. In a power supply it is used to eliminate unwanted AC components present in the output of a rectifier. Filters are classified as :

- **Inductor Filter**

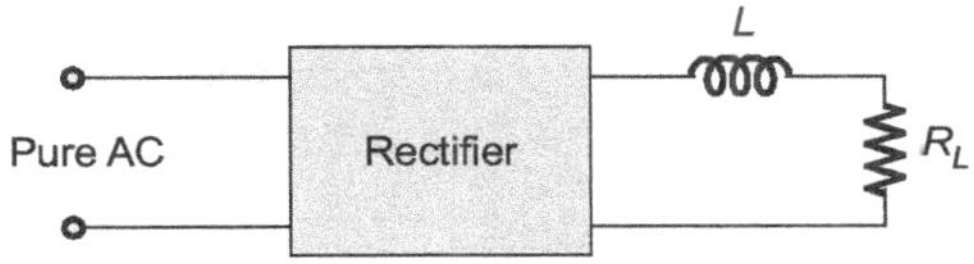

Ripple factor of inductor filter for full wave rectifier, $r = \dfrac{R_L}{3\sqrt{2}\,\omega_o L}$

Ripple factor of inductor filter for half wave rectifier, $r = \dfrac{\pi R_L}{2\sqrt{2}\,\omega_o L}$

where, $\omega_o \rightarrow$ Source frequency

- ***L-C* Filter**

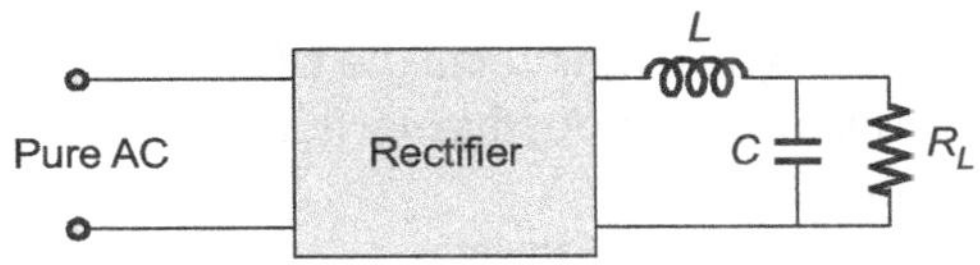

Ripple factor of LC filter for full wave rectifier, $r = \dfrac{\sqrt{2}}{3}\dfrac{|X_C|}{|X_L|}$

where, $X_C \rightarrow$ Reactance of capacitor calculated at $\omega = 2\omega_0$;
$X_L \rightarrow$ Reactance of inductor calculated at $\omega = 2\omega_0$

Critical inductance (L_C) is the value of inductance used in LC filter at which peak AC current in the circuit is equal to DC current

$$L_C = \frac{R_L}{3\omega_o}$$

For half wave rectifier, $r = \dfrac{\pi}{2\sqrt{2}}\dfrac{|X_C|}{|X_L|}$; $L_C = \dfrac{\pi R_L}{2\omega_o}$

where, X_C, X_L are calculated at $\omega = \omega_0$

- **Capacitor Filter**

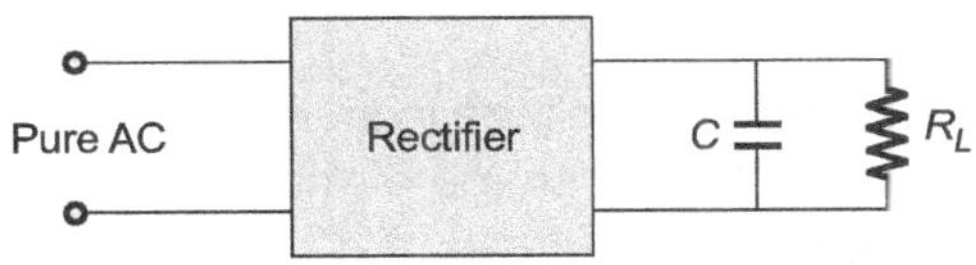

For half wave rectifier, ripple factor, $r = \dfrac{1}{2\sqrt{3}\, f_o R_L C}$

Peak to peak ripple voltage, $V_r = \dfrac{I_{dc}}{f_o C}$

For full wave rectifier, $r = \dfrac{1}{4\sqrt{3}\, f_o R_L C}$; $V_r = \dfrac{I_{dc}}{2 f_o c}$

- **C-L-C Filter**

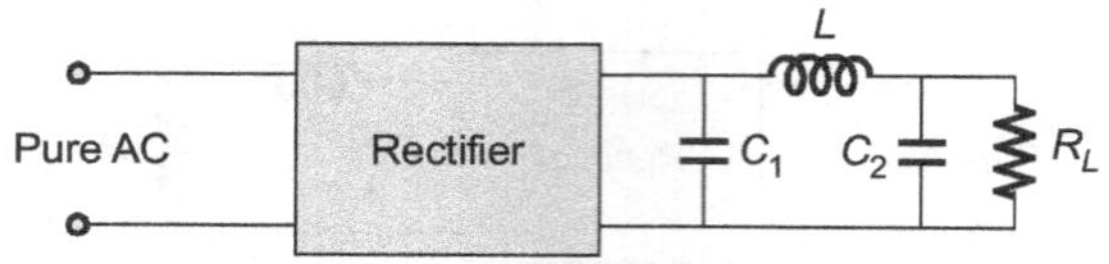

For half wave rectifier and full wave rectifier, ripple factor is

$$r = \frac{\sqrt{2}}{R_L}\frac{|X_{C_1}||X_{C_2}|}{|X_L|}$$

where, $X_{C_1}, X_{C_2} \rightarrow$ Reactance of capacitors calculated at $\omega = \omega_0$ for HWR and $\omega = 2\omega_0$ for FWR ; $X_L \rightarrow$ Reactance of inductor

Regulator

An electronic circuit which maintains the DC output voltage of a power supply stable or constant irrespective in fluctuation in AC supply and variations in load current.

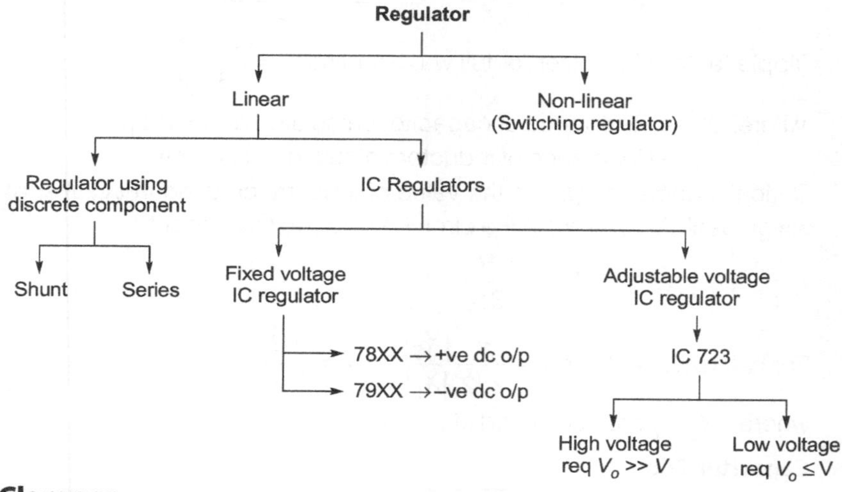

Clamper

- A circuit which can add DC voltage to an AC waveform.
- A clamper does not change shape of the waveform, it only shift the waveform up or down.

Application

A clamper is used to reinsert or restore DC component into a waveform which has lost DC component after passing through a processing circuit such as amplifier.

Clamper is also called DC reinserter or DC restorer.

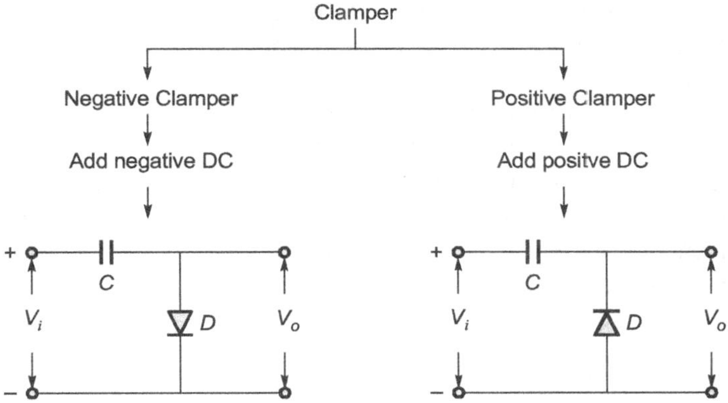

Voltage Multiplier

A circuit whose input is an AC waveform of peak V_m and output is DC voltage which is an integer multiple of peak AC input. $V_c = 2\ V_m$, $3\ V_m$, $4\ V_m$ etc.

If V_o = DC voltage of 2 V_m ; Voltage doubler
V_o = DC voltage of 3 V_m ; Voltage tripler
V_o = DC voltage of 4 V_m ; Voltage quadrupler

Voltage doubler

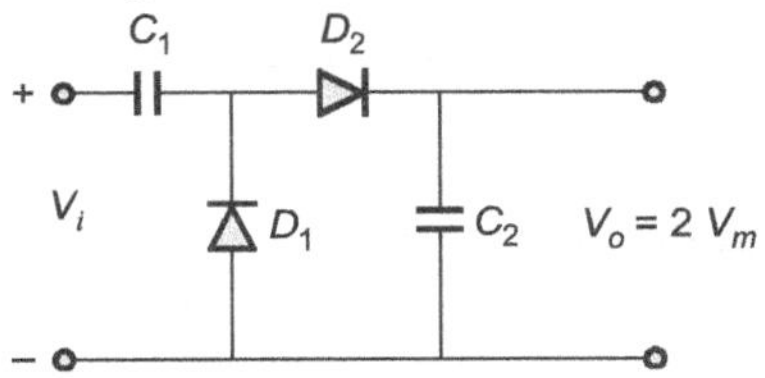

Voltage Tripler/Quadrupler

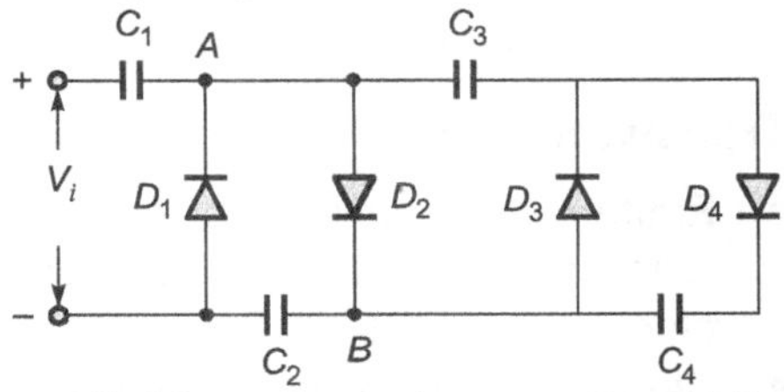

- DC voltage of 4 V_m can be obtained if V_o is measured across C_4 and C_2.
- DC voltage of 3 V_m can be obtained if V_o is measured across C_3 and C_1 combination.

III BJT Characteristics

Introduction

- A transistor is 3 layer, 3 terminal, 2 junction semiconductor device.
- It consist of either two *n*-type and one *p*-type layers of material called *npn* transistor or two *p*-type and one *n*-type material called *pnp* transistor.
- In transistor current flows due to both majority as well as minority carrier that's why called a bipolar device.
- It is a current controlled device.

- Its input impedance is low and output impedance is high.
- Thermal stability is less because of leakage current or reverse saturation current.
- In transistor all the major currents are diffusion currents.
- Base current is recombination current.
- A transistor represented by two diode connected back to back cannot work as transistor as there is no bonding between base and collector.
- Current conduction in *pnp* transistor is carried out by hole whereas in *npn* transistor it is due to electrons.

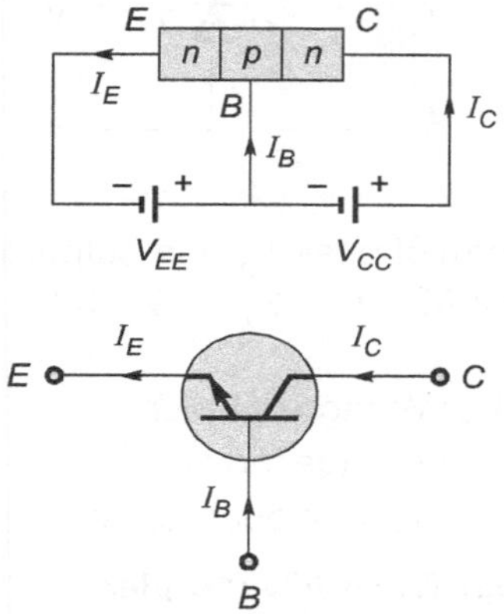

(a) *npn* transistor

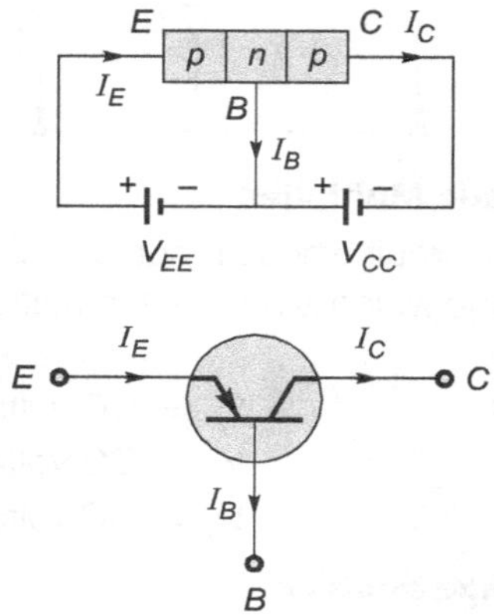

(b) *pnp* transistor

Transistor Sections

- **Emitter:** It is heavily doped as compared to other regions. The main function of this region is to supply majority charge carrier i.e. either electrons or holes.
- **Base:** This section is very lightly doped and very thin as compared to other sections. Its main function is to transfer the majority carriers from emitter to collector.
- **Collector:** This section is moderately doped and has largest area compared to other two regions because it collects the charge carriers from emitter and base. There is a large amount of heat liberation and so it is provided with large area to dissipate the heat.

BJT Configuration

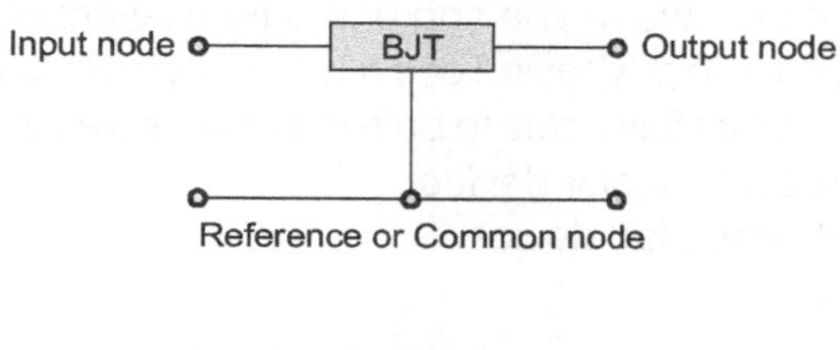

Based on the reference node a BJT can be used in 3 configuration as given in table below:

Configuration	Input node	Output node
Common Base (CB)	E	C
Common Emitter (CE)	B	C
Common Collector (CC)	B	E

Common Base Configuration

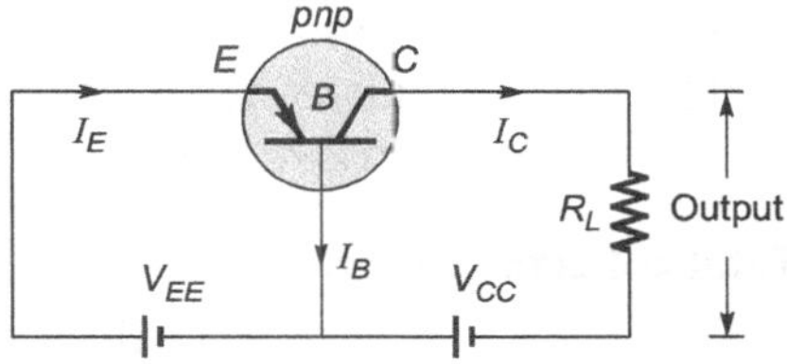

Common base *pnp* transistor

$I_E = I_B + I_C + I_{CBO}$

or $I_E \approx I_B + I_C$

where, $I_E \rightarrow$ Emitter current

$I_B \rightarrow$ Base current

$I_C \rightarrow$ Collector current

$I_{CBO} \rightarrow$ Collector to base leakage current

Note:

☑ $I_B \approx 2\%$ of I_E ; $I_C \approx 98\%$ of I_E

Current amplification factor (α)

When no signal is applied then the ratio of the collector current to emitter current is called current amplification factor or current gain.

$$\alpha = \frac{I_C}{I_E}$$

Note:

α of a transistor is a measure of the quality of transistor. Higher is the value of α better is the transistor as I_C approaches I_E.

Collector current can also be expressed as $I_C = \alpha I_E + I_{CBO}$

Common Emitter Configuration

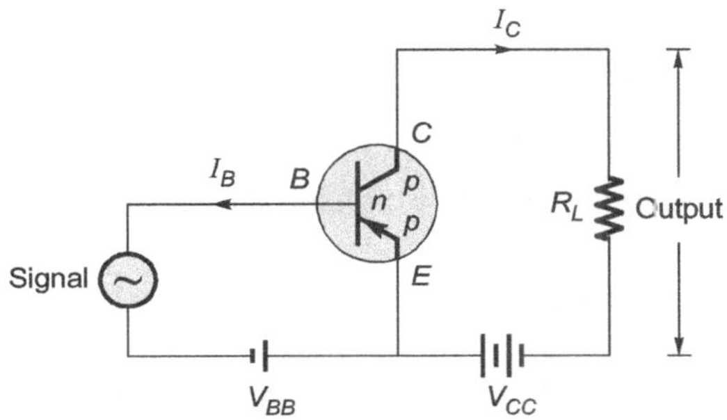

Common emitter *pnp* transistor

Base current amplification factor (β)

When no signal is applied, then the ratio of collector current to the base current is called β.

$$\beta = \frac{I_C}{I_B} \quad ; \quad \text{Also} \quad \alpha = \frac{\beta}{1+\beta}$$

Note:

- ☑ β ranges from 20 to 500.
- ☑ This configuration is used when appreciable current gain as well as voltage gain is required.

Total collector current

$$I_C = \beta I_B + I_{CEO} \quad \text{(or)} \quad I_C = \beta I_B + (1+\beta)I_{CBO}$$

where, $I_{CEO} \rightarrow$ Collector to emitter leakage current ; $I_{CEO} = (1+\beta)I_{CBO}$

Common Collector Configuration

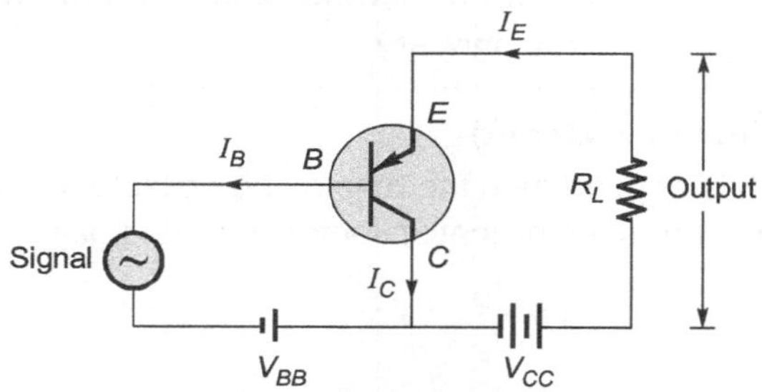

Common collector *pnp* transistor

Current amplification factor (γ)

When no signal is applied then the ratio of emitter current to the base current is called γ of the transistor.

$$\gamma = \frac{I_E}{I_B} \quad ; \quad \gamma = 1+\beta = \frac{1}{1-\alpha}$$

- Total emitter current, $I_E = (1+\beta)I_B + (1+\beta)I_{CBO}$

Note:

- ☑ This configuration has very high input resistance ≈ 750 kΩ and very low output resistance.
- ☑ Voltage gain is always less than one so never used for amplification purpose.
- ☑ This configuration is used for impedance matching i.e. driving a low impedance load from a high impedance source.

Characteristics	CB Configuration	CE Configuration	CC Configuration
Input resistance	Very low (40 Ω)	Low (50 kΩ)	Very high (750 kΩ)
Output resistance	Very high (1 MΩ)	High (10 kΩ)	Low (50 Ω)
Current gain	Less than unity	High (100)	High (100)
Voltage gain	Small (150)	High (500)	Less than unity
Leakage current	Very small	Very large	Very large
Applications	For high frequency applications	For audio frequency applications	For impedance matching
Phase shift between input and output	0°	180°	0°

Input and Output Characteristics of Different Configuration

The collector current I_C is completely determined by the input current I_E and V_{CB} voltage.

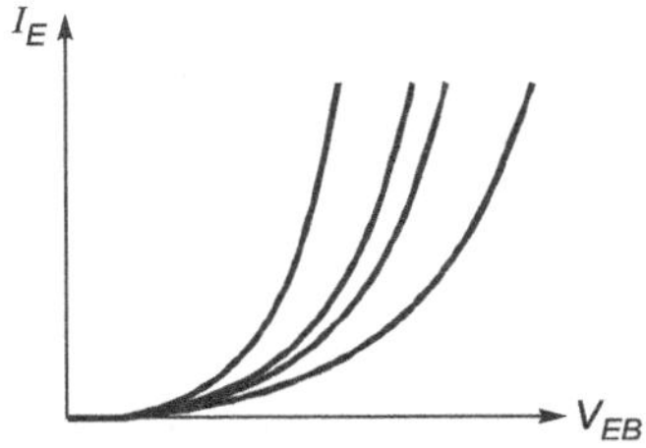

Input characteristics of CB configuration

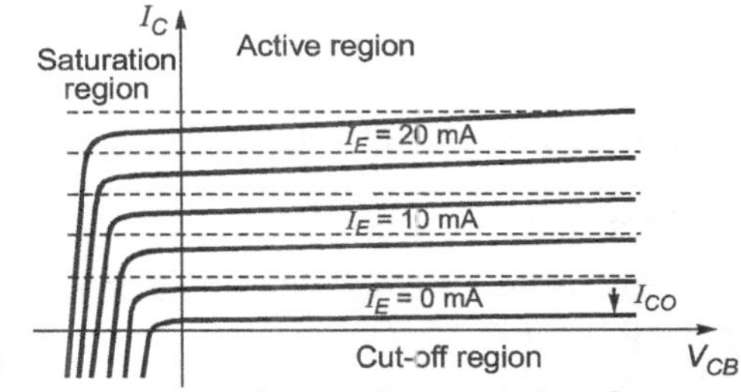

Output characteristics of CB configuration

The curve between I_B and V_{BE} for different values of V_{CE} are shown in figure.

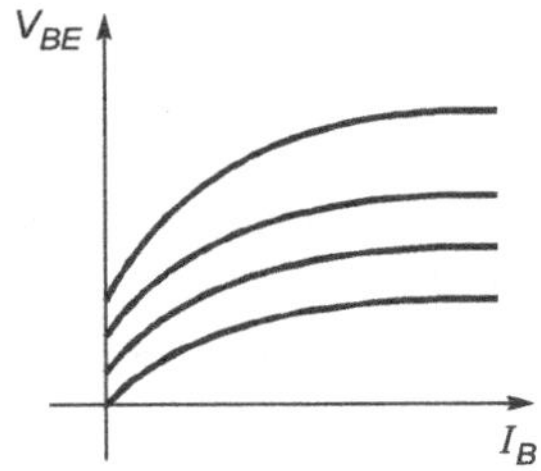

Input characteristics of CE configuration

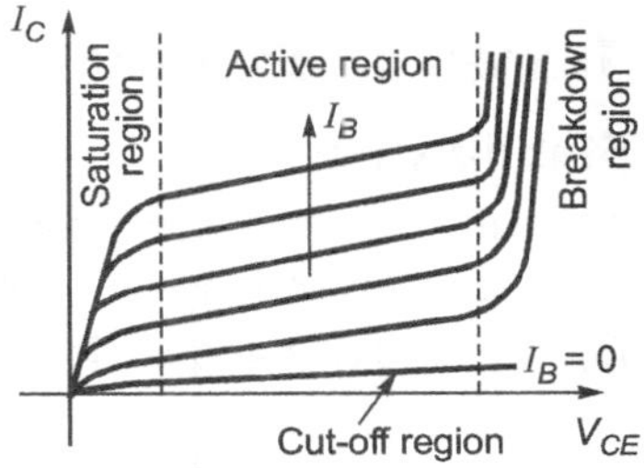

Output characteristics of CE configuration

The output characteristics are the curve between V_{CE} and I_C for various values of I_B.

DC Load Lines

- DC load line is the locus of all possible operating point at which BJT remains in active region.

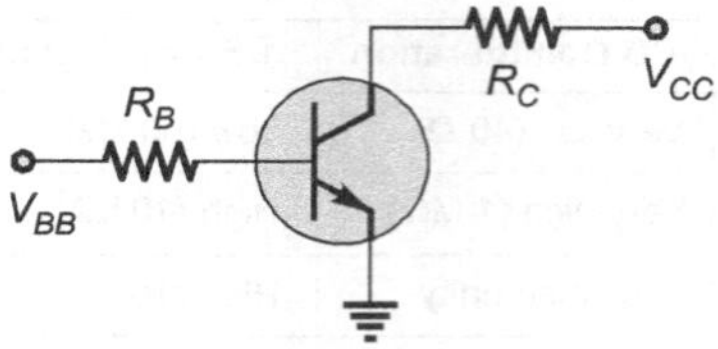

- If base current $I_B < \frac{I_{C\,sat}}{\beta}$ then transistor operate in active region.

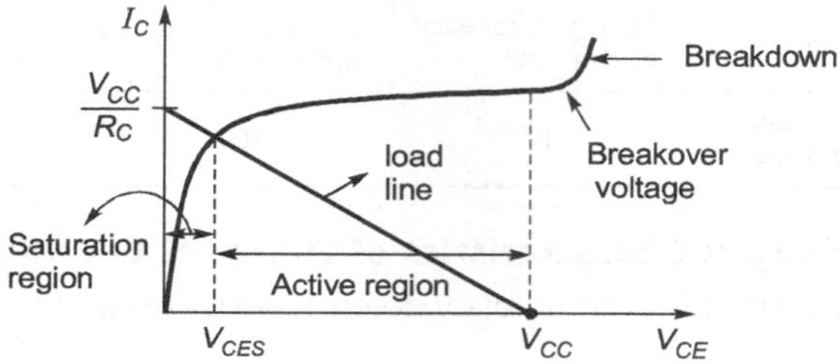

- If $I_B > \frac{I_{C\,sat}}{\beta}$ then it operates in saturation region.
- Maximum value of current, $I_{C\max} = \frac{V_{CC}}{R_C}$ [Taking $V_{CE} = 0$ at saturation]

  This is the ideal case.
- ***Power dissipated*** in a transistor, $P_D = I_C V_{CE}$

Remember:

☑ Power dissipation is maximum in active region and minimum in cutoff region and saturation region.

IV Transistor Biasing Circuits

Bias Stabilization

- Stabilization is a process of making Q-point independent of changes in temperature and changes in transistor parameter.
- If I_{CO}, V_{BE} and β changes simultaneously then net change in I_C.

$$\Delta I_C = \frac{\partial I_C}{\partial I_{CO}}\Delta I_{CO} + \frac{\partial I_C}{\partial V_{BE}} \times \Delta V_{BE} + \frac{\partial I_C}{\partial \beta}\Delta\beta$$

where, $\frac{\partial I_C}{\partial I_{CO}} = S \rightarrow$ Current stability factor

$\frac{\partial I_C}{\partial V_{BE}} = S' \rightarrow$ Voltage stability factor

$\frac{\partial I_C}{\partial \beta} = S_\beta = S'' \rightarrow$ Amplification stability factor

Note:

☑ Out of three stability factor S is most significant reason being, if S is within tolerable limit then other S' and S'' are guaranteed to remain within tolerable limit.

$$S_{\text{ideal}} = 1$$

☑ Practically S should be less than 20.

$$\text{Alternate evaluation of } S = \frac{1+\beta}{1-\beta\frac{\partial I_B}{\partial I_C}}$$

BJT Biasing

Biasing refers to providing appropriate DC voltage and DC current to an electronics device to operate it in a desired way.

A BJT is biased:

(i) To operate the BJT in active region so that it can be used as amplifier.

(ii) To maintain I_C stable so that the operating point does not drift and thermal run away does not happen.

Commonly Used BJT Biasing Circuit are:

(i) Fixed biased circuit; (ii) Collector to base bias circuit; (iii) Self bias circuit

Fixed biased circuit

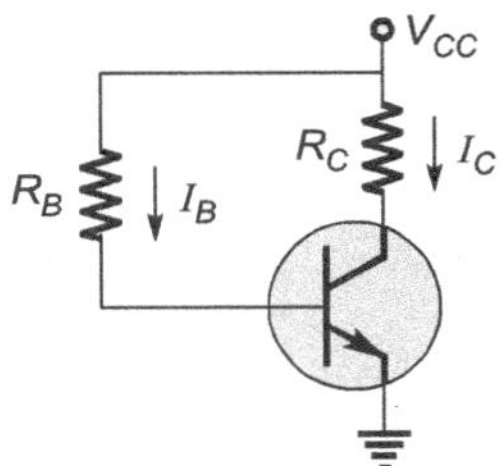

$$I_B = \frac{V_{CC}-V_{BE}}{R_B} \;;$$

$$I_C = \frac{V_{CC}-V_{CE}}{R_C}$$

I_B is constant or fixed. Hence circuit is called fixed biased circuit.

$S = 1+\beta$; since S is very large I_C will be unstable.

Collector to Base bias circuit

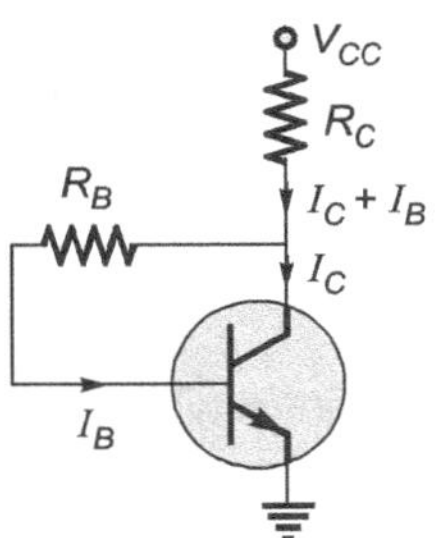

$$I_C = \frac{V_{CC}-V_{BE}}{R_C} \quad \text{(Assuming } \beta \text{ to be large or } I_B \approx 0\text{)}$$

$$I_C = \left[\frac{V_{CC}-V_{BE}}{R_C}\right]\times\left[\frac{\beta}{\beta+1}\right]$$

$$S = \frac{1+\beta}{1+\beta\left(\frac{R_C}{R_B+R_C}\right)}$$

Advantage: It gives more stable I_C than fixed biased circuit.

Drawback: Stability factor depend upon R_C, if R_C becomes smaller or zero, stability factor becomes very large and I_C does not remain stable.

Self-Bias Circuit (Voltage Divider Circuit)

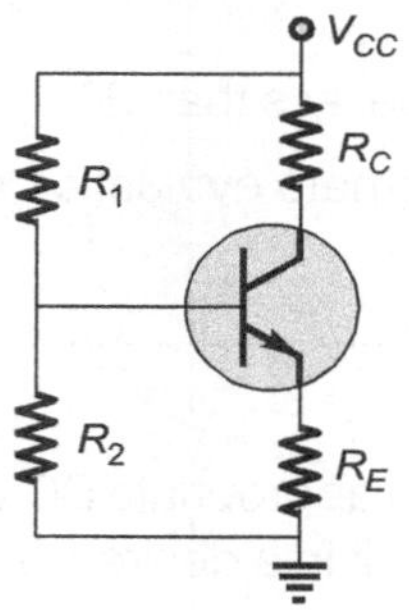

Thevenin Equivalent

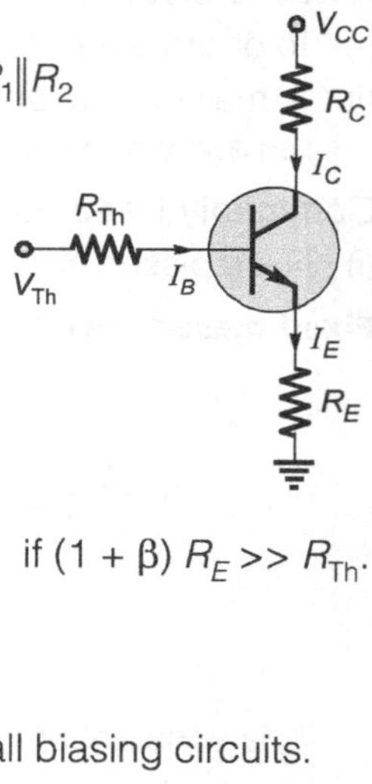

$$V_{Th} = \frac{R_2}{R_1 + R_2} V_{CC} \ ; \ R_{Th} = R_1 \| R_2$$

$$I_C \cong \frac{V_{CC} - V_{CE}}{R_C + R_E}$$

$$I_E = \frac{V_{Th} - V_{BE}}{\left(R_E + \frac{R_{Th}}{\beta + 1}\right)}$$

Assuming β to be large

$$S = \frac{1+\beta}{1+\beta\left(\frac{R_E}{R_{Th} + R_E}\right)}$$

if $(1 + \beta) R_E >> R_{Th}$.

Advantages:

(i) Self bias circuit has least stability factor among all biasing circuits.

(ii) Stability factor is independent of R_C.

(iii) Self bias circuit can be used to bias BJT for any configuration.

Drawback:

(i) Resistance R_E creates negative feedback due to which voltage gain $\frac{V_o}{V_i}$ or $\frac{V_o}{V_s}$ becomes smaller.

Thermal Run-away

It is the process of self destruction or damage of a transistor where an increase in temperature increase the collector current, causing a further increase in the temperature.

Thermal resistance (θ)

$$\theta = \frac{T_j - T_A}{P_D} \quad (°C/Watt \text{ or } °K/Watt)$$

where, $T_j \rightarrow$ Junction temperature (collector junction) ; $T_A \rightarrow$ Ambient temperature in Kelvin ; $P_D \rightarrow$ Power dissipated across collector junction

Condition to avoid thermal run away

A transistor will be thermally stable if

$$\frac{\partial P_C}{\partial T_j} \leq \frac{\partial P_D}{\partial T_j} = \frac{1}{\theta}$$

where, $\frac{\partial P_C}{\partial T_j} \rightarrow$ Rate at which heat is released

$\frac{\partial P_D}{\partial T_j} \rightarrow$ Rate at which heat is dissipated

- For thermal stability $V_{CE} < \frac{V_{CC}}{2}$

Early Effect

It is also known as base width modulation. Due to Early effect :

(i) Common base current gain increases.

(ii) Emitter current increases.

(iii) Punch through or reach through occurs.

The output resistance, $r_o = \frac{V_{CE} + V_A}{I_C}$ where, V_A = Early voltage

Compensation Method

- In this method collector current is made stable by connecting temperature dependent devices such as diode, thermistor, sensistor.
- Thermistor and sensistor are temperature dependent resistors.

Thermistor	Sensistors
1. Temperature coefficient is negative	1. Temperature coefficient is positive
2. Fabricated from pure semiconductor	2. Fabricated from heavily doped semiconductor

<table>
<tr><th>Method</th><th>Diode compensation for I_{CO}</th><th>Thermistor compensation for I_{CO}</th><th>Compensation in self bias circuit</th></tr>
<tr><td>Circuit Diagram</td><td>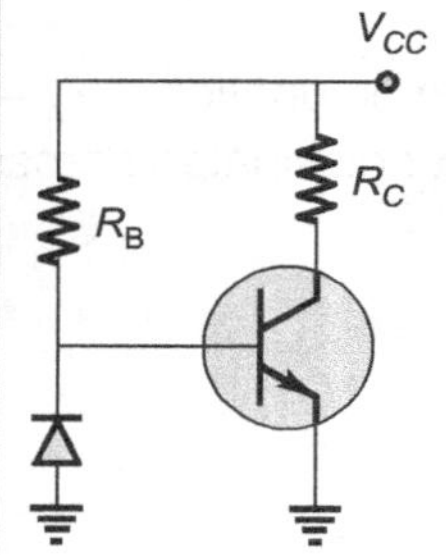
</td><td>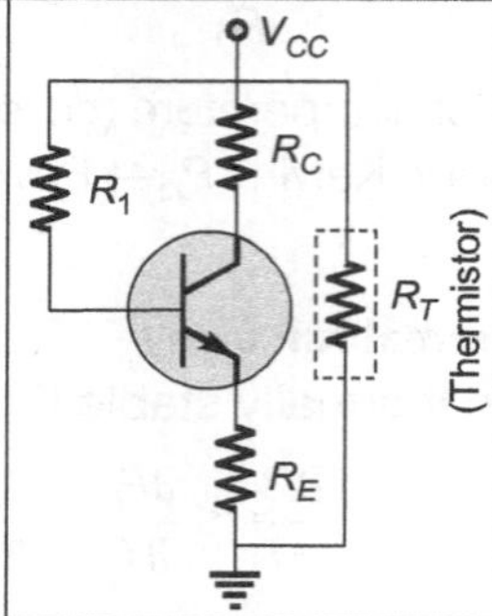
</td><td>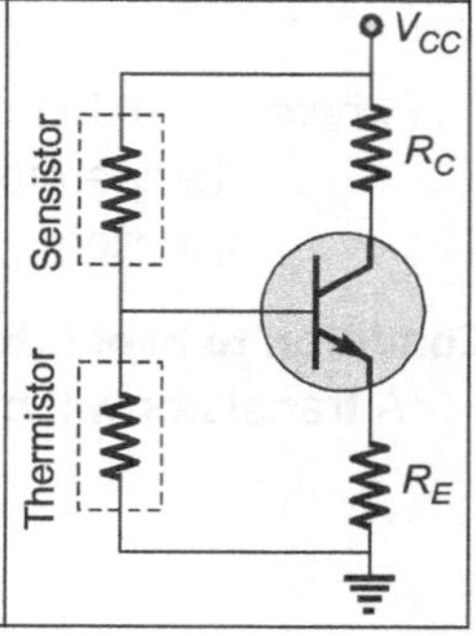
</td></tr>
</table>

BJT Biasing in Integrated Circuits

- Self bias circuit cannot be used to bias BJT in amplifier because R_1 and R_2 are large resistances and they require greater area on silicon chip.
- Therefore BJT is biased in IC amplifier using special circuit known as current mirror.

Basic Current Mirror

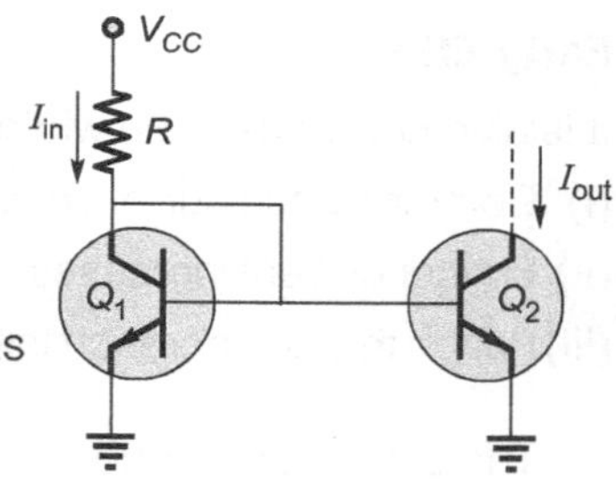

$$I_{out} = \frac{I_{in}}{1+\frac{2}{\beta}}$$

Current mirror is symbolically shown as constant current source.

Applications of current mirror

(i) A current mirror is used to bias BJT in IC amplifiers.

(ii) Current mirror can be used as active load in IC amplifier.

Advantage of current mirror biasing

(i) It require smaller chip area than self bias circuit.

(ii) A current mirror is equivalent to constant current source hence the DC collector current of bias transistor remains highly stable.

V BJT as an Amplifier

Common Emitter Amplifier

Output voltage	$V_{out} \cong -I_E R_C$
Voltage gain	$A = \dfrac{V_{out}}{V_{in}} \cong -\dfrac{R_C}{r'_e}$
Input impedance	$Z_{in} \cong R_1 \parallel R_2 \parallel \beta r'_e$
Input impedance of base	$Z_{in(base)} \cong \dfrac{\beta I_b r'_e}{I_b} = \beta r'_e$
Output impedance	$Z_{out} \cong R_C$

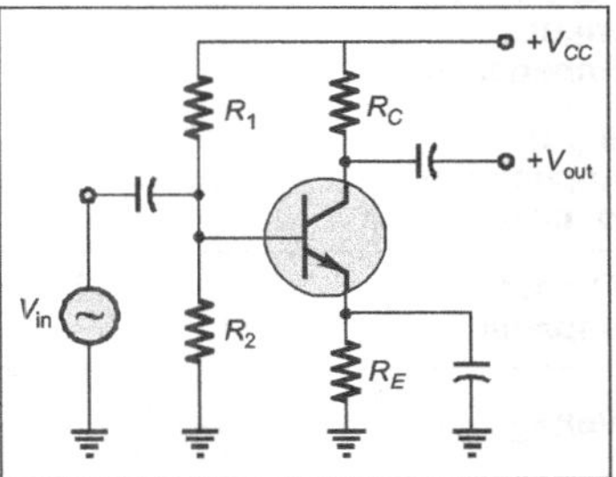

Swamped Amplifier

Effect of swamping on voltage gain	$A \cong \dfrac{R_C}{R_E + r'_e}$
Input impedance of base	$Z_{in(base)} \cong \beta(R_E + r'_e)$

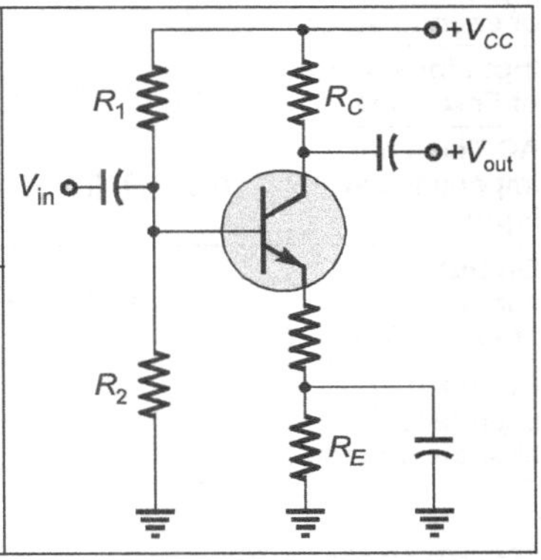

Common Base Amplifier

Input impedance	$Z_{in} \cong r'_e$
Voltage gain	$A = \dfrac{V_{out}}{V_{in}} \cong \dfrac{R_C}{r'_e}$
Output impedance	$Z_{out} \cong R_C$

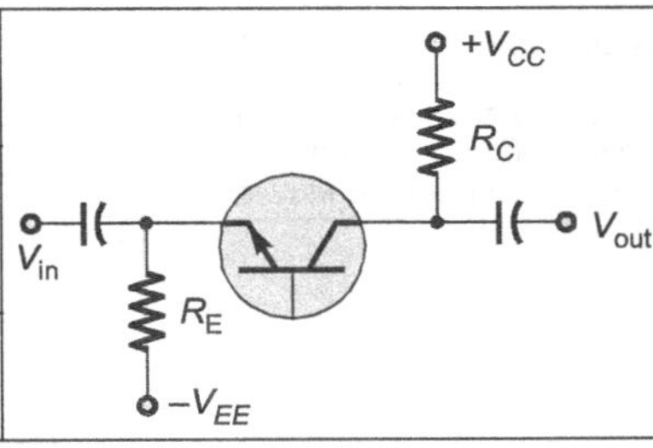

Common Collector Amplifier (or Emitter Follower)

DC output voltage	$V_{out} = V_{in} - V_{BE}$
Input impedance	$Z_{in} = R_1 \parallel R_2 \parallel \beta R_E$
Input impedance of base	$Z_{in(base)} = \beta(R_E + r_e') \cong \beta R_E$
Output impedance	$Z_{out} = r_e' + \dfrac{R_s \parallel R_1 \parallel R_2}{\beta}$
Voltage gain	$A = \dfrac{R_E}{R_E + r_e' + (R_s \parallel R_1 \parallel R_2)/\beta} \cong 1$

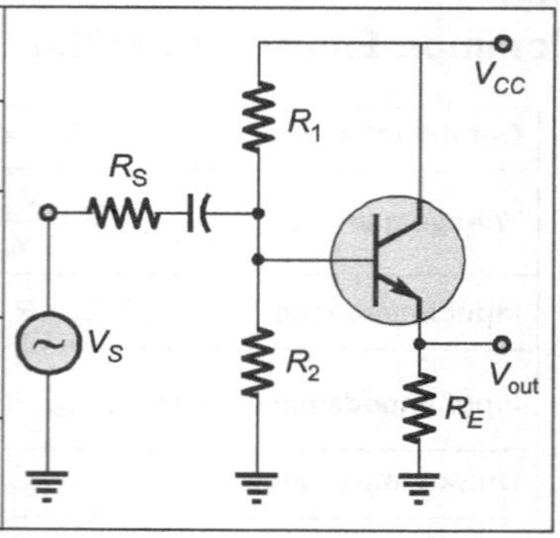

Darlington Amplifier

Input impedance of second stage	$Z_{in(2)} \cong \beta_2 R_E$
Input impedance of first stage	$Z_{in(1)} \cong \beta_1\beta_2 R_E$
AC Thevenin impedance at input	$R_{Th} = R_s \parallel R_1 \parallel R_2$
Output impedance of first stage	$Z_{out(2)} = r_{e1}' + \dfrac{R_{Th}}{\beta_1}$
Output impedance of second stage	$Z_{out(2)} = r_{e2}' + \dfrac{r_{e1}' + R_{Th}/\beta_1}{\beta_2}$

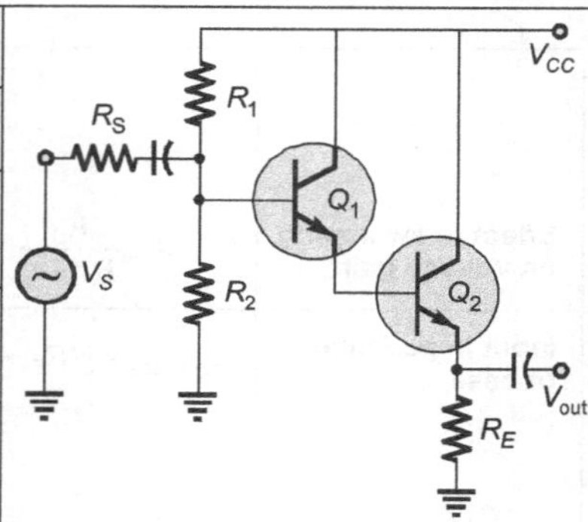

Cascode amplifier

Voltage gain	$A_v = -g_{m1}(R_C \,\Pi\, R_L)$
Current gain	$A_I \approx 1$
Cascode amplifier eliminates miller effect thus contribute to a much higher bandwidth	

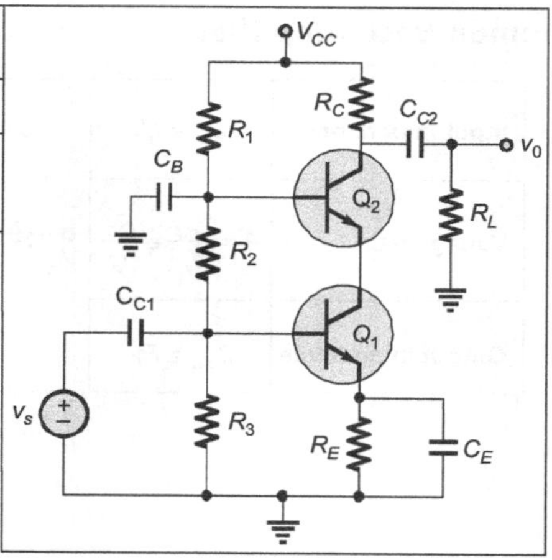

Note:

- ☑ Input impedance of common collector amplifier is highest among CE, CB and CC.
- ☑ Output impedance of common collector is lowest among CE, CB and CC.
- ☑ Darlington pair is known as super b circuit.
- ☑ Darlington pair is used where very high input impedance and high current gain is required.
- ☑ Voltage gain of CB is highest and current gain of CC is highest.
- ☑ CE has highest power gain.

VI Junction Field Effect Transistors

- FET is a unipolar and voltage controlled device.
- The terminals drain, gate and source of a FET are identical to collector base and emitter of a BJT.
- Since the input function is reverse biased in JFET, the current drawn is very small and so input impedance is very high.
- Less noisy device due to absence of minority carriers.
- Excellent thermal stability due to absence of leakage current.
- FET is considered as excellent signal chopper because of zero offset voltage.

Circuit Diagram and Symbol

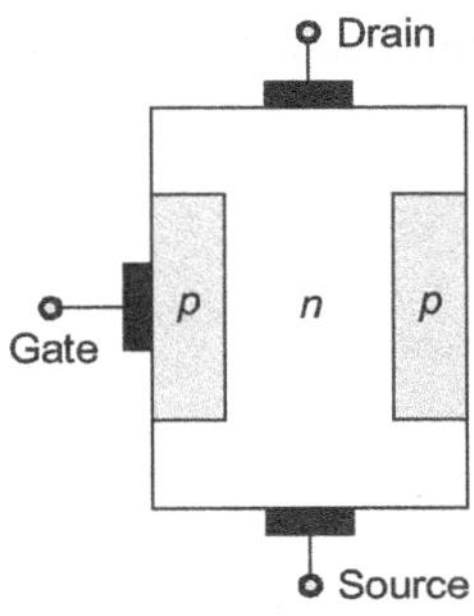

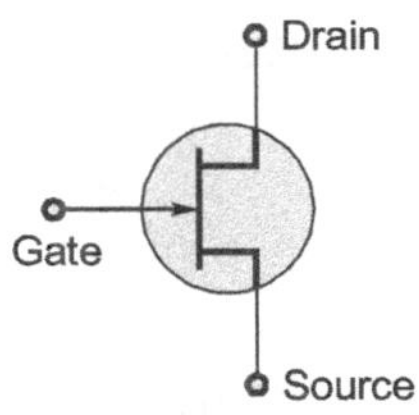

Parameters

Drain current

$$I_D = I_{DSS}\left[1-\frac{V_{GS}}{V_p}\right]^2 \qquad \text{In saturation region}$$

where, $I_D \rightarrow$ Drain current

$I_{DSS} \rightarrow$ Maximum value of current when $V_{GS} = 0$

$V_p \rightarrow$ Pinch off voltage

$V_{GS} \rightarrow$ Gate to source voltage

Drain resistance

$$r_d = \left.\frac{\Delta V_{DS}}{\Delta I_{DS}}\right|_{V_{GS} = \text{constant}}$$

Note:

r_d ranges from 100 kΩ to 500 kΩ.

Transconductance

$$g_m = \left.\frac{\Delta I_D}{\Delta V_{GS}}\right|_{V_{DS\text{-constant}}}$$

Note:

g_m ranges from 0.1 mS to 10 mS.

In saturation region

$$g_m = -\frac{2I_{DSS}}{V_p}\left(1 - \frac{V_{GS}}{V_p}\right) = -\frac{2}{V_p}\sqrt{I_D . I_{DSS}}$$

Also, $$g_m = g_{mo}\left(1 - \frac{V_{GS}}{V_p}\right)$$

where, $$g_{mo} = -\frac{2I_{DSS}}{V_p} = \text{maximum value of transconductance}$$

Amplification Factor

$$\mu = \left.\left(\frac{\Delta V_{OS}}{\Delta V_{GS}}\right)\right|_{I_D = \text{Constant}}$$

Note:

☑ μ ranges from 2.5 to 150.

☑ Relation between μ, r_d and g_m is $\mu = r_d \cdot g_m$.

Remember:

If two FETs are connected in parallel having transconductance g_{m_1} and g_{m_2}, drain resistance r_{d_1} and r_{d_2}, amplification factor μ_1 and μ_2 then,

Effective transconductance $g_m = g_{m_1} + g_{m_2}$.

Effective drain resistance $r_d = r_{d_1} \| r_{d_2}$.

Effective amplification factor $\mu = \dfrac{\mu_1 r_{d_2} + \mu_2 r_{d_1}}{\mu_1 + \mu_2}$.

Characteristics of JFET

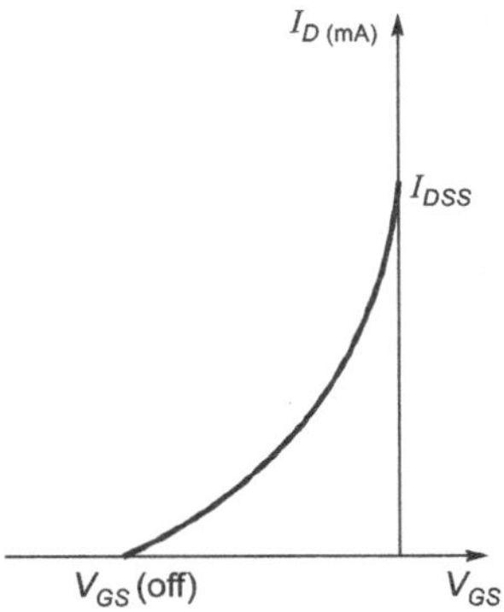

Transfer characteristic

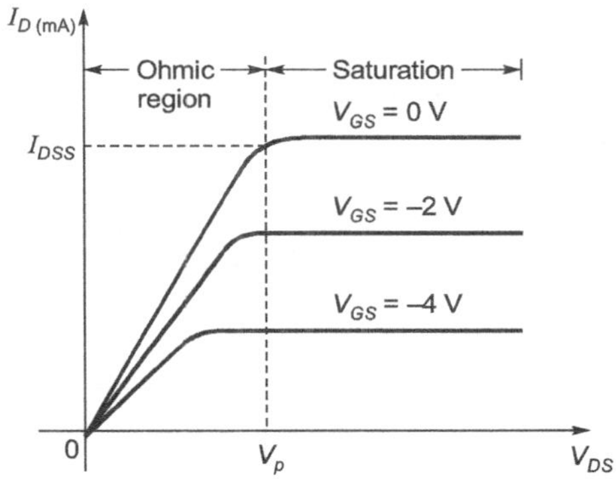

Drain characteristic

Remember:

When FET is operated below pinch-off voltage (V_p), it acts as a voltage variable resistor.

FET Amplifiers

Common Source Amplifier

AC output voltage	$V_{out} = -g_m V_{GS} R_D$
Unloaded voltage gain	$A = -g_m R_D$

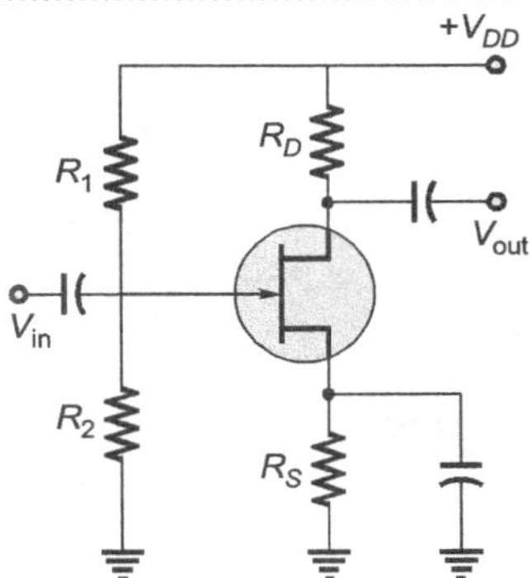

Common Drain (CD) Amplifier

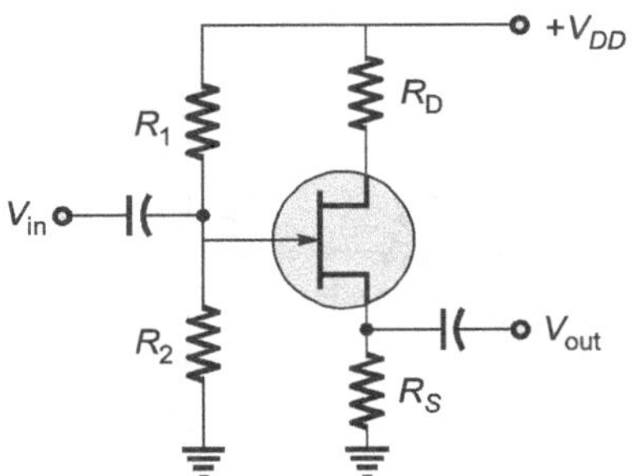

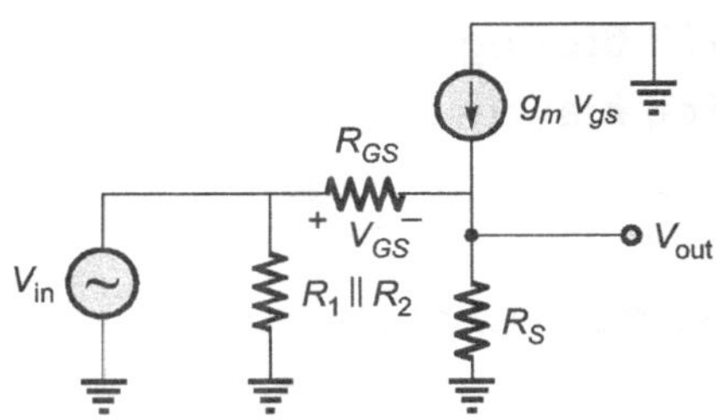

AC input voltage	$V_{in} = (1+g_m R_S)V_{GS}$
AC output voltage	$V_{out} = g_m V_{GS} R_S$
Unloaded voltage gain	$A = \frac{R_S}{R_S + \frac{1}{g_m}} \cong 1 \quad \ldots\left(\text{as } R_S \; ? \; \frac{1}{g_m}\right)$
Output impedance	$Z_{out(source)} = \frac{1}{g_m}; \; Z_{out} = R_S \, \Pi \frac{1}{g_m}$

Common Gate (CG) Amplifier

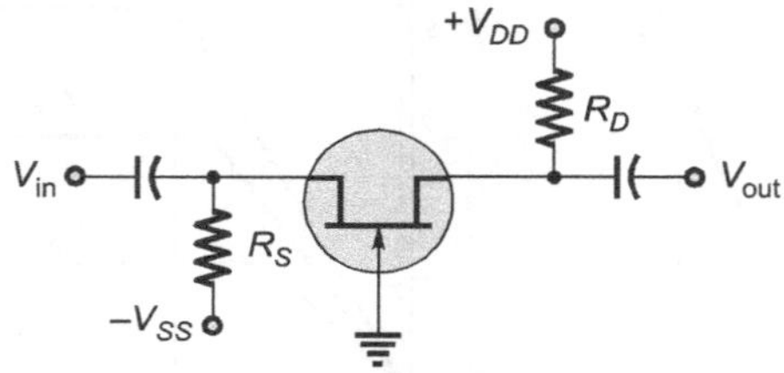

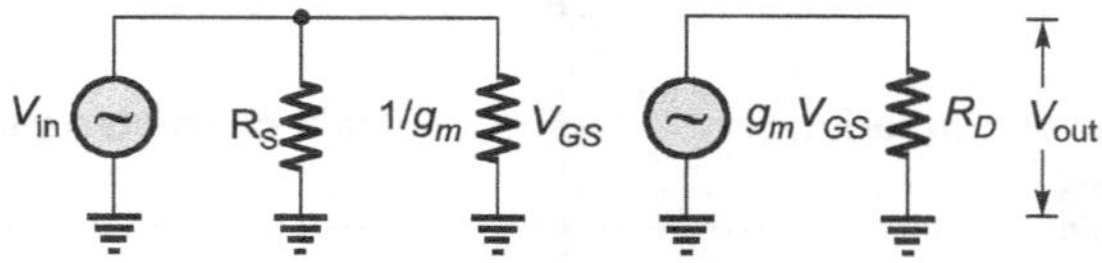

AC equivalent circuit

AC input voltage	$V_{in} = V_{GS}$
AC output voltage	$V_{out} = g_m V_{GS} R_D$
Unloaded voltage gain	$A = g_m R_D$
Input impedance	$Z_{in} = \frac{1}{g_m}$
DC on-state resistance	$r_{DS(on)} = \frac{V_{DS}}{I_D}$

where, $r_{DS(on)} \rightarrow$ DC resistance in saturation region

$V_{DS} \rightarrow$ DC drain-source voltage

$I_D \rightarrow$ DC drain current

FET Biasing Circuit

Self bias circuit

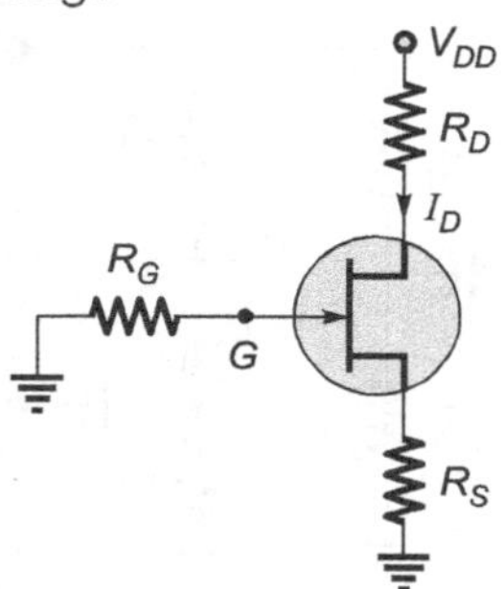

Gate to source voltage

$$V_{GS} = -I_D R_s$$

Source resistance

$$R_s = -V_p/I_{DSS}$$

Note:

Q-point is the interaction between tranconductance and self bias line.

Voltage divider/Source bias circuit

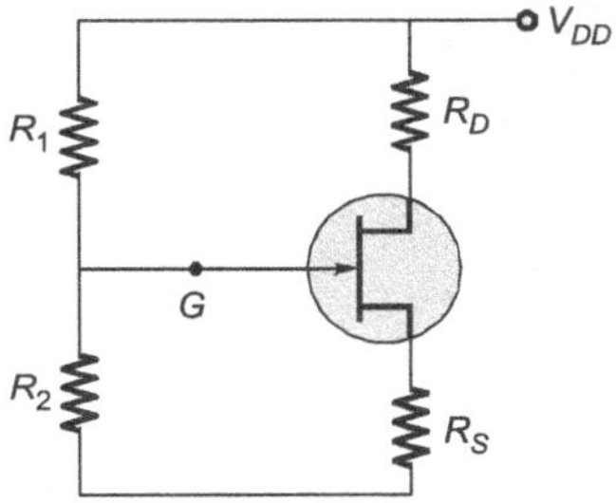

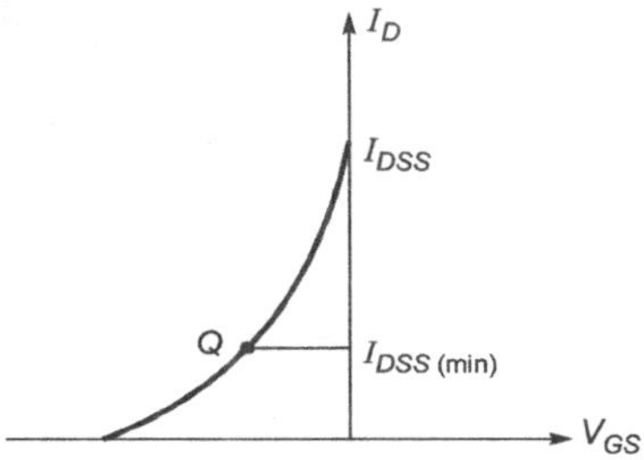

Drain current : $I_D = \dfrac{V_{Th} - V_{GS}}{R_s}$ where, $V_{Th} = \left(\dfrac{R_2}{R_1 + R_2}\right)V_{DD}$

Current source bias circuit

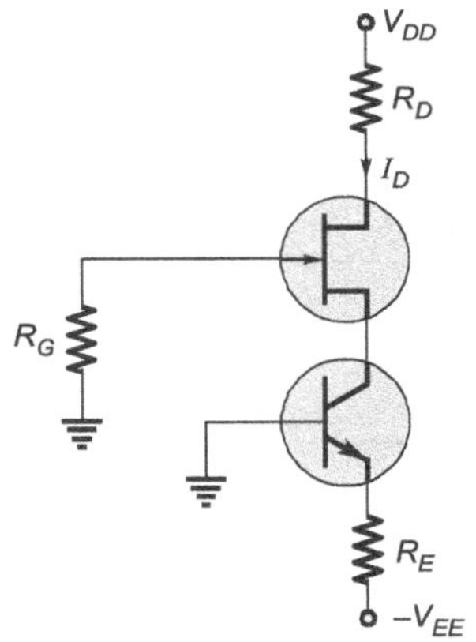

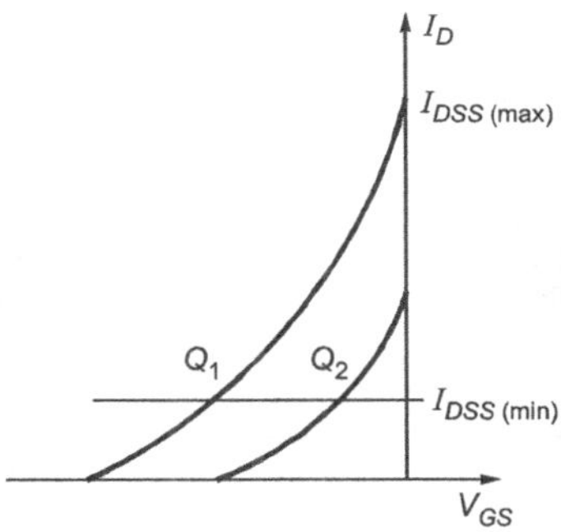

Collector current : $I_C = I_D = \dfrac{V_{EE} - V_{BE}}{R_E}$

VII Metal Oxide Semiconductor Field Effect Transistor

- MOSFET is an integrated device, fabricated by VLSI using planner technology.
- It is a voltage controlled device.
- In the MOSFET the plate and semiconductor channel will be working as the plate of capacitor and SiO_2 as dielectric.
- The large input impedance of MOSFET is due to the SiO_2.
- MOSFET is very sensitive to static electrical noise and static electrical disturbances.

Schematic Symbol

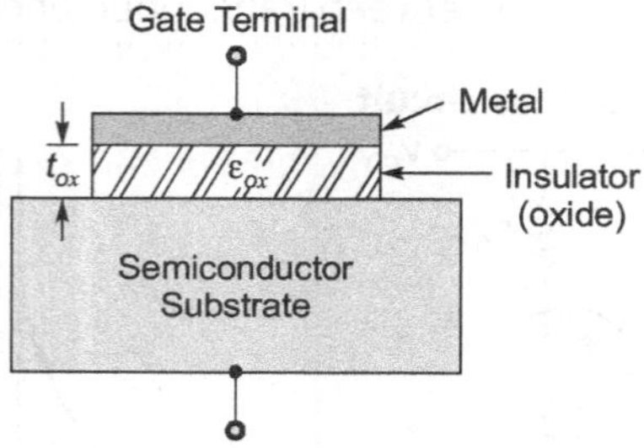

<table>
<tr><td>p-channel Depletion MOSFET</td><td>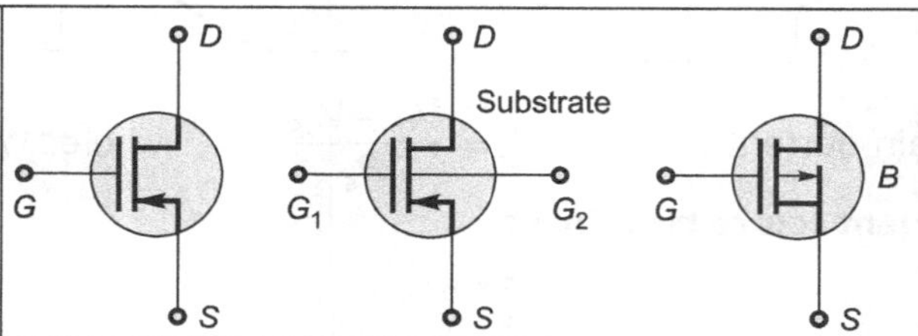
</td></tr>
<tr><td>n-channel Depletion MOSFET</td><td>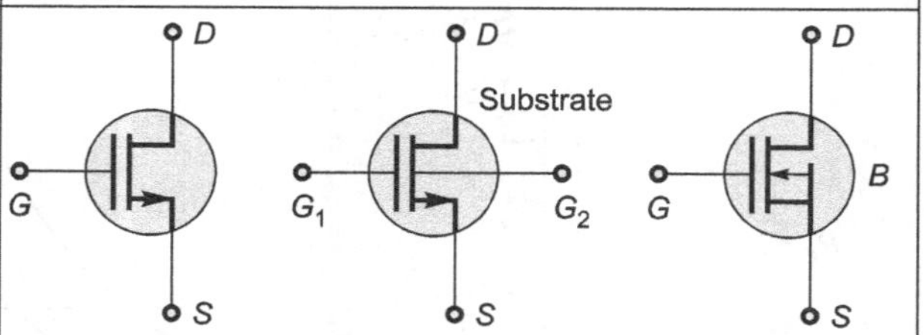
</td></tr>
<tr><td>p-channel Enhancementn MOSFET</td><td>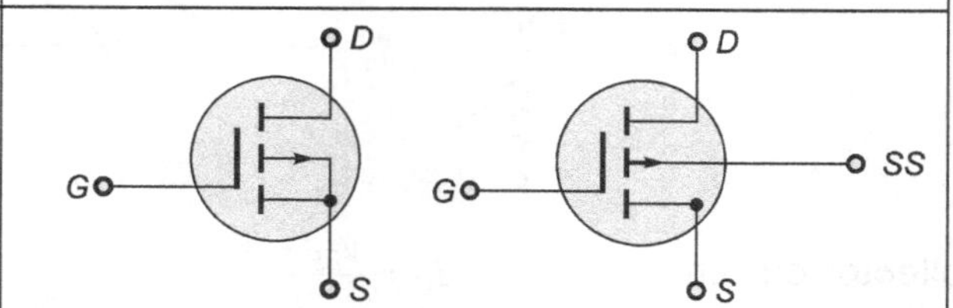
</td></tr>
<tr><td>n-channel Enhancementn MOSFET</td><td>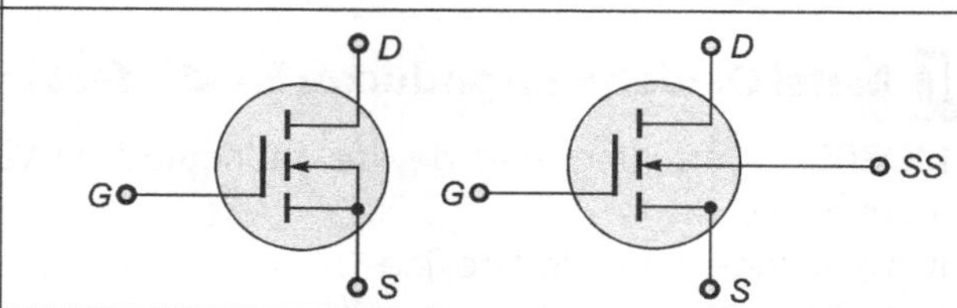
</td></tr>
</table>

Characteristics of MOSFET

For depletion MOSFET characteristics are same as JFET characteristics.

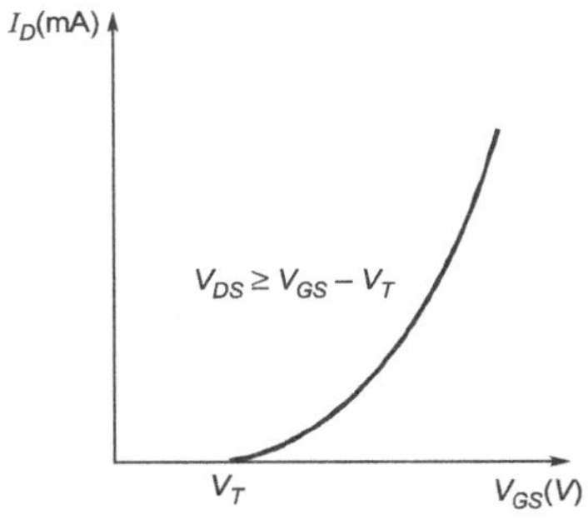

Transfer characteristic of N-MOSFET

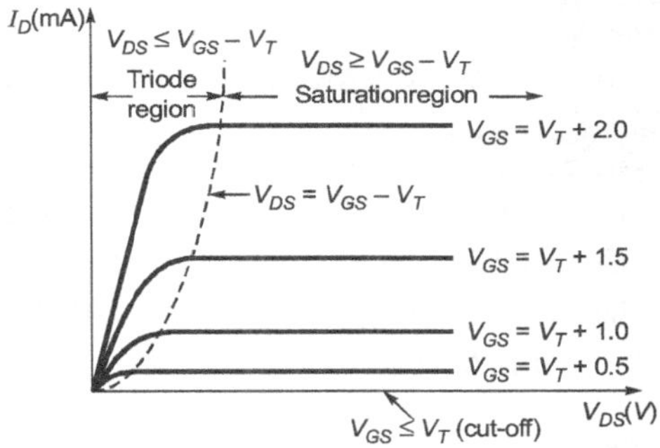

Drain characteristic of N-MOSFET

Remember:

- ☑ Threshold voltage (V_T) is defined as the "Minimum gate to source voltage required for MOSFET turn into ON state".
- ☑ For better performance of the MOSFET, threshold voltage must be as small as possible.
- ☑ The threshold voltage of MOSFET can be graphically obtained from the transfer characteristics.
- ☑ Threshold voltage V_T should be smaller.

Parameters

For depletion MOSFET

Drain current, $$I_D = I_{DSS}\left(1 - \frac{V_{GS}}{V_p}\right)^2 \text{ where, } I_{DSS} = I_{D\max}$$

Note:

In depletion MOSFET channel is pre-existing.

For enhancement type *n*-channel MOSFET

- $V_{GS} > V_T$ for 'ON' or saturation region.
  $V_{GS} < V_T$ for 'OFF' or cut-off region; where V_T = Threshold voltage
- $V_{DS} < (V_{GS} - V_T) \rightarrow$ Triode region of operation.
  $V_{DS} \geq (V_{GS} - V_T) \rightarrow$ Saturation region of operation.
- For saturation region : $I_D = K(V_{GS} - V_T)^2$
- For *p*-channel MOSFET

$$I_D = K(V_{GS} + V_T)^2$$

Note:

In enhancement MOSFET channel is diffused channel.

Remember:

- ☑ MOSFET is less noisy as compared to JFET.
- ☑ MOSFET is faster than JFET but BJT is faster than MOSFET.
- ☑ When compare to BJT, MOSFET is more suitable for high frequency application.

VIII Transistor Hybrid Model

Introduction

The h-parameter model of BJT is defined by two-port network as

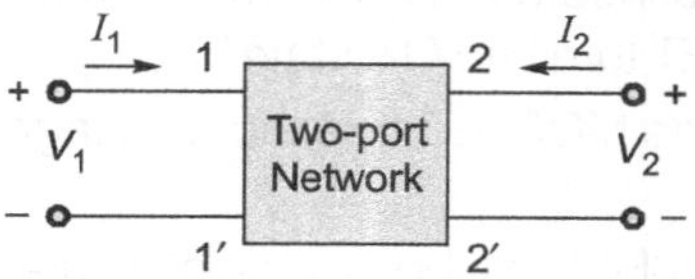

$$V_1 = h_{11} I_1 + h_{12} V_2$$
$$I_2 = h_{21} I_1 + h_{22} V_2$$

Parameters	Formula
Input impedance h_{11}/h_i	$h_{11} = \left.\frac{V_1}{I_1}\right\|_{V_2=0}$
Current gain h_{12}/h_f	$h_{21} = \left.\frac{I_2}{I_1}\right\|_{V_2=0}$
Reverse voltage gain h_{12}/h_r	$h_{12} = \left.\frac{V_1}{V_2}\right\|_{I_1=0}$
Output Admittance h_{22}/h_o	$h_{22} = \left.\frac{I_2}{V_2}\right\|_{I_1=0}$

Typical h-paramter Values for Transistors

Parameters	*CE*	*CC*	*CB*
$h_{11} = h_i$	1.1 kΩ	1.1 kΩ	21.6 Ω
$h_{12} = h_r$	2.5×10^{-4}	; 1	2.9×10^{-4}
$h_{21} = h_f$	50	–51	–0.98
$h_{22} = h_o$	24 μA/V	24 μA/V	0.49 μA/V
$1/h_o$	40 kΩ	40 kΩ	2.04 MΩ

Approximated Conversion Formulaes for h-paramters

CE to CC	CE to CB
$h_{ic} = h_{ie}$	$h_{ib} = \frac{h_{ie}}{1 + h_{fe}}$
$h_{rc} = 1$	$h_{rb} = \frac{h_{ie} h_{oe}}{1 + h_{fe}} - h_{re}$
$h_{fc} = -(1 + h_{fe})$	$h_{fb} = \frac{-h_{fe}}{1 + h_{fe}}$
$h_{oc} = h_{oe}$	$h_{ob} = \frac{h_{oe}}{1 + h_{fe}}$

Hybrid Model

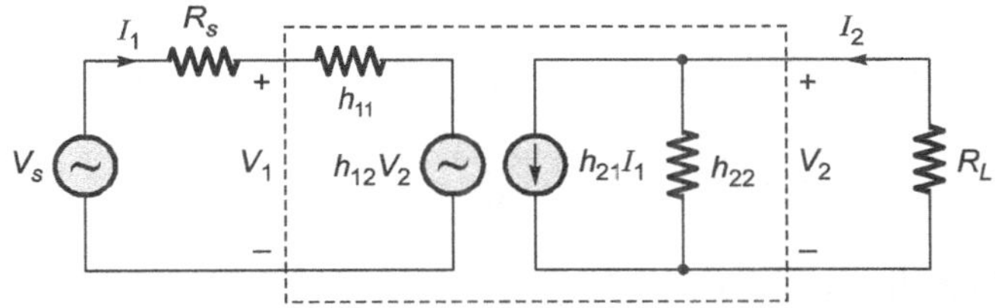

Current gain

$$A_I = \frac{I_2}{I_1} = \frac{-h_{21}}{1 + h_{22} R_L} = \frac{-h_f}{1 + h_o R_L}$$

where, $I_1 \rightarrow$ AC input current; $I_2 \rightarrow$ AC output current

Voltage gain

$$A_V = \frac{V_2}{V_1}$$

where, $V_1 \rightarrow$ AC input voltage; $V_2 \rightarrow$ AC output voltage

$$A_V = \frac{-h_{21} R_L}{h_{11} + (h_{11}h_{22} - h_{12}h_{21})R_L} = \frac{-h_f R_L}{h_i + (h_i h_o - h_f h_r)R_L}$$

Input impedance

$$Z_{in} = \frac{V_1}{I_1} = h_{11} - \frac{h_{12} h_{21} R_L}{1 + h_{22} R_L} = h_i - \frac{h_f h_r R_L}{1 + h_o R_L}$$

Output impedance

$$Z_{out} = \frac{V_2}{I_2} = \frac{R_s + h_{11}}{(R_s + h_{11})h_{22} - h_{12}h_{21}} = \frac{R_s + h_i}{(R_s + h_i)h_o - h_r h_f}$$

$$Y_{out} = h_o - \frac{h_r h_f}{h_f + R_s}$$

Multistage Amplifier

- Upper cutoff frequency of overall configuration is

  $f_H^* = f_H\sqrt{2^{1/n}-1}$ (For identical amplifiers in cascade)

  where, $n \rightarrow$ number of amplifiers in cascade

  $f_H \rightarrow$ upper cutoff frequency of one amplifier

- Lower cutoff frequency of overall configuration

  $f_l^* = \dfrac{f_l}{\sqrt{2^{1/n}-1}}$ where, n = number of identical amplifiers in cascade

  Approximate BW of amplifier in cascade is f_H^*.

- When amplifiers are non-identical then

  $f_l^* = 1.1\sqrt{f_{l1}^2 + f_{l2}^2 + f_{l3}^2 + ... + f_{ln}^2}$ and $\dfrac{1}{f_H^*} = 1.1\sqrt{\dfrac{1}{f_{H1}^2} + \dfrac{1}{f_{H2}^2} + ... + \dfrac{1}{f_{Hn}^2}}$

- Rise time $t_r^* = 1.1\sqrt{t_{r1}^2 + t_{r2}^2 + ... + t_{rn}^2}$

IX Feedback Amplifiers

Basic Feedback Concept

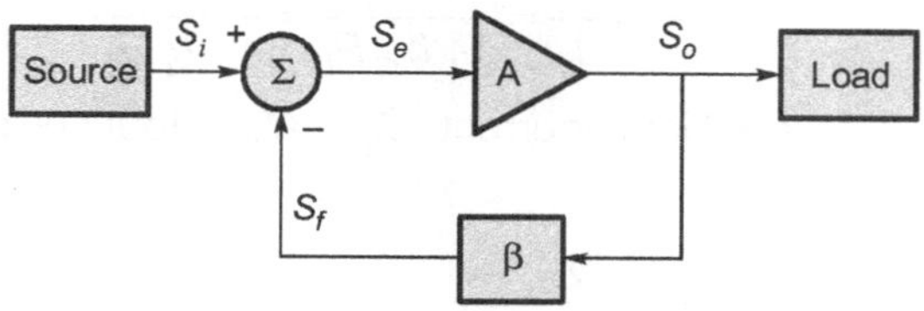

where, $A \rightarrow$ Open loop gain ; $\beta \rightarrow$ Feedback factor ; $S_i \rightarrow$ Input signal
$S_e \rightarrow$ Error signal ; $S_f \rightarrow$ Feedback signal

Gain with feedback

$A_f = \dfrac{A}{1+A\beta}$ for –ve feedback. $A_f = \dfrac{A}{1-A\beta}$ for +ve feedback.

Remember:

☑ The amount of feedback introduced into amplifier is often expressed in dB and is given by

$$N(\text{Feedback in dB}) = 20\log\left|\frac{1}{1+A\beta}\right|$$

Here, $N \rightarrow$ Negative for negative feedback and
$N \rightarrow$ Positive for positive feedback

Loop gain or return ratio

$$\text{Loop gain} = -A\beta$$

Return difference

Difference between unity and loop gain is called return difference (D)

$$D = 1 + A\beta$$

D is also called desensitivity.

Sensitivity

In the case of negative feedback, if

$\frac{\partial A}{A} \rightarrow$ % change without feedback ; $\frac{\partial A_f}{A_f} \rightarrow$ % change with feedback

Reduction in sensitivity of amplifier : $\frac{\partial A_f}{A_f} = \frac{\partial A / A}{1 + A\beta}$

Feedback Amplifier

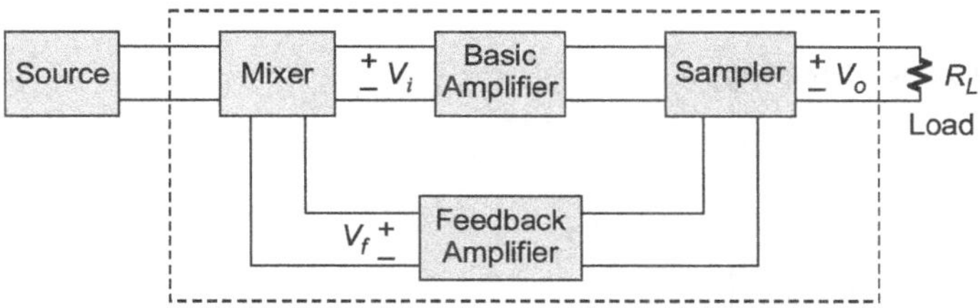

Sample Circuit	Mixer (Comparator)
1. Voltage/Shunt Sampler	**1. Series/Voltage Mixer (Comparator)**
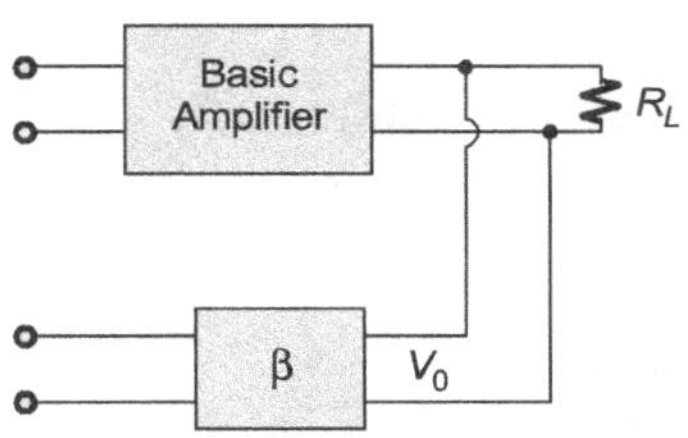	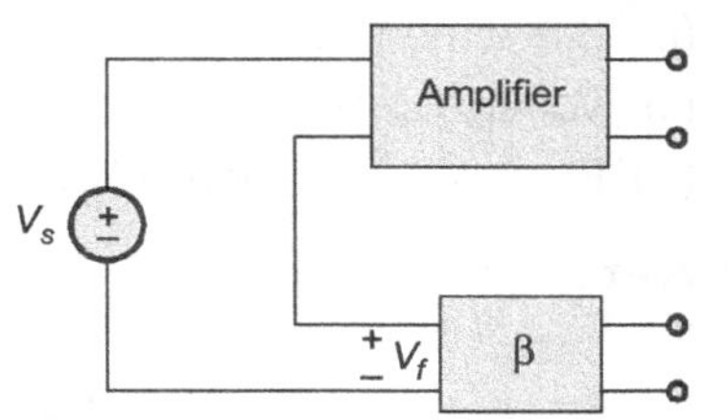
2. Current/Series Sampler	**2. Shunt/Current Mixer (Comparator)**
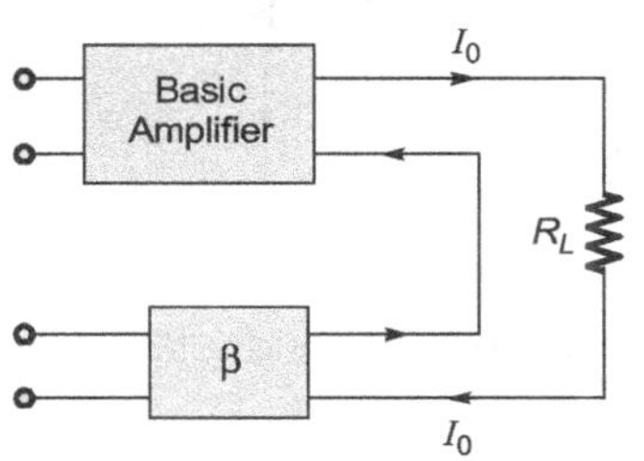	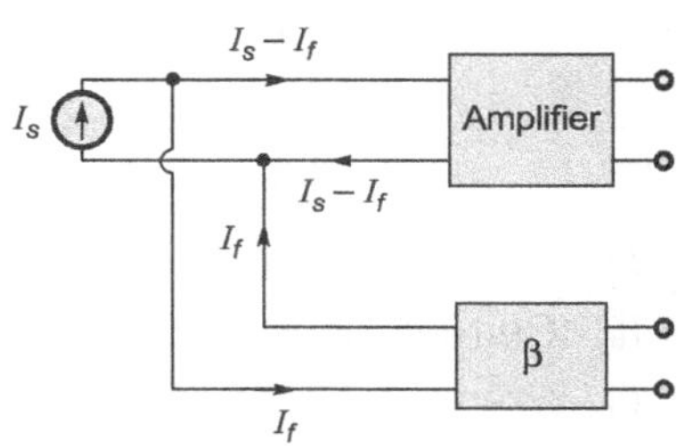

Effect of Feedback

- Reduction in noise, $N_{of} = \dfrac{N_o}{1+A\beta}$

  where, $N_o \rightarrow$ Noise without feedback ; $N_{of} \rightarrow$ Noise with feedback

- Lower cut-off frequency, $f_{lf} = \left(\dfrac{f_l}{1+A\beta}\right)$

  where, $f_{lf} \rightarrow$ Lower 3-dB frequency with feedback ; $f_l \rightarrow$ Lower 3-dB frequency without feedback

- Upper cut-off frequency, $f_{Hf} = f_H(1 + A\beta)$

  where, $f_{Hf} \rightarrow$ Upper 3-dB frequency with feedback ; $f_H \rightarrow$ Upper 3-dB frequency without feedback

Note:

☑ Bandwidth increases with negative feedback.

Voltage series topology

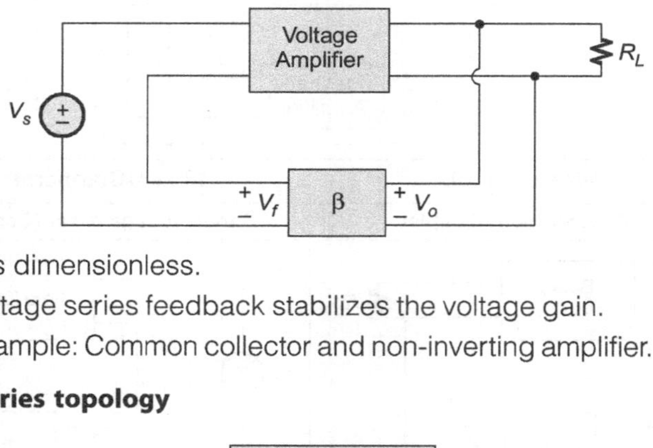

(i) β is dimensionless.

(ii) Voltage series feedback stabilizes the voltage gain.

(iii) Example: Common collector and non-inverting amplifier.

Current series topology

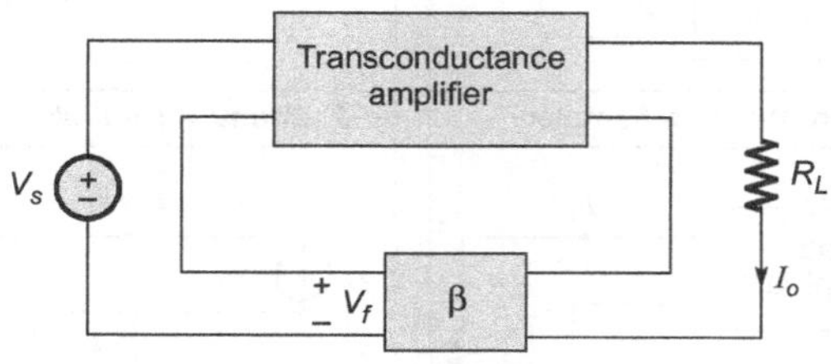

(i) Unit of β is ohm.

(ii) Current series feedback stabilizes the transconductance gain.

(iii) Example: Common emitter.

Voltage shunt topology

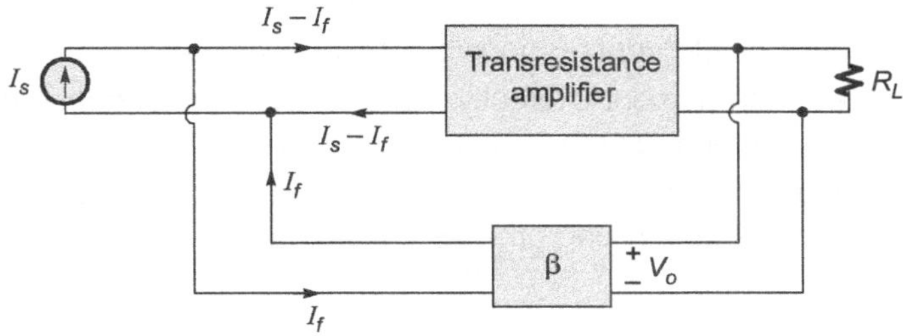

(i) β has unit mho.
(ii) It stabilizes the transresistance gain of the amplifier.
(iii) Example: Collector to base biasing.

Current shunt topology

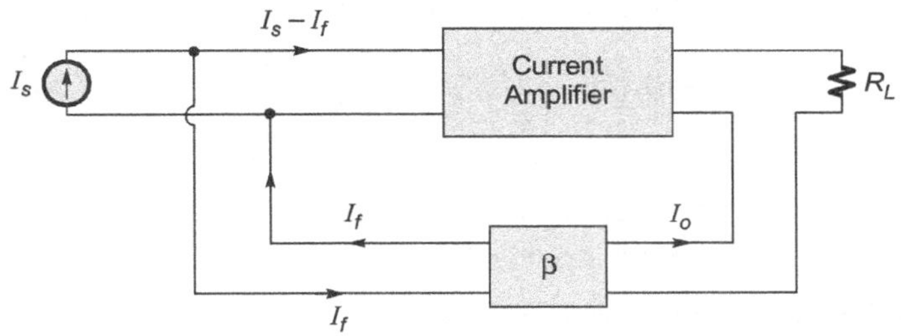

(i) β has no unit.
(ii) It stabilizes the current gain of amplifier.
(iii) Example: Cascading amplifier.

- For different feedback configuration effect on input and output impedance and gain.

Feedback	Input	Output	Gain	R_i	R_o
Voltage series	Voltage	Voltage	A_V	↑	↓
Current series	Voltage	Current	G_m	↑	↑
Voltage shunt	Current	Voltage	R_m	↓	↓
Current shunt	Current	Current	A_I	↓	↑

Note:

☑ Voltage series feedback is the best arrangement in terms of R_i and R_o.

☑ Current shunt is the worst for the same.

Remember:

Procedure to identify the type of feedback:

- ☑ Identify the element responsible for the feedback.
- ☑ If feedback element is directly connected to output node, it indicates voltage sampling otherwise it indicates current sampling.
- ☑ If feedback element is directly connected to the input node, it indicates shunt mixing otherwise it indicates series mixing.

X Operational Amplifier

Differential Amplifier

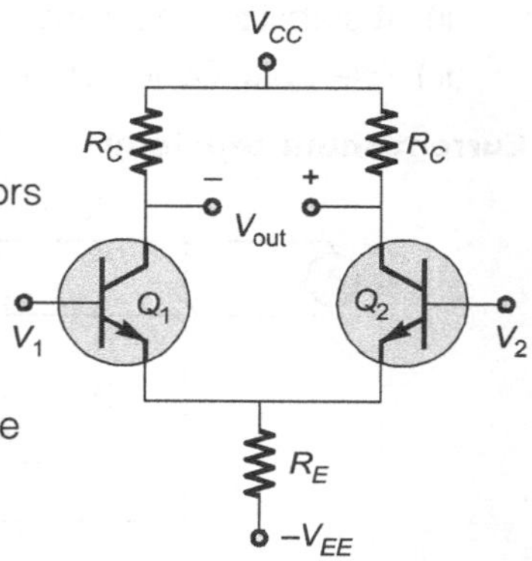

- Here,

$$V_{out} = A(V_1 - V_2)$$

where, $V_{out} \rightarrow$ Voltage between collectors

$$A = \frac{R_c}{r_e} = g_m R_c$$

where, $r_e \rightarrow$ Emitter resistance

$V_1 \rightarrow$ Non-inverting input voltage

$V_2 \rightarrow$ Inverting input voltage

DC analysis of a differential amplifier

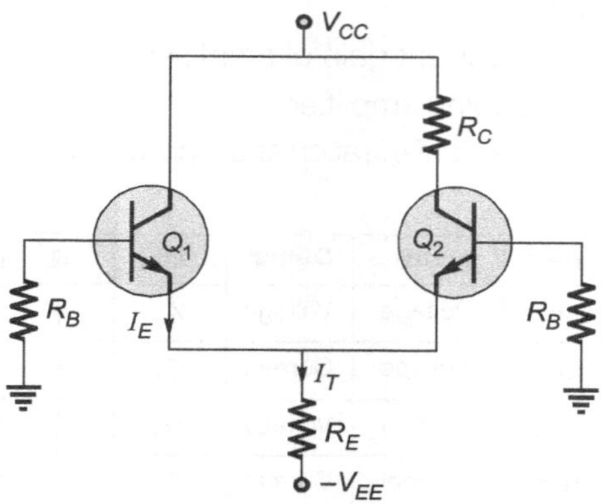

- Emitter bias current, $I_E = \dfrac{V_{EE} - V_{BE}}{2R_E}$

- Tail current, $I_T = \dfrac{V_{EE} - V_{BE}}{R_E}$

AC analysis of a differential amplifier

Non inverting input:

- AC emitter current, $i_e = \frac{V_1}{2r_e}$; where, $r_e \to$ Emitter resistance
- AC output voltage, $V_{out} = i_C R_C = \frac{V_1}{2r_e} R_C$
- Voltage gain for non-inverting input, $\frac{V_{out}}{V_1} = \frac{R_c}{2r_e}$

Inverting input:

- AC emitter current, $i_e = \frac{V_2}{2r_e}$
- AC output voltage, $V_{out} = -i_C R_C = \frac{-V_2}{2r_e} R_C$
- Voltage gain for non-inverting input, $\frac{V_{out}}{V_2} = \frac{R_c}{2r_e}$
- Differential voltage gain, $A = \frac{R_C}{2r_e}$
- Input impedance, $r_{in} = 2\beta r_e$

Common mode voltage gain

$$A_{CM} = \frac{V_{out}}{V_{in(CM)}} = \frac{-R_C}{r_e + 2R_E} = \frac{-R_C}{2R_E}$$

$R_E >> r_e$

$$\text{CMRR} = \left|\frac{A}{A_{CM}}\right| = 2g_m R_E$$

where, $A_{CM} \to$ Common mode gain
$A \to$ Differential mode gain

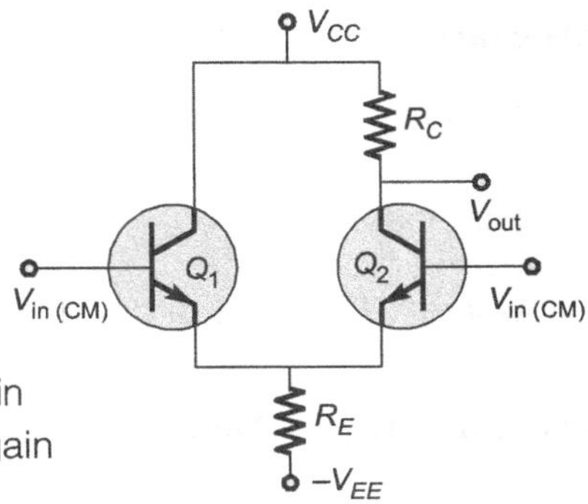

Operational Amplifier

An op-amp is basically a very high gain, direct-coupled amplifier with high input impedance and low output impedance.

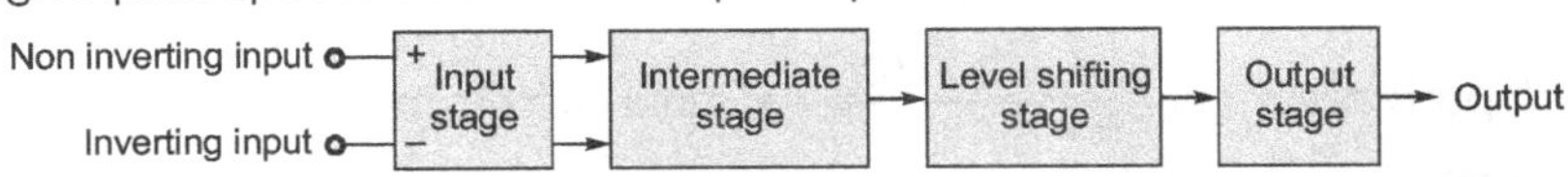

Equivalent circuit of an op-amp

Op-amp amplifies the between the two input signals applied at non-inverting and inverting input terminals.

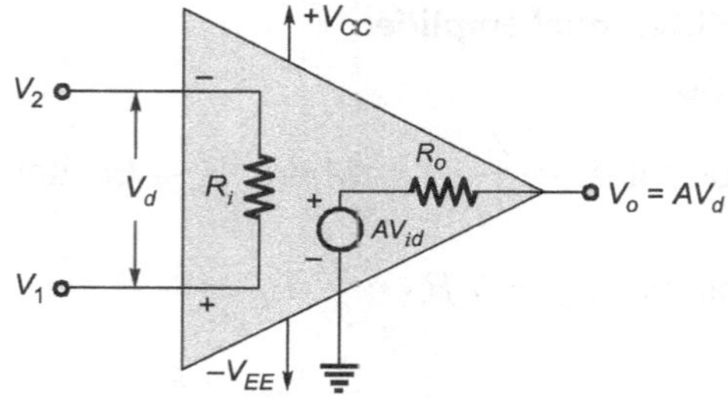

Equivalent circuit of op-amp

where, Difference input voltage → $V_d = V_1 - V_2$; A → Open loop gain of op-amp ; Output voltage → $V_o = AV_d$; R_i → Differential input resistance ; R_o → Output resistance

Properties of Op-amp

Parameters	Ideal value	Practical value
Voltage gain	∞	10^6
Input resistance	∞	10^6 Ω or 1 MΩ
Output resistance	0	10 Ω to 100 Ω
B.W.	∞	10^6 Hz or 1 MHz
CMRR	∞	10^6 or 120 dB
Slew rate	∞	80V/μ sec.

Slew rate

Condition for distortionless output, $SR \geq \dfrac{2\pi f V_m A_{CL}}{10^6}$

For input $V_m \sin \omega t$, $f_{max} = \dfrac{SR}{2\pi A_{CL} V_m}$

where, SR → Slew rate ; f_{max} → Maximum frequency of operation ; A_{CL} → Closed loop gain of OP-AMP.

Bias Currents and Voltage

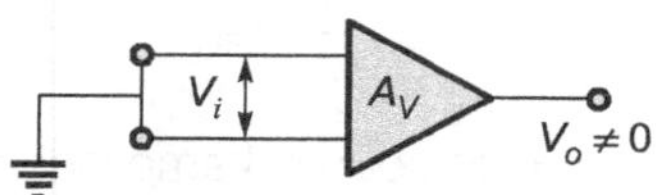

$V_0 \neq 0$ → Called O/P offset voltage.

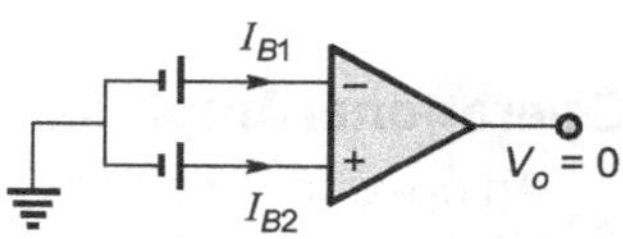

Balanced condition.

- Input bias current, $I_B = \dfrac{I_{B_1} + I_{B_2}}{2}$
- Input offset current, $I_{io} = (I_{B1} - I_{B2})$

Note: Input offset voltage is the voltage which must be applied between input terminals to balance amplifier ($V_0 = 0$).

Linear Op-Amp Circuits

Inverting Amplifier

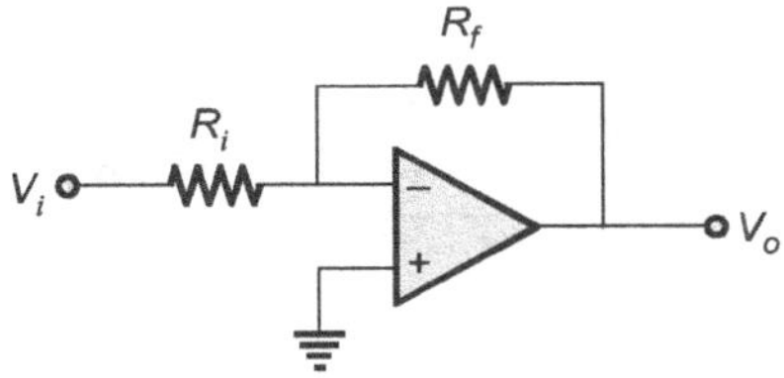

$$\text{Gain} = A_v = \frac{V_o}{V_i} = \frac{-R_f / R_1}{1+\left(\dfrac{1+R_f / R_1}{A_{OL}}\right)}$$

where, $A_{OL} \rightarrow$ Open loop gain of op-amp

ideally; $A_{OL} = \infty$

$$A_V = -\frac{R_f}{R_1}$$

Non-inverting Amplifier

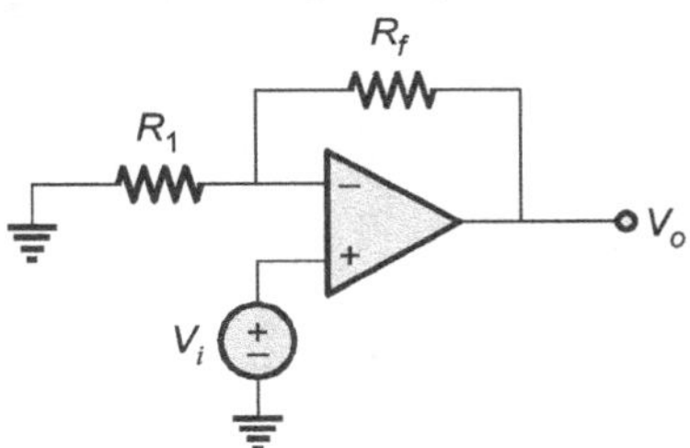

$$\text{Gain} = A_v = \frac{V_o}{V_i} = \frac{\left(1+\dfrac{R_f}{R_1}\right)}{1+\left(\dfrac{1+R_f / R_1}{A_{OL}}\right)}$$

where, $A_{OL} \rightarrow$ Open loop gain of op-amp

ideally; $A_{OL} \rightarrow \infty$

$$A_V = \left(1+\frac{R_f}{R_1}\right)$$

Difference Amplifier

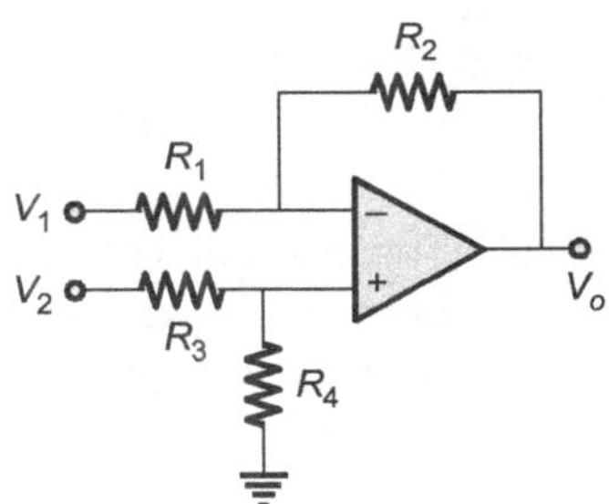

$$V_o = -\frac{R_2}{R_1}V_1 + \left(\frac{R_4}{R_3+R_4}\right)\left(1+\frac{R_2}{R_1}\right)V_2$$

if $\dfrac{R_2}{R_1} = \dfrac{R_4}{R_3}$

then, $V_0 = \dfrac{R_4}{R_3}[V_2 - V_1] = \dfrac{R_2}{R_1}(V_2 - V_1)$

Note:

☑ In this case when $\dfrac{R_2}{R_1} = \dfrac{R_4}{R_3}$,

☑ $A_c = 0 \rightarrow$ Common mode gain and CMRR $= \infty$

Nonlinear Op-Amp circuits

Integrator

$$V_0 = -\frac{1}{RC}\int V_i\,dt + V_c(0^+)$$

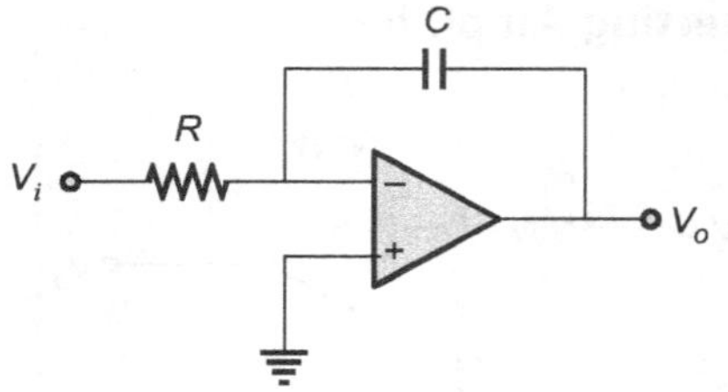

Note:

☑ This integrator becomes unstable at low frequency due to finite gain.

☑ The drawback of instability can be eliminated by connecting a resistor R_f in feedback path and such integrator circuit is called practical integrator.

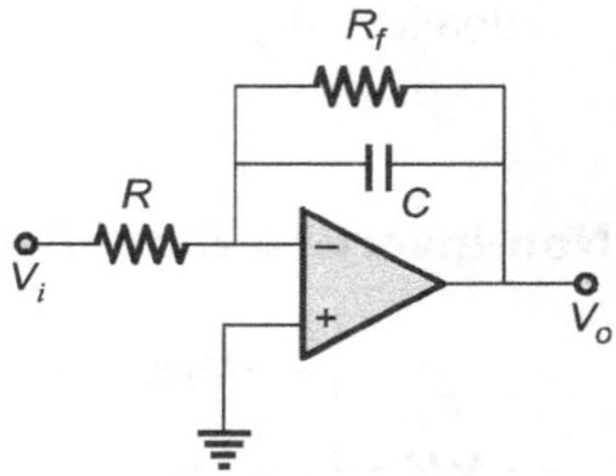

Differentiator

$$V_0 = -RC\frac{dV_i}{dt}$$

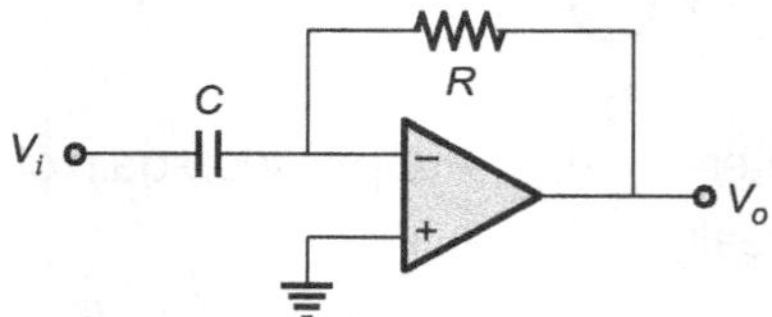

Note:

☑ This differentiator becomes unstable at high frequency due to finite gain.

☑ The drawback of instability can be eliminated by connected R_f in input path and such differentiator is called practical differentiator.

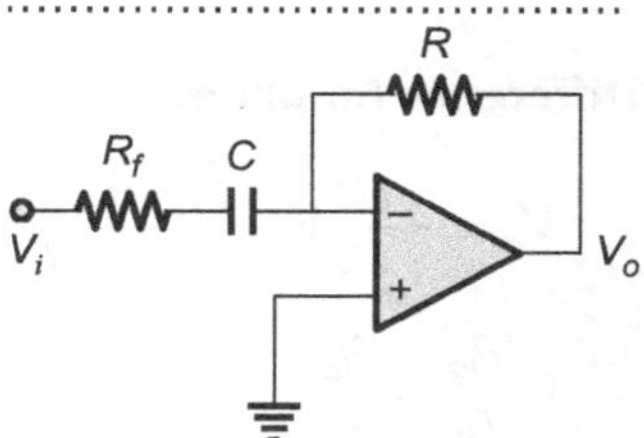

Instrumentation Amplifier

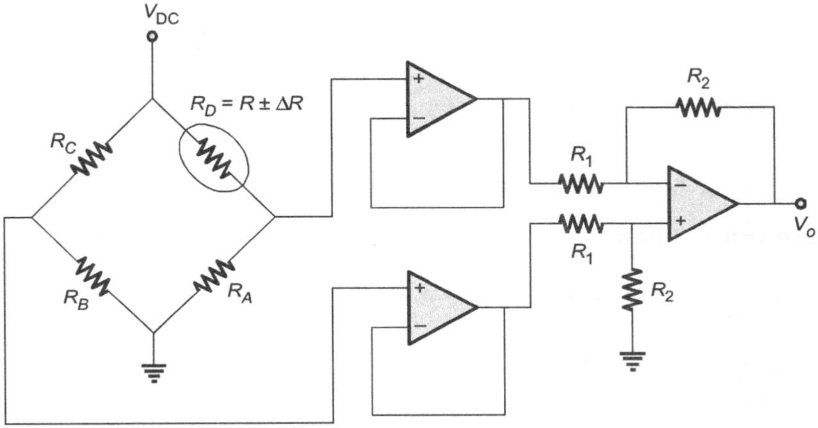

If $R_A = R_B = R_C = R$

$$V_o \simeq \frac{R_2}{R_1}\frac{V_{DC}}{4R} \times (\pm \Delta R)$$

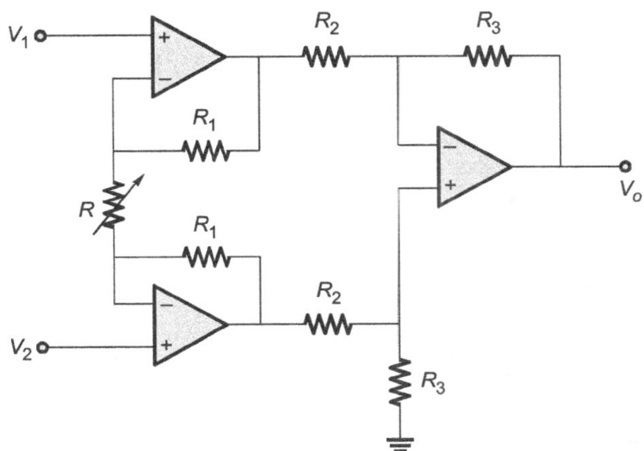

$$\boxed{V_o = \frac{R_3}{R_2}\left(1 + \frac{2R_1}{R}\right)(V_2 - V_1)}$$

Logarithmic Amplifier

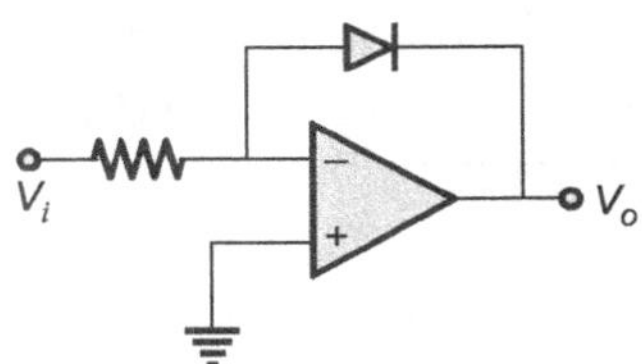

$$V_0 = -\eta V_T \ln \frac{V_i}{I_0 R}$$

where, $\eta \rightarrow$ Recombination factor

$V_T \rightarrow$ Thermal voltage

$I_o \rightarrow$ Reverse saturation current of diode

Antilog Amplifier

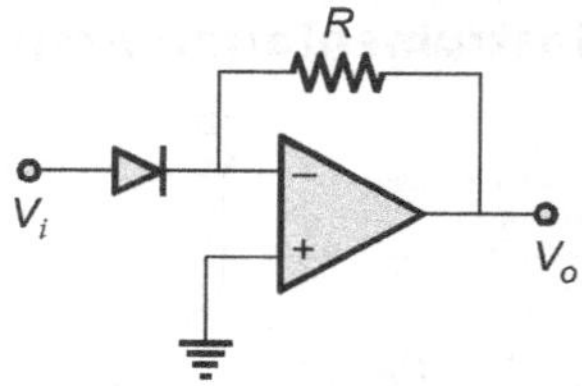

$$V_o = -I_o R \text{ antilog} \frac{V_i}{\eta V_T}$$

where, $\eta \rightarrow$ Recombination factor

$V_T \rightarrow$ Thermal voltage

$I_0 \rightarrow$ Reverse saturation current of diode

Grounded Load

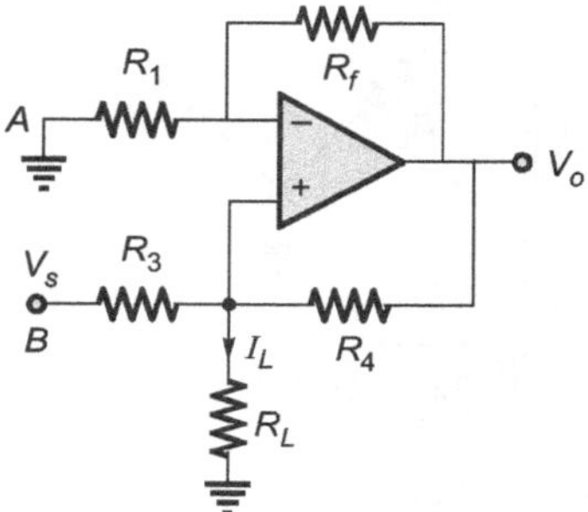

If $\frac{R_f}{R_1} = \frac{R_4}{R_3}$ then $I_L = \frac{V_s}{R_3}$

If V_s is applied at point A and B is grounded and

$\frac{R_f}{R_1} = \frac{R_4}{R_3}$ then $I_L = \frac{-V_s}{R_3}$

Schmitt Trigger

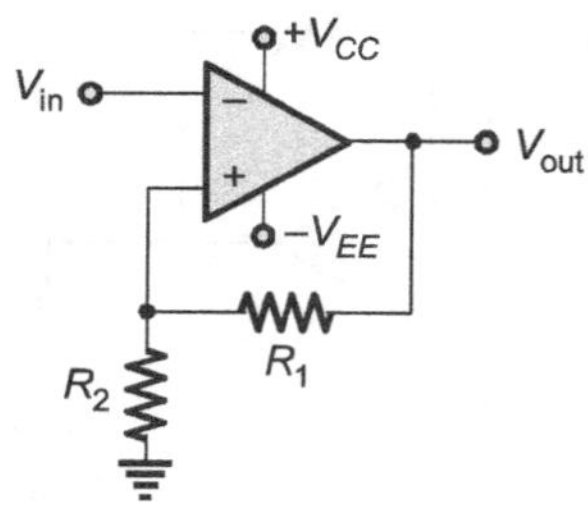

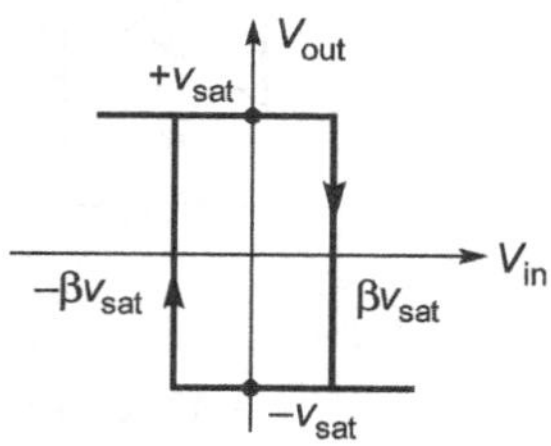

$$V_{Th} = \beta V_{sat} = \left(\frac{R_2}{R_1 + R_2}\right) V_{sat} \; ; \; V_{TL} = -\beta V_{sat} = \left(\frac{-R_2}{R_2 + R}\right) V_{sat}$$

Remember:

☑ The complete transfer characteristics of Schmitt trigger shows a hysteresis effect.

☑ The width of hysteresis is the difference between the two cross-over voltage V_{Th} and V_{TL}.

Triangular Wave Generator

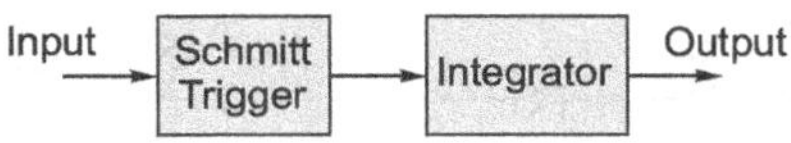

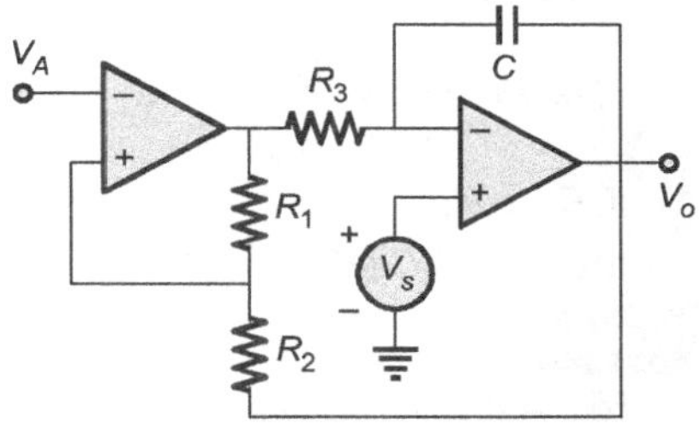

where, $V_s \rightarrow$ Control the slope ; $V_A \rightarrow$ Control the amplitude output
$V_s = 0 \rightarrow$ Symmetrical triangular wave is obtained

where, $$T_1 = \frac{2R_2\,R_3\,CV_{sat}}{R_1(V_{sat}+V_s)}$$

$$T_2 = \frac{2R_2\,R_3\,CV_{sat}}{R_1(V_{sat}-V_s)}$$

$$T_1 + T_2 = T = \frac{4R_2\,R_3\,C}{R_1}$$

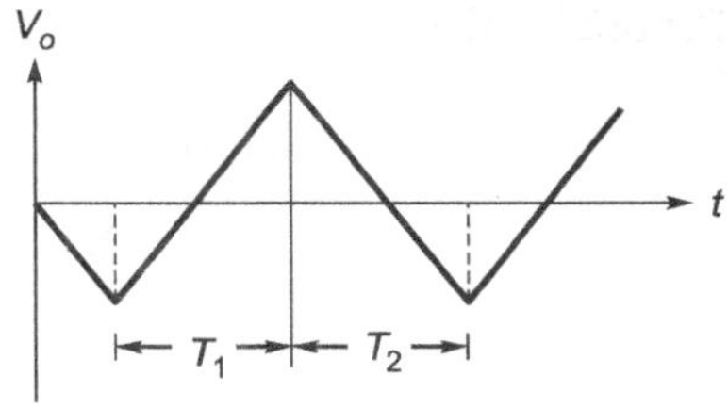

Op-Amp as a Multivibrator

Square Wave Generator or Astable Multivibrator

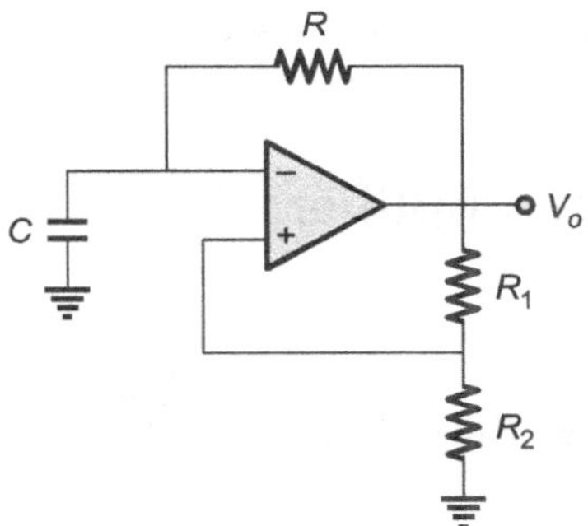

Time period of output waveform generated is

$$T = 2RC\ln 3 \qquad (\text{For } R_1 = R_2)$$

$$T = 2RC\ln\left[\frac{1+\beta}{1-\beta}\right] \qquad \left(\text{where } \beta = \frac{R_2}{R_1+R_2}\right)$$

Monostable Multivibrator

Width of pulse generated at output,

$$T = RC\ln\left(1+\frac{R_2}{R_1}\right)$$

if $R_2 = R_1$

then $T = RC\ln 2$

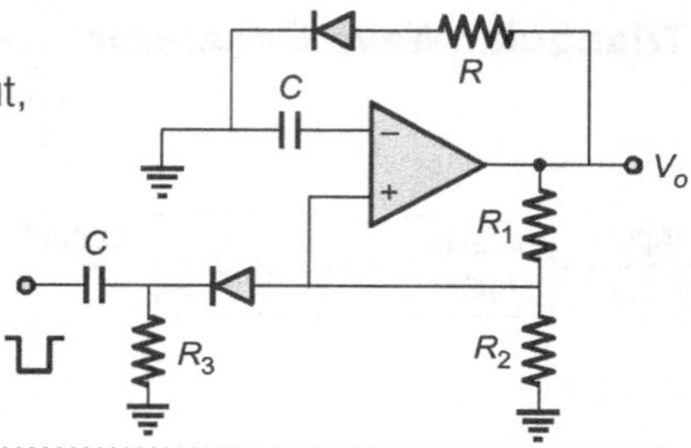

Note:

- ☑ Astable multivibrator is a square wave generator.
- ☑ Monostable multivibrator is used as pulse stretcher and missing pulse detector.

Bistable Multivibrator

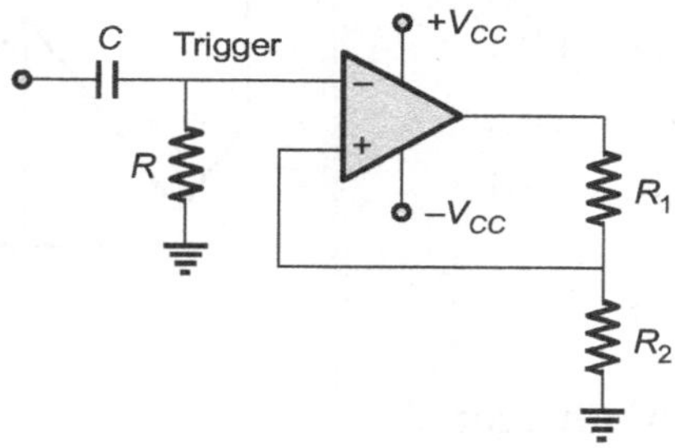

Note:

- ☑ Astable multivibrator is a square wave generator.
- ☑ Monostable multivibrator is used or pulse stretcher and missing pulse detector.

XI Large Signal Amplifiers

Load Line Analysis

For Common Emitter Amplifier

where,

I_{CQ} = Quiescent collector current

V_{CEQ} = Quiescent collector-emitter voltage

Thevenin resistance driving base

$$r_B = R_s \parallel R_1 \parallel R_2$$

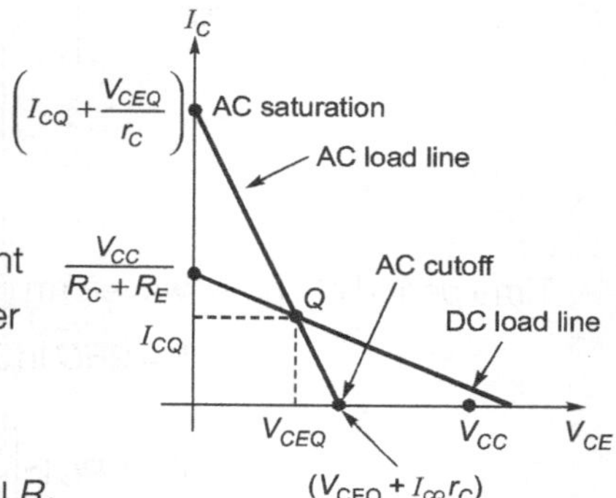

AC load line

AC load resistance seen by collector

$$r_C = R_C \parallel R_L$$

Equation of AC load line

$$I_C = I_{CQ} + \frac{V_{CEQ}}{r_C} - \frac{V_{CE}}{r_C}$$

where, $I_{CQ} \rightarrow$ DC collector current
$V_{CEQ} \rightarrow$ DC collector emitter voltage
$r_c \rightarrow$ AC resistance seen by collector

AC saturation current

$$I_{C(sat)} = I_{CQ} + \frac{V_{CEQ}}{r_C}$$

Note:

When the transistor goes into saturation, $V_{CE} = 0$.

AC cutoff voltage

$$V_{CE(cut)} = V_{CEQ} + I_{CQ} r_C$$

AC output compliance of a CE amplifier

$$PP \cong \min.(2I_{CQ} r_C, 2V_{CEQ})$$

Note:

AC output compliance is the maximum unclipped peak to peak ac voltage that an amplifier can produce.

For Emitter Follower Amplifier

Parameters	Formula
Effective AC load resistance	$r_E = R_E \parallel R_L$
AC saturation current	$I_{C(sat)} = I_{CQ} + \frac{V_{CEQ}}{r_E}$
AC cut-off voltage	$V_{CE(cut)} = V_{CEQ} + I_{CQ} r_E$
AC output compliance of an emitter follower	$PP \cong \min.(2I_{CQ} r_E, 2V_{CEQ})$

For Common Base Amplifier

AC load resistance of CB amplifier

$$r_C = R_C \parallel R_L$$

Note:

AC load line and ac output compliance of CB amplifier are same as that of CE amplifier.

For Swamped Amplifier

AC saturation current

$$I_{C(sat)} = I_{CQ} + \frac{V_{CEQ}}{r_C + r_E}$$

AC cut-off voltage

$$V_{CE(cut)} = V_{CEQ} + I_{CQ}(r_C + r_E)$$

AC output compliance of a swamped amplifier

$$PP \cong \min.\left(2I_{CQ}r_C, 2V_{CEQ}\frac{r_C}{r_C + r_E}\right)$$

For maximum AC output compliance

$$I_{CQ}r_C = V_{CEQ} \qquad \text{....... (CE stage)}$$

$$I_{CQ}r_E = V_{CEQ} \qquad \text{....... (CC stage)}$$

$$I_{CQ}r_C = V_{CEQ}\frac{r_C}{r_C + r_E} \qquad \text{....... (Swamped)}$$

Power Amplifiers

It is a large signal amplifier which has greater AC output voltage and greater AC output current hence, it can provide greater AC output power to load.

Conversion Efficiency

It is the ability of power amplifier to convert DC power into AC power:

$$\eta = \frac{P_{AC}}{P_{DC}} \times 100\%$$

Harmonic Distortion

The harmonic distortion means the presence of the frequency components in the waveform, which are not present in the input signal.

If $I_C = I_o + B_0 + B_1\cos\omega t + B_2\cos 2\omega t + \cdots + B_n\cos n\omega t$

then $$\%D_n = \frac{|B_n|}{|B_1|} \times 100\% \quad ; \quad \text{where, } n = 1, 2, 3, \text{ and so on}$$

Here, fundamental frequency component has an amplitude B_1 and n^{th} harmonic component has an amplitude of B_n.

Thus, total harmonic distortion is

$$\%D = \sqrt{D_2^2 + D_3^2 + D_4^2 + \cdots + D_n^2} \times 100\%$$

XII The Signal Generators and Wave Shaping Circuit

Oscillators

The oscillators are positive feedback amplifier in which the part of output is feedback to the input via feedback circuit.

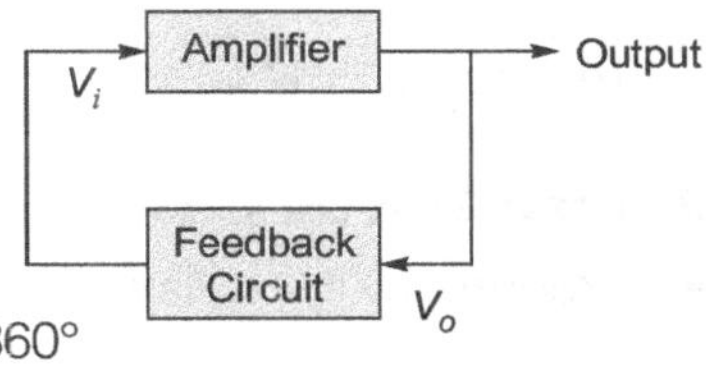

The gain, $\dfrac{V_o}{V_i} = \dfrac{A}{1-A\beta}$

$\because$ $V_i = 0$ (Noise)

$A\beta = 1$ or $1\angle 0° \cong \angle 360°$

Note:

- ☑ $|A\beta| = 1$, the amplitude of sinusoidal output remains constant.
- ☑ If $|A\beta| < 1$, the amplitude of oscillation will decrease with time and eventually or finally the oscillation will disappear.
- ☑ If $|A\beta| > 1$, the amplitude of oscillation will increase with time practically the amplitude becomes constant after reaching maximum value due to supply voltage limitation and non-linearity of transistor.

Oscillator Types

Type of Component	Frequency of oscillation	Waveform generated
RC oscillator	Audio frequency	Sinusoidal
LC oscillator	Radio frequency	Square wave
Crystal oscillator	Radio frequency	Triangular wave sawtooth wave etc.

RC Phase Shift Oscillator

RC phase shift oscillator using BJT

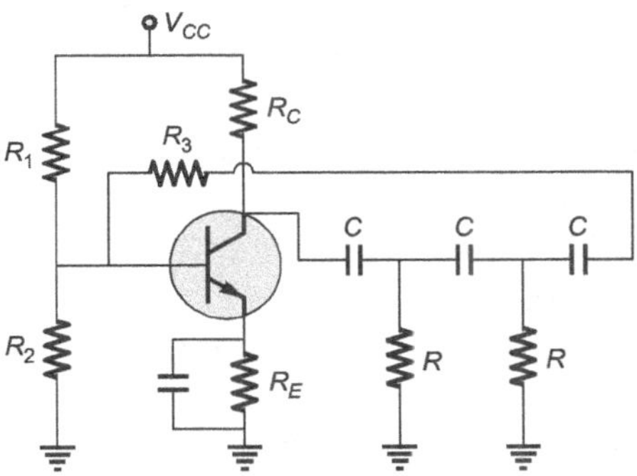

- Frequency of oscillation is

$$\omega = \frac{1}{RC\sqrt{4K+6}}; \quad f = \frac{1}{2\pi RC\sqrt{4K+6}} \quad \text{where, } K = \frac{R_C}{R}$$

Note:

☑ For transistor to work as oscillator

$$h_{fe} \geq \left(4K + 23 + \frac{29}{K}\right)$$

☑ $h_{fe\,(min)} = 44.5$.

RC Phase Shift Oscillator using FET

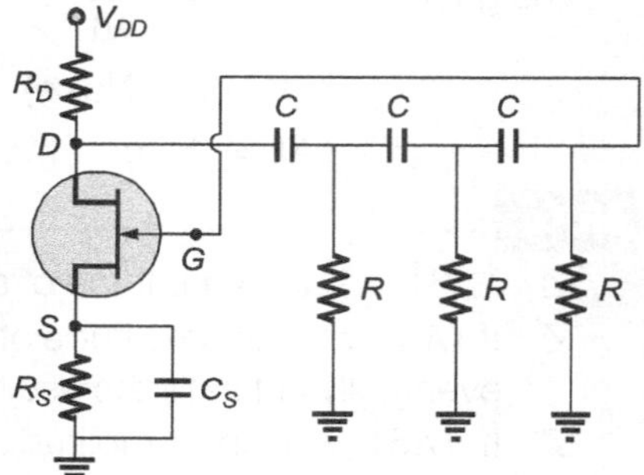

- Frequency of Oscillation

$$f = \frac{1}{2\pi RC\sqrt{6}}$$

$$\beta = -\frac{1}{29} \text{(feedback factor)}$$

Note:

☑ Amplification factor for FET to work as an Oscillator is $\mu \geq 29$.

RC Phase Shift Oscillator using Op-Amp

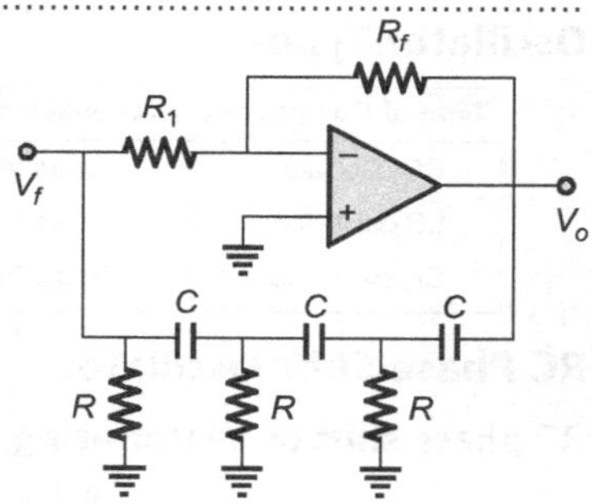

- Feedback factor

$$\beta = \frac{V_f}{V_o} = -\frac{1}{29} \text{(for oscillation)}$$

- Frequency of oscillation

$$f = \frac{1}{2\pi RC\sqrt{6}}$$

Note:

☑ $R_f \geq 29R_1 \rightarrow$ for oscillation.

☑ In *RC* phase shift oscillator minimum three *RC* circuits are needed to create a phase shift of 180°.

Wein Bridge Oscillator

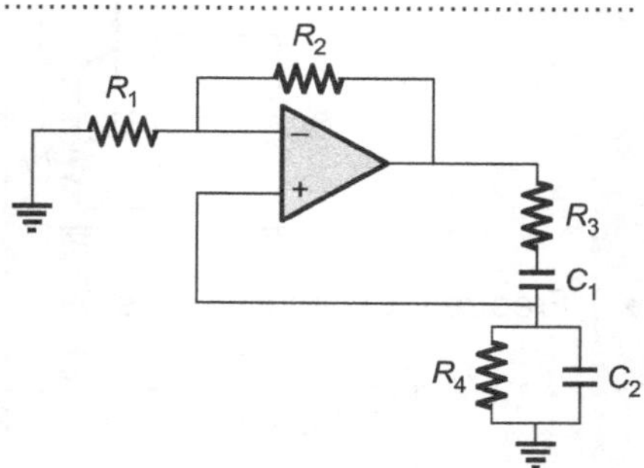

- Frequency of oscillation

$$f = \frac{1}{2\pi\sqrt{R_3R_4C_1C_2}}$$

if $R_3 = R_4 = R$ and $C_1 = C_2 = C$

then $$f = \frac{1}{2\pi RC}$$

Remember:

☑ $\beta = \frac{1}{3}$ for oscillation. ☑ $R_2 \geq 2R_1$ for oscillation.

LC Oscillator

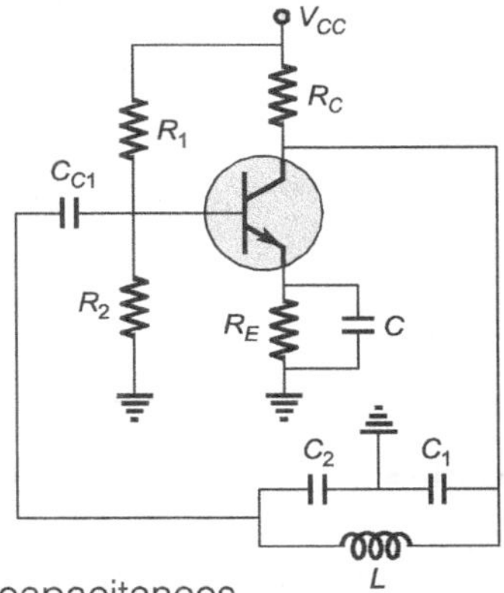

Colpitts Oscillator

- Frequency of oscillation

$$f = \frac{1}{2\pi\sqrt{L\left(\frac{C_1C_2}{C_1+C_2}\right)}}$$

and $g_m R_C \geq \frac{C_2}{C_1}$

Drawback of Colpitts oscillator

f_o is unstable due to the effect of internal capacitances.

Net input capacitance C_i appears in parallel with C_1 and net output capacitance C_o appears in parallel with C_2. As a result

$$f_o = \frac{1}{2\pi}\sqrt{\frac{1}{L}\left(\frac{1}{C_1'}+\frac{1}{C_2'}\right)}$$

where, $C_1' = C_1 + C_i$; $C_2' = C_2 + C_o$.

Since C_i and C_o vary with time and temperature, f_o becomes unstable. To eliminate this drawback an additional capacitance C_3 can be connected in series with inductor. Such circuit is called clapp oscillator.

Clapp Oscillator

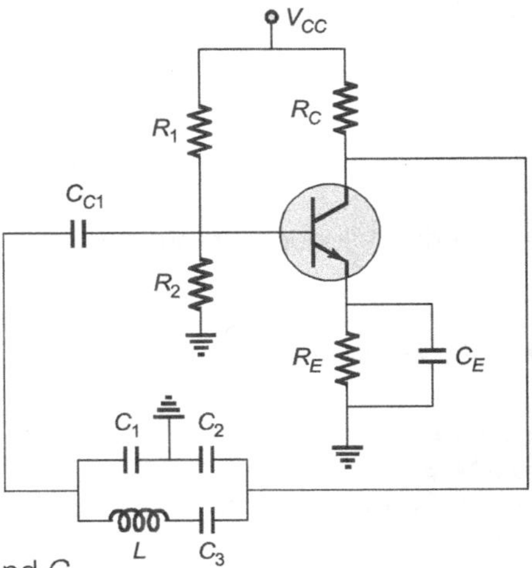

Frequency of oscillation

$$f_o = \frac{1}{2\pi}\sqrt{\frac{1}{L}\left(\frac{1}{C_1}+\frac{1}{C_2}+\frac{1}{C_3}\right)}$$

If C_3 is selected such that it is much smaller than C_1' and C_2' then

$$\frac{1}{C_3} >> \frac{1}{C_1}+\frac{1}{C_2}$$

Thus, $f_o = \frac{1}{2\pi}\sqrt{\frac{1}{L}\times\frac{1}{C_3}}$

f_o becomes independent of C_1 and C_2.

Hartley Oscillator

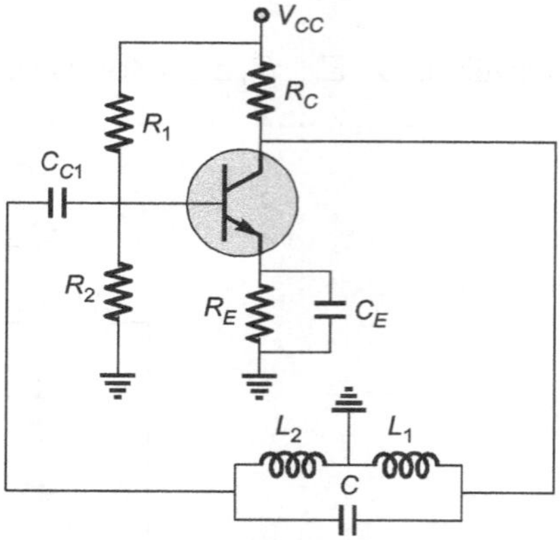

- Frequency of oscillation for Hartley Oscillator

$$f = \frac{1}{2\pi\sqrt{C(L_1 + L_2)}} \quad \text{and} \quad g_m R_C \geq \frac{L_1}{L_2}$$

Crystal Oscillator

- It uses a quartz-crystal which is a piezoelectric material.
- Quartz has excellent physical characteristics which do not get affected by time and temperature, hence crystal oscillator provides highly stable f_o.

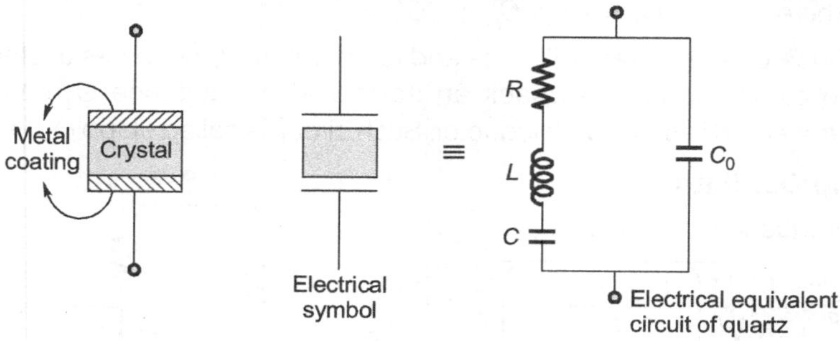

- Series resonance frequency, $f_s = \dfrac{1}{2\pi\sqrt{LC}}$
- Parallel resonance frequency, $f_p = \dfrac{1}{2\pi\sqrt{LC_{eq}}}$ where, $C_{eq} = \dfrac{CC_0}{(C + C_0)}$

Note:

☑ Frequency of oscillation is chosen between f_p and f_s.

Relaxation Oscillator

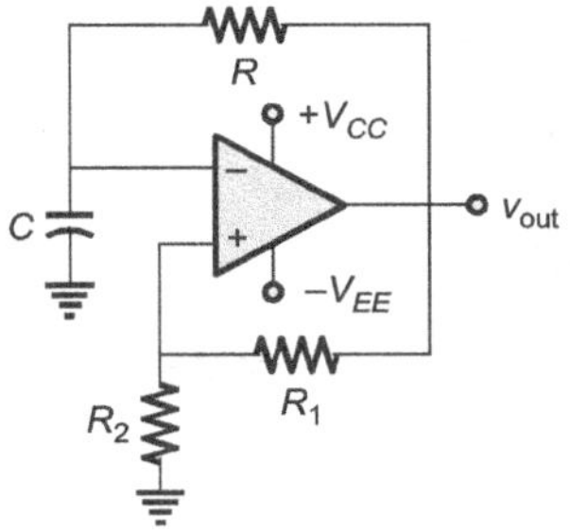

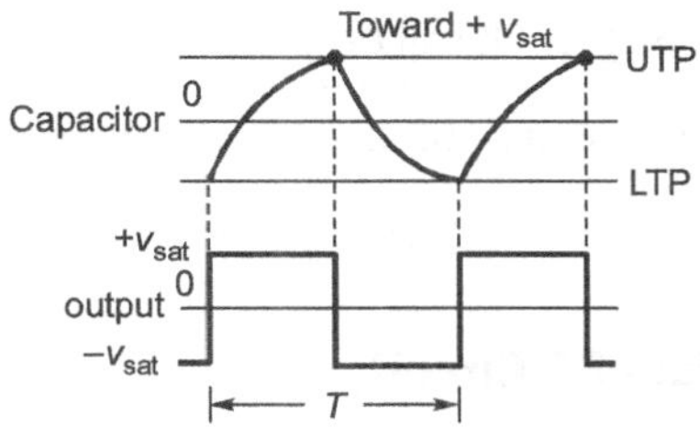

Time period of output signal, $T = 2RC\ln\left(\frac{1+B}{1-B}\right)$

where, $R \rightarrow$ Feedback resistance ; $B \rightarrow$ Feedback fraction

where, $B = \frac{R_2}{R_1 + R_2}$

Note:

- ☑ Linear oscillator generates sinewaves utilizing resonance phenomenon.
- ☑ Non linear oscillators or function generators generate square-wave, triangular wave, pulses etc. waveforms. They uses function blocks as multivibrators.
- ☑ Relaxation oscillator and UJT oscillator are non-linear oscillator.
- ☑ The oscillators which uses a tunnel diode or UJT possesses a current voltage characteristic curve of negative slope within some range of operation are known as negative resistance oscillator.
- ☑ Wein's Bridge oscillator is a variable frequency oscillator.

Linear Wave Shaping Circuits

High Pass RC Circuit

Pulse response of high pass R.C circuit

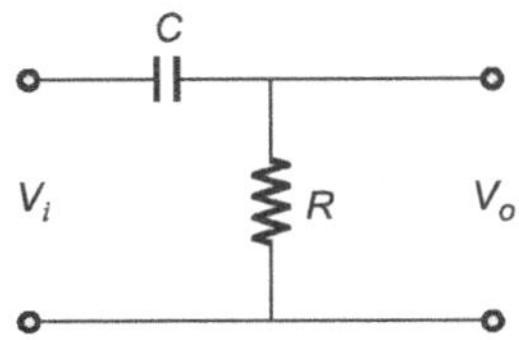

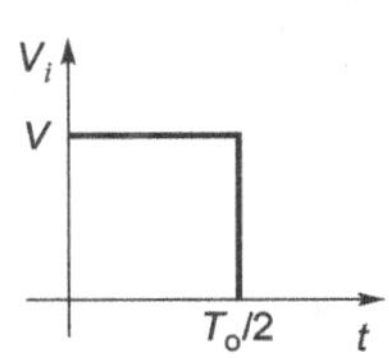

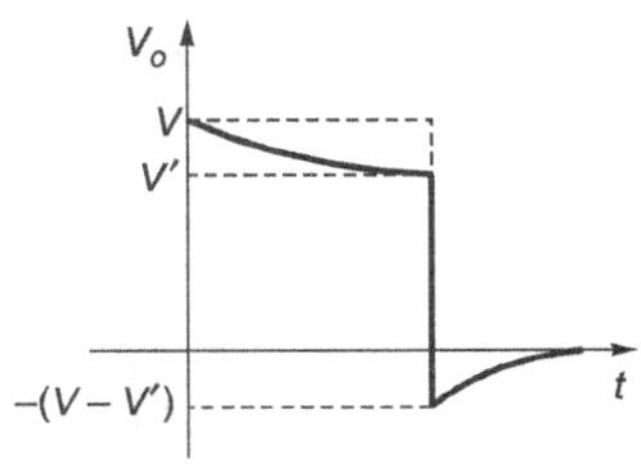

$(V - V')$ is called tilt or sag.

$$\%\,\text{tilt} = \frac{V - V'}{V} \times 100\%$$

Note:

☑ Sag or tilt will be obtained if $RC >> T_o/2$.

- If input is a square wave of time period T then

$$\% \text{ tilt} = \frac{\pi f_L}{f} \times 100 \quad \text{where, } f_L = \frac{1}{2\pi RC} \text{ and } f = \frac{1}{T_o}$$

Low Pass RC Circuit

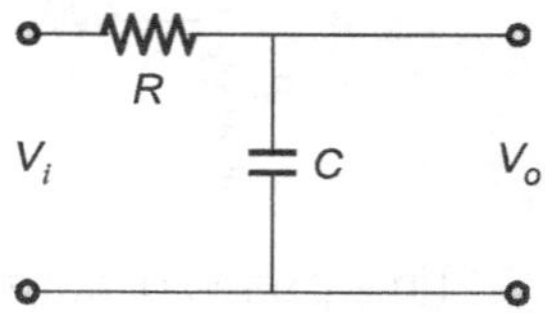

- Rise time

$$t_r = \frac{0.35}{f_H} \text{ (rise from 10\% to 90\%)}$$

where, $f_H = \dfrac{1}{2\pi RC}$

555 Timer

The 555 is a timing circuit that can produce accurate and highly stable time delays or oscillation.

555 Timer Pin Configuration

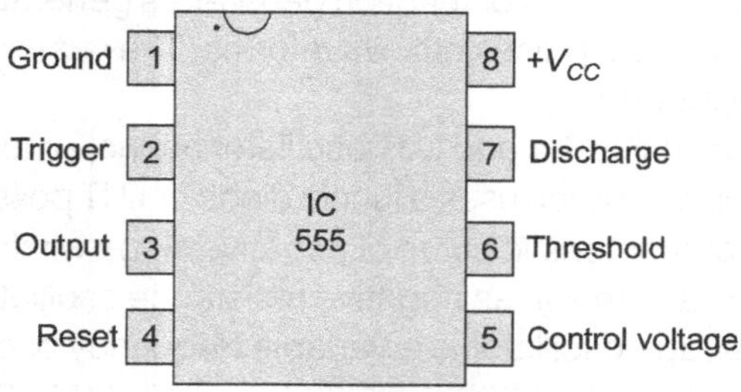

Monostable Multivibrator

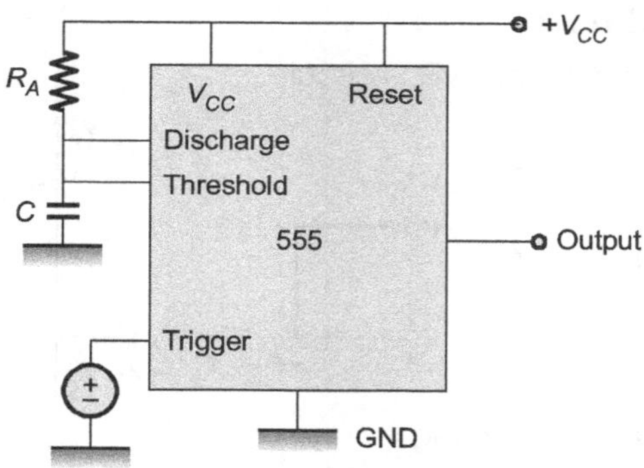

$$T = RC \ln(3) = 1.1\, RC$$

Note:

☑ The main applications of monostable multivibrator are pulse width modulator, linear ramp generator, frequency divider, pulse stretcher etc.

Astable Multivibrator

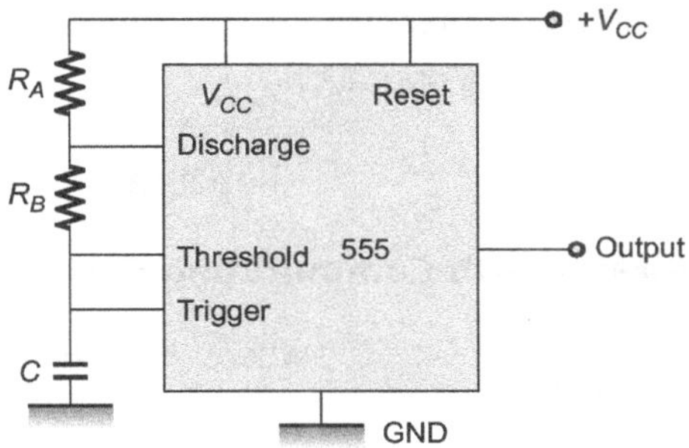

$$T = \tau \ln(2)$$

- Charging time constant, $T_C = 0.693(R_A + R_B)C$
- Discharging time constant, $T_D = 0.693\, R_B C$
- Total time, $T = T_C + T_D = 0.693\,(R_A + 2R_B)C$
- Duty cycle, $D = \dfrac{T_C}{T_C + T_D} \times 100\% = \dfrac{R_A + R_B}{R_A + 2R_B} \times 100\%$

Note:

☑ The main applications of astable are square-wave generator, voltage control oscillator, FSK generator, free running oscillator etc.

Remember:

☑ Duty cycle (D) > 50%. (Because $T_C > T_D$)

☑ By placing a diode parallel to R_B and making $R_A + R_f = R_B$, we get square wave output of duty cycle 50%.

Astable multivibrator

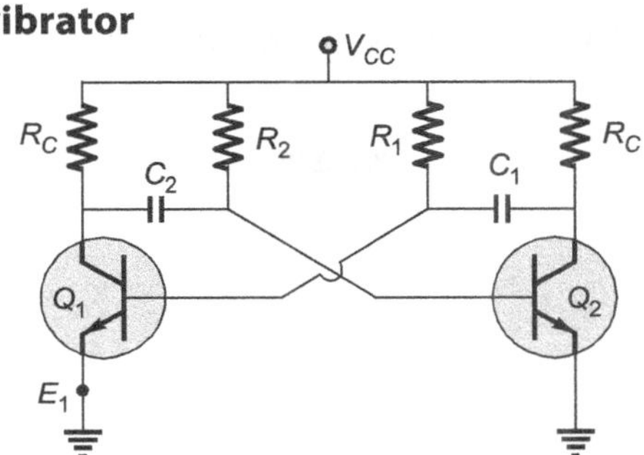

- Time period of output wave form generated is $T = 0.69[R_1C_1 + R_2C_2]$

  output frequency, $f = \frac{1}{T} = \frac{1.45}{(R_1C_1 + R_2C_2)}$
- Resolution time for bistable multivibrator is

  $\tau_s \rightarrow$ Smallest allowable interval between trigger.

  So, maximum frequency of operation

$$f_{max} = \frac{1}{\tau_s} \qquad \text{[Practical value]}$$

Bistable Multivibrator with Commutating Capacitor

- $f_{max} = \frac{R_1 + R_2}{2CR_1R_2} = \frac{1}{2CR}$ where, $R = R_1 \parallel R_2$ and $\tau_s = CR$

  where, $R_1 \rightarrow$ Resistance between collector of one transistor and base of other transistor.

  $R_2 \rightarrow$ Resistance between base and ground of each transistor.

  $f_{max} \rightarrow$ maximum frequency of operation
- In case of monostable multivibrator, width of pulse generated is given by

$$T = 0.69\ RC$$

where, $C \rightarrow$ Capacitance between collector of one transistor and base of other transistor.

$R \rightarrow$ Resistance between V_{CC} and base of one transistor.

Note:

☑ Astable multivibrator is nothing but square wave generator.

☑ Monostable multivibrator is used for missing pulse detection, in PWM, PPM.

☑ Bistable multivibrator is nothing but a flip flop.

■■■■

Digital Electronics

9

I Number System and Codes

Introduction

A digital system is a combination of devices designed to manipulate physical quantities or information that are represented in digital form, i.e. they can take on only discrete values. Examples of digital system are digital computer, calculator, telephone etc.

Advantages of Digital Techniques

1. Size and cost is less.
2. Power dissipation is less.
3. Digital circuits are less affected by noise.
4. Accuracy and precision are greater.
5. Information storage is easy.

Digital Number System

Many number systems are used in digital technology. The most common are the binary, octal, decimal and hexadecimal system.

Note:

A number system with base '*b*' will have '*b*' different digits from 0 to (*b* - 1).

Number Representation

$$N_b = \underbrace{d_{n-1}\, d_{n-2} \ldots d_i \ldots d_0}_{\text{Integer Portion}} \underset{\substack{\uparrow \\ \text{Radix point}}}{\bullet} \underbrace{d_{-1}\, d_2 \ldots d_{-f} \ldots d_{-m}}_{\text{Fraction Portion}}$$

where, $N \rightarrow$ Number ; $b \rightarrow$ Base or radix

Weighted number system

- It is a positional weightage system.
- **Ex:** Binary, octal, hexadecimal, BCD, 2421 etc.

Unweighted number system

- It is non positional weightage system.
- **Ex:** Gray code, Excess 3-code etc.

Number System	Base (b)	Digits
Binary	2	0, 1
Octal	8	0, 1, 2, 3, 4, 5, 6, 7
Decimal	10	0, 1, 2, 3, 4, 5, 6, 7, 8, 9
Hexadecimal	16	0, 1, 2, 3, 4, 5, 6, 7, 8, 9 *A, B, C, D, E, F*

Decimal	Hexadecimal	Binary
0	0	0000
1	1	0001
2	2	0010
3	3	0011
4	4	0100
5	5	0101
6	6	0110
7	7	0111
8	8	1000
9	9	1001
10	*A*	1010
11	*B*	1011
12	*C*	1100
13	*D*	1101
14	*E*	1110
15	*F*	1111

Note:

☑ In positional weightage system, the position of each digit of a number has some positional weightage.

☑ In non-positional weightage system, a digit of a number does not indicate any significance in a position and weight.

Codes

Binary Coded Decimal Code (BCD)

Each digit of a decimal number is represented by binary equivalent.

In 4-bit binary formats

Total number of possible representation = $2^4 = 16$

Valid BCD codes = 10, Invalid BCD codes = 6

In 8-bit binary formats

Valid BCD codes = 100, Invalid BCD codes = 256 – 100 = 156

Note:

☑ During arithmetic operation if invalid *BCD* is present then add (0110) to get correct result.

Excess-3 code

- It is a 4-bit code.
- It can be derived from BCD code by adding "3" to each coded number.
- It is a "self-complementing code".

Gray code

- Also called "minimum change code" or "unit distance code" in which only one bit in the code group changes when going from one step to the next.
- Gray code is a minimum error code.

Binary-to-Gray conversion

- 'MSB' in the gray code is same as corresponding digit in binary number.
- Starting from "Left to Right", add each adjacent pair of binary digits to get next and gray code digit. (Discard the carry if generated).

Gray-to-Binary conversion

- "MSB" of Binary is same as that of gray code .
- Add each binary digit to the generated gray digit in the next adjacent position (discard the carry if generated).

Remember:

☑ In self-complemented code, the sum of weightage = 9.

☑ The largest number that can be represented by using N-bits is $(2^N - 1)_{10}$.

Code Converters

BCD to Excess-3 code

- Let input variables of ***BCD* code** is A, B, C, D and output variables of excess-3 is W, X, Y, Z.
- Required truth table: (BCD + 0011 = excess-3-code)

Input (BCD)				Output (Excess-3 code)			
A	*B*	*C*	*D*	*W*	*X*	*Y*	*Z*
0	0	0	0	0	0	1	1
0	0	0	1	0	1	0	0
0	0	1	0	0	1	0	1
0	0	1	1	0	1	1	0
0	1	0	0	0	1	1	1
0	1	0	1	1	0	0	0
0	1	1	0	1	0	0	1
0	1	1	1	1	0	1	0
1	0	0	0	1	0	1	1
1	0	0	1	1	1	0	0

Minimized Boolean function:

$$Z = \bar{D}$$
$$Y = CD + \bar{C}\bar{D}$$
$$X = \bar{B}C + \bar{B}D + B\bar{C}\bar{D}$$
$$W = A + BC + BD$$

Binary-to Gray Code Converter

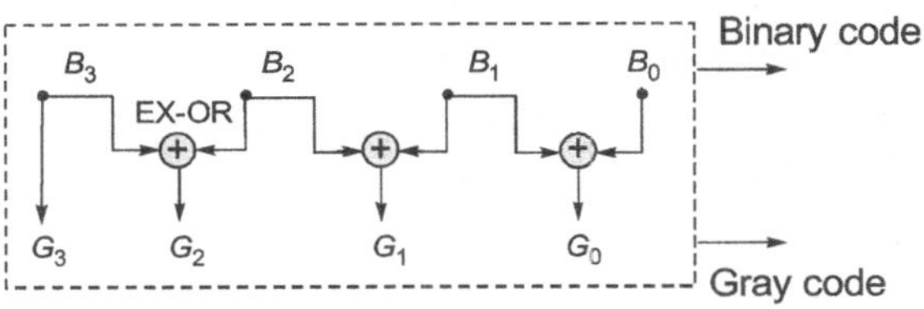

Equivalent logical gate diagram:

$G_3 = B_3$

$G_2 = B_3 \oplus B_2$

$G_1 = B_2 \oplus B_1$

$G_0 = B_1 \oplus B_0$

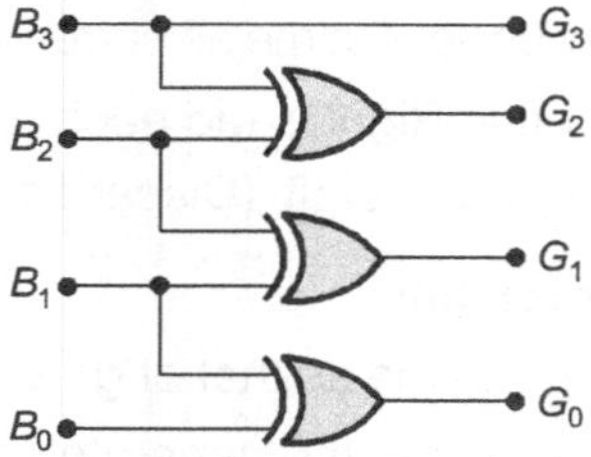

Gray to Binary Converter

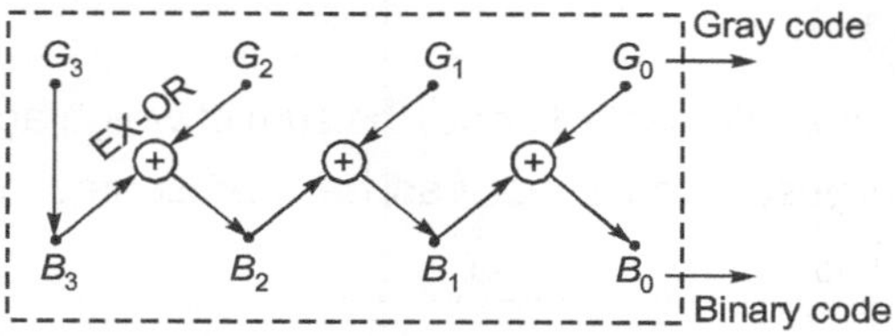

Equivalent Logical Gate Diagram:

$B_3 = G_3$

$B_2 = B_3 \oplus G_2$

$B_1 = B_2 \oplus G_1$

$B_0 = B_1 \oplus G_0$

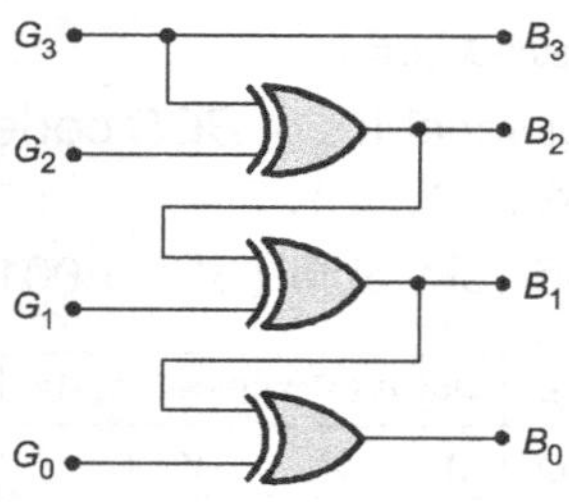

Conversion of Number System

Decimal Number System to Other Number System Conversion

Steps for integer part

- **Step-1:** Divide the decimal number to be converted by the value of new base.
- **Step-2:** Get the remainder from step-1 as the rightmost digit (LSB) of new base number.
- **Step-3:** Divide the quotient of the previous divide by the new base.
- **Step-4:** Record the remainder from step-3 as the next digit (to the left) of the new base number.

Example: $(29)_{10} \rightarrow (\ )_2$

$\therefore\ (29)_{10} \rightarrow (11101)_2$.

Operation	Result	Remainder
29/2	14	1
14/2	7	0
7/2	3	1
3/2	1	1
1/2	0	1

Fraction Part

Example: $(0.125)_{10} \rightarrow (\ )_2$

$\therefore\ (0.125)_{10} \rightarrow (0.001)_2$.

Operation	Result	Carry
0.125 × 2	0.25	0
0.25 × 2	0.5	0
0.5 × 2	1.0	1 ↓

Other Number System to Decimal Number System Conversion

Steps

- **Step-1:** Determine the column (positional value of each digit).
- **Step-2:** Multiply the obtained column value by the digits in the corresponding column.
- **Step-3:** Sum the products calculated in step-2. The total is equivalent value in decimal.

$$(x_2\, x_1\, x_0 \cdot y_{-1}\, y_{-2}) \rightarrow (x_2 \times r^2 + x_1\, r^1 + x_0\, r^0 + y_{-1}\, r^{-1} + y_{-2}\, r^{-2})_{10}$$

Octal to Binary

Each digit is represented with 3 bit binary equivalent.

Hexadecimal to Binary

Each digit is represented with 4 bit binary equivalent.

Octal to Hexadecimal

(i) First convert octal to binary and then binary to Hexadecimal.

(ii) From Hexadecimal to octal conversion first convert Hexadecimal to binary and then binary to octal number.

Digital Number Representation

- In unsigned magnitude representation with 'n' bits, the possible integer values are [0 to $2^n - 1$].
- Extra bit → sign bit → MSB

  if MSB = 0 → +ve number ; if MSB = 1 → –ve number
- In a sign magnitude representation the range of number.

$$-(2^{n-1} - 1) \text{ to } + (2^{n-1} - 1)$$

Compliment: For base 'r' → (r – 1)'s Complement; r's Complement

- Binary → 1's, 2's
- Octal → 7's, 8's
- Decimal → 9's, 10's
- Hexadecimal → F's, 16's

1. To determine $(r-1)$'s complement, the given number is subtracted from the maximum possible number in given base.
2. To determine r's complement, first write $(r-1)$'s complement then add 1 to the LSB (Least Significant Bit).

- In 1's complement representation, the positive number are represented similar to positive number in sign magnitude, but for representing negative number, first write positive number and then take 1's complement of that.
- Range of 1's complement number $-(2^{n-1}-1)$ to $+(2^{n-1}-1)$
- Range of 2's complement representation number -2^{n-1} to $+(2^{n-1}-1)$

Binary Arithmetic

- When both number have same sign then we add only magnitudes and use the sign as MSB.

2's complement addition

In 2's complement addition if any carry is present, then discard it.

Note:

☑ In 2's complement there is only one way to represent '0'.
☑ 2's complement of 2's the complement number is equal to number itself.
☑ In 2's complement representation to extend the number of bits MSB bit is is always copied.

II Logic Gates

Introduction

- The Boolean '0' and '1' represents the "logic level".

Logic 0	Logic 1
False	True
OFF	ON
Low	High
No	Yes
Open switch	Closed switch

- A "Truth table" is a means of describing how a logic circuits output depends on the logic levels present at circuits input.

Note:

The number of input combinations will equals to '2^N' for an "N-input" truth table.

- The logic Gates can be classified as
  (a) **Basic Gate:** NOT, AND, OR.
  (b) **Universal Gate:** NAND, NOR.
  (c) **Special Purpose Gates:** EX-OR and EX-NOR. They are used in arithmetic circuit, comparators, code conversion, parity generators and parity checkers etc.

NOT Gate

It is also referred as "inversion" or "complementation".

Symbol and Truth Table

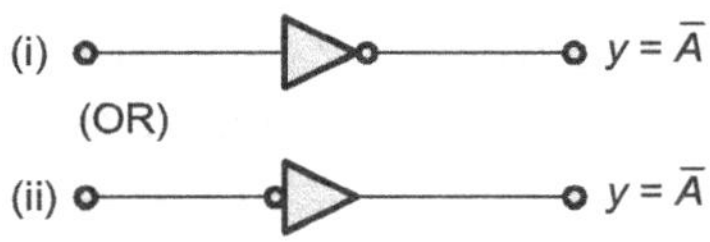

Input	Output
A	$y = \bar{A}$
0	1
1	0

Switching Circuit

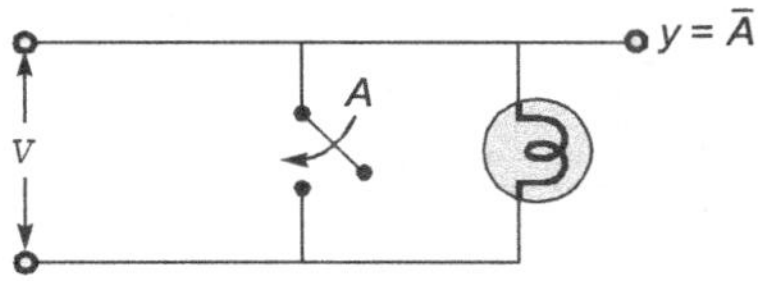

Transistor Circuit

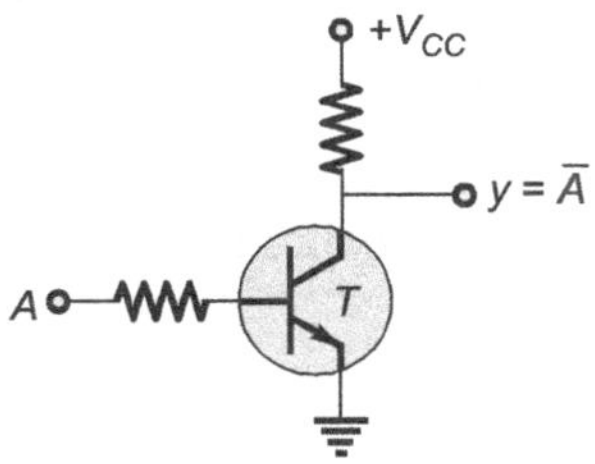

A	T	y
0	OFF	$+V_{CC} \approx 1$
1	0	Ground ≈ 0

- When even number of NOT Gates are connected in series then it acts like Buffer Circuit.

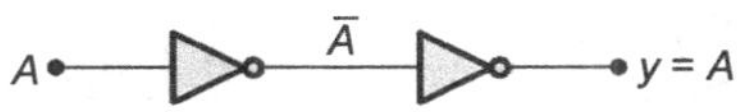

- When even number of NOT Gates are connected with feedback then it acts like a "Bistable multivibrator". It is also a basic memory element.

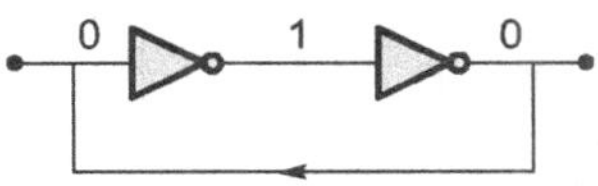

- When odd number of NOT Gates are connected with feedback, then it acts like an astable multivibrator (AMV) or square-wave generator or clock generator or ring oscillator.

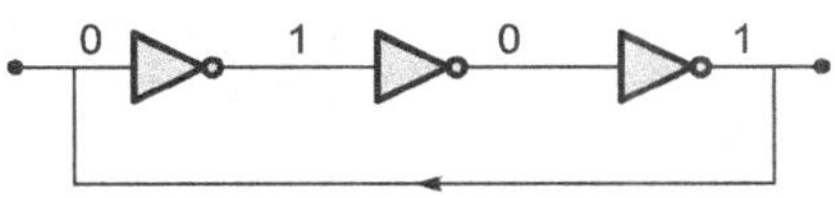

- All inverter take some time to get the response 'Y', this time is called propagation delay time (t_{pd}).

For an Astable Multivibrator (AMV)

Time period of Square Wave Generated by AMV: $T = 2nt_{pd}$; $f = \frac{1}{T}$

where, $n \rightarrow$ Number of inverters (NOT Gates) connected in feedback ; $t_{pd} \rightarrow$ Propagation delay time of each inverter ; $T \rightarrow$ Time period of a square wave generated by AMV or Ring oscillator ; $f \rightarrow$ Frequency of oscillation

AND Gate

Symbol, Truth Table and Switching Circuit

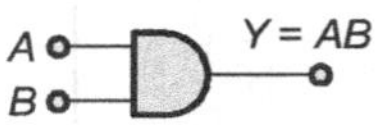

Inputs		Output
A	B	$Y = AB$
0	0	0
0	1	0
1	0	0
1	1	1

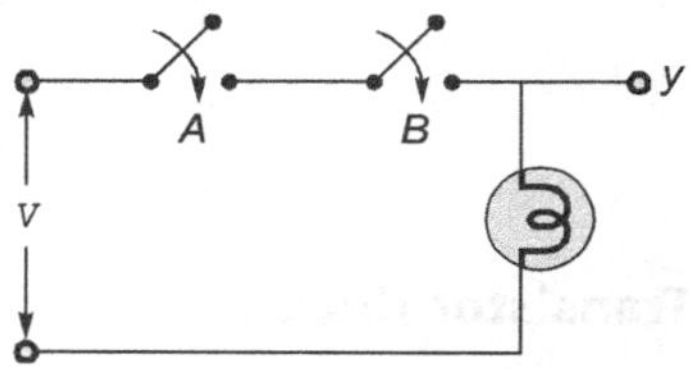

Note:

In AND operation

☑ ENABLE INPUT ⇒ Logic '1' ☑ DISABLE INPUT ⇒ Logic '0'

Transistor circuit

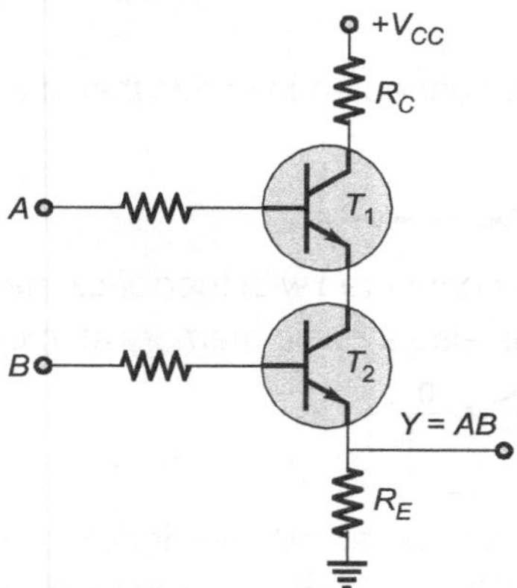

A	B	T_1	T_2	Y
0	0	OFF	OFF	0
0	1	OFF	ON	0
1	0	ON	OFF	0
1	1	ON	ON	1

Diode circuit diagram

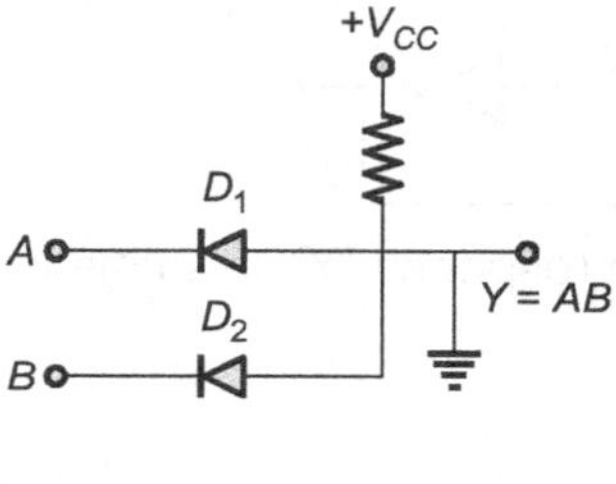

A	B	D_1	D_2	Y
0	0	ON	ON	0
0	1	ON	OFF	0
1	0	OFF	ON	0
1	1	OFF	OFF	1

Remember:

- ☑ In AND gate operation, any unused inputs (Floating inputs) may be connected as: Logic '1' for TTL circuit and Logic '0' for ECL circuit
- ☑ AND gate is also known as detector logic

OR Gate

Symbol, Truth Table and Switching Circuit

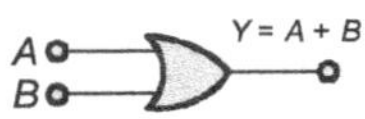

Inputs		Output
A	B	Y = A + B
0	0	0
0	1	1
1	0	1
1	1	1

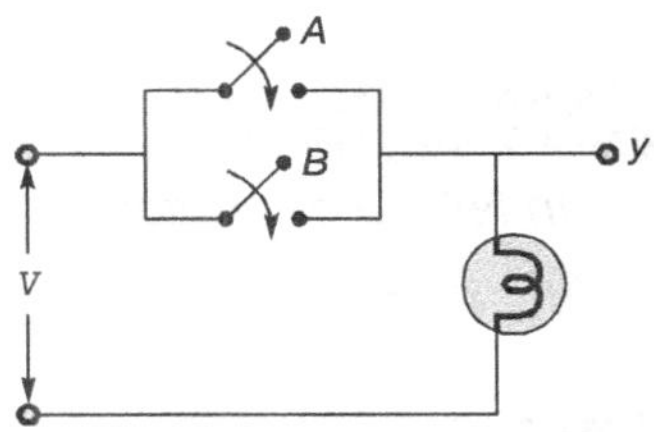

Remember:

In OR operation

- ☑ ENABLE INPUT ⇒ Logic '0'
- ☑ DISABLE INPUT ⇒ Logic '1'

Transistor Circuit

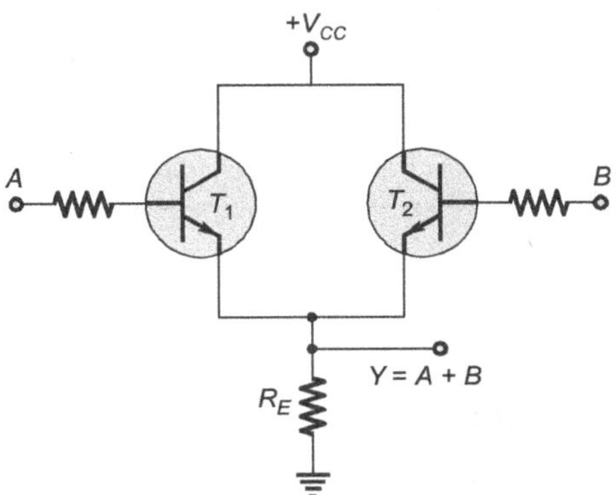

A	B	T_1	T_2	Y
0	0	OFF	OFF	0
0	1	OFF	ON	1
1	0	ON	OFF	1
1	1	ON	ON	1

Diode Circuit Diagram

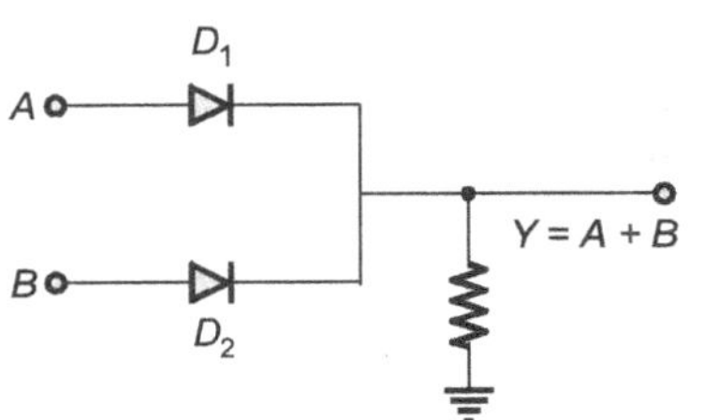

A	B	D_1	D_2	Y
0	0	OFF	OFF	0
0	1	OFF	ON	1
1	0	ON	OFF	1
1	1	ON	ON	1

NAND Gate

Symbol, Truth Table and Switching Circuit

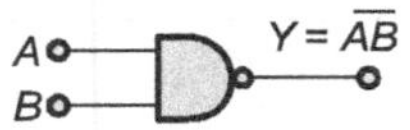

Inputs		Output
A	B	$Y = \overline{AB}$
0	0	1
0	1	1
1	0	1
1	1	0

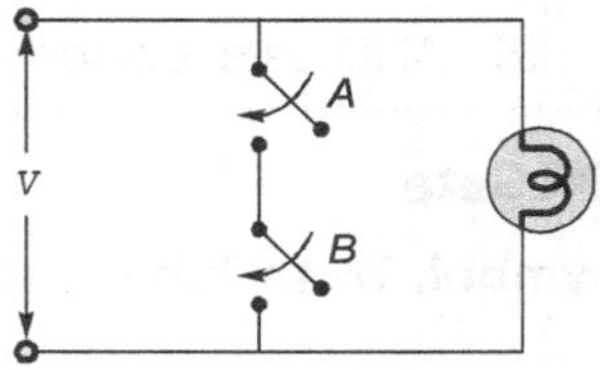

Remember:

In NAND operation

☑ ENABLE INPUT ⇒ Logic '1' ☑ DISABLE INPUT ⇒ Logic '0'

Transistor Circuit

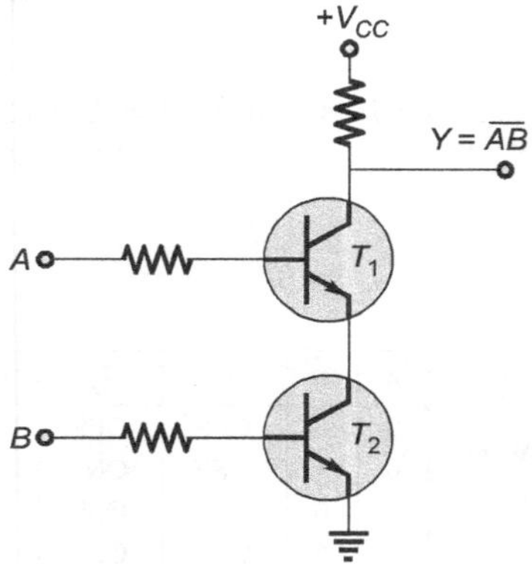

A	B	T_1	T_2	Y
0	0	OFF	OFF	1
0	1	OFF	ON	1
1	0	ON	OFF	1
1	1	ON	ON	0

NOR Gate

Symbol, Truth Table and Switching Circuit

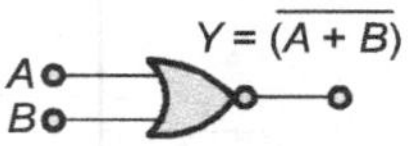

Inputs		Output
A	B	$Y = \overline{(A+B)}$
0	0	1
0	1	0
1	0	0
1	1	0

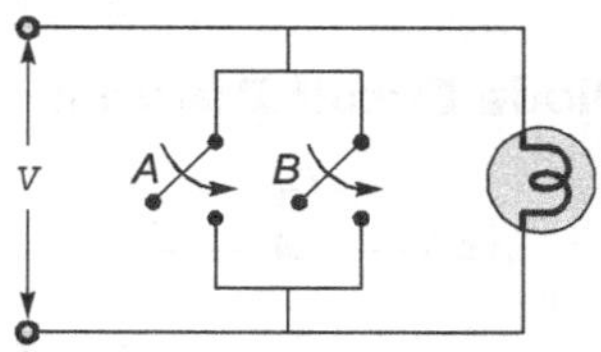

Remember:

In NOR operation

☑ ENABLE INPUT ⇒ Logic '0' ☑ DISABLE INPUT ⇒ Logic '1'

Transistor circuit

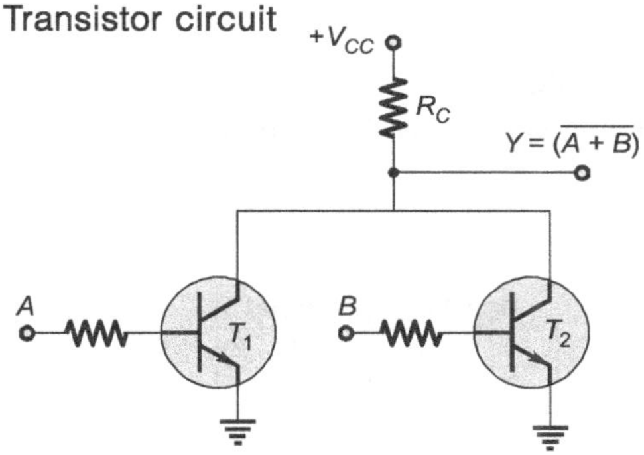

A	B	T_1	T_2	Y
0	0	OFF	OFF	1
0	1	OFF	ON	0
1	0	ON	OFF	0
1	1	ON	ON	0

EX-OR Gate

It is also called "stair case switch".

Symbol and Truth Table

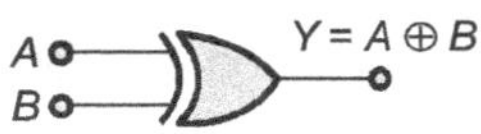

Inputs		Output
A	B	$Y = A \oplus B$
0	0	0
0	1	1
1	0	1
1	1	0

Boolean function of 2-input EX-OR operation , $Y = A \oplus B = \bar{A}B + A\bar{B}$

Remember:

☑ It acts as "odd number of 1's detector in the input".

☑ It is mostly used in "parity generation and detection".

☑ When both the inputs are same, then output becomes LOW or Logic '0'.

☑ When both the inputs are different, then output becomes HIGH or Logic '1'.

☑ In EX-OR operation

(i) For BUFFER CIRCUIT ⇒ Logic '0'

(ii) For INVERSION CIRCUIT ⇒ Logic '1'

Note:

$$A \oplus A = 0, \quad A \oplus 0 = A$$

$$A \oplus \bar{A} = 1, \quad A \oplus 1 = \bar{A}$$

☑ $A \oplus A \oplus A \oplus$..........upto n terms = 0, when n = even

= A, when n = odd

EX-NOR Gate

- It acts as "even number of 1's detector".
- It is also called "Gate of equivalence" or "coincidence logic".

Symbol and Truth Table

Inputs		Output
A	B	$Y = A \odot B$
0	0	1
0	1	0
1	0	0
1	1	1

- **Boolean function of 2-input EX-NOR operation**

$$Y = A \odot B = \overline{A \oplus B} = \overline{(\bar{A}B + A\bar{B})} = AB + \bar{A}\bar{B}$$

Remember:

☑ When both the inputs are same, then output becomes HIGH or Logic '1'.

☑ When both the inputs are different, then output becomes LOW or Logic '0'.

☑ In EX-NOR operation
  (i) For BUFFER CIRCUIT ⇒ Logic '1'
  (ii) For INVERSION CIRCUIT ⇒ Logic '0'

Note:

$$A \odot A = 1, \quad A \odot 1 = A$$
$$A \odot \bar{A} = 0, \quad A \odot 0 = \bar{A}$$

☑ $A \odot A \odot A \odot$ .... upto n terms = 1, when n = even
= A, when n = odd

☑ $\bar{A} \oplus B = A \odot B$ and $A \oplus \bar{B} = A \odot B$
$\bar{A} \odot B = A \oplus B$ and $A \odot \bar{B} = A \oplus B$

☑ $A \oplus B \oplus AB = A + B$.

Alternative Symbols of Gates

Bubbled-OR gate ≡ NAND gate :

Bubbled-NAND gate ≡ OR gate :

Bubbled-NOR gate ≡ AND gate :

Bubbled-AND gate ≡ NOR gate :

NAND and NOR Gate as Universal Gate

NAND and NOR gates are called universal gates because combinations of them can be used to accomplish any of the basic operations and thus can produce inverter, OR gate or an AND gate.

Logic gates	No. of NAND gate required	No. of NOR gate required
NOT	1	1
AND	2	3
OR	3	2
EX-OR	4	5
EX-NOR	5	4

III Boolean Algebra and Reduction Techniques

Boolean Algebraic Laws

Commutative Law : $A + B = B + A$ **and** $A \cdot B = B \cdot A$

Associative Law : $A + (B + C) = (A + B) + C = A + B + C$

and $A \cdot (B \cdot C) = (A \cdot B)\, C = A\,B\,C$

Distributive Law : $A(B + C) = AB + AC$

$(A + B)(C + D) = AC + AD + BC + BD$

Boolean Algebraic Theorems

AND-Operation Theorem: $A \cdot A = A$; $A \cdot 0 = 0$

$A \cdot 1 = A$; $A \cdot \bar{A} = 0$

Involution Theorem: $(A')' = \bar{\bar{A}} = A$

OR-Operation Theorem: $A + A = A$; $A + 0 = A$

$A + 1 = 1$; $A + \bar{A} = 1$

De Morgan's Theorem:

$$\overline{(A_1 \cdot A_2 \cdot A_3 \cdot ....A_n)} = \bar{A}_1 + \bar{A}_2 + \bar{A}_3 + ... + \bar{A}_n$$

$$\overline{(A_1 + A_2 + A_3 + \cdots + A_n)} = \bar{A}_1 \cdot \bar{A}_2 \cdot \bar{A}_3 \cdots \bar{A}_n$$

Transposition Theorem: $(A + B)(A + C) = A + BC$

Distribution Theorem: $A + BC = (A + B)(A + C)$

Consensus Theorem

- Used to eliminate redundant term.
- It is applicable only when,
  1. A Boolean function contains 3 variables.
  2. Each variable is used 2-times
  3. Only one variable is in complemented and uncomplemented form. Then the related terms to that complemented or uncomplemented variable is the answer.

  Ex: $AB + \bar{A}C + BC = AB + \bar{A}C$

Boolean Algebraic Theorems

Theorem No.	Theorem
1.	$(A+B)\cdot(A+\bar{B}) = A$
2.	$AB + \bar{A}C = (A+C)(\bar{A}+B)$
3.	$(A+B)(\bar{A}+C) = AC + \bar{A}B$
4.	$AB + \bar{A}C + BC = AB + \bar{A}C$
5.	$(A+B)(\bar{A}+C)(B+C) = (A+B)(\bar{A}+C)$
6.	$\overline{A \cdot B \cdot C \cdot ...} = \bar{A} + \bar{B} + \bar{C} + ...$
7.	$\overline{A + B + C + ...} = \bar{A} \cdot \bar{B} \cdot \bar{C} ...$

Duality Theorem

"Dual expression" is equivalent to writing a negative logic of the given boolean relation. The procedure is,

1. Change each OR sign by an AND sign and vice-versa.
2. Complement any '0' or '1' appearing in expression.
3. Keep literals as they are

Note:

- ☑ If a dual of an expression results the expression itself then it is called "self dual expression".
- ☑ For N-variables, maximum possible Self-Dual Function = $(2)^{2^{(n-1)}}$.
- ☑ With N-variables, maximum possible distinct logic functions = 2^{2^n}.

Complementary Theorem

For obtaining complement expression we

1. Change each OR sign by AND sign and vice-versa.
2. Complement any '0' or '1' appearing in expression.
3. Complement the individual literals.

Boolean Function Representation

Canonical Form

All the terms contains each literal.

Ex.: $F(A,B,C) = \overline{A}BC + ABC + \overline{A}B\overline{C}$

Minimal Form

All the term do not have each literal.

Ex.: $F(A,B,C) = \overline{A} + BC + A\overline{B}C$

Standard Form

All the term do not have each literal.

Ex.: $F(A, B, C) = \Sigma m(1, 2, 3, 6)$

Note:

A binary-variable is called "LITERALS".

Minterms and Maxterms

- n-binary variables have 2^n possible combinations and each of these possible combination is called "Minterm or Standard Product form".
- "Maxterm" is the complement of corresponding "Minterm" i.e. $M = \overline{m}$.

Reduction Techniques

SOP (Sum of Product)

In SOP form each product term is known as min term. SOP form is used when output is logic 1.

e.g. $\underbrace{AB\overline{C}}_{\text{Minterm}} + A\overline{B}C + ABC$

Ex.: $F(A, B, C) = \Sigma m(5, 6, 7)$

POS (Product of SUM)

In POS form each product term is called as max term. POS form is used when output is 0 logic '0'.

e.g. $\underbrace{(A+B+C)}_{\text{Max term}}+(A+\bar{B}+C)\cdot(\bar{A}+B+C)$

Ex.: $F(A, B, C)=\pi M(0, 4, 5, 7)$

Note:

- ☑ AND-OR Logic ≡ NAND-NAND Logic and is used in SOP.
- ☑ OR-AND Logic ≡ NOR-NOR Logic and is used in POS.

Venn Diagram

It is a diagram that shows all possible relations between a finite collection of different sets.

Here, the AND operation is considered as an intersection and the OR operation is considered as a union.

Ex. :

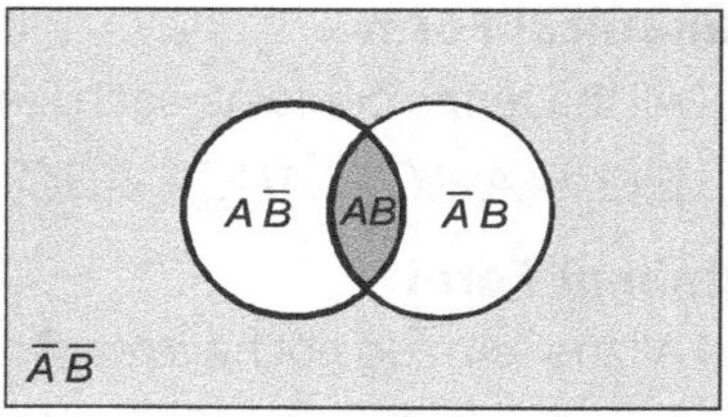

K-Map

Implicant

Each individual min-term in canonical SOP form is called implicant.

Prime Implicant

Prime implicant is an implicant, which is obtained by combining maximum possible adjacent cell in K-map.

Essential Prime Implicant

An essential prime implicant is a prime implicant in which one or more min term are unique i.e. which is obtain by combining only one way in K-map.

Ex:

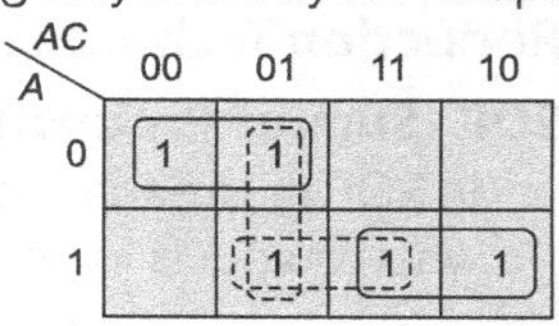

Here,

Implicants: $\bar{A}\bar{B}\bar{C}, \bar{A}\bar{B}C, A\bar{B}C, ABC, AB\bar{C}$.

Prime implicants: $\bar{A}\bar{B}, AB, AC, \bar{B}C$.

Essential prime implicants: $\bar{A}\bar{B}, AB$.

Note:

The *QM* method gives the prime implicants but not the essential prime implicants.

IV Digital Logic Circuits

Introduction

Digital logic circuits are classified as:

(a) Combinational circuit

(b) Sequential circuit

Combinational Circuit	Sequential Circuit
1. Present output depends on present input only.	1. Present output depends on present input as well as previous output
2. No feedback is present.	2. Feedback is present.
3. No memory is present.	3. Due to feedback, memory is present
4. Example: Comparator, half adder, full adder, half subtractor, full-subtractor, multiplexer, demultiplexer,encoder, decoder.	4. Example: Flip-flop, counter, register.

Combinational Circuits

Arithmetic Circuit

Half Adder (H.A)

A logic circuit for the addition of two one-bit numbers is referred to as an "HALF ADDER (H.A)".

Logic circuit and Truth table:

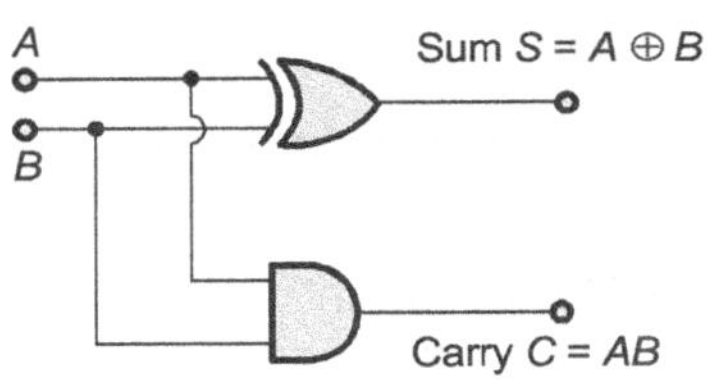

Inputs		Outputs	
A	B	Sum(S)	Carry(C)
0	0	0	0
0	1	1	0
1	0	1	0
1	1	0	1

Logical expression:

Sum $S = \bar{A}B + A\bar{B} = A \oplus B$; Carry $C = AB$

Remember:

- ☑ Total number of NAND-gates required to implement half adder = 5
- ☑ Total number of NOR-gates required to implement half adder = 5
- ☑ To implement the half adder circuit by minimum number of logic gates (if we have all gates except EX-OR and EX-NOR) is "3".
- ☑ Total number of 2 : 1 MUX required to implement half adder = 3.

Full Adder (F.A.)

It performs the arithmetic sum of the three input bits i.e. addend bit, augend bit and carry bit.

Logical expression:

Sum, $S = A \oplus B \oplus C$

Carry, $C = AB + BC + CA = AB + C(A \oplus B)$

Remember:

- ☑ A F.A. can be implemented by two H.A. and one OR-gate
- ☑ Total number of NAND-gate/NOR-gate required to implement a F.A is equals to "9".
- ☑ Total number of 2 : 1 MUX required to implement full adder = 7.

Half Subtractor (H.S)

Logic Circuit

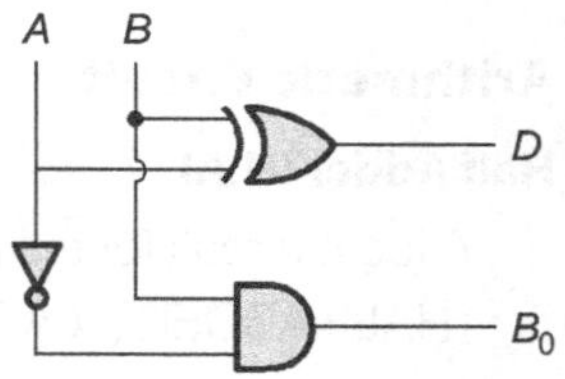

Logical expression:

When B is subtracted from A.

Difference, $D = \bar{A}B + A\bar{B} = A \oplus B$

Borrow, $B_0 = \bar{A}B$

Remember:

- ☑ Total number of NAND/NOR gates required to implement the H.S is equal to "5".
- ☑ Total number of 2 : 1 MUX required to implement half subtractor = 2.

Full Subtractor (F.S)

It is a circuit which performs subtraction between two bits taking into account that a '1' may have been borrowed by a lower significant stage.

Logical expression:

Difference, $D = A \oplus B \oplus C$

Borrow, $B_o = \bar{A}B + \bar{A}C + BC = \bar{A}B + \left(\overline{A \oplus B}\right) \cdot C$

Remember:

- ☑ A full subtractor can be implemented with two half subtractor and one OR Gate.
- ☑ Number of NAND/NOR gates required to implement the full subtractor is equal to '9'
- ☑ In parallel adder n full adder or {$(n-1)$ F.A. and 1 H.A.} or {$(2n-1)$ H.A. and (n – 1) OR Gate} are required to add two n-bit numbers.
- ☑ Total number of 2 : 1 MUX required to implement full subtractor = 7.

Look Ahead Carry Adder

- Look ahead carry adder makes the addition faster.
- It is based on the two additional signal that must be computed.

  Generator signal, $G_i = A_i B_i$

  Propagation signal, $P_i = A_i \oplus B_i$
- The carry signal generated, $C_{i+1} = P_i C_i + G_i$
- Number of AND gates required in carry circuit for n-bit adder is

$$N = \frac{n(n+1)}{2}$$

- Number of OR gates required in carry circuit of n-bit adder is $N = n$.

Non Arithmetic Circuits

Multiplexer : Multiplexer is a combinational circuit which have many data input and single output depending on select or control input, one of the input line is transferred to the output, hence it is known as "many to one circuit" or "universal logic circuit" or "data selector circuit".

- The selection of a particular input line is controlled by a set of select lines.
- There are 2^n input lines where 'n' is the number select line.

4 : 1 MUX

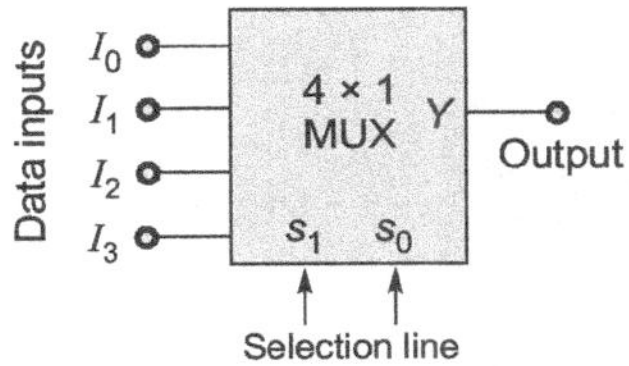

S_1	S_0	Y
0	0	I_0
0	1	I_1
1	0	I_2
1	1	I_3

$$Y = \text{output} = \bar{S}_1 \bar{S}_0 I_0 + \bar{S}_1 S_0 I_1 + S_1 \bar{S}_0 I_2 + S_1 S_0 I_3$$

Note:

☑ The size of the MUX is specified by the 2^n of its input line and the single output line.

☑ MUX contains AND gate followed by an OR-gate.

2 : 1 MUX

Higher order MUX using Lower order MUX:

Given MUX	To be Implemented MUX	Required number of MUX
2 : 1	4 : 1	3
4 : 1	16 : 1	4 + 1 = 5
4 : 1	64 : 1	16 + 4 + 1 = 21
8 : 1	64 : 1	8 + 1 = 9
8 : 1	256 : 1	32 + 4 + 1 = 37

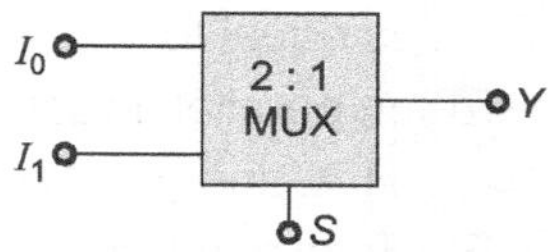

$$Y = \bar{S}I_0 + SI_1$$

Remember:

- To implement 2^n : 1 MUX by using 2 : 1 MUX, the total number of 2 : 1 MUX required is $(2^n - 1)$.
- MUX is a Universal Logic gate.
- Using one 2^n : 1 MUX
  → All 'n' variable logic functions can be implemented.
  → Some 'n + 1' variable logic function can be implemented.

  Using one 2^n : 1 MUX and 1 NOT gate.
  → All 'n' and 'n + 1' variable logic functions can be implemented.

MUX as Universal Logic Circuit

- AND gate using MUX

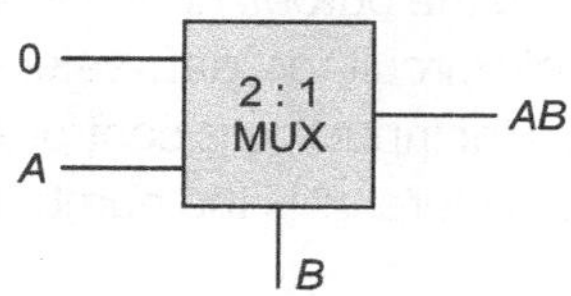

- OR gate using MUX

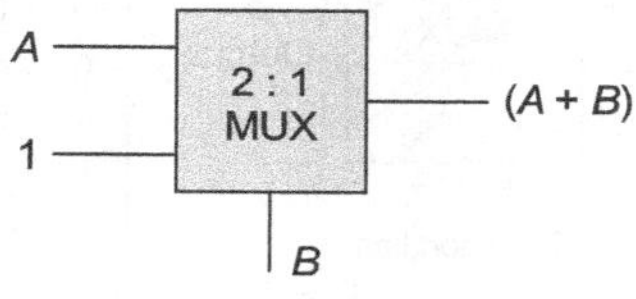

- NOT gate using MUX

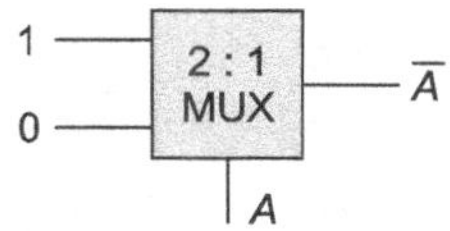

- NAND gate using MUX

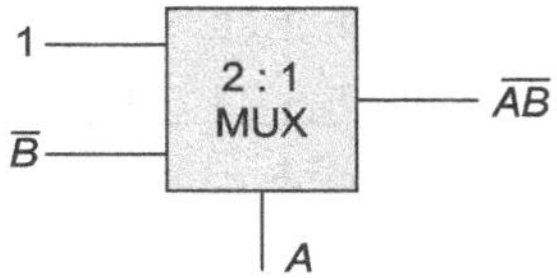

Requires 2, 2 : 1 MUX

- NOR gate using MUX

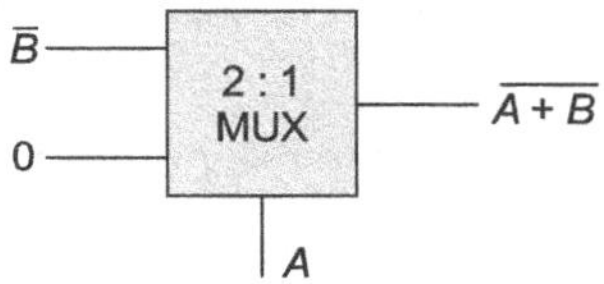

Requires 2, 2 : 1 MUX

- EXOR gate using MUX

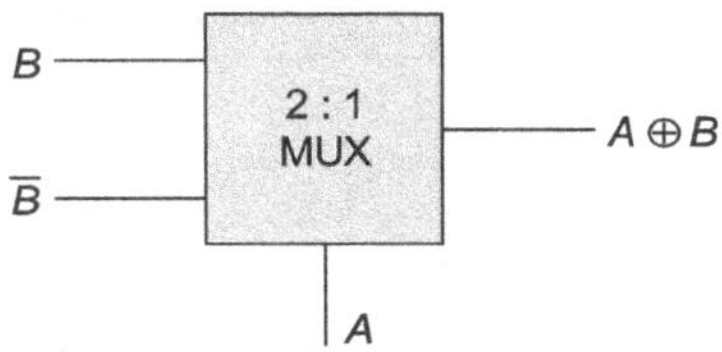

It requires 2, 2 : 1 MUX

- EXNOR gate using MUX

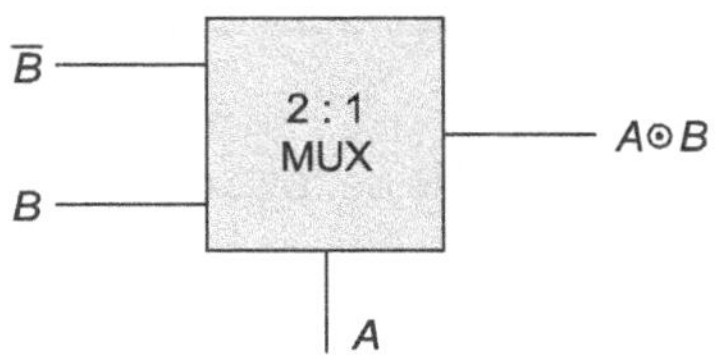

Requires 2, 2 : 1 MUX

Demultiplexer (DEMUX)

- It receives information on a single line and transmits it on one of 2^n possible output lines.
- A "Decoder" with an enable input can function as a "Demultiplexer".

1 : 2 Demux

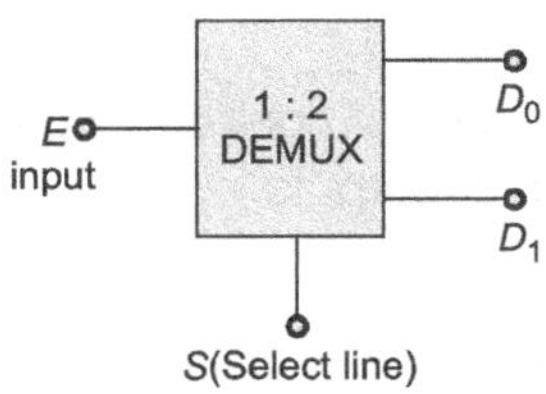

$D_0 = \overline{S}\,E$ and $D_1 = SE$

1 : 4 DEMUX

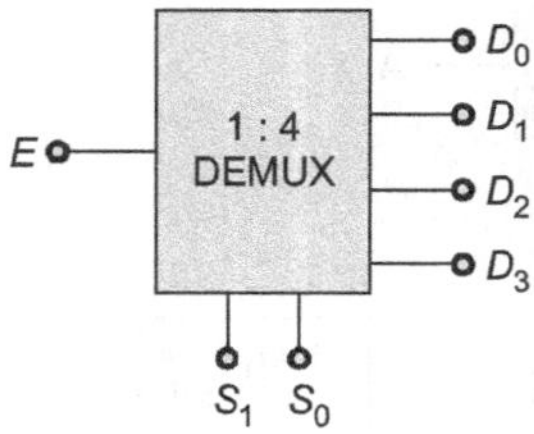

Select lines		Outputs			
S_1	S_0	D_3	D_2	D_1	D_0
0	0	0	0	0	E
0	1	0	0	E	0
1	0	0	E	0	0
1	1	E	0	0	0

$$D_0 = \bar{S}_1\bar{S}_0 E, D_1 = \bar{S}_1 S_0 E, D_2 = S_1\bar{S}_0 E, D_3 = S_1 S_0 E$$

Remember:

☑ Other name of demultiplexer are "one to many circuit" or "data distributor circuit".

☑ Number of select line is $\log_2 n$ (where n is number of output line).

Higher Order DEMUX Using Lower Order DEMUX:

Given DEMUX	To be Implemented MUX	Required number of MUX
1 : 2	1 : 4	1 + 2 = 3
1 : 2	1 : 8	1 + 2 + 4 = 7
1 : 2	1 : 16	1 + 2 + 4 + 8 = 15
1 : 2	1 : 64	1 + 2 + 4 + 8 + 16 + 32 = 63
1 : 4	1 : 16	1 + 4 = 5

Decoder

- Decoder is a combinational circuit that is used to convert binary to other codes such as
  - (a) Binary to octal
  - (b) Binary to decimal
  - (c) Binary to hexadecimal
- In decoder, depending on binary data applied, one of the output line will become logic 1.
- A "Decoder" has many inputs and many outputs line.
- It is a combinational circuit that converts binary information from n input lines to a maximum 2^n unique output lines.
- If the n-bit decoded information has unused or don't care combinations, the decoder output will have less than 2^n outputs.
- **Total number of output lines**

$$m \le 2^n \quad \text{where,} \quad n \to \text{Total number of input lines}$$

2 × 4 Decoder

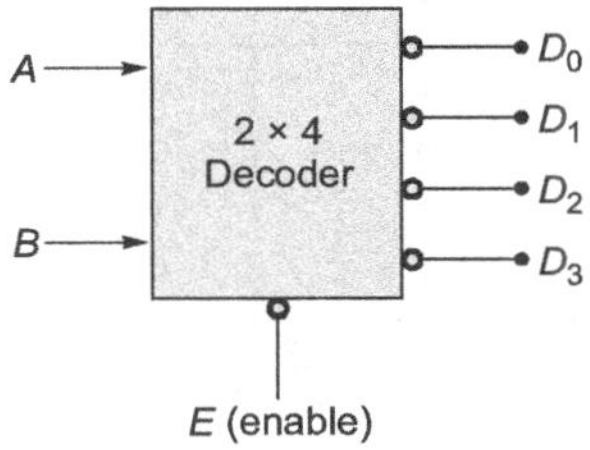

E	A	B	D_0	D_1	D_2	D_3
1	X	X	1	1	1	1
0	0	0	0	1	1	1
0	0	1	1	0	1	1
0	1	0	1	1	0	1
0	1	1	1	1	1	0

Note:

- ☑ 2 × 4 - Decoder may act like 1 : 4 DE MUX and vice-versa.
- ☑ Decoder and Demux circuits are almost same.
- ☑ Decoder contains AND-gates or NAND-gates.

Higher Order Decoder Using Lower Order Decoder:

Given Decoder	To be Implemented Decoder	Required number of Decoder
2 : 4	4 : 16	1 + 4 = 5
2 : 4	3 : 8	2 + a NOT Gate
4 : 16	8 : 256	1 + 16 = 17

Encoder

- Encoder is used to convert other codes to binary such as
  - (a) Octal to binary
  - (b) Hexadecimal to binary
  - (c) Decimal to binary
  - (d) Decimal to *BCD*
- In encoder one of the input line is high and corresponding binary is available at the output.
- In priority encoder any number of inputs can be high but based on priority highest priority input corresponding binary is available at the output.

Magnitude Comparator

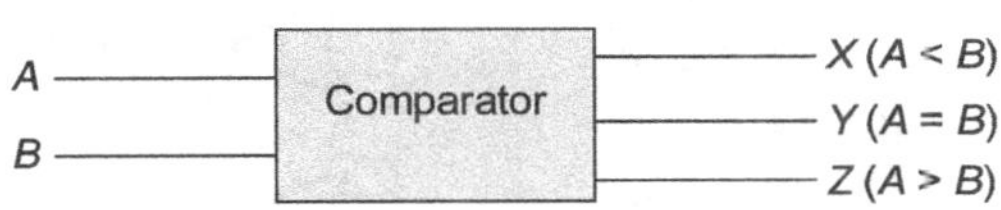

Truth table:

A	B	X	Y	Z
0	0	0	1	0
0	1	1	0	0
1	0	0	0	1
1	1	0	1	0
Output expression		$\bar{A}B$	$\bar{A}\bar{B} + AB = A \odot B$	$A\bar{B}$

V Sequential Circuits

Flip-Flops

- Flip-flop is a basic memory element.
- It can store "one bit of information".
- A flip-flop has 2 outputs, which are always complementary to each other.
- A flip-flop has 2 stable state hence it is known as bi-stable multivibrator.
- A simplest form of flip-flop is called "latch". It can be constructed with two cross coupled NAND gates or NOR gates.

Triggering

Triggering is used to initiate the operation of latches or flip-flops.

(i) **Level trigger:** Input signal affects the flip-flop only when the clock is at logic '1' or logic '0'.

(ii) **Edge trigger:** Input signal affects the flip-flop only if they are present at the positive or negative going edge of the clock pulse.

Setup Time

It is minimum time required to keep input at proper level before applying clock.

Hold Time

Hold time is minimum time required to keep input at same level after applying clock.

Note:

If set-up time and hold time is not properly given, the output may go to meta stable state.

Clocked S-R Flip-Flop

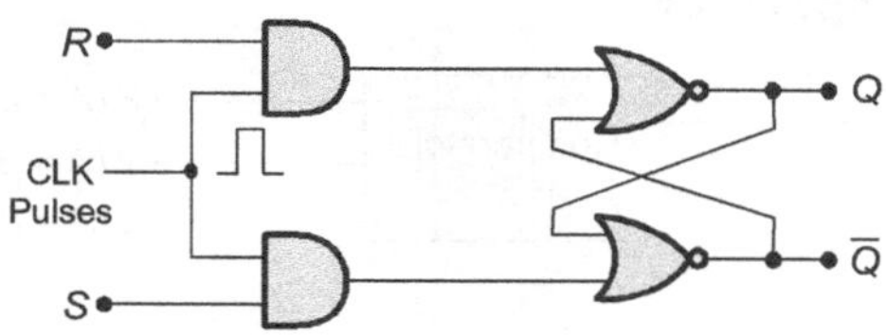

Truth table

Clock	S_n	R_n	Q_{n+1}	
0	X	X	Q_n	
1	0	0	Q_n	→ HOLD state
1	0	1	0	→ RESET state
1	1	0	1	→ SET state
1	1	1	Invalid	→ FORBIDDEN state

- S_n and R_n denotes the inputs and Q_n the output during the bit time 'n'.
- 'Q_{n+1}' denotes the output Q after CLK passes, i.e. in bit time $(n + 1)$.

Characteristic table

S	R	Q_n	Q_{n+1}	
0	0	0	0	→ Q_n
0	0	1	1	
0	1	0	0	→ 0
0	1	1	0	
1	0	0	1	→ 1
1	0	1	1	
1	1	0	X	→ Invalid
1	1	1	X	

Characteristic Equation

$$Q_{n+1} = S + \bar{R}Q_n$$

Remember:

Disadvantage of SR flip-flop is invalid state will be present when both inputs are 1 or high. To avoid this J-K flip-flop is used.

J-K Flip-Flop

JK-FF is called universal FF because the FFs like *D*-FFs, *SR*-FF and *T*-FF can be derived from it.

$$S = J\bar{Q} \text{ and } R = KQ$$

Logic diagram

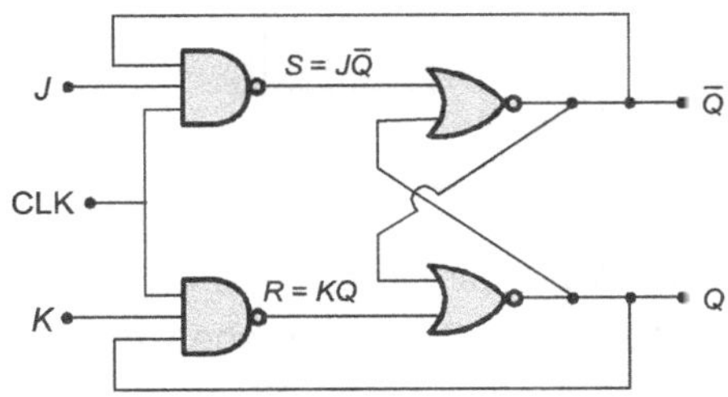

Truth table

Clock	J	K	Q_{n+1}	
0	X	X	Q_n	
1	0	0	Q_n	→ HOLD state
1	0	1	0	→ RESET state
1	1	0	1	→ SET state
1	1	1	$\bar{Q}_n$	→ TOGGLE state

Characteristic table

J	K	Q_n	Q_{n+1}	
0	0	0	0	→ Q_n
0	0	1	1	
0	1	0	0	→ 0
0	1	1	0	
1	0	0	1	→ 1
1	0	1	1	
1	1	0 →	1	→ $\bar{Q}_n$
1	1	1 →	0	

Characteristic equation

$$Q_{n+1} = J\bar{Q}_n + \bar{K}Q_n$$

Race Around Condition

- The Race around condition will occur when
  (i) $J = K = 1$
  (ii) $t_{pd(FF)} < t_{pw}$
  (iii) Flip-flop is level triggered.
- For the duration "t_{pw}" of the clock pulse, the output Q will oscillate back and forth between 0 and 1. At the end of the clock pulse, the value "Q" is uncertain. This situation is referred as "race around condition".
- To avoid this, the condition is $t_{pw} < t_{pd(FF)} < T$

Remember:

Race around condition does not occur in edge triggered flip-flop.

Master-Slave J-K Flip-Flop

To avoid race around condition master slave FF is used.

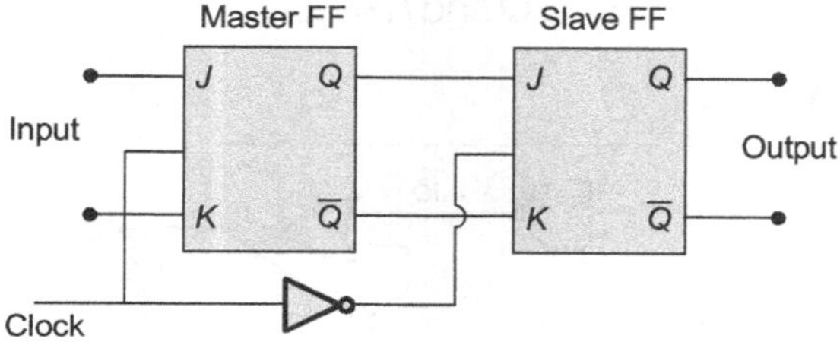

- In master slave flip-flop master is applied with input clock and slave is applied with inverted clock. Due to this when master output is changing slave output remains in previous state.

- In master slave flip-flop, output will change only when slave output changes.
- Output of master can change many times but, slave output can change only 1 time so master FF act as level triggered and slave FF act as edge triggered. Therefore there is no "race around condition" at the output of the slave.

D-Flip-Flop

It is a FF with a delay equal to exactly one cycle of CLK.

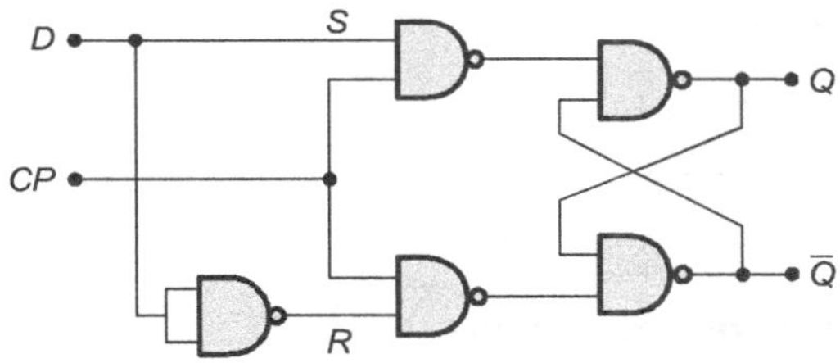

Graphical diagram

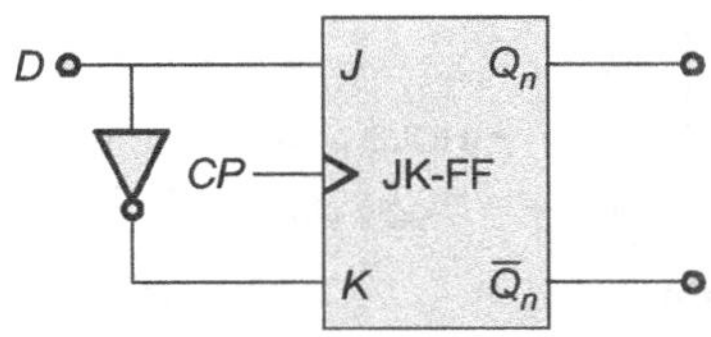

$$J = D \text{ and } K = \bar{D}$$

Truth table and characteristic table

CLK	D	Q_{n+1}
0	X	Q_n
1	0	0
1	1	1

Truth Table

D	Q_n	Q_{n+1}
0	0	0
0	1	0
1	0	1
1	1	1

Characteristic Table

Characteristic equation

$$Q_{n+1} = D$$

Note:

It is also known as "Transparent Latch" because $Q_{n+1} = D$.

Toggle Flip-Flop (T-FF)

- The T-FF is a single input version of the J-K FF.
- T-FF can be obtained from a J-K FF if J and K are tied together as in figure shown below.
- The designation 'T', comes from the ability of the FF to "toggle" or "change state".

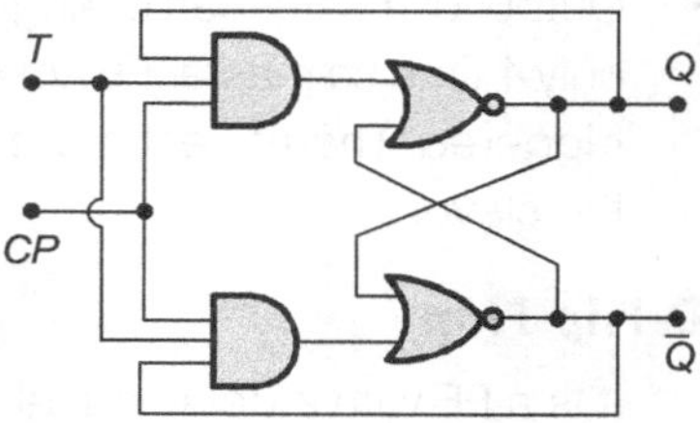

Graphical symbol of T-FF

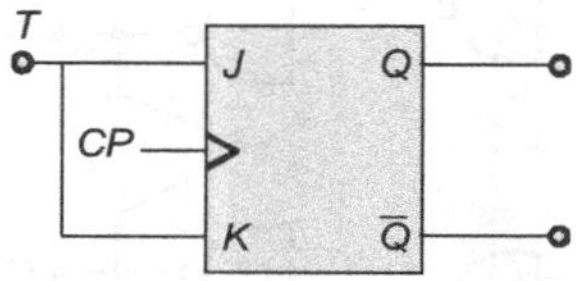

$$J = K = T$$

Truth table and characteristic table

CLK	T	Q_{n+1}
0	X	Q_n
1	0	Q_n → HOLD state
1	1	$\bar{Q}_n$ → Toggle state

Truth Table

T	Q_n	Q_{n+1}
0	0	0
0	1	1
1	0	1
1	1	0

Characteristic Table

Characteristic equation

$$Q_{n+1} = \bar{T}\,Q_n + T\,\bar{Q}_n = T \oplus Q_n$$

Conversion of Flip-Flops

SR FF to other FFs	JK FF to other FFs	D-FF to other FFs	T-FF to other FFS
$S = J\bar{Q}_n$ & $R = KQ_n$	$J = S$ & $K = R$	$D = S + \bar{R}Q_n$	$T = S\bar{Q}_n + RQ_n$
$S = D$ & $R = \bar{D}$	$J = D$ & $K = \bar{D}$	$D = J\bar{Q}_n + \bar{K}Q_n$	$T = J\bar{Q}_n + KQ_n$
$S = T\bar{Q}_n$ & $R = TQ_n$	$J = K = T$	$D = T \oplus Q_n$	$T = D \oplus Q_n$

Excitation Table of Flip-Flops

Q_n	Q_{n+1}	S	R	J	K	D	T
0	0	0	X	0	X	0	0
0	1	1	0	1	X	1	1
1	0	0	1	X	1	0	1
1	1	X	0	X	0	1	0

VI Registers

Shift Registers (SRs)

- Registers are used to store group of bits.
- In register to store *N*-bit, it requires *N* flip-flop.
- The "serial input" determines what goes into the left most FF during the shift.
- Depending upon input and output, registers can be classified into 4 types.:
  - (a) SISO: Serial In Serial Out
  - (b) SIPO: Serial In Parallel Out
  - (c) PISO: Parallel In Serial Out
  - (d) PIPO: Parallel In Parallel Out

SISO (Serial In Serial Out)

4-bit Right-Shift SISO Register

- In right shift SISO register, LSB data is applied at the MSB FF'(D-FF).
- In '*n*' bit register, to enter '*n*' bit data, it requires 'n' clock pulses in serial form.
- If '*n*' bit data is stored in SISO register then output s taken serially for this $(n-1)$ clock pulses are required.
- SISO register is used to provide '*n*' clock pulse delay to the input data.
- If '*T*' is the time period of clock pulse, then delay provided by SISO is nT.

4-bit Left-Shift SISO Register

- In this above SISO register MSB data is applied to the LSB FF(D-FF).
- To enter the '*n*' bit data in serial form we require '*n*' clock pulse.
- To exit or getting output of 'n' bit data as serially we require $(n-1)$ clock pulse.

SIPO (Serial In Parallel Out)

- For '*n*' bit- serial input data to be stored the number of CLK pulse required = n.
- For '*n*' bit-parallel output data to be stored the number of CLK pulse required = 0 (there is no need of CLK pulse).

PISO (Parallel In Serial Out)

- To store parallel in data, if we store '*n*' bit then the number of CLK pulse required = 1 CLK pulse.
- To store serial out data if we store '*n*' bit then the number of CLK pulse required = $(n-1)$.

Note:

To convert temporal code into spatial code, we use SIPO register, whereas to convert spatial code into temporal code we use PISO register.

PIPO (Parallel In Parallel Out)

- For parallel in data, the number of CLK pulse required = 1 CLK pulse.
- For parallel out data, the number of CLK pulse required = 0 CLK pulse.

Remember:

☑ All shift registers are made of *JK* flip-flops.

☑ "PIPO" register is a storage register made up with D-FFs.

☑ Clock pulses required by various shift registers.

	Input	Output
SISO	n	$n-1$
SIPO	n	0
PISO	1	$n-1$
PIPO	1	0

☑ If 'n' shift left operations are performed then data will be multiplied 2^n times.

☑ If 'n' shift right operation are performed then data will be divided 2^n times.

Time delay

A SISO SRs may be used to introduce time delay "Δt" in digital signals

$$\Delta t = N \times T = N \times \frac{1}{f_c}$$

where, $N \rightarrow$ Number of FFs ; $T \rightarrow$ Time period of CLK pulse ;
$f_C \rightarrow$ CLK frequency

- The amount of delay can be controlled by the "f_c" or number of FFs in the SR.

VII Counters

Introduction

- It is a sequential circuit formed by the cascading of FFs.
- Counter are basically used for
  (i) Counting of the number of clock pulses applied

(ii) Frequency division
(iii) Timers
(iv) Frequency measurement
(v) Waveform generation

- Counters are classified as:
  (i) Asynchronous counter
  (ii) Synchronous counter

Remember:

☑ If N = total number of states and 'n' = number of FFs then

$$N \leq 2^n$$

$$n \geq 3.32 \log_{10} N \quad \text{or} \quad n \geq \log_2 N$$

☑ If $N = 2^n$, then we get BINARY COUNTER.

☑ If $N < 2^n$, then we get NON-BINARY COUNTER.

MOD number

- The "MOD-number" indicates the number of states in counting sequence.
- For n-FFs, counter will have 2^n different states and then this Counter is said to be "MOD-2^n Counter".
- MOD number indicates the frequency division factor obtained from the Last FF.
- It would be capable of counting upto $(2^n - 1)$ before returning to zero state.

Note:

☑ In "MOD-N Counter", if applied input frequency is "f", then output frequency is f/N.

☑ If two counters are cascaded with MOD-M followed by MOD-N, then number of overall states of combined counter is $(M \times N)$ and counter is called "MOD-MN" counter.

Asynchronous (Series) Counter

Binary Ripple Counter

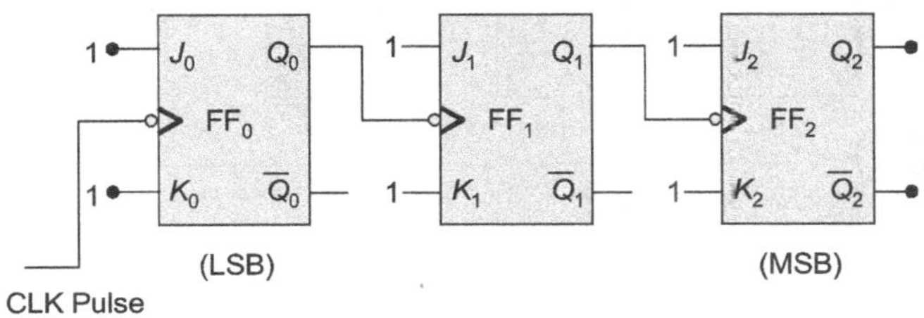

- In ripple counter with n-FFs there are 2^n possible states.
- With n-FFs the maximum count that can be counted by this counter is $(2^n - 1)$.
- If input frequency is 'f' then output frequency is $\frac{f}{2^n}$.
- It is also known as $2^n : 1$ scalar divider.

For proper operation of the ripple counter:

$$T_{CLK} \geq nt_{pd(FF)} \quad ; \quad f_{CLK} \leq \frac{1}{nt_{pd(FF)}}$$

Maximum CLK frequency:

$$f_{CLK,m} = \frac{1}{nt_{pd(FF)}}$$

Disadvantage of Ripple Counter:

Decoding error is present due to propagation delay of FFs i.e. $t_{pd(FF)}$.

Up/Down Counter

An "Up/Down Counter" can count in any direction depending upon the control input.

For determination of Up/Down Counter:

Triggering with	CLK connection in	Access as
(–ve) edge	Q	UP Counter
(–ve) edge	$\overline{Q}$	Down Counter
(+ve) edge	Q	Down Counter
(+ve) edge	$\overline{Q}$	UP Counter

Non-Binary Ripple Counter

Decade Counter or Mod-10 Counter

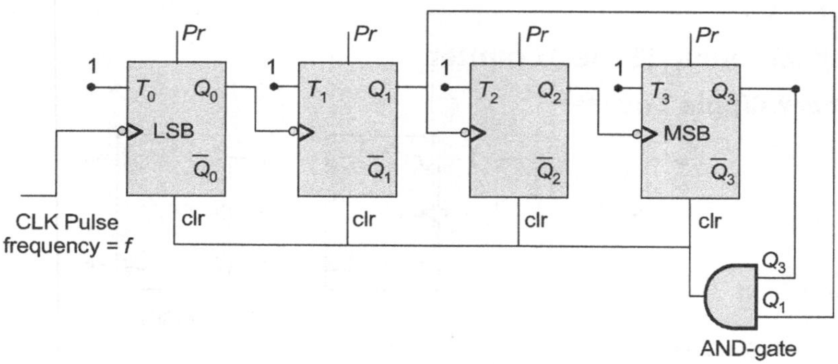

- Used states = 10 and unused states = 6
- Output frequency of MOD-10 counter = $f/10$.
- Clear and preset is used in non binary ripple counter.
- Clear is used to reset counter without clock.
- Preset is used to set counter.
- When decade counter counts from 0 to 9 then it is known as BCD counter.

Synchronous (Parallel) Counters

Ring Counter

It is SISO shift register.

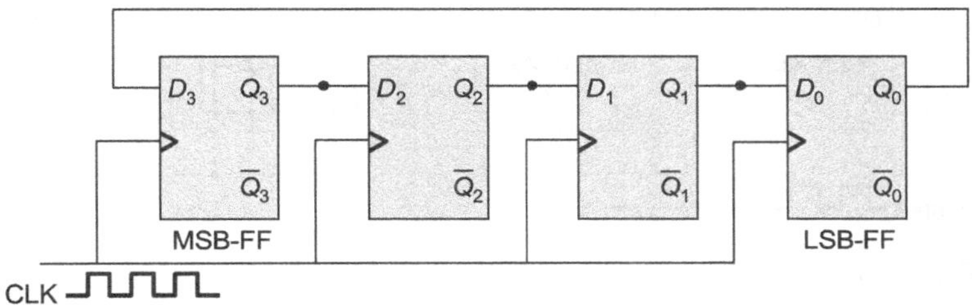

Remember:

- ☑ Ring counter is a non-self starting counter.
- ☑ With 'n' FFs, there are n-states present in ring counter.
- ☑ With 'n' FFs, maximum count possible in ring counter is (2^{n-1}).
- ☑ Decoding is very easy in Ring counter, because there is no aid of extra circuit.

In a 4-bit ring counter:

Used states = 4 ; Unused states = $2^4 - 4 = 12$

- In a ring counter if CLK frequency is "f" the FFs output frequency is "f/N" (where N = Number of states = modulus of the counter).

Twisted-Ring Counter

Also known as Johnson Counter or Switch Tail Ring Counter

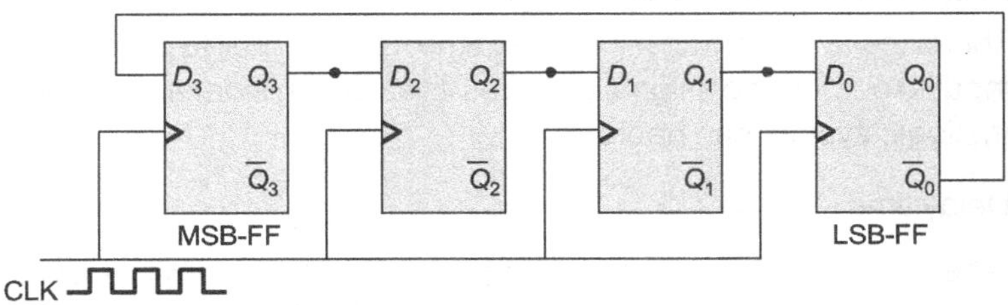

Remember:

☑ With 'n' FFs there are '2*n*' states in this counter.

☑ With 'n' FFs the maximum count by this counter is $(2^n - 1)$.

☑ In normal "Johnson Counter" with 'n' FFs and the input frequency is '*f*' then output frequency of FFs is "f/2*n*"

Synchronous-Series Carry Counter

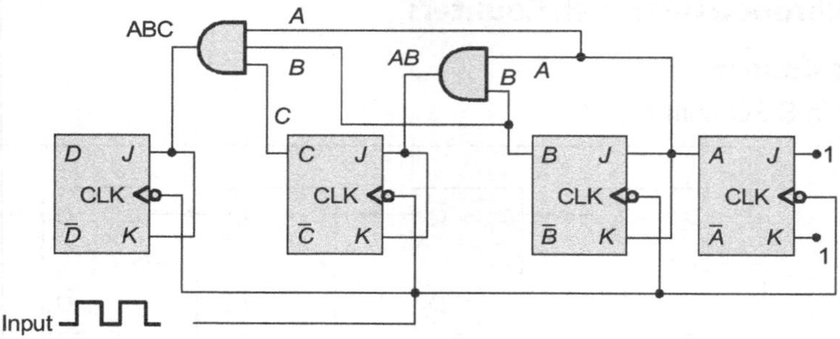

Clock frequency ; $f_{CLK} \leq \dfrac{1}{t_{pd(FF)} + (n-2)\, t_{pd(\text{AND gate})}}$

Synchronous-Parallel Carry Counter

It is the "Fastest Counter".

Clock frequency: $f_{CLK} = \dfrac{1}{t_{pd(FF)} + t_{pd(\text{AND gate})}}$

VIII Digital ICs Family

Logic Family

A group of compatible ICs with the same logic levels and supply voltages for performing various logic functions fabricated using a specific circuit configuration is called "logic family".

Characteristics of Logic Families

- **Propagation Time Delay (t_{pd})**

  This is the average transition delay time for the signal to propagate from input to output when signal changes its value. This determines how fast the logic system can operate.

  Delay time: $t_{pd} = \dfrac{t_{PHL} + t_{PLH}}{2}$ ns

  where, $t_{PLH} \rightarrow$ Delay time in going form LOW logic to HIGH logic
  $t_{PHL} \rightarrow$ Delay time in going form HIGH logic to LOW logic

Remember:

- ☑ The delay times are measured between the 50% points on the input and output transitions.
- ☑ In BJT, $t_{PLH} > t_{PHL}$ due to reverse recovery time.
- ☑ In FET, $t_{PLH} < t_{PHL}$ due to large capacitance formed.

- **Power Dissipation (P_D)**

  It is the amount of power dissipated in an *IC*.

  It is determined by the current "I_{CC}" which draws from the V_{CC} supply.

  Collector current: $I_{CC} = \dfrac{I_{CCH} + I_{CCL}}{2}$

  where, $I_{CCH} \rightarrow$ current value when all inputs are HIGH ; $I_{CCL} \rightarrow$ current value when all inputs are LOW.

  Power dissipation: $P_{D(avg)} = I_{CC} \times V_{CC}$ mW

- **Figure of Merit (FOM)**

  $$FOM = t_{pd} \times P_{D(avg)} \quad \text{pico Joules}$$

  For the best operation of ICs, figure of merit (FOM) should be as small as possible.

- **Fan-in and Fan-out**

  The maximum number of inputs which can be applied to a logic gate is known as fan-in. On the other hand, fan out of a logic gate is the number of gates that can be driven by it.

  $$\text{Fan out} = \left(\frac{I_{OH}}{I_{IH}}\right) \text{ or } \left(\frac{I_{OL}}{I_{IL}}\right) \text{; whichever is minimum}$$

- **Noise-Margin**

  It is the property of a logic circuit to withstand unwanted noise voltage at input or the maximum value of noise signal that a system can reject with performance unaffected.

  $V_{OH} > V_{IH} > V_{IL} > V_{OL}$

  High state N.M. $= V_{NH} = V_{OH(min)} - V_{IH(min)}$

  Low state N.M. $= V_{NL} = V_{IL(max)} - V_{OL(max)}$

  Noise margins $= (V_{NH}, V_{NL})_{min}$

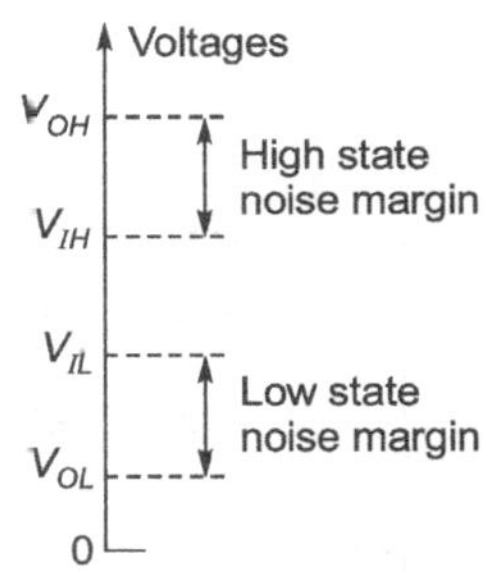

Bipolar Logic Families

Resistor Transistor Logic (*RTL*)

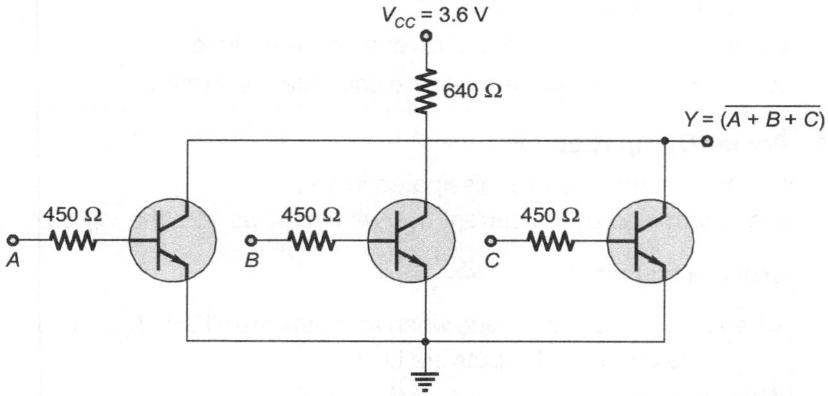

RTL provides wired AND logic.

Direct Coupled Transistor Logic (DCTL)

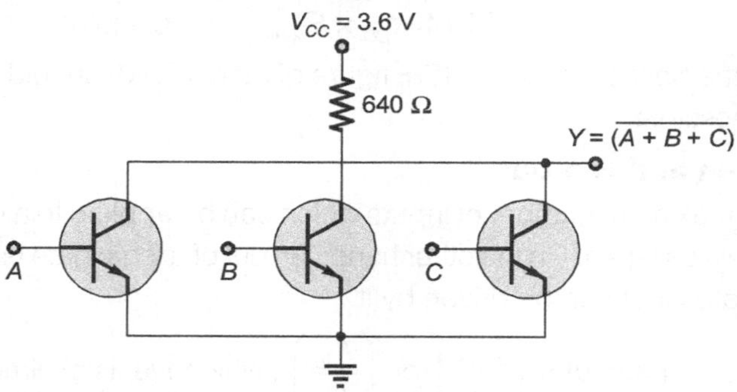

- DCTL provides wired AND logic.
- It suffers from "current Hogging".

Diode Transistor Logic (DTL)

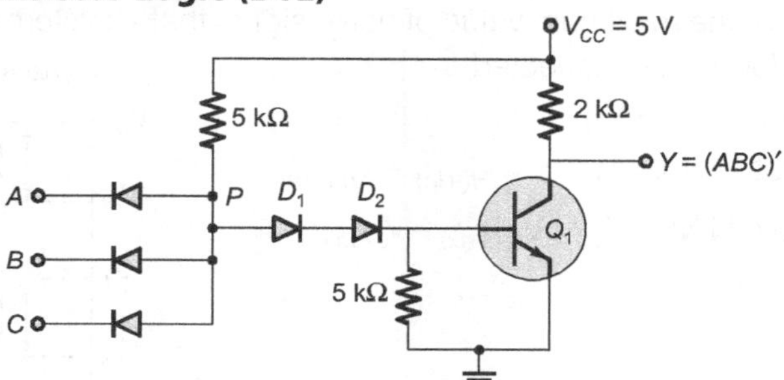

DTL provides wired AND logic.

High Threshold Logic (HTL)

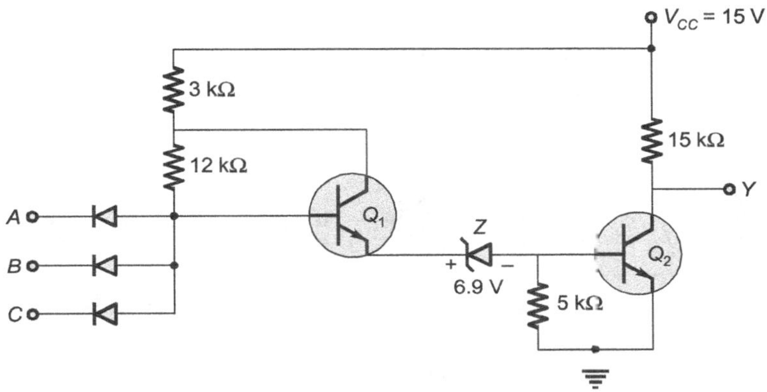

- HTL circuits has excellent noise margin and largest voltage swing.
- It provides wired AND operation.

Transistor Transistor Logic (TTL)

- TTL circuits are classified as:
  - (i) Tristate logic (high impedance logic)
  - (ii) Totem pole logic (active pull-up)
  - (iii) Open collector logic (passive pull-up)

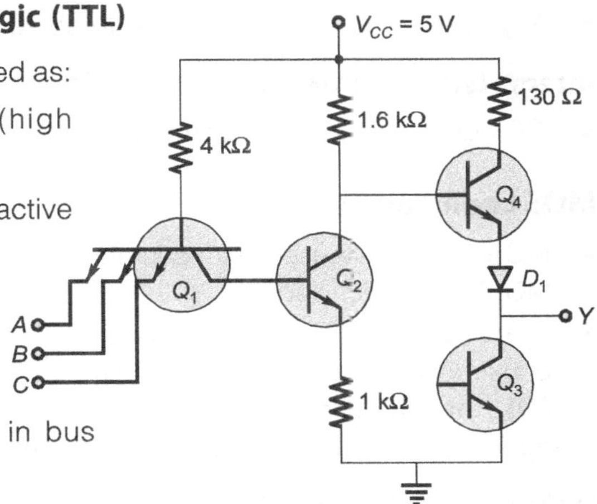

TTL gate with totem-pole output

- Tristate logic is used in bus oriented system.
- Totem pole logic does not provide wired logic.
- The advantage of active pull-up over passive pull-up is, reduced power dissipation and increased speed of operation.
- Any floating input TTL is considered as logic '1'.

Emitter Coupled Logic (ECL)

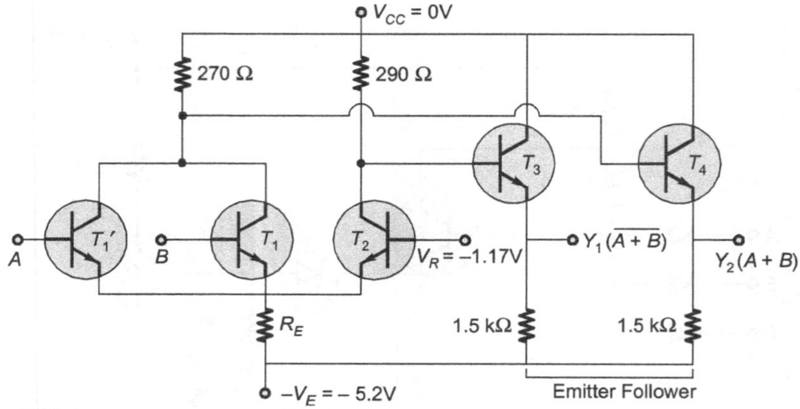

- ECL is a non saturated logic.
- It is the fastest logic among the all hence called "current mode logic".
- It provides wired 'OR' logic.
- It uses negative power supply to avoid noise and glitches.
- Any floating input in ECL is considered as logic '0'.

Integrated Injection Logic (IIL or I^2L)

- I^2L logic has best figure of merit.

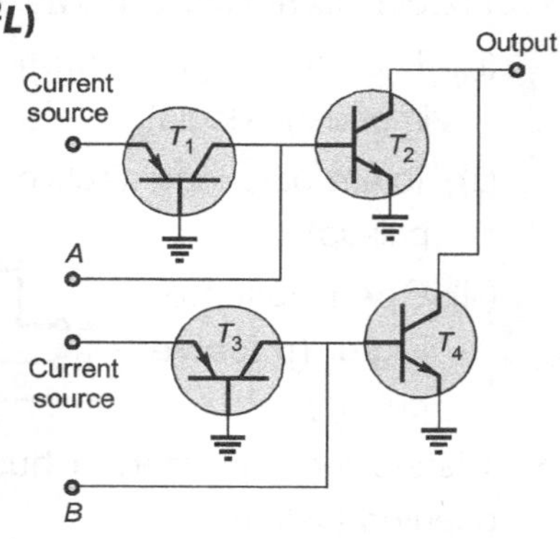

MOS Logic Family

- MOS digital ICs use enhancement MOSFETs exclusively.
- Because of the symmetrical construction of source (S) and drain (D), the MOS transistor can be operated as a bilateral device.

N-Channel MOS (NMOS)

- It is faster than PMOS.
- NMOS conducts whenever input is HIGH.

Symbol

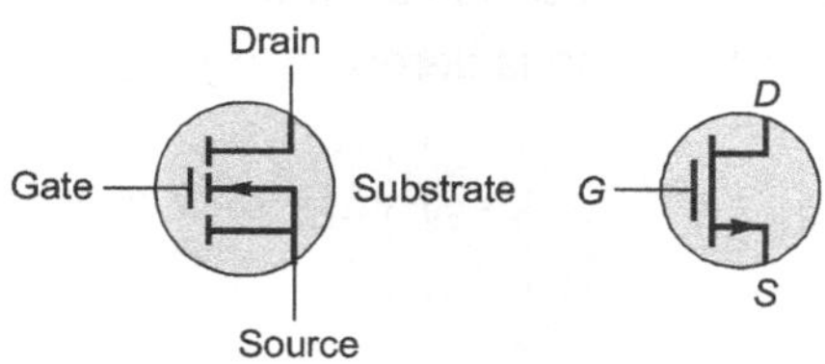

Different Logic Gates by NMOS

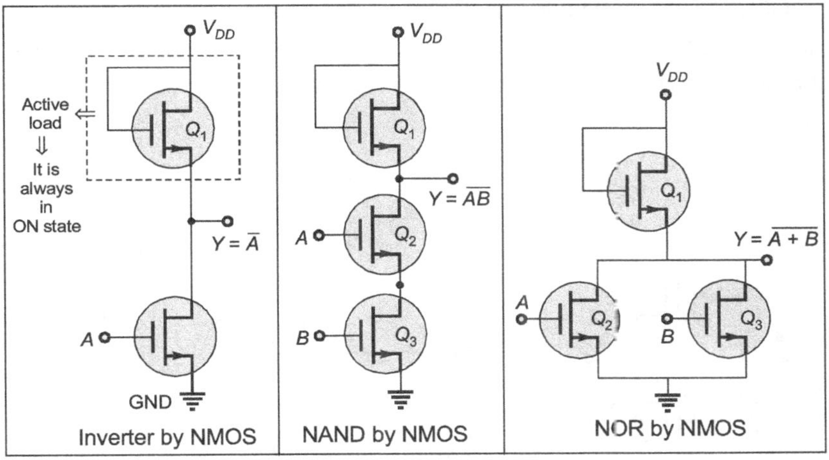

P-channel MOS (PMOS)

- Also called "Pull-Up Network".
- PMOS uses FETs having heavily doped P-channel.
- PMOS conducts whenever input is LOW.

Symbol

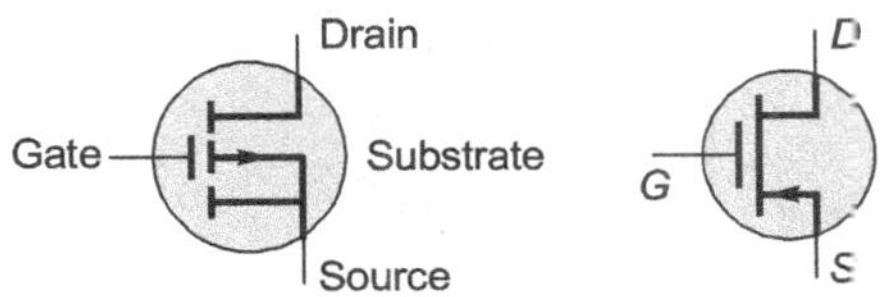

Different logic gates by PMOS

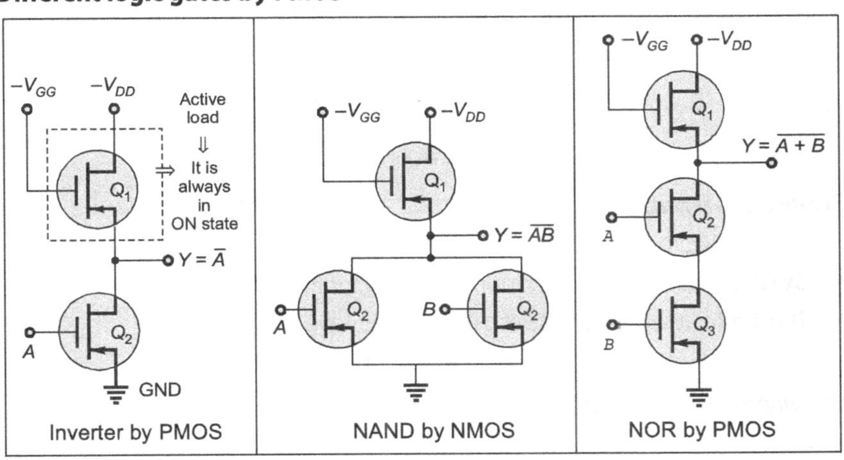

Complementary MOS (CMOS)

CMOS is a combination of NMOS and PMOS and is fabricated as twin tub. Various logic functions can be realized using CMOS. They are characterized by low power dissipation and excellent noise immunity.

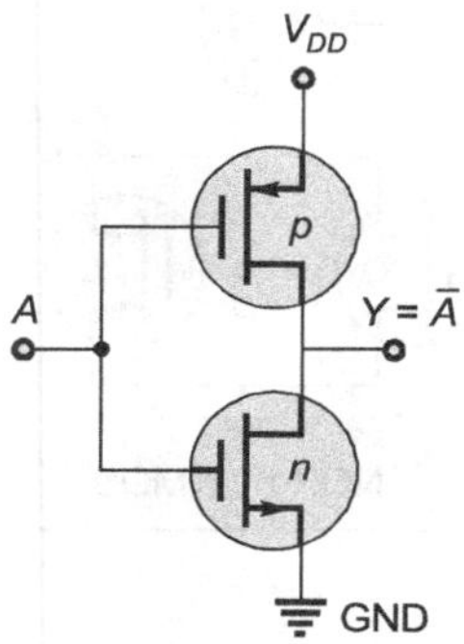

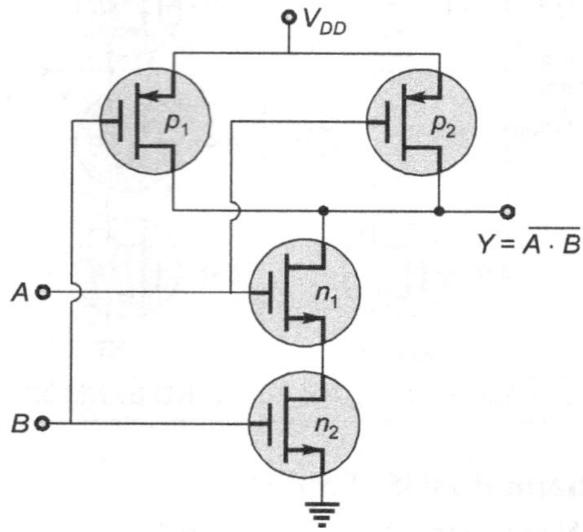

NOR by CMOS

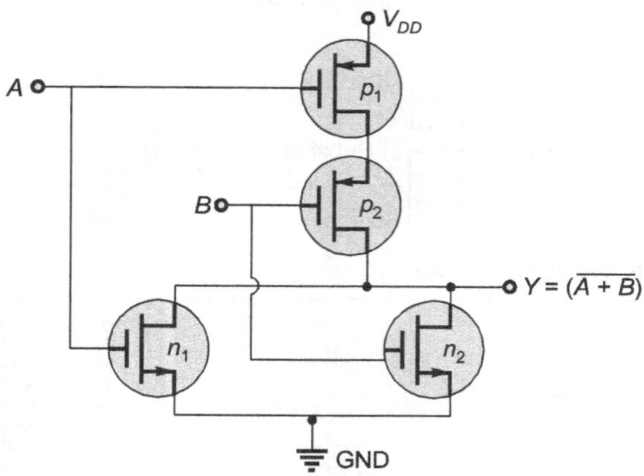

Power Dissipation

Power dissipation is CMOS is of two types:

- **Static:** It will be present during logic '0' or logic '1' input.
- **Dynamic:** It will be present during logic level transition

$$P_d(\text{dynamic}) = Cf\,V_{DD}^2$$

where, $C \to$ Capacitance ; $f \to$ Frequency ; $V_{DD} \to$ Supply voltage

Comparison of Various Logic Families

Logic parameter	Basic GATE	Fan-out	Power dissipation (mW)	t_{PD} (nsec)	F.O.M. (pJ)	N.M. (volt)
RTL	NOR	5	12	50	500	0.2
DTL	NAND	8	8-12	30	240	0.75
HTL	NAND	10	55	90	5000	4-5
TTL	NAND	10	12-22	12-6	144-132	0.4
ECL	OR/NOR	25	40-55	4-1	55-160	0.3
MOS	NAND/NOR	20	0.2-10	300	60	1.5
CMOS	NAND/NOR	>50	0-0.01	70	0.7	$\frac{V_{DD}}{2}$

IX DACs and ADCs

Digital to Analog Converter (DAC)

Specification of DAC

- **Resolution**

  Resolution of DAC is change in analog voltage corresponding to LSB bit increment at the input.

$$\text{Resolution} = \frac{V_r}{2^n - 1}$$

where, $V_r \rightarrow$ Reference voltage corresponding to logic 1 ;

$n \rightarrow$ Number of bit

- **Analog Output Voltage**
- Resolution × Decimal equivalent of binary data = Output voltage
- Analog output voltage of an N-bit straight binary DAC

$$V_0 = K\left[2^{N-1}\, b_{N-1} + 2^{N-2} b_{N-2} + \dots\dots\dots + 2^2\, b_2 + 2^1 b_1 + b_0\right]$$

where, $K \rightarrow$ Proportionality factor.

$b_n \rightarrow$ 1; if the nth bit of digital input is '1'.

$\rightarrow$ 0; if the nth bit of digital input is '0'.

- Full scale output (V_{FS}) voltage is maximum output of DAC

$$V_{FS} = \text{Resolution} * \text{Maximum decimal}$$

$$= \frac{V_r}{2^n - 1} * (2^n - 1) = V_r$$

- **% Resolution**

$$\% \text{ Resolution} = \frac{\text{Resolution}}{V_{FS}} \times 100 = \frac{1}{2^n - 1} \times 100$$

- **Error/Accuracy**

  Maximum error acceptable in ADC/DAC is 1 LSB bit which is equal to resolution.

Weighted-Resistor DAC

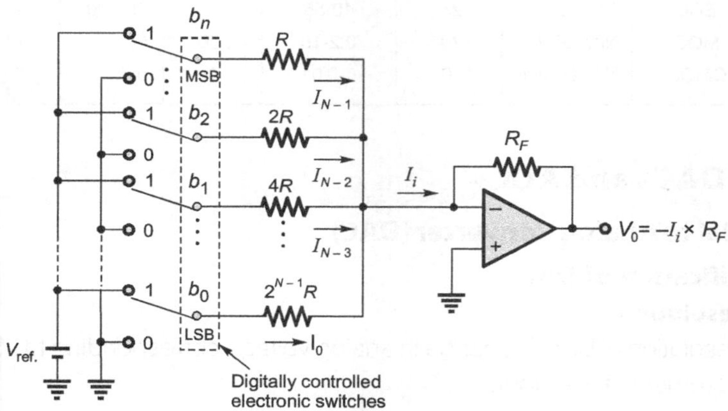

Output voltage:

$$\boxed{V_o = -\frac{R_F}{R}\left[\frac{b_0}{2^{N-1}} + \frac{b_1}{2^{N-2}} + \ldots + \frac{b_n}{2^{N-N}}\right] V_R}$$

Resistance: LSB Resistance = $(2^{N-1}) \times$ MSB Resistance

Remember:

☑ The resistance values are weighted in accordance with the binary weight.

☑ Here "OP-AMP" is employed as a "summing amplifier".

☑ OP-AMP is used in negative feedback mode to work as a "current to voltage converter".

☑ For N-bit DAC

(i) Number of different levels = 2^N (ii) Number of steps = $2^N - 1$

R-2R Ladder DAC by Using Non-inverting OP-AMP

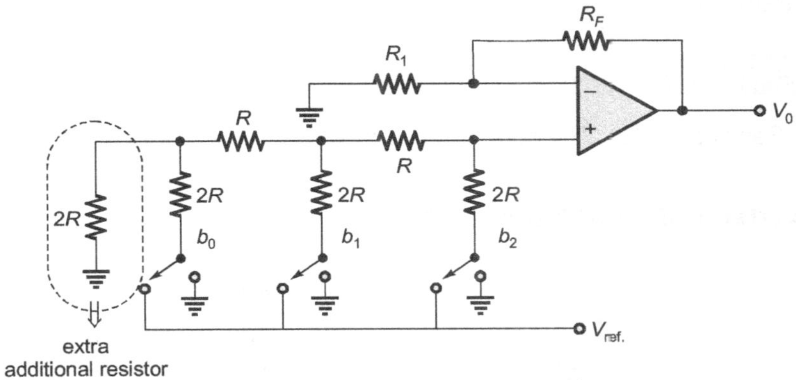

3-bit ($b_2\, b_1\, b_0$) R-$2R$ ladder DAC

Gain of OP-AMP : $\dfrac{V_0}{V_{ref.}} = \left(1 + \dfrac{R_F}{R_1}\right)$

where, $V_{ref.} \rightarrow$ Reference voltage or Input voltage

$V_0 \rightarrow$ Output voltage

Output analog voltage:

$$V_0 = \frac{V_{ref.}}{2^N}\left[\sum_{i=0}^{N-1} 2^i b_i\right] \times \left[1 + \frac{R_F}{R_1}\right]$$

R-2R Ladder DAC by Using Inverting OP-AMP

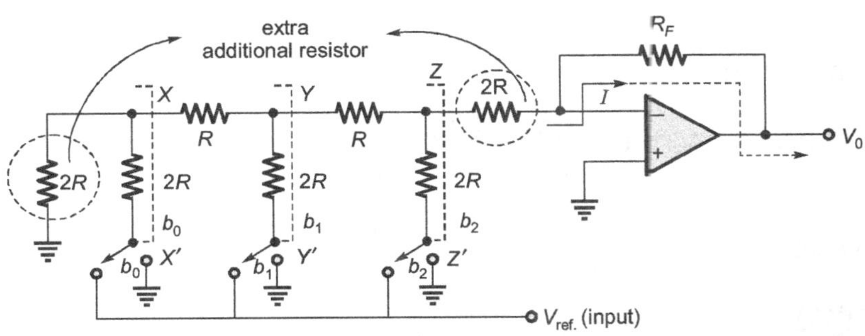

3-bit ($b_2\, b_1\, b_0$) R-$2R$ ladder DAC

Output voltage:

For N-bit DAC , $V_0 = \dfrac{V_{ref.}}{2^N} \times \left[\displaystyle\sum_{i=0}^{N-1} 2^i b_i\right] \times \left[\dfrac{-R_F}{3R}\right]$

Output current: $I = \frac{V_{ref.}}{2^N} \times \left[\sum_{i=0}^{N-1} 2^i b_i\right] \times \left[\frac{1}{3R}\right]$

Remember:

Resolution of R-2R ladder network is $\frac{V_r}{2^n}$.

Inverted Ladder (R-2R)type DAC

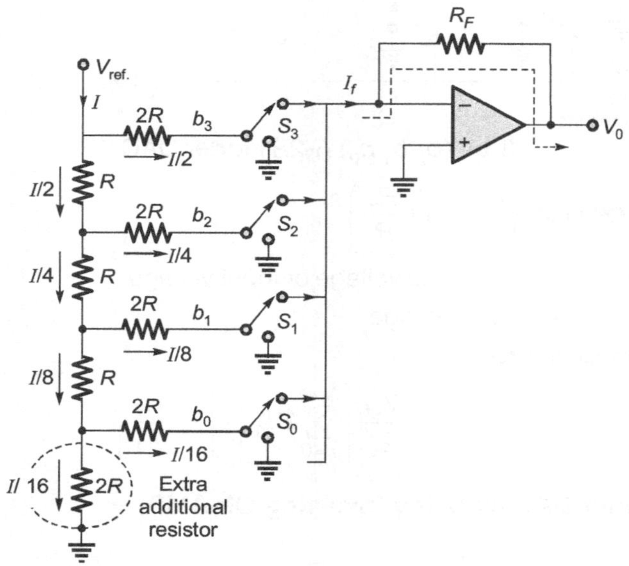

Output voltage:

For N-bit DAC , $V_0 = \frac{V_{ref.}}{2^N} \times \left[\sum_{i=0}^{N-1} 2^i b_i\right] \times \left(\frac{-R_F}{R}\right)$

Forward current : $I_f = \frac{V_{ref.}}{2^N} \times \left[\sum_{i=0}^{N-1} 2^i b_i\right] \times \frac{1}{R}$

Note:

☑ ON-OFF Switches (S_0, S_1, S_2, S_3) are at the same potential.

☑ We always consider the bit as MSB, where the input reference voltage or supply to be given.

☑ The bit stream $(b_3\ b_2\ b_1\ b_0)_2$ has MSB = b_3 and LSB = b_0.

Analog to Digital Converter

Specifications of ADC

$$\text{Voltage range} = V_{max} - V_{min}$$

$$\text{Resolution} = \frac{V_{range}}{2^n - 1} = \frac{1}{2^n - 1} \times 100$$

$$\text{Dynamic range} = (1.8 + 6n) \text{ dB}$$

Types of ADCs are

- Counter type
- Dual slope type
- Successive Approximation Resistor type (SAR)
- Flash type

Counter type ADC

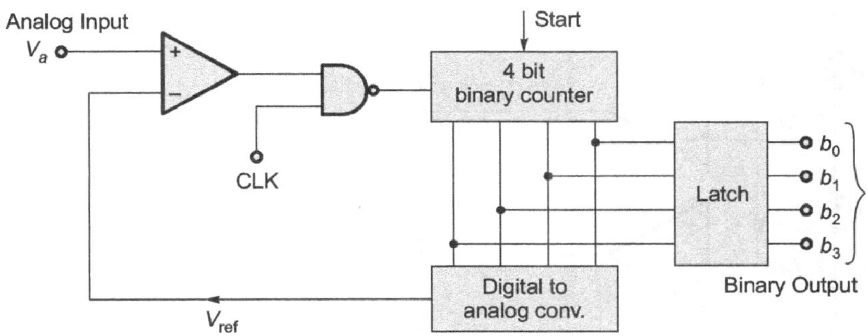

- Number of clock pulse required for n-bit conversion = $2^n - 1$
- Maximum conversion time = $(2^N - 1)\ T_{clk}$.
- Counter type ADC also known as 'ramp type' ADC.
- In this ADC, conversion time depends on input analog voltage.

Flash type ADC

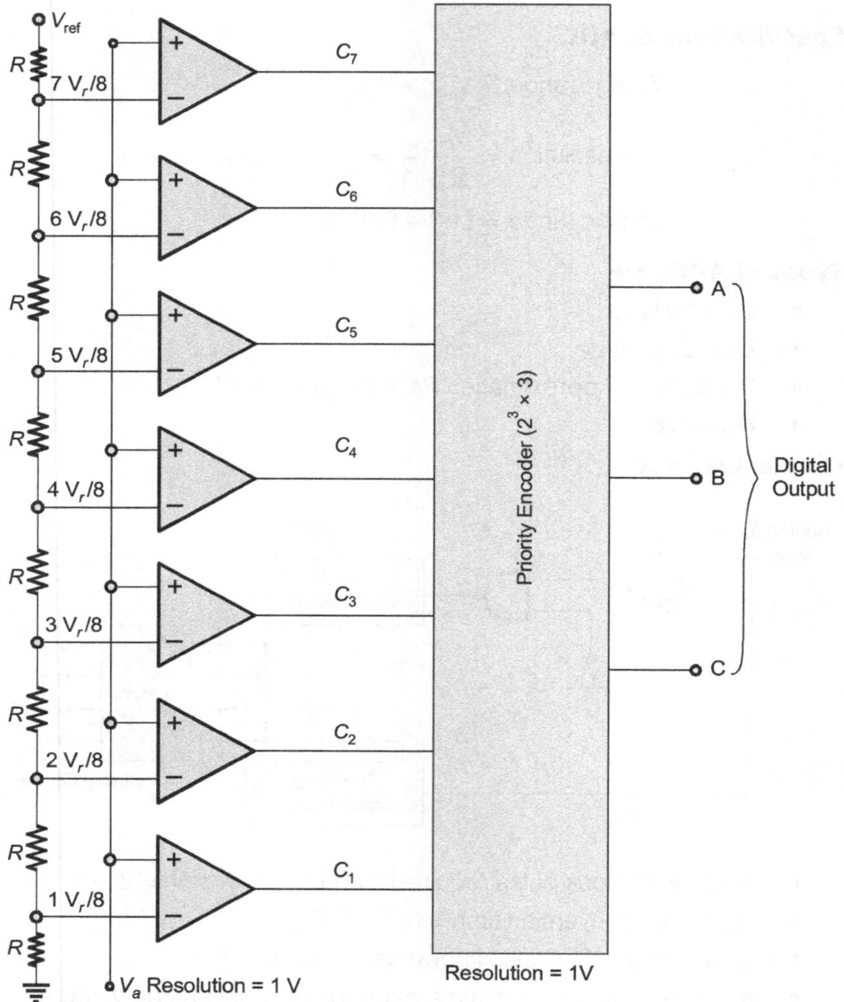

- For n bit conversion flash type ADC requires
  - (i) $2^n - 1$ comparators
  - (ii) 2^n resistors
  - (iii) One $2^n \times n$ priority encoder
- It is the fastest ADC among all.
- It is also known as parallel comparator type ADC.

SAR type ADC

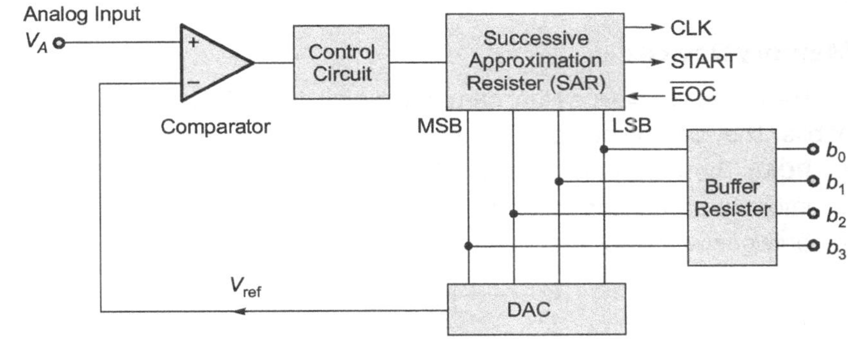

- Number of clock pulse required for n-bit conversion = n.
- Maximum conversion time = $n * T_{clk}$.
- In SAR ADC, conversion time is independent of input analog voltage.

Dual slope integrating type ADC

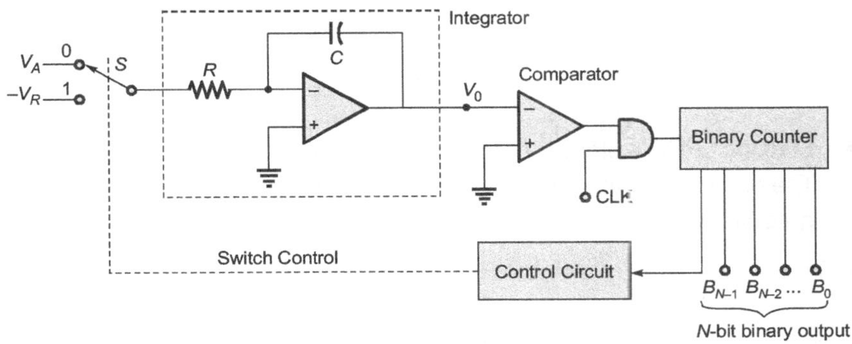

- Total number of clock pulse required = $2^n + N \approx 2^{n+1}$.
- It is most accurate ADC therefore mostly used in digital voltmeter.
- It is slowest ADC.

Remember:

	No. of clocks	Time
Counter type	2^{n-1}	$(2^{n-1})\ T_{clk}$
SAR	n	nT_{clk}
Counter type	1	T_{clk}
Counter type	2^{n+1}	$2^{n+1}\ T_{clk}$

where, $T_{clk} \rightarrow$ Clock time period

X Miscellaneous

Memory Devices

Memories are devices that can store digital data or information in terms of bits. They can be categorised as:

- **ROM:** The read only memory (ROM) is also called mask ROM. The information is inscribed in the form of presence or absence of a link between word (access) line and the bit (sense) line.

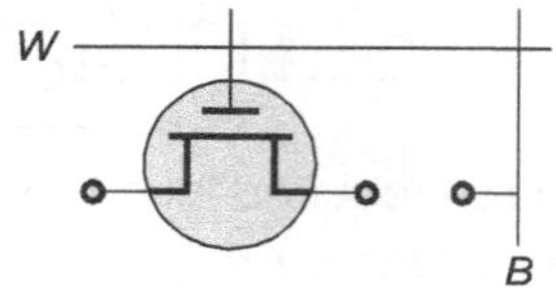

- **PROM:** The programmable read only memory (PROM) are ROMs which are field programmable but lack "erase" capability.

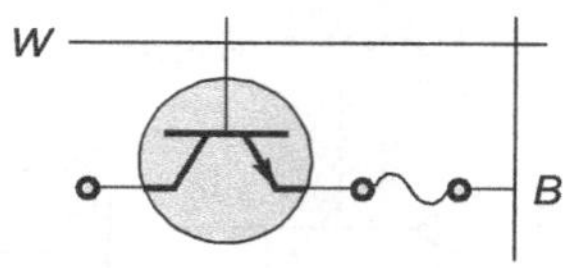

- **EPROM:** EPROM stands for erasable programmable read only memory. The presence or absence of charge in the floating gate of a double poly gate MOSFET determines the logic state.

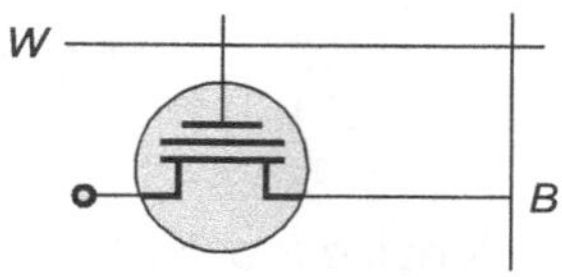

- **EEPROM:** EEPROM stands for electrically erasable PROM. Programming and erasing in EEPROM is achieved by forcing the channel current to flow between the gate and substrate.

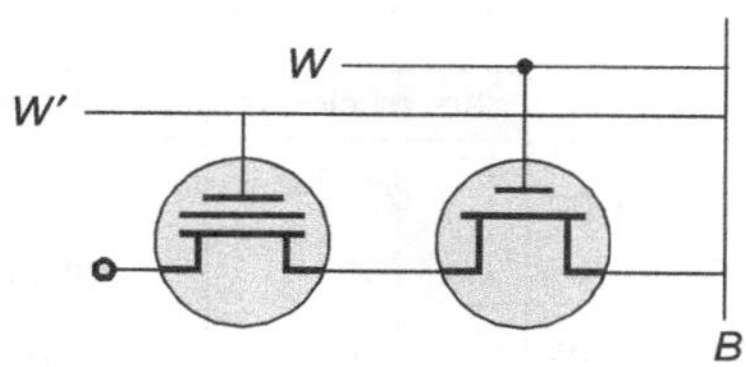

- **SRAM:** Static random access memory can retain stored data indefinitely and it can be implemented as a flip-flop circuit to store one bit of information.

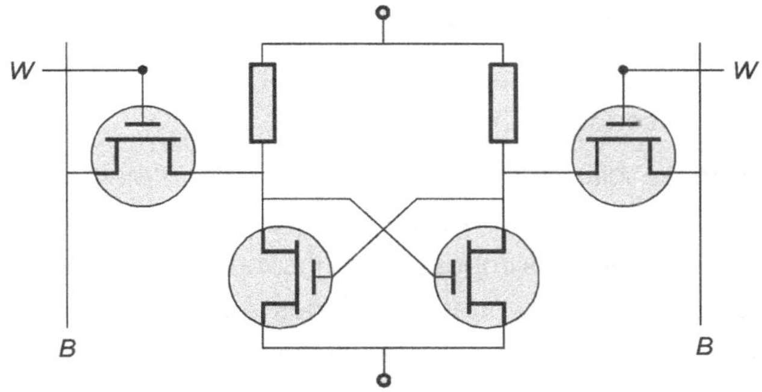

- **DRAM:** Dynamic RAM uses a single transistor to access the capacitor's charge. Since, the stored charge is gradually lost because of space-charge generation-recombination process, cells have to the read and refreshed intermittently.

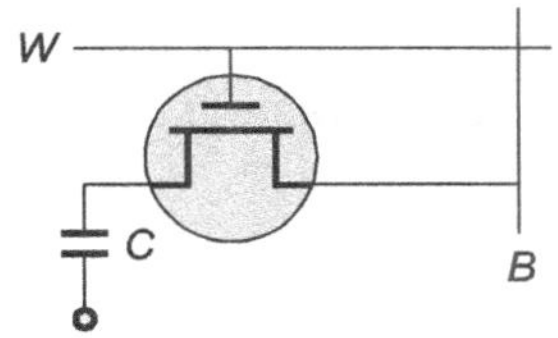

Difference between RAM and ROM

RAM	ROM
1. It is read/write memory.	1. It read only memory.
2. Random access (time required to access any memory is same).	2. Random access.
3. Volatile memory (temporary data storage).	3. Permanent data storage.

RAM

- Static RAM
- Dynamic RAM

Static RAM	Dynamic RAM
1. Data is stored like FF.	1. Data is stored in MOS capacitor.
2. BJT, MOSFET is used.	2. MOSFET is used.
3. Faster	3. Slower
4. Power dissipation is more.	4. Power dissipation is low.
5. Density is less.	5. Density is high.
6. Used as cache memory.	6. Used as main memory.
7. No refreshing is required.	7. Refreshing is required.

Finite State Model/Machine

- The finite state machine is an abstract model that describes the synchronous sequential machine.
- In a sequential circuit, the output depends upon the present input as well as past history and therefore it would need an infinite capacity for storing them. Since it is impossible to implement machines with infinite storage capacity, finite state machines are used. Thus finite state machines are sequential circuits whose past history can affect their future behaviour in only a finite number of ways.
- That means finite state machines are machines with a fixed number of states. Every finite state machine has a finite number of memory devices.
- The block diagram of a finite state model is shown in Figure-1. Here x_1, $x_2 \cdots x_n$ are the inputs $y_1, y_2 \cdots y_m$ are the outputs, $z_1, z_2 \cdots z_K$ are state variables and $Z_1, Z_2 \cdots Z_K$ represents the next state.

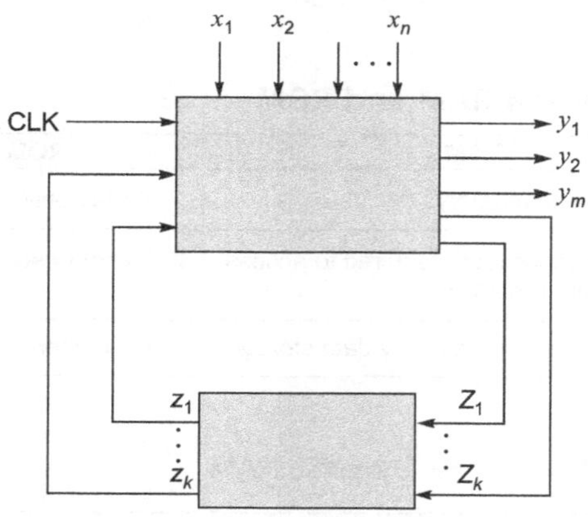

- With an n-state machine, we can generate a periodic sequence of less than or equal to n-states.
- Finite machines are of two types. The major difference between them is the way the output is generated. They are
  - (i) Moore type model
  - (ii) Mealy type model.

Mealy machine

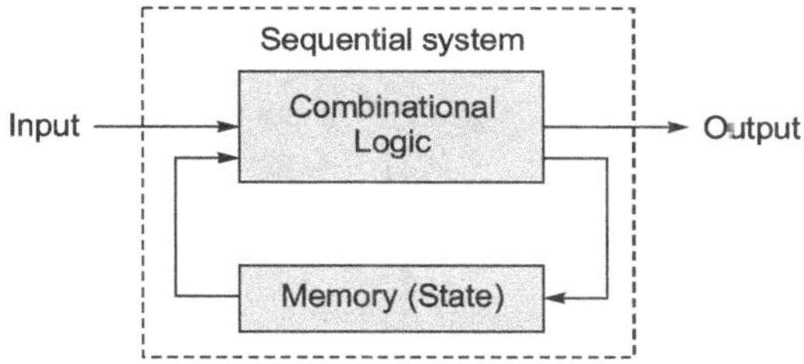

- The output depends on current input and state.
- Less number of states are involved.
- Complex.

Moore machine

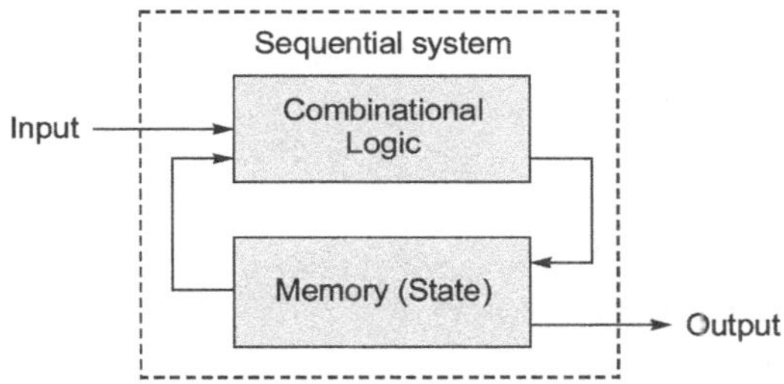

- Output depends only on current state.
- More number of states are involved.
- It is relatively less complex to implement.

Moore Machine	Mealy Machine
1. The output is a function of present state only.	1. The output is a function of present state as well as present input.
2. The change in input does not affect the output of the circuit.	2. The change in input may affect the output of the circuit.
3. For implementing same function, it requires more number of states.	3. It requires less number of states for implementing same function.
4. The designing of Moore model is easy.	4. The designing of Mealy model is complex.

State Diagram

A state diagram represents a state machine and contains:

- **Circles:** They represent machine states labelled with a binary encoded number.

- **Directed arcs:** They represent transition between states labelled with input/output for that state transition.

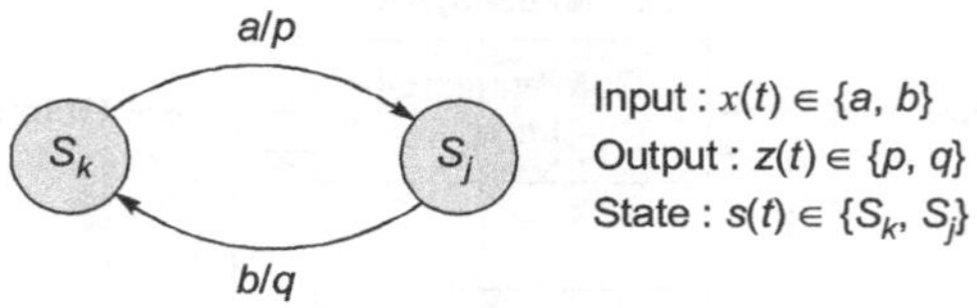

State Table

A state table expresses systems behaviour and consists of

- **Present state:** It is typically given by binary encoded form or with S_k.
- **Inputs:** The external inputs used to cause state transitions.
- **Next state:** The state which occurs after transition, expressed in binary encoded form of S_j.
- **Outputs:** Outputs which occur other than the state. Note that there would be no outputs in Moore machine.

Application

- In sequence detection.
- Vending machine designing.
- Electronic design automation.
- Communication protocol design.

Note: A sequence detector is a sequential machine which produces an output '1' every time when the desired sequence is deleted and an output '0' at all the other time.

Electrical Materials

10

I Dielectric Properties of Materials

Insulators

The insulating material is a substance which prevents the leakage of electric current in unwanted directions. In other words, when the main function of nonconducting materials is to provide electrical insulation they are called insulators.

Dielectrics

Dielectrics is a nonconducting materials which can be polarized in the presence of electric field. In other words materials which form charged dipole instantaneously under the influence of electric field are known as dielectrics i.e. they have ability to store energy when external electric field is applied. The dielectrics are of two types i.e. polar and non-polar.

Remember:

A dielectric is always an insulator.

Dielectric Parameters

Permittivity ($\in$)

The term permittivity or dielectric constant is the measurement of electrostatic energy store within it and therefore depends on the material.

$$\in = \in_r \in_0$$

where, $\in_r \rightarrow$ Relative permittivity ;

$\in_0 \rightarrow$ Permittivity of vacuum $= 8.854 \times 10^{-12}$ Farad meter^{-1}

Note:

- ☑ Unit of permittivity is farad meter^{-1}.
- ☑ Relative permittivity of vacuum is equal to 1.
- ☑ Materials used for capacitor (ceramic) have higher dielectric constant than polymer.

Dipole Moment

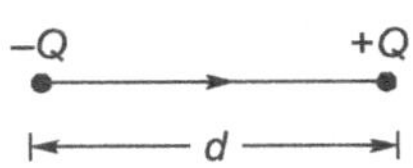

$$\bar{p} = Q\bar{d} \text{ Debye}$$

Two charge of equal magnitude but opposite in polarity placed distance d apart constitute a dipole moment.

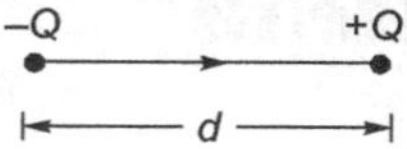

$$\bar{p} = Q\bar{d} \text{ Debye}$$

Note:

- ☑ Dipole moment is a vector quantity and pointing from the negative charge to the positive charge.
- ☑ 1 Debye = 3.33×10^{-30} Coulomb-meter

Polarization

The electric dipole moment per unit volume is called as polarization.

$$P = Np \text{ Coulomb/m}^2$$

where, $N \rightarrow$ Number of dipoles per unit volume.

In many dielectric materials the polarization vector is proportional to the electric field E, as

$$P = \epsilon_0 \chi_e E \text{ and } P = \epsilon_0 (\epsilon_r - 1)E$$

where, $E \rightarrow$ Electric field intensity ; $\chi_e \rightarrow$ Electrical susceptibility

$$\chi_e = \epsilon_r - 1$$

Note:

χ_e is a measure of the ability of the material to become polarized.

$$\chi_e = \frac{\text{Bound charge density}}{\text{Free charge density}}$$

Polarizability

The elemental dipole moment is proportional to electric field strength E, that acts on the particle so that,

$$p = \alpha E$$

where, $\alpha \rightarrow$ Polarizability, Farad-m^2 ; $E \rightarrow$ Electric field intensity, V/m

$p \rightarrow$ Dipole moment, Coulomb-m

$$\alpha = \frac{\epsilon_0 \chi_e}{N} = \frac{\epsilon_0 (\epsilon_r - 1)}{N}$$

Note:

α represents the extent of degree of polarization.

Electric Flux Density

$$\bar{D} = \epsilon \bar{E} \text{ ... for isotropic materials}$$

Remember:

- ☑ In isotropic materials $\in$ is independent of the direction.
- ☑ Electric flux density when an electric field is applied.

$$D = \in_0 E + P$$

- ☑ The stored energy per unit volume in a dielectric medium due to polarization.

$$W = \frac{1}{2} PE$$

Mechanism of Polarization

Electronic polarization

The electronic polarizability of a molecule can be defined as the dipole moment induced per unit field strength resulting only from shift of the electron clouds relative to nuclei.

$$P_e = N\alpha_e E \quad \text{where, } \alpha_e \rightarrow \text{Electronic polarizability}$$

$$\alpha_e = 4\pi \in_0 R^3 \quad \text{where, } R \rightarrow \text{Radius of atom}$$

$$\alpha_e = \frac{\in_0 (\in_r - 1)}{N} \; ; \; \in_r = 1 + 4\pi N R^3$$

Ionic polarization

When in a molecule some of the atoms have excess positive and negative charges resulting from ionic character of the bonds, an electric field will tend to shift positive ions relative to nuclei.

$$P_i = N\alpha_i E \quad \text{Where, } \alpha_i \rightarrow \text{ionic polarizability}$$

Note:

- ☑ $\alpha_i = \frac{1}{10} \times \alpha_e$.
- ☑ Ionic polarization is independent of temperature.

Orientational polarization

When an internal field is applied to a molecule carrying a permanent dipole moment, the external field will tend to align the dipole along the direction of external field by applying a torque.

$$P_0 = \frac{NP_P^2}{3KT} E$$

where, $N \rightarrow$ Number of permanent dipoles ; $P_p \rightarrow$ Permanent dipole moment ; $E \rightarrow$ Applied external electric field intensity
$K \rightarrow$ Boltzmann constant ; $T \rightarrow$ Temperature in Kelvin

Orientational polarizability, $\alpha_0 = \dfrac{P_P^2}{3KT}$ ... Curie law

Total polarization of a polyatomic gas

$$P = P_e + P_i + P_0$$

$$P = N\left(\alpha_e + \alpha_i + \frac{P_P^2}{3KT}\right)E$$

$$\epsilon_0 \chi_e = N(\alpha_e + \alpha_i) + \left(\frac{NP_p^2}{3K}\right)\frac{1}{T}$$

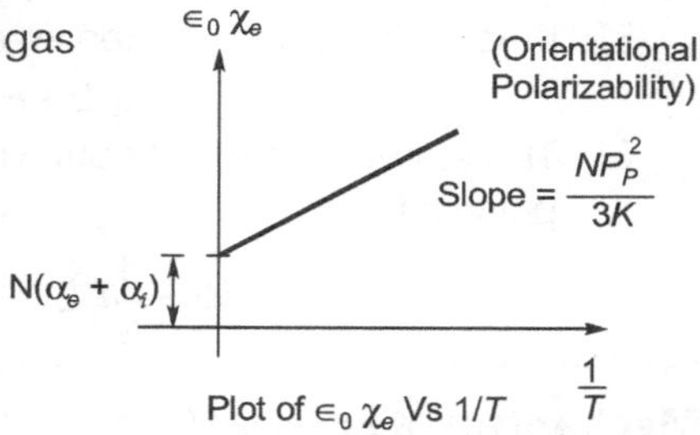

Plot of $\epsilon_0 \chi_e$ Vs $1/T$

Note:

- ☑ Orientational polarization is dependent on temperature and frequency.
- ☑ Orientational polarization is applicable only to polar dielectrics.

Interfacial Polarization or Space charge polarization

This type of polarization occurs due to accumulation of charges at the interface in a multiphase material.

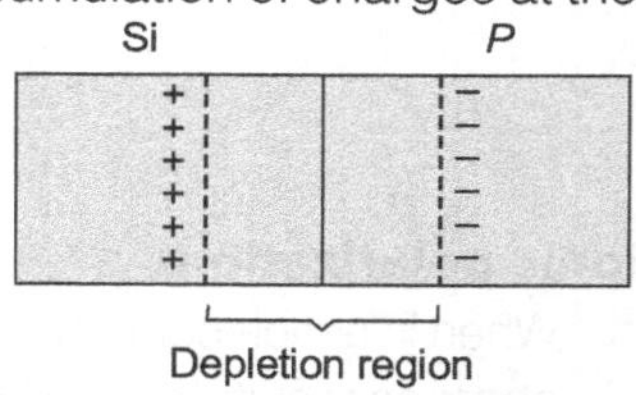

Net polarization

$$P = P_e + P_i + P_0 + P_s$$

where, $P_s \rightarrow$ Interfacial polarization

Note:

For single phase material, value of P_s become zero.

Internal Field in Solids and Liquids

Internal field, $E_i = E_{\text{ext}} + \dfrac{\gamma}{\epsilon_0}P$

where, $P \rightarrow$ Dipole moment per unit volume ; $\gamma \rightarrow$ Internal field constant

$\because$ γ, ϵ_0, P are positive quantities. Thus $E_i > E_{ext}$

Note:

For cubic lattice $\gamma = 1/3$ and internal field is called Lorentz's internal field

and given by : $E_{i(\text{Lorentz's})} = E_{ext} + \dfrac{P}{3\epsilon_0}$

Types of Dielectric Materials

Elemental Solid Dielectrics

These are the materials consisting of single type of atoms. Such materials contains neither ions nor permanent dipoles.

$$\epsilon_r = (\chi_e + 1) = \frac{1-(1-\gamma)\dfrac{N\alpha_e}{\epsilon_0}}{1-\dfrac{\gamma N\alpha_e}{\epsilon_0}}$$

Remember:

- ☑ In elemental solid dielectric only electronic polarization is present.
- ☑ Diamond, sulphur, germanium etc. are example of elemental dielectric materials.

Ionic Nonpolar Solid Dielectrics

These solids contains more than one type of atoms but no permanent dipoles. In ionic crystals, total polarization is electronic and ionic in nature.

Polar Solid

In these solids molecules possess permanent dipole moments. The total polarization has all the three components, i.e., ionic, electronic and orientational polarization.

Clausius-Mossotti relation , $\dfrac{N\alpha_e}{3\epsilon_0} = \dfrac{\epsilon_r - 1}{\epsilon_r + 2}$

where, $N \rightarrow$ Number of molecules per unit volume; $\alpha_e \rightarrow$ Electronic polarizability

Maxwell's relation for index of refraction

$$\epsilon_{re} = n^2 \qquad \text{(assuming } \mu = \mu_0\text{)}$$

where, $\epsilon_{re} \rightarrow$ Dielectric constant at optical frequencies ; $n \rightarrow$ Refractive index ; $\mu \rightarrow$ Magnetic permeability of material ; $\mu_0 \rightarrow$ Magnetic permeability of vacuum

In Clausius-Mossotti relations, for gases at low pressure

$$\epsilon_r = 1 \quad \text{or} \quad \epsilon_r + 2 = 3$$

So,
$$\frac{N\alpha_e}{\epsilon_0} = \epsilon_r - 1 = \chi_e$$

Debye's generalization of Clausius-Mossotti relation

- Polarizability per kilogram molecule

$$\pi = \frac{N_A\alpha}{3\epsilon_0} = \frac{\epsilon_r - 1}{\epsilon_r + 2}\cdot\frac{M}{\rho}$$

where, $\pi \rightarrow$ Molar polarizability ;
$N_A \rightarrow$ Avogadro's number = 6.023×10^{23} ; $\alpha \rightarrow$ Polarizability which includes effect of orientational polarization ; $M \rightarrow$ Molecular weight of material, kg ; $\rho \rightarrow$ Density, kg/m^3

- Maxwell's relation for optical frequencies or Lorentz-Lorentz equation

$$\pi = \frac{N_A \alpha}{3\epsilon_0} = \frac{n^2-1}{n^2+2}\cdot\frac{M}{\rho} \quad \text{where, } n \rightarrow \text{Refractive index}$$

Note:

Lorentz-Lorentz equation will apply only for the case $\alpha = \alpha_e$.

- Condition for spontaneous polarization : $\dfrac{N\alpha\gamma}{\epsilon_0} = 1$

where, $N \rightarrow$ Number of molecules per unit volume, m^{-3}
$\alpha \rightarrow$ Polarizability, Farad-m^2 ; $\gamma \rightarrow$ Internal field constant

- Curie-Weiss Law, $\chi_e = \epsilon_r - 1 = \dfrac{C}{T-\theta}$

where, $C \rightarrow$ Curie constant ; $\theta \rightarrow$ Characteristic temperature which is usually a few degree smaller than the ferroelectric Curie temperature θ_f ; $T \rightarrow$ Temperature

Note:

Above temperature θ, the spontaneous polarization vanishes and the material becomes piezoelectric from ferroelectric.

Classification of Dielectric Material Based on the Dielectric Behavior in the Presence of Electric Field

Piezoelectric Material

Piezoelectric materials are those which get polarized when they are subjected to mechanical stress. Piezoelectric material also get strained when subjected to electrical stress.

Example: Quartz crystal, $BaTiO_3$ crystal, $BaTiO_3$ ceramic modified lead zirconate titanate ceramic, Rochelle salt.

Piezoelectricity

- **Direct Effect :** The application of stress to a crystal produces a strain which results in a net polarization.
- **Inverse Effect :** The application of an electric field produces a strain whose sign depends on the field direction
- These are both linear effects. Such materials obey the following equations:

$$\left.\begin{aligned} T &= cS \\ S &= sT \end{aligned}\right] \qquad \text{...(i)}$$

$$P = dT \qquad \text{...(ii)}$$

- A stress T results in a strain S in a material and c is an elastic stiffness constant; s is reverse of c. The stress T will produce a polarization; d is called piezoelectric strain constant.

- The dielectric displacement, in the presence of stress.

$$D = \in E + dT \quad \text{...(iii)}$$

and the inverse effect is given by the relation

$$S = sT + dE \quad \text{...(iv)}$$

Ferroelectric Material

A ferroelectric material is one which exhibits an electric dipole moment and is said to be spontaneously polarised even in the absence of an electric field. Ferroelectric materials shows a hysteresis in polarization.

Example: Rochelle salt, KDP, Barium titanate, sodium nitrite, lead titanate.

Note:

☑ The direction of polarization can be reversed by applying an external field in reverse direction of spontaneous polarization.

☑ Curie-Weiss law is applicable only for ferroelectric material.

Pyroelectric Material

Pyroelectric material are those which exhibits spontaneous polarization in the absence of external electric field and changes its polarization on heating.

$$\Delta P = \lambda \Delta T$$

where, $\Delta P \rightarrow$ Change of polarization ; $\lambda \rightarrow$ Pyroelectric constant
$\Delta T \rightarrow$ Change in temperature

Example: Tourmaline and polyvinylidene fluoride.

Anti-ferroelectric Material

These are materials in which the dipole moment are aligned in anti-parallel direction therefore net spontaneous polarization is zero.

Example: ADP, Lead Zirconate, Sodium Niobate.

Remember:

☑ All ferroelectric can be piezoelectric and pyroelectric

☑ All pyroelectric are piezo electric

Quartz

- Quartz is a crystalline material.
- Structure of quartz is "silicate tetrahedral (SiO_4)".
- It has shape of hexagonal prism with pyramid attached at both the ends.
- Quality factor range for quartz crystal is $10^4 - 10^6$.

Crystal Cuts

AT cut crystal

- It is most commonly used cut.

- It has high degree of frequency stability over a large temperature range.
- Crystal plate makes an angle 35° 15′ with respect to z-axis.

***BT* cut crystal**

- Crystal plate makes an angle of 49° with respect to z-axis.
- It can be used for high frequency application.

Equivalent Circuit

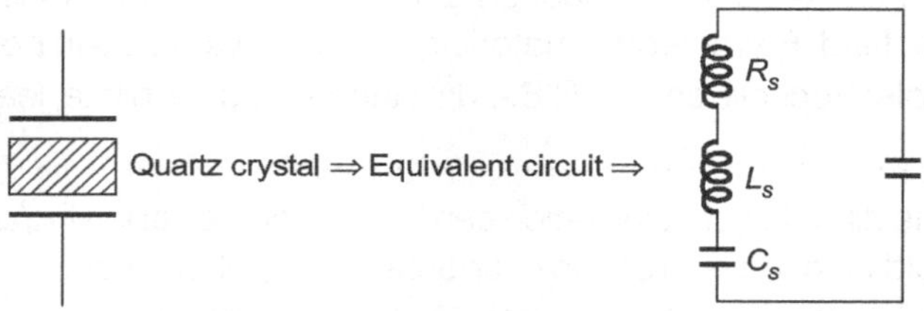

where, $R_s \rightarrow$ Represents damping constant ; $L_s \rightarrow$ Represents mass of true crystal ; $C_s \rightarrow$ Represents spring/force constant
$C_p \rightarrow$ Represents electrostatic capacitance between opposite surface of the crystal

- Series resonance frequency, $\omega_S = \dfrac{1}{\sqrt{L_s C_s}}$
- Impedance at series resonance frequency, $Z|_{\omega=\omega_S} = R_s$
- Parallel resonance frequency, $\omega_p = \sqrt{\dfrac{1}{L_s}\left(\dfrac{1}{C_s}+\dfrac{1}{C_p}\right)}$
- Impedance at parallel resonance frequency, $z|_{\omega=\omega_p} = \dfrac{R_s^2 + \left(\dfrac{1}{\omega_p C_p}\right)^2}{R_s}$

Dielectric in AC fields

Dielectric constant which is real for static fields breaks into real and imaginary part in case of AC fields.

$$\epsilon_r = \epsilon_r' - j\epsilon_r''$$

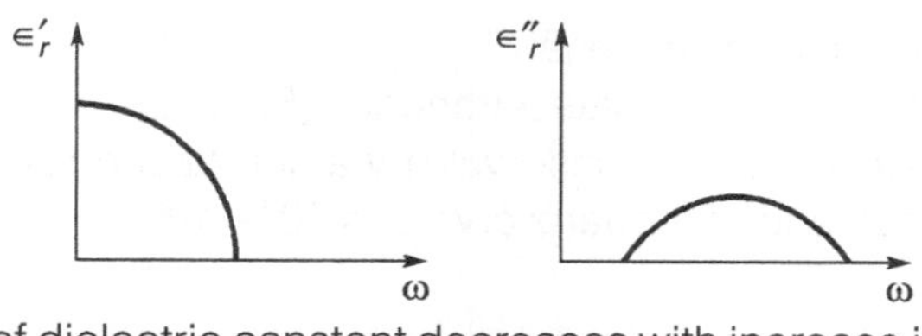

- Real part of dielectric constant decreases with increase in frequency.
- Imaginary part of dielectric constant first increases, attains a maximum value and then decreases with frequency.

Dielectric loss

- Absorption of electrical energy by dielectric material subjected to an AC field is called dielectric loss. The result is dissipation of energy in form of the heat. It occurs due to two reasons.
  1. Oscillation of ions 2. Continuous change in or entation of dipoles

  Energy absorbed per m^3, $W = \frac{1}{2} \epsilon_o \epsilon_r'' E_o^2 \omega$

  $$\boxed{W = \frac{E^2 f \epsilon_r \tan\delta}{1.8 \times 10^{12}}} \text{ W/cm}^2 \quad \text{where, } E \rightarrow \text{RMS value in V/cm.}$$

  Loss tangent $\tan\delta = \frac{\epsilon_r''}{\epsilon_r'}$; Loss factor = $\epsilon_r' \tan\delta = \epsilon_r''$

  $$0.0001 \le \tan\delta \le 0.001$$

II Dielectric Breakdown

Dielectric Breakdown of Gases

- **Average velocity of the charge carrier in a gas**

  $$v = \mu E$$

  where, $\mu \rightarrow$ Mobility of the charge carrier; $E \rightarrow$ Applied electric field

  Condition of Ionization

  $$E\lambda = V_i$$

  where $V_i \rightarrow$ Ionization potential ; $\lambda \rightarrow$ Mean free path

- **Electron ionization coefficient (Townsend breakdown process)**

  $$\alpha = \frac{1}{\lambda} e^{-(V_i/E\lambda)}$$

  where, α = Townsend's first ionization coefficient or ionization coefficient of electrons

  $$\lambda \propto \frac{1}{\text{Pressure}} \quad \text{.....at constant temperature}$$

  $$n_d = n_o e^{\alpha d}$$

  where, n_d = Number of electrons striking the anode per second

- **Secondary ionization coefficient**

  $$n = \frac{n_0 e^{\alpha d}}{1 - \gamma\left(e^{\alpha d} - 1\right)}$$

  where, $n \rightarrow$ Total number of electrons arriving at the anode

  $n_0 \rightarrow$ No. of primary photoelectrons per second emitted from cathode

$d \rightarrow$ Distance between anode and cathode

$\gamma \rightarrow$ Townsend's second ionization coefficient

Townsend criterion for spark breakdown

$$\gamma\left(e^{\alpha d} - 1\right) = 1$$

Note:

☑ The avalanche breakdown develops over relatively long period of time over 1 μs and does not generally occur with impulse voltage.

Dielectric Breakdown of Liquids

- Colloidal theory

$$\frac{\epsilon - \epsilon'}{\epsilon + 2\epsilon'} \cdot r^3 E^2 \geq \frac{kT}{4}$$

where $r \rightarrow$ Radius of insulating particle
$\epsilon \rightarrow$ Permittivity of insulating material
$\epsilon' \rightarrow$ Permittivity of oil ; $E \rightarrow$ Field strength
$k \rightarrow$ Boltzmann's constant ; $T \rightarrow$ Absolute temperature

- Bubble theory

$$E_i = \frac{3\epsilon E_0}{2\epsilon + 1}$$

where $E_i \rightarrow$ Electric field in a gas bubble which is immersed in a liquid
$\epsilon \rightarrow$ Permittivity of liquid
$E_0 \rightarrow$ Electric field in the liquid in absence of the bubble

- **Break down due to liquid globules**

$$E = 487.7\sqrt{\frac{\sigma}{R\epsilon_1}} \text{ V/cm}$$

where $E \rightarrow$ Critical field at which the globule looses its stability
$\epsilon_1 \rightarrow$ Permittivity of the liquid medium
$\sigma \rightarrow$ Pressure due to surface tension, dyne/cm
$R \rightarrow$ Radius of globule

Dielectric Breakdown of Solids

- Theory of Von Hippel

$$E_c = \frac{2\pi^2 vem}{h}\left[\frac{1}{n_\infty^2} - \frac{1}{\epsilon}\right] \text{ V/m}$$

where $E_c \rightarrow$ Critical field at which the breakdown occurs
$v \rightarrow$ Optical frequency of lattice

$n_\infty \rightarrow$ The index of refraction for infinite wave length
$\in \rightarrow$ Dielectric constant; $m \rightarrow$ Effective mass of electron
$e \rightarrow$ Electron charge; $h \rightarrow$ Plank's constant

Thermal breakdown

- Heat generated per unit volume in unit time

$$W = E^2 f.\frac{\in_r \tan\delta}{1.8 \times 10^{12}} \text{ Watts/cm}^3$$

where $E \rightarrow$ Uniform electric field
$f \rightarrow$ Frequency of applied voltage
$\delta \rightarrow$ Loss angle; $\in_r \rightarrow$ Relative permittivity

For direct voltage

Power dissipated per volume is given by

$$W = \frac{E^2}{\rho} \text{ Watts/m}^3$$

where $\rho \rightarrow$ Resistivity of the insulation, ohm-m

- Power loss

$$\in \tan\delta = \in' \tan\delta' \cdot e^{c(T-T')}$$

where $\in \tan\delta \rightarrow$ Loss factor of the dielectric at a temperature
$\in' \tan\delta' \rightarrow$ Loss factor of dielectric at initial temperature T'
$c \rightarrow$ Coefficient depending on the properties concerned

III Magnetic Properties of Materials

Magnetic field

- **Force on a current Element**

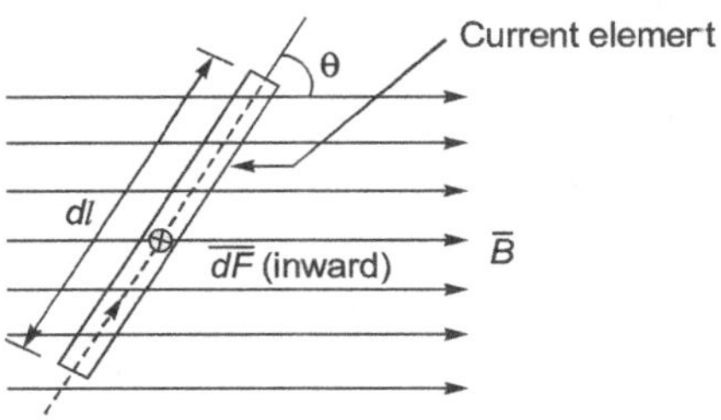

$$\overline{dF} = I(\overline{dl} \times \overline{B})$$

where, $B \rightarrow$ Flux density, Wb/m^2; $dF \rightarrow$ Force on conductor, Newtons
$dl \rightarrow$ Length of current carrying conductor, metres
$I \rightarrow$ Current in conductor

For linear conductor of length l in a uniform field B, $\overline{F} = I\left(\overline{l} \times \overline{B}\right)$

- **Biot savart law**

$$\overline{dB} = \frac{\mu}{4\pi}\frac{I(\overline{dl}\times\overline{r})}{|r|^3}$$

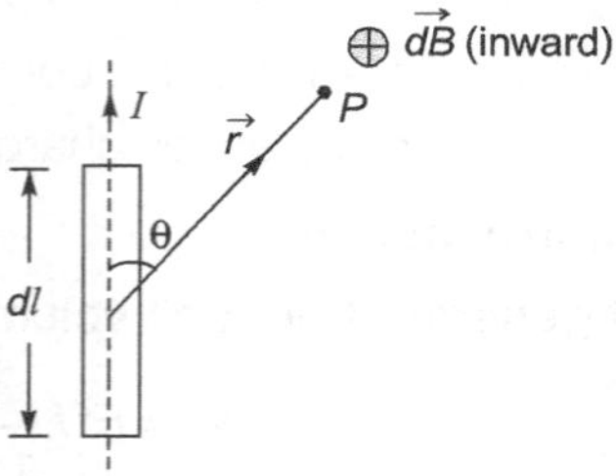

where, $\overrightarrow{dB}$ → Magnetic flux density produced by a current carrying element
$\vec{r}$ → Radius vector

- Flux density produced by a infinitely long current carrying wire at a distance R

$$B_P = \frac{\mu I}{2\pi R}$$

- Torque on a current carrying loop

$$T = BINA\sin\alpha$$

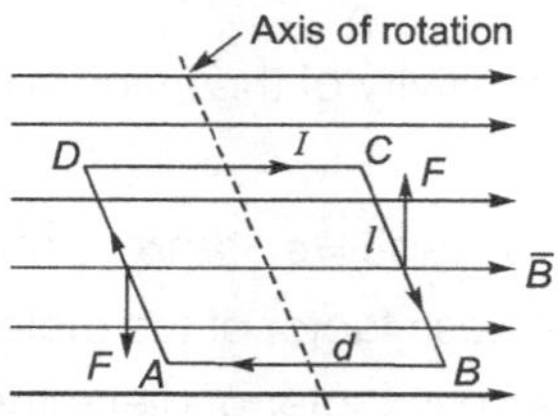

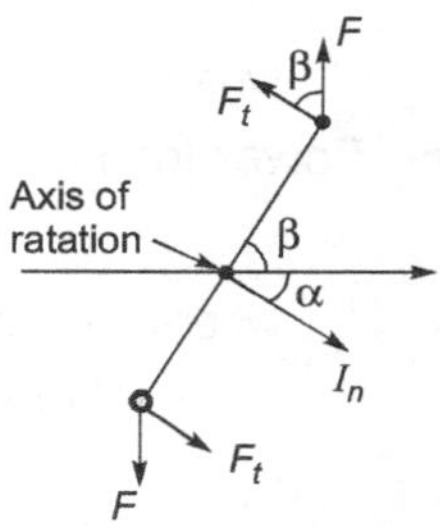

where, A → Area of the loop = ld ; α → Angle between normal to the plane of the loop and direction B ; N → Number of loops

- Magnetic dipole moment, $p_m = I\overline{A}$
- Torque, $\overline{T} = \overline{p}_m \times \overline{B}$

Magnetic Parameters

Permeability of the Medium

$$\mu = \mu_r\,\mu_0$$

where, μ_r → Relative permeability
μ_0 → Permeability of vacuum = $4\pi \times 10^{-7}$ henry-metre^{-1}

Magnetic Moment

Magnetic dipole moment

$$p_m = (\text{Current} \times \text{Area of the coil})\ \text{Ampere-metre}^2$$

For a permanent bar magnet the dipole moment is defined as the product of pole strength and distance between them.

Magnetization

$$\bar{M} = \frac{\bar{p}_m}{\text{Volume}} \quad ; \quad \bar{M} = N\bar{p}_m$$

where, $N \rightarrow$ Number of magnetic dipoles per unit volume

Magnetic Flux Density

$$B = \mu H \quad \text{where, } H \rightarrow \text{Magnetic field intensity}$$

The above equation is valid for the materials in which factor m is constant.

- Magnetic flux density when a magnetic field is applied

$$B = \mu_0 H + \mu_0 M = \mu_0 \mu_r H$$

- Magnetization , $M = (\mu_r - 1)H = \chi_m H$
- Magnetic susceptibility , $\chi_m = \mu_r - 1$

Remember:

☑ For diamagnetic materials χ_m is negative

☑ For paramagnetic materials χ_m is very small and positive

☑ For ferromagnetic materials χ_m is large and positive

Magnetic Moment

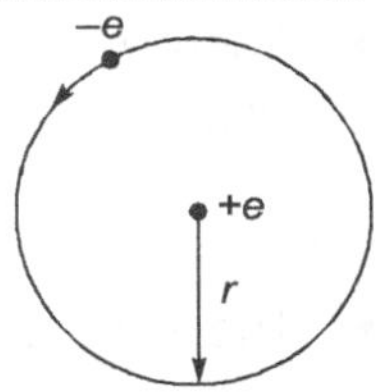

Orbital Dipole Moment

Magnetic dipole moment of the orbit

$$p_m = \frac{1}{2} e\omega_0 r^2 \quad \text{.......(in absence of fields)}$$

where, $\omega_0 \rightarrow$ Angular velocity of electron, $r \rightarrow$ Radius of the orbit

Angular Momentum

- **Orbital angular moment**

$$|M_a| = m\omega_0 r^2 \; ; \quad \frac{p_m}{M_a} = -\frac{e}{2m} \quad \text{where, } m \rightarrow \text{Mass of the electron}$$

Remember:

☑ Ratio of p_m to M_a is independent of the angular velocity and the radius of the orbit.

☑ Atomic unit of magnetic moment is called Bohr magneton.

☑ 1 Bohr magneton = 9.27×10^{-24} ampere-m^2.

- Larmor angular frequency, $\omega_L = \frac{e}{2\,m} B$
- Angular velocity for any value of B, $\omega = \omega_0 + \frac{e}{2m} B$

- Magnetic dipole moment induced by the field, $p_{m,ind} = -\frac{e^2}{4m} \cdot r^2 B$
- Bohr's Magneton, $\beta = \frac{eh}{4\pi m}$ ampere-metre^2

Electron Spin Dipole Moment

It is generated due to spin of electron

$$p_{m,spn} = (\text{No. of unpaired } e^{-}s) \times \beta \;;\; |\bar{p}_{m,\text{spin}}| > |\bar{p}_{m,\text{orb}}|$$

Diamagnetic Materials

These materials do not have permanent dipole moment

$$\bar{p}_m \cong \bar{p}_{m,\text{orb}} + \bar{p}_{m,\text{spin}} = 0$$

Remember:

☑ For perfect diamagnetic material, $\chi_m = -1$.

☑ In general, χ_m comes out to the order of -10^{-5} to -10^{-6}.

Para Magnetic Material

- Spontaneous magnetization for para magnetic material is zero
  Magnetic susceptibility

$$\chi_m = -\frac{N\beta^2\mu_0}{kT}$$

- The magnetic field inside the material, when the effect of internal field is considered : $H_i = H + \gamma M$
  where, γ = Internal or molecular field constant

Curie law

Magnetic susceptibility, $\chi_m = \frac{C}{T}$

where, $C \rightarrow$ Curie constant ; $T \rightarrow$ Temperature

Curie-Weiss law

Magnetic susceptibility, $\chi_m = \frac{C}{T - \theta}$

where, $\theta \rightarrow$ Paramagnetic curie temperature

Ferromagnetism

- Ferromagnetism is chracterized by the presence of parallel alignment of magnetic dipole moment.
- Common characteristic of these materials is partly filled inner electron shells.
- They follow ***Curie-Weiss Law***.

- Below Curie temperature they exhibit ferromagnetism but above curie temperature they behave as ***paramagnetic*** materia s.
  Total internal field which exists in these materials is
  $$H_i = H + \gamma M$$
  where, $\gamma \rightarrow$ Molecular field constant ; $M \rightarrow$ Magnetization
  $H \rightarrow$ Applied field
- Spontaneous magnetization dies out as temperature increases.

Magnetic Anisotropy

In some materials like iron magnetic properties depends on the direction in which they are measured, this phenomenon is called ***magnetic anisotropy***.

Magnetostriction

When a ferromagnetic material is magnetized, its physical dimension changes, this phenomenon is called ***magnetostriction***.

Villari Effect

This is converse of magnetostriction, that is when strain is applied there is change in the magnetic property of the material.

Antiferromagnetism

- These materials have a small susceptibility at all temperatures.
- As the temperature decreases χ_m increases but at a critical temperature T_N called Neel temperature, it goes maximum and if temperature is further lowered it further decreases.
- The variation of susceptibility with temperature of an antiferromagnetic material above critical temperature T_N is given by
  $$\chi_m = \frac{C}{T - \theta}$$
  where, $C \rightarrow$ Curie constant ; $\theta \rightarrow$ Paramagnetic curie temperature

Classification of Magnetic Materials

Type	Susceptibility χ_m	χ_m vs T relation	Examples
1. Diamagnetic	$\sim -10^{-n}$(negative)	Independent	Atoms of solids having closed shells and some metals Au, Ge, etc.
2. Paramagnetic	$\sim 10^{-5}$ (positive)	$\chi_m = C/T$ Curie law or or $\chi_m = C/T - \theta$ Curie Weiss law	Atoms possessing odd number of electrons, ionic crystals etc.
3. Ferromagnetic	Very large & positive	$(\chi_m \rightarrow \infty)$	Iron, cobalt, nickel, godolinium.
4. Antiferromagnetic	Small and positive	χ_m decreases with temperature	Salts and oxides of transition metals, e.g. NiO,. MnF_2.
5. Ferrimagnetic	Large and positive	$\chi_m \rightarrow \infty$	Ferrites, e.g. Fe_3O_4.

IV Conductive Materials

Electrical Conductivity

A very important electrical property of a material is its resistivity.

$$R = \rho \cdot \frac{l}{A}$$

Electrical resistivity is reciprocal of electrical conductivity and is denoted by σ

where, $R \rightarrow$ Resistance of conductor, Ω ; $\rho \rightarrow$ Resistivity of material, Ω-m
$l \rightarrow$ Length of conductor, m ; $A \rightarrow$ Area of cross section, m^2

Ohm's Law

$$J = \sigma E = \frac{I}{A} \text{ A/m}^2 \qquad \text{... Point form}$$

where, $J \rightarrow$ Current density, A/m^2 ; $\sigma \rightarrow$ Conductivity of material, Ω^{-1}-m^{-1}
$E \rightarrow$ Applied electric field, v/m ; $I \rightarrow$ Current, A

Joule's Law

Volume density of heat developed per second, $W = \sigma E^2 = JE$ Watts/m^3

Remember:

This is the energy which the electrons transfer to the lattice in the collision process and is converted into heat.

Mobility and Conductivity

It is the magnitude of the average drift velocity per unit field.

$$\mu_e = \frac{e\tau_c}{m} \text{ also } \sigma = \frac{ne^2\tau_c}{m}$$

Mobility and conductivity have the relation, $\sigma = ne\mu_e$

where, $n \rightarrow$ Number of electrons/unit volume ; $\tau_c \rightarrow$ Collision time, sec;
$e \rightarrow$ Charge of electron ; $m \rightarrow$ Mass of electron ;
$\mu_e \rightarrow$ Mobility of electron, m^2 volt^{-1} sec^{-1}

Drift Velocity of Electron

This velocity is associated with the electric field and is called drift velocity (v_d)

$$v_d = \frac{e\tau_c E}{m} = \mu_e E \text{ (m/sec)}$$

Mean Free Path (*d*)

It is average distance travelled by electron before the collision takes place

$d = v\tau_c$ where, $v \rightarrow$ Average electron velocity

Velocity of an electron

$$v_p = \sqrt{\frac{2E_F}{m}} \quad \text{where, } E_F \rightarrow \text{Fermi energy}$$

Note: At absolute zero, all energy levels below a certain value E_F are filled, and all those above E_F being empty: E_F is the Fermi level of electron.

Relaxation time (τ_r)

It is defined as the time at which the drift velocity of electrons reduces to 37% of its initial value after the removal of the field.

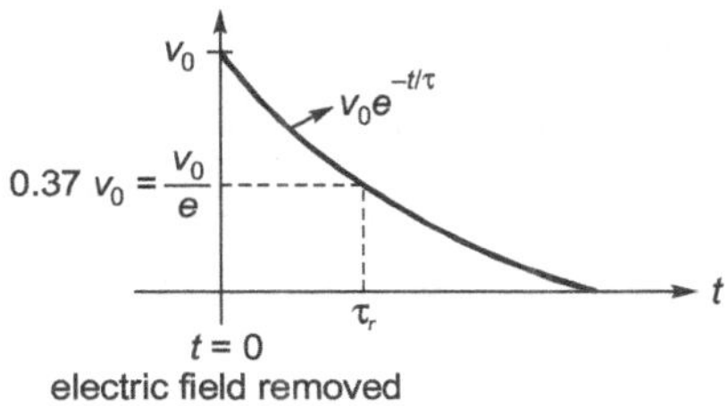

Note: For isotropic materials, the mean time of collision is the same as the relaxation time.

- Mean free path at Fermi level, $d_F = \tau_c \sqrt{\frac{2E_F}{m}}$

Thermal conductivity

- Flux of thermal energy, $Q = -K\left(\frac{dT}{dx}\right)$ Watts/m^2

  where, $K \rightarrow$ Thermal conductivity, Watts/m°C ; $\frac{dT}{dx} \rightarrow$ Temperature gradient, °C/m

- Thermal conductivity, $K = \frac{1}{3}\frac{n\pi^2 k^2 \tau T}{m}$ Watt/mK

  where, $n \rightarrow$ Number of conduction electrons per m^3
  $T \rightarrow$ Temperature
  $K \rightarrow$ Boltzmann's constant

Note: τ varies as T^{-1} above Debye temperature.

Super Conductors

In the state of super conductivity material exhibit zero resistivity and perfect diamagnetism.

Superconductivity appears at low temperature. Example: Hg, Pb, Zn, PbAu, $PbTl_2$, ZrC, CuS

Transition temperature (T_c)

The critical temperature (T_c) is the temperature at which there is change of state from normal to super conducting and vice-versa.

Meissner's Effect

Magnetic susceptibility in a super conductor is negative. This is referred to as perfect diamagnetism. This phenomenon is called Meissner effect.

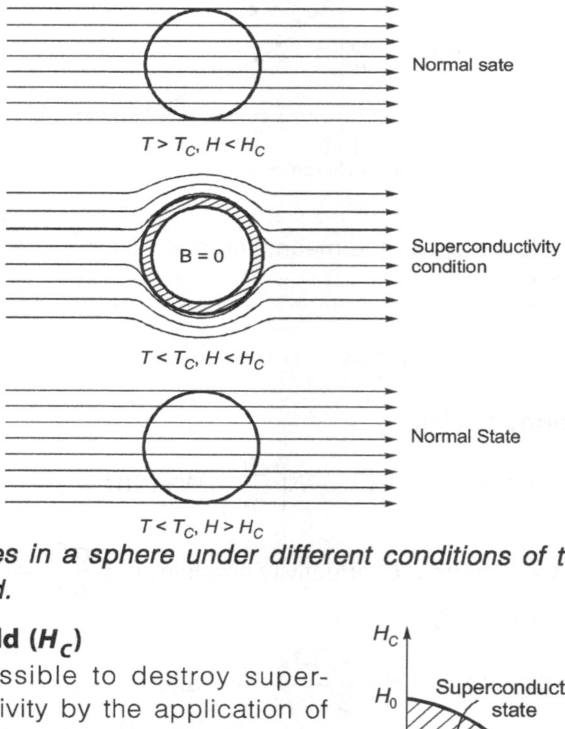

Flux lines in a sphere under different conditions of temperature and field.

Critical Field (H_C)

It is possible to destroy superconductivity by the application of sufficiently strong magnetic field.

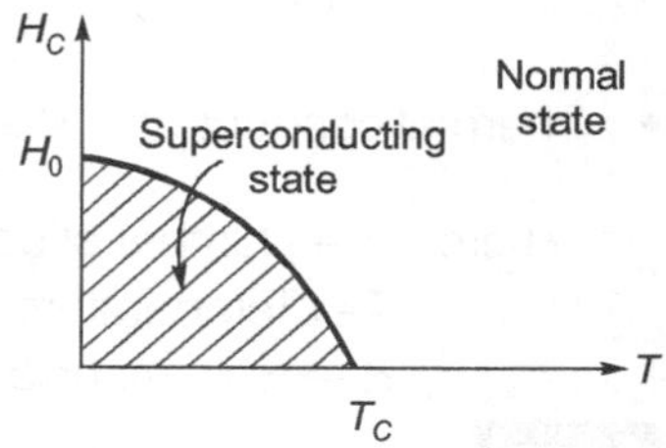

$$H_C = H_0\left[1-\left(\frac{T}{T_C}\right)^2\right]$$

where, $H_0 \rightarrow$ Critical field at absolute zero temperature ; $H_C \rightarrow$ Critical field at any temperature T ; $T_C \rightarrow$ Transition temperature.

Silsbee's Rule

In a long superconductor wire of radius R, the superconductivity may be destroyed when a current I exceeds the critical current value I_C

$$I_C = 2\pi R H_C$$

Factors Affecting the Super Conductivity

Frequency Effect : When frequency increases above 10^{13} Hz (infrared region); material loses its super conductivity.

Entropy Effect : Increase in entropy results in, change in state from super conducting to normal.

Isotope Effect : It has been observed that the critical temperature of a super conductor varies with isotopic mass as

$$T_C \propto \frac{1}{\sqrt{M}} \qquad \text{where, } M \rightarrow \text{Mass of isotope}$$

Types of superconductor

Type-I

- These are also called soft or ideal superconductor.
- Their critical temperature and critical fields are low.
- The change of state from normal to superconducting and vice-versa is abrupt.
- They follow Silsbee rule and exhibit Meissner's effect.

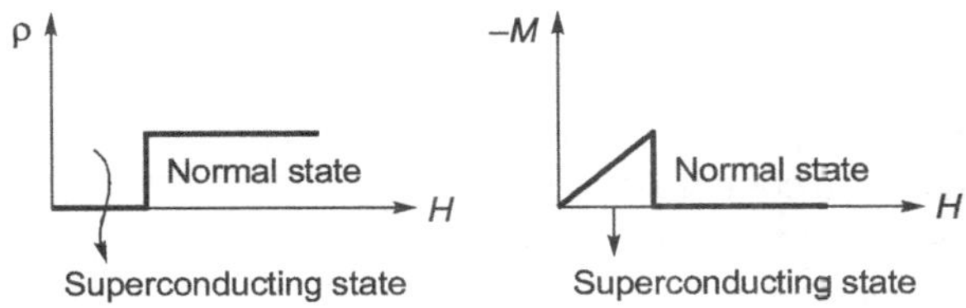

Ex: Pd, Hg, Zn, Al etc.

Type-II

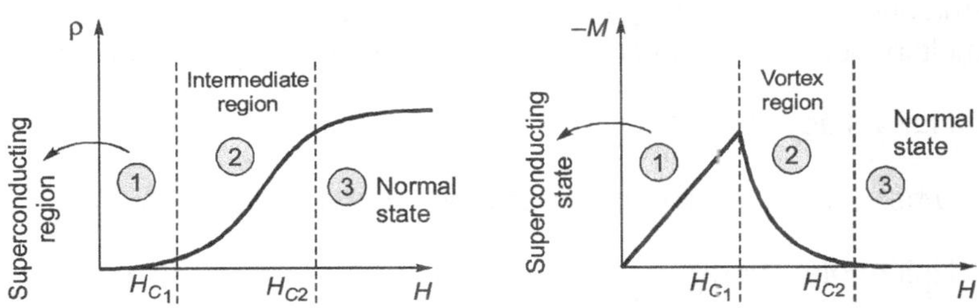

- These superconductor are also known as hard or non ideal superconductor.
- Their critical temperature and critical fields are high.

- The change of state from normal to superconducting and vice-versa is gradual.
- They exhibit incomplete Meissner's effect and Silsbee rule.

Ex: Nb_3 Al (Niobium aluminium), NbTi (Niobium titanate) and Nb_3S_n etc.

V Semiconductors

Conductivity of Semiconductor

- The current flowing through a pure semiconductor is carried by two kinds of carriers, i.e., electrons and holes.
- Conductivity of intrinsic semiconductor

$$\sigma = \frac{n_e e^2 \tau_e}{m_e} + \frac{n_h e^2 \tau_h}{m_h}$$

Where, $\tau_e, \tau_h \rightarrow$ Relaxation times for electrons and holes respectively.
$m_e, m_h \rightarrow$ Effective mass of electrons and holes respectively.
$n_e, n_h \rightarrow$ Number of conduction electrons and holes respectively.

- Current density, $J = ne\,v_d$; where, $v_d \rightarrow$ Drift velocity

Mobility

- Mobility of electron, $\mu_e = \dfrac{e\tau_e}{m_e}$
- Mobility of Hole, $\mu_h = \dfrac{e\tau_h}{m_h}$

Hall Effect

If a specimen (metal or semiconductor) carrying a current-I is placed in a transverse magnetic field B, an electric field E is induced in the direction perpendicular to both I and B.

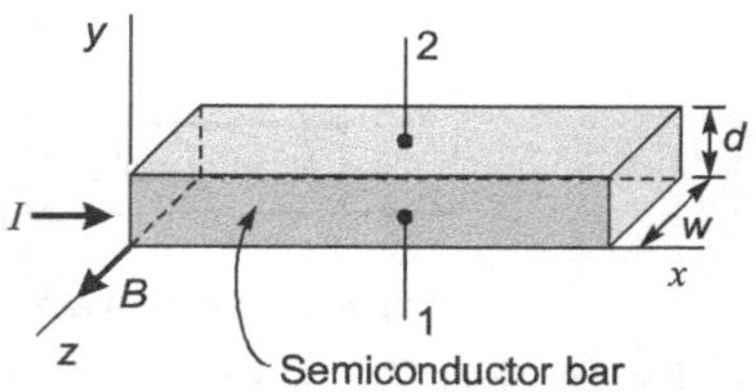

This phenomenon, known as Hall effect, is used to determine whether a semiconductor is n-or p-type and to find the carrier concentration. Also, by simultaneously measuring the conductivity, the mobility μ can be calculated.

- Hall Voltage, $V_H = \dfrac{BI}{\rho w}$

where, $w \rightarrow$ Width of the specimen ; $\rho \rightarrow$ the charge density

- Hall Angle, $\tan\theta_H = \dfrac{E_y}{E_x}$; $\theta_H = \tan^{-1}(B\mu)$
- Hall Coefficient, $R_H = \dfrac{E}{BJ} = \dfrac{1}{en_h} = \dfrac{-1}{en_e}$

where, $E \rightarrow$ Applied electric field ; $B \rightarrow$ Applied magnetic field
$J \rightarrow$ Current density in the

If the conductivity is measured together with the Hall coefficient, the mobility can be determined from

$$\mu = \sigma R_H$$

Note: ....

For intrinsic semiconductor

$$R_H = \frac{1}{n_i e}\left[\frac{\mu_p - \mu_n}{\mu_p + \mu_n}\right]$$

....

Einstein Relation (Electrical Mobility Equation)

$$D_h = \left(\frac{kT}{e}\right)\mu_h \ , \quad D_e = \left(\frac{kT}{e}\right)\mu_e$$

where $D_h, D_e \rightarrow$ Diffusion constant for holes and electron respectively

$$\frac{D_h}{\mu_h} = \frac{D_e}{\mu_e} = \frac{kT}{e} = V_T$$

VI Insulating Materials

Complex Permittivity

When an alternating field is applied to a dielectric, the relative dielectric constant become a complex quantity whose value is given by

$$\epsilon_r^* = \epsilon_r' - j\epsilon_r''$$

Where, ϵ_r' = real part of dielectric constant

ϵ_r'' = imaginary part of dielectric constant

Remember: ....

☑ When a time varying electric field is applied to a dielectric material, the response is not entirely instantaneous.

$$D = \epsilon'' E$$

☑ The imaginary part is responsible for energy loss in the material.

☑ The real part represents the relative permittivity.

....

Debye Equations

It gives the variation of both the inphase and out of phase component of ϵ_r as a function of angular frequency ω with relaxation time τ

$$\epsilon_r' = \epsilon_\infty + \frac{\epsilon_s - \epsilon_\infty}{1 + \omega^2\tau^2}$$

$$\epsilon_r'' = \frac{(\epsilon_s - \epsilon_\infty)\omega\tau}{1+\omega^2\tau^2}$$

where, ϵ_∞ = dielectric constant at infinite frequency

ϵ_s = dielectric constant under static field

τ = relaxation time

Dielectric losses

The absorption of electric energy by the dielectric material subjected to an alternating electric field is known as the dielectric losses. This results in dissipation of the electrical energy as heat in a material.

Note:

☑ Dielectric loss occurs due to two reason
- Oscillation of the ion
- Continuous change in orientation of dipoles

Energy absorbed per m³

$$W = \frac{1}{2}\omega \epsilon_0 \epsilon_r'' E_0^2$$

Dielectric loss tangent

It represents how lossy the material is for electrical AC signals.

$$\tan\delta = \frac{\epsilon_r''}{\epsilon_r'}$$

Loss Factor

$$\epsilon_r' \tan\delta = \epsilon_r''$$

Lossy Capacitor

Equivalent circuits

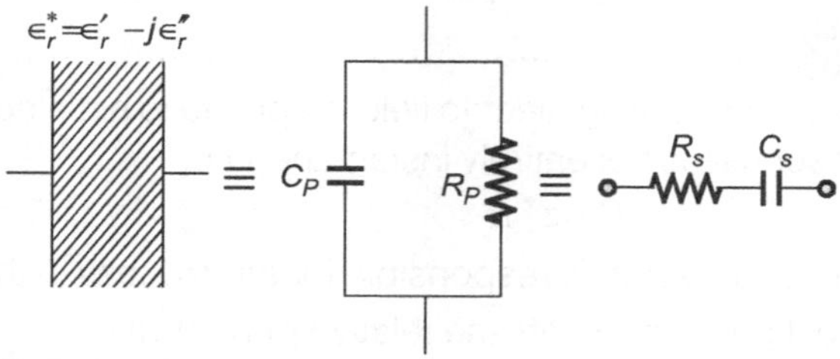

For parallel circuit

$$R_P = \frac{1}{\omega\epsilon_r'' C_0}$$

where C_0 = capacitance; $C_0 = \dfrac{\epsilon_0 A}{d}$; $C_P = \epsilon_r' C_0$

For series circuit

$$R_s = \frac{R_p}{1+(\omega C_P R_P)^2}$$

$$C_s = \frac{1+(\omega C_P R_P)^2}{\omega^2 C_P R_1^2}$$

Loss tangent

For series circuit : $\tan\delta = \omega C_s R_s$

For parallel circuit : $\tan\delta = \dfrac{1}{\omega C_P R_P}$

Quartz

Quartz is a piezoelectric material, used for high frequency oscillation

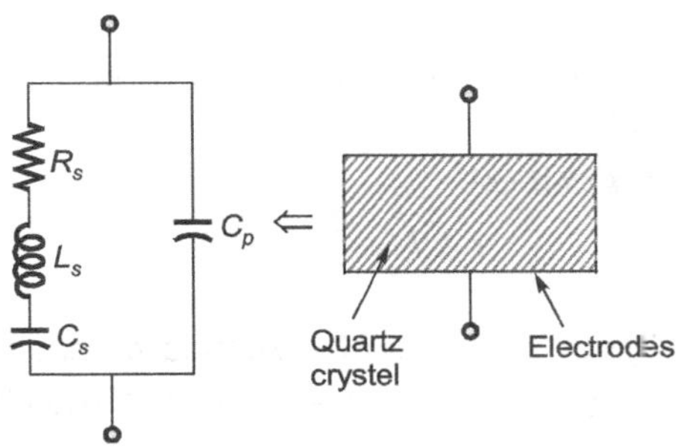

$$z = \left(\frac{1}{j\omega C_S} + j\omega L + R_1\right) \Bigg\| \left(\frac{1}{j\omega C_p}\right)$$

- Q-factor

$$Q = \frac{\omega_S L_S}{R_S} = \frac{1}{\omega R_S C_S}$$

- Series resonance frequency

$$\omega_S = \frac{1}{\sqrt{L_S C_S}}$$

- Impedance at series resonance frequency

$$Z\big|_{\omega=\omega_S} = R_S$$

- Parallel resonance frequency

$$\omega_p = \sqrt{\frac{1}{L_s}\left(\frac{1}{C_s} + \frac{1}{C_P}\right)} = \omega_s\sqrt{1+\frac{C_s}{C_p}}$$

- Impedance at parallel resonance frequency

$$Z\big|_{\omega=\omega_p} = \frac{R_s^2 + \left(\omega_p L_s - \dfrac{1}{\omega_p C_s}\right)}{R_s}$$

$$Z\big|_{\omega=\omega_p} > Z\big|_{\omega=\omega_s}$$

Note:

☑ Quartz crystal can be modelled as an electrical network with a low impedance (series) and a high impedance (parallel).

VII Structure of Materials

Unit Cell

The smallest unit of crystal structure which completely defines the behaviour of a crystal.

Number of atoms per unit cell

$$n = \frac{\rho a^3 N_A}{M}$$

where, $a \rightarrow$ Edge of unit cell ; $M \rightarrow$ Atomic weight of element
$\rho \rightarrow$ Density of material ; $N_A \rightarrow$ Avogadro's number $= 6.023 \times 10^{23}$

Note:

☑ Unit cell is cubic in shape.

☑ Unit cell is repeated to form a crystal. So most important characteristic of crystal is periodicity of unit cell inside the crystal.

Simple Cubic (SC)

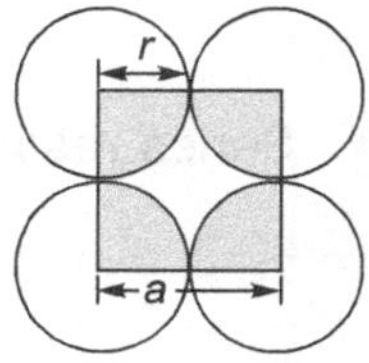

- Distance between adjacent atom : $d_{sc} = 2r = a$
  where, $r \rightarrow$ Radius of an atom ;
  $a \rightarrow$ Edge length of cube
- Effective number of atoms in a unit cell

$$Z_{sc} = 8 \times \frac{1}{8} = 1$$

- Atomic packing fraction : $APF = \dfrac{\text{Volume occupied by atoms}}{\text{Volume of the cell}}$

$$APF_{SC} = \frac{\pi}{6} = 0.523$$

- Coordination number of simple cubic is 6.

Body Centered Cubic (BCC)

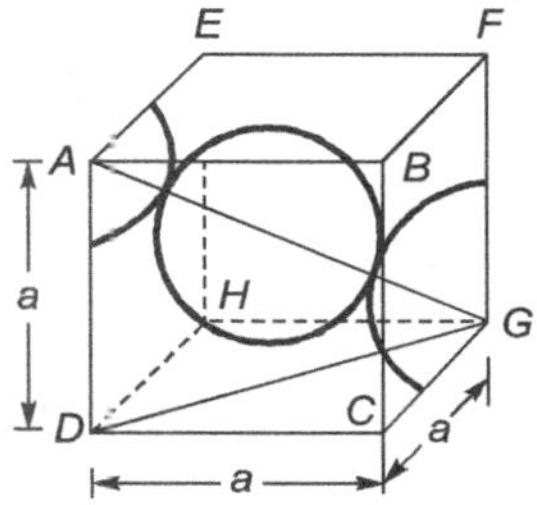

- The distance between adjacent atoms

$$d_{BCC} = 2r = \frac{\sqrt{3}}{2}a$$

- Effective number of atom in unit cell

$$Z_{BCC} = 8 \times \frac{1}{8} + 1 = 2$$

- Atomic packing fraction

$$APF_{BCC} = \frac{\sqrt{3}}{8}\pi = 0.68$$

- The coordination number of Body Centered Cubic is 8.

Face Centered Cubic (FCC)

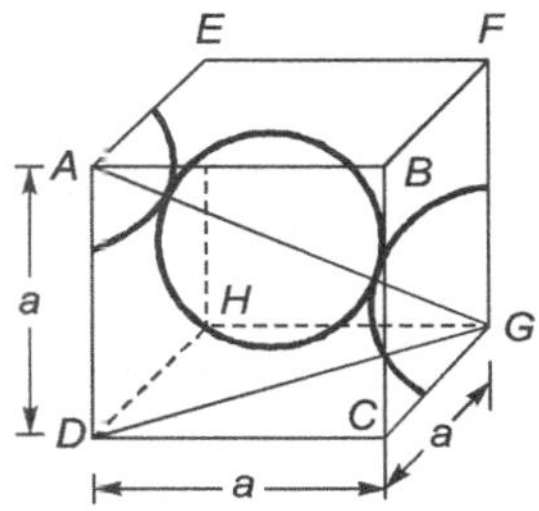

- The distance between adjacent atoms

$$d_{FCC} = 4r = a\sqrt{2}$$

- Effective number of atoms in a unit cell

$$Z_{FCC} = 4$$

- Atomic packing fraction

$$APF_{FCC} = \frac{\pi}{3\sqrt{2}} = 0.7404$$

- The coordination number of face centered cubic is 12.

Miller Index , $MI = \dfrac{\text{Cell edge length}}{\text{Intercept by plane}}$

Comparison between Crystal Structures

Structures	Number of atoms/ Unit cell	Co-ordination Number	APF	Atomic Radius	Examples
Diamond Cubic	8	4	0.34	$\frac{a\sqrt{3}}{8}$	C, Si, Ge Diamond, GaAs etc.
Simple Cubic	1	6	0.52	$\frac{a}{2}$	Mn, Fluorspar, Polonium etc.
Body Centered Cubic	2	8	0.68	$\frac{a\sqrt{3}}{4}$	Li, Na, K Cr, Fe (α, δ)
Face Centered Cubic	4	12	0.74	$\frac{a\sqrt{2}}{4}$	Au, Ag, Ni, Cu, Al, Pt, Pb, Nacl, Fe(γ)
Hexagonal Closed Packing	6	12	0.74	$\frac{c}{a} = \frac{\text{Max. dimension}}{\text{Min. dimension}} = 1.63$	Be, Cd, Zn, Mg

where, $a \rightarrow$ Cell width

Comparison between Graphite and Diamond

Properties	Diamond	Graphite
Structure	Tetrahedral	Hexagonal
Electrical properties	Insulator	Conductor
Hardness	Extremely hard	Soft
Transparency	Transparent	Opaque

Bravais Crystal Structure

Crystal structure	Cell dimension	Angle between axis	Bravais lattice
Orthorhombic	$a \neq b \neq c$	$\alpha = \beta = \gamma = 90°$	P, I, C, F
Monoclinic	$a \neq b \neq c$	$\alpha = \beta = 90° \neq \gamma$	P, C
Triclinic	$a \neq b \neq c$	$\alpha \neq \beta \neq \gamma \neq 90°$	P
Hexagonal	$a = b \neq c$	$\alpha = \beta = 90°$ $\gamma = 120°$	P
Teteragonal	$a = b \neq c$	$\alpha = \beta = \gamma = 90°$	P, I
Trigonal/ Rhombohedral	$a = b = c$	$\alpha = \beta = \gamma \neq 90°$	P
Cubic	$a = b = c$	$\alpha = \beta = \gamma = 90°$	P, I, F

where, $P \rightarrow$ Simple cubic (Primitive); $I \rightarrow$ Body center cubic
$F \rightarrow$ Face center cubic; $C \rightarrow$ Side center

Structural imperfections

- Point defect
  - → Vacancy defect: is a point defect which involves a missing atom within crystal structure
  - → Impurity defect
    - → Substitutional
    - → Interstitial
- Line defect
- Surface defect
- Volume defect

VIII Ceramic Materials

- These are the materials containing the oxides of metals and non-metals. They are characterised by high melting point, big resistivity brittle, high resistance to a chemical action, large dielectric strength, simple to manufacture and economically cheap.

Classifications :

- Abrasives: Eg: Al_2O_3, Mgo, Corundum (Sic)
  powdered used to polish and grinding the substrate.
- Pure Oxide Ceramics : Al_2O_3, Ago, SiO_2
- Fire clay products : Porcelain, Bricks, tiles.
- Glass Ceramics
- Rocks : Minerals
- Refractories : Magnesita

Ceramics are also distinguished based upon relative permittivity as

Material	Relative permittivity (ε_r)
Glass	5.9
Porcelain	4.1 - 11.0
Fused silica	4
Backelite	4.4
Mica (dielectric ceramic)	5.7
Rochell's salt	500
$BaTiO_3$	1500
PZT	1500
80% $BaTiO_3$ + 20% $PbTiO_3$ (ferroelectric ceramic)	850

- $\epsilon_r < 12$: Ceramics having relative permittivity ($\epsilon_r < 12$) used as insulators. Ex.: Porcelain, steatite, alumina.
- $\epsilon_r > 12$: This ceramics used in capacitor applications due to high dielectric constant. Ex.: TiO_2 (Kutile)

Applications of Ceramics:

- For low and high voltage insulation
- In high frequency applications
- As insulators
- As piezo electric transducers
- As metal cutting tools (Tungsten carbide and glass free Al_2O_3)
- As substrate in the IC chip fabrication.

■■■■■

Electromagnetic Theory

11

I Vector Calculus

Vector Analysis

Vector algebra

- **Addition**

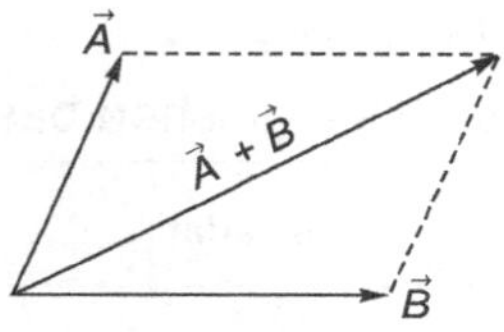

$$\vec{A}+\vec{B} = \vec{B}+\vec{A}$$

Note:

☑ It follow commutative law: $\vec{A}+\vec{B}=\vec{B}+\vec{A}$

☑ It also obey associative law: $(\vec{A}+\vec{B})+\vec{C}=\vec{A}+(\vec{B}+\vec{C})$

- **Subtraction**

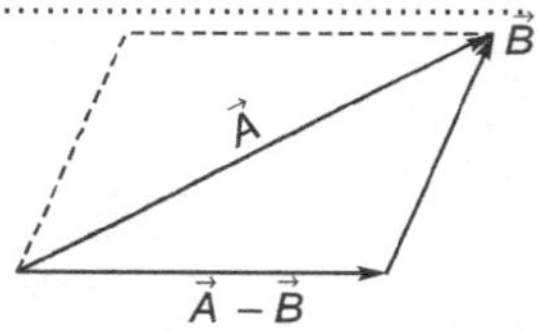

$$\vec{A}-\vec{B} = \vec{A}+(-\vec{B})$$

–ve sign shows reversal of direction

Note:

☑ It does not follow cumulative law : $\vec{A}-\vec{B}\neq\vec{B}-\vec{A}$

But follow associative law.

- **Multiplication**
  - Scalar or Dot product

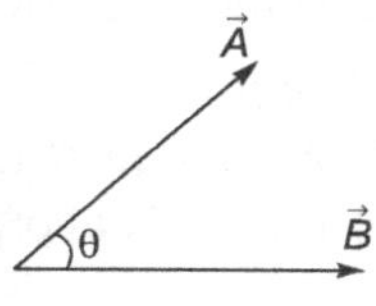

$$\vec{A}\cdot\vec{B} = |A||B|\cos\theta$$

Note:

☑ $\bar{A}\cdot\bar{B} = \bar{B}\cdot\bar{A}$

☑ Dot product of vector is a scalar.

  - **Vector or cross product**

$$\vec{A}\times\vec{B} = |A||B|\sin\theta\,\hat{i}_n$$

Where, $\hat{i}_n$ = unit vector in the direction of perpendicular to $\vec{A}$ and $\vec{B}$

Note:

☑ To find direction $\hat{i}_n$, use right hand screw rule.

☑ Cross product of two vector is always a vector.

☑ $\vec{A}\times\vec{B} = (\vec{B}\times\vec{A})$

☑ $(\vec{A}\times\vec{B})\times\vec{C} = \vec{A}\times(\vec{B}\times\vec{C})$

- Scalar triple product

$$\vec{A}\times(\vec{B}\times\vec{C}) = \vec{B}\cdot(\vec{C}\times\vec{A})-\vec{C}\cdot(\vec{A}\times\vec{B})$$

- **A vector multiplied/divided by a scalar**
  - $S\cdot\vec{V}$ = it increases the magnitude by S unit
  - $\dfrac{\vec{V}}{S}$ = it decreases the magnitude by S unit
  - $\dfrac{\vec{A}}{\vec{B}}$ = it does not exist

- **Differentiation of vectors**

If $\vec{A} = A_x i_x + A_y i_y + A_z i_z$

$$d\vec{A} = dA_x i_x + dA_y i_y + dA_z i_z$$

For cylindrical and spherical coordinate system:

$$d\vec{A} = dA_r i_r + dA_\phi i_\phi + dA_z i_z$$

and $$d\vec{A} = dA_r i_r + dA_\theta i_\theta + dA_\phi i_\phi$$

II Cartesian Coordinate System & Vector Calculus

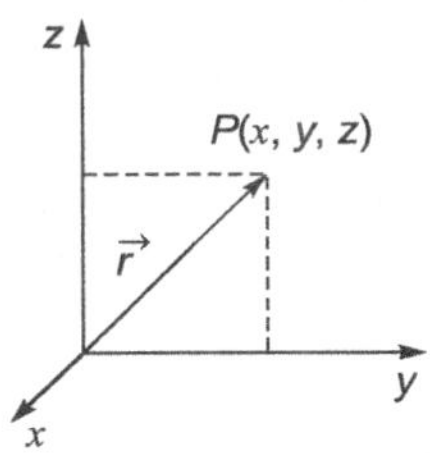

Range

$-\infty < x < \infty$

$-\infty < y < \infty$

$-\infty < z < \infty$

where $x, y, z \rightarrow$ Perpendicular distances from the axis;

$P \rightarrow$ Arbitrary point

Remember:

- ☑ From the definition of dot product

$$\hat{i}_x \cdot \hat{i}_x = \hat{i}_y \cdot \hat{i}_y = \hat{i}_z \cdot \hat{i}_z = 1$$

$$\hat{i}_x \cdot \hat{i}_y = \hat{i}_y \cdot \hat{i}_z = \hat{i}_z \cdot \hat{i}_x = 0 \quad \text{[orthogonality]}$$

- ☑ From the definition of cross product

$$\hat{i}_x \times \hat{i}_x = \hat{i}_y \times \hat{i}_y = \hat{i}_z \times \hat{i}_z = 0$$

$$\hat{i}_x \times \hat{i}_y = \hat{i}_z \,;\, \hat{i}_y \times \hat{i}_z = \hat{i}_x \,;\, \hat{i}_z \times \hat{i}_x = \hat{i}_y \quad \text{[orthonormality]}$$

where $\hat{i}_x, \hat{i}_y, \hat{i}_z$ are the unit vectors along the x, y, z direction respectively.

General Parameters	
Position Vector of P	$\vec{r} = x\hat{i}_x + y\hat{i}_y + z\hat{i}_z$
Displacement vector	$\vec{dl} = dx\, \hat{i}_x + dy\, \hat{i}_y + dz\, \hat{i}_z$ $\lvert\vec{dl}\rvert = \sqrt{(dx)^2 + (dy)^2 + (dz)^2}$
Unit vector along $\vec{r}$	$\dfrac{\vec{r}}{\lvert\vec{r}\rvert} = \dfrac{x\hat{i}_x + y\hat{i}_y + z\hat{i}_z}{\sqrt{x^2 + y^2 + z^2}}$
Differential length elements	$dx\, \hat{i}_x$, $dy\, \hat{i}_y$, $dz\, \hat{i}_z$
Differential area ($\vec{ds}$)	$dy\, dz\, \hat{i}_x$, $dz\, dx\, \hat{i}_y$, $dx\, dy\, \hat{i}_z$
Differential volume (dV)	$dx\, dy\, dz$

Cylindrical Coordinate System

where, $\rho \rightarrow$ Shortest distance from reference z-axis

$\phi \rightarrow$ Orientation angle (azimuth) around the reference

$z \rightarrow$ Height of the point P along the reference z-axis

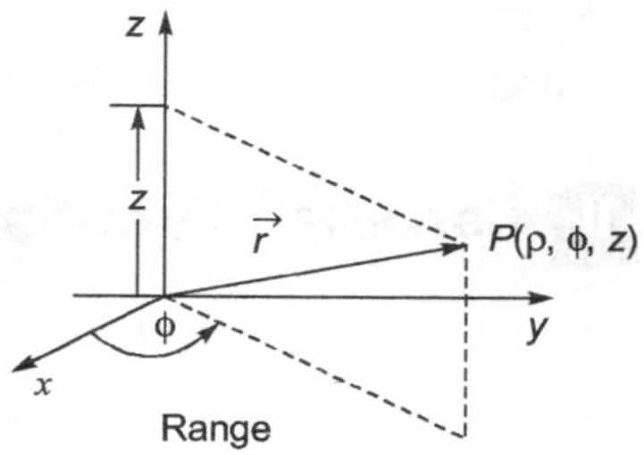

Range

$$\begin{cases} 0 \le \rho < \infty \\ 0 \le \phi < 2\pi \\ -\infty < z < \infty \end{cases}$$

Remember:

$\hat{i}_\rho, \hat{i}_\phi, \hat{i}_z$ follow orthogonality and orthonormality conditions, (where, $\hat{i}_\rho, \hat{i}_\phi, \hat{i}_z$ are unit vectors along ρ, ϕ, z direction).

General Parameters	
Position Vector	$\vec{r} = \rho\hat{i}_\rho + z\hat{i}_z$
Displacement vector	$\vec{dl} = d\rho\hat{i}_\rho + \rho d\phi\hat{i}_\phi + dz\hat{i}_z$ $\lvert\vec{dl}\rvert = \sqrt{(d\rho)^2 + (\rho d\phi)^2 + (dz)^2}$
Differential length	$d\rho\hat{i}_\rho, \rho d\phi\hat{i}_\phi, dz\hat{i}_z$
Differential area $\vec{ds}$	$\rho d\phi\, dz\, \hat{i}_\rho, d\rho dz\hat{i}_\phi, \rho d\phi d\rho\hat{i}_z$
Differentiated volume (dV)	$\rho d\rho\, d\phi\, dz$

Spherical Coordinate System

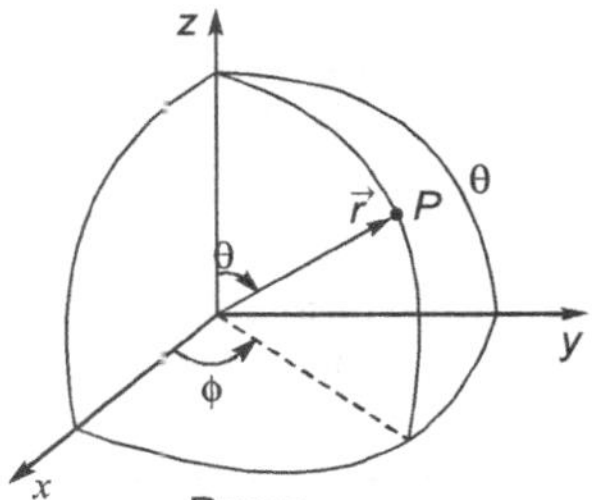

where, $r \to$ Shortest distance of P from origin

$\theta \to$ Angle between the z-axis and position vector, known as colatitudes

$\phi \to$ Angle measured from x-axis in xy plane, known as azimuth.

Range

$0 \le r < \infty$
$0 \le \theta \le \pi$
$0 \le \phi < 2\pi$

Remember: ....

$\hat{i}_r, \hat{i}_\theta, \hat{i}_\phi$ are orthogonal and orthonormal.

....

General Parameters	
Position Vector	$\vec{r} = r\,\hat{i}_r$
Displacement vector	$\vec{dl} = dr\hat{i}_r + rd\theta\hat{i}_\theta + r\sin d\phi\hat{i}_\phi$ $\lvert\vec{dl}\rvert = \sqrt{(dr)^2 + (rd\theta)^2 + (r\sin\theta d\phi z)^2}$
Differential length	$dr\ \hat{i}_r, rd\theta\ \hat{i}_\theta, r\sin\theta\ d\phi\ \hat{i}_\phi$
Differential area $\vec{ds}$	$r^2\sin\theta d\theta\ d\phi\ \hat{i}_r, r\sin\theta\ drd\phi\ \hat{i}_\theta, rdr\ d\theta\ \hat{i}_\theta$
Differentiated volume (dV)	$r^2\sin\theta drd\theta\ d\phi$

Note: ....

Parameter			Scaling Factor		
u	**v**	**w**	h_1	h_2	h_3
x	y	z	1	1	1
ρ	ϕ	z	1	ρ	1
r	θ	ϕ	1	r	$r\sin\theta$

- ☑ The table can be used to calculate general parameters of any coordinate system, like,
- ☑ Differential length $(\vec{dl}) = h_1\, du\, \hat{i}_u + h_2 dv \hat{i}_v + h_3 dw \hat{i}_w$
- ☑ Surface $(\vec{ds}) = h_2\, h_3 dv\, dw\, \hat{i}_u + h_3 h_1\, dw\, du\, \hat{i}_v + h_1\, h_2\, du\, dv\, \hat{i}_w$
- ☑ Volume $(dv) = h_1\, h_2\, h_3\;\; du\, dv\, dw$

Relationship Between Different Sets

	Cartesian (x, y, z)	**Cylindrical** (ρ, ϕ, z)	**Spherical** (r, θ, ϕ)
Cartesian (x, y, z)		$x = \rho\cos\phi$ $y = \rho\sin\phi$ $z = z$	$x = r\sin\theta\cos\phi$ $y = r\sin\theta\sin\phi$ $z = r\cos\theta$
Cylindrical (ρ, ϕ, z)	$\rho = \sqrt{x^2 + y^2}$ $\phi = \tan^{-1}\dfrac{y}{x}$ $z = z$		$\rho = r\sin\theta$ $\phi = \phi$ $z = r\cos\theta$
Spherical (r, θ, ϕ)	$r = \sqrt{x^2 + y^2 + z^2}$ $\theta = \tan^{-1}\dfrac{\sqrt{x^2 + y^2}}{z}$ $\phi = \tan^{-1}\dfrac{y}{x}$	$r = \sqrt{\rho^2 + z^2}$ $\theta = \tan^{-1}\dfrac{\rho}{z}$ $\phi = \phi$	

Relationship Between Component Vectors

- Cartesian and cylindrical

$$\begin{bmatrix} \hat{i}_\rho \\ \hat{i}_\phi \\ \hat{i}_z \end{bmatrix} = \begin{bmatrix} \cos\phi & \sin\phi & 0 \\ -\sin\phi & \cos\phi & 0 \\ 0 & 0 & 1 \end{bmatrix} \begin{bmatrix} \hat{i}_x \\ \hat{i}_y \\ \hat{i}_z \end{bmatrix}$$

- Cartesian and spherical

$$\begin{bmatrix} \hat{i}_r \\ \hat{i}_\theta \\ \hat{i}_\phi \end{bmatrix} = \begin{bmatrix} \sin\theta\cos\phi & \sin\theta\sin\phi & \cos\theta \\ \cos\theta\cos\phi & \cos\theta\sin\phi & -\sin\theta \\ -\sin\phi & \cos\phi & 0 \end{bmatrix} \begin{bmatrix} \hat{i}_x \\ \hat{i}_y \\ \hat{i}_z \end{bmatrix}$$

- Cylindrical and spherical

$$\begin{bmatrix} \hat{i}_r \\ \hat{i}_\theta \\ \hat{i}_\phi \end{bmatrix} = \begin{bmatrix} \sin\theta & 0 & \cos\theta \\ \cos\theta & 0 & -\sin\theta \\ 0 & 1 & 0 \end{bmatrix} \begin{bmatrix} \hat{i}_\rho \\ \hat{i}_\phi \\ \hat{i}_z \end{bmatrix}$$

Scalar Triple Product

Geometrically, $\vec{A}\cdot(\vec{B}\times\vec{C}) = \vec{B}\cdot(\vec{C}\times\vec{A}) = \vec{C}\cdot(\vec{A}\times\vec{B})$

Vector Triple Product

Geometrically, $\vec{A}\times(\vec{B}\times\vec{C}) = \vec{B}(\vec{A}\cdot\vec{C}) - \vec{C}(\vec{A}\cdot\vec{B})$

Gradient

Gradient of a scalar field (V) is a vector that represents the magnitude and direction of maximum spatial rate of change of scalar field.

$$\vec{E} = \nabla V \qquad \text{(here } \vec{E} \text{ is a vector)}$$

Generally,

$$\nabla V = \frac{1}{h_1}\frac{\partial V}{\partial u}\hat{i}_u + \frac{1}{h_2}\frac{\partial V}{\partial v}\hat{i}_v + \frac{1}{h_3}\frac{\partial V}{\partial w}\hat{i}_w$$

Cartesian coordinates,

$$\nabla V = \frac{\partial V}{\partial x}\hat{i}_x + \frac{\partial V}{\partial y}\hat{i}_y + \frac{\partial V}{\partial z}\hat{i}_z$$

Cylindrical coordinates

$$\nabla V = \frac{\partial V}{\partial \rho}\hat{i}_\rho + \frac{1}{\rho}\frac{\partial V}{\partial \phi}\hat{i}_\phi + \frac{\partial V}{\partial z}\hat{i}_z$$

Spherical coordinates

$$\nabla V = \frac{\partial V}{\partial r}\hat{i}_r + \frac{1}{r}\frac{\partial V}{\partial \theta}\hat{i}_\theta \frac{1}{r\sin\theta}\frac{\partial V}{\partial \phi}\hat{i}_\phi$$

Divergence

Generally, if $\vec{A} = A_u\hat{i}_u + A_v\hat{i}_v + A_w\hat{i}_w$

then $\nabla\cdot\vec{A} = \frac{1}{h_1 h_2 h_3}\left[\frac{\partial}{\partial u}(h_2 h_3 A_u) + \frac{\partial}{\partial v}(h_3 h_1 A_v) + \frac{\partial}{\partial w}(h_1 h_2 A_w)\right]$

Co-ordinate System	Divergence
(i) Cartesian	$\nabla.\vec{A} = \frac{\partial A_x}{\partial x} + \frac{\partial A_y}{\partial y} + \frac{\partial A_z}{\partial z}$
(ii) Cylindrical	$\nabla.\vec{A} = \frac{1}{\rho}\frac{\partial}{\partial \rho}(\rho A_\rho) + \frac{1}{\rho}\frac{\partial}{\partial \phi}(A_\phi) + \frac{\partial A_z}{\partial z}$
(iii) Spherical	$\nabla.\vec{A} = \frac{1}{r^2}\frac{\partial}{\partial r}(r^2 A_r) + \frac{1}{r\sin\theta}\frac{\partial}{\partial \theta}(\sin\theta A_\theta) + \frac{1}{r\sin\theta}\frac{\partial}{\partial \phi}(A_\phi)$

Properties of Divergence Operator

(i) $\nabla(A + B) = \nabla\cdot A + \nabla\cdot B$

(ii) $\nabla(VA) = V\nabla \cdot A + A \cdot \Delta V$

(iii) A vector $\vec{A}$ is rotational vector with magnitude being that of the maximum rotation of $\vec{A}$ per unit area as the area tends to zero.

Curl

The curl of a vector $\vec{A}$ is rotational vector with magnitude being that of the maximum rotation of $\vec{A}$ per unit area as the area tends to zero.

Remember:

- Divergence of Curl is always zero, i.e., $\left[\vec{\nabla}\cdot(\vec{\nabla}\times\vec{A})\right]=0.$
- Curl of gradient is always zero, i.e., $\left[\vec{\nabla}\times(\nabla A)\right]=0.$
- A vector $\vec{A}$ is solenoidal if its divergence is zero.
- A vector $\vec{A}$ is irrotational if its curl is zero.

Generally, if $\vec{A} = A_u\,\hat{i}_u + A_v\,\hat{i}_v + A_w\,\hat{i}_w$

$$\nabla\times\vec{A} = \frac{1}{h_1 h_2 h_3}\begin{bmatrix} h_1\,\hat{i}_u & h_2\,\hat{i}_v & h_3\,\hat{i}_w \\ \dfrac{\partial}{\partial u} & \dfrac{\partial}{\partial v} & \dfrac{\partial}{\partial w} \\ h_1 A_u & h_2 A_v & h_3 A_w \end{bmatrix}$$

Properties of Curl Operator

(i) $\nabla \cdot (\nabla \times A) = 0$; $\nabla \times \nabla A = 0$

(ii) $\nabla \times (A + B) = \nabla \times A + \nabla \times B$

(iii) $\nabla \times (VA) = V(\nabla \times A) + \nabla V \times A$

(vi) $\nabla \times \nabla \times A = \nabla(\nabla \cdot A) - \nabla^2 A$

(v) A vector $\vec{A}$ is irrotational if $\nabla \times A = 0$

Laplacian Operator

Laplacian of Scalar

Laplacian operation can be performed both on scalar and vector function.

$$\text{Laplacian } V = \nabla^2 V = \nabla.\nabla V$$

A scalar field V is said to be harmonic in a given region if its Laplacian vanishes in that region.

$$\nabla^2 V = 0$$

(i) In Cartesian coordinate system

$$\nabla^2 V = \frac{\partial^2 V}{\partial x^2} + \frac{\partial^2 V}{\partial y^2} + \frac{\partial^2 V}{\partial z^2}$$

(ii) In Cylindrical coordinate system

$$\nabla^2 V = \frac{1}{\rho}\left[\frac{\partial}{\partial \rho}\left(\frac{\rho dV}{\partial \rho}\right)\right] + \frac{1}{\rho^2}\left[\frac{\partial^2 V}{d\phi^2}\right] + \frac{\partial^2 V}{\partial z^2}$$

(iii) In Spherical coordinate system

$$\nabla^2 V = \frac{1}{r^2}\frac{\partial}{\partial r}\left(r^2 \frac{\partial V}{\partial r}\right) + \frac{1}{r^2 \sin\theta}\frac{\partial}{\partial \theta}\left(\sin\theta \frac{\partial V}{\partial \theta}\right) + \frac{1}{r^2 \sin^2\theta}\left(\frac{\partial^2 V}{\partial \phi^2}\right)$$

Laplacian of vector

The Laplacian of vector A is, $\nabla^2 \vec{A} = \nabla(\nabla \cdot \vec{A}) - \nabla \times \nabla \times \vec{A}$

Remember:

- ☑ The Laplacian of a scalar field V is the divergence of the gradient of V
- ☑ Laplacian of a scalar results in a scalar

Divergence Theorem

"The volume integral of the divergence of vector field A is taken over any volume V is equal to the surface integral of A taken over the closed surface (total outward flux of the vector through the surface) surrounding the volume V."

$$\int_V (\nabla \cdot \vec{A}) dV = \oint_S \vec{A} \cdot d\vec{S}$$

Note:

The divergence theorem applies to any volume V that is bounded by surface S.

Stoke's Theorem

"The circulation of vector field A around a closed path L is equal to the surface integral of the Curl of A over the open surface S bounded by L, provided A and $\nabla \times A$ are continuous on S."

$$\int_S (\nabla \times \vec{A}) \cdot d\vec{S} = \oint_L \vec{A} \cdot d\vec{l}$$

III Electrostatics

Coulomb's Law

- The force acting between two point charges is directly proportional to the products of charges and inversely proportional to the square of the distance between them

$$F \propto \frac{Q_1 Q_2}{R^2}$$

$$F = K \cdot \frac{Q_1 Q_2}{R^2} = \frac{Q_1 Q_2}{4\pi \epsilon_0 R^2} \quad : \text{Newton}$$

Where, $K = \dfrac{1}{4\pi \epsilon_0} = 9 \times 10^9$ in S.I. system

- If two positive charges are placed, then they experience a repulsive force.

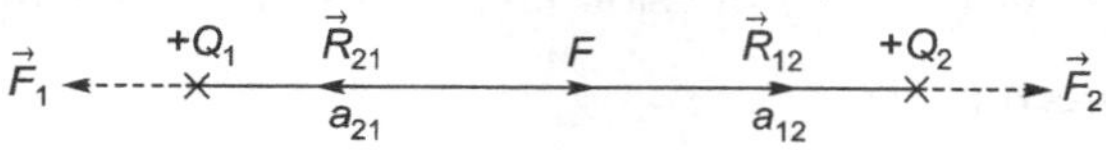

$$\vec{a}_{12} = \frac{\vec{R}_{12}}{|\vec{R}_{12}|} \quad \text{and} \quad \vec{a}_{21} = \frac{\vec{R}_{21}}{|\vec{R}_{21}|}$$

$$\vec{R}_{12} = -\vec{R}_{21} \quad \text{and} \quad \vec{a}_{12} = -\vec{a}_{21}$$

$$\vec{F}_1 = \frac{Q_1 Q_2}{4\pi \epsilon_0 R_{21}^2}\vec{a}_{21}, \quad \vec{F}_2 = \frac{Q_1 Q_2}{4\pi \epsilon_0 R_{12}^2}\vec{a}_{12}$$

$$\vec{F}_1 = -\vec{F}_2$$

Note:

☑ Coulomb's force obeys law of superposition.

☑ Coulomb's forces are mutual and linear.

Charge Distributions

- **Point Charge**

  If charge is put at a point only • Q

- **Line charge density**

  It is charge per unit length.

$$\frac{dQ}{dL} = \rho_L \ \text{C/m}$$

$$Q = \int_L \rho_L dL \ \text{C}$$

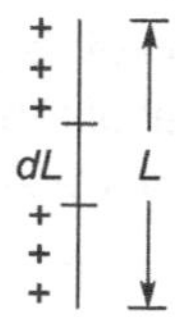

- **Surface charge density**

  It is the charge per unit area

$$\frac{dQ}{dS} = \rho_s \ \text{C/m}^2$$

$$Q = \iint_s \rho_S dS \ \text{C}$$

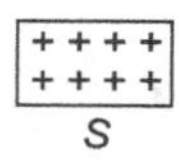

- **Volume charge density**

  It is the charge per unit volume

$$\frac{dQ}{dV} = \rho_v \ \text{C/m}^3$$

$$Q = \iiint\limits_V \rho_V dV \ \text{C}$$

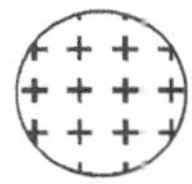

V

Charge in terms of Various Charge Distributions

$$Q = \int\limits_L \rho_L dL = \iint\limits_S \rho_s dS = \iiint\limits_V \rho_v \, dV$$

where, ρ_L, ρ_S and ρ_V are line charge, surface charge and volume charge density respectively.

Electric/Displacement Flux Density

The electric flux density is always tangential to electric flux lines. Electric flux lines originates from a positive charge and ends on a negative charge.

$$D = \frac{\text{Flux}}{\text{Unit area}}$$

$$\vec{D} = \frac{Q}{4\pi r^2}\hat{i}_r \ \text{ and } \ \vec{E} = \frac{Q}{4\pi \epsilon \, r^2}\hat{i}_r \ ; \ \vec{D} = \epsilon \vec{E}$$

where, $\vec{D} \rightarrow$ Electric flux density C/m^2

$\vec{E} \rightarrow$ Electric field intensity V/m

$\epsilon \rightarrow$ Electrical permittivity of medium

$$\epsilon = \epsilon_0 \epsilon_r$$

$\epsilon_0 \rightarrow$ Free space permittivity

$$= 8.854 \times 10^{-12} \text{ F/m} = \left(\frac{1}{36\pi}\right) \times 10^{-9} \ \text{F/m}$$

$\epsilon_r \rightarrow$ Relative permittivity or Dielectric constant of medium

$\epsilon_r = 1$ (For air (or) free space)

$\epsilon_r > 1$ (For rest of the material)

Note:

Electric field $\vec{E}$ depends upon the medium whereas $\vec{D}$ is independent of the medium.

Gauss's Law

Gauss law states that flux leaving any closed surface is equal to the charge enclosed by that surface.

$$\psi = \oint_S \vec{D} \cdot d\vec{S} = Q = \int_V \rho_V \cdot dV$$ .... Gauss law in integral form

$\rho_V = \nabla \cdot \vec{D}$ .... Gauss law in differential or point form.

Note:

☑ The Gauss law is applicable for time varying as well as static fields.

☑ The equation is valid irrespective of the shape of closed surface area 'S'.

Electric Field Intensity

The Electric Field Intensity or Electric Field Strength (E) is the force per unit charge when placed in an electric field

$$E = \lim_{Q \to 0} \frac{\vec{F}}{Q} \text{ Newton/C}$$

Electric field due to continuous charge distribution

(i) Line charge, $E = \int_L \frac{\rho_L dL}{4\pi \epsilon_0 R^2} \hat{i}_r$

(ii) Surface charge, $E = \int_S \frac{\rho_S dS}{4\pi \epsilon_0 R^2} \hat{i}_r$

(iii) Volume charge, $E = \int_V \frac{\rho_V dV}{4\pi \epsilon_0 R^2} \hat{i}_r$

Electric field intensity due to infinitely long line charge

$$E = \frac{\rho_L}{2\pi \epsilon_o r} \hat{i}_r$$

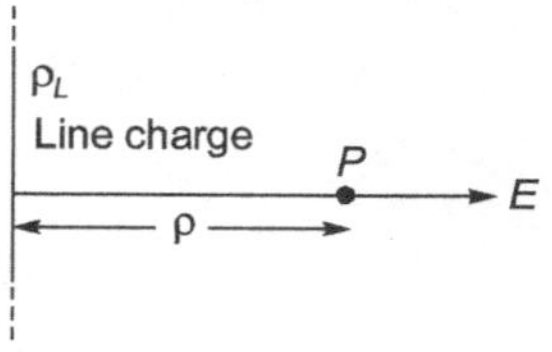

where, $\rho_L \to$ Charge per unit length

= Line charge density C/m

Electric field intensity due to infinite charge sheet

$$E = \frac{\rho_s}{2 \epsilon_o} \hat{i}_n$$

where, $\rho_s \to$ Surface charge density, C/m^2

If sheet is on xy plane $z = z_o$

$$E = \begin{cases} \frac{\rho_s}{2\epsilon_o} \hat{i}_z & \text{for } z > z_o \\ -\frac{\rho_s}{2\epsilon_o} \hat{i}_z & \text{for } z < z_o \end{cases}$$

Remember:

In a parallel plate capacitor, the electric field existing between the two plates having equal and opposite charges is : $E = \frac{\rho_s}{\epsilon_0} \hat{i}_n$

Electric field intensity due to uniformly charged sphere

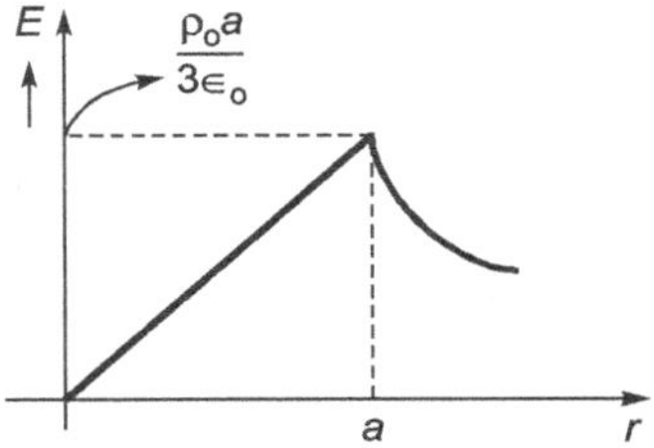

$$E = \begin{cases} \dfrac{r}{3\epsilon_o}\rho_o \hat{i}_r \, ; & 0 < r \leq a \\ \dfrac{a^3}{3\epsilon_o r^2}\rho_o \hat{i}_r \, ; & r \geq a \end{cases}$$

Where $\rho_o \rightarrow$ Volume charge density ; $a \rightarrow$ Radius of sphere

Electric Potential

Potential difference

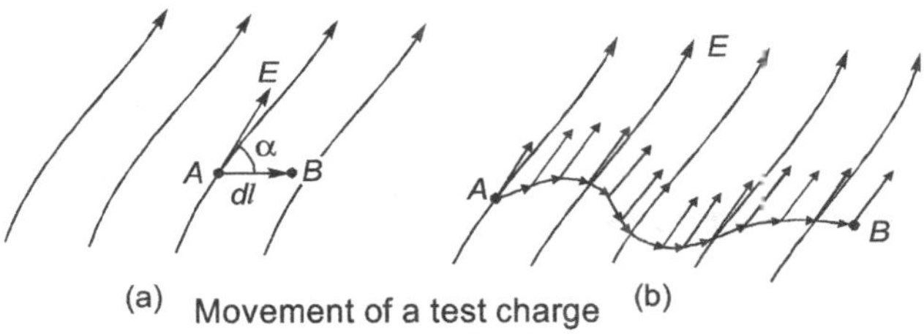

Movement of a test charge in an electric field

$$V_{AB} = \frac{Q}{4\pi\epsilon_0}\left[\frac{1}{r_B} - \frac{1}{r_A}\right] \quad ; \quad V_{AB} = \frac{W_{AB}}{q} = -\int_A^B \vec{E}\cdot d\vec{l}$$

where, $V_{AB} \rightarrow$ Potential difference between the points A and B
$W_{AB} \rightarrow$ Work done by the field in moving a test charge q from A to B
$dl \rightarrow$ Infinitesimal length of segments

Remember:

☑ The negative sign in above equation indicates that the work is being done by an external agent.

☑ If V_{AB} is negative, there is a loss in potential energy in moving charge Q from A to B i.e. the field does the work.

Potential at a point *P* due to charge distribution

(i) Point charge, $V(r) = \int_r^\infty \vec{E}\cdot d\vec{l} = -\int_\infty^r \vec{E}\cdot d\vec{l}$ Volts

where, $V(r) \rightarrow$ Potential at a distance r from the point charge
$r \rightarrow$ distance of point P from point charge

$$V = \frac{Q}{4\pi\epsilon_0 r}$$

(ii) Line charge, $V = \frac{\rho_L}{2\pi\epsilon} ln\left(\frac{1}{\rho}\right)$ Volt

where, $\rho_L \rightarrow$ Line charge density, $\rho \rightarrow$ Distance from line charge

(iii) Cylindrical sheet, $V = \frac{\rho_L}{2\pi\epsilon} ln\frac{b}{a}$ Volt

where, $\rho_L \rightarrow$ Line charge density ; $a \rightarrow$ Inner radius ; $b \rightarrow$ Outer radius

(iv) Concentric spheres, $V = \frac{Q}{4\pi\epsilon}\left(\frac{1}{a} - \frac{1}{b}\right)$ Volt

where, $Q \rightarrow$ Charge in inner sphere; $a \rightarrow$ Inner radius; $b \rightarrow$ Outer radius

(v) Conducting Sphere

$$V = \frac{Q}{4\pi\epsilon_o a} \quad \text{for } r \leq a$$

$$= \frac{Q}{4\pi\epsilon_o r} \quad \text{for } r > a$$

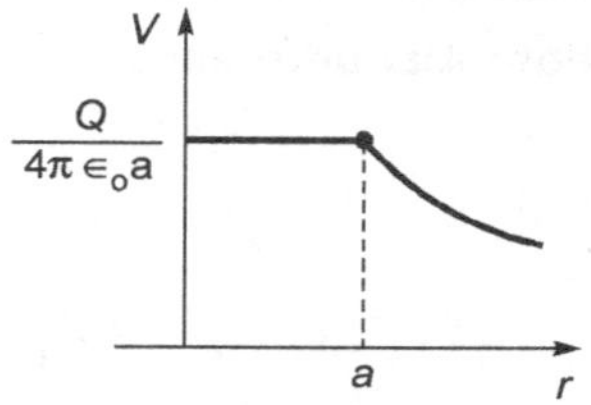

For static electric field

$$\oint \vec{E} \cdot \vec{dl} = 0 \quad ; \quad \nabla \times \vec{E} = 0$$

Remember:

The potential at any point is the potential difference between that point and a chosen point (or reference point) at which the potential is zero.

Electric Dipole

An electric dipole is formed when two point charges of equal magnitude but opposite sign are separated by small distance

Dipole moment

$$\vec{p} = Q\vec{d}$$

–Q ●←—— d ——→● Q

Direction of dipole moment is from negative charge to positive charge.

Electric field intensity due to electric dipole

$$\vec{E} = \frac{Q\vec{d}}{4\pi\epsilon_o r^3}(2\cos\theta\,\hat{i}_r + \sin\theta\,\hat{i}_\theta)$$

$$\vec{E} = \frac{\vec{p}}{4\pi\epsilon_o r^3}(2\cos\theta\,\hat{i}_r + \sin\theta\,\hat{i}_\theta)$$

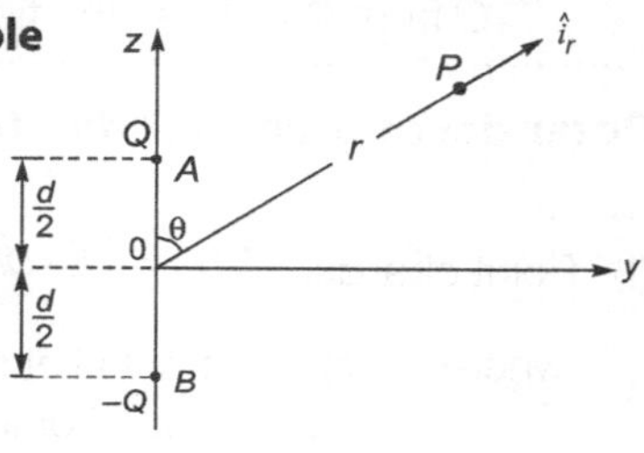

Conductor

A perfect conductor ($\sigma = \infty$) cannot contain an electrostatic field within it.

Inside a conductor

$$\vec{E}=0, \rho_v=0, V_{ab}=0$$

where, $V_{ab}\rightarrow$ Potential difference between points *a* and *b* in the conductor.

Power

Joule's law, $P=\int_v \vec{E}\cdot\vec{J}\,dV$ where, $J\rightarrow$ Current density, A/m^2

Continuity Equation

$$\nabla\cdot\vec{J}=-\frac{\partial\rho_v}{\partial t}$$

Note:

☑ For steady current $\frac{\partial\rho_0}{\partial t}=0.$

☑ For lossless region $\rho_v=0$; $\nabla\cdot\vec{J}=0$.

Boundary Conditions

Dielectric-Dielectric Boundary Condition

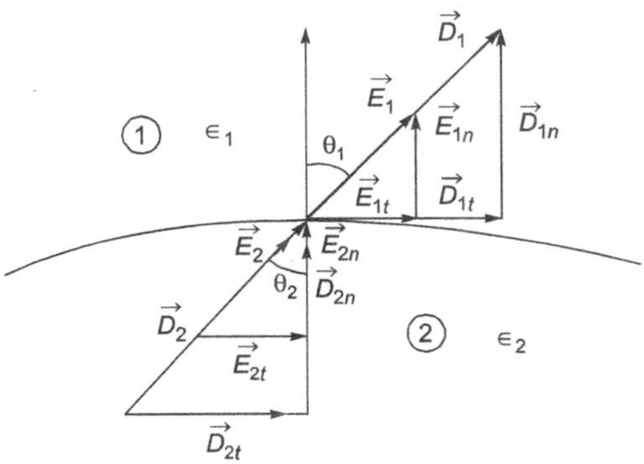

$$\vec{E}_1=\vec{E}_{1t}+\vec{E}_{1n} \quad \text{and} \quad \vec{E}_2=\vec{E}_{2t}+\vec{E}_{2n}$$

where, $\vec{E}_1, \vec{E}_2\rightarrow$ Fields in media 1 and 2 respectively

$\vec{E}_t, \vec{E}_n\rightarrow$ Tangential and normal components of E

Tangential component relation

$$\vec{E}_{1t}=\vec{E}_{2t} \text{ or } \frac{\vec{D}_{1t}}{\in_1}=\frac{\vec{D}_{2t}}{\in_2}$$

Normal component relation

$$\vec{D}_{1n}-\vec{D}_{2n}=\rho_s$$

where, $\in_1$ and $\in_2\rightarrow$ Permittivity of dielectric 1 and 2

$\rho_s\rightarrow$ Free charge density placed deliberately at the boundary

Remember:

- ☑ The tangential component of $\vec{E}$ is continuous while that of $\vec{D}$ is discontinuous at boundary.
- ☑ The normal component of $\vec{D}$ is continuous while that of $\vec{E}$ is discontinuous at boundary.

If no free charges exists at the interface

$$\vec{D}_{1n} = \vec{D}_{2n} \quad \text{or} \quad \in_1 \vec{E}_{1n} = \in_2 \vec{E}_{2n}$$

$$\frac{\tan\theta_1}{\tan\theta_2} = \frac{\in_{r1}}{\in_{r2}}$$

Conductor-Dielectric Boundary Conditions.

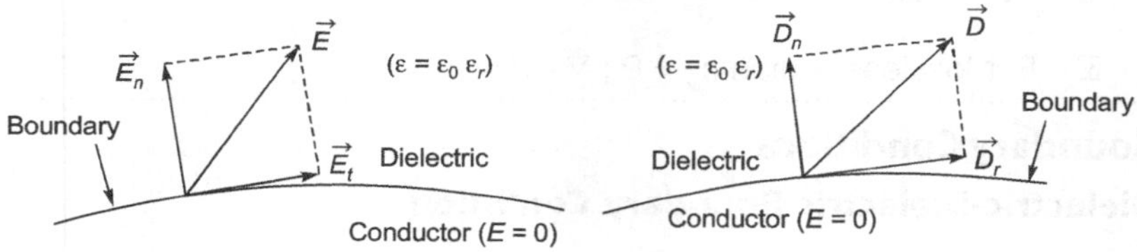

$$\vec{D}_t = \in \vec{E}_t = 0 \quad \text{and} \quad \vec{D}_n = \in \vec{E}_n = \rho_s$$

Remember:

- ☑ Since $\vec{E} = -\nabla V = 0$, there can be no potential difference between any two points in the conductor (i.e. a conductor is an equipotential body).
- ☑ An electric field E must be external to the conductor and must be normal to its surface.

$$\vec{D}_t = \in \vec{E}_t = 0 \quad \text{and} \quad \vec{D}_n = \in \vec{E}_n = \rho_s$$

Electrical Energy Density

It is total electrical energy per unit volume : $W_e = \frac{1}{2}\in \vec{E}^2 = \frac{1}{2}\vec{D}\cdot\vec{E}$ J/m³

Total electrical energy stored : $W_e = \frac{1}{2}CV^2 = \frac{1}{2}QV = \frac{1}{2}\frac{Q^2}{C}$

Energy Density in Electrostatic Field

- Electrostatic energy density, $W_E = \frac{dW_E}{dv} = \frac{1}{2}\vec{D}\cdot\vec{E} = \frac{1}{2}\in_0 E^2 = \frac{D^2}{2\in_0}$
- Total electrostatic energy, $W_E = \frac{1}{2}\int_v \vec{D}\cdot\vec{E}\, dV = \frac{1}{2}\int_v \in_0 E^2\, dV$

IV Magnetostatics

Biot-Savart's Law

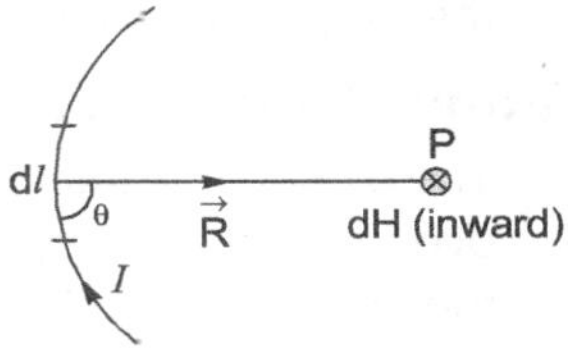

The Biot-Savart's law states that the differential magnetic field intensity $\overrightarrow{dH}$ produced at a point P, by the differential current element $I\overrightarrow{dl}$ is proportional to the product $I\overrightarrow{dl}$ and the sine of angle θ between the element and the line joining P to the element and is inversely proportional to the square of the distance R between P and the element.

$$\overrightarrow{dH} \propto \frac{I\overrightarrow{dl}\sin\alpha}{R^2} \text{ A/m} \;;\; \overrightarrow{dH} = \frac{I\overrightarrow{dl}\times\hat{i}_r}{4\pi R^2} = \frac{I\overrightarrow{dl}\times\vec{R}}{4\pi|\vec{R}|^3} \;;\; \vec{H} = \oint\frac{I\overrightarrow{dl}\times\hat{i}_r}{4\pi R^2} \text{ A/m}$$

where, $\hat{i}_r \rightarrow$ Unit vector pointing from the different element of current to the point of interest.

Magnetic Field Intensity for Various Current Distribution

(i) Line current, $\vec{H} = \int_L \frac{I\overrightarrow{dl}\times\hat{i}_r}{4\pi R^2}$ A/m

(ii) Surface current, $\vec{H} = \int_S \frac{K\cdot dS\times\hat{i}_r}{4\pi R^2}$ (A/m²); where, $K \rightarrow$ Surface current density

(iii) Volume current, $\vec{H} = \int_V \frac{J\cdot dV\times\hat{i}_r}{4\pi R^2}$ (A/m³); where $J \rightarrow$ Volume current density

Magnetic Field Intensity due to straight current carrying conductor

(i) Finite length AB,

$$\vec{H} = \frac{I}{4\pi\rho}(\cos\alpha_2 - \cos\alpha_1)\hat{i}_\phi$$

(ii) Semi-infinite length

$$\vec{H} = \frac{I}{4\pi\rho}\hat{i}_\phi$$

(iii) Infinite length

$$\vec{H} = \frac{I}{2\pi\rho}\hat{i}_\phi \quad ; \qquad \hat{i}_\phi = \hat{i}_l \times \hat{i}_\rho$$

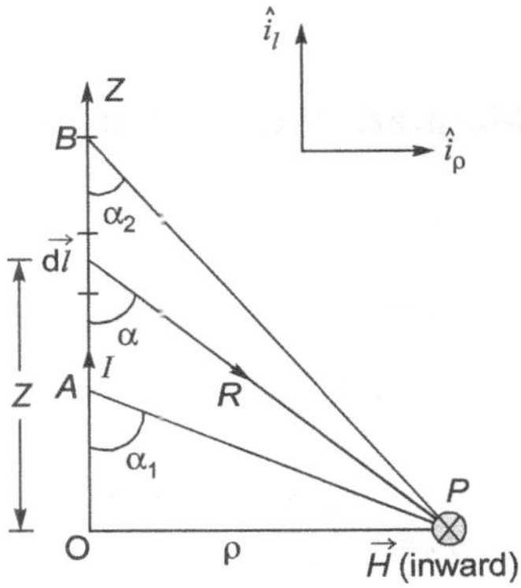

where, $\hat{i}_l$ is unit vector along the line current and $\hat{i}_\rho$ is unit vector along the perpendicular line from the line current to the field point.

Ampere's Circuit Law

The closed line integral of static magnetic field intensity $\vec{H}$, integrated over any closed curve '*C*' is always equal to total current enclosed within the closed curve '*C*'.

$$\boxed{\oint \vec{H}\cdot\vec{dl} = I_{enc}} \text{ A}$$

$$\oint \vec{H}\cdot\vec{dl} = \iint \vec{J}\cdot\vec{dS} \quad \text{.... Ampere's circuit law in the integral form}$$

The Curl of static magnetic field intensity $\vec{H}$ at any point in the electromagnetic region is equal to volume current density $\vec{J}$ present at that point.

$\nabla\times\vec{H} = \vec{J}$ .... Ampere's circuit law in point form or differential form

Note:

- ☑ $\nabla\times\vec{H} = \vec{J}$ is the also known as Maxwell's third equation.
- ☑ Ampere's circuit law can be applied irrespective of shape of the closed curve '*C*'.
- ☑ The magnetostatic field is not conservative as $\nabla\times\vec{H} = \vec{J} \neq 0$.

Magnetic Field Intensity due to Infinite Line Current

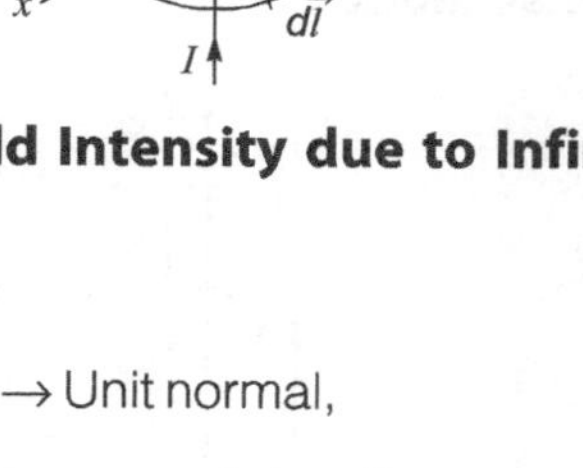

$$\vec{H} = \frac{I}{2\pi\rho}\hat{i}_\phi$$

Magnetic Field Intensity due to Infinite Sheet of Current

$$\vec{H} = \frac{1}{2}\vec{K}\times\hat{i}_n$$

where, $\hat{i}_n \rightarrow$ Unit normal,

$\vec{K} \rightarrow$ Uniform current density

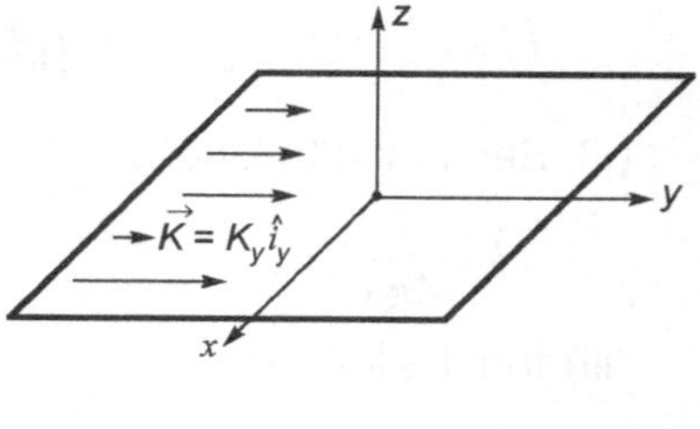

$$H = \begin{cases} \frac{1}{2}K_y\,\hat{i}_x; & z > 0 \\ -\frac{1}{2}K_y\,\hat{i}_x; & z < 0 \end{cases}$$

Magnetic Flux Density

$\boxed{\vec{B} = \mu_0 \vec{H}}$ Wb/m^2 where, $\mu_0 \rightarrow$ Permeability of free space.

Magnetic flux through a surface S

$$\boxed{\psi = \int_S \vec{B} \cdot d\vec{S}} \text{ Wb}$$

Remember:

An isolated magnetic charge does not exist. Thus the total flux through a closed surface in a magnetic field must be zero.

$$\oint \vec{B} \cdot d\vec{S} = 0 \quad \text{or} \quad \nabla \cdot \vec{B} = 0$$

Magnetic Energy Density (W_m)

The magnetic energy density represents magnetic energy stored at a point in the electro magnetic region and gives total magnetic energy per unit volume of the given configuration.

$$W_m = \frac{1}{2}\mu \vec{H}^2 = \frac{1}{2}\vec{B}\,\vec{H}$$

Magnetic Energy

$$W_m = \iiint_V W_m dV = \frac{1}{2}LI^2 \text{ Joule}$$

Remember:

The magnetic energy density depends upon

(i) Magnetic field due to given current distribution in the configuration.

(ii) Permeability of the magnetic medium.

Magnetic Scalar and Vector Potentials

The magnetic potential could be scalar V_m or vector $\vec{A}$.

Magnetic scalar potential (V_m)

The magnetic scalar potential is only defined in a region where $J = 0$.

$$\vec{H} = -\nabla V_m$$

Vector Magnetic Potential $\vec{A}$

$$\vec{B} = \vec{\nabla} \times \vec{A}$$

For line current, $\vec{A} = \int_L \frac{\mu_0 I dl}{4\pi R}$ Wb/m

For surface current, $\vec{A} = \int_S \frac{\mu_0 \vec{K} dS}{4\pi R}$

For volume current, $\vec{A} = \int_V \frac{\mu_0 \vec{J} dV}{4\pi R}$

Remember:

Important identities which must always hold for any scalar field V and vector field $\vec{A}$.

$$\nabla \times (\nabla V) = 0 \quad \text{and} \quad \nabla \cdot (\nabla \times \vec{A}) = 0$$

Forces due to Magnetic Field

Force on a Charged Particle

(i) Magnetic force : $\vec{F}_m = Q(\vec{v} \times \vec{B})$ Newton

where, $\vec{v} \rightarrow$ Velocity of moving charge Q; $\vec{B} \rightarrow$ Magnetic field

(ii) Electric force : $\vec{F}_e = Q\vec{E}$

where, $Q \rightarrow$ Electric charge ; $\vec{E} \rightarrow$ Electric field intensity

Lorentz force Equation

For a moving charge Q in the presence of both electric and magnetic fields

$$\vec{F} = \vec{F}_e + \vec{F}_m = Q(\vec{E} + \vec{v} \times \vec{B})$$

Force on a Current Element

Force on a current carrying conductor

The magnetic field is defined as the force per unit current element.

$$\vec{F} = \oint_L I\vec{dl} \times \vec{B}$$

where $I\vec{dl} \rightarrow$ Current element of current carrying conductor

Magnetic Torque and Magnetic Moment

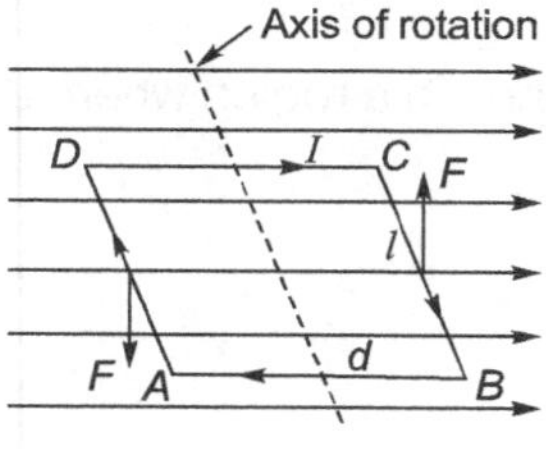

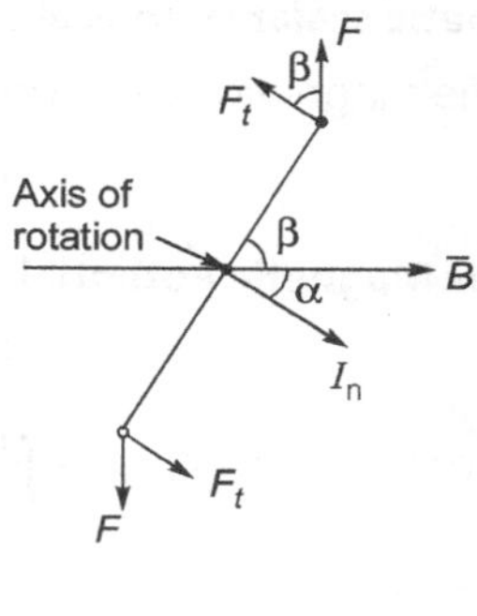

- Magnetic Torque, $T = BIS\sin\alpha$ where, $S \rightarrow$ area of the loop
  $\vec{T} = \vec{m} \times \vec{B}$
- Magnetic Dipole Moment, $\vec{m} = IS\hat{i}_n$

Magnetic Boundary Conditions

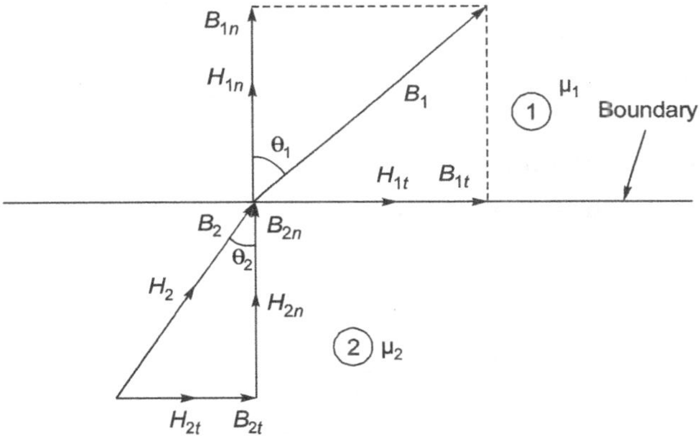

Tangential component relation

$$\vec{H}_{1t} = \vec{H}_{2t} \quad \text{or} \quad \frac{\vec{B}_{1t}}{\mu_1} = \frac{\vec{B}_{2t}}{\mu_2}$$

Normal component relation

$$\vec{B}_{1n} = \vec{B}_{2n} \text{ or } \mu_1 \vec{H}_{1n} = \mu_2 \vec{H}_{2r}$$

Remember:

☑ The tangential component of Magnetic field intensity (H) is continuous while that of Magnetic flux density (B) is discontinuous at boundary.

☑ The normal component of B is continuous while that of H is discontinuous at boundary.

$$\frac{\tan\theta_1}{\tan\theta_2} = \frac{\mu_1}{\mu_2}$$

V Time-Varying Electromagnetic Fields

Introduction

- Stationary charges $\rightarrow$ Electrostatic field
- Steady currents $\rightarrow$ Magnetostatic field
- Time-varying currents $\rightarrow$ Electromagnetic field (or waves)

Faraday's Law

- The induced emf (V_{emf}), in any closed circuit is equal to the time rate of change of the magnetic flux linkage by the circuit

$$V_{emf} = -\frac{d\lambda}{dt} = -N\frac{d\psi}{dt}$$

where, $\lambda \rightarrow$ Flux linkage ; $N \rightarrow$ Number of turns in the circuit
$\psi \rightarrow$ Flux through each turn

- The negative sign shows that the induced voltage acts in such a way as to oppose the flux producing it.

Transformer and motional Electromotive Forces

Stationary loop in time-varying magnetic field

(i) Transformer emf

$$V_{emf} = \oint_L \vec{E}\cdot d\vec{l} = -\frac{d}{dt}\int_S \vec{B}\cdot d\vec{S}$$

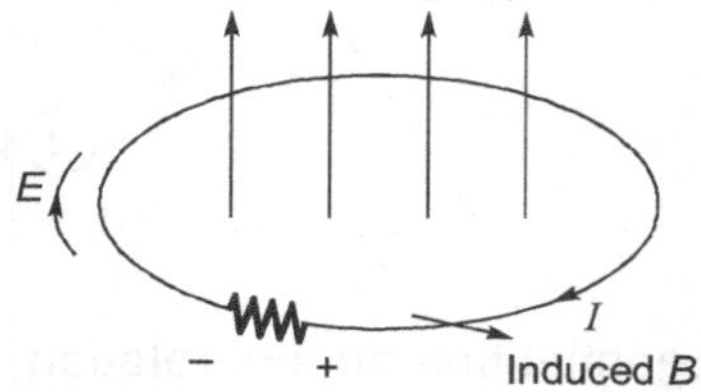

(ii) Maxwell's equation for time-varying field

$$\nabla\times\vec{E} = -\frac{\partial\vec{B}}{\partial t}$$

Note:

Time varying electric field E is not conservative ($\nabla\times\vec{E} \neq 0$).

Moving loop in static magnetic field

(i) Motional electric field, $\vec{E}_m = \frac{\vec{F}_m}{Q} = \vec{v}\times\vec{B}$

where, $F_m \rightarrow$ Force on charge moving with uniform velocity $\vec{v}$, in a magnetic field

(ii) Motional emf or flux-cutting emf, $V_{emf} = \oint \vec{E}_m\cdot d\vec{l} = \oint_L (\vec{v}\times\vec{B})\cdot dl$

Moving loop in time-varying field

$$V_{emf} = \oint_L \vec{E}\cdot d\vec{l} = -\int_S \frac{\partial\vec{B}}{\partial t}\cdot d\vec{S} + \oint_L (\vec{v}\times\vec{B})\cdot d\vec{l}$$

$$\nabla\times\vec{E} = -\frac{\partial\vec{B}}{\partial t} + \nabla\times(\vec{v}\times\vec{B})$$

Displacement Current

Displacement current density, $\vec{J}_d = \dfrac{\partial \vec{D}}{\partial t}$

Maxwell's Equations in Final Forms

Maxwell's equation for static electric and magnetic fields

Differential or Point form	Integral Form	Remarks
$\vec{\nabla} \cdot \vec{D} = \rho_v$	$\oint_S \vec{D} \cdot d\vec{S} = \int_V \rho_v dv$	Gauss's law
$\vec{\nabla} \cdot \vec{B} = 0$	$\oint_S \vec{B} \cdot d\vec{S} = 0$	Non existence of magnetic monopole
$\vec{\nabla} \times \vec{E} = 0$	$\oint_L \vec{E} \cdot d\vec{l} = 0$	Conservative nature of electrostatic field
$\vec{\nabla} \times \vec{H} = \vec{J}$	$\oint_L \vec{H} \cdot d\vec{l} = \int_S \vec{J} \cdot d\vec{S}$	Ampere's Law

Maxwell's equation for time-varying fields

Differential Form	Integral Form	Remarks
$\vec{\nabla} \cdot \vec{D} = \rho_v$	$\oint_S \vec{D} \cdot d\vec{S} = \int_V \rho_v dv$	Gauss's law
$\vec{\nabla} \cdot \vec{B} = 0$	$\oint_S \vec{B} \cdot d\vec{S} = 0$	Non existence of isolated magnetic charge
$\vec{\nabla} \times \vec{E} = -\dfrac{\partial \vec{B}}{dt}$	$\oint_L \vec{E} \cdot d\vec{l} = -\dfrac{\partial}{\partial t} \int_S \vec{B} \cdot d\vec{S}$	Faraday's law
$\vec{\nabla} \times \vec{H} = \vec{J} + \dfrac{\partial \vec{D}}{\partial t}$	$\oint_L \vec{H} \cdot d\vec{l} = \int_S \left(\vec{J} + \dfrac{\partial \vec{D}}{\partial t} \right) d\vec{S}$	Ampere's circuit law

Time-Varying Potentials

$$\vec{E} = -\nabla V - \frac{\partial A}{\partial t}$$

Lorentz condition for potentials, $\nabla \cdot \vec{A} = -\mu \in \dfrac{\partial V}{\partial t}$

Time Harmonic Field

- A time harmonic field is one that varies periodically or sinusoidally with time.
- Phasor form of vector A is $A_s(x, y, z)$

$$A = \text{Re}\left\{A_s e^{j\omega t}\right\}$$

Maxwell's equation for time harmonic field

Point Form	Integral Form
$\vec{\nabla}\cdot\vec{D}_s = \rho_{vs}$	$\oint \vec{D}_s \cdot d\vec{S} = \int \rho_{vs} dv$
$\vec{\nabla}\cdot\vec{B}_s = 0$	$\oint \vec{B}_s \cdot d\vec{S} = 0$
$\vec{\nabla}\times\vec{E}_s = -j\omega\vec{B}_s$	$\oint \vec{E}_s \cdot d\vec{l} = -j\omega\int \vec{B}_s \cdot d\vec{S}$
$\vec{\nabla}\times\vec{H}_s = \vec{J}_s + j\omega\vec{D}_s$	$\oint \vec{H}_s \cdot d\vec{l} = \int (\vec{J}_s + j\omega\vec{D}_s)\cdot d\vec{S}$

VI Electromagnetic Wave Propagation

Introduction

Waves are means of transporting energy or information. Some examples of EM waves are radio waves, TV signals, radar beams and light rays.

Remember:

Characteristics of EM medium:

(a) Free space : $\sigma = 0$, $\in = \in_0$, $\mu = \mu_0$.

(b) Perfect dielectric : $\sigma = 0$, $\in$ and μ can have any value.

(c) Good dielectrics : $\sigma \simeq 0$, $\in = \in_r \in_0$, $\mu = \mu_r \mu_0$ or $\sigma << \omega\in$.

(d) Perfect conducting : $\sigma = \infty$, $\in$ and μ can have any value.

(e) Good conductor : $\sigma \simeq \infty$, $\in = \in_0$, $\mu = \mu_r \mu_0$ or $\sigma >> \omega\in$.

where, $\sigma \rightarrow$ Conductivity ; $\omega \rightarrow$ Angular frequency of wave

Wave propagation in Lossy Dielectrics

A lossy dielectric is a medium in which an EM wave, as it propagates, loses power owing to imperfect dielectric. A lossy dielectric is a partially conducting medium.

Vector wave equation or vector Helmholtz's equation

(i) For $\vec{E}$-field, $\nabla^2 \vec{E}_s - \gamma^2 \vec{E}_s = 0$; **(ii)** For $\vec{H}$-field, $\nabla^2 \vec{H}_s - \gamma^2 \vec{H}_s = 0$

where, $\gamma \rightarrow$ Propagation constant of medium ; E_s, $H_s \rightarrow$ Source electric and magnetic fields respectively

$$\gamma = \alpha + j\beta \text{ and } \gamma^2 = j\omega\mu(\sigma + j\omega\in)$$

where, α = Attenuation constant (Neper/m)

β = Phase constant (rad/m)

$$\alpha = \omega\sqrt{\frac{\mu\in}{2}\left[\sqrt{1+\left[\frac{\sigma}{\omega\in}\right]^2} - 1\right]} \quad ; \quad \beta = \omega\sqrt{\frac{\mu\in}{2}\left[\sqrt{1+\left[\frac{\sigma}{\omega\in}\right]^2} + 1\right]}$$

Field equation of EM wave in s-domain

$$E_S(z) = E_0 e^{-\alpha z} e^{j(\omega t - \beta z)}\, \hat{i}_x \text{ and } H_S(z) = H_0 e^{-\alpha z} e^{j(\omega t - \beta z)}\, \hat{i}_y$$

Field equation of EM wave in time domain

$$E(z, t) = E_0 e^{-\alpha z} \cos(\omega t - \beta z)\, \hat{i}_x \text{ and } H(z, t) = H_0 e^{-\alpha z} \cos(\omega t - \beta z)\, \hat{i}_y$$

Intrinsic Impedance (η)

- It is the ratio of Electric Field Intensity to the corresponding Magnetic Field Intensity for an electromagnetic wave.

  where, $\eta = \dfrac{E}{H}\,\Omega$

  It is found to be independent of the location of measurement.
- It is a complex quantity, $\eta = |\eta|\, e^{j\theta_\eta}$

For general lossy medium

$$\eta = \sqrt{\frac{j\omega\mu}{\sigma + j\omega\epsilon}} = |\eta|\angle\theta_n$$

$$\text{where, } |\eta| = \frac{\sqrt{\mu/\epsilon}}{\left[1+\left(\dfrac{\sigma}{\omega\epsilon}\right)^2\right]^{1/4}} \quad ; \quad \theta_n = \frac{1}{2}\tan^{-1}\left(\frac{\sigma}{\omega\epsilon}\right)$$

Remember:

- $\mu = \dfrac{|\gamma \cdot \eta|}{\omega}$
- $\sigma = \text{Re}(\gamma \times \eta)$
- $\epsilon = \dfrac{\text{Img}(\gamma/\eta)}{\omega}$

Phase Velocity

$$v = \frac{\omega}{\beta} = \frac{1}{\text{wave number}} = \frac{c}{\sqrt{\mu_r \epsilon_r}} \text{ m/s}$$

where, $v \rightarrow$ Wave velocity ; $c \rightarrow$ Free space velocity

also, $\beta = \dfrac{2\pi}{\lambda}$ where, $\lambda \rightarrow$ Wavelength

Remember:

Phase velocity decreases as loss tangent increases from 3×10^8 m/s in free space to few m/s in good conductor.

Loss tangent/Dissipation factor

It is the ratio of conduction current density to the displacement current density and given by

$$\tan\delta = \left|\frac{\vec{J}_C}{\vec{J}_D}\right| = \frac{\sigma}{\omega\epsilon} \qquad \text{where,} \quad \delta \to \text{loss angle}$$

Note:

If θ_η is intrinsic angle then the loss tangent is given by $\tan 2\theta_\eta$.

Complex Permittivity

$$\epsilon_c = \epsilon' - j\epsilon'' = \epsilon\left[1 - \frac{\sigma}{\omega\epsilon}\right]$$

where, $\epsilon' = \epsilon \to$ Real component of complex permittivity

$\epsilon'' = \sigma/\omega \to$ Imaginary component of permittivity

Loss tangent, $\tan\delta = \dfrac{\epsilon''}{\epsilon'} = \dfrac{\sigma}{\omega\epsilon}$

Plane Waves in Free Space

In free space, $\sigma = 0, \epsilon = \epsilon_0, \mu = \mu_0$

- $\alpha = 0, \beta = \omega\sqrt{\mu_0 \epsilon_0} = \dfrac{\omega}{c}$
- $v = \dfrac{1}{\sqrt{\mu_0 \epsilon_0}} = c; \lambda = \dfrac{2\pi}{\beta}$
- $\eta_0 = \sqrt{\dfrac{\mu_0}{\epsilon_0}} \simeq 377\ \Omega = 120\pi\ \Omega$

$$\hat{i}_E \times \hat{i}_H = \hat{i}_K$$

where, $\hat{i}_K \to$ Direction of wave propagation

Plane Waves in Lossless Dielectrics/Perfect Dielectric

In lossless dielectric, $\sigma << \omega\epsilon$

$$\sigma \simeq 0, \ \epsilon = \epsilon_0\epsilon_r, \ \mu = \mu_0\mu_r$$

- $\alpha = 0, \quad \beta = \omega\sqrt{\mu\epsilon}$
- $v = \dfrac{\omega}{\beta} = \dfrac{1}{\sqrt{\mu\epsilon}}, \quad \lambda = \dfrac{2\pi}{\beta}$
- $\eta = \sqrt{\dfrac{\mu}{\epsilon}}\angle 0° = \eta_0\sqrt{\dfrac{\mu_r}{\epsilon_r}} = 120\pi\sqrt{\dfrac{\mu_r}{\epsilon_r}}$
- *E* and *H* are in time phase with each other

Plane waves in Good Conductors

In good conductor, $\sigma >> \omega\epsilon$

$$\sigma \simeq \infty, \ \epsilon = \epsilon_0, \ \mu = \mu_0\mu_r$$

- $\alpha = \beta = \sqrt{\frac{\omega\mu\sigma}{2}} = \sqrt{\pi f \mu\sigma}$
- $v = \frac{\omega}{\beta} = \sqrt{\frac{2\omega}{\mu\sigma}}, \lambda = \frac{2\pi}{\beta}$
- $\eta = \sqrt{\frac{\omega\mu}{\sigma}} \angle 45°$

Remember:

Electric and magnetic fields are out of phase by 45° in time in good conductors.

Skin Depth

The skin depth is a measure of the depth to which the amplitude of an EM wave will reduce to 36.8% of initial value.

$$\delta = \frac{1}{\alpha} = \sqrt{\frac{1}{\pi f \mu\sigma}} \text{ meter}$$

Surface or Skin Resistance

$$R_s = \frac{1}{\sigma\delta} = \sqrt{\frac{\pi f \mu}{\sigma}} \ \Omega / m^2$$

Remember:

☑ The skin depth is useful in calculating the *ac resistance* due to skin effect.

☑ $\delta \propto \frac{1}{\sqrt{f}}$.

☑ The perfect dielectric medium behaves like a perfect transmitter of EM wave.

☑ The perfect conducting medium behaves like a perfect reflector of EM wave.

Poynting's Theorem

The theorem states that "the net power flowing out of a given volume V is equal to the sum of the Ohmic losses and the time rate of decrease in the energy stored with in volume V."

$$\underbrace{-\oint_S (\vec{E}\times\vec{H})\cdot d\vec{S}}_{\text{Total Power leaving the volume}} = \underbrace{\frac{\partial}{\partial t}\int_V \left[\frac{1}{2}\in \vec{E}^2 + \frac{1}{2}\mu\vec{H}^2\right] dV}_{\text{Rate of decrease in energy stored in electric and magnetic fields}} + \underbrace{\int_V \sigma\vec{E}^2 dV}_{\text{Ohmic Power dissipated}}$$

Poynting Vector

$$\vec{P} = \vec{E} \times \vec{H} \quad \text{Watt/m}^2$$

Poynting vector represents the following parameter associated with the EM wave.

(i) Power density associated in EM wave in Watt/m^2

(ii) The power flow for the given EM wave and this is the direction of unit vector normal to the plane containing $\vec{E} \times \vec{H}$, according to right hand system. So this is also the direction of propagation of EM wave.

Instantaneous and Average Power Density

Parameters	Formula
Instantaneous power density	$\vec{P} = \frac{E_0^2}{\eta}\cos^2(\omega t - \beta z)\cdot \vec{i}_z$ Watt/m^2
Average power density	$\vec{P}_{avg} = \frac{1}{2}\text{Re}\{\vec{E} \times \vec{H}^*\}$ Watt/m^2
Average power density in lossless medium	$\vec{P}_{avg} = \frac{1}{2}\frac{E_x^2}{\eta_0}\vec{i}_z = \frac{1}{2}\eta_0 \cdot H_y^2 \vec{i}_z = \frac{1}{2}E_x H_y \vec{i}_z$ Watt/m^2 $\vec{P}_{avg} = \frac{E_{rms}^2}{\eta}\vec{i}_z = \eta H_{rms}^2 \vec{i}_z = E_{rms}H_{rms}\vec{i}_z$ Watt/m^2

Incidence of EM Wave

Normal Incidence

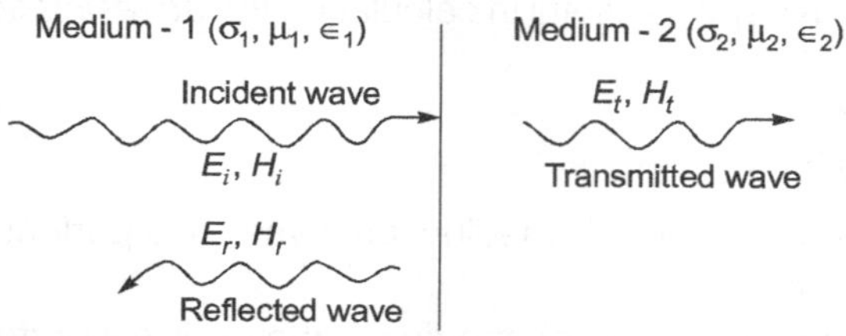

Reflection Coefficient

It is ratio of electric field of reflected wave to electric field of incident wave

$$\Gamma = \frac{E_r}{E_i} = \frac{\eta_2 - \eta_1}{\eta_2 + \eta_1} = \frac{\sqrt{\epsilon_1} - \sqrt{\epsilon_2}}{\sqrt{\epsilon_1} + \sqrt{\epsilon_2}}$$

where, η_1, $\eta_2 \rightarrow$ Intrinsic impedance of medium-1 and medium-2 respectively

Note:

Reflection coefficient is a complex quantity and can be written as

$$\Gamma = |\Gamma|e^{j\theta} \quad \text{where, } |\Gamma| = \rho$$

Transmission Coefficient (T)

It is defined as the ratio of transmitted electric field to the incident electric field

$$T = \frac{E_t}{E_i} \qquad\qquad T = 1 + \Gamma$$

$$T = \frac{2\eta_2}{\eta_2 + \eta_1} = \frac{2\sqrt{\epsilon_1}}{\sqrt{\epsilon_1} + \sqrt{\epsilon_2}}$$

Standing Wave Ratio

$$S = \frac{1+|\Gamma|}{1-|\Gamma|} \quad \text{or} \quad S = \frac{1+\rho}{1-\rho}$$

Note:

- ☑ $0 \le |\Gamma| \le 1$
- ☑ $1 \le S \le \infty$
- ☑ Total power reflected = $P_{\text{reflected}} = \Gamma^2 \times P_{\text{incident}}$
- ☑ Total power transmitted = $P_{\text{transmitted}} = (1 - \Gamma^2)\, P_{\text{incident}}$

Oblique Incidence

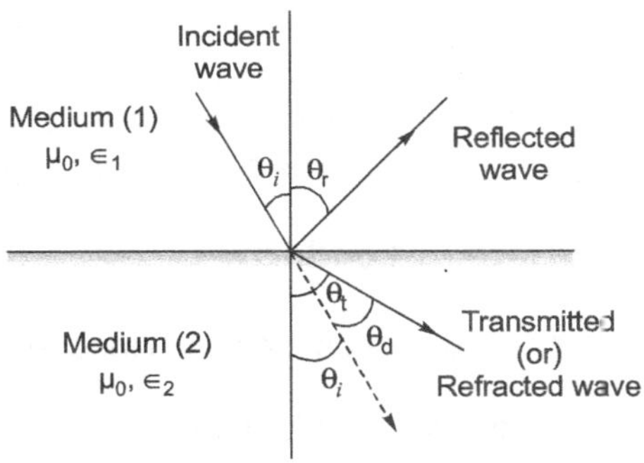

$$\theta_d = \theta_t - \theta_i$$

Snell's Law

$$n_1 \sin\theta_i = n_2 \sin\theta_t$$

where, $n_1 = c\sqrt{\mu_1 \epsilon_1} = c\sqrt{\epsilon_1}$; $n_2 = c\sqrt{\mu_2 \epsilon_2} = c\sqrt{\epsilon_2}$

are the refractive indices of the media.

Snell's law for EM wave, $\dfrac{\sin\theta_t}{\sin\theta_i} = \sqrt{\dfrac{\epsilon_1}{\epsilon_2}}$, where ϵ_1, ϵ_2 are relative permittivity.

Total internal Reflection of EM wave

$$\theta_c = \sin^{-1}\sqrt{\frac{\epsilon_2}{\epsilon_1}}$$ where, $\theta_c \rightarrow$ Critical angle for total internal reflection

Note:

θ_c is the minimum angle of incidence at which total internal reflection just starts.

Condition for total internal reflections

(a) $n_1 > n_2$

(b) $\theta_i \geq \theta_c$ or $\theta_i \geq \sin^{-1}\sqrt{\frac{\epsilon_2}{\epsilon_1}}$

(c) $T = 0 \quad \Rightarrow \Gamma = (T - 1) = -1$

Brewster's Angle (θ_B)

This is the angle of incidence for which complete transmission of EM wave occurs.

$$\tan\theta_B = \sqrt{\frac{\epsilon_2}{\epsilon_1}} = \frac{n_2}{n_1}$$

Note:

When a circularly or elliptically polarized wave is incident at Brewster's angle then the reflected and transmitted wave is linearly polarized. Therefore this angle is also known as polarization angle.

Reflection at the Surface of Conductors

- $\Gamma = \frac{\eta_2 - \eta_1}{\eta_2 + \eta_1} = -1 \qquad T = 0$ because $\eta_2 = 0$ for conductors.
- At the surface of conductor, electric field is minimum while magnetic field is maximum.

Polarization

This is the orientation of electric vector at a fixed position in space with respect to time.

Linear Polarization

Condition: ($\phi = 0°$ or $180°$)

$$E_x = E_{x0}\cos\omega t \text{ and } E_y = E_{y0}\cos(\omega t + \phi)$$

- **Parallel polarization**

$$E_x = E_{x0}\cos\omega t \text{ and } E_y = 0$$

- **Perpendicular polarization**

$$E_x = 0 \text{ and } E_y = E_{y_0} \cos\omega t$$

Circular Polarization

Condition: $(\phi = \pm 90°)$ and $(E_{x0} = E_{y0} = E_0)$

$$E_x = E_{x_0} \cos\omega t \text{ and } E_y = E_{y_0} \cos(\omega t + \phi) = \pm E_0 \sin\omega t$$

Note:

☑ Wave is assumed to be propagating in $+z$ direction.

☑ If $\phi = +90°$, the rotation is in clockwise direction then the wave is said to be left hand circularly polarized.

☑ If $\phi = -90°$ the rotation is in anti-clockwise direct on then the wave is said to be right hand circularly polarized.

Elliptical Polarization

Condition: $(\phi = \pm 90°)$ and $(E_{x0} \neq E_{y0})$

$$E_x = E_{x_0} \cos\omega t \text{ and } E_y = E_{y_0} \cos(\omega t + \phi) = \pm E_{y_0} \sin\omega t$$

Note:

☑ If $\phi = +90°$, the rotation is in clockwise direction then the wave is said to be left hand elliptically polarized.

☑ If $\phi = -90°$, the rotation is in anti-clockwise direction then the wave is said to be right hand elliptically polarized.

S and *P* Polarized Waves

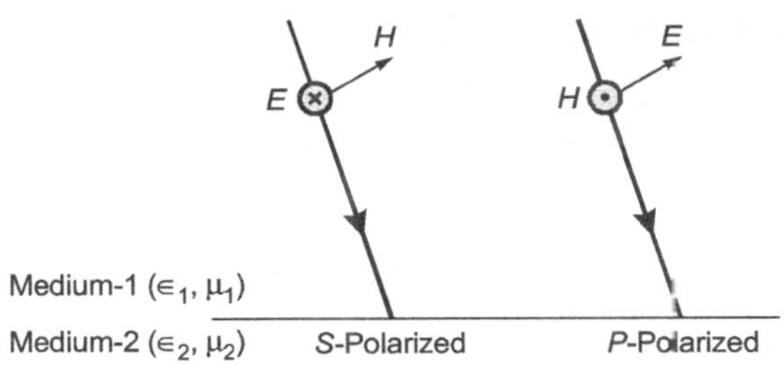

- In *S*-polarization, the electric field lies in the plane perpendicular to the at of interface. The reflection coefficient is given as

$$\Gamma_s = \frac{\eta_2 \sec\theta_t - \eta_1 \sec\theta_i}{\eta_2 \sec\theta_t + \eta_1 \sec\theta_i}$$

where, $\theta_i \rightarrow$ Incidence angle ; $\theta_t \rightarrow$ Transmission angle

$\eta_1, \eta_2 \rightarrow$ Intrinsic impedance

- In P-polarization, the $\vec{E}$ field lies in the plane of interface. The reflection coefficient is given as

$$\Gamma_P = \frac{\eta_2 \cos\theta_t - \eta_1 \cos\theta_i}{\eta_2 \cos\theta_t + \eta_1 \cos\theta_i}$$

Note:

Brewster's angle is valid for P-polarized waves only.

VII Transmission Lines

Introduction

These are the structures used for transmitting EM power from source or generator to the load.

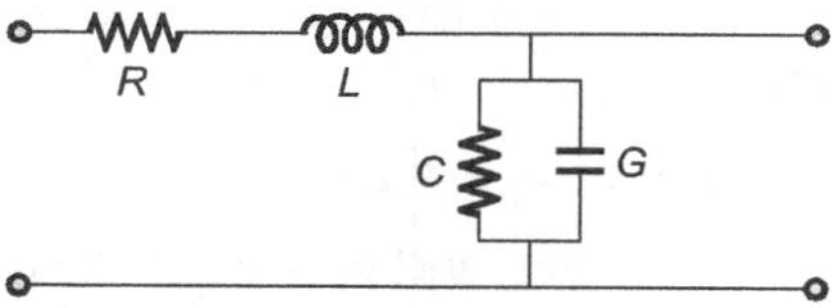

(Electrical equivalent circuit of transmission line in distributed from l length of line)

Voltage and Current in Transmission Line

$$V = \underbrace{\underbrace{V^+e^{-\gamma z}}_{\text{incident wave}} + \underbrace{V^-e^{\gamma z}}_{\text{Reflected wave}}}_{\text{standing wave}} \quad ; \quad I = \frac{1}{Z_0}[V^+e^{-\gamma z} - V^-e^{\gamma z}]$$

where, V_0^+ and V_0^- are wave amplitudes and Z_0 = characteristic impedance.

Characteristic Impedance (Z_0)

Characteristic impedance (Z_0) of the line is the ratio of the positively travelling voltage wave to the current wave at any point on the line

$$Z_0 = \frac{V^+}{I^+} = \sqrt{\frac{R + j\omega L}{G + j\omega C}}\ \Omega$$

Remember:

- Characteristic impedance of coaxial cable is $Z_0 = 60\sqrt{\frac{\mu_r}{\epsilon_r}} \ln\left(\frac{b}{a}\right)$.
- Characteristic impedance of parallel line is $Z_0 = 120\sqrt{\frac{\mu_r}{\epsilon_r}} \ln\left(\frac{D}{r}\right)$.

Propagation Constant

$$\gamma = \alpha + j\beta = \sqrt{(R + j\omega L)(G + j\omega C)} \text{ m}^{-1}$$

where, $\gamma \rightarrow$ propagation constant ; $\alpha \rightarrow$ Attenuation constant

$\beta \rightarrow$ phase constant

Lossless Transmission Line

A transmission line is said to be lossless if the conductors of the line are perfect and the dielectric medium separating them is loss less.

$$R = G = 0 \; ; \; \gamma = j\omega\sqrt{LC}$$

$\therefore \; \alpha = 0$ and $\beta = \omega\sqrt{LC}$

$$V = \frac{\omega}{\beta} = \frac{1}{\sqrt{LC}} \quad ; \quad Z_0 = R_0 = \sqrt{\frac{L}{C}}$$

Note: Every lossless line is distortionless but vice versa is not true.

Distortionless Transmission Line

A distortionless line is one in which the attenuation constant α is frequency independent while the phase constant β is linearly dependent on frequency.

$$\frac{R}{L} = \frac{G}{C} \; ; \; \alpha = \sqrt{RC} \text{ and } \beta = \omega\sqrt{LC}$$

$$V = \frac{\omega}{\beta} = \frac{1}{\sqrt{LC}} = 2\pi f \cdot \delta \quad ; \quad Z_C = \sqrt{\frac{R}{G}} = \sqrt{\frac{L}{C}}$$

Input Impedance, Standing Wave Ratio and Power

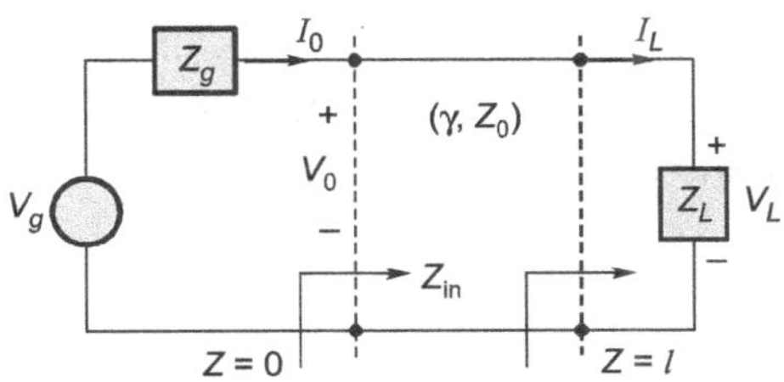

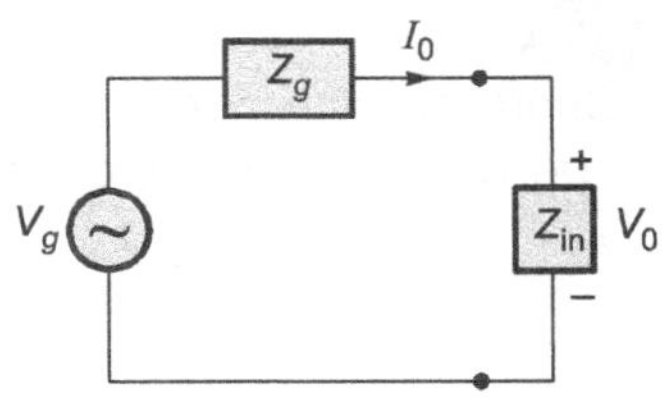

Input Impedance

$$Z_{in} = Z_0\left[\frac{Z_L + Z_0 \tanh\gamma l}{Z_0 + Z_L \tanh\gamma l}\right] \quad \text{.. lossy medium}$$

$$Z_{in} = Z_0\left[\frac{Z_L + jZ_0 \tan\beta l}{Z_0 + jZ_L \tan\beta l}\right] \quad \text{... lossless medium}$$

Voltage reflection coefficient

The voltage reflection coefficient at any point on the line is the ratio of the reflected voltage wave to that of the incident wave.

$$\Gamma_L = \frac{Z_L - Z_0}{Z_L + Z_0} \qquad 0 \le \Gamma_L \le -1$$

Current reflection coefficient

The current reflection coefficient at any point on the line is the negative of the voltage reflection coefficient at that point i.e. $-\Gamma_L$.

Standing wave ratio

It is the ratio of maximum voltage of standing wave to the minimum voltage of standing wave.

$$S = \frac{V_{max}}{V_{min}} = \frac{1+|\Gamma_L|}{1-|\Gamma_L|} \qquad 1 \le S < \infty$$

Power carried by forward wave

$$P_{avg} = \frac{|V^+|^2}{2Z_0}(1-|\Gamma|^2)$$

Behaviour of Transmission Line due to Variation in Load Impedance (Z_L)

(i) Short circuited line ($Z_L = 0$)

$$Z_{in} = Z_{sc} = jZ_0 \tan\beta l \qquad ; \quad \Gamma_L = -1, \quad S = \infty$$

(ii) Open circuited line ($Z_L = \infty$)

$$Z_{in} = Z_{oc} = -jZ_0 \cot\beta l \qquad ; \quad \Gamma_L = 1, \quad S = \infty$$

Note:

Characteristic impedance (Z_0) is geometric mean of short circuited and open circuited line input impedance.

$$Z_{sc} Z_{oc} = Z_0^2$$

$$Z_0 = \sqrt{Z_{sc} Z_{oc}}$$

(iii) Matched line ($Z_L = Z_0$)

$$Z_{in} = Z_0 \quad ; \quad \Gamma_L = 0 \quad ; \quad S = 1$$

Behaviour of Transmission Line due to variation in Length of line 'l'

(i) Infinitely long line ($l = \infty$) : $Z_{in} = Z_0$

(ii) For length $l = \lambda$: $Z_{in} = Z_L$

(iii) For length $l = \lambda/2$ (half wave line) : $Z_{in} = Z_L$

(iv) For length $l = \lambda/4$ (quarter wave line)

$$Z_{in} = \frac{Z_0^2}{Z_L} \quad \text{or} \quad Z_0 = \sqrt{Z_{in} \cdot Z_L}$$

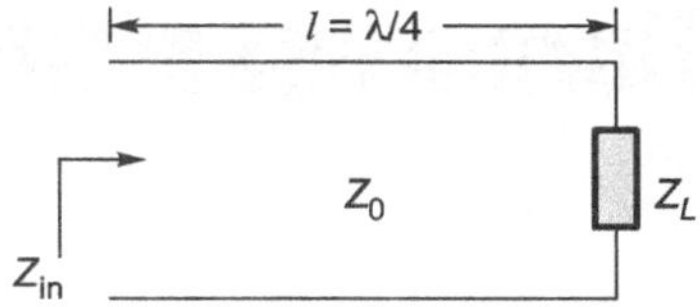

(v) For length $l = \lambda/8$

$$Z_{in} = Z_0\left[\frac{Z_L + jZ_0}{Z_0 + jZ_L}\right] \quad ; \quad |Z_{in}| = Z_0$$

Remember:

The magnitude of Z_{in} for $\lambda/8$ line and $l = \infty$ length line are always equal but phase is different hence Z_{in} for $l = \lambda/8$ and $l = \infty$ are not equal.

■■■■

12 Microprocessors

I Introduction

Definition

Microprocessor is a programmable integrating device having computing, storing, retrieving and decision making capability.

Note:

☑ Computer with microprocessor as its CPU is known as microcomputer.

☑ Microcomputer on a single chip is known as microcontroller.

History of Microprocessor

Microprocessor	Word length	Memory capacity
Intel4004 (PMOS)	4-bit	640 B
Intel 8008	8-bit	16 kB
Intel 8080 (NMOS)	8-bit	64 kB
Intel 8085 (NMOS)	8-bit	64 kB
Intel 8086 (HMOS)	16-bit	1 MB
Intel 8088	8/16-bit	1 MB
Intel 80186	16-bit	1 MB
Intel 80286	16-bit	16 MB real, 4 GB virtual
Intel 80386	32-bit	4 GB real, 4 GB virtual
Intel 80486	32-bit	4 GB real, 64 TB virtual
Pentium-II	64-bit	64 GB real
Z-80	8-bit	64 kB
Z-800	8-bit	500 kB

A brief review of various microprocessors

Computer language

Mnemonic : A combination of letters to suggest the operation of an instruction.

Program : A set of instruction written in a specific sequence for the computer to accomplish a given task.

Machine language : A set of instructions executed directly by computer's CPU.

Assembly language : It is a low level programming language in which there is a strong correspondence between the language and architecture's machine code instructions.

Low level language : Machine specific languages are known as low level language Ex: Machine language, Assembly language.

High level language : A programming language that enables a programmer to write programs that are independent of particular type of computer.

Assembler : It is a computer program that translates an assembly language program from mnemonics to the binary machine code of a computer.

Compiler : A program that translates high level language program into machine level language by translating the entire program at a time and executing it.

Interpreter : It translates high level language program in object codes statementwise.

Remember:

High level language programming requires a translator like a compiler or an interpreter to translate the program written in a high level language to binary form.

Bit: A binary digit, 0 or 1

Nibble: A group of four bits

Byte: A group of eight bits

Word: A group of byte the computer recognizes and processor at a time.

II Architecture of 8085

Introduction

There are two types of architecture depending upon storage of program and data in memory

(i) **Von Neumann architecture** : Both data and program are in the same memory. Intel 8085 Intel 8086

(ii) **Harvard architecture:** Program and data are in different memories. TMS 32010, Intel 8051, e.g. Intel's Pentium etc.

Pin Configuration

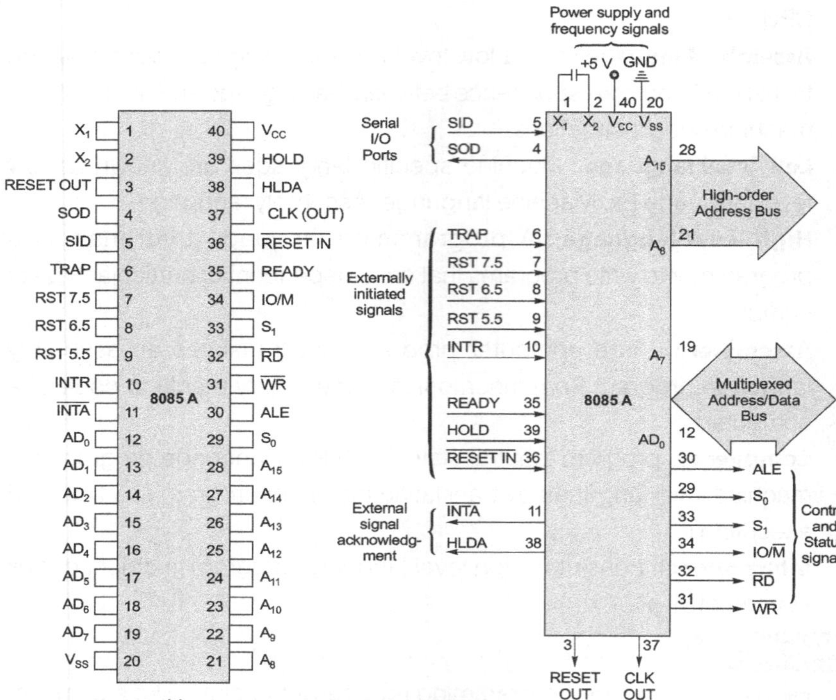

(a) 8085 Configuration (b)

Key features of 8085

(i) It is a 8-bit processor.

(ii) It has 8 data bus lines, which is the bit capacity of the microprocessor.

(iii) It has total 16 address lines with addressing capacity of 64 kB.

(iv) The crystal frequency of processor is 6 MHz and the clock frequency is 3.07 MHz (= 3 MHz), which is approximately half the crystal frequency.

(v) Low order address bus ($AD_0 - AD_7$) is multiplexed with data bus in order to reduce the number of pins.

(vi) To de-multiplex address from data, ALE (Address Latch Enable) signal is used.

ALE = 1, Address transfer to bus

ALE = 0, Data transfer to bus.

(vii) Disadvantage of multiplexing is that speed will be reduced.

(viii) There are 5 hardware interrupts available for 8085.

(ix) 8085 has 74 basic instructions with 246 Opcodes.

Sr. No.	Signals	Pins
1.	Power supply signals	V_{CC}, V_{SS}
2.	Clock signals	X_1 X_2, CLK OUT
3.	Reset signals	RESETIN, RESET OUT
4.	Interrupt signals	TRAP, Restart interrupts (RST 7.5, REST 6.5 and RST 5.5), INTR, INTA
5.	Address bus and data bus	Address bus ($A_8 - A_{15}$) Multiplexed address/data bus ($AD_0 - AD_7$)
6.	Status signals and control signals	Address latch enable (ALE), Input output/ memory ($IO/\overline{M}$), Status signals (S_1 and S_0), Read ($\overline{RD}$), Write ($\overline{WR}$), READY
7.	Serial input/output signals	HOLD, HLDA
8.	DMA request signals	SID, (Serial input data), SOD (Serial output data)

System Bus

Bus is a group of wires or conductors used for communication between processor, memory and I/O device.

- **Address Bus**
  - (i) It is used to transfer address of memory or I/O from the processor. It is unidirectional.
  - (ii) Address Bus defines maximum memory that can be connected to a processor related by $2^n = N$.

  where, $n \rightarrow$ Number of address lines ; $N \rightarrow$ Number of address/memory location

- **Data Bus**
  - (i) It is used to transfer data between processor, memory or peripheral devices. It is bidirectional.
  - (ii) In 8085, it is multiplexed with lower order address bus.

  ALE = 1 ; All lines are address bus

  ALE = 0 ; A_{15} - $A_8 \rightarrow$ Address bus

  AD_7 - $AD_0 \rightarrow$ Data bus

- **Control Bus**
  - (i) It is a group of control signals required for various operations of the processor.
  - (ii) It is considered to the unidirectional if signals from processor are considered. It is partially unidirectional and bidirectional if the control signals are considered.

Control and status signal

(i) $\overline{RD}$ (Read) : When the signal is low on this pin, the microprocessor performs memory reading or I/O reading operation.

(ii) $\overline{WR}$ (Write) : When the signal is low on this pin, the microprocessor performs memory writing or I/O writing operation.

(iii) IO/$\overline{\text{M}}$: This status signal is used to give information of operation to be performed with memory or I/O device.

IO/$\overline{\text{M}}$	$\overline{\text{RD}}$	$\overline{\text{WR}}$	Description
0	0	1	Memory read ($\overline{\text{MEMR}}$)
0	1	0	Memory write ($\overline{\text{MEMW}}$)
1	0	1	IO Read (IO$\overline{\text{R}}$)
1	1	0	IO Write (IO$\overline{\text{W}}$)

S_1 **and** S_0 : These two status signal are used to indicate the status of the operation.

S_1	S_0	Microprocessor operation
0	0	Halt (No operation)
0	1	Write operation
1	0	Read operation
1	1	Opcode fetch (reading instruction)

Power supply and frequency signals

V_{CC} **:** Power supply pin of +5 V

X_1 **and** X_2 **:** A crystal is connected across these pins. The frequency is internally divided by two.

CLK (out) : This pin, there will be synchronization between the different peripherals and microprocessor.

Serial I/O Port

(i) SID (Serial Input Data) : This pin is used for receiving the data into microprocessor serially.

(ii) SOD (Serial output data) : This pin is used for sending the data from the microprocessor serially.

Externally initiated signals

- **Hardware interrupts :** TRAP, RST-7.5, RST-6.5, RST-5.5, INTR are called Hardware interrupts. It is also used to accept external interrupts to provide acknowledgment (Ack) to the external device.

Remember:

In priority order, interrupts are as

TRAP > RST-7.5 > RST-6.5 > RST-5.5 > INTR

- **READY :** It is used to interface with the slow peripheral devices (Memory or I/O devices) to the microprocessor.
- **Reset :**

(a) $\overline{\text{RESET IN}}$: It is an active low signal, when this signal gets activated, microprocessor is reset.

(b) RESET OUT : This signal is used to reset the other peripheral devices.

- **HOLD and HLDA :**
  (a) HOLD : A peripheral like DMA controller sends the HOLD request to the microprocessor through this pin to leave the data bus which is being used by microprocessor.
  (b) HLDA : HOLD acknowledge.

Internal Architecture of 8085 Microprocessor

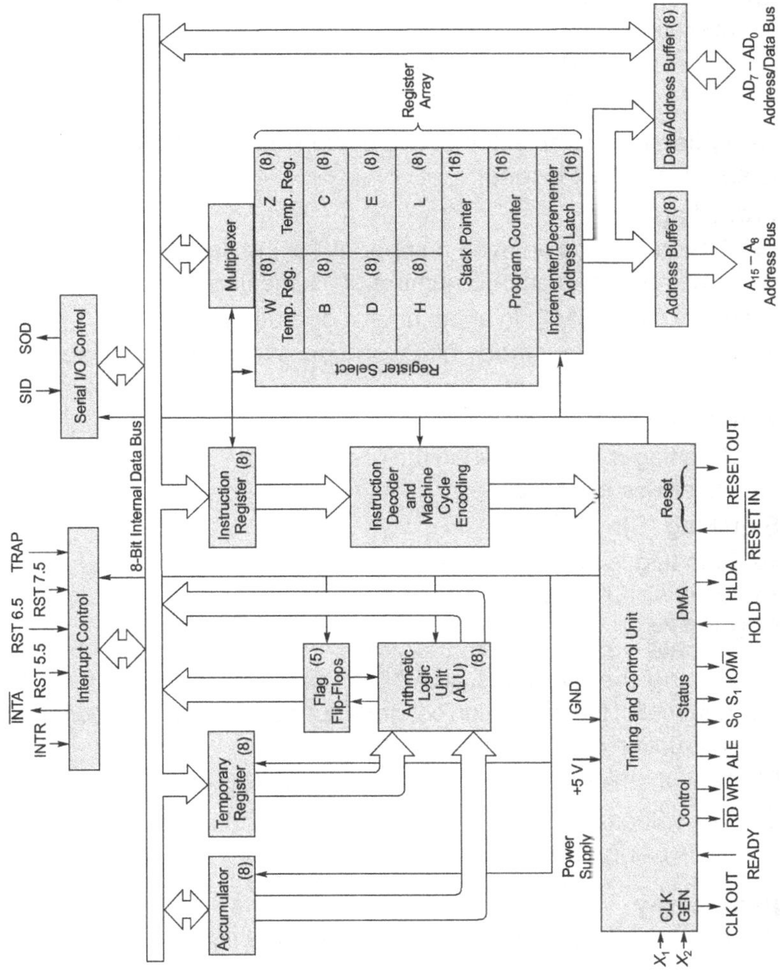

The architecture is divided into following functional units:

Arithmetic and Logical Unit (ALU)

The ALU performs arithmetical and logical operations. It includes :

(a) Accumulator : It is a 8 bit programmable register. All the arithmetic and logical operations performed with contents of accumulator and the results are stored in accumulator only.

(b) Temporary Register : It is an 8 bit non-programmable register used to hold data during an arithmetic and logic operation.

(c) Arithmetic and logic circuits : This unit performs the actual numerical and logical operation.

(d) Flags : The flags generally reflect the status of arithmetic or logical operation.

D_7	D_6	D_5	D_4	D_3	D_2	D_1	D_0
S	Z	×	AC	×	P	×	CY

Carry flag (CY): If an arithmetic operation results in a carry, the CY flag is **set** otherwise it is **reset.**

Parity flag (P):

⇒ If the result has an even number of 1's, the flag is **set.**

⇒ If the result has an odd number of 1's the flag is **reset.**

Auxiliary Carry (AC): In an arithmetic operation.

⇒ If carry is generated by bit D_3 and passed to D_4, flag is **set.**

⇒ Otherwise it is **reset.**

Zero Flag (Z):

⇒ Zero flag is **set** to 1, when the result is zero.

⇒ Otherwise it is **reset.**

Sign Flag (S):

⇒ Sign flag is **set** if bit D_7 of the result is 1.

⇒ Otherwise it is **reset.**

Remember:

☑ Among the five flags, the AC flags is used internally for BCD arithmetic; the instruction set does not include any conditional Jump instruction based on the AC flag.

☑ 'x' in the flag register indicate the unused flip-flops.

☑ The values of D_1, D_3 and D_5 bits should be taken as '0' in programs while using PSW instruction.

Register array

The architecture of 8085 consist of following registers.

Temporary registers :

(i) Temporary data register : It is also called as operand register (8-bit). It provides operands to the ALU.

(ii) Temporary registers (W and Z) : This registers are not available to the user.

Remember:

W and Z registers are used by 8085 for swap instruction.

General purpose registers :

The 8085 microprocessor consists of 6 general purpose registers i.e. B, C, D, E, H, and L of 8 bits each. These registers are available to the user.

Remember:

- ☑ The general purpose registers put together is called scratch and memory.
- ☑ The valid register pairs available are BC, DE and HL to store 16-bit of data.
- ☑ The HL register pair functions as default data pointer. If used as memory pointer it holds the address of a 16 bit address at a memory location.
- ☑ Accumulator can also be used along with status register to form a 16 bit programmable register called program status word (PSW).

Special purpose registers

There are two 16-bit special purpose register i.e. program counter (PC) and stack pointer (SP).

(i) Program counter : It is a 16 bit register used to hold memory addresses.

(ii) Stack pointer : It is a 16 bit register used as a memory pointer.

Remember:

- ☑ When the microprocessor is reset, the PC sets to zero.
- ☑ Stack grows from bottom to top following last in first out (LIFO) structure and the SP contents keep·decreasing as stack grows.

Timing and control unit

It controls all internal and external circuits in the microprocessor system.

Remember:

The microprocessor uses a quartz crystal (LC or RC circuits) to determine the clock frequency, so that other timing and control signals are developed.

Instruction Register and Decoder

It is the part of ALU.

- **Instruction register :** The instruction register holds the opcode of the instruction that is decoded and executed.

- **Instruction decoder:** The output of instruction register connected to the decoder. The decoder decodes the instruction and establishes the sequence of events to follow.

Interrupt Control Unit

The interrupt control units job is to service the interrupt and after completing the interrupt service routine return back the control to the main program where it was interrupted.

III Instruction Set and Data Formats

Timing Diagram

Instruction cycle

The CPU fetches one instruction from the memory at a time and executes it. One instruction cycle can consists of 1-6 machine cycle.

Note:

An instruction cycle consists of a fetch cycle and execute cycle.

Machine Cycle

The time required by the microprocessor to complete the operation of accessing memory or I/O device is called as a machine cycle. One machine cycle can consists of 3-6 T-states.

Note:

☑ An instruction cycle consists of several machine cycles.

☑ Different types of machine cycles are opcode fetch (4T-6T), memory read (3T), memory write (3T), I/O read(3T), I/O write (3T), INTR Acknowledge and BUS Idle(3T).

T-State

Microprocessor perform an operation in a specific time period i.e. specific clock cycles. Each clock cycle is called as T-state.

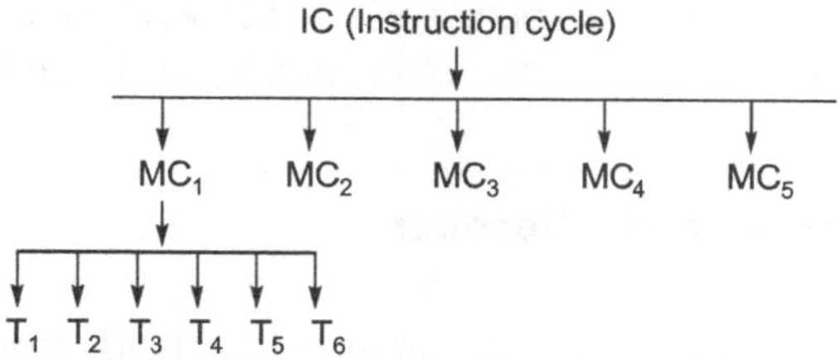

Classification of Instructions Set of 8085 Microprocessor

Instructions set

- Based on Length of wordsize
  - One-byte
  - Two-byte
  - Three-byte
- Based on addressing modes
  - Direct
  - Register
  - Register indirect
  - Immediate
  - Implicit/Implide
- Based on operations
  - Data transfer
  - Arithmetic
  - Logical
  - Branch control
  - Stack, I/O and machine control

An instruction is a binary pattern designed inside a microprocessor to perform a specific function.

Based on Length of Word Size

(i) **One-Byte Instructions:** The one-byte instructions specify the operation to be performed and who is going to perform it. These nstructions require single memory location.

(ii) **Two-Byte Instructions:** The Two-byte instruction uses first byte to specify the operation i.e. opcode and second byte to specify the operand. These instructions required two successive memory locations in the memory.

(iii) **Three-Byte instructions:** First byte stores opcode, second byte stores lower order 8-bit of 16-bit operand or address and third byte stores higher order 8-bit of 16-bit operand or address.

Based on Addressing Modes

(i) **Register addressing mode:** In register addressing mode the source and the destination are general purpose registers.

(ii) **Immediate addressing mode:** In immediate addressing mode the data (8/16 bit) is specified in the instruction itself.

(iii) **Direct addressing mode:** In direct addressing mode 16 bit address of the operand is given within the instruction itself.

(iv) **Indirect addressing mode:** In indirect addressing mode the instructions reference the memory through a register pair, i.e. the memory address where the operand is located is specified by the contents of a register pair.

(v) **Implicit addressing mode:** The implicit mode of addressing does not require any operand. The data is specified within the opcode itself.

Based on Operations

Abbreviations used in the description of the instruction set:

R = Register   Rs = Register source

Rd = Register destination   M = Memory

Rp = Register pair
XX = Random information
Z = Zero flag
P = Parity flag
( ) = Contents of
S = Sign flag
AC = Auxiliary carry flag
CY = Carry flag

Data Transfer Instruction

MOV : Move – Copy from Source to Destination

Opcode	Operand	Bytes	M-Cycles	T-States
MOV	Rd/M, Rs/M	1	1/2	4T/7T

Description: This instruction copies the contents of the source register into the destination register; the contents of the source register are not altered.

MOV Rd, Rs requires 1 M-cycle and 4T. MOV R_d, M and MOV M, Rs requires 2M-cycle and 7T.

MVI : Move Immediate 8-Bit

Opcode	Operand	Bytes	M-Cycles	T-States
MVI	R/M, 8-bit data	2	2/3	7T/10T

Description: The 8-bit data are stored in the destination register or memory.

LXI : Load Register Pair Immediate

Opcode	Operand	Bytes	M-Cycles	T-States
LXI	Rp, 16-bit data	3	3	10T

Description: The instruction loads 16-bit data in the register pair designated in the operand.

Comments:

This is the only instruction that can directly load a 16-bit address in the stack pointer register.

LDA : Load Accumulator Direct

Opcode	Operand	Bytes	M-Cycles	T-States
LDA	16-bit address	3	4	13T

Description: The contents of a memory location, specified by a 16-bit address in the operand, are copied to the accumulator. The contents of the source are not altered.

STA : Store Accumulator Direct

Opcode	Operand	Bytes	M-Cycles	T-States
STA	16-bit address	3	4	13T

Description: The contents of the accumulator are copied to a memory location specified by the operand.

LHLD : Load H and L Registers Direct

Opcode	Operand	Bytes	M-Cycles	T-States
LHLD	16-bit address	3	5	16T

Description: The instruction copies the contents of the memory location pointed out by the 16-bit address in register L and copies the contents of the next memory location in register H.

SHLD : Store H and L Registers Direct

Opcode	Operand	Bytes	M-Cycles	T-States
SHLD	16-bit address	3	5	16T

Description: The contents of register L are stored in the memory location specified by the 16-bit address in the operand, and the contents of H register are stored in the next memory location by incrementing the operand. The contents of registers HL are not altered.

LDAX : Load Accumulator Indirect

Opcode	Operand	Bytes	M-Cycles	T-States
LDAX	Rp	1	2	7T

Description: The contents of the designated register pair point to a memory location. This instruction copies the contents of that memory location into the accumulator.

Flags : No flags are affected.

STAX : Store Accumulator Indirect

Opcode	Operand	Bytes	M-Cycles	T-States
STAX	Rp	1	2	7T

Description: The contents of the accumulator are copied into the memory location specified by the contents of the operand (register pair).

Flags : No flags are affected.

XCHG : Exchange H and L with D and E

Opcode	Operand	Bytes	M-Cycles	T-States
XCHG	None	1	1	4T

Description: The contents of register H are exchanged with the contents of register D, and the contents of register L are exchanged with the contents of register E.

Arithmetic Instructions

ADD : Add Register to Accumulator

Opcode	Operand	Bytes	M-Cycles	T-States
ADD	R/M	1	1/2	4T/7T

Description: The contents of the operand (register or memory) are added to the contents of the accumulator and the result is stored in the accumulator.

ADC : Add Register to Accumulator with Carry

Opcode	Operand	Bytes	M-Cycles	T-States
ADC	R/M	1	1/2	4T/7T

Description: The contents of the operand (register or memory) and the Carry flag are added to the contents of the accumulator and the result is placed in the accumulator.

ACI : Add Immediate to Accumulator with Carry

Opcode	Operand	Bytes	M-Cycles	T-States
ACI	8-bit data	2	2	7

Description: The 8-bit data (operand) and the Carry flag are added to the contents of the accumulator, and the result is stored in the accumulator.

ADI : Add Immediate to Accumulator

Opcode	Operand	Bytes	M-Cycles	T-States
ADI	8-bit data	2	2	7T

Description: The 8-bit data (operand) is added to the contents of the accumulator, and the result is placed in the accumulator.

Flags: All flags are modified to reflect the result of the addition.

SBB : Subtract Source and Borrow from Accumulator

Opcode	Operand	Bytes	M-Cycles	T-States
SBB	R/M	1	1/2	4T/7T

Description: The contents of the operand (register or memory) and the Borrow flag are subtracted from the contents of the accumulator and the results are placed in the accumulator.

SUB : Subtract Register or Memory from Accumulator

Opcode	Operand	Bytes	M-Cycles	T-States
SUB	R/M	1	1/2	4T/7T

Description: The contents of the register or the memory location specified by the operand are subtracted from the contents of the accumulator, and the result is placed in the accumulator.

SBI : Subtract Immediate with Borrow

Opcode	Operand	Bytes	M-Cycles	T-States
SBI	8-bit data	2	2	7T

Description: The 8-bit data (operand) and the borrow are subtracted from the contents of the accumulator, and the result is placed in the accumulator.

SUI : Subtract Immediate from Accumulator

Opcode	Operand	Bytes	M-Cycles	T-States
SUI	8-bit data	2	2	7T

Description: The 8-bit data (the operand) is subtracted from the contents of the accumulator, and the result is placed in the accumulator.

INR : Increment Contents of Register/Memory by 1

Opcode	Operand	Bytes	M-Cycles	T-States
INR	R/M	1	1/3	4T/10T

Description: The contents of the designated register/memory is incremented by 1 and the result is stored in the same place.

DCR : Decrement Source by 1

Opcode	Operand	Bytes	M-Cycles	T-States
DCR	R/M	1	1/3	4T/10T

Description: The contents of the designated register/memory is decremented by 1 and the result is stored in the same place.

Note:

In INR and DCR operation except carry, all flags are affected.

INX : Increment Register Pair by 1

Opcode	Operand	Bytes	M-Cycles	T-States
INX	Rp	1	1	6T

Description: The contents of the specified register pair are incremented by 1. The instruction views the contents of the two registers as a 16-bit number.

DCX : Decrement Register Pair by 1

Opcode	Operand	Bytes	M-Cycles	T-States
DCX	Rp	1	1	6T

Description: The contents of the specified register pair are decremented by 1. This instruction views the contents of the two registers as a 16-bit number.

Flags: No flags are affected for INX and DCX.

DAA : Decimal Adjust Accumulator

Opcode	Operand	Bytes	M-Cycles	T-States
DAA	None	1	1	4T

Description:

1. If the value of the low-order four bits ($D_3 - D_0$) in the accumulator is greater than 9 or if AC flag is set, the instruction adds 6(06) to the low-order four bits.
2. If the value of the high-order four bits ($D_7 - D_4$) in the accumulator is greater than 9 or if the Carry flag is set, the instruction adds 6(60) to the high-order four bits.

Flags: S, Z, AC, P, CY flags are altered to reflect the results of the operation.

DAD : Add Register Pair to H and L Registers

Opcode	Operand	Bytes	M-Cycles	T-States
DAD	Rp	1	3	10T

Description: The 16-bit contents of the specified register pair are added to the contents of the HL register and the sum is saved in the HL register. The contents of the source register pair are not altered.

Flags: If the result is larger than 16-bits the CY flag is set. No other flags are affected.

Note:

After the execution of the instruction, the contents of the stack pointer register are not altered.

Logical Instructions

ANA : Logical AND with Accumulator

Opcode	Operand	Bytes	M-Cycles	T-States
ANA	R/M	1	1/2	4T/7T

Description: The contents of the accumulator are logically ANDed with the contents of the operand (register or memory), and the result is placed in the accumulator.

ANI : AND Immediate with Accumulator

Opcode	Operand	Bytes	M-Cycles	T-States
ANI	8-bit data	2	2	7T

Description: The contents of the accumulator are logically ANDed with the 8-bit data (operand) and the result is placed in the accumulator.

Flags: AC = 1, CY = 0 after execution

ORA : Logically OR with Accumulator

Opcode	Operand	Bytes	M-Cycles	T-States
ORA	R/M	1	1/2	4T/7T

Description: The contents of the accumulator are logically ORed with the contents of the operand (register of memory).

ORI : Logically OR Immediate

Opcode	Operand	Bytes	M-Cycles	T-States
ORI	8-bit data	2	2	7T

Description: The contents of the accumulator are logically ORed with the 8-bit data in the operand and the result is placed in the accumulator.

Flags: AC = 0, CY = 0 after execution.

XRA : Exclusive OR with Accumulator

Opcode	Operand	Bytes	M-Cycles	T-States
XRA	R/M	1	1/2	4T/7T

Description: The contents of the operand (register or memory) are Exclusive ORed with the contents of the accumulator.

XRI : Exclusive OR Immediate with Accumulator

Opcode	Operand	Bytes	M-Cycles	T-States
XRI	8-bit data	2	2	7T

Description: The 8-bit data (operand) are Exclusive ORed with the contents of the accumulator, and the result is placed in the accumulator.

Flags : Z, S, P are altered to reflect the results of the operation. CY and AC are reset.

CMA : Complement Accumulator

Opcode	Operand	Bytes	M-Cycles	T-States
CMA	None	1	1	4T

Description: The contents of the accumulator are complemented.

Flags: No flags are affected.

CMP : Compare with Accumulator

Opcode	Operand	Bytes	M-Cycles	T-States
CMP	R/M	1	1/2	4T/7T

Description:

- The contents of the operand (register or memory) are compared with the contents of the accumulator. Both contents are preserved and the comparison is shown by setting the flags as follows:
  - ⇒ If (A) < (R/M) : Carry flag is set and Zero flag is reset.
  - ⇒ If (A) = (R/M) : Zero flag is set and Carry flag is reset.
  - ⇒ If (A) > (R/M) : Carry and Zero flags are reset.
- The comparison of two bytes is performed by subtracting the contents of the operand from the contents of the accumulator; however, neither contents are modified.

Flags : S, P, AC are also modified in addition to Z and CY to reflect the results of the operation.

CPI : Compare Immediate with Accumulator

Opcode	Operand	Bytes	M-Cycles	T-States
CPI	8-bit	2	2	7T

Description:

- The second byte (8-bit data) is compared with the contents of the accumulator.
  - ⇒ If (A) < Data : Carry flag is set and Zero flag is reset.
  - ⇒ If (A) = Data : Zero flag is set and Carry flag is reset.
  - ⇒ If (A) > Data : Carry and Zero flags are reset.
- The comparison of two bytes is performed by subtracting the data byte from the contents of the accumulator; however, neither contents are modified.

Flags: S, P, AC are also modified in addition to Z and CY to reflect the result of the operation.

CMC : Complement Carry

Opcode	Operand	Bytes	M-Cycles	T-States
CMC	None	1	1	4T

Description: The Carry flag is complemented.

Flags: The Carry flag is modified, no other flags are affected.

STC : Set Carry

Opcode	Operand	Bytes	M-Cycles	T-States
STC	None	1	1	4T

Description: The Carry flag is set.

Flags: No other flags are affected.

RLC : Rotate Accumulator Left

Opcode	Operand	Bytes	M-Cycles	T-States
RLC	None	1	1	4T

Description: Each binary bit of the accumulator is rotated left by one position. Bit D_7 is placed in the position of D_0 as well as in the Carry flag.

Flags: CY is modified according to bit D_7. S, Z, P, AC are not affected.

RAL : Rotate Accumulator Left through Carry

Opcode	Operand	Bytes	M-Cycles	T-States
RAL	None	1	1	4T

Description: Each binary bit of the accumulator is rotated left by one position through the Carry flag. Bit D_7 is placed in the bit in the Carry flag and the Carry flag is placed in the least significant position D_0.

Flags: CY is modified according to bit D_7. S, Z, AC, P are not affected.

RRC : Rotate Accumulator Right

Opcode	Operand	Bytes	M-Cycles	T-States
RRC	None	1	1	1

Description: Each binary bit of the accumulator is rotated right by one position. Bit D_0 is placed in the position of D_7 as well as in the Carry flag.

Flags: CY is modified according to bit D_0. S, Z, P, AC are not affected.

RAR : Rotate Accumulator Right through Carry

Opcode	Operand	Bytes	M-Cycles	T-States
RAR	None	1	1	4T

Description : Each binary bit of the accumulator is rotated right by one position through the Carry flag. Bit D_0 is placed in the Carry flag and the bit in the Carry flag is placed in the most significant position D_7.

Flags: CY is modified according to bit D_0. S, Z, P, AC are not affected.

Branch Control Instructions

JMP : Jump Unconditionally

Opcode	Operand	Bytes	M-Cycles	T-States
JMP	16-bit	3	3	10T

Description : This instruction is equivalent to a 1-byte unconditional Jump instruction. The program sequence is transferred to the memory location specified by the 16-bit address.

Conditional Jump

Op Code		Operand	Bytes	M-cycles	T-States
Carry =	1 JC	16-bit	3	2/3	7/10
	0 JNC	16-bit	3	2/3	7/10
Zero =	1 JZ	16-bit	3	2/3	7/10
	0 JNZ	16-bit	3	2/3	7/10
Sign =	1 JM	16-bit	3	2/3	7/10
	0 JP	16-bit	3	2/3	7/10
Parity =	1 JPE	16-bit	3	2/3	7/10
	0 JPO	16-bit	3	2/3	7/10

Description : Jump execution of program immediately at the given address if condition is true. The machine cycles and 10 T-states are consumed if condition is true while two machine cycles and 7 T-states are consumed if condition is false.

PCHL : Load Program Counter with HL Contents

Opcode	Operand	Bytes	M-Cycles	T-States
PCHL	None	1	1	6T

Description : The contents of registers H and L are copied into the program counter.

Flags : No flags are affected.

CALL : Unconditional Subroutine Call

Opcode	Operand	Bytes	M-Cycles	T-States
CALL	16-bit address	3	5	18T

Description : The program sequence is transferred to the address specified by the operand. Before the transfer, the address of the next instruction to CALL (the contents of the program counter) is pushed on the stack.

Flags : No flags are affected.

Conditional Call

Op Code		Operand	Bytes	M-cycles	T-States
Carry =	1 CC	16-bit	3	2/5	9/18
	0 CNC	16-bit	3	2/5	9/18
Zero =	1 CZ	16-bit	3	2/5	9/18
	0 CNZ	16-bit	3	2/5	9/18
Sign =	1 CM	16-bit	3	2/5	9/18
	0 CP	16-bit	3	2/5	9/18
Parity =	1 CPE	16-bit	3	2/5	9/18
	0 CPO	16-bit	3	2/5	9/18

Description : Call immediately the subroutine stored at given 16-bit address if condition is true. Five machine cycles and 18 T-states are consumed if condition is true while two machine cycles and 9-T states are consumed if condition is false.

RET : Return from Subroutine Unconditionally

Opcode	Operand	Bytes	M-Cycles	T-States
RET	None	1	3	10T

Description : The program sequence is transferred from the subroutine to the calling program. The instruction is equivalent to POP program counter.

Flags : No flags are affected.

Conditional Return

Op Code		Operand	Bytes	M-cycles	T-States
Carry =	1 RC	—	1	1/3	6/12
	0 RNC	—	1	1/3	6/12
Zero =	1 RZ	—	1	1/3	6/12
	0 RNZ	—	1	1/3	6/12
Sign =	1 RM	—	1	1/3	6/12
	0 RP	—	1	1/3	6/12
Parity =	1 RPE	—	1	1/3	6/12
	0 RPO	—	1	1/3	6/12

Description : Return to main program if condition is true. Three machine cycles and 12 T-states are consumed if condition is true while one machine and 6T-states are consumed if condition is false.

XTHL : Exchange H and L with Top of Stack

Opcode	Operand	Bytes	M-Cycles	T-States
XTHL	None	1	5	16T

Description : The contents of the L register are exchanged with the stack location pointed out by the contents of the stack pointer register. The contents of the H register are exchanged with the next stack location (SP + 1).

Flags : No flags are affected.

Stack, I/O and Machine Control Instructions

NOP : No Operation

Opcode	Operand	Bytes	M-Cycles	T-States
NOP	None	1	1	4T

Description : No operation is performed.

Comments:

The instruction is used to fill in time delays or to delete and insert instructions while troubleshooting.

HLT : Halt and Enter Wait State

Opcode	Operand	Bytes	M-Cycles	T-States
HLT	None	None	2 or more	5T or more

Description : The microprocessor finishes executing the current instruction and halts any further execution.

Note:

NOP and HLT both are same as no operation. However, when NOP is executed the PC is increased by 1 time where as for HLT PC remains the same.

PUSH : Push Register Pair onto Stack

Opcode	Operand	Bytes	M-Cycles	T-States
PUSH	Rp	1	3	12T

Description : The contents of the register pair designated in the operand are copied into the stack in the following sequence. The stack pointer register is decremented and the contents of the high-order register (B, D, H, A) are copied into that location. The stack pointer register is decremented again and the contents of the low-order register (C, E, L, flags) are copied to that location.

Flags : No flags are modified.

POP : Pop off Stack to Register Pair

Opcode	Operand	Bytes	M-Cycles	T-States
POP	Rp	1	3	10T

Description : The contents of the memory location pointed out by the stack pointer register are copies to the low-order register (such as C, E, L and flags) of the operand. The stack pointer is incremented by 1 and the contents of that memory location are copied to the high-order register (B, D, H, A) of the operand. The stack pointer register is again incremented by 1.

Flags : No flags are modified.

Note:

The contents of the source, stack locations, are not altered after the POP instruction.

RSTn : Restart

Opcode	Operand	Bytes	M-Cycles	T-States
RSTn	None	1	3	12T

Description : This instruction is equivalent to 1-byte CALL instruction. Here n = 0 to 7 are 8 software interrupts.

Flags : No flags are affected.

SPHL : Copy H and L Registers to the Stack Pointer

Opcode	Operand	Bytes	M-Cycles	T-States
SPHL	None	1	1	6T

Description : The instruction loads the contents of the H and L registers into the stack pointer register.

Flags : No flags are affected.

Note:

- ☑ This instruction performs the same function as MOV A, M except this instruction uses the contents of BC or DE as memory pointers.
- ☑ This instruction is used in conjunction with CALL or conditional call instructions.

IN : Input Data to Accumulator from a Port with 8-bit Address

Opcode	Operand	Bytes	M-Cycles	T-States
IN	8-bit port address	2	3	10T

Description : The contents of the input port designated in the operand are read and loaded into the accumulator.

Flags : No flags are affected.

OUT : Output Data from Accumulator to a Port with 8-Bit Address

Opcode	Operand	Bytes	M-Cycles	T-States
OUT	8-bit port address	2	3	10T

Description : The contents of the accumulator are copied into the output port specified by the operand.

Flags : No flags are affected.

EI : Enable Interrupts

Opcode	Operand	Bytes	M-Cycles	T-States
EI	None	1	1	4T

Description : The Interrupt Enable flip-flop is set and all interrupts are enabled.

Flags : No flags are affected.

Comments:

After a system reset or the acknowledgment of an interrupt, the Interrupt Enable flip-flop is reset, thus disabling the interrupts. This instruction is necessary to enable the interrupts (except TRAP).

DI : Disable Interrupts

Opcode	Operand	Bytes	M-Cycles	T-States
DI	None	1	1	4T

Description : The Interrupt Enable flip-flop is reset and all the interrupts except the TRAP (8085) are disabled.

Flags : No flags are affected.

Comments:

This instruction is commonly used when the execution of a code sequence cannot be interrupted.

RIM : Read Interrupt Mask

Opcode	Operand	Bytes	M-Cycles	T-States
RIM	None	1	1	4T

Description : This is a multipurpose instruction used to read the status of interrupts 7.5, 6.5, 5.5 and to read serial data input bit.

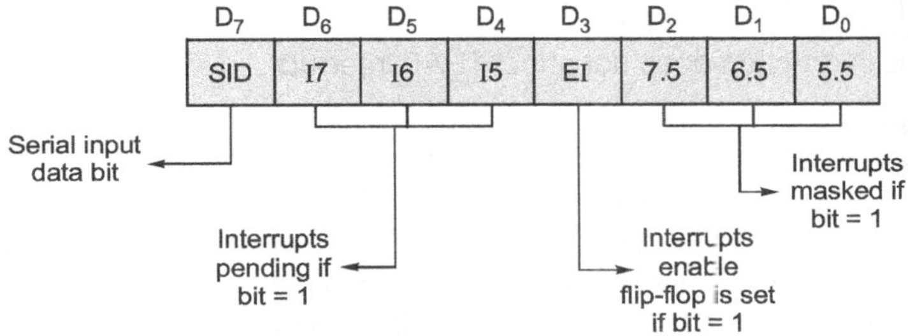

Flags : No flags are affected.

SIM : Set Interrupt Mask

Opcode	Operand	Bytes	M-Cycles	T-States
SIM	None	1	1	4T

Description : This is a multipurpose instruction and used to implement the 8085 interrupts (RST 7.5, 6.5 and 5.5) and serial data output.

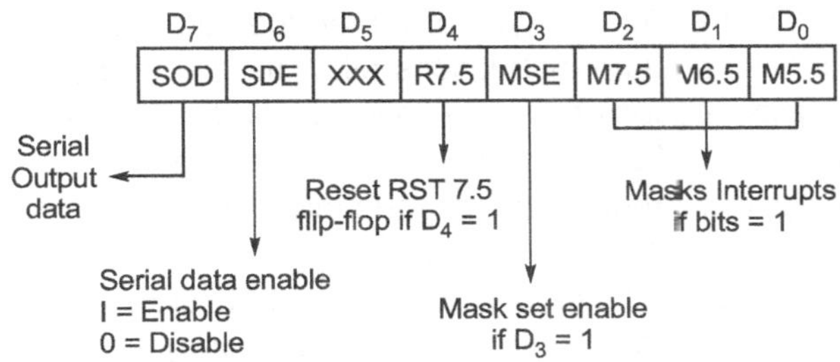

⇒ **SOD (Serial Output Data):** Bit D_7 of the accumulator is latched into the SOD output line and made available to a serial peripheral if bit $D_6 = 1$.

⇒ **SDE (Serial Data Enable):** If this bit = 1, enables the serial output. To implement serial output, this bit needs to be enabled.

⇒ **XXX:** Don't care

⇒ **R7.5 (Reset RST 7.5):** If this bit = 1, RST 7.5 flip-flop is reset. This is an additional control to reset RST 7.5.

⇒ **MSE (Mask Set Enable):** If this bit is high, it enables the functions of bits D_2, D_1, D_0. This is a master control over all the interrupt masking bits. If this bit is low, bits D_2, D_1 and D_0 do not have any effect on the masks.

⇒ **M7.5:** D_2 = 0, RST 7.5 is enabled.
= 1, RST 7.5 is masked or disabled

⇒ **M6.5:** D_1 = 0, RST 6.5 is correct.
= 1, RST 6.5 is masked or disabled.

⇒ **M6.5:** D_0 = 0, RST 5.5 is enabled.
= 1, RST 5.5 is masked or disabled.

Comments:

This instruction does not affect TRAP interrupt.

Remember:

- ☑ Data copy instructions do not affect the flags.
- ☑ Operand PSW (Program Status Word) represents the contents of the accumulator and the flag register; the accumulator is high-order register and the flags are low-order register.
- ☑ Add and subtract are performed in relation to contents of the accumulator, however the increment or decrement operation can be performed in any register.
- ☑ Out of all instructions PUSH , CALL, RET, RSTn, INX, DCX, SPHL and PCHL uses 6 T-states for fetch machine cycle.
- ☑ DAD instruction uses bus idle machine cycle.
- ☑ NOT operation does not affect any flags.
- ☑ JMP: Instruction is a type of immediate addressing.
- ☑ Conditional jump instructions allow the microprocessor to make decision, based on certain conditions indicated by the flags.
- ☑ INX and DCX does not affect any of the flags.
- ☑ Compare instruction work like subtraction but content of accumulator and register does not change.
- ☑ Stack is a set of memory locations used to store binary information (byte) temporarily during execution of a program.
- ☑ Stack and stack pointer are two different things.
- ☑ Subroutine is also a program written outside main program.
- ☑ CALL and RET instructions are used for execution of subroutine and after execution, return to main program.

IV Interrupts

Hardware Interrupts

The 8085 microprocessor has five interrupt signals that can be used to interrupt a program execution.

Interrupt	Triggering	Vectored Address	Maskable/ Non-maskable	Instructions
TRAP	Edge and level	0024H	Non-maskable	Independent of EI and DI
RST 7.5	Edge	003CH	Maskable	Controlled by EI and DI unmasked by SIM
RST 6.5	Level	0034H	Maskable	Controlled by EI and DI unmasked by SIM
RST 5.5	Level	002CH	Maskable	Controlled by EI and DI unmasked by SIM
INTR	Level	Non vectored	Maskable	Controlled by EI and DI

$\overline{\text{INTA}}$:

It is the active low interrupt acknowledgment signal which is only used with INTR.

Note:

To calculate vector address of RSTn

$$n \times 8 = (X)_{10} = (Y)_{16}$$

Example : For RST 4.5,

So, $(4.5 \times 8)_{10} = (36)_{10} \xrightarrow{\text{Heaxadecimal}} (24)_H = (0024)_H$

Software Interrupt

There are 8 software interrupts which are used either in instructions or alongwith INTR interrupt. They are defined as RSTn where n $\rightarrow$ 0 to 7.

Software interrupt (RSTn)

Vectored Address	
RST 0	0000 H
RST 1	0008 H
RST 2	0010 H
RST 3	0018 H
RST 4	0020 H
RST 5	0028 H
RST 6	0030 H
RST 7	0038 H

Remember:

- ☑ TRAP is also called RST 4.5 interrupt.
- ☑ In interrupts priority order is
  TRAP > RST 7.5 > RST 6.5 > RST 5.5 > INTR
- ☑ Hold has higher priority than TRAP.
- ☑ RIM and SIM instructions are not only used for interrupt process but also used for serial I/O process.

☑ 8259A is a programmable interrupt controller and is used to implement and extend the capability of the 8085 interrupt. It manages 8 interrupt requests.

V Interfacing with Microprocessor

Interfacing

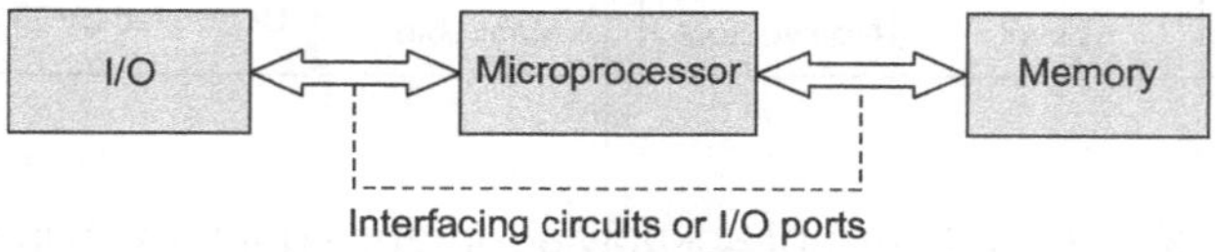

- Designing logic circuits and writing program to make the processor communicate either with memory or I/O is known as interfacing.
- The logic circuits used are known as interfacing circuits or I/O ports.

Characteristics of Memory

- **Capacity** : Memory capacity depends upon the amount of data that can be stored.
- **Memory size** $= 2^A \times D$

  where, $A \rightarrow$ Address lines ; $D \rightarrow$ Data lines
- **Number of chips required**

$$n = \frac{\text{Size of } M_2}{\text{Size of } M_1}$$

  where, $n \rightarrow$ Number of chips required ; $M_1 \rightarrow$ Available capacity

  $M_2 \rightarrow$ Memory to be designed
- If initial address and memory size is given then formulae for last address

  last address = [Initial address in hexadecimal + hexadecimal equivalent of memory size – $(1)_H$]
- If memory range is given then formulae for calculating memory size

  Memory size = [$(\text{last address})_H - (\text{Initial address})_H + (1)_H$] byte

Interfacing with I/O Ports

There are two ways by which I/O port can be connected to the microprocessor:

1. Memory mapped I/O scheme
2. I/O mapped I/O scheme

Comparison between Memory Mapped I/O and I/O Mapped I/O Scheme

Characteristic	Memory-mapped I/O	I/O Mapped I/O
Device Address	16-bit	8-bit
Control signals for Input/Output	$\overline{MEMR}$ / $\overline{MEMW}$	$\overline{IOR}$ / $\overline{IOW}$
Instruction available	Memory-related instructions such as LDA; STA; LDAX; STAX; MOV M, R; ADD M, SUB M; ANA M; etc.	IN and OUT
Data Transfer	Between any register and I/O	Only between I/O and the accumulator.
Maximum number of input/output devices possible	The memory map (64 K) is shared between I/O devices and system memory.	The I/O map is independent of the memory map; 256 input devices and 256 output devices can be connected.
Execution speed	13 T-states (STA, LDA) 7 T-states (MOV M.R)	10 T-states
Hardware requirements	More hardware is needed to decode 16-bit address.	Less-hardware is needed to decode 8-bit address.
Other features	Arithmetic or logical operations can be directly performed with I/O data.	Not available

Interfacing Devices

Intel 8155 : Programmable Peripheral Interface

The 8155 includes 256 bytes of RD/WR memory i.e. RAM, 3 I/O ports and a 16-bit timer.

Format of 8155 Timer

LSB Timer =	T_7	T_6	T_5	T_4	T_3	T_2	T_1	T_0
MSB Timer =	M_2	M_1	T_{13}	T_{12}	T_{11}	T_{10}	T_9	T_8

Where, M_2 M_1 Modes

M_2	M_1	Modes		
0	0	Mode 0	⇒	Single square wave cycle
0	1	Mode 1	⇒	Square wave
1	0	Mode 2	⇒	Single pulse after terminal count
1	1	Mode 3	⇒	Pulse every terminal count

Intel 8255 : Programmable Peripheral Interface/Adopter

- The 8285 is widely used programmable parallel I/O device. It can be programmed to transfer data under various conditions from I/O to interrupt I/O.
- Operating modes of 8255

1. Mode 0 ⇒ Simple input/output
2. Mode 1 ⇒ Strobed input/output
3. Mode 2 ⇒ Bidirectional Strobed

Control Word for BSR Mode

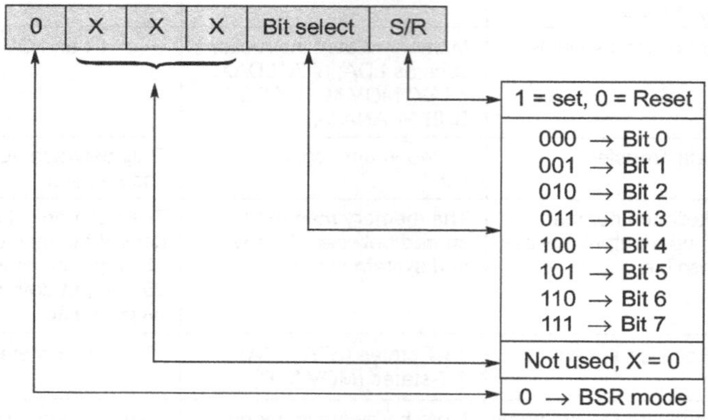

Structure of Control Word

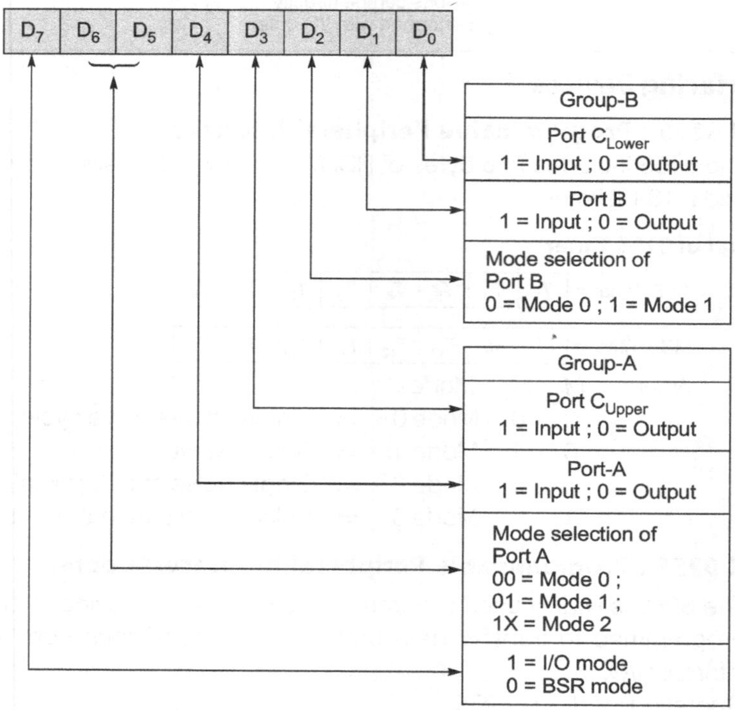

Intel 8251 : Programmable Communication Interface

Intel 8251 is also known as universal synchronous/asynchronous receiver/transmitter (USART) used to transmit serial data.

Intel 8253 : Programmable Interval Timer

- The programmable counter/interval time is used in real time application of timing and counting such as BCD/binary counting, generation of time delay etc.
- The 8253 uses NMOS technology and operates in any of the following six modes:
  1. Mode 0 ⇒ Interrupt on terminal count
  2. Mode 1 ⇒ Programmable one shot
  3. Mode 2 ⇒ Rate generator
  4. Mode 3 ⇒ Square wave generator
  5. Mode 4 ⇒ Software trigger strobe
  6. Mode 5 ⇒ Hardware trigger strobe

Intel 8257 : Programmable DMA Controller

In DMA data transfer scheme, data are directly transferred from an I/O device to RAM or from RAM to an I/O device thus it is capable of performing three operations i.e. read, write and verify.

Intel 8259 : Programmable Interrupt Controller

The programmable interrupt controller is used when several I/O devices transfer data using interrupt and they are to be connected to the same interrupt level of the microprocessor.

Intel 8272 : Floppy Disk Controller

The function of floppy disk controller is to interface a floppy disk system to a microprocessor.

Intel 8275 : Programmable CRT Controller

The Intel 8275 is a single chip device. Its function is to interface CRT raster scan display with the microcomputer.

Intel 8279 : Programmable Keyboard/Display Interface

The Intel 8279 is a programmable keyboard interfacing device. It has two sections namely keyboard section and display section. The function of keyboard section is to interface the keyboard and the function of the display section is to drive alphanumeric displays or indicator lights.

Counter and Time Delays

- Counters are used primarily to keep track of events.
- Time delays are important in setting up reasonably accurate timing between two events.

- Time delay can be introduced using a loop and total delay = time to execute instructions outside loop + time to execute loop instructions.

$$T_D = T_O + T_{LA}$$

where T_D = Total delay ; T_O = Time to execute instructions outside loop. ; T_{LA} = Time to execute loop instructions.

- Normally T_O is very small and neglected in most of the cases.
- The accuracy of time delay depends on the accuracy of the system's clock.
- Intel 8254 is a programmable timer chip that can be interfaced with microprocessor and programmed to provide with considerable accuracy.

VI Introduction to 8086

Important Features of 8086

- 8086 is a 16 bit processor having 20 address lines.
  ∴ a maximum of 2^{20} = 1 Mega Byte can be connected.
- All registers are 16 bit in length.
- There are 8 8-bit general data registers.

AX		BX		CX		DX	
AH	AL	BH	BL	CH	CL	DH	DL

- It can be operated in 2 modes with the help of control signal

$$MN/\overline{MX} = 0;\ \text{maximum/multi processor mode}$$
$$= 1;\ \text{minimum/single processor mode.}$$

- When processor is reset, control of the program is transferred to **FFFF0H.**
- It contains 6 bytes of pre-decoding instruction byte queue by which 6 bytes of instruction can also be prefetched into the processor.
  At the same time interval more than one instruction can be at different stages of execution i.e. parallel execution of instruction. This process is known as PIPELINING.
  Technique involved in 6 bytes of queue is First In First Out.
- There are 9 flags in 8086

D_{15}	D_{14}	D_{13}	D_{12}	D_{11}	D_{10}	D_9	D_8	D_7	D_6	D_5	D_4	D_3	D_2	D_1	D_0
X	X	X	X	O	D	I	T	S	Z	X	AC	X	P	X	CY

Overflow ← O; Direction ← D; Interrupt ← I; Trap ← T

Similar to 8085 ← D_7–D_0

(i) Status flags → S, Z, A, C, P, CY, O.

(ii) Control flags → D, I, T.

- It follows VON NEUMANN architecture for memory but it is segmented memory.
  1 MB is segmented into 16 logical segment.
  $$1\text{ MB} \rightarrow 2^4 \times 2^{16} = 2^{20}$$
  16 (64 kB)
- **Interrupts:**

Hardware	Software
(i) NMI	(i) INTn n = 0 to 255
(ii) INTR	Type 0 to type 255.

Architecture of 8086

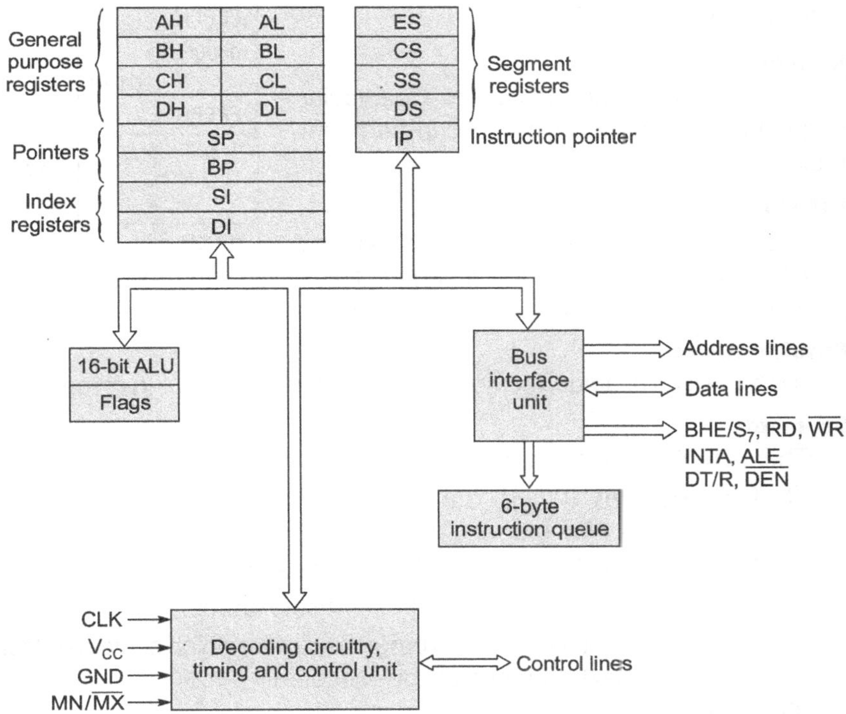

Functional Units

Bus Interface Unit

It contains address conversion mechanism in which 2 16-bit registers are used to find out or calculate 20-bit effective or physical address. It contains 6-bytes of instruction byte queue used for pre fetching the instructions for parallel execution known as pipelining technique. It has different segment registers to hold base address of **code, data, stack** and

extra segments. It makes the processor to communicate with external devices including memory.

Execution Unit

It is responsible for decoding the prefetched instruction in BIU. It generates various control signals for BIU and also for operations of processor.

It contains different registers like

(i) general data registers,

(ii) pointer and index registers,

(iii) flag register

Memory Segmentation

1 MB of memory is divided into 16 logical segments each of 64 kB.

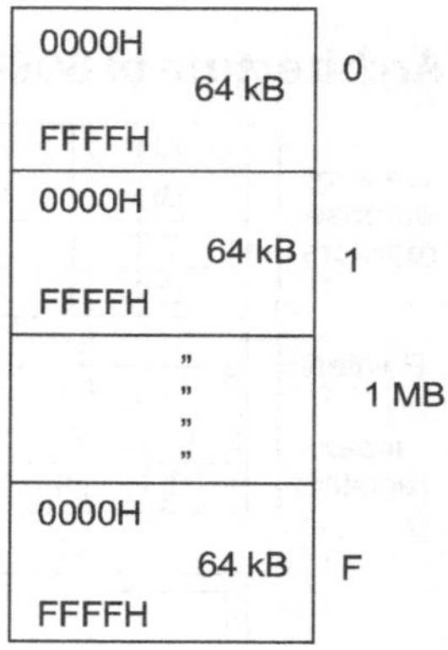

Base Address

It indicates starting location of a particular segment which may be present in segment register.

Offset Address

It is the distance of actual memory location from base address in a particular segment. It is present in pointer or index registers or BX register.

Physical Address

It is a position of actual memory location from the beginning of memory.

Segments

There are 4 types of segments:

(i) Code Segment :

Contains code/program,

Code Segment Register CS → Base Address.

IP → Offset Address.

Instruction pointer is similar to program counter which contains offset address of next instruction to be executed.

(ii) Data Segment :

Contains data

DS → Base Address

BP/SI/DI/BX → Offset address

(iii) Stack Segment :

Contains temporary data

SS → Base Address

SP → Offset address

(iv) Extra Segment :

Contains data

ES → Base Address

DI → Offset address

Remember:

☑ **Range of Segment Address** can be from 000H - F000H

☑ **Range of Offset Address** can be from 0000H - FFFFH

Register Organisation

(i) General Data Register

AX	AH	AL
BX	BH	BL
CX	CH	CL
DX	DH	DL

(ii) Special Purpose Registers

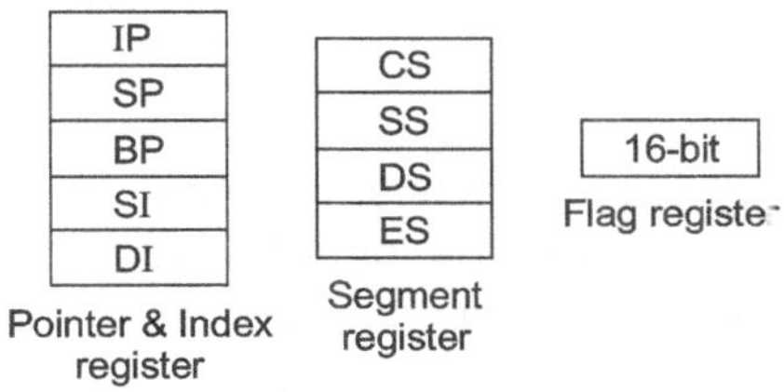

General Data Registers

AX: It is also called as Accumulator.

One of the operands in multiplication & division operation must be present in accumulator.

BX : It is used to hold offset address in Data Segment.

CX : Used as default counter in instruction like LOOP, LOOPZ, LOOPNZ.

DX : Used in multiplication and division operations and also as implicit register.

Flag Registers

TRAP : TF = 1; processor enters into single step execution mode in which execution steps after every instruction such that contents of registers can be verified.

Interrupt Flag : IF = 1; maskable interrupts are recognized by processor.

Direction flag : It is used in string manipulation instructions

D = 0; Auto increment mode

= 1; Auto decrement mode.

Overflow flag: It is SET (1) if result of a signed operation (8-bit on 16-bit) is not possible to be accommodated in destination register. It is 0 otherwise.

Addressing Modes

Sequential flow instructions: There are 8 addressing modes

(i) Immediate addressing mode
(ii) Direct addressing mode
(iii) Register addressing mode
(iv) Register indirect addressing mode
(v) Register Relative addressing mode
(vi) Indexed Relative addressing mode
(vii) Based Indexed addressing mode
(viii) Relative based addressing mode

Control transfer instructions

(i) Intra Segment — Direct / Indirect

(ii) Inter Segment — Direct / Indirect

Intra Segment

If control of the program is transferred within the same segment, it is known as intra segment which can be **direct** or **indirect addressing** modes.

- If an 8-bit signed displacement is used, there can be +127 to –128 locations possible known as short jump.
- If a 16-bit signed displacement is used, there can be +32,767 to –32,768 locations possible known as long jump.

Inter Segment

If control of the program is transferred from one segment to another, it is known as intersegment which can be direct or indirect addressing modes. Unconditional jump instructions are used in this type.

VII Microcontroller

Microprocessor	Microcontroller
1. It includes ALU, register arrays and control circuits on a single chip.	1. It includes microprocessor, RAM, ROM and I/O signal lines on a single chip.
2. Addition of external ROM, RAM and I/O devices makes the microprocessor based circuitry bulky and costly.	2. As there is no need of such external devices, the microcontroller based circuitry becomes simple and cheap.
3. As microprocessors support large amount of external memory, these are designed with large number of instructions.	3. As microcontrollers operate with small amount of fixed internal memory, these are designed with small number of instructions.

4. Microprocessors are used for designing general purpose digital computing systems like personnel computers.	4. Microcontrollers are used for designing dedicated (or single purpose) digital computing systems like automatic traffic light control system.
5. Microprocessor based system consumes more power	5. Microcontroller based system consumes less power

A brief history of the 8051

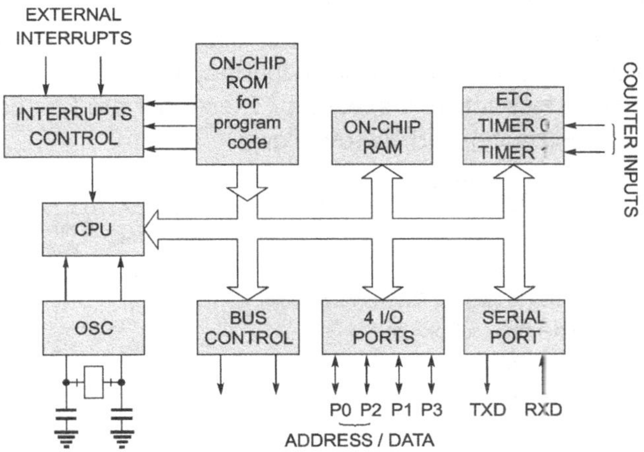

Inside the 8051 Microcontroller Block Diagram

Comparison of 8051 Family Members

Feature	8051	8052	8031
ROM (on-chip program space in bytes)	4 K	8 K	0 K
RAM (bytes)	128	256	128
Timers	2	3	2
I/O pins	32	32	32
Serial Port	1	1	1
Interrupt sources	6	8	6

Inside 8051 : In this section we examine the major registers of the 8051.

Registers :

- In the CPU, registers are used to store information temporarily. That information could be a byte of data to be processed, or an address pointing to the data to be fetched. The vast majority of 8051 registers are 8-bit registers. In the 8051 there is only one data type : 8 bits.

- The most widely used registers of the 8051 are A (accumulator), B, R0, R1, R2, R3, R4, R5, R6, R7, DPTR (data pointer) and PC (program counter). All of the above registers are 8-bits, except DPTR and the program counter.
- In the 8051, register A must be involved and be the destination for all arithmetic operations. There are two 16-bit registers in the 8051 : PC (program counter) and DPTR (data pointer).

PSW (program status word) register

The program status word (PSW) register is an 8-bit register. Four of the flags are called conditional flags, meaning that they indicate some conditions that result after in instruction is executed. These four are CY (carry), AC (auxiliary carry), P (parity), and OV (overflow).

CY	AC	F0	RS1	RS0	OV	--	P

CY	PSW.7	Carry flag.
AC	PSW.6	Auxiliary carry flag.
F0	PSW.5	Available to the user for general purpose.
RS1	PSW.4	Register Bank selector bit 1.
RS0	PSW.3	Register Bank selector bit 0.
OV	PSW.2	Overflow flag.
--	PSW.1	User-definable bit.
P	PSW.0	Parity flag. Set / cleared by hardware each instruction cycle to indicate an odd/even number of 1 bits in the accumulator.

RS1	RS0	Register Bank	Address
0	0	0	00 H - 07 H
0	1	1	08 H - 0 FH
1	0	2	10 H - 17 H
1	1	3	18 H - 1 FH

CY, the carry flag : This flag is set whenever there is a carry out from the D7 bit. This flag bit is affected after an 8-bit addition or subtraction. It can also be set to 1 or 0 directly by an instruction such as "SETB C" and "CLR C" where "SETB C" stands for "set bit carry" and "CLR C" for "clear carry".

AC, the auxiliary carry flag : If there is a carry from D3 to D4 during an ADD or SUB operation, this bit is set; otherwise, it is cleared. This flag is used by instructions that perform BCD (binary coded decimal) arithmetic.

P, the parity flag : The parity flag reflects the number of 1s in the A (accumulator) register only. If the A register contains an odd number of 1s, then $P = 1$. Therefore, $P = 0$ if A has an even number of 1s.

OV, the overflow flag : This flag is set whenever the result of a signed number operation is too large, causing the high-order bit to overflow into the sign bit. In general, the carry flow is used to detect errors in unsigned arithmetic operations. The overflow flag is only used to detect errors in signed arithmetic operations.

8051 Register Banks and Stack

The 8051 microcontroller has a total of 128 bytes of RAM. In this section we discuss the allocation of these 128 bytes of RAM and examine their usage as registers and stack.

RAM memory space allocation in the 8051 : There are 128 bytes of RAM in the 8051 (some members, notably the 8052, have 256 bytes of RAM). The 128 bytes of RAM inside the 8051 are assigned addresses 00 to 07 FH. As we will see in Chapter 5, they can be accessed directly as memory locations. These 128 bytes are divided into three different groups as follows.

- As total of 32 bytes from locations 00 to 1 F hex are set aside for register and the stack.
- A total of 16 bytes from locations 20 H to 2 FH are set aside for bit-addressable read/write memory.
- A total of 80 bytes from locations 30 H to 7 FH are used for read and write storage, or what is normally called a scratch pad. These 80 locations of RAM are widely used for the purpose of storing data and parameters by 8051 programmers. We will use them in future chapters to store data brought into the CPU via I/O ports.

Register banks in the 8051

A total of 32 bytes of RAM are set aside for the register banks and stack. These 32 bytes are divided into 4 banks of registers in which each bank has 8 register, R0 - R7. RAM locations from 0 to 7 are set aside for bank 0 of R0 - R7 where R0 is RAM location 0, R1 is RAM location 1, R2 is location 2, and so on, bank 1 uses the same RAM space as the stack. This is a major problem in programming the 8051. We must either not use register bank 1, or allocate another area of RAM for the stack.

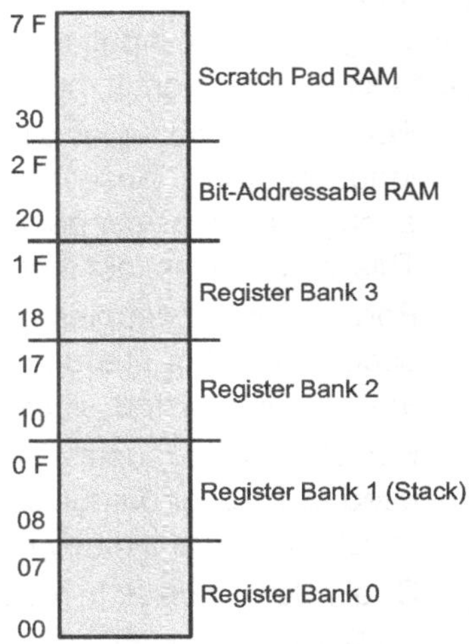

RAM Allocation in the 8051

How to switch register banks : We can switch to other banks by use of the PSW (program status word) register. Bits D4 and D3 of the PSW are used to select the desired register bank as shown in table.

How stack are accessed in the 8051 : The register used to access the stack is called the SP (stack pointer) register. The stack pointer in the 8051 is only 8 bits wide, which means that it can take values of 00 to FFH. When the 8051 is powered up, the SP register contains value of 07.

Pushing onto the stack: Stack pointer (SP) points to the last used location of the stack. As we push data onto the stack, the stack pointer (SP) is incremented by one, this is different from many microprocessors, notably x86 processors in which the SP is decremented when data is pushed onto the stack.

The upper limit of the stack : If in a given program we need more than general application programmes (08 to 1 FH = 24 bytes) of stack, we can change the SP to point to RAM location 30 - 7 FH. This is done with the instruction "MOV SP, # xx".

8051 I/O Programming

In the 8051 there are a total of four ports for I/O operations. Examining figure 4-1, note that of the 40 pins, a total of 32 pins are set aside for the four ports P0, P1, P2 and P3, where each port takes 8 pins. The rest of the pins are designated at V_{CC}, GND, XTAL1, XTAL2, RST, $\overline{EA}$, ALE / $\overline{\text{PROG}}$ and $\overline{\text{PSEN}}$.

I/O port pins and their functions

The four ports P0, P1, P2 and P3 each use 8 pins, making them 8-bits ports. All the ports upon RESET are configured as inputs, ready to be used as input ports. When the first 0 is written to a port, it becomes an output. To reconfigure it as an input, a 1 must be sent to the port. To use any of these ports as an input port, it must be programmed,

Port 0: Port 0 occupies a total of 8 pins (pins 32 - 39). It can be used for input or output. To use the pins of port 0 as both input and output ports, each pin must be connected externally to a 10 K - ohm pull-up resistor. This is due to the fact that P0 is an open drain, unlike P1, P2 and P3.

Port 1: Port 1 occupies a total of 8 pins (pins 1 through 8). It can be used as input or output. In contrast to port 0, this port does not need any pull-up resistors since it already has pull-up resistor internally.

Port 2: Port 2 occupies a total of 8 pins (pins 21 through 28). It can be used as input or output. Just like P1, port 2 does not need any pull-up resistors since it already has pull-up resistors internally.

Dual role of port 2: In many system based on the 8051, P2 is used as simple I/O. However, in 8031-based system, port 2 must be used along with P0 to provide the 16-bit address for external memory.

Port 3: Port 3 occupies a total of 8 pins, pins 10 through 17. It can be used as input or output. P3 does not need any pull-up resistors, just as P1 and P2 did not. The 8051 provides a total of five distinct addressing modes.

Addressing Modes

The CPU can access data in various ways. The data could be in a register, or in memory, or be provided as an immediate value. These various ways of accessing data are called addressing modes. The 8051 provides a total of five distinct addressing modes.

Immediate addressing mode : In this addressing mode, the sources operand is a constant. In immediate addressing mode, as the name implies, when the instruction is assembled, the operand comes immediately after the opcode.

MOV A, #79 H ; load 79H into A

MOV R4, #56 ; load the decimal value 56 into R4

MOV B, #84 H ; load 84H into B

MOV DPTR, #5678 H; DPTR = 5678H

Register addressing mode : Register addressing mode involves the use of registers to hold the data to be manipulated. Examples of register addressing mode follow.

MOV A, R0 ; copy the contents of R0 into A

MOV R2, A ; copy the contents of A into R2

ADD A, R5 ; add the contents of R5 to contents of A

ADD A, R7 ; add the contents of R7 to contents of A

Direct addressing mode : The RAM has been assigned addresses 00 to 7FH.

- RAM locations 00 - 1FH are assigned to the register banks and stack.
- RAM locations 20 - 2FH are set aside as bit-addressable space to save single-bit data.
- RAM locations 30 -7FH are available as a place to save byte-sized data.

Although the entire 128 bytes of RAM can be accessed using direct addressing mode, it is most often used to access RAM location 30 -7FH.

MOV R0, 40H ; save content of RAM location 40H in R0

MOV 56H, A ; save content of A in RAM location 56H

MOV R4, 7FH ; move contents of RAM location 7FH to R4

Register indirect addressing mode : In the register indirect addressing mode, a register is used as a pointer to the data. If the data is inside the CPU, only registers R0 and R1 are used for this purpose.

R2-R7 cannot be used to hold the address of an operand located in RAM when using this addressing mode. '@" sign, as shown below :

```
MOV  A, @R0   ;  move contents of RAM location whose
              ;  address is held by R0 into A
MOV  @R1, B   ;  move contents of B into RAM location
              ;  whose address is held by R1
```

Indexed addressing mode and on-chip ROM access : Indexed addressing mode is widely used in accessing data elements of look-up tables entries located in the program ROM space of the 8051. The instruction used for this purpose is "MOVC A, @A+DPTR". The 16-bit register DPTR and register A are used to form the address of the data element stored in on-chip ROM.

8051 Special Function Register (SFR) Address

Symbol	Name	Address
ACC*	Accumulator	0E0H
B*	B register	0F0H
PSW*	Program status word	0D0H
SP	Stack pointer	81H
DPTR	Data pointer 2 bytes	
DPL	Low byte	82H
DPL	High byte	83H
P0*	Port 0	80H
P1*	Port 1	90H
P2*	Port 2	0A0H
P3	Port 3	0B0H
IP*	Interrupt priority control	0B8H
IE*	Interrupt enable control	0A8H
TMOD	Timer / counter mode control	89H
TCON*	Timer / counter control	88H
T2CON*	Timer / counter 2 control	0C8H
T2MOD	Timer / counter mode control	0C9H
TH0	Timer / counter 0 high byte	8CH
TL0	Timer / counter 0 low byte	8AH
TH1	Timer / counter 1 high byte	8DH
TL1	Timer / counter 1 low byte	8BH
TH2	Timer / counter 2 high byte	0CDH
TL2	Timer / counter 2 low byte	0CCH
RCAP2H	T/C 2 capture register high byte	0CBH
RCAP2L	T/C 2 capture register low byte	0CAH
SCON*	Serial control	98H
SBUF	Serial data buffer	99H
PCON	Power control	87H

For this reason the 8051 has another 64K bytes of memory space set aside exclusively for data storage. This data memory space is referred to as external memory and it is accessed only by the MOVX instruction. In other words, the 8051 has a total of 128K bytes of memory space since 64K bytes of code added to 64K bytes of data space gives us 128K bytes.

Programming 8051 Timers

The 8051 has two timers: Timer 0 and Timer 1. These can be used either as timers or as event counters. In this section we first discuss the timers' register and then slow how to program the timers to generate time delays.

Basic registers of the timer

Both Timer 0 and Timer 1 are 16 bit bits wide. Since the 8051 has an 8-bit architecture, each 16-bit timer is accessed as two separate registers of low byte and high byte.

Timer 0 registers : The 16-bit register of Timer 0 is accessed as low bytes and high byte. The low byte register is called TL0 (Timer 0 low byte) and the high byte register is referred to as TH0 (Timer 0 high byte). These registers can be accessed like any other register, such as A, B, R0, R1, R2, etc.

Timer 1 registers : Timer 1 is also 16 bits, and its 16-bit register is split into two bytes, referred to as TL1 (Timer 1 low byte) and TH1 (Timer 1 high byte). These registers are accessible in the same way as the registers of Timer 0.

TMOD (timer mode) register

Both timers 0 and 1 use the same register, called TMOD, to set the various timer operation modes. TMOD is an 8-bit register in which the lower 4 bits are set aside for Timer 0 and the upper 4 bits for Timer 1. In each case, the lower 2 bits are used to set the timer mode and the upper 2 bits to specify the operation.

TH0								TL0							
D15	D14	D13	D12	D11	D10	D9	D8	D7	D6	D5	D4	D3	D2	D1	D0

Fig. Timer 0 Registers

TH1								T1							
D15	D14	D13	D12	D11	D10	D9	D8	D7	D6	D5	D4	D3	D2	D1	D0

Fig. Timer 1 Registers

(MSB) (LSB)

GATE	C/T	M1	M0	GATE	C/T	M1	M0
Timer 1				Timer 0			

GATE Gating control when set. The timer/counter is enabled only while the INTx pin is high and the TRx control pin is set. When cleared, the timer is enabled whenever the TRx control bit is set.

C/T Timer or counter selected cleared for timer operation (input from internal system clock). Set for counter operation (input from Tx input pin).

M1 Mode bit 1

M0 Mode bit 0

M1	M0	Mode	Operating Mode
0	0	0	13-bit timer mode
			8-bit timer/counter THx with TLx as 5-bit prescaler
0	1	1	16-bit timer mode
			16-bit timer/counters THx and TLx are cascaded; there is no prescaler
1	0	2	8-bit auto reload
			8-bit auto reload timer/counter; THx holds a value that is to be reloaded into TLx each time it overflows.
1	1	3	Split timer mode

Fig. TMOD Register

C/T (clock/timer) : This bit in the TMOD register is used to decide whether the timer is used as a delay generator or an event counter. If C/T = 0, it is used as a timer for time delay generation. The clock source for the time delay is the crystal frequency of the 8051.

GATE : The other bit of the TMOD register is the GATE bit. Notice in the TMOD register of figure 9.3 that both Timers 0 and 1 have the GATE bit. What is its purpose? Every timer has a means of starting and stopping. Some timers do this by software, some by hardware, and some have both software and hardware controls. The timers in the 8051 have both. The start and stop of the timer are controlled by way of software by the TR (timer start) bits TR0 and TR1. This is achieved by the instructions "SETB TR1" and CLR TR1" for Timer 1, and "SETB TR0" and "CLR TR0" for Timer 0. The SETB instruction starts it, and it is stopped by the CLR instruction. These instructions start and stop the times as long as GATE = 0 in the TMOD register.

The 8051 has serial communication capability built into it, thereby making possible fast data transfer using only a few wires.

■■■■

Communication Systems 13

I Basics

Communication System

A communication system is used to transfer or exchange the information between two points.

Basic Elements of Communication System

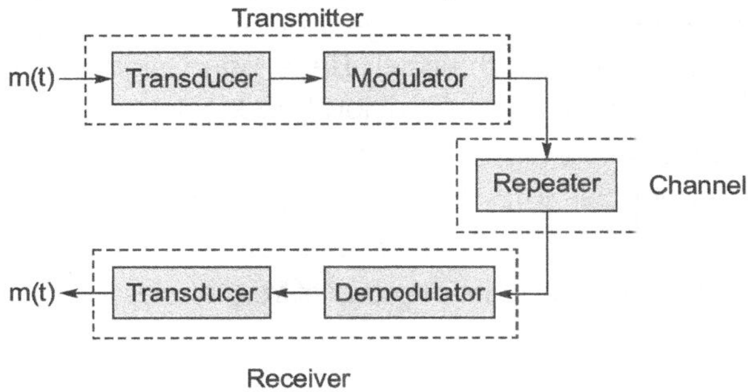

Communication system basically consists transmitter, channel and receiver.

- **Transducer:** It is a device which converts one form of energy to another form. The input transducer convert the message signal into a time varying electrical signal.
- **Modulator:** It is used to perform the modulation.
- **Channel:** It is a physical medium that carries the e ectrical signal. It is used for connection between transmitter and receiver.
- **Repeaters:** These produce a fresh copy of transmitted signals.
- **Demodulator:** It performs reverse operation of modulator i.e. it converts high frequency signal to low frequency signal.

Modulation

Modulation is a process that causes a shift in the range of frequencies in a signal.

Need of Modulation

- To decrease the length of transmitting and receiving antenna.
- The low frequencies are attenuated fast and therefore low frequency

signal can not be transmitted over a large distance. By translating the low frequency component to high frequency component, long distance communication is possible.

- By varying the signal power which is being transmitted, required signal to noise ratio (S/N) can be obtained.
- Frequency division multiplexing (FDM) is possible and therefore large number of signals can then be transmitted with different carrier frequencies over a common communication channel.

Note:

☑ Message signal is modulating signal and it modulates carrier signal.

☑ In modulation some properties of carrier signal are varied in accordance with the message signal.

Base Band Signal : The message signal generated from the information source is called base band signal. Base band signal has significant frequency component near to zero or low frequencies.

Band Pass Signal : It is a signal having significant frequency components for range of frequencies away from zero frequency or low frequency.

Spectrum : It is frequency domain representation of a signal.

Bandwidth : It is defined as band or group of frequencies for which amplitude of signal is not zero.

Note:

☑ Carrier frequency is much higher than message signal frequency.

☑ Noise is mainly added to the signal in the channel.

II Fourier Series, Energy and Signals

Fourier Series

Fourier Representation of Periodic Function

$$f(t) = a_0 + 2\sum_{n=1}^{\infty}[a_n \cos n\omega_0 t + b_n \sin n\omega_0 t]$$

where $f(t)$ = Periodic function

$$\omega_0 = \frac{2\pi}{T_0} = \text{Fundamental frequency}$$

T_0 = Fundamental period of function $f(t)$

$$a_0 = \frac{1}{T_0}\int_{-T_0/2}^{T_0/2} f(t)dt \quad \text{...mean value of function}$$

$$a_n = \frac{1}{T_0} \int_{-T_0/2}^{T_0/2} f(t) \cdot \cos n\omega_0 t \, dt$$

$$b_n = \frac{1}{T_0} \int_{-T_0/2}^{T_0/2} f(t) \cdot \sin n\omega_0 t \, dt$$

Complex Exponential Fourier Series

$$f(t) = \sum_{n=-\infty}^{\infty} c_n \exp(-jn\omega_0 t)$$

where, $$c_n = \frac{1}{T_0} \int_{-T_0/2}^{T_0/2} f(t) \cdot \exp(-jn\omega_0 t) \, dt \text{ for } n = 0, \pm 1, \pm 2 \ldots$$

Remember:

☑ Fourier series is generally used to represent a periodic signal where as Fourier transform is used to represent a non-periodic signal.

- For distortionless transmission

$$y(t) = k\, g(t - t_d)$$

where $y(t)$ = Output ; $g(t)$ = Input ; k = Gain ; t_d = Time delay

- Transfer function required for distortionless transmission

$$H(\omega) = ke^{-j\omega t_d}$$

Paley-Wiener criterion

The necessary and sufficient condition for amplitude response $|H(\omega)|$ to be realizable

$$\int_{-\infty}^{\infty} \frac{\ln|H(\omega)|}{1+\omega^2} d\omega < \infty$$

- **Energy contained in a given signal f(t)**

$$E = \int_{-\infty}^{\infty} f^2(t) dt$$

Parseval's theorems

$$E = \frac{1}{2\pi} \int_{-\infty}^{\infty} |F(\omega)|^2 \, d\omega$$

- **Energy spectral density**

$$G_E(\omega) = |F(\omega)|^2, \text{ J/Hz}$$

It is the energy contained in the signal per unit bandwidth

- Auto correlation function (ACF)

$$R(\tau) = \int_{-\infty}^{\infty} f(t) \cdot f(t-\tau) d\omega$$

III Analog Modulation

Amplitude Modulation

Amplitude of the carrier signal (high frequency) is varied with the amplitude of modulating signal keeping frequency and the phase of the carrier fixed.

Single Tone Modulation (Sinusoidal Signal)

- Modulating Signal : $v_m(t) = V_m \cos\omega_m t$
  where $\omega_m \rightarrow$ Modulating frequency; $V_m \rightarrow$ Amplitude of modulating signal
- Carrier Signal : $v_c(t) = V_c \cos\omega_c t$
  where, $\omega_c \rightarrow$ Carrier frequency; $v_c \rightarrow$ Amplitude of carrier signal

Remember:

☑ $\omega_c >> \omega_m$

☑ f_m = 50 Hz - 15 kHz ; f_c = 535 kHz - 1605 Hz

☑ AM broadcast range is 535 MHz - 1605 MHz.

Amplitude-Modulated (AM) Signal

$$v_{AM}(t) = [V_c + K_a V_m \cos\omega_m t]\cos\omega_c t$$

or

$$v_{AM}(t) = V_c[1 + m_a \cos\omega_m t]\cos\omega_c t$$

$$m_a = K_a \frac{V_m}{V_c}$$

where, $m_a \rightarrow$ Modulation index = Depth of modulation
= Percentage modulation

$K_a \rightarrow$ Constant of proportionality, and depends upon circuit from where AM signal has been generated

Note:

☑ Unless specified take $K_a = 1$ then $m_a = \frac{V_m}{V_c}$.

☑ Practically $0 < m_a < 1$, generally take value of m_a = 0.45 to 0.6.

☑ If $m_a = 0$ then unmodulated signal.
If $m_a = 1$ then 100% modulated signal.

Representation of Various AM Signal

AM-DSB/FC

It is a standard signal

$$v_{AM}(t) = \underbrace{V_c \cdot \cos\omega_c t}_{\text{Free carrier}} + \frac{1}{2}m_a V_c \left[\underbrace{\cos(\omega_c + \omega_m)t}_{\text{Upper side band}} + \underbrace{\cos(\omega_c - \omega_m)t}_{\text{Lower side band}}\right]$$

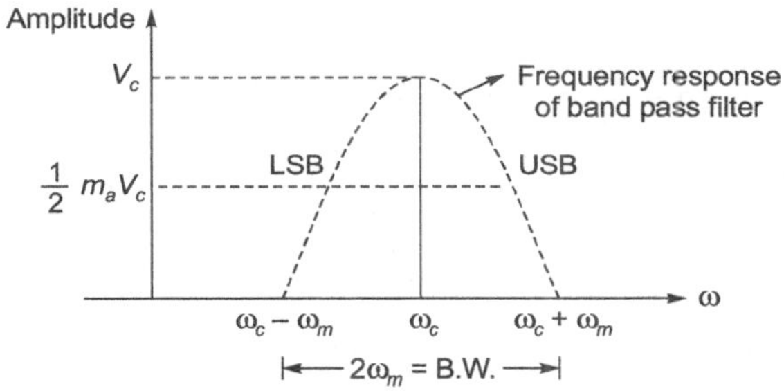

Spectrum of AM-DSB/FC sigal

Important Points:

- AM-DSB/FC signal represents standard AM signal which requires maximum amount of power and maximum bandwidth for its transmission.
- Such system is used for broadcast purpose since its receiver circuit configuration is simplest.
- Carrier carries no information. USB and LSB carry equal information since both have same frequency component with reference to the centre frequency ω_c.
- For the transmission of such signal, a band pass f lter is used, having centre frequency of ω_c and a bandwidth of $2\,\omega_m$.
- **Total power required for transmission of AM-DSB/FC**

$$P_{\text{Total}} = P_{\text{carrier}} + P_{\text{LSB}} + P_{\text{USE}}$$

$$P_t = P_c\left(1 + \frac{m_a^2}{2}\right)$$

where, $P_c \rightarrow$ Carrier power ; $P_{\text{USB}} \rightarrow$ Upper side band power
$P_{\text{LSB}} \rightarrow$ Lower side band power

Note:

☑ If carrier is modulated by several modulating signals then over all modulation index:

$$m_a = \sqrt{m_{a1}^2 + m_{a2}^2 + \cdots m_{an}^2}$$

AM-DSB/SC

$$v_{AM}(t) = \frac{1}{2} m_a V_c [\cos(\omega_c + \omega_m)t + \cos(\omega_c - \omega_m)t]$$

AM-SSB/FC

$$v_{AM}(t) = V_c \cdot \cos\omega_c t + \frac{1}{2} m_a V_c [\cos(\omega_c \pm \omega_m)t]$$

(+ → USB; − → LSB)

AM-SSB/SC

$$v_{AM}(t) = \frac{1}{2} m_a V_c [\cos(\omega_c \pm \omega_m)t]$$

Different AM System: At $K_a = 1$

System	P_t	P_s (saved power)	BW	% power saving	Complexity	SNR
AM-DSB/FC	$P_c\left(1+\frac{m_a^2}{2}\right)$	–	$2\omega_m$	–	Min.	Max.
AM-DSB/SC	$P_c \times \frac{m_a^2}{2}$	P_C	$2\omega_m$	67%	↓	Low
AM-SSB/FC	$P_c\left(1+\frac{m_a^2}{4}\right)$	$P_c + \frac{m_a^2}{4}$	ω_m	16%	↓	High
AM-SSB/SC	$P_c \times \frac{m_a^2}{4}$	$P_c\left(1+\frac{m_a^2}{4}\right)$	ω_m	83%	Max.	Min.

Note:

☑ AM-DSB/FC system is the best from the circuit configuration point of view. It has least complexity in the circuit but requires maximum bandwidth as well as maximum power.

☑ AM-SSB/SC system is the worst from the circuit complexity point of view. It has maximum complexity in the circuit but power and bandwidth requirement are minimum.

☑ AM-DSB/FC systems are used for broadcast purpose whereas AM-SSB/SC system is used for point to point communication.

- **Transmission Efficiency**

$$\eta = \frac{\text{Useful power}}{\text{Total power}} \quad ; \quad \eta = \frac{P_s}{P_T} \times 100\%$$

Case-1: AM-DSB/FC

$$\eta = \frac{m_a^2}{2 + m_a^2} \times 100\%$$

If $m_a = 1$ then $\eta = 33\%$.

Case-2: AM-DSB/SC

$$\eta = 100\%$$

- **Percentage power savings**

$$\% \text{ Power saving} = \frac{\text{Power saved}}{\text{Total power}}$$

Case-1: AM-DSB/SC

$$\% \text{ Power saving} = (1-\eta) = \frac{2}{2+m_a^2}$$

Case-2: AM-SSB/SC

$$\% \text{ Power saving} = \frac{4+m_a^2}{2(2+m_a^2)}$$

Note:

- ☑ For full carrier system the transmission efficiency depends upon modulation index (m_a).
- ☑ For full carrier system the maximum efficiency is only 33% for maximum modulation index of unity.
- ☑ For suppressed carrier system the transmission efficiency is always independent of modulation index and is always 100%.

Multi Tone Modulation (Non Sinusoidal)

Representation of Various AM-Signal

AM-DSB/FC

Standard signal

In time domain

$$v(t) = (A + f(t))\left[\frac{1}{2}\left(e^{j\omega_c t} + e^{-j\omega_c t}\right)\right]$$

In frequency domain

$$V(\omega) = \frac{1}{2}\left[\{F(\omega-\omega_c) + F(\omega+\omega_c)\} + \pi A\{\delta(\omega-\omega_c) + \delta(\omega+\omega_c)\}\right]$$

AM-DSB/SC

Standard signal

In time domain

$$v(t) = f(t)\left[\frac{1}{2}\left(e^{j\omega_c t} + e^{-j\omega_c t}\right)\right]$$

In frequency domain

$$V(\omega) = \frac{1}{2}[F(\omega - \omega_c) + F(\omega + \omega_c)]$$

Important Points:

- In suppressed carrier signal, the free carrier is absent.
- Modulation represents frequency translation. Therefore original spectrum centered about zero frequency is translated to frequency spectrum centered at ω_c.
- Using modulation, any low pass signal is modulated to band pass signal.
- Negative frequencies in the original spectrum have no physical meaning. They are represented since
    - The frequency spectrum of any signal is always an even function of ω and therefore negative and positive frequencies occur in pairs.
    - Any negative frequency can be made a positive frequency by frequency translation or using modulation.

AM Modulators and Demodulators

- Product modulation is used for generation of AM waves using non linear devices.
- Switching modulator is also used for generation of AM waves.
- Balanced modulator and ring modulators are used for generation of DSB-SC waves.
- SSB wave is generated using analog multiplier and band pass filter.
- Filter method and phase shift methods are also used for generation of SSB wave.

- **Envelope Detector**

  It is used for detection of AM wave.

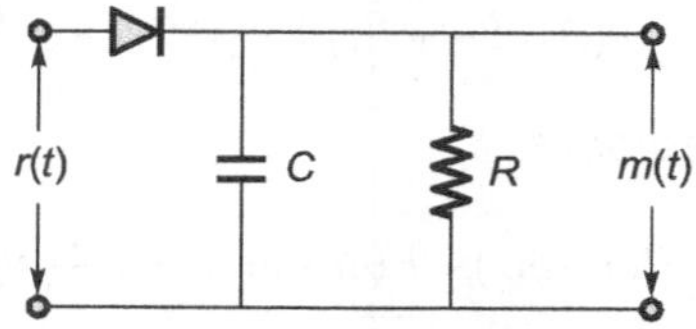

$r(t)$ is received signal and $m(t)$ is message signal and for better reception RC must be selected such as

$$\frac{1}{f_c} << RC << \frac{1}{\omega}$$

where, ω is bandwidth of message signal (in Hz).

Note:

- ☑ Demodulation of DSB-SC and SSB waves is done using coherent carrier signal at receiver.
- ☑ Envelope detector can also be used to recover message by passing received VSB signal through it.

Angle Modulation

Angle Modulation
- Phase modulation (PM)
- Frequency modulation (FM)

Phase Modulation (PM)

In phase modulation the phase of the carrier is varied in accordance with instantaneous value of the amplitude of the modulating signal keeping the amplitude and frequency of the carrier fixed.

General Expression

$$v_{PM}(t) = A\cos[\omega_c t + K_p f(t)]$$

where, $f(t) \rightarrow$ Modulating signal; $K_p \rightarrow$ Phase sensitivity or phase deviation

Frequency Modulation (FM)

In frequency modulation, the frequency of the carrier is varied in accordance with the instantaneous value of the amplitude of modulating signal keeping amplitude and phase of the carrier fixed.

General Expression

$$V_{FM}(t) = A\cos\left[\omega_c t + K_f \int f(t)dt\right]$$

where, $f(t) \rightarrow$ Modulating signal; $K_f \rightarrow$ Frequency sensitivity

If $f(t) = V_m \cos\omega t$, then $V_{FM}(t) = A\cos[\omega_c t + m_f \sin\omega_{mt}]$

where, $m_f \rightarrow K_f \dfrac{V_m}{\omega_m}$ = Modulation index

Note:

- ☑ A pure PM signal or a pure FM signal cannot be obtained since the phase change or the frequency change are inter-related.
- ☑ In PM the modulation index is directly proportional to the amplitude of the modulating signal where as the modulation index in the FM depends upon the amplitude as well as the frequency of modulating signal.

- **Total Power Required for FM Signal**

$$v_{FM}(t) = A\cos\left[\omega_c t + K_f \int f(t)dt\right]$$

$$P = \frac{A^2}{2}$$

- **Frequency Deviation**

$$\delta = m_f\ \omega_m \text{ rad/sec.} \quad \text{or} \quad \delta = m_f\ f_m \text{ Hz}$$

δ represents total bandwidth require for the transmission of FM signal.

Note:

- ☑ Bandwidth required for FM signal is always more than the bandwidth required for the AM signal.
- ☑ ω_m control the spacing between two successive sidebands.
- ☑ The modulation index (m_f) controls the number of significant sidebands.

FM Standards

Remember:

- ☑ Carrier frequency = 88 MHz – 108 MHZ
- ☑ Maximum frequency deviation = ±75 kHz
- ☑ Intermediate frequency (I.F.) fi = 10.7 MHz

- **Bandwidth**

Practical bandwidth required is given by Carson's rule

$$B.W. \cong 2\omega_m(m_f + 1) \text{ rad/sec}$$

or

$$B.W. \cong 2(\delta + \omega_m) \text{ rad/sec.}$$

Case-1: Narrow band FM (NBFM)

$$m_f << 1$$

$$B.W. \cong 2W_m$$

Case-2: Wide band FM (WBFM)

$$m_f >> 1$$

$$B.W. \cong 2\delta$$

It is used in FM radio broadcast.

Note:

- ☑ Same bandwidth is required for AM-DSB system.
- ☑ NBFM used in mobile radio.

Remember:

- ☑ The modulation index determines the number of sidebands. Components have significant amplitudes.
- ☑ In FM, the total transmitted power always remains constant, but with increased depth of modulation, the required bandwidth, is increased.
- ☑ Deviation ratio is the shift in carrier frequency from its resting point compared to the amplitude of the modulating voltage.
- ☑ A deviation ratio of 5 is the maximum allowed in commercially broadcast FM.

Angle Modulators and Demodulators

- FM wave can be generated using VCO caused direct method.
- FM wave can be generated using varactor diode.
- FM and PM both can be generated using reactance tubes.
- Another method is first generate narrow band FM then pass through it frequency multiplier.
- Demodulation is normally done using PLL.

Remember:

The FM system requires larger bandwidth and therefore the information is transmitted through larger number of frequency components or larger number of sidebands. Therefore the quality of recovered signal using FM system is much better as compared to that using AM system. Therefore FM system is Hi-Fi system and therefore has the ability to reproduce same quality of signal which was actually transmitted.

Note:

- ☑ If two tuned circuit are connected in cascade then $\alpha = \alpha_2 \cdot \alpha_2$.
- ☑ The image signal rejection represents the ratio of desired signal amplitude to the amplitude of undesired signal.
- ☑ For any good radio receiver, the parameter α must have high value.

IV Pulse Modulation

Pulse Analog Modulation

- **Pulse amplitude modulation (PAM):** The amplitude of pulse is varied in accordance with the height of the modulating signal, keeping the width and the position of the pulse constant.

- **Pulse width modulation (PWM):** The width of the pulses is varied in accordance with the height of modulating signal keeping amplitude and position of pulse constant.
- **Pulse position modulation (PPM):** The position or the location of the pulses is varied in accordance with the height of the modulating signal keeping the amplitude and the width of the pulses constant.

Note:

☑ Speed control of DC motor is done using PWM.
☑ PWM is generated using 555 timer in monostable multivibrator mode.
☑ PPM is generated using 555 timer by using PWM as a trigger signal in monostable multivibrator mode.

Comparison of various pulse modulation signals

	Circuit Complexity	BW	SNR
PAM	Minimum	$\sim 2\omega_m$	Minimum
PWM	↓	$\sim 10\omega_m$	↓
PPM	Maximum	$\sim 1/t_r$	Maximum

where, t_r = Rise time of each pulse

Note:

☑ PPM is most effective pulse analog modulator scheme in forms of SNR.
☑ SNR should be high, it is an advantage.
☑ The aperture effect occurs only when flat topped sampling is used. This error voltage is acceptable since the generation of flat top sampling is least complex.

Time Division Multiplexing (TDM)

- In TDM entire time interval is divided into smaller time slots and corresponding to each time slot, the sample from a specified signal is transmitted over a common communication channel.
- A common sampling frequency is used for transmission of various signal.
- Simplex circuit and less costly.
- Used for discrete signals such as pulse modulated signals or pulse code modulated signals (PCM).

Frequency Division Multiplexing (FDM)

- In FDM, entire frequency band available is divided into smaller frequency bands and corresponding to each frequency band, the signals are transmitted with different carrier frequencies at all times.

- A separate transmitter is required for the transmission of each signal.
- Complex circuit and therefore more costly.
- FDM system is used for continuous signals such as AM and FM signals.

Pulse Digital Modulation

Sampling Theorem

In order to recover the original modulating signal from its sampled version the signal must be having a sampling frequency of greater than or equal to twice of highest modulating frequency component contained in the given signal, i.e.,

$$f_s \geq 2f_m; \quad \omega_s \geq 2\omega_m \qquad \text{where, } f_s = \text{Sampling rate}$$

The Nyquist rate of sampling represents the minimum rate of sampling so that the original signal can be recovered from its sampled version.

$$f_{s(\text{min})} = f_{s(\text{Nyquist})} = 2f_m \quad \text{Samples/sec}$$

where, f_m = Maximum modulating frequency component

Pulse Code Modulation (PCM)

- The signal to noise ratio of PCM signal is very high.
- Only two level signal and therefore the noise can always be minimised by passing the PCM signal through a limiter circuit.
- For detecting the PCM signal, a threshold level is fixed and incoming voltage is detected on the basis of whether the incoming voltage is more than or less than this threshold level. The detection does not depend upon the absolute value of the voltage received but depends only upon the relative voltage receive with reference to the threshold value.
- Very low probability of error. Error in the detection takes place only when
  - The noise exceeds the threshold value.
  - The signal is detected at the time instant where the noise exceeds the threshold value.

Note:

The PCM signal cannot be transmitted through antenna since the sampling is done at a rate comparable to highest modulating frequency component. Therefore the PCM signal is transmitted through a transmission line or co-axial cable and therefore the range of transmission is limited.

- **Quantization levels**

$$L = 2^n$$

where, $n \rightarrow$ Number of bits/sample ; $L \rightarrow$ Number of quantization levels

- **Step size**

$$\delta = \frac{2V_m}{L} = \frac{2V_m}{2^n}$$

where, $V_m \rightarrow$ Maximum amplitude of sinusoidal signal

- **Quantization Noise**

$$N_q = \frac{\delta^2}{12}$$

- **Signal to Noise Ratio (SNR)**

$$\frac{S_0}{N_q} = \frac{3}{2}L^2 = \frac{3}{2}2^{2n} \; ; \quad \left(\frac{S_0}{N_q}\right)_{\text{dB}} \cong 1.76 + 6n$$

- **Bit rate**

$R_b = nf_s$ where, $f_s \rightarrow$ Number of samples per sec.

- **Min-bandwidth**

$$BW_{\min} = \frac{1}{2}R_{b(\min)} = nf_m$$

Note:

The bit coading parameter 'n' is adjusted depending upon.

☑ Availability of B.W. of transmission.

☑ Required value of the signal to noise ratio.

☑ The complexity of circuit which we can afford for the transmission of given signal.

V Digital Carrier Modulation

- In digital carrier modulation system, the modulating signal is digital which is the output of PCM or DM system and carrier signal is high frequency sinusoidal carrier.
- In digital carrier modulation system, one of the properties of the carrier, namely amplitude, frequency or the phase is varied at a time in accordance with the binary modulating signal.
  - **Amplitude shift keying:** The generation of ASK signal is simplest, amplitude dependent but maximum noise is introduced and therefore ASK system has lowest signal to noise ratio.
    - ASK signal uses ON-OFF signaling.
    - In ASK probability of error (P_e) is high.
    - The ASK system is used for telegraphy.

- **Frequency shift keying:** The circuit configuration of FSK system is most complex, requires large bandwidth for its transmission but has relatively high S/N ratio.
  - FSK signal uses NRZ signaling.
  - In case of FSK, P_e is less.
  - Multiplexing is difficult.
  - Used in MODEM.
- **Phase shift keying:** The PSK system has relatively high SNR, relatively complex circuit and requires lesser bandwidth as compare to FSK system.
  - PSK signal uses NRZ signaling.
  - This system has lowest probability of error
  - The PSK in its modified form is used for satellite communication or for the mobile communication.

- **Comparison between ASK, PSK, FSK**

Parameter	ASK	PSK	FSK
E	$\frac{1}{2}A^2T$	$\frac{1}{2}A^2T$	$E' \equiv 2E$
P_s	$\frac{1}{2}\left(\frac{E}{T}\right)$	$\left(\frac{E}{T}\right)$	$\left(\frac{E}{T}\right)$
$P_{e(min)}$	$\text{erfc}\sqrt{\frac{P_sT}{\eta}}$	$\text{erfc}\sqrt{\frac{2P_sT}{\eta}}$	$\text{erfc}\sqrt{\frac{P_s\bar{}}{\eta}}$
BW	$\frac{2}{T_b}$	$\frac{2}{T_b}$	$(f_2 - f_1) + \frac{2}{T_b}$

Binary Phase Shift Keying (BPSK)

- Waveform $b(t)$ is a NRZ (non-return-to-zero) binary waveform
- **Transmitted Signal**

$$V_{BPSK}(t) = b(t)\sqrt{2P_s}\cos\omega_0 t$$

where, $P_s \rightarrow$ Signal power

$b(t) \rightarrow 1$ V for Logic level 1

$\rightarrow -1$ V for Logic level 0

- **Received signal**

$$V_{BPSK}(t) = b(t)\sqrt{2P_s}\cos(\omega_0 t + \theta)$$

where, θ = Phase shift corresponding to time delay θ/ω_0

Phase shift depends on the length of the path from transmitter to receiver

- **Power spectral density of the BPSK signal**

$$G_{BPSK}(f) = \frac{P_s T_b}{2}\left\{\left[\frac{\sin \pi(f-f_0)T_b}{\pi(f-f_0)T_b}\right]^2 + \left[\frac{\sin \pi(f+f_0)T_b}{\pi(f+f_0)T_b}\right]^2\right\}$$

where, $T_b \rightarrow$ Bit duration

- **Energy contained in a bit duration**

$$E_b = P_s T_b$$

- **Bandwidth**

$$BW = \frac{2}{nT_b}$$; where, n = Number of input bits

Remember:

☑ Differential phase-shift keying (DPSK) and differential encoded PSK (DEPSK) are modifications of BPSK.

☑ DPSK avoids the need to provide the synchronous carrier required at the demodulator for detecting a BPSK signal.

Quadrature Phase Shift Keying (QPSK)

- Four quadrature signals

$$V_m(t) = \sqrt{2P_s}\cos\left[\omega_0 t + (2m+1)\frac{\pi}{4}\right] \quad .... m = 0, 1, 2, 3$$

- Bandwidth

$$BW = \frac{2}{2T_b} = R_b$$

M-ARY PSK

$$V_m(t) = \sqrt{2P_s}\cos(\omega_0 t + \phi_m) \quad ....m = 0, 1, 2, ...(M-1)$$

- Phase Angle

$$\phi_m = (2m+1)\frac{\pi}{M} \quad ....m = 0, 1, 2, ...(M-1)$$

Binary Frequency Shift Keying

$$V_{BFSK}(t) = \sqrt{2P_s}\cos[\omega_0 t + d(t)\,\Omega t]$$

where, $d(t)$ = 1 for logic levels 1 of data waveform

= −1 for logic levels 0 of data waveform

Ω = Constant offset from nominal carrier frequency

Note:

$$BW_{(BFSK)} = 2 \times BW_{(BPSK)}$$

Minimum Shift Keying

- Transmitted signal

$$V_{MSK}(t) = \sqrt{2P_s}\left[\left\{b_e(t)\sin 2\pi\left(\frac{1}{4T_b}\right)\right\}\cos\omega_0 t + \left\{b_0(t)\cos 2\pi\left(\frac{1}{4T_b}\right)\right\}\sin\omega_0 t\right]$$

- Most important and useful feature of MSK is its phase continuity

$$\int_0^{T_b}\sin\omega_H t\cdot\sin\omega_L t\,dt = 0$$

- Probability of error in the detection of any signal.

$$P_{e(\min)} = \text{erfc}\left(\frac{a}{\sqrt{\eta E/2}}\right)$$

where, $\text{erfc}(u)$ = Complementary error function; a = Threshold level

VI Random Variables and Noise

Theory of Random Signals and Process

- If $p(x)$ is a probability density function

$$\int_{-\infty}^{\infty} p(x)dx = 1$$

- Cumulative distribution function (CDF)

$$P(X) = \begin{cases}\int_{-\infty}^{x} p(x)dx & , X \le x\\ 1 & , X < \infty\end{cases}$$

- Avg. or dc value of signal/mean

$$\bar{X} = E(X) \equiv m = \int_{-\infty}^{\infty} xp(x)dx$$

- DC power contained in the given signal

$$P_{dc} = m^2$$

- Total power contained in the given signal/mean square value

$$\bar{X}^2 = E\left(X^2\right) \equiv \int_{-\infty}^{\infty} x^2 p(x)dx$$

- AC power contained in the signal/variance

$$P_{AC} = \sigma_x^2 = \left[E\left(X^2\right) - m^2\right]$$

where, $\sigma_x \rightarrow$ Standard deivation

- RMS voltage

$$V_{rms} = \sqrt{\overline{x}^2} = \sqrt{E(x^2)}$$

Noise

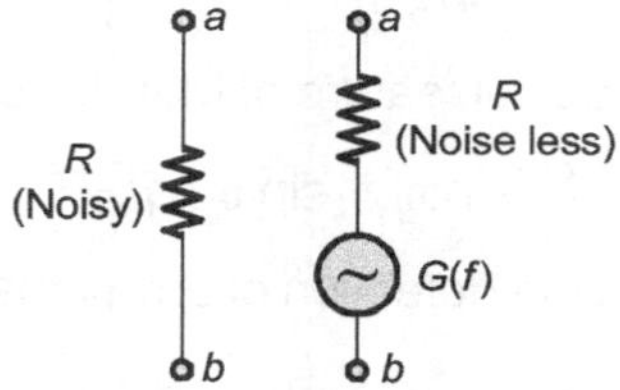

- **Power spectral density of thermal noise**

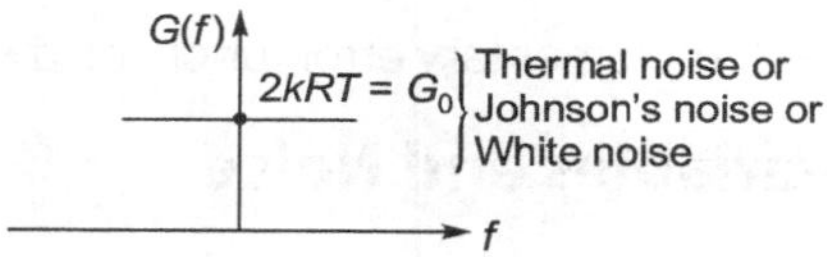

$$G(f) = 2kRT$$

where $k \rightarrow$ Boltzmann constant = 1.38×10^{-23} J/°K

- **Noise power**

$$N_0 = \int_{-\infty}^{\infty} G(f)df = \int_{-\infty}^{\infty} 2kRTdf = \infty$$

- The thermal noise requires infinite amount of power for its generation and infinite amount of BW for its transmission.
- Band limited thermal noise.

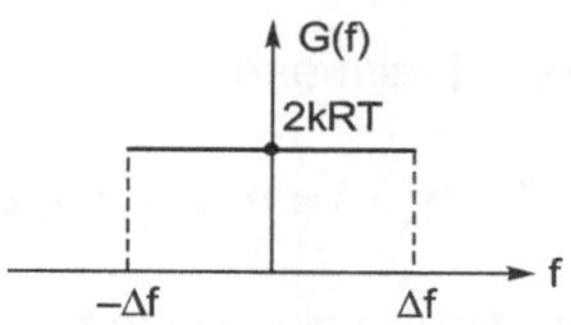

$$G(f) = 2kRT \text{ W/Hz} \quad \ldots\ldots -\Delta f < f < \Delta f$$

$$N_0 = \int_{-\Delta f}^{+\Delta f} G(f)df = 4kRT\Delta f \text{ W}$$

$$V_{rms} = \sqrt{N_0} = \sqrt{4kRT\Delta f} \text{ V}$$

Noise Figure of a Communication System

- The noise figure of any communication system is a measure of total noise power contributed by that communication system.

- For an ideal communication system or for a noise-less system the noise figure has a minimum value of unity.

$$F = \frac{\text{Input SNR}}{\text{Output SNR}}$$

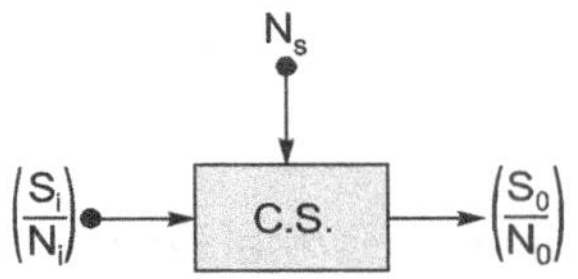

$$F = 1 + \frac{N_s}{AN_i} = 1 + \frac{T_{eq}}{T_0} = 1 + \frac{R_{eq}}{R_s}$$

where, $A \rightarrow$ Power gain ; $N_s \rightarrow$ System noise power
$N_i \rightarrow$ Input noise power ; $T_{eq} \rightarrow$ Equivalent noise temperature
$T_0 \rightarrow$ Room temperature ;
$R_{eq} \rightarrow$ Equivalent input noise resistance
$R_s \rightarrow$ Source resistance

- Noise figure of cascaded Communication Systems

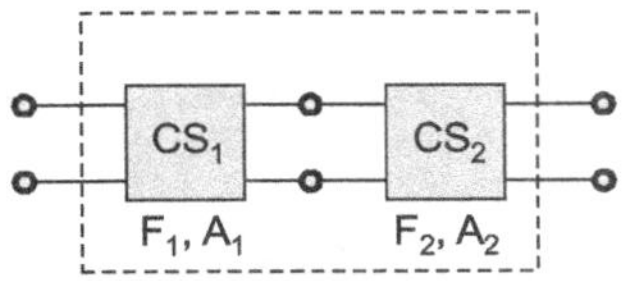

$$F = F_1 + \frac{F_2 - 1}{A_1} + \frac{F_3 - 1}{A_1 A_2} + \ldots$$

Note:

☑ Higher the value of noise figure, higher is the contribution of the noise by that system.

☑ The noise figure of any system does not depend upon signal power at the input. Therefore the noise figure remains same irrespective of the variation in the input signal power.

☑ By cascading, the power gain of the overall system increases to a very large value but contribution of the noise due to cascading is only marginal. Therefore cascading of the communication system is always preferred.

- **Maximum signal to noise ratio**

$$\rho_{max} = (SNR)_{max} = \frac{1}{\pi\eta} \int_{-\infty}^{\infty} |S(\omega)|^2 \, d\omega$$

$$\rho_{max} = \frac{2E}{\eta}$$

where, $S(\omega) \rightarrow$ Fourier transform of message signal s(t)

Note:

☑ Maximum SNR at output depends upon the spectrum and energy contained in the input signal.

- **Power spectral density of Thermal Noise**

$$G_V(f) = 2kRT \;\; (V^2/Hz)$$

- **Thermal Noise power**

$$T_N = \frac{G_V(f)}{4R} = \frac{kT}{2} \;\; (W/Hz)$$

- **Power spectral density of white noise**

$$G_{WN}(f) = \frac{\eta}{2} \; ; \qquad \text{where, } \eta \rightarrow \text{constant}$$

- **Noise equivalent Band-width**

$$B_N = \frac{\pi}{2} B_{eff.} \quad \text{where, } B_N \rightarrow \text{Width of white noise}$$

■■■■

Computer Fundamentals 14

I Data Representation

- If a number system has base b then we can find its b's complement and $(b-1)$'s complement. To determine $(b-1)$'s complement subtract given number from maximum number possible to the given base, i.e., $(r^n - 1)$. To determine b's complement, first determine $(b-1)$'s complement then add 1 to get b's complement of number.
- In unsigned magnitude representation with 'n' bits, the possible integer values are $[0 \text{ to } (2^n - 1)]$.
- In a sign magnitude representation the range of number $-(2^{n-1} - 1)$ to $+(2^{n-1} - 1)$.
- Range of 1's complement number $-(2^{n-1} - 1)$ to $+(2^{n-1} - 1)$.
- Range of 2's complement representation number : -2^{n-1} to $+(2^{n-1} - 1)$.
- In 2's complement representation there is only one representation for zero while in 1's complement representation there are two zeros +0 and –0.
- Fixed Point Signed Numbers: The fixed point signed numbers are denoted using sign magnitude notation, 1's and 2's complement notation.

  (i) **Signed Magnitude Form :**
  - Sign bit is considered explicitly, hence additional hardware is required for resultant sign of arithmetic.
  - Addition and subtraction are performed on separate hardware.
  - 0 has 2 representations i.e., +0 : 0000 0000
    – 0 : 1000 0000

  (ii) **1's Complement Form :**
  - Sign bit is not considered explicitly hence no additional hardware required.
  - 0 has 2 representations i.e., +0 : 0000 0000
    – 0 : 1111 1111
  - Non weighted system.

  (iii) **2's Complement Form :**
  - Sign bit is not considered explicitly.
  - Addition and subtractions are performed by using adder only.

$$-B = \overline{B} + 1$$

$$A - B = A + \overline{B} + 1$$

- 0 has only 1 representation.
- Only code that assigns negative weight to the sign bit.

- The floating numbers are stored in mantissa exponent form. Any floating point number is represented in the form $\pm\, m \times r^{\pm e}$.
- Any floating point number is represented in "normalised" form. In a normalised floating point number the MSB of mantissa must be non-zero.
- Zero cannot be normalized.
- The value of expression is given by :
  (i) Explicit Normalization : $V = (-1)^{\text{sign}}\,[0.M]_2 \times 2^{E-\text{Bias}}$
  (ii) Implicit Normalization : $V = (-1)^{\text{sign}}\,[1.M]_2 \times 2^{E-\text{Bias}}$
- The biased exponent is an unsigned number which can represent signed exponent of original number.
- True Exponent = Biased Exponent – Bias

Sign	$E(K)$ Biased Exponent	Mantissa

Bias = 2^{K-1}

$0 \le E \le 2^K - 1$, Bias = 2^{K-1}

- In IEEE 754 the base of the system is 2, there is a provision for the values ±0 and ±∞, the floating point number is stored either with single precision or with double precision, certain mantissa exponent combinations do not represent any number representing NAN, and The value represented in single precision.

Single Precision (32 bits)

S	E	M
1	8	23

$0 \le E \le 255$

Double Precision (64 bits)

S	E	M
1	11	52

$0 \le E \le 2047$

II Computer Architecture

- **Computer organization:** Interconnection of hardware to form the computer system
- **Computer architecture:** the structure and behaviour of the computer perceived by the user.
- **Main Memory:** Area of the computer that temporarily holds data waiting to be processed, stored, or output. *Example:* Cache and Main memory

- **Secondary Storage:** Area of the computer that holds data on a permanent basis when it is not immediately needed for processing. Example: Disk, Floppy, etc.

CISC and RISC Architectures

CISC (Complex Instruction Set Computers)

- Large instruction set.
- Instruction formats are of different lengths.
- Instructions perform both elementary and complex operations.
- Control unit is micro-programmed.
- Not pipelined or less pipelined.
- Single register set.
- Numerous memory addressing options for operands.
- Emphasis on hardware.
- Includes multi-clock complex instructions.
- Memory-to-memory: "LOAD" and "STORE" incorporated in instructions.
- Small code sizes, high cycles per second.
- Transistors used for storing complex instructions.
- **Examples of CISC processors :**
    - VAX
    - PDP-11
    - Motorola 68000 family
    - Intel x86 architecture based processors.

RISC (Reduced Instruction Set Computers)

- Compact instruction set.
- Instruction formats are all of the same lengths.
- Instructions perform elementary operations.
- Control unit is simple and hardwired.
- Pipelined.
- Multiple register set.
- Compiler and IC developed simultaneously.
- Emphasis on software.
- Single-clock, reduced instruction only.
- Register to registe: "LOAD" and "STORE" are independent instructions.

- Low cycles per second, large code sizes.
- Spends more transistors on memory registers.
- **Examples of RISC processors :**
    - Apple iPods (custom ARM7TDMI SoC)
    - Apple iPhone (Samsung ARM1176JZF)
    - Nintendo Game Boy Advance ARM7)
    - Sony Network Walkman Sony inhouse ARM based chip)

General CPU Architecture (8086 Microprocessor)

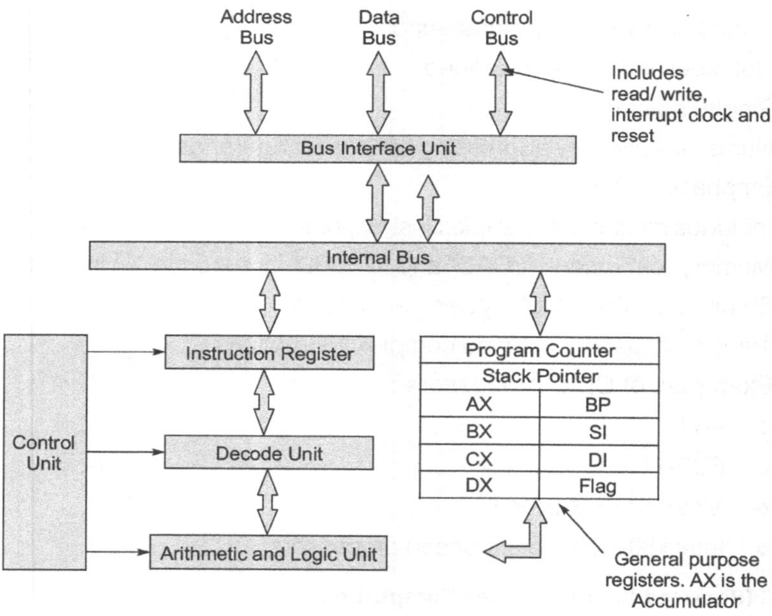

Computer Architecture Vs Computer Organization

Architecture and organization are independent, you can change the organization of a computer without changing its architecture.

1. The architecture indicates its hardware whereas the organization reveals its performance.
2. For designing a computer, its architecture is fixed first and then its organization is decided.

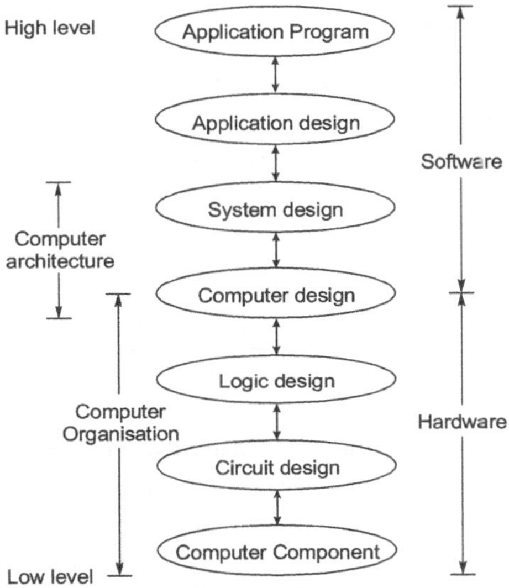

Computer System

Computer Organization	Computer Architecture
• Computer organization deals with structural relationships that are not visible to the programmer like clock frequency or the size of the physical memory).	• Computer architecture deals with the functional behaviour of a computer system as viewed by a programmer like the size of a data type – 32 bits to an integer).
• A computer's organization expresses the realization of the architecture.	• A computer's architecture is its abstract model and is the programmer's view in terms of instructions, addressing modes and registers.
• Organization describes how it does it.	• Architecture describes what the computer does.

Von Neumann Architecture Vs Harvard Architecture

Von Neumann Architecture

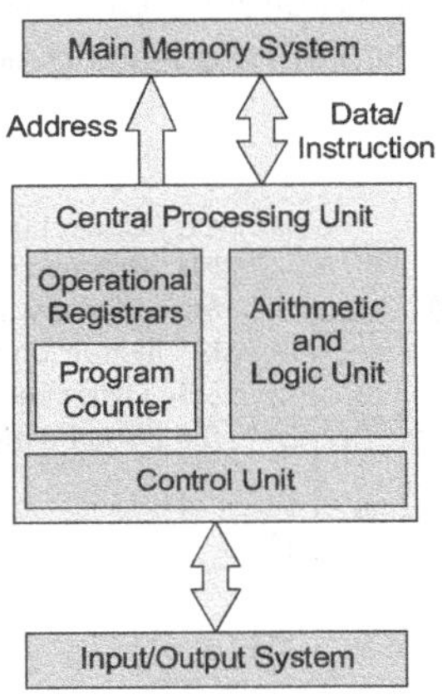

Harvard Architecture

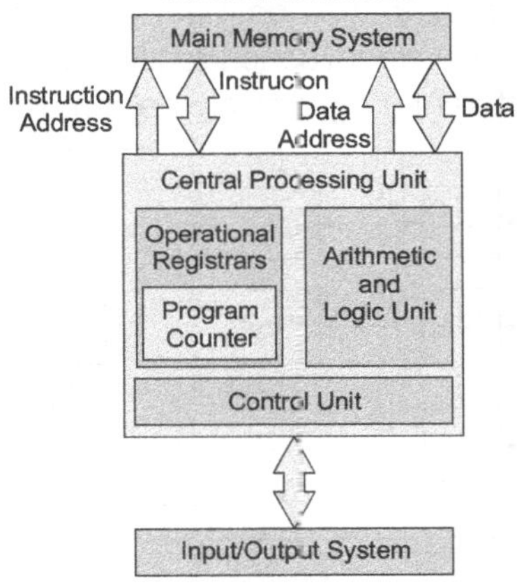

Flynn's Classification

Single Instruction Stream, Single Data Stream (SISD)

A computer with a single processor is called a Single Instruction Stream, Single Data Stream (SISD) Computer. It represents the organization of a single computer containing a control unit, a processor unit, and a memory unit.

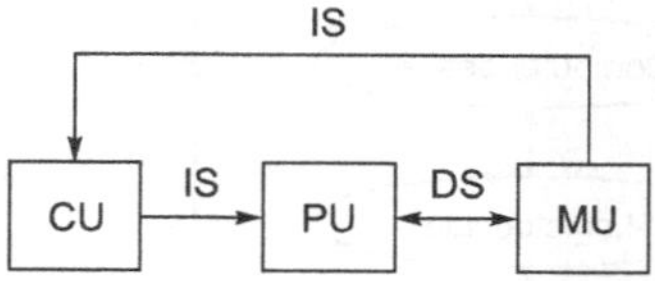

CU : Control unit
PU : Processor unit
MU : Memory unit
IS : Instruction stream
DS : Data stream

Single Instruction Stream, Multiple Data Stream (SIMD)

It represents an organization of computer which has multiple processors under the supervision of a common control unit. All processors receive the same instruction from the control unit but operate on different items of the data.

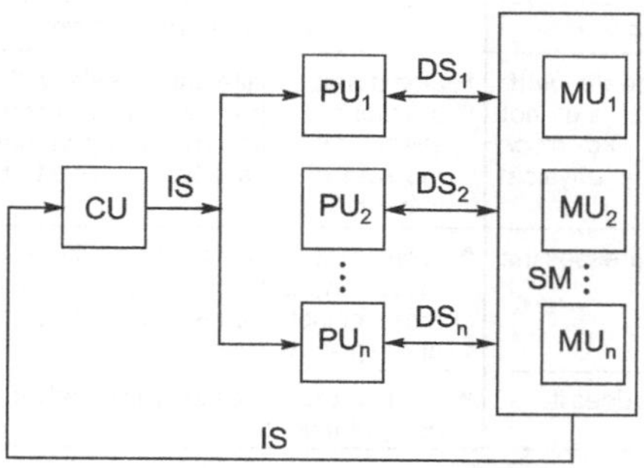

CU : Control unit
PU : Processor unit
MU : Memory unit
IS : Instruction stream
DS : Data stream
SM : Shared memory

Multiple Instruction Stream, Single Data Stream (MISD)

It refers to the computer in which several instructions manipulate the same data stream concurrently. In the structure different processing element run different programs on the same data.

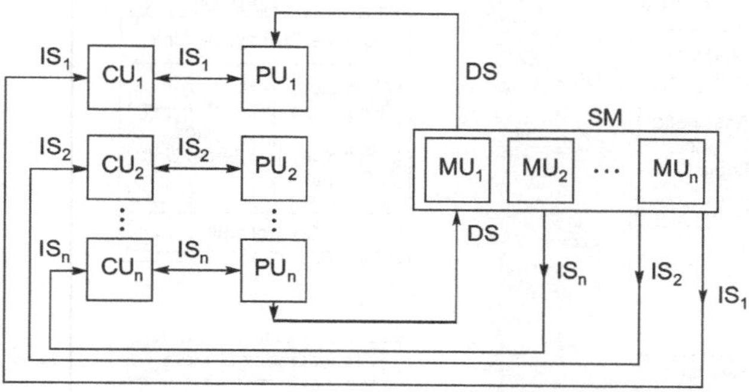

CU : Control unit
PU : Processor unit
MU : Memory unit
IS : Instruction stream
DS : Data stream
SM : Shared memory

Multiple Instruction Stream, Multiple Data Stream (MIMD)

MIMD computers are the general purpose parallel computers. Its organization refers to a computer system capable of processing several programs at a same time.

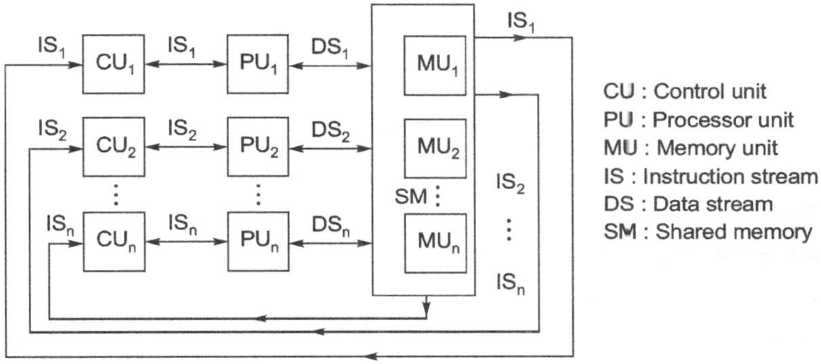

III Memory Organization

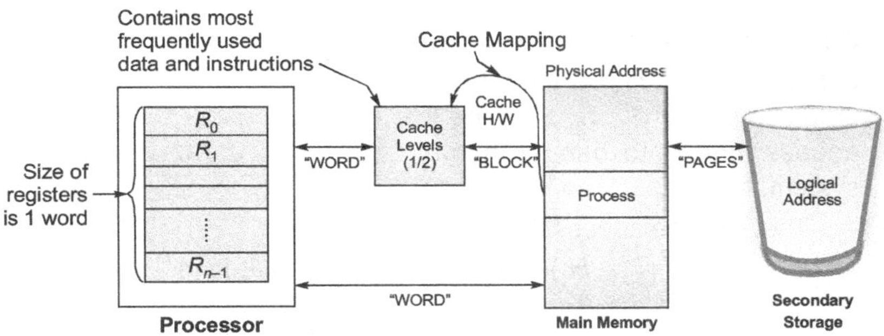

Memory Hierarchy

- The memory hierarchy shows the accessing order of the existing memory system.
- $$\text{Average Memory Access Time} = \frac{\text{All Instruction Fetch Time + Write Time}}{\text{Total instruction}}$$
- Smaller cache block $\rightarrow$ $\downarrow$ Cache miss penalties

  $\uparrow$ Larger TAG $\rightarrow$ $\uparrow$ Cache TAG overhead

  $\uparrow$ Spatial locality
- The purpose is to bridge the speed mismatch between fastest processor to the slowest memories at reasonable cost.

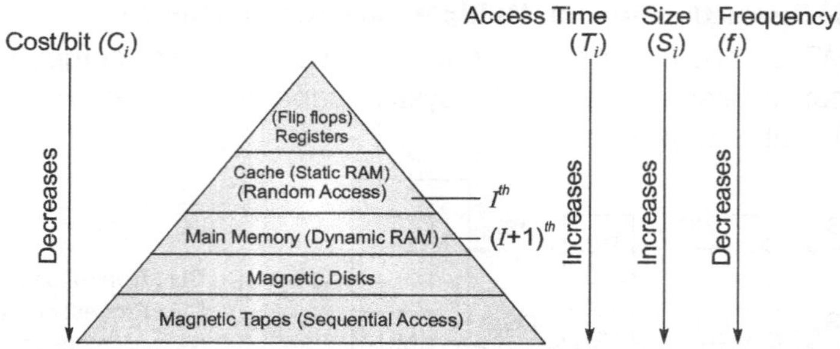

- When the processor refers to I^{th} level memory if it is found then "hit" else "miss".

Types of Memory Organisation

Based on the order of accessing the memory system, the memory organisation is divided into two types:

Simultaneous Access

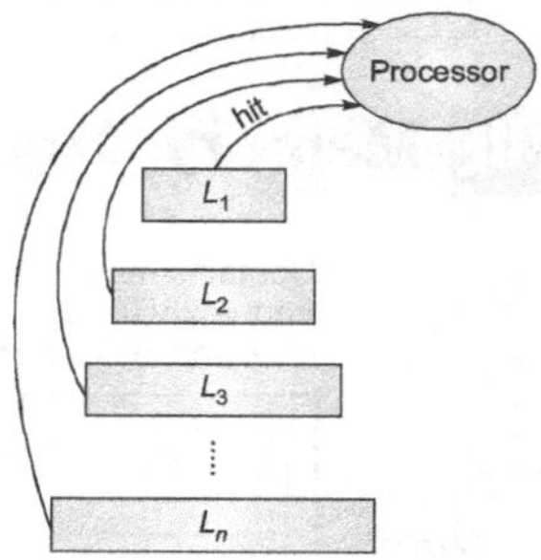

- In this the CPU directly communicates with all the levels of memories.
- Consider a '*n*' level memory system. Let H_1, H_2, ... , H_n be the hit ratios upto *n* level and T_1, T_2, ... T_n be the access time upto *n* level memory system.
  - The average access time of the memory is;
    $T_{avg} = H_1T_1 + (1 - H_1)H_2T_2 + ... + (1 - H_1)(1 - H_2)...(1 - H_{n-1})H_nT_n$
  - Throughput of the memory $= \frac{1}{T_{avg}}$ word/sec
  - $C_{avg}/\text{bit} = \frac{C_1 S_1 + C_2 S_2 + ... + C_n S_n}{S_1 + S_2 + ... + S_n}$

Hierarchical Access

- In this the data from 2nd level is transferred to 1st level and CPU accesses the data from 1st level.
- Consider a '*n*' level memory system.
- Let H_1, H_2, ..., H_n be the hit ratios upto *n* level memory system and T_1, T_2, ... T_n be the access time upto *n* level memory system.

$$T_{avg} = H_1T_1 + (1 - H_1)H_2(T_1 + T_2) + (1 - H_1)(1 - H_2)H_3(T_3 + T_2 + T_1) + ... + ((1 - H_1)(1 - H_2)...(1 - H_{n-1})H_n(T_n + T_{n-1} + ... + T_2 + T_1))$$

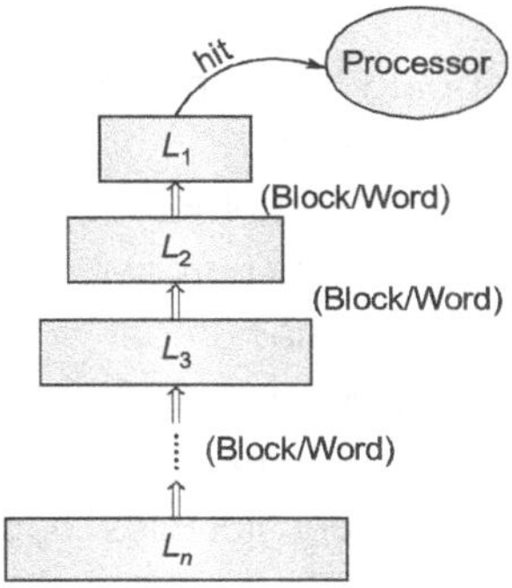

Cache Memory

- Cache is used as the intermediate memory between CPU and main memory.
- It is small and fast memory.
- By placing most frequently used data and instructions in cache, the average access time can be minimized.
- When miss occurs the processor directly obtains data from main memory and a copy of it is send to cache for future references.
- The cache memory works on the principle of Locality of Reference (LOR). LOR states that "the references to memory at any given interval of time tends to be confined to within localized areas of memory".
- **The performance of cache depends on:**
  - Cache size (small)
  - Cache block size
  - Number of levels of cache
  - Cache mapping technique
  - Cache replacement policy
  - Cache updation policy
- Cache is divided into equal parts called as cache blocks/cache lines.
- Data is transferred from main memory to cache in form of blocks, thus both are divided into blocks of equal size.
- Cache line size = Main memory block size

  $$\text{Number of cache lines} = \frac{\text{Cache size}}{\text{Block size}}$$

Mapping Techniques

The process of transferring the data from the main memory to cache memory is called as mapping. There are three different mapping techniques:

- Direct mapping
- Associative mapping
- Set associative mapping

Direct Mapping

- In this technique, mapping function is used to transfer the data from main memory to cache memory.
  The mapping function is K **Mod** $N = i$
  where K = Main memory block number ; N = Number of cache lines
  i = Cache memory line number
- **The address representation is:**

← Physical address (in bits) →

Tag	Block	Word

- Accessing same block from different pages is always a miss. The replacement is done when "conflict miss" occurs.
- Hit ratio is less.
- The higher order bits of the address is compared simultaneously with each cache tag. If a match is not found it is miss. The delay of tag comparator is called hit latency or hit delay. So, the reference is forward to the main memory.
- The main memory control logic interprets the request into its known format.

Tag	Word

  The tag field is directly connected with the address logic of the main memory. Therefore, the corresponding block is enabled.
- By using the mapping function the main memory block is transferred to corresponding cache line along with the complete tag. Later the CPU accesses the data from the cache.
- The number of bits in the tag is tag size or tag length.
  The maximum number of tag comparisons = 1.

Associative Mapping

- Any main memory block can be mapped to any location in cache. Thus, the cache address is not in use.
- The data is transferred from main memory to cache along with the complete tag. The address representation is

← Physical address (in bits) →

Tag	Word

- The replacement is done only when the cache is full or when "capacity miss" occurs.
- The block number is updated in the tag field. The higher order bits of the address compared sequentially with each cache tag, if a match is not found then it is "miss".
- The maximum number of tag comparisons = N (Number of cache blocks)

- The complexity of tag comparator is more.
- Cache size = Tag memory size + Data memory size

  Data memory size = Number of cache lines × Number of bits in the line

Set Associative Mapping

- In this mapping the cache memory is organised into sets, each set is capable to hold multiple main memory blocks.

  $$\text{Number of sets} = \frac{N}{\text{P-way}}$$

  where N = Number of cache lines

  P = Number of MM blocks possible on each cache line
- The address interpretation is:
- The mapping function used is

  $K \bmod S = i$

  where, K = Main memory block number

  S = Number of cache sets

  i = Cache memory set number

← Physical address (in bits) →

Tag	Set offset	Word offset

- The replacement is done when the set is full.
- Hit ratio and complexity of tag comparator is optimal.
- Maximum number of tag comparisons = P.

  When P =1 set associative is direct mapping.
- There is a need of multiplexers to compare the existing tags in the set one by one with respect to CPU generated tag.
- Tag memory size = Number of sets in the cache × Number of blocks in the set × Number of tag bits in each block

Replacement Algorithms

When the cache is full, the existing block is replaced with new block. There are two replacement algorithms used

- **FIFO (First-In-First-Out):** It replaces the cache block with new block which is having longest time stamp.
- **LRU (Least Recently Used):** It replaces the cache block with new block which is having less number of references with longest time stamp.

Updating Techniques

Coherence: The same address contains different values at different places (cache is updated while main memory is not updated). This property is called "coherence" which creates loss of data.

There are two updating techniques used: Write through and write back.

Write Through

In this technique, the CPU performs the simultaneous updation in the cache memory and main memory. Thus, there is no coherence.

- Word updation (T_W) Time = $\max\left(\begin{array}{c|c}\text{Word updation time in cache} & \text{Word updation time in main memory}\end{array}\right)$
- $T_{avg_{read}} = H_r \; T_c + (1 - H_r) \cdot (T_m + T_c)$

  (H_r: read hit; T_c: read data; $(1 - H_r)$: read miss; T_m: read allocate; T_c: read data)
- $T_{avg_{write}} = H_w \; T_w + (1 - H_w) \cdot (T_m + T_w)$

  (H_w: write hit; T_w: word updation time; $(1 - H_w)$: write miss; T_m: write allocate; T_w: word updation time)
- Total access time of memory, when both read and write cycle is considered is:

  $T_{avg} = f_{read} \times T_{avg_{read}} + f_{write} \times T_{avg_{write}}$
- Throughput $\eta_{WT_{cache}} = \dfrac{1}{T_{avg_{WT}}}$ word/sec
- The hit ratio for the write operation is always 1,
- $T_{avg_{write}} = H_w T_m$; where T_m = Main memory access time

Write Back

The CPU performs the write operation only on the cache, still coherence is present but that doesn't create problem because before replacing the cache block with new block the updation are taking back into the main memory.

- Each cache line contains one extra bit, called modified bit (update bit), when the block is updated the corresponding modified bit is set.
- 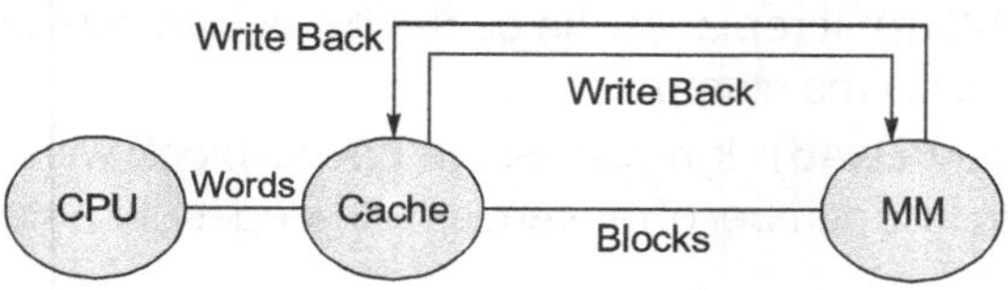

Modified bit → 1 (Dirty bit) [Hierarchical: ($T_{read} = T_{write}$)]

[$T_{read} = H.\, T_c + (1 - H)\,(T_{cache} + T_{mem\text{-}block} + \%(\text{dirty}) \,.\, T_{mem\text{-}block})$]

Modified bit → 0 (Clean bit) [Simultaneous: $T_{read} = H.\, T_c + (1 - H)\,(T_m + \%(\text{dirty}) - T_{mem\text{-}block})$]

- $T_{avg_{read}} = \underset{\text{read hit}}{\underset{\downarrow}{H_r}} \; \underset{\text{read data}}{\underset{\downarrow}{T_c}} + \underset{\text{read miss}}{\underset{\downarrow}{(1 - H_r)}} \; [\% \text{ dirty bits } (\underset{\text{write back}}{\underset{\downarrow}{T_m}} + \underset{\text{read allocate}}{\underset{\downarrow}{T_m}} + \underset{\text{read data}}{\underset{\downarrow}{T_c}}) + \% \text{ clean bits } (\underset{\text{read allocate}}{\underset{\downarrow}{T_m}} + \underset{\text{read data}}{\underset{\downarrow}{T_c}})]$

- $T_{avg_{write}} = H_w T_c + (1 - H_w) \cdot [\% \text{ dirty bits } (\underset{\text{write back}}{\underset{\downarrow}{T_m}} + \underset{\text{write allocate}}{\underset{\downarrow}{T_m}} + \underset{\text{write data}}{\underset{\downarrow}{T_c}}) + \% \text{ clean bits } (\underset{\text{write allocate}}{\underset{\downarrow}{T_m}} + \underset{\text{write data}}{\underset{\downarrow}{T_c}})]$

- $T_{avg_{write\ back}} = f_{read} \times T_{avg_{read}} + f_{write} \times T_{avg_{write}}$

- Throughput $\eta_{write\ back} = \dfrac{1}{T_{avg_{write\ back}}}$ word/sec

Multilevel Caches

- To reduce the miss penalty, multilevel caches are used in the system design.

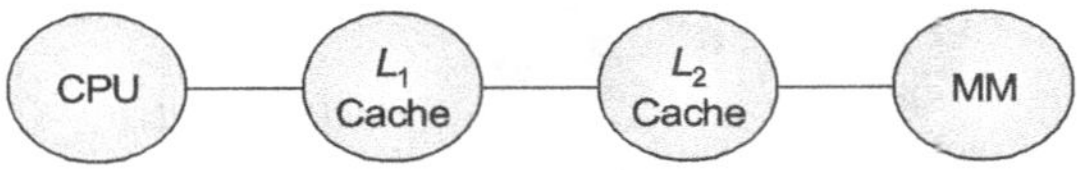

- In multilevel caches, two kinds of miss rates will be calculated

    (*i*) **Local miss rate:**

    $$\text{Local miss rate} = \frac{\text{No. of misses in the cache}}{\text{Total No. of references to that cache}}$$

    (*ii*) **Global miss rate:**

    $$\text{Global miss rate} = \frac{\text{No. of misses in the cache}}{\text{Total No. of CPU generated references}}$$

- Average access time can be calculated as

    $T_{avg} = \text{Hit time}_{L_1} + \text{Miss rate}_{L_1}(\text{Local miss rate}) \times \text{Miss penalty}_{L_1}$

    $\text{Miss penalty}_{L_1} = \text{Hit time}_{L_2} + \text{Miss rate}_{L_2} \times \text{Miss penalty}_{L_2}$

    $\text{Miss penalty}_{L_2} = \text{MM Access time}$

- Average memory stall cycles = (No. of miss$_{L1}$/Instruction × Hit time$_{L2}$)+ (No. of misses$_{L2}$/Instruction × Miss penalty$_{L2}$)

Note: ....

☑ There are three types of caches based on whether the index or tag correspond to physical or virtual addresses.

☑ **Physically indexed, physically tagged:** Caches use the physical address for both the index and the tag.

☑ **Virtually indexed, physically tagged:** Caches use the virtual address for the index and the physical address in the tag.

☑ **Virtually indexed, virtually tagged:** Caches use the virtual address for both the index and the tag.

Secondary Memory

- It is often required to have larger programs which exceeds the size of main memory.
  Example: Operating System.
- A magnetic disk is a thin circular metal plate, the data on the disk surface is organised as

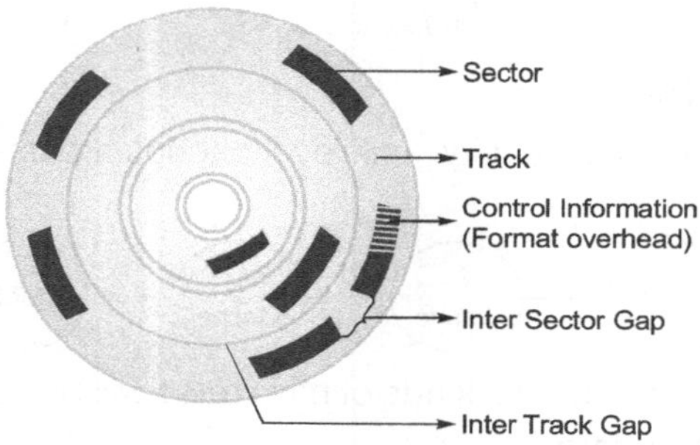

- A set of concentric circles is called "tracks".
- Each track holds same number of manageable units called sectors. The universal size of a sector is 512 Bytes.
  The basic unit of transfer from the disk is a "sector".
- The disk space without control information is called formatted disk space.
- The disk is a semi random access memory and data is distributed across circular tracks and sectors. Two types of disk construction:
  - **Constant Track Capacity Disk:** In this case variable recording density is used across all the tracks and the disk has to move with angular velocity.
  - **Variable Track Capacity Disk:** In this case constant recording density is used.
- Each rotation of the disk covers 1 track.
- The data transfer rate depends on rpm.
- **Seek Time (t_s):** The time to move Read/Write (R/W) head to the desired track.
- **Rotational Latency (t_r):** The time to move Read/Write (R/W) head to the desired sector beginning.

By default $t_r = \frac{1}{2} \times \text{rotation time}$

$$t = \text{access time of the disc} = t_s + t_r$$

$$t_{avg} = t_s + t_L + t_{data\ transfer} + t_{overhead}$$

- The storage density **ρ = Number of Bytes/cm**
  ρ is maximum at innermost track.
- The data transfer rate of the disc
  D = **Number of Bytes/sec**
- Multiple disks can be organised as one above another.

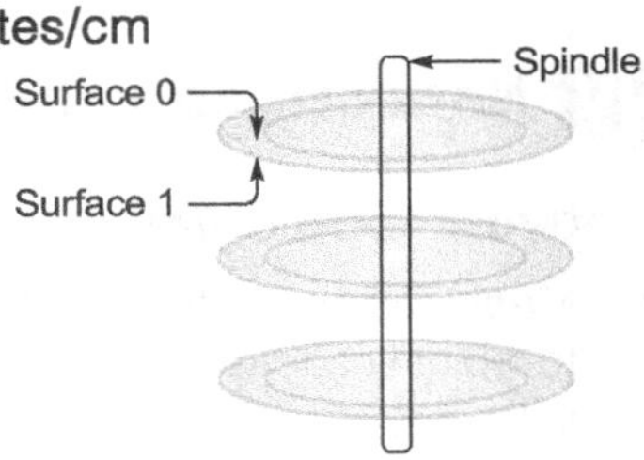

- **Characteristics of disks:**
  – **Single sided/Double sided disks:** By default double sided disks are used, in which recording is done on both the surfaces.
  – Fixed Read/Write (R/W) head or Movable R/W head

Fixed R/W Head	Movable R/W Head
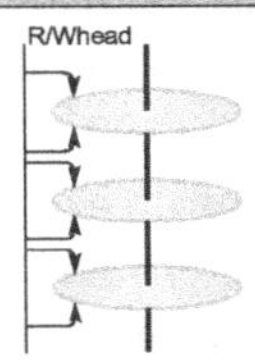	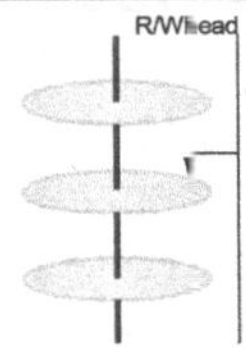
(a) One R/W head per surface (b) Each rotation covers '*n*' tracks *n* = no. of surfaces (c) Hardware increases and hence cost is more	(a) One R/W head for all surface (b) Each rotation covers 1 track (c) Less hardware required but access time increases

(*iii*) Single platter (or) Multiple platter disks

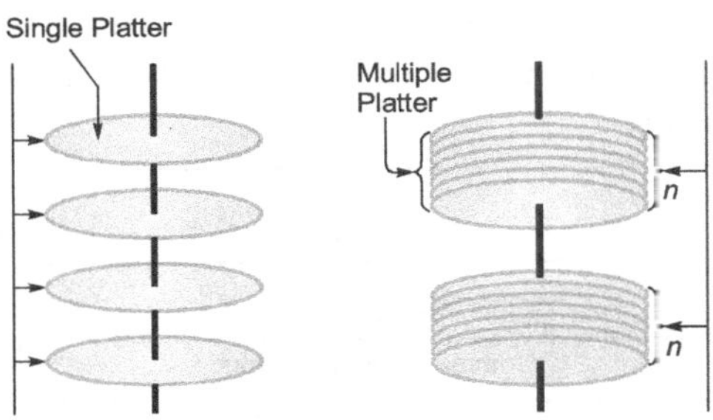

- A vertical set of all the tracks from the same position in a disk pack forms a cylinder.
  Number of cylinders = Number of tracks.

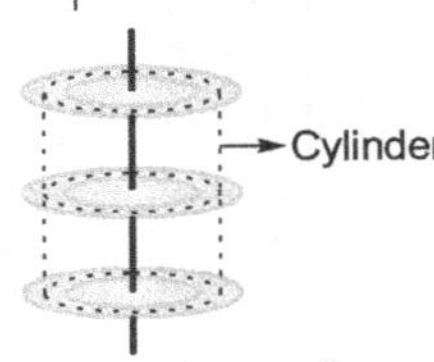

Note:

☑ Capacity = Track × Sectors/Track × Bytes/Sector.

☑ $\text{Capacity}_{\text{formatted}}$ = Tracks × Sectors/Track × (Bytes/Sector – Format overhead).

IV Networking Fundamental

Network Topologies

- **Mesh Topology:** In mesh topology, every device has a dedicated point-to-point link to every other device. The term *dedicated* means that the link carries traffic only between the two devices it connects. Node 1 must be connected to n–1 nodes, node 2 must be connected to n–1 nodes, and finally node n must be connected to n – 1 nodes. Hence in total we need n(n–1) physical links. Every node must have n –1 input/output (I/O) ports.
- **Star Topology:** In a star topology, each device has a dedicated point-to-point link only to a central controller, usually called a **hub**. The devices are not directly linked to one another. Unlike a mesh topology, a star topology does not allow direct traffic between devices. In a star, each device needs only one link and one I/O port to connect it to any number of others.
- **Bus Topology:** A bus topology is multipoint. One long cable acts as **backbone** to link all the devices in a network. A drop line is a connection running between the device and the main cable.
- **Ring Topology:** In a ring topology, each device has a dedicated point-to-point connection with only the two devices on either side of it. A signal is passed along the ring in one direction, from device to device, until it reaches its destination. Each device in the ring incorporates a repeater. Each device is linked to only its immediate neighbours.

Protocol Layering

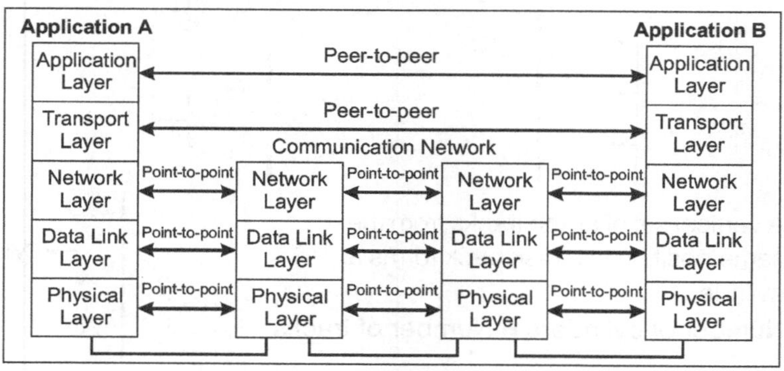

OSI layer vs TCP/IP layer

OSI (Open System Interconnection)	TCP (Transmission Control Protocol/ Internet Protocol)
• OSI provide layer function and also defines functions of all layers.	• TCP/IP model the transport layer does not guarantees delivery of packets.
• In OSI model the transport layer guarantees the delivery of packets.	• Follows vertical approach.
• Follows horizontal approach.	• TCP/IP does not have a separate presentation layer.
• OSI model has a separate presentation layer.	• TCP/IP model cannot be used in any other application.
• OSI is a general model.	• The network layer in TCP/IP model provides connectionless services.
• Network layer of OSI model provide both connection oriented and connectionless service.	• TCP/IP model does not fit any protocol.
• Protocols are hidden in OSI model and are easily replaced as the technology changes.	• In TCP/IP replacing protocol is not easy.
• OSI model defines services, interfaces and protocols very clearly and makes clear distinction between them.	• In TCP/IP it is not clearly separated its services, interfaces and protocols.
• It has 7 layer.	• It has 5 layers.

V Programming Elements

- **'Auto' Storage Class:** The auto storage class is implicitly the default storage calls used simply specifies a normal local variable which is visible within its own code block only and which is created and destroyed automatically upon entry and exit respectively from the code block.
- **'Static' Storage Class:** The static storage class causes a local variable to become permanent within its own code block i.e. it retains its memory space and hence its value between function calls.
- **'Register' Storage Class:** The register storage class also specifies a normal local variable but it also requests that the compiler store a variable so that it may be accessed as quickly as possible, possible from a CPU register.
- **'Extern' Storage Class:** An extern class which informs the compiler for the existence of the global variable which enables it can be accessed in more that one source file.
- **Control Statements:** 1. **if** statement, 2. Nested **if** statement, 3. Conditional operator?,4. **switch** statement, 5. **for** statement, 6. **while** statement, 7. **do-while** statement, 8. **break** statement and 9. **continue** statement.

- **Call by value:** "Actuals copied to formal", but not formals to actuals.
- **Call by reference:** "Actuals and formals uses same address space".
- **Call by value result (Restore):** "Actuals copied to formals and formals copied to actuals".
- **Call by Result:** "Actuals not copied to formals", but formals copied to actuals.
- **Call by Constant:** Formal values are never changed.
- **Call by Name:** The textual substitution of every occurrence of a formal parameter in the called routines body by the corresponding actual parameter.
- **Call by Text:** Same as call by name, but if variables are named same as local variables then it prefers local.
- **Call by Need:** Memory allocates only if formals are used in function.
- **Static Scoping:** The method of binding names to non local variables called static scoping. Scope of variable can be statically determined prior to execution.
- **Dynamic Scoping:** It is based on the calling sequence of subprograms. Scope can be determined at runtime.
- **Pointer:** A pointer is a variable that is used to store a memory address of another variable in memory. If one variable holds the address of another then it is said to the second variable.

VI Operating System Concepts

Scheduling Criteria

- **CPU Utilization:** CPU should be as busy as possible
- **Throughput:** The number of processes that are completed per unit time
- **Turnaround Time:** Completion Time – Arrival Time
- Waiting Time = Turnaround Time – Burst Time
- Response Time = First Response Time – Arrival Time

CPU Scheduling Algorithms

The following algorithms are priority/non-priority based and preemptive/ non-preemptive based.

- **First Come First Serve (FCFS)/ First-in First-out (FIFO):** It assigns the processor in the order of arrival of requests. This algorithm is non-preemptive. In this policy, the disadvantage is shorter jobs can get stacked behind the long running jobs.

  Convoy Effect: If the first process is having large burst time, then it will have major impact on average waiting time of other processes.

- **Shortest Job First (SJF)/Shortest Job Next (SJN):** It favours shorter jobs, it is used as a benchmark in order to compare its performance with other process. This algorithm is non-preemptive. In this policy, the advantage is "turnaround time can be minimised". The disadvantage is it can lead to starvation of processes.

  "Aging": It is a technique of gradually increasing the priority of processes that wait in the system for a long time. It avoids starvation.
- **Shortest Job Remaining Time First (SRTF) scheduling:** Preemptive SJF scheduling is called Shortest Remaining Time First (SRTF).
- **Round Robin Scheduling:** Designed for *time sharing* system CPU is allocated to each process for certain time interval called time *quantum* (time slice). It is preemptive algorithm. If time quantum decreases, no of context switch increases, response time decreases.
- **Priority based scheduling** can be preemptive or non preemptive.

Internal Fragmentation

In fixed size partitioning if process size is smaller than partition size some unused free space is left within the partition and these are called memory fragments/holes/waste memory/unused free space and their sum is called internal fragmentation.

External Fragmentation

In variable size partitioning initially processes are contiguous but as processes leave their space these partitions may not suit the new processes for allocation. Those unused partitions is called External Fragmentation.

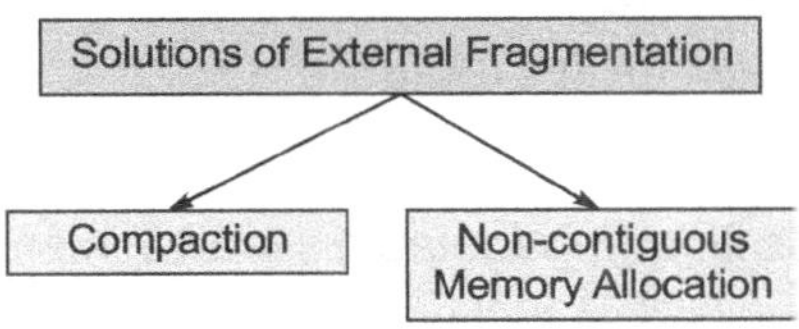

Compaction

- Processes are on one side and free space is on other side in the memory.
- During memory compaction:
  - Block processes for compaction
  - Move them towards lower address space
  - Allocate memory for pending request
  - Resume blocked process w.r.t. compaction
- It consumes CPU time (overhead)
- Program need to support the dynamic allocation.

Non-contiguous Allocation

Size of Logical Address Space (LAS) generated by CPU is greater than or equal to size of Physical Address Space (PAS) generated by Main Memory. So that there will be efficient use of multiprogramming, can run larger programs in smaller memory area.

Paging

Logical Address Space

- Logical Address Space is divided into equal size pages.
- Page size (PS) is generalisation of power of 2.
- Number of pages in LAS (N) = $\frac{\text{LAS}}{\text{PS}}$
- Page No (P) bits = $\lceil \log_2 N \rceil$
- Page offset depends on page size
- Page No dependents on number of pages
- Page offset (d) bits = $\lceil \log_2 \text{PS} \rceil$

Physical Address Space

- Physical Address Space is divided into equal size frames (Page frames)
- Frame size = Page size
- Number of frames (M) = $\frac{\text{PAS}}{\text{PS (FS)}}$
- Frame No (f) bits = $\lceil \log_2 M \rceil$
- Frame offset = Page offset = d

Page Table

- Each process has its own pagetable stored in memory.
- Pagetable is organised as series of entries.
- Number of entries in page table = Number of pages in logical address space (single level paging)
- Page table entry contain frame number.
- Page table implementation maintains a four-bit per page table entry.
  - **Use bit:** Set when a page is referenced, cleared by the clock algorithm
  - **Modified bit:** Set when page is modified, cleared when a page is written to disk
  - **Valid bit:** Set when a program can make legitimate use of this page table entry
  - **Read-only:** Set for a program to read the page, but not to modify it.

- Pagetable (PT) size in bytes = $N \times e$; e = page table entry size in bytes ($e \geq 1$)
- 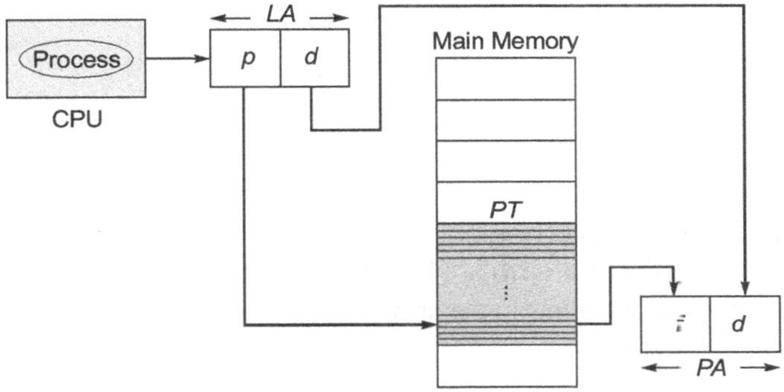

Performance of Paging

- Paging without TLB:
  - Main Memory Access Time = 'm'
  - Effective Memory Access Time (EMAT) = '$2m$'
- Paging with TLB:
  - TLB Access Time = 'c' [$c <<< m$]
  - TLB Hit Ratio(x) = $\dfrac{\text{Number of hits}}{\text{Total References}}$
  - Effective Memory Access Time = $x(c + m) + (1 - x)(c + 2m)$

Multilevel Paging

Larger pagetable is not desirable because they take lot of memory.

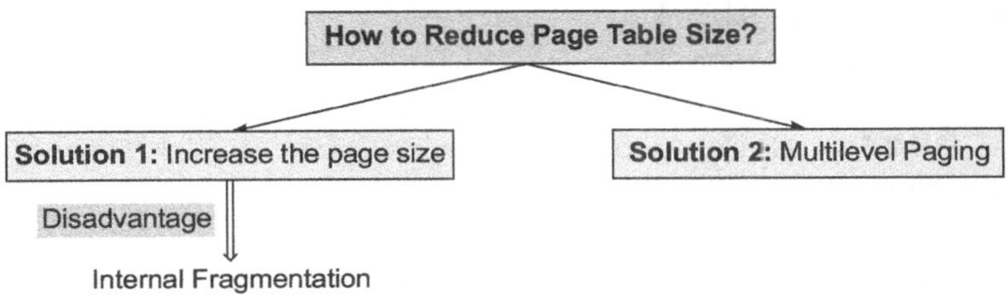

Optimal Page Size

- LAS = 'S' bytes
- PT entry size = 'e' bytes
- Page size (PS) = P
- Page Table (PT) size = $\dfrac{S}{P} \times e$
- Internal Fragmentation (IF) = $P/2$

- Total overhead = PT size + IF = $\left(\frac{P}{2}+\frac{S}{P}\times e\right)$ = Should be minimum
- Differentiating w.r.t. to P

$$\frac{d}{dP}\left(\frac{P}{2}+\frac{S}{P}\times e\right) \Rightarrow \frac{1}{2}+\left(\frac{-1}{P^2}\times Se\right)=0$$

$$\Rightarrow \frac{Se}{P^2}=\frac{1}{2} \Rightarrow P=\sqrt{2Se}$$

- Performance of multilevel paging:

$$EMAT = x(c+m)+(1-x)(c+(n+1)m)$$

where n is number of levels in paging, c is cache access time and m is memory access time.

Difference between Paging and Segmentation

Paging	Segmentation
(a) Paging suffers from Internal Fragmentation only in the last page.	(a) There is no Internal Fragmentation
(b) Paging does not preserve users view of memory allocation	(b) Segmentation is visible to the programmer
(c) There is no External Fragmentation	(c) There is External Fragmentation

Virtual Memory

- Virtual memory gives an illusion to the programmer that a huge amount of memory is available for waiting and executing programs greater than size of available physical address space.
- **Page Fault Service:**
  - Process gets blocked
  - Change the mode bit
    Virtual Memory Manager (VMM) runs in Kernel mode
  - VMM → Device Manger → Device Driver → Disk (Device)
  - **Cases:**
    (i) Empty frame may be available
    (ii) No empty frame is available ⇒ Use page replacement algorithm
- More number of page faults, Page Fault Service Time (PFST) is more, execution time increases and throughput decreases.
- Virtual memory is implemented by demand paging.

Performance of Virtual Memory

Effective Memory Access Time (EMAT): $EMAT_{DP} = P\times s+(1-P)m$
Main Memory Access Time = 'm' [in nsec/μsec]

PFST = ‘s’ [$s >>> m$] (in msec)

Page fault rate = ‘P’ ; Page hit ratio = $1 - P$

Demand Paging

- It allows the pages that are being referenced actively to be stored in memory.
- It allows bigger virtual address spaces and provides the illusion of infinite physical memory.
- It allows more processes to fit in the physical memory, and to be running at the same time.
- Demand paging means that page tables need to sometimes point to the disk locations as opposed to memory locations.
- Page table needs an additional *present* or *valid* bit.
  - If present, the page table entry points to a page in memory.
  - If not present, a reference to an invalid page ***(page fault)*** will cause the hardware to trap, and the OS performs the following steps while running another process in the meantime.
    (i) Choose an old page in memory to replace
    (ii) If the old page has been modified, write contents back to disk
    (iii) Change the corresponding page table entry and TLB entry
    (iv) Load new page into memory from disk
    (v) Update page table entry
    (vi) Continue the process

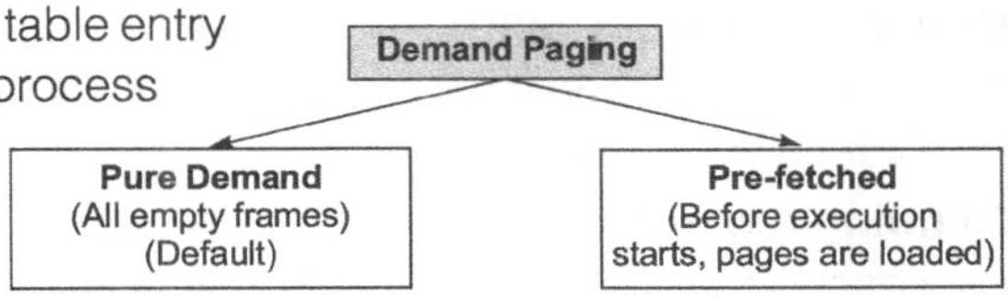

Page Replacement

- **Page reference string:** Set of successively unique pages referred in given list of logical address or virtual address.
- **Frame Allocation Policies:** All frames are not allocated, less than maximum demand are allocated.
  Total number of frames = ‘m’
  Total number of processes = ‘n’ [$P_1, P_2, \ldots, P_n$]
  Demand of each process = s_i

  Total demand of all processes = $S = \sum_{i=1}^{n} s_i$ ($s >>> m$)

Note:

☑ If process requires ‘*n*’ frames and operating system allocate ‘*n*’ frames to it at once, then there is no need of page replacement.

Page Replacement Algorithms

First-In First-Out (FIFO)

- In general within the increase in page frames the page fault rate of the process should decrease (natural characteristic).
- Sometimes with increase in number of page frames of the process, the page fault rate also increases. This unnatural behavior is called **Belady's Anamoly**.
- FIFO and FIFO based page replacement algorithms (Second Chance/ Clock Algorithm as it degenerates to FIFO) are bound to suffer from Belady's Anamoly.

Optimal

- Replace the page that will not be used for the longest period of time.
- It does not follow Beladys Anamoly.
- It is not implementable. It is used as a Benchmark to measure the performance of other page replacement algorithms.

Least Recently Used (LRU)

- Replace the page that has not been used for the longest period of time.
- Performance of LRU is closer to optimal as it is practical approximation to optimal page replacement.
- Criteria is Time of Reference.

Most Recently Used (MRU)

- Replace the page that is used most recently.
- Criteria is Time of Reference.

Counting Algorithms

- Least Frequently Used (LFU): Replace the page with the smallest count.
- Most Frequently Used (MFU).
- Implementation is expensive.

Variation of LRU/LRU Approximation

- LRU approximation are not exactly LRU, they approximation to the behaviour of LRU (i.e. they work like LRU).
- Reference bit (R)
  - 0 → Page is not referred so far during the current epoch
  - 1 → Page is referred atleast once during the current epoch (a period of time)
- **Second Chance/Clock Algorithm:** It degenerates to FIFO. Criteria is Arrival Time and Reference Bit.
- **Enhanced Second Chance/Not Recently Used:** Criteria is Reference Bit and Modify/Dirty Bit

Modify Bit
- 0 → Page is clean
- 1 → Page is modified

Thrashing

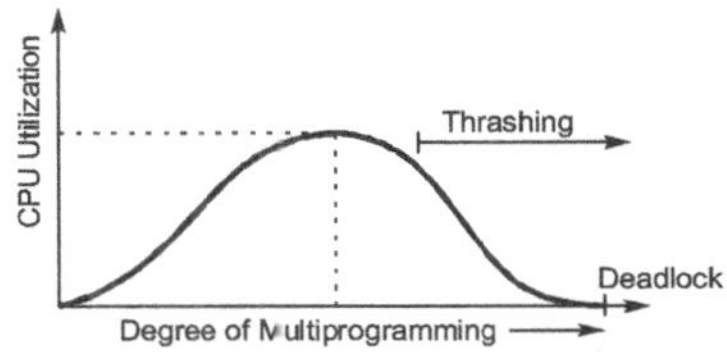

- High paging activity is called thrashing. A process is thrashing if it is spending more time paging than executing.
- **Thrashing** occurs when the memory is over committed, and pages are tossed out while still needed.
- CPU utilization is very low when the page fault rate is high.
- **Reasons:**
  - **Primary Reasons:**
    - Lack of memory frames
    - High degree of multiprogramming
  - **Secondary Reasons:**
    - Page replacement policy
    - 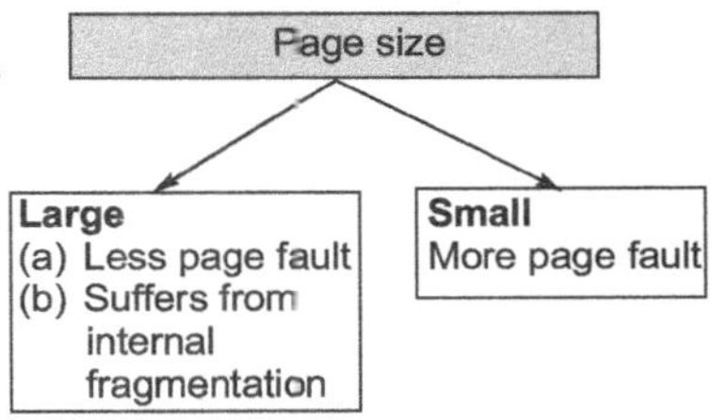

- **Control strategies:**

  Prevention: Controlling the degree of multiprogramming.

  Detection:
  - Low CPU utilization
  - High degree of multiprogramming
  - High paging disk utilization

  Recovery:
  - Process suspension
  - Increase RAM

Inverted Page Table

- An inverted page table has one entry for each real page (or frame) of memory.
- Each entry consists of the virtual address of the page stored in that real memory location, with information about process that owns that page.
- Only one page table is in the system and it has only one entry for each page of physical memory.
- Each inverted page table entry is a pair < process-id, page-number> where the process id assumes the role of the address space identifier.

■■■■